Related titles

Geothermal Energy: An Alternative Resource for the 21st Century
(ISBN 978-0-44452-875-9)
Geothermal Reservoir Engineering
(ISBN 978-0-12383-880-3)
Geothermal Power Plants
(ISBN 978-0-08098-206-9)

Woodhead Publishing Series in Energy:
Number 97

Geothermal Power Generation

Developments and Innovation

Edited by

Ronald DiPippo

AMSTERDAM • BOSTON • CAMBRIDGE • HEIDELBERG
LONDON • NEW YORK • OXFORD • PARIS • SAN DIEGO
SAN FRANCISCO • SINGAPORE • SYDNEY • TOKYO
Woodhead Publishing is an imprint of Elsevier

WOODHEAD
PUBLISHING

Woodhead Publishing is an imprint of Elsevier
The Officers' Mess Business Centre, Royston Road, Duxford, CB22 4QH, UK
50 Hampshire Street, 5th Floor, Cambridge, MA 02139, USA
The Boulevard, Langford Lane, Kidlington, OX5 1GB, UK

British Library Cataloguing-in-Publication Data
A catalogue record for this book is available from the British Library

Library of Congress Cataloging-in-Publication Data
A catalog record for this book is available from the Library of Congress

ISBN: 978-0-08-100337-4 (print)
ISBN: 978-0-08-100344-2 (online)

For information on all Woodhead Publishing publications
visit our website at https://www.elsevier.com/

Working together
to grow libraries in
developing countries

www.elsevier.com • www.bookaid.org

Publisher: Joe Hayton
Acquisition Editor: Cari Owen
Editorial Project Manager: Alex White
Production Project Manager: Omer Mukthar
Designer: Mark Rogers

Typeset by TNQ Books and Journals
Transferred to Digital Printing in 2016

Geothermal Power Generation

Contents

Many of the images used in the book are historic photographs. They are the only available best quality pictures.

Woodhead Publishing Series in Energy

Author biographies

Lucien Y. Bronicki, co-founder and past chairman and chief technology officer for Ormat Technologies Inc., holds a BSME and an MS in physics from the University of Paris and a postgraduate degree in nuclear engineering from the Conservatoire National des Arts et Métiers, Paris. He worked in the Nuclear Research Center in Saclay (France) designing equipment for particle research at CERN. While at the National Physical Laboratory (Israel), his development of solar-powered turbines evolved into geothermal power plants. Besides publishing over 30 articles, Lucien holds more than 70 patents on the design of thermodynamic devices, turbines, and controls. He received the Pioneers Award from the Geothermal Resources Council and an honorary doctorate from the Weizmann Institute of Science.

Patrick (Pat) Brown, a Chartered Accountant Fellow and director of Strettons, Accountants, Taupo, New Zealand, has experience in geothermal development mainly with Maori Land Trusts. Pat developed and managed the Mokai geothermal field with the Tuaropaki Trust Energy Park comprising a 110 MWe power plant, greenhouses, and a powdered milk plant. He advises the Maori Trusts including acquisition and management of the world's largest geothermal direct heat supply to the forestry industry at Kawerau. Semi-retired, Pat has an ongoing interest in economic development and the establishment of a local community foundation.

Louis E. Capuano, Jr., a petroleum engineering graduate of the University of Southwestern Louisiana, worked as a drilling engineer in the Gulf of Mexico until 1974 when he moved to Santa Rosa, CA, to work in the company's geothermal division. Louis has extensive experience in geothermal drilling having worked on many geothermal projects, including for his own firm, ThermaSource, since 1980. He has been aggressively involved in the geothermal industry, serving on the Geothermal Resource Council Board for 15 years and as GRC president for 1999–2000 and 2013–14. The GRC awarded him its *Joseph W. Aidlin Award* in 2004.

Raffaele Cataldi is former manager at the ENEL Geothermal Department and consultant to the Italian Government, the United Nations, and the European Union. He promoted and co-founded the International Geothermal Association and the Italian Geothermal Union, of which he is now the honorary president. Raffaele is the author of over 200 geothermal papers and three books on the history of science and technology. He has been involved in research, exploration, exploitation, and management of geothermal fields in Italy and around the world, having 57 years of activity. He is the recipient of several national and international recognitions in the geothermal sector.

Andrew Chiasson is a faculty member in the Department of Mechanical & Aerospace Engineering, University of Dayton, OH, where he teaches courses and conducts research in the areas of thermo-fluid sciences and renewable and clean energy. He has academic and professional practical experience in a wide range of geothermal applications related to small-scale electrical power generation, direct use heating, borehole thermal energy storage, and geothermal heat pumps (GeoExchange), in addition to hydrogeological site evaluations and groundwater modeling.

William (Bill) Cumming is an independent consultant who provides technical and management services for geophysical surveys, geothermal resource risk assessment, and geophysical research and training in the geothermal industry. His 35 years of geothermal experience include over 20 years with Unocal Corporation (now Chevron) in positions from geophysicist to chief geoscientist. Since 2000, he has provided consulting services to geothermal industry, academic, and government clients at over 40 geothermal fields and 100 prospects in the Americas, SE Asia, Europe, and Africa.

Surya Darma has over 30 years' experience in the geothermal, power, and energy sectors. From 2006 to 2009, he served as director of operations of Pertamina Geothermal Energy, the largest geothermal company in Indonesia. In his role as past-president of Indonesia Geothermal Association, he was responsible for hosting the 2010 World Geothermal Congress in Bali. Surya is now chairman of the Indonesian Renewable Energy Society and is on the Expertise Board of the Indonesia Electric Power Society. Since 2012, he has been a lecturer in the Geothermal Exploration Master's Program at the University of Indonesia.

Agnes C. de Jesus is chief sustainability officer of First Philippine Holdings Corporation. Previously she was senior vice president for environment and external relations and the compliance officer of the Energy Development Corporation in the Philippines. Her extensive experience spans a number of renewable energy technologies (geothermal, hydropower, wind and solar), specifically on environmental protection, watershed management, corporate governance, corporate social responsibility, public engagement, and conflict management. As her way of giving back to the energy industry, Agnes shares her knowledge as a reviewer of social and environmental safeguards guidelines for multilateral agencies and as an international lecturer.

Ronald DiPippo has 40 years' experience in geothermal power and has served on international advisory boards in five countries where he participated in the development of 22 geothermal fields. His book *Geothermal Power Plants* is in its fourth edition. Ron has written over 120 technical articles and contributed to and/or edited nine handbooks dealing with geothermal power. He received the *Ben Holt Geothermal*

Power Plant Award from the Geothermal Resources Council. See also the back cover of this book.

Wilfred A. Elders, professor of geology emeritus at the University of California, Riverside, has studied the geology of high-temperature geothermal systems in the USA, Mexico, Japan, New Zealand, and Iceland. From 1983 to 1988, he was chief scientist of the Salton Sea Scientific Drilling Project. He is the co-principal investigator of the Iceland Deep Drilling Project that in 2009 drilled the world's hottest (440°C) flowing well. In 2016, the IDDP will drill to 5.0 km depth at Reykjanes to continue exploring for supercritical geothermal energy: see http://www.iddp.is.

Steven (Steve) L. Enedy is a reservoir engineer for Calpine Corporation. Previously he held the positions of Steamfield Superintendent and Operating Supervisor for the Northern California Power Agency at The Geysers geothermal field in California.

Luis Carlos A. Gutiérrez-Negrín is a Mexican geologist retired in 2008 from the geothermal-electric division of the Comisión Federal de Electricidad (CFE). He is currently an independent consultant and executive director of the company Geocónsul and member of the Steering Group of the Mexican Innovation Center for Geothermal Energy (CEMIE-Geo). He is also a member of the BoD of the International Geothermal Association (IGA) and chair of its Information Committee and a former president of the Mexican Geothermal Association.

Jill Robinson Haizlip is president and principal geochemist of Geologica Geothermal Group Inc. She has over 30 years of experience in geothermal resource evaluation, including all aspects of geothermal exploration, geochemistry, development strategy, production, well testing, environmental compliance, and due diligence for financial and environmental risk. Her work involves the integration of the resource characteristics with the design and operation of geothermal power plants to maximize efficiency within environmental constraints, as well as economic and social settings. Jill holds a BS from Middlebury College and an MS from Columbia University, both in geology.

William (Bill) Harvey is a mechanical engineer with POWER Engineers that has served in all project phases for flash and binary plants, including projects commissioned in Africa, Turkey, the Americas, and Asia. His roles span detailed design, owner's engineering and independent engineering. Bill writes and lectures extensively on geothermal and renewable energy topics and is a guest professor at Reykjavik University. He holds a PhD from the University of Minnesota and a BS and MS from the University of Wisconsin.

Paul von Hirtz is president of Thermochem, Inc. and PT. Thermochem Indonesia. Paul studied chemistry at UC Berkeley and Sonoma State University, with a BS from SSU. He trained and worked as a chemical engineer with 30 years' experience in the geothermal industry. He did research for the US Department of Energy and the California Energy Commission, chairs the ASTM committees on Materials and

Geothermal Fluid Sampling and Analysis, and is an associate editor for the international journal *Geothermics*.

Roland N. Horne is the Thomas Davies Barrow Professor of Earth Sciences at Stanford University, and Senior Fellow in the Precourt Institute for Energy. He holds BE, PhD, and DSc degrees from the University of Auckland, NZ. He is best known for his work in well test interpretation, production optimization, and analysis of fractured reservoirs. He is an honorary member of the Society of Petroleum Engineers (SPE) and a member of the US National Academy of Engineering. He was the 2010–13 president of the International Geothermal Association (IGA).

Ernst Huenges is head of Geothermal Research at the German Research Center for Geosciences. He has an engagement as lecturer at the Technical University Berlin. His research interests are related to the development of new technologies for an economical exploration and utilization of geothermal energy for power, heating, or cooling. Ernst studied process engineering, physics, and geology and obtained a PhD in natural sciences (1987) at the University Bonn. He is spokesman on Geothermal Energy Systems within the program Renewable Energies of the Helmholtz Research Association.

Francesco Lazzeri works in Enel Green Power and is responsible for the operation of the Italian geothermal plants that generate 5800 TWh of electricity per year. Until May 2014, he held the position of country manager and general manager of the EGP subsidiaries in Romania, where they developed around 500 MW of wind plants. He developed his technical background mainly inside the geothermal O&M field and in wind and hydroelectric fields. He achieved the degree in mechanical engineering at the University of Pisa in 2000.

Marcelo J. Lippmann is a retired geological staff scientist, Lawrence Berkeley National Laboratory. He holds a master's in geology, University of Buenos Aires, and a PhD in engineering science, University of California, Berkeley. His research mainly focuses on the characterization and environmentally safe utilization of groundwater aquifers and in the evaluation of geothermal systems with emphasis on determining their commercial longevity and in designing appropriate exploitation strategies. Marcelo was also editor-in-chief of the journal *Geothermics* from 2004 to 2008 and is an instructor at geothermal courses and workshops.

Joseph (Joe) N. Moore is a research professor in the College of Engineering at the University of Utah. Since joining the Energy & Geoscience Institute in 1977, his career has focused on the geology and hydrothermal alteration of geothermal resources in the United States, Mexico, and Central and South America, Indonesia and the Philippines. Joe has served as lead scientist on numerous U.S. Department of Energy sponsored projects and as a consultant to government and private organizations throughout the world. He was also an associate editor for *Geothermics*.

Michal C. Moore is professor of energy economics, Senior Fellow at the University of Calgary and a professor of economics and systems engineering at Cornell University. His research is focused on energy regulatory and market oversight issues. Michal, a California native, was a former regulator of the energy industry in California and served as the chief economist for the U.S. National Renewable Energy Laboratory in Golden, Colorado. He read for his PhD at the University of Cambridge and is a member of Darwin College.

Gregory (Greg) Mines is a retired engineer from the Idaho National Laboratory who has conducted geothermal research for the U.S. Department of Energy since the mid 1970s. The emphasis of this research has been on both reducing power generation costs and improving the performance of binary cycles. It has included field testing to validate project cost and performance benefits in commercial operating plants, as well in a prototype binary research facility. Greg graduated from Montana State University with a BS in mechanical engineering.

Paul Moya Rojas is a Costa Rican civil engineer who earned his MS from the University of Kansas and did some geothermal studies at the University of Kyushu (Japan), the University of Braunschweig (Germany), and the University of Auckland (New Zealand). Paul participated in the geothermal development in Costa Rica (Miravalles and Las Pailas fields) mainly as steamfield manager. He has worked in geothermal energy for 30 years and currently works for West Japan Engineering Consultants as adviser to the general manager of the Geothermal Department.

John P. O'Sullivan is a lecturer in the Department of Engineering Science, University of Auckland. He completed his PhD and his undergraduate degree in engineering science at Auckland and his master's in aeronautics and astronautics at Stanford. His main area of research is geothermal reservoir modeling, and he teaches the subject at postgraduate level. He also works as a consultant for geothermal projects around the world developing models to assist with strategic decision making.

Michael J. (Mike) O'Sullivan is a professor in the Department of Engineering Science, University of Auckland. As an undergraduate at Auckland, he completed degrees in civil engineering and mathematics and then went on to complete a PhD in applied mechanics at Caltech. After a brief spell at New York University he returned to the University of Auckland. His main research interest is in geothermal modeling, recognized in 2011 by the Geothermal Resources Council with the *Henry J. Ramey Geothermal Reservoir Engineering Award.*

Roberto Parri is a mechanical engineer with more than 33 years of professional experience in the geothermal energy sector. Starting in 1981 he has worked in the fields of drilling technology, pipeline design, and installation and commissioning of geothermal power plants. He was responsible for the operations activity of Technical Services and then head of the Operation Unit of Enel Green Power where he was managing the operations of all the Italian geothermal plants. Now in retirement, Roberto is an energy consultant.

Kenneth Phair is a registered professional mechanical engineer. During 34 years of engineering geothermal power stations, Ken has contributed to the engineering design or upgrade of power plants that represent nearly 40% of the worldwide installed geothermal generation capacity. His work includes engineering two of the world's most efficient geothermal power plants. He received the Geothermal Resources Council *Ben Holt Geothermal Power Plant Award* in 2013 for outstanding achievements in geothermal power plant engineering, design, and construction.

Emilia Rodríguez Arias is a Costa Rican lawyer who works as an independent consultant on environmental legal issues. Emilia was a member of the Congress of the Republic of Costa Rica during 2002−06 and the first congressperson in the country to introduce a bill that would permit geothermal development in national parks. She has been a professor at the University of Costa Rica and the National University of Costa Rica. Currently, she is working on her thesis for a master's degree in renewable energies.

Subir K. Sanyal is a senior technical advisor to the Geothermal Resource Group in California. His area of expertise includes geothermal reservoir engineering, feasibility studies, financial assessment, and risk analysis. Prior to this, he was president of GeothermEx Inc., a Schlumberger company, for 35 years. Over his career, he has assisted clients worldwide in geothermal project development, power sales negotiations, and financing and provided due diligence for project financing in numerous countries leading to the development of some 7000 MW of capacity.

Ian K. Smith holds BSc and PhD degrees in mechanical engineering and a postgraduate diploma in gas turbine technology. His industrial experience includes 3 years in the aircraft gas turbine industry and 35 years as a consultant. Ian also has an academic position as professor of applied thermodynamics at City University, London. His research interests are power recovery from low-grade heat and the use of alternative working fluids and screw expanders in such systems. He has 162 refereed publications, five patents, and three monographs and has been recognized with 13 prizes and professional awards.

Ian A. Thain (Fellow IME) is a mechanical engineer with over 35 years' experience in operation, maintenance and development of geothermal power plants, and in the design and construction of geothermal direct use projects utilizing downhole heat energy recovery systems. Ian was involved in the establishment of the International Geothermal Association in 1988 and served as IGA vice president in 1998−2001. Although now retired, he is a lifetime member of the New Zealand Geothermal Association.

Kevin Wallace is the renewable generation lead for POWER Engineers, Inc. He has been a project manager, project engineer, process engineer, and commissioning engineer for more than 750 MW of new geothermal plant capacity around the world. Before joining POWER, he worked for PG&E in California. Primary assignments included projects at The Geysers and Diablo Canyon Nuclear Power Plant. He holds a BS in chemical/nuclear engineering from UC Berkeley and an MS in chemical/environmental engineering from San Jose State University.

Preface

Nearly 40 years ago, I became involved with geothermal power generation as a member of the organizing committee and editorial team at Brown University (Providence, RI) that produced a collaborative book titled *Sourcebook on the Production of Electricity from Geothermal Energy*. That 1000-page tome was supported by funds from the U.S. Department of Energy and featured chapters written by more than 40 specialists from companies, universities, and national laboratories across the United States.

The editor-in-chief was Dr. Joseph Kestin, a renowned thermodynamicist, who, like a number of the book's contributors, was a newcomer to geothermal energy. In 1976, when that project began, only a handful of countries had commercially operating geothermal power plants, with a total installed capacity of about 1000 MW, roughly the equivalent of one nuclear power station. That book and a companion volume, *Geothermal Energy as a Source of Electricity: A Worldwide Survey of the Design and Operation of Geothermal Power Plants*, which I wrote, appeared in 1980 and were widely circulated around the globe. They helped inform scientists, engineers, and policy-makers regarding the prospects of developing geothermal energy as a dependable, renewable, and environmentally benign source of electric power.

Jumping ahead to the present, we find that more than 25 countries now have or have had operating geothermal power plants, and as of 2015 the total worldwide capacity has grown to about 12,000 MW. Growth in development, however, has been uneven, driven and constrained by the price of oil and its availability. A wake-up call was sounded to the nations of the world by the two oil shocks of 1973 and 1979, both caused by political events, the first being the 1973 Arab—Israeli War (or Yom Kippur War) and the ensuing OPEC Oil Embargo, and the second being the Iranian Revolution and the overthrow of the shah. The growth rate of geothermal installed power leaped to 17% annually from 1977 to 1985 but has declined to about 3% since then.

Owing to the fact that geothermal energy is unique in many ways relative to other renewable sources of energy, it is the least understood among the public, legislators, and corporate leaders. This new volume will address the need for a better-informed populace, and it is hoped it will lead to policies that encourage the growth of this reliable and benign resource.

The authors of the chapters were invited and chosen for their expertise and stature in the geothermal community. Many are senior individuals with decades of experience. Others are relatively young but with stellar records of accomplishment in their respective fields. All have the desire to share with our readers what they have learned through study and practice.

The book is organized into four sections: Part One—Resource Exploration, Characterization, and Evaluation; Part Two—Energy Conversion Systems; Part Three—Design and Economic Considerations; and Part Four—Case Studies.

In the first part, the reader will learn about the preliminary work that must be done before a power plant can be properly designed, namely, the geology, geophysics, and hydrogeochemistry of prospective sites, drilling and well logging, and reservoir modeling. The geoscientific studies help define the geothermal system at a particular site, leading to confirmation of the resource by drilling deep wells at locations indicated as most promising by the geoscience. The data obtained from all these activities lead to a preliminary model of the system and the reservoir, which will be continually updated as the development of the field takes place over a number of years.

The second part deals comprehensively with the various energy conversion systems that can be brought to bear to exploit the energy contained in the geothermal fluids found in the reservoir. Since geothermal energy is extremely site specific, the fluids found in various sites around the world have dramatically different technical characteristics, creating engineering challenges that demand innovation. The first geothermal power plants from the early 1900s tapped into steam reservoirs at the Larderello field in Italy but that experience was not sufficient to permit development of the hot water resources that far outnumber steam systems. Thus a whole new approach was needed and new designs were developed that led to a wide expansion of geothermal power. Most recently, systems have been designed to make use of low-temperature resources previously thought unfeasible to exploit. Currently, a large variety of power systems are available to handle the wide spectrum of geothermal fluids found around the world, including those with high concentrations of dissolved solids and noncondensable gases once deemed too aggressive to be commercialized.

In the third part, we deal with specific aspects that influence the design and economics of geothermal projects, including means of discharging waste heat, how to cope with fluids that produce scale and corrosion, ways to mitigate environmental effects, and meeting regulations, obtaining permits, and financing.

The last part takes the reader to selected geothermal plant sites including Larderello, The Geysers (California, USA), and Los Azufres (Mexico) to learn the details of development and to see the diversity of resources represented. Detailed expositions are also presented on developments in Indonesia, New Zealand, and Latin America. The topic of enhanced geothermal systems (EGS) is included here because it may hold the key to truly worldwide use of geothermal energy. And the book concludes with a chapter on the environmental law that applies internationally and that must be followed to allow both geothermal development and preservation of the living conditions on our planet.

The idea for this book originated with Woodhead Publishers in May 2014 when I was contacted by Ginny Mills, Acquisitions Editor, asking if I was interested in being the editor for a comprehensive volume on geothermal power generation. About a month later, Sarah Hughes, Senior Acquisitions Editor, returned from maternity leave and took over for Ginny. Sarah and I fashioned the outline of the book and recruited authors. In October 2014, Alex White, Editorial Project Manager, came on-board and has shepherded the work ever since. Omer Mukthar Moosa and Jayanthi Bhaskar handled the design and production of the volume; the credit for its excellent appearance goes to them. A great deal of thanks goes to these people for their foresight and steady, efficient management.

Of course we would not have this volume without the contributions of the authors, most of whom had to fit their writing into their hectic work and travel schedules. A few authors are retired, including me, which allowed for somewhat more time, albeit at the expense of our families and other retirement activities. To all of the authors, we are most appreciative for your efforts and the outstanding chapters you created that will enlighten the next generation of geothermal scientists, engineers, and policy-makers.

Ronald DiPippo, PhD
Dartmouth, Massachusetts, USA
November 3, 2015

Introduction to geothermal power generation

1

L.Y. Bronicki
Ormat Technologies Inc., Reno, Nevada, United States

Geothermal energy is the only alternative energy source that can supply base-load or dispatchable power, independent of the climate. Although not evenly distributed geographically, geothermal energy potential is very important, particularly if research and development into technology such as engineered geothermal systems (EGS) will be actively pursued.

Chapters included in this publication cover subjects dealing with the nature of geothermal energy resources, their utilization, conversion technologies, as well as its future development. The chapters also highlight the greatest challenge in geothermal development, namely, the geothermal resource. Power conversion is the least uncertain part of a geothermal project but requires designs to ensure an economic exploitation of the resource over the designed life of the project. The risks and challenges are related to exploration, drilling, and managing the resource. Optimization depends on the manner in which the power station configuration is adapted to the available resource.

The **distribution of geothermal resources** is irregular due to unequal distribution of volcanoes, hot springs, and heat manifestations at specific locations over the Earth's surface. Geothermal resources are a reflection of the underlying global, local geological, and hydrological frameworks. The most thermally rich resources tend to concentrate in environments with abundant volcanic activity and tend to be controlled by plate tectonic processes or spreading centers evident as volcanic chains associated with subduction zones and hot spots. The local geological characteristics that favor useful resources include relatively shallow resource depths to high permeability in the rocks surrounding the resource, and adequate resource fluids.

Exploration starts with the analysis of available geological information to identify the potential target. Once the target is identified, geochemical studies and core drilling are undertaken. These studies are complimented or sometimes preceded by geophysical surveys including aeromagnetic or resistivity studies and remote infrared and hyper spectral techniques.

Hydrothermal systems have differing types of chemical properties which in turn impact the choice of materials and design of the power plant. The source of heat is usually a magma chamber a few kilometers below the surface. Fluid origin is meteoric, namely rain water, which infiltrates the ground to depths of a few kilometers. The permeability and degree of fracturing of this cap rock vary from site to site according to the intensity and abundance of the surface hydrothermal systems manifestations, such as hot springs, steam vents, geysers, etc.

Geothermal Power Generation. http://dx.doi.org/10.1016/B978-0-08-100337-4.00001-2

In the early stages of exploration of a hydrothermal system when there is only sur-face evidence, the aim of a geochemical survey is the generation of a model that eval-uates the temperature and chemical conditions of the fluid at depth.

Drilling of the wells for extracting geothermal energy resources is a niche within the larger drilling services industry that focuses primarily on oil, gas, and minerals. In particular, deep drilling required in most exploration programs for geothermal po-wer generation projects will likely use large drill rigs typical in oil and gas extraction.

There are several aspects unique to geothermal drilling. Mainly, geothermal forma-tions, by their nature, involve elevated temperatures, which are usually significantly higher than those experienced when drilling for oil and gas. The rock that hosts these formations are typically harder (granite, granodiorite, quartzite, basalt, volcanic tuff), more abrasive, highly fractured, and underpressured. Caustic elements may be present that can cause corrosion and scaling in the wellbore.

These unique characteristics present challenges in dealing with geothermal wells, which, unlike oil and gas wells, do not produce economically until used through elec-tric generation or direct uses. For power production, geothermal wells must be of a larger diameter than oil and gas wells to produce sufficiently high flow rates for commercial production. Depths of geothermal wells vary according to location and can reach over 3000 m and even more for EGS projects.

Reservoir engineering is the comprehensive integration of all available surface and underground information regarding geology, geophysics, geochemistry, well drilling-testing, exploitation data, information concerning the geothermal developer, and objectives of a geothermal development (eg, market targets, costs, and finance), so it is the most powerful tool to evaluate the feasibility of a project. As in any scien-tific or engineering activity, results derived from reservoir engineering depend on the quantity and quality of the information, as well as the associated processing and inter-pretation of the information. Reservoir engineering is not limited to the final numerical tool but also includes acquisition of information which allows prediction of the impacts on a geothermal resource 20−30 years into the future.

Geothermal reservoir monitoring is the means for maintaining a sustainable geothermal field during operation. Using techniques such as downhole monitoring, surface monitoring, and introducing the collected data into the numerical model of the reservoir, the impact on the long-term sustainability of a geothermal field can be closely evaluated to ensure that the resource is not prematurely cooling and that any cooling or adverse impact can be mitigating by drilling makeup wells correctly located.

The promising **engineered geothermal systems (EGS)** aims to exploit hot rock not accessible via conventional geothermal technology. Commercialization of this tech-nology could unlock many thousands of megawatts of power. For example, the esti-mate of the technical potential for EGS in the United States is estimated at 100 GWe, which is more than 30 times the total current installed geothermal capacity in the United States from current sources. As of the end of 2014 there has been some success in particular on engineered injection wells of existing plants, but no actual electricity production using both engineered production and injection wells on the same plant. Once EGS becomes commercialized, the system needs to demonstrate sustainability.

As with any other geothermal energy source, EGS development involves some impact on the environment. Geothermal resources are environmentally important as natural thermal features. Typically, the most significant environmental impacts are associated with the exploitation of high-temperature, liquid-dominated geothermal systems for electric power generation. However, the majority of these impacts can be avoided or minimized with appropriate techniques. Moreover, as geothermal energy generally offsets use of fossil fuels, the use of geothermal resources is more likely to improve air quality and overall water quality.

Geothermal power conversion are the techniques used for the conversion of thermal energy content of geothermal fluid into mechanical power to drive a generator and produce electrical power. Power conversion is the most predictable part of a geothermal project, as it consists of well-established and straightforward engineering design with work executed by experienced manufacturers, engineering firms, and contractors.

Today, more than 10 GW of geothermal power plants are in operation in the world, and a majority of them use steam turbines that operate on dry steam or steam produced by single- or double-flash with about 1.5 GW using organic Rankine cycles (ORC) or geothermal combined cycles. However, to widen the range of resources suitable for power generation beyond dry steam and flashed steam plants, ORC cycles have been implemented in the last 30 years and will probably continue to grow as a common technology driving future development of geothermal resources.

Operational experience confirms the advantages of ORC power stations, not only for low-temperature, liquid-dominated resources but also for certain high-temperature resources where the brine is aggressive or the fluid contains a high percentage of noncondensable gas. The higher installation cost of these systems, where economically feasible, is justified by environmental and long-term resource management considerations.

From the concept of **sustainability and renewability** of geothermal systems and the relationship between renewable and sustainable capacities, it is possible to estimate the commercial, sustainable, and renewable capacities of a geothermal system. **Sustainability** is defined as the ability to economically install and maintain power capacity over the amortized life of a power plant. This is done by taking practical steps, such as drilling "makeup" wells as required to compensate for resource degradation. **Renewability** is defined as the ability to maintain an installed power capacity indefinitely without encountering any resource degradation. Typically, the renewable power capacity at a geothermal site is generally too small for commercial development of electrical power capacity but may be adequate for district heating or other direct uses of the geothermal energy.

The **cost of production** for geothermal electric generation is important. In particular, the levelized cost of power is the applicable measurement for the cost of geothermal energy. Unlike fossil fuel power plants, most of the capital costs are incurred upfront in the development of the resource. Power cost is an objective criterion that favors geothermal solutions compared to other alternative energy sources. However, the costs are heavily tied to the resource and the need for makeup well drilling to maintain full generation capacity over the planned period of operation to provide an adequate return on investment.

Part One

Resource exploration, characterization and evaluation

Geology of geothermal resources

W.A. Elders[1], J.N. Moore[2]
[1]University of California, Riverside, CA, United States; [2]University of Utah, Salt Lake City, UT, United States

2.1 Introduction

Geology plays an essential role in the exploration and development of geothermal resources. The geologist is responsible for targeting potential resources, developing conceptual models of geothermal systems, assessing permeability and the nature and distribution of subsurface rocks, providing essential information to develop quantitative reservoir models, and supporting civil works required for drilling and plant operations. Because geologic and subsurface data are commonly limited during the early phases of a geothermal program, geologists must rely on conceptual models of geothermal systems in similar geologic environments. In this chapter, we present the general characteristics of geothermal systems, describe representative occurrence models, and discuss the geologic techniques utilized during exploration and development.

2.2 Heat flow and plate tectonics

Estimates of the total heat flow from the Earth's interior to surface average 47 TW (TW = terawatt or 10^{12} W) but span a range of 43−49 TW. The average crustal heat flow (heat flux per unit area) is 91.6 mW/m^2, but it varies according to the geologic environment concerned. For example, the mean heat flows of continental and oceanic crust are 70.9 and 105.4 mW/m^2, respectively. There are two different sources that make roughly equal contributions to this heat flow: first, radiogenic heat, due to the decay of radioactive elements such as ^{238}U, ^{232}Th, and ^{40}K, and second, residual primordial heat generated during the initial accretion of the Earth and its subsequent gravitational differentiation into core, mantle, and crust. In stable areas of the Earth's crust, this heat flow results in geothermal gradients in the shallow crust of about 30°C/km but can be up to 500°C/km in tectonically active zones.

Geothermal systems can be classified on the basis of their reservoir temperature, fluid type, and mechanisms of heat transfer. Three broad groups of conventional resources can be recognized. Low-temperature resources with temperatures of less than 150°C provide energy for heat pumps and direct use applications. Moderate-temperature geothermal systems, with temperatures between 150 and 200°C, can be used for electric generation, but the wells must be pumped to produce the quantities of water required for large-scale electric production. Wells with temperatures exceeding 225°C are currently

Geothermal Power Generation. http://dx.doi.org/10.1016/B978-0-08-100337-4.00002-4

the chief class of geothermal resources developed to generate electricity and can be produced by thermo-artesian flow, without the need for pumping.

Unconventional geothermal systems are found at either greater depths or higher temperatures than the geothermal resources currently being exploited. These include deep sedimentary basins, magma systems, and enhanced geothermal systems (EGS). The characteristics of these different geothermal environments are discussed next.

2.2.1 Geothermal systems and plate tectonics

There is a strong correlation between heat flow and plate tectonic settings. Geothermal systems are concentrated at plate boundaries (Fig. 2.1). Divergent and convergent plate boundaries are loci of volcanic activity, high heat flow, and stress regimens that are favorable for the development of high-temperature geothermal systems [53]. There is, for example, a concentration of andesite volcanoes and geothermal fields around the margins of the Pacific Ocean along a series of convergent plate margins.

There is also a close relationship between high heat flow and active basaltic volcanism along divergent margins where spreading is active, such as the Mid-Atlantic Ridge, the East Pacific Rise, the Red Sea, and the East African Rift system. High-temperature geothermal systems are also abundant along complex continental plate boundaries (ie, areas of recent thrust faulting and orogenesis), with their associated volcanism, for example, in the zones stretching from northern Italy, through the Caucasus, and into the Himalayas.

Table 2.1 summarizes the diversity of geothermal resource types and presents a simple classification of them. Geothermal resources occur in geological environments so diverse that each requires individual study to derive a realistic conceptual model.

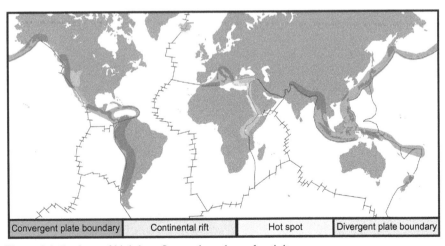

Figure 2.1 Regions of high heat flow and geothermal activity.
Adapted from Moeck IS. Catalog of geothermal play types based on geologic controls. Renew Sustain Energy Rev 2014;37:867−82.

Table 2.1 Simplified classification of geothermal systems with economic potential

Geothermal system	Surface features and topography	Tectonic setting and heat source	Reservoir depth and rock type	Surface features and hydrology	Examples
Andesitic volcanoes	Stratovolcanoes; high relief	Volcanic arcs at subduction zones; thin dikes and sills within the cones; larger bodies at >5-km depth	Andesitic lava flows, volcaniclastic deposits; resource depth is moderate to deep	Fumaroles (above upflow zones) and minor hot springs (distal to main upflow); rare sinter deposits; extensive lateral outflows; fumaroles only if vapor dominated	Indonesia, Philippines geothermal systems
Silicic volcanic systems	Calderas, dome fields; low relief	Continental environments; large silicic magma chambers at >5-km depth	Dominantly rhyolite to dacite pyroclastic deposits; Mesozoic and Paleozoic basement rocks; depth to resource is moderate to deep	Hot springs and minor fumaroles; sinter deposits common; outflows of limited extent; fumaroles if vapor dominated	Taupo volcanic zone, New Zealand; Coso Hot Springs, USA
Sedimentary hosted	Low relief	Rift basin; dikes and regionally elevated heat flow	Recent sedimentary deposits; depth to resource is moderate	Few surface features; outflows of limited extent	Salton Trough, USA
Extensional tectonic systems	Low relief	Continental rift zones and back-arc basins; regionally elevated heat flow and deep circulation along faults	Mesozoic and Paleozoic basement rocks; resource is deep	Sparse hot springs and fumaroles; outflows of limited extent	Basin and Range, USA
Oceanic spreading center	Shield volcanoes and rift zones; low to moderate relief	Ocean spreading centers; basaltic sills and dikes	Basalt; resource at moderate depths	Sparse hot springs and fumaroles; outflows of limited extent	Iceland

Depths: shallow (<1 km); moderate (1–2 km); deep (2–3 km). Basement rocks include sedimentary, metamorphic, and intrusive rocks.

Heat from the mantle can be transferred to the upper crust by convection or conduction so that geothermal systems can have a wide variety of geothermal gradients (ie, conductive, convective, or a combination of the two).

Because of their importance as sources of geothermal electricity, emphasis here is placed on convective systems where heat transfer occurs via the circulation of liquid water or steam. Conductively heated waters are characteristic of many deep sedimentary basins and the geopressured systems of the Gulf of Mexico where fluids are trapped within permeable sands between impermeable clay layers.

The convecting fluids may be liquid, steam, or two-phase mixtures of steam and liquid. Liquid-dominated systems are characterized by pressure profiles that follow hydrostatic or hot hydrostatic profiles and maximum temperatures that follow the boiling point to depth curve (Fig. 2.2). Boiling is common in the upflow zones of geothermal systems and many systems contain two-phase regions of liquid plus vapor or shallow steam caps. Because of their diversity no single conceptual model fits the broad range of economic geothermal systems.

Fig. 2.2 shows a range of representative temperature profiles found in diverse geothermal systems. The maximum temperatures are defined by gas-free boiling point to depth curves. For reference, curves for 0 and 10 wt% total dissolved solids (TDS) are shown as curves VI and VII, respectively. However, the salinities of most geothermal systems are relatively low, with TDS contents from about 0.5 to 3 wt%. In contrast, even the presence of only minor amounts of CO_2 significantly affects the boiling point of the fluid, reducing the depth at which boiling can occur (curve V, Fig. 2.2). In contrast to liquid-dominated resources, steam is the pressure-controlling phase in vapor-dominated geothermal systems (curve III, Fig. 2.2).

2.3   Geologic techniques

Geologic studies are needed throughout the exploration, development, and production phases of a geothermal project and should be regarded as an integral part of any program. As projects proceed, different techniques can be applied and integrated with ongoing geochemical, geophysical, and engineering studies. The result of these efforts is the development of a conceptual model that can guide exploration and development.

Table 2.2 presents the techniques frequently used by geologists. During the initial stages of the project, these rely heavily on literature reviews, satellite images, and geologic mapping to locate and characterize surface manifestations, faults or porous rocks that may control fluid flow within the reservoir. These relationships can be incorporated into relatively simple conceptual models similar to those discussed next. Site visits and detailed mapping of relatively small areas, on the order of 10×10 km, typically follow. Samples of hot spring and fumarole discharges are collected for geochemical analyses and rock samples can be selected for characterization and dating. $^{40}Ar/^{39}Ar$ spectrum dating of potassium-bearing minerals can be particularly valuable for determining the ages and thermal history of young volcanic rocks and adularia veins, which are common in the reservoir rocks. ^{14}C dating by accelerator mass spectrometry (AMS) can be used to determine the ages of pollen

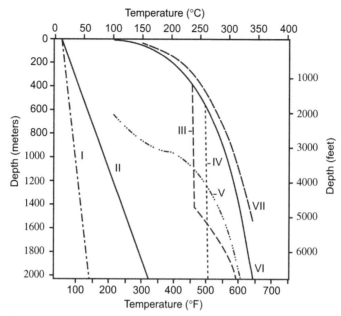

Figure 2.2 Temperature—depth relationships for geothermal fluids. I — low thermal gradient, normal conductive heat flow; II — average thermal gradient, conductive heat flow; III — a vapor-dominated system, with pressures considerably lower than hydrostatic. Below 1400-m depth, the curve represents a convecting brine that boils forming the vapor-dominated zone above, that condenses to form a vapor—liquid mixture along the boiling point curve (VI); IV — a low-salinity, moderate-temperature, water-dominated system, boiling at 600 m depth; V — boiling point to depth curve for pure water with 1 mol% CO_2; VI — boiling point to depth curve for pure water; VII — boiling point to depth curve for water with 10 wt% NaCl. Compiled from Haas Jr JL. The effect of salinity on the maximum thermal gradient of a hydrothermal system at hydrostatic pressure. Econ Geol 1971;66:940—6; Henley RW, Ellis AJ. Geothermal systems ancient and modern: a geochemical review. Earth Sci Rev 1983;19:1—50; Sasada M. CO_2 — bearing fluid inclusions from geothermal fields. Geotherm Resour Counc Trans 1971;5:351—6; White DE. Characteristics of geothermal systems. In: Kruger P, Otte C, editors. Geothermal energy: resources, production, stimulation. Stanford University; 1973. p. 69—94; ISBN-10: 0804708223; ISBN-13: 978-0804708227 [chapter 4].

and plant fragments encased in hot spring deposits and lake beds that are less than about 57,000 years old [56].

Once the most prospective areas are identified, geophysical surveys and the drilling of thermal gradient wells can proceed. The primary objective of these wells is to provide evidence of the nature and distribution of heat. Slim holes, because of their narrow diameters, typically do not produce fluids. However, petrographic analyses of the core and cuttings samples from these wells can assist in determining the stratigraphy, hydrology and structure of the geothermal system. Examination of the samples in thin section is used to identify rock types, the alteration assemblages and the sequence of alteration events.

Table 2.2 **Geologic techniques frequently used during exploration and development of a geothermal system**

Technique	Information obtained	Purpose
Satellite imagery	Locations of possible faults and surface manifestations;	Determine areas for geologic mapping; locate areas of potential geohazards (eg, unstable slopes, debris flows)
Geologic mapping	Types and locations of surficial manifestations, rock types, locations of faults	Identify potential reservoir lithologies, caprocks, and aquicludes; develop a preliminary conceptual model of system and initial drill targets; provide input into civil works
Petrography	Rock type, alteration mineralogy, relative ages of secondary minerals, distribution of fractures and porosity	Develop geologic history, locate zones of upflow and recharge, assess permeability distributions
X-ray diffraction	Determine rock mineralogy and percentage of primary and secondary minerals	Identify minerals that cannot be characterized petrographically (eg, mixed layer clays, zeolites); assess degree of alteration, chemistry, and temperature of the geothermal fluids, assist in developing well casing programs
Fluid inclusion analyses	Homogenization temperature, apparent salinities and the presence of daughter minerals	Characterize temperatures and compositions of fluids that have circulated through the system; assess fluid processes (boiling, mixing, conductive cooling)
Electron microprobe, SEM-EDS analyses	Mineral compositions and textures	Identification and compositions of individual minerals; mineral paragenesis and textural data
Radiometric dating	Age of rocks and hot spring deposits	Evidence of young heat sources

Drill core samples, although more expensive, are preferred by geologists because lithologies and mineral and fracture relationships can be directly observed; cutting samples are not a substitute for core. Nevertheless, careful systematic studies of drill cuttings can still provide important information on the hydrothermal alteration and vein distributions. For example, the abundance of minerals with euhedral shapes indicating formation in open spaces can provide a measure of the rocks porosity and permeability. Intervals that show little evidence of open-space filling typically show little evidence of permeability.

2.4   Hydrothermal alteration

An important role for geologists is the study of petrology of geothermal systems. This is necessary to determine the lithology of subsurface reservoir and cap rocks as an essential step in interpreting wireline logs and in determining the geologic structure of geothermal systems by correlation between multiple wells. Petrology can supplement borehole geophysics and reservoir engineering to investigate the physical characteristics of the reservoir rocks, their geothermometry and the direction of temperature change, the patterns of fluid flow that operated in the undisturbed state, the shape, size, and boundaries of the reservoir, discharge and recharge zones, and inferences about the location and nature of heat sources (see, for example, Ref. [18]).

The types of alteration minerals that are present provide a record of the temperatures and type of fluid that have circulated through the rocks (eg, acidic, neutral pH, or bicarbonate rich). X-ray diffraction analyses of the samples can provide additional detailed information on the mineral assemblages, particularly of clay minerals and zeolites present and on the overall degree of hydrothermal alteration. Although the intensity of alteration has been found to vary from system to system, studies of individual geothermal reservoirs suggest that hydrothermal alteration within the reservoir is most intense within the upflow zones and decreases in the recharge system. Additional information on past reservoir temperatures, salinities, and fluid processes can be provided by measurements of fluid inclusions, especially in vein minerals [32].

During the exploration phase of geothermal development, petrology is used to identify the nature, age and history of surface rocks, and the presence of hydrothermal discharge zones. Thus, petrological studies should begin with outcrop samples as an aid in siting exploratory wells and then continue during drilling with investigation of cuttings and cores. During the drilling phase, petrology is necessary to identify the protolith of the rocks, the characteristics of the hydrothermal alteration (precipitation, leaching, and replacement), and likely production zones and casing depths. If laboratory facilities are available, this information can be obtained from petrologic studies conducted while drilling, before a well comes to thermal equilibrium. During the production phase, petrology is used to study ejecta and wellbore scales to help reservoir engineers make inferences about questions such as "What is likely to happen if we inject fluid of a particular composition in a specific location?"

Factors affecting the formation of hydrothermal minerals are temperature, pressure, rock type, permeability, fluid composition, and duration of hydrothermal activity [11]. Assemblages of hydrothermal minerals thus record the processes and conditions that have occurred in a geothermal reservoir [17]. Because hydrothermal alteration changes the mineralogy of rocks, it also can strongly affect their physical properties. Water–rock reactions can both increase or decrease permeability and porosity by mineral

Table 2.3 Occurrence of common hydrothermal minerals as a function of temperature and pH

pH	<225°C	>225°C
Acid environment	Chalcedony, cristobalite, kaolinite, alunite	Diaspore, pyrophyllite, quartz, pyrite
Neutral to alkaline environment	Smectite, interstratified clay minerals, feldspars, zeolites (eg, laumontite), quartz, calcite	Epidote, prehnite, wairakite, illite, chlorite, biotite, actinolite, clinopyroxene, tourmaline, garnet

Based on Browne PRL. Hydrothermal alteration in active geothermal fields. Annu Rev Earth Planet Sci 1978;6:229−50; Henley RW, Ellis AJ. Geothermal systems ancient and modern: a geochemical review. Earth Sci Rev 1983;19:1−50; Reyes AG. Petrology of Philippine geothermal systems and the application of alteration mineralogy to their assessment. J Volcanol Geotherm Res 1990;43:279−309.

deposition or dissolution. These interactions can modify thermal and hydraulic stresses, leading to faulting and fracturing.

Two broadly different hydrothermal assemblages are commonly recognized [11,34,62]. The most common is produced by interactions with neutral pH to basic chloride fluids. Less commonly, acid alteration occurs as fluids of low pH react with the rocks—usually sulfuric acid, but hydrochloric acid may also be present. Table 2.3 presents a summary of the hydrothermal minerals that are most typical of different ranges of temperature and pH.

Overall, temperature, fluid composition, pressure, and water−rock ratios have the greatest effects on the mineralogy of the alteration assemblages. Because of limited thermodynamic data and the effects of solid solution, researchers have tended to rely on empirical relationships based on measured well temperatures to determine the thermal ranges of common geothermal minerals. This approach assumes the minerals are in thermal equilibrium and that no heating and cooling have occurred since mineral formation. Despite these uncertainties and the individual characteristics of geothermal systems, there is considerable consistency in the mineral assemblages and temperatures estimated by different researchers in many different geothermal systems (Fig. 2.3).

Fluid composition and pressure are strongly interrelated because pressure plays a major role in controlling the gas contents of the fluid and its pH. Where pressures are reduced, resulting in boiling, gas loss can lead to mineral deposition through the change in pH and solute concentrations. Condensation of the gases and steam into overlying groundwaters produces the acidic fluids that can cause intense alteration of the host rocks. These mineral assemblages are distinct from those produced by neutral pH fluids (Fig. 2.3). The absolute concentrations of the components in the fluids, in contrast, may have only minor influence on the compositions of the mineral assemblages. For example, alteration assemblages in the Salton Sea geothermal field in California, United States, which has a salinity of approximately 250,000 mg/kg TDS [50] are similar to those in many volcanic systems with salinities of 10,000 mg/kg, despite the differences in rock type.

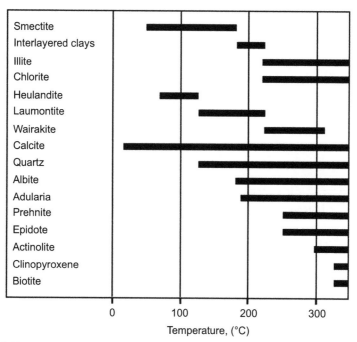

Figure 2.3 Temperature ranges of common hydrothermal minerals formed at neutral pH. Compiled from Bird DK, Schiffman P, Elders WA, Williams AE, McDowell SD. Calc-silicate mineralization in active geothermal systems. Econ Geol 1984;79:671−95; Browne PRL. Hydrothermal alteration in active geothermal fields. Annu Rev Earth Planet Sci 1978;6: 229−50; Henley RW, Ellis AJ. Geothermal systems ancient and modern: a geochemical review. Earth Sci Rev 1983;19:1−50; Reyes AG. Petrology of Philippine geothermal systems and the application of alteration mineralogy to their assessment. J Volcanol Geotherm Res 1990;43:279−309.

Rock type, permeability, and duration of activity, while affecting the intensity of hydrothermal alteration, generally have only a minor influence on the actual mineral assemblages formed. With the exceptions of reservoir rocks with unusual compositions, such as limestone, the chemistry of most host rocks varies within relatively narrow limits.

Permeability can influence alteration mineral assemblages by controlling the water−rock ratios and locations where boiling can occur but overall has only second-order effects on the assemblages of hydrothermal minerals formed.

Similarly, the duration of hydrothermal activity primarily influences the overall degree of alteration, and whether equilibrium has been achieved. Little is known about the longevity of geothermal systems although there is evidence that some may persist, at least episodically, for 200,000−500,000 years [11]. For example, dating of adularia from the Tiwi, Philippines, indicates the system has undergone two periods of heating: the earliest from about 280 to greater than 315 ka and a younger period of renewed heating during the last 10−50 ka [55].

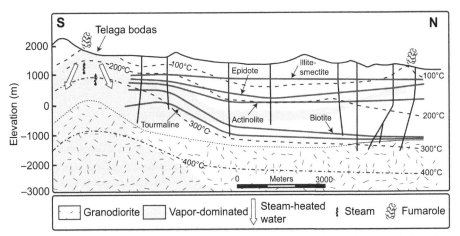

Figure 2.4 North—south section of the Karaha—Telaga Bodas geothermal system, Indonesia, showing surfaces representing the first appearance of key alteration minerals, present-day isotherms (*dash-dot* pattern where inferred), and the top of the granodiorite (*dotted* where inferred); smectite occurs above the first appearance of interlayered illite-smectite. The system is partially vapor dominated. Telaga Bodas is an acid lake.
Modified from Moore JN, Allis RG, Nemčok M, Powell TS, Bruton CJ, Wannamaker PE, et al. The evolution of volcano-hosted geothermal systems based on deep wells from Karaha—Telaga Bodas, Indonesia. Am J Sci 2008;308:1—48.

Fig. 2.4 provides one example, of many, of mineral successions formed at progressively greater temperatures [56]. Two features are noteworthy. First, comparison of the measured temperatures with those defined by the mineral distributions (refer to Fig. 2.4) indicates cooling has occurred in the system. It is not uncommon to find mineral assemblages that appear to be out of equilibrium with the present-day conditions when cooling has occurred as higher-temperature alteration minerals tend to persist when temperatures fall. Understanding the relationships between the mineral distributions and temperature allows reconstruction of the evolution of the system. Second, the surfaces showing first occurrences of minerals are generally flat lying, suggesting the top of the heat source (in this case, the top of the granodiorite) is also relatively flat lying.

As shown in Fig. 2.3, at the lowest temperatures, the common secondary minerals include the clay minerals smectite, interlayered illite-smectite and chlorite-smectite, quartz, carbonates, and zeolites. The most common low- to moderate-temperature zeolites are heulandite, stilbite, clinoptilolite, and laumontite. This assemblage persists up to about 220°C. Smectite is the first clay mineral to form and is common in the temperature range from about 50 to 180°C; at higher temperatures, it is replaced by interlayered clays. The smectite content of the interlayered clays typically decreases with increasing temperature up to 220°C, when it is replaced by illite or chlorite. Because smectite and smectite-rich interlayered clays are electrically conductive and are commonly found in the low-permeability cap rocks, electrical resistivity (primarily magnetotelluric) surveys are commonly used to map their distributions above the underlying reservoir rocks.

Epidote typically becomes abundant at temperatures above about 250°C, and the appearance of persistent epidote often coincides with the top of the economic geothermal reservoir [6]. Actinolite appears at about 290°C, followed by clinopyroxene at 300°C, biotite at 325°C, and garnet at about 350°C [18]. Illite, chlorite, K-feldspar, and albite are characteristic additional phases. Tourmaline may be present, particularly where temperatures have exceeded 400°C [59], reflecting the influx of boron-bearing magmatic fluids and proximity to the heat source.

Acidic fluids are characterized by distinctive mineral assemblages (see Table 2.3). These fluids are of concern to power plant development because of the corrosion they can cause to wells and pipelines. There are several mechanisms that produce acid fluids. The most common is through the oxidation of H_2S, released by boiling from the reservoir fluids, to H_2SO_4. This reaction requires the presence of atmospheric oxygen and can produce fluids with a pH as low as about 2–3 [28,59,63]. Such surficial acid waters are found in both volcanic and nonvolcanic geothermal systems.

Fluids with pH less than 2 require reactions involving magmatic gases (eg, SO_2, HCl, HF, and HBO_3) with groundwater or steam condensate. These acidic fluids can form at depth, with no surface expression of their presence or reach the surface through vapor-chimneys to form acidic crater lakes [12,28,64]. The discharge of magmatic gases is common within the central upflow zones of stratovolcanoes above the cooling magma bodies.

In contrast to the hot spring deposits produced by chloride and bicarbonate-rich waters, surficial deposits associated with fumaroles, acid mud pots, and crater lakes are associated with intense cation leaching of the surrounding rocks. The resulting residues consist dominantly of silica and minor amounts of aluminum, iron, and titanium [69]. These leached rocks may superficially resemble sinter deposits and their recognition has important implications. True siliceous sinters are indicative of past or present reservoir temperatures above 210°C [23], whereas siliceous residues formed by acid leaching provide no information on conditions within the reservoir.

Acid alteration is common above the upflow zones of geothermal systems. Descending acidic fluids may migrate long vertical distances along faults or permeable horizons, such as the contacts between volcanic units [63], as the ability of the rocks to neutralize the fluids progressively decreases and anhydrite is deposited on the pathway walls. Both anhydrite and calcite will deposit in response to heating, as both minerals have retrograde solubility, or to mixing with hot reservoir fluids [37]. In this way, the acid fluids and accompanying acid alteration may be isolated from the surrounding neutral pH waters, making understanding of the fluid regimen more difficult.

Veins filled with various proportions of calcite and anhydrite, but lacking silicate phases, are common in the marginal parts of many geothermal systems where they seal open fractures. Their presence can be used to characterize the locations of recharge zones and regions of enhanced cooling due to the influx of downflowing fluids [18].

Zones of boiling are frequently regions of enhanced permeability, and their recognition can be important in locating drilling targets and mapping the upflow zones of geothermal systems. They can be recognized by the presence of hydrothermal breccias, veins filled with intergrown calcite and quartz, bladed calcite, adularia, and vapor-rich fluid inclusions [10,68].

2.5   Volcanic-hosted systems

2.5.1   Liquid-dominated geothermal systems

The most favorable conditions for the formation of moderate- and high-temperature geothermal systems are found in magmatic, extensional, and transtensional environments characterized by active tectonism, which maintains permeable fractures capable of sustained fluid flow (eg, [53]). The majority of high enthalpy geothermal systems occur in volcanic arcs above subduction zones. Numerous examples of these systems are found in Japan, the Philippines, Indonesia, the Andes, and New Zealand, within the Ring of Fire that surrounds the Pacific Ocean (Fig. 2.1). Less commonly, volcanic-hosted geothermal systems are related to local heating at mantle hot spots or to oceanic or continental type rift systems. The geothermal system at Puna, Hawaii, is an example of this first type and Reykjanes, Iceland, of the second.

Fig.2.5(a) illustrates the characteristic features of geothermal systems typical of arc volcanism. These systems are characterized by a deep water table, a chloride reservoir fluid at depth, two-phase conditions within the central upflow zone, and marginal acid-sulfate and bicarbonate-rich waters. Although isotopic data suggest the reservoir fluids are dominated by waters that were originally meteoric in origin, their compositions can be strongly influenced by interactions with magmatic gases.

The distribution of surficial features and their origins is of particular importance to the geologist. The most conspicuous features are found on the upper part of the volcanic edifice, above the main upflow zone. Here, a crater lake, fumaroles, acid-sulfate waters, boiling mud pots, and surficial acid alteration are commonly present (Fig. 2.5(a)). Because the gases rise vertically, they provide information on the locations of permeable zones where boiling is occurring. In some places, acid-sulfate springs may also discharge from perched aquifers marginal to the main upflow zone [29].

Bicarbonate-rich waters form where steam and gas condense into groundwaters that are not in contact with atmospheric oxygen [47]. These acidic waters can be highly corrosive to geothermal wells [31] and readily react with the volcanic host rocks to produce low permeability "umbrella"-shaped regions of argillic alteration characterized by smectite and interlayered clay minerals. Bicarbonate-rich waters often discharge at moderate to low elevations on the stratovolcano, producing bubbling warm springs and travertine ($CaCO_3$). These deposits are interpreted to be indicative of temperatures of less than about 150°C [59].

Because of the steep hydraulic gradients and the formation of marginal clay-rich cap rocks, there is often little evidence of the deep chloride reservoir waters near the upflow zones (Fig. 2.5(a)). However, outflow plumes of chloride water are common, with some extending up to 20 km from the upflow zones along permeable horizons and fault zones. Discharges of chloride waters produce hot springs and geysers that deposit sinter (amorphous silica). Even though the main upflow zone of the system is located beneath the central cone of the stratovolcano, geothermal development typically occurs on the lower slopes of the volcano's flanks. These regions lack direct input of acidic magmatic fluids and can provide access to larger volumes of permeable rocks than sites at higher elevations.

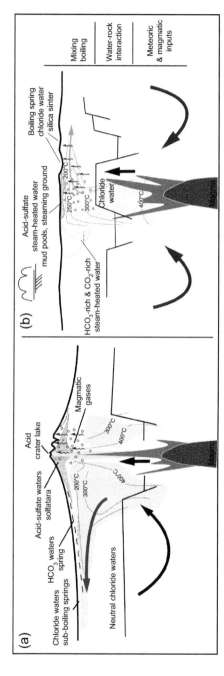

Figure 2.5 Conceptual models of geothermal systems related to volcanic activity (*Modified from Henley RW, Ellis AJ. Geothermal systems ancient and modern: a geochemical review. Earth Sci Rev 1983;19:1—50.*). (a) An andesite volcano. (b) A silicic volcanic system.

The host rocks are typically andesitic to basaltic lava flows, pyroclastic deposits, and volcaniclastic deposits derived from them, although the underlying sedimentary and metamorphic sequences as well as subvolcanic intrusions may also be present. Volcaniclastic deposits, which frequently include thick sequences of debris flows composed of fine-grained ash and rock fragments, are common in many volcanic-hosted geothermal systems. At low to moderate grades of alteration, these deposits are poor reservoir rocks. With more advanced alteration at higher temperatures, where epidote is present, volcaniclastic rocks can maintain open fractures and sustain fluid flow.

Temperatures in the range of 225°C to about 350°C are optimal, as wells at these temperatures will be self-flowing. The corresponding depths can be estimated from boiling point to depth curves for the relevant fluid composition [30]. Typically, the tops of reservoirs range in depth from about 500 to 1000 m, depending on the depth of the overlying conductive regimen.

At temperatures greater than 350−370°C, pressures in excess of hydrostatic may be encountered. These temperature and pressure conditions have been observed in the low permeability regions of deep wells at The Geysers, Nesjavellir (Iceland), and Larderello [24]. Two factors are thought to contribute to these conditions: sealing of fractures by mineral deposition and changes in the behavior of the reservoir rocks from brittle to quasi-plastic [24,25]. The low permeabilities of these high-temperature zones precludes their development as conventional geothermal resources; however, they may be suitable for permeability enhancement (EGS).

The features of geothermal systems developed in silicic volcanic terrains, illustrated in Fig. 2.5(b), are somewhat different. These systems are characterized by a silicic magmatic heat source at depth, a reservoir in which there is convective upflow of chloride reservoir fluids, formation of shallow, peripheral fluids, and a shallow outflow plume. Their differences from stratovolcanic complexes reflect the high relief of the stratovolcanoes, the depth to the heat source, and the mobility of magmatic gases.

Magma-driven geothermal systems developed in continental environments within back-arc rifts (Taupo Volcanic Zone, New Zealand) and continental rift zones (East African Rift Zone) are characterized by silicic volcanism that forms domes and calderas. Host rocks include locally derived pyroclastic deposits and silicic lava flows, although in some systems these rocks are absent and the reservoirs are developed in the underlying basement rocks.

2.5.2 Vapor-dominated geothermal systems

Although vapor-dominated geothermal systems are not common, they are important resources because of their size and ability to produce dry steam. The largest of these systems overlie young plutons at The Geysers, United States, with a current output of 850 MWe and Larderello, Italy, which produces nearly 600 MWe [57] (see also chapters: Larderello: 100 years of geothermal power plant evolution in Italy and Fifty-five years of commercial power generation at the Geysers geothermal field, California: the lessons learned). Smaller volcanic-hosted vapor-dominated systems have been described in West Java and include Dieng, Darajat, Kamojang, Karaha-Telaga Bodas, and Wayang Windu [9].

Vapor-dominated systems can be modeled as heat pipes [60,67]. Within the reservoir, steam and gas migrate upward along fractures and condense at the base of the cap rock. Condensate that drains down the fracture walls is re-boiled. No deep boiling water table has yet been encountered at The Geysers, despite drilling to depths of nearly 4000 m. This suggests all of the liquid originally in the fractures has boiled off [71]. Fumaroles and acid alteration occur where H_2S and steam reach the surface.

The main features of The Geysers are shown in Fig. 2.6. The reservoir is developed primarily in metagraywacke of the Mesozoic Franciscan Complex beneath a low permeability cap rock of argillite and serpentinite [51]. $^{40}Ar/^{39}Ar$ dating suggests the hydrothermal system initially developed 1.5−2 My ago in response to the intrusion of a hypabyssal granitic pluton [14]. Although wells produce only steam, mineralogical, fluid inclusion and isotope data demonstrate the present vapor-dominated conditions evolved from an earlier, more-extensive liquid-dominated system [54].

The vapor-dominated regimen consists of two hydraulically connected steam reservoirs [72]. Within the upper normal-temperature reservoir, pressures are vapor static (about 3.5 MPa) and temperatures are near 240°C. These conditions are close to the maximum enthalpy of saturated steam and typical of many vapor-dominated systems. At greater depths, in the northern third of The Geysers, temperatures range from 240 to greater than 400°C [45]. The high temperatures suggest heat is supplied by a recent intrusion. Wells drilled into this hotter, high-temperature reservoir produce corrosive chloride-bearing steam.

The persistence of vapor-dominated conditions requires that discharge exceeds recharge [76]. Because vapor-dominated systems are inherently unstable due to the lower steam pressure of the reservoir relative to the surrounding liquid-saturated

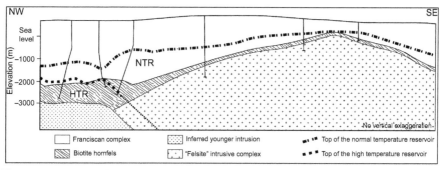

Figure 2.6 Cross section of The Geysers geothermal field showing the vapor-dominated reservoirs and major rock types. *NTR*, normal-temperature reservoir; *HTR*, high-temperature reservoir.

Modified from Lowenstern JB, Janik CJ. The origins of reservoir liquids and vapors from the Geysers geothermal field, California. In: Simmons SF, Graham I, editors. Volcanic, geothermal, and ore-forming fluids: rulers and witnesses of processes within the earth, vol. 10. Society of Economic Geologists, Special Publication; 2003. p. 181−95; Lutz SJ, Walters M, Moore JN, Pistone S. New insights into the high-temperature reservoir, northwest Geysers: Geotherm Resour Counc Trans 2012;36:907−16.

regimen, flooding will eventually occur, leading to the formation of liquid-dominated conditions. Mineral sealing of marginal fractures by calcite and anhydrite can inhibit recharge rates.

2.6   Sediment-hosted geothermal systems

Although many geothermal reservoirs are hosted in volcanic rocks, there are also important occurrences of moderate- and high-temperature systems in sedimentary basins. In North America, the geothermal systems of the Salton Trough of southern California and northern Mexico are examples of this resource type (Fig. 2.7).

The Salton Trough is a 150- to 300-km-wide structural depression, with surface elevations mostly about 40 m below sea level, that trends northwest–southeast, between the transform tectonic regimen of the San Andreas fault system to the north and the *trans*tensional tectonic regimen of the Gulf of California to the south. In the last 5 million years (Ma), 5–6 km of sediments have been deposited by the Colorado River forming a delta that has isolated the Salton Trough from the marine environment of the Gulf [16,43].

The installed capacity of the geothermal systems in the Salton Trough currently exceeds 1500 MWe, with the Salton Sea and Cerro Prieto fields being among the

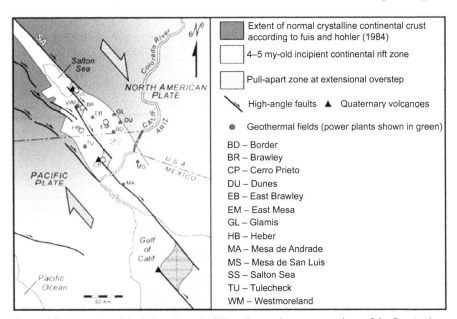

Figure 2.7 Tectonics of the Salton Trough (ST) at the southeastern terminus of the San Andreas transform fault zone (SA) and the northern end of the Gulf of California. Geothermal systems shown by red and green dots.
Modified from Hulen JB, Kasperit D, Norton DL, Osborn W, Pulka FS. Refined conceptual modeling and a new resource estimate for the Salton Sea geothermal field, Imperial Valley, California. Geotherm Resour Counc Trans 2002;26:29−36.

hottest and most productive of the world's liquid-dominated geothermal reservoirs. The Salton Sea field has an installed nameplate capacity of 437 MWe with another 396 MWe in the planning stage. Estimated thermal energy outputs are of the order of 10^3-10^4 MWt, and one well (production casing of $13^3/_8$-inch diameter) is capable of producing about 50 MWe [38]. Cerro Prieto has an installed nameplate capacity of 720 MWe, but due to pressure declines after 42 years of production, it currently produces only about 515 MWe (see chapter: Enhanced geothermal systems: review and status of research & development). These high-temperature fields are located in pull-apart zones formed where a series of active, right-lateral, transform faults step to the right, Outside these pull-apart zones moderate-temperature geothermal fields, typically with temperatures of about 200°C at less than 3000-m depth, such as Heber, with 180 MWe, and East Mesa, with 115 MWe, are associated with regional faulting.

The geothermal fluids of the Salton Trough range from relatively dilute to hypersaline chloride brines. Most have about 30,000 mg/kg TDS but the most extreme concentration is the hypersaline brine of the Salton Sea, where production fluids can contain up to 30 wt% TDS, primarily NaCl, $CaCl_2$, and KCl. The Salton Sea brines have a Cl/Br ratio of 1600, identical to that of the Colorado River, which has a TDS of only 800 mg/kg. These relationships suggest the Salton Sea brines originated by dissolution of nonmarine evaporates formed in lakes during periods of flooding by the Colorado River [61]. The deeper and hotter fluids are rich in Fe, Mn, Zn, Li, and Pb [49]. Pilot studies have shown high-purity Zn and Li can be extracted commercially. The fluids at Cerro Prieto, with 28,000 mg/kg TDS, have Cl:Br ratios of about 300, similar to that of seawater, and are interpreted to have formed in high-salinity coastal lagoons [42].

The reservoir rocks in these developed geothermal fields consist of sandstones, siltstones, and shales that are overlain by lake sediments deposited after the basin was isolated from the marine environment of the Gulf of California [35,58]. These lake sediments form effective seals over the geothermal systems, and except for the Cerro Prieto and the Salton Sea, the geothermal systems have no surface expressions. They were discovered during regional oil and gas surveys and by the presence of positive residual gravity anomalies due to densification produced by hydrothermal alteration [5].

Despite the difference in fluid salinity, reservoir rocks at Cerro Prieto and Salton Sea have similar alteration mineralogy. Fluid production comes primarily from hydrothermally altered sandstone aquifers separated by discontinuous altered mudstones. The unaltered sandstones are cemented with calcite and, less commonly, anhydrite. With increasing temperatures, a series of decarbonation reactions occur that create secondary porosity. At temperatures of 290−310°C, calcite, illite, and quartz react with the brine to form Fe-epidote and K-feldspar with further release of CO_2. Three distinct metamorphic zones have been recognized: (1) a chlorite-calcite zone between about 190 and 300°C; (2) a biotite zone between about 325°C and about 360°C; and (3) a deeper, hotter, garnet-clinopyroxene zone [18,48]. At the highest temperatures, the clastic textures are completely destroyed and recrystallization has formed hornfelsic textures; there is a transition from porosity to fracture-dominated permeability.

Detailed petrologic studies at Cerro Prieto have aided investigation into the patterns of fluid flow before development of the field. Based on estimates of water: rock ratios and the heat emplaced during the approximately 50 ka that the system appears to have been active, the heat source was modeled as a basaltic intrusion at least 4−5 km wide emplaced at depths of 5−6 km, to the northeast of the producing field. This model is consistent with the presence of gravity and magnetic anomalies in that location [18].

At the Salton Sea, the heat source is clearly related to five small rhyolite domes occurring in an arc along the northern margin of the producing field. Rhyolitic intrusive and extrusive rocks are also encountered during drilling [38]. Densification of the sediments and the presence of igneous intrusions have produced a residual Bouguer gravity anomaly of +23 mGal [19]. U-Th-Pb dating of zircons in both intrusive and extrusive rhyolites indicates a range of zircon crystallization ages from 30 to 9 ka, and U-series disequilibrium in the brines suggests similar ages for the geothermal system [66].

2.7   Extensional tectonic geothermal systems

In contrast to the high-temperature geothermal systems found on the flanks of young volcanoes and intrusive centers, many geothermal systems with temperatures less than about 225°C show no apparent connection to young volcanism. This is the situation in the Basin and Range Province of the western United States (Fig. 2.8), a broad region of extension between stable cratonic blocks to the east and west [13]. This extension has produced fault bounded northerly trending mountain ranges and valleys [15]. Similar tectonically active extensional environments are found in western Turkey and the upper Rhine Graben [21,53].

More than 20 geothermal power plants with a total nameplate capacity of nearly 1000 MWe are currently operating in the Basin and Range Province, and of these, only three, Coso Hot Springs, Long Valley, and Roosevelt Hot Springs, can be directly linked to recent magmatic activity. The reservoirs are developed primarily in Paleozoic and Mesozoic intrusive, sedimentary, and metasedimentary rocks. Permeability and fluid flow are controlled by recently active faults (Fig. 2.9), particularly in the vicinity of complex fault geometries [39]. Many of these systems have fossil sinter and travertine deposits, but the flow rates of active springs are characteristically low. Although most of the valleys where the geothermal systems are found receive little precipitation, the hot spring deposits suggest higher flow rates in the past. The oldest ^{14}C ages of pollen and plant fragments trapped in hot spring deposits at Steamboat Hot Springs, Beowawe, and McGinness Hills range from about 11 to 15 ka. [46] (Wannamaker and Moore, unpublished data, 2014). These dates are interpreted to indicate early spring activity was related to an elevated water table occurring during the last glacial period, a conclusion consistent with stable isotopic data suggesting the geothermal waters originated during a colder climate [3].

Numerous thermal springs (>35°C), warm mineral exploration holes, and hot deep oil and gas wells suggest the currently producing systems may represent only a small

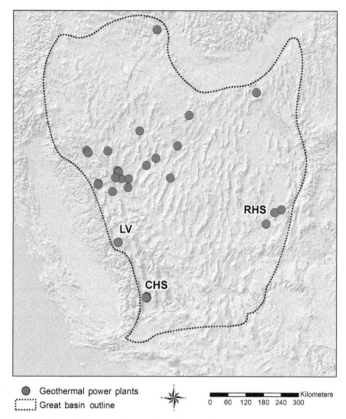

Geothermal power plants
Great basin outline

Kilometers
0   60   120 180 240 300

Figure 2.8 Distribution of power plants of geothermal systems in the Basin and Range. *CHS*, Coso Hot Springs; *LV*, Long Valley; *RHS*, Roosevelt Hot Springs.
Modified from Faulds JE, Hinz NH. Favorable tectonic and structural settings of geothermal systems in the Great Basin region, western USA: proxies for discovering blind geothermal systems: Proceedings World Geothermal Congress 2015, 2015, Melbourne, Australia.

fraction of the geothermal potential [1,22]. Measurements from these oil and gas wells indicate temperatures above 150°C exist at depths of 3−4 km in the carbonate and sandstone sequences beneath the valleys and may constitute a large, untapped resource.

Regionally, the Basin and Range Province is characterized by high strain rates and high heat flows, which typically range from 75 to 100 mW/m^2. In general, strain rates of 10^{-15} s^{-1} and higher characterize areas with electric grade resources, whereas lower strain rates of 10^{-16} s^{-1} are typical of less productive areas [7,8,41].

Kennedy and van Soest used geochemical data to demonstrate that high strain rates could be correlated with regions of enhanced permeability. They measured higher concentrations of mantle-derived ^{3}He in geothermal fluids from systems related to high strain rates than those where strain rates are lower [40]. These ^{3}He enrichments were interpreted to reflect the presence of deep, permeable structures that allowed

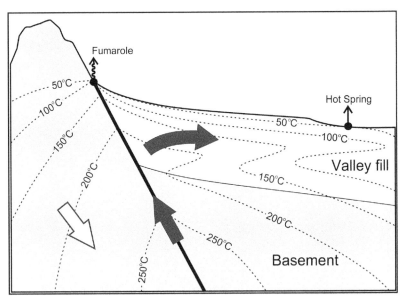

Figure 2.9 Idealized conceptual model of a fault-controlled geothermal system. Hot fluids move upward along the fault and then laterally through the valley fill as an outflow plume.

both the upward movement of mantle gases and the circulation of meteoric fluids to deeper, hotter portions of the crust.

The existence of the pathways interpreted from the He isotope geochemistry is corroborated by steeply dipping zones of low resistivity that have been mapped to middle and lower crustal levels [74]. These zones connect to quasi-tabular regions of low resistivity in the deep crust inferred to represent modern magmatic underplating and high-temperature fluid release. This underplating is interpreted to be the result of upper mantle adiabatic upwelling, fusion, and melt egress to the crust during regional extension and the regionally elevated heat flow values [52,73].

2.8   Unconventional geothermal resources

Among geothermal resources that we term "unconventional" are enhanced geothermal systems (EGS), supercritical geothermal resources, and resources in ductile, high-enthalpy, geological reservoirs, all of which are currently in different stages of development but have the potential to make considerable contributions in future.

Enhanced geothermal resources, which may be found at any depth, have temperatures potentially within the economic range but lack the natural permeability necessary for the production of commercial quantities of geothermal fluids. The concept is to create permeability by hydrofracturing the hot rocks, inject water into the fractures created, and then drill production wells into the newly created reservoir. Low-permeability hot rocks suitable for EGS are believed to be widespread [70]. Despite efforts during the past four decades in several countries, large-scale EGS development has proved to be elusive. However, there are a number of successful small experimental plants operating, the oldest of which is at Soultz-sous-Forêts in the Rhine Graben of France, where a binary plant has been producing 3 MWe from a 200°C EGS granite basement at a depth of 5 km [33]. EGS hold significant promise for future electric generation if and when the technological and economic problems are solved. Chapter "Enhanced geothermal systems: review and status of research & development" covers EGS in depth.

Supercritical geothermal resources require identifying and producing geothermal fluids at pressures and temperatures above the critical point. With increasing temperature and pressure, the properties of liquid water and vapor phases converge until at the critical point of 374°C and 22.2 MPa (for pure water) only a single, high-enthalpy supercritical phase exists. Supercritical water at 400°C has five times the available energy of water at 250°C. However, dissolved solids and gases affect both the boiling point and critical point of water. For example, the critical point of seawater (3.3% TDS) occurs at 407°C and 29.8 MPa. The first practical attempts to investigate the technical and the economic feasibility of producing electricity from supercritical geothermal resources is being carried out by the Iceland Deep Drilling Project (IDDP, see www.iddp.is). In 2009, the IDDP attempted to drill a 5-km-deep supercritical well in the Krafla basaltic caldera in northeast Iceland. However, drilling had to be terminated at only 2.1 km depth when 900°C rhyolite magma flowed into the well. The well was highly productive and capable of generating more than 35 MWe from superheated steam at a well-head temperature of about 450°C [20,27]. In 2016, the IDDP will drill a 4.5-km-deep well in a high-temperature geothermal field in southwest Iceland on the Reykjanes Peninsula, the landward extension of the Mid-Atlantic Ridge [26]. This well will penetrate the roots of a hydrothermal system similar to those that produce black smokers on mid-ocean ridges. Perhaps in the future, it will be economically feasible to develop the extensive submarine geothermal systems associated with ocean spreading centers [4,36].

The Japan Beyond the Brittle Project (JBBP), a very ambitious endeavor being developed in northeast Japan, is an extension of both the concepts of EGS and producing deep supercritical geothermal resources. JBBP is aimed at creating an EGS reservoir in greater than 500°C rocks. The plan is to drill beyond the brittle zone into ductile greater than 500°C neogranite, fracture the hot rock by thermo-hydroshearing, and so create a self-contained, supercritical enhanced geothermal system. The hope is to use a combination of government and industry funding to commence drilling in 2016−2017 [2].

2.9 Conclusions

During the past decade, 2005−2015, considerable effort has been directed toward the development and application of new geophysical and geochemical tools for geothermal assessment. These investigations have not overshadowed the application of basic geologic techniques, such as field mapping and investigations of secondary minerals, in the evaluation of geothermal systems. Instead, it has become even more important that geologists work as part of a team including geochemists, geophysicists, and reservoir engineers (see chapters: Geophysics and resource conceptual models in geothermal exploration and development and Application of geochemistry to resource assessment and geothermal development projects).

A broad spectrum of geological techniques can be applied to geothermal systems, including remote sensing, dating, analysis of the mechanical properties of the rocks and stress fields, and the coupling of thermomechanical and hydrological interactions. In this chapter, we have emphasized the study of the formation and interpretation of the mineral assemblages formed through interactions between the rocks and geothermal fluids, because of the proven utility of that approach. Despite differences in the individual characteristics of geothermal systems, consistent patterns of hydrothermal minerals occur in diverse geothermal systems with different host rocks and fluid chemistries. Secondary minerals, particularly those deposited in veins and open pore space, provide information on fluid temperature, chemistry, permeability, and fluid processes. Microthermometric measurements on samples of fluids trapped in fluid inclusions during mineral deposition and later fracturing can further elucidate the temperatures, salinities, gas contents, and evolution of the circulating waters.

These geological techniques should be integrated with complementary geochemical and geophysical investigations. For example, water—rock reactions have a strong effect on the physical and electrical properties of rocks and thus influence application of geophysical surveys and drilling programs. Even where secondary mineralization is weak, variations in mineral abundances can assist in developing a reservoir model. At the earliest stages of an exploration project, knowledge of the surficial deposits can be integrated with simple conceptual models to guide and interpret geophysical data and target thermal gradient wells. As the project progress, these conceptual models can then be updated to incorporate new geoscientific, drilling, and reservoir engineering data.

The development of unconventional geothermal resources, including those in low permeability rocks (EGS), deep sedimentary basins, and around still-cooling magma bodies will require an expansion of current geological techniques. In these environments, the mechanical properties of the reservoir rocks and the effects of the in situ stress field will be of critical importance. Rock mineralogy will still play an important role, as it can affect both permeability formation and destruction.

References

[1] Allis G, Gwynn M, Hardwick C, Mines G, Moore J. Will stratigraphic reservoirs provide the next big increase in U.S. geothermal power generation? Geotherm Resour Counc Trans 2015;39.

[2] Asanuma H, Tuschiya N, Muraoka H, Ito H. Japan Beyond-Brittle Project: development of EGS beyond brittle-ductile transition. In: Proceedings World Geothermal Congress 2015, Melbourne, Australia; April 19—25, 2015.

[3] Ayling B, Moore JN. Fluid geochemistry at the Raft River geothermal field, Idaho, USA: new data and hydrogeological implications. Geothermics 2013;47:116−26.

[4] Baker ET, German CR. On the global distribution of hydrothermal vent fields. In: German CR, Parsons LM, editors. Mid-ocean ridges: hydrothermal interactions between the lithosphere and Oceans. Geophysical monograph series, vol. 148; 2004. p. 245−66.

[5] Biehler S, Kovach RL, Allen CR. Geophysical framework of the northern end of the Gulf of California structural province. In: van Andel TH, Shor GG, editors. Marine geology of the Gulf of California. American Association of Petroleum Geologists, Memoir 3; 1964. p. 126−43.

[6] Bird DK, Schiffman P, Elders WA, Williams AE, McDowell SD. Calc-silicate mineralization in active geothermal systems. Econ Geol 1984;79:671−95.

[7] Blackwell DD. Heat flow in the northern Basin and Range province. In: The role of heat in the development of energy and mineral resources in the Northern Basin and Range province. Geothermal Resources Council Special Report, vol. 13; 1983. p. 81−92.

[8] Blewitt G, Coolbaugh MF, Holt W, Kreemer C, Davis JL, Bennett RA. Targeting of potential geothermal resources in the Great Basin from regional relationships between geodetic strain and geological structures. Geotherm Resour Counc Trans 2002;26:523−5.

[9] Bogie I, Ussher G, Lovelock B, Mackenzie K. Finding the productive sweet spots in the vapour and transitional vapour-liquid dominated geothermal fields of Java, Indonesia. In: Proceedings world geothermal Congress 2015, Melbourne Australia; 2015. 10 p.

[10] Browne PRL, Ellis AJ. The Ohaki-Broadlands hydrothermal area, New Zealand: mineralogy and related geochemistry. Am J Sci 1970;269:97−131.

[11] Browne PRL. Hydrothermal alteration in active geothermal fields. Annu Rev Earth Planet Sci 1978;6:229−50.

[12] Christenson BW. Gases in volcanic lake environments. In: Rouwet D, Christenson B, Tassi F, Vandemeulebrouck J, editors. Volcanic Lakes. Springer-Verlag Berlin Heidelberg; 2015, ISBN 978-3-642-36832-5. p. 125−53. http://dx.doi.org/10.1007/978-3-642-36833-2.

[13] Coolbaugh M, Zehner R, Kreemer C, Blackwell D, Oppliger G, Sawatzky D, et al. Geothermal potential map of the Great Basin, western United States. Nev Bureau Mines Geol Map 2005;151.

[14] Dalrymple GB, Grove M, Lovera OM, Harrison TM, Hulen JB, Lanphere MA. Age and thermal history of the Geysers plutonic complex (felsite unit), Geysers geothermal field, California: a $^{40}Ar/^{39}Ar$ and U-Pb study. Earth Planet Sci Lett 1999;173:285−98.

[15] Eaton GP. The Basin and Range province: origin and tectonic significance. Annu Rev Earth Planet Sci 1982;10:409−40.

[16] Elders WA, Rex RW, Meidav T, Robinson PT, Biehler S. Crustal spreading in southern California. Science 1972;178:15−24.

[17] Elders WA. Petrology as a practical tool, in geothermal studies. Geotherm Resour Counc Trans 1977;1:85−7.

[18] Elders WA, Williams AE, Bird DK, Schiffman P. Hydrothermal flow regime and magmatic heat source of the Cerro Prieto geothermal system, Baja California, Mexico. Geothermics 1984;13:27−47.

[19] Elders WA, Sass JH. The Salton Sea Scientific Drilling Project. J Geophys Res 1988;93: 12953−68.

[20] Elders WA, Friðleifsson GÓ, Albertsson A. Drilling into magma and the implications of the Iceland Deep Drilling Project (IDDP) for high-temperature geothermal systems worldwide. Geothermics 2014;49:111−8.

[21] Faulds JE, Bouchot V, Moeck I, Oguz K. Structural controls of geothermal systems in western Turkey: a preliminary report. Geotherm Resour Counc Trans 2009;33:375−83.

[22] Faulds JE, Hinz NH. Favorable tectonic and structural settings of geothermal systems in the Great Basin region, western USA: proxies for discovering blind geothermal systems. In: Proceedings World Geothermal Congress 2015, Melbourne, Australia; 2015.

[23] Fournier RO. The behavior of silica in hydrothermal solutions. In: Berger BR, Bethke PM, editors. Geology and geochemistry of epithermal systems: reviews in economic geology, 1985;2:45−62.

[24] Fournier RO. The transition from hydrostatic to greater than hydrostatic fluid pressure in presently active continental hydrothermal systems in crystalline rock. Geophys Res Lett 1991;18:955−8.

[25] Fournier RO. Hydrothermal processes related to movement of fluid from plastic into brittle rock in the magmatic-epithermal environment. Econ Geol 1999;94:1193−211.

[26] Friðleifsson GÓ, Elders WA, Bignall G. A plan for a 5 km-deep borehole at Reykjanes, Iceland, into the root zone of a black smoker on land. Sci Drill 2013;16:73−9.

[27] Friðleifsson GÓ, Elders WA, Albertsson A. The concept of the Iceland deep drilling project. Geothermics 2014;49:2−8.

[28] Giggenbach WF. The origin and evolution of fluids in magmatic-hydrothermal systems. In: Barnes HL, editor. Geochemistry of hydrothermal ore deposits. 3rd ed. New York: John Wiley & Sons, Inc.; 1997. p. 737−96.

[29] Giggenbach WF, Garcia PN, Londono CA, Rodriquez VLA, Rojas GN, Calvache VML. The chemistry of fumarolic vapor and thermal spring discharges from the Nevado del Ruiz volcanic-magmatic hydrothermal system, Colombia. J Volcanol Geotherm Res 1990;42: 13−39.

[30] Haas Jr JL. The effect of salinity on the maximum thermal gradient of a hydrothermal system at hydrostatic pressure. Econ Geol 1971;66:940−6.

[31] Hedenquist JW, Stewart MK. Natural CO_2-rich steam-heated waters in the Broadlands-Ohaaki geothermal system, New Zealand. Geotherm Resour Counc Trans 1985;9:245−50.

[32] Hedenquist JW, Reyes AG, Simmons SF, Taguchi S. The thermal and geochemical structure of geothermal and epithermal systems: a framework for interpreting fluid inclusion data. Eur J Mineral 1992;4:989−1015.

[33] Held S, Genter A, Kohl T, Kobel TH, Schoenball M. Economic evaluation of geothermal reservoir performance through modeling the complexity of the operating EGS in Soultz-sous-Forêts. Geothermics 2014;51:270−80.

[34] Henley RW, Ellis AJ. Geothermal systems ancient and modern: a geochemical review. Earth Sci Rev 1983;19:1−50.

[35] Herzig CT, Mehegan JM, Stelting CE. Lithostratigraphy of the state 2-14 borehole, Salton Sea Scientific Drilling Project. J Geophys Res 1988;93:12969−80.

[36] Hiriart G, Hernández I. Electricity production from hydrothermal vents. Geotherm Resour Counc Trans 2010;34:1033−7.

[37] Holland HD, Malinin SD. The solubility of occurrence of non-ore minerals. In: Barnes HL, editor. Geochemistry of hydrothermal ore deposits. 2nd ed. New York: John Wiley & Sons; 1979. p. 461−508.

[38] Hulen JB, Kasperit D, Norton DL, Osborn W, Pulka FS. Refined conceptual modeling and a new resource estimate for the Salton Sea geothermal field, Imperial Valley, California. Geotherm Resour Counc Trans 2002;26:29−36.

[39] Jolie E, Moeck I, Faulds JE. Quantitative structural-geological exploration of fault-controlled geothermal systems-A case study from the Basin and Range Province Nevada (USA). Geothermics 2015;54:54−67.

[40] Kennedy BM, van Soest MC. Flow of mantle fluids through the ductile lower crust: helium isotope trends. Science 2007;318:1433−6.

[41] Lachenbruch AH, Sass JH. Heat flow of the United States and thermal regime of the crust. In: Heacock JG, editor. The nature and physical properties of the Earth's crust, vol. 20. American Geophysical Union Monograph; 1977. p. 626−75.

[42] Lippmann MJ, Truesdell AH, Halfman SE, Mañon A. A review of a hydrogeological and geochemical model for Cerro Prieto. Geothermics 1991;20:39−52.

[43] Lonsdale P. Geology and tectonic history of the Gulf of California. In: Winterer EL, Husong DM, Decker RW, editors. The geology of North America, v. N, the Eastern Pacific Ocean and Hawaii. Geological Society of America; 1989. p. 499−521.

[44] Lowenstern JB, Janik CJ. The origins of reservoir liquids and vapors from the Geysers geothermal field, California. In: Simmons SF, Graham I, editors. Volcanic, geothermal, and ore-forming fluids: rulers and witnesses of processes within the earth, vol. 10. Society of Economic Geologists, Special Publication; 2003. p. 181−95.

[45] Lutz SJ, Walters M, Moore JN, Pistone S. New insights into the high-temperature reservoir, northwest Geysers. Geotherm Resour Counc Trans 2012;36:907−16.

[46] Lynne BY, Campbell KA, Moore JN, Browne PRL. Origin and evolution of the Steamboat Springs siliceous sinter deposit, Nevada, U.S.A. J Sediment Geol 2008. http://dx.doi.org/ 10.1016/j.sedgeo.2008.07.006.

[47] Mahon WAJ, Klyen LE, Rhode M. Natural sodium-bicarbonate-sulphate hot waters in geothermal systems. Chinetsu J Jpn Geotherm Energy Assoc 1980;17:11−24.

[48] McDowell SD, Elders WA. Authigenic layer silicate minerals in borehole Elmore No 1, salton sea geothermal field, California, USA. Contrib Mineral Petrol 1980;74:293−310.

[49] McKibben MA, Elders WA. Fe-Zn-Cu-Pb mineralization in the Salton Sea geothermal system. Econ Geol 1985;80:511−23.

[50] McKibben MA, Hardie LA. Ore forming brines in active continental rifts. In: Barnes HL, editor. Geochemistry of hydrothermal ore solutions. 3rd ed. New York: John Wiley; 1997. p. 877−935.

[51] McLaughlin RJ. Tectonic setting of pre-Tertiary rocks and its relation to geothermal resources in the Geysers-Clear Lake area. In: McLaughlin RJ, Donnelly-Nolan JM, editors. Research in the Geysers-Clear Lake geothermal area, Northern California. USGS Prof. Pap. 1141; 1981. p. 25−45.

[52] Meqbel NM, Egbert GD, Wannamaker PE, Kelbert A, Schultz A. Deep electrical resistivity structure of the Pacific NW derived from 3-D inversion of Earthscope US Array magnetotelluric data. Earth Planet Sci Lett 2014. http://dx.doi.org/10.1016/j.epsl. 2013.12.026.

[53] Moeck IS. Catalog of geothermal play types based on geologic controls. Renew Sustain Energy Rev 2014;37:867−82.

[54] Moore JN, Gunderson RP. Fluid inclusion and isotopic systematics of an evolving magmatic-hydrothermal system. Geochim Cosmochim Acta 1995;59:3887−907.

[55] Moore JN, Powell TS, Heizler MT, Norman DI. Mineralization and hydrothermal history of the Tiwi geothermal system, Philippines. Econ Geol 2000;95:1001−23.

[56] Moore JN, Allis RG, Nemčok M, Powell TS, Bruton CJ, Wannamaker PE, et al. The evolution of volcano-hosted geothermal systems based on deep wells from Karaha − Telaga Bodas, Indonesia. Am J Sci 2008;308:1−48.

[57] Moore JN, Simmons SF. More power from below (invited perspective). Science 2013;380: 933. http://dx.doi.org/10.1126/science.1235640.

[58] Muffler LJP, Doe BR. Composition and mean age of detritus of the Colorado River delta in the Salton Trough, southeastern California. J Sediment Petrol 1968;38:384−99.

[59] Nicholson K. Geothermal fluids: chemistry and exploration techniques. Berlin: Springer Verlag; 1993. 263 p.

[60] Pruess K. A quantitative model of vapor dominated geothermal reservoirs as heat pipes in fractured porous rock. Geotherm Resour Counc Trans 1985;9:353−61.

[61] Rex RW. The origin of brines of the Imperial Valley, California. Trans Geotherm Resour Counc 1983;7:321−4.

[62] Reyes AG. Petrology of Philippine geothermal systems and the application of alteration mineralogy to their assessment. J Volcanol Geotherm Res 1990;43:279−309.

[63] Reyes AG. Mineralogy, distribution and origin of acid alteration in Philippine geothermal systems. In: Matsuhisa Y, Masahiro A, Hedenquist J, editors. High-temperature acid fluids and associated alteration and mineralization. Geological survey of Japan Report, vol. 277; 1991. p. 59−65.

[64] Reyes AG, Giggenbach WF, Saleras JRM, Salonga ND, Vergara MC. Petrology and geochemistry of Alto Peak, a vapor-cored hydrothermal system, Leyte Province, Philippines. Geothermics 1993;22:479−519.

[65] Sasada M. CO_2 − bearing fluid inclusions from geothermal fields. Geotherm Resour Counc Trans 1985;5:351−6.

[66] Schmitt AK, Vazquez JA. Alteration and remelting of nascent oceanic crust during continental rupture: evidence from zircon geochemistry of rhyolites and xenoliths from the Salton Trough, California. Earth Planet Sci Lett 2006;252:260−74.

[67] Shook GM. Development of a vapor-dominated reservoir with a "high-temperature" component. Geothermics 1995;24:489−505.

[68] Simmons SF, Christenson BW. Origins of calcite in a boiling geothermal system. Am J Sci 1994;294:361−400.

[69] Stoffregen RE. Genesis of acid-sulfate alteration and Au-Cu-Ag mineralization at Summitville, Colorado. Econ Geol 1987;82:1575−91.

[70] Tester JW, Anderson BJ, Batchelor AS, Blackwell DD, DiPippo R, Drake EM, et al. Impact of enhanced geothermal systems on US energy supply in the twenty-first century. Philos Trans R Soc Lond A 2007;365:1057−94.

[71] Truesdell AH, Walters M, Kennedy M, Lippmann M. An integrated model for the origin of the Geysers geothermal field. Geotherm Resour Counc Trans 1993;9:273−80.

[72] Walters MA, Haizlip JR, Sternfeld JN, Drenick AF, Combs J. A vapor-dominated reservoir exceeding 600°F at the Geysers, Sonoma County California. In: Proceedings 13th workshop on geothermal reservoir engineering. Stanford University; 1988. p. 73−81.

[73] Wannamaker PE, Hasterok DP, Doerner WM. Possible magmatic input to the Dixie Valley geothermal field, and implications for district-scale resource exploration, inferred from magnetotelluric (MT) resistivity surveying. Geotherm Resour Counc Trans 2006;30: 71−475.

[74] Wannamaker PE, Hasterok DP, Johnston JM, Stodt JA, Hall DB, Sodergren TL, et al. Lithospheric dismemberment and magmatic processes of the Great Basin-Colorado Plateau transition, Utah, implied from magnetotellurics. Geochem Geophys Geosystems 2008;9: Q05019. http://dx.doi.org/10.1029/2007GC001886.

[75] White DE. Characteristics of geothermal systems. In: Kruger P, Otte C, editors. Geothermal energy: resources, production, stimulation. Stanford University; 1973. p. 69−94. ISBN-10: 0804708223; ISBN-13: 978-0804708227 [chapter 4].

[76] White D, Muffler LPJ, Truesdell AH. Vapor-dominated hydrothermal systems compared with hot-water systems. Econ Geol 1971;66:75.

Geophysics and resource conceptual models in geothermal exploration and development

3

W. Cumming
Cumming Geoscience, Santa Rosa, CA, United States

3.1 Introduction

In a manner analogous to the use of geophysics in other contexts like petroleum and mineral exploration, geophysical applications to geothermal resource exploration had an initial phase that emphasized targeting anomalous data, and that evolved into approaches to assessing target and capacity risk using conceptual models [61]. In the 1980s and 1990s, parts of the geothermal industry increasingly emphasized targeting geothermal resource conceptual models supported by geoscience data, rather than targeting the data itself, along with an increasing geophysical emphasis on magnetotelluric (MT) and time domain electromagnetic (TEM) resistivity surveys. A conceptual model approach using combined MT-TEM surveys was promoted as an international best practice for geothermal exploration geophysics in 2000 [16]. Geothermal resource risk assessment approaches adapted from the petroleum industry [61] were more routinely integrated with geothermal conceptual model assessments in the late 1980s. Standards for reporting geothermal reserves to investors in Australia and Canada were adopted in 2010 that emphasized a frequentist statistical process instead of a conceptual model approach, but, after serious shortcomings became apparent [31], best practices for estimation of geothermal reserves have been under review through 2015. However, most large geothermal field developers have continued to base high-value geothermal resource decisions on risk assessments supported by resource conceptual models, and, when well test data are available, by numerical reservoir simulation models [32]. MT remains the dominant geophysical method used to support exploration resource decisions for volcano-hosted [21] and most sediment-hosted geothermal resources [17]; reflection seismic is crucial to targeting deep stratigraphic reservoirs [76]; and precision gravity [6] and seismic monitoring [63] provide important constraints on numerical reservoir simulation models of developed fields.

Geothermal prospects that best fit the paradigm for assessment based on conceptual models and particular geophysical methods like MT are given higher priority for drilling. As a result, the remaining prospects are necessarily poorer fits to this paradigm. For most regions, this implies a trend toward increasing uncertainty, more pessimistic resource capacity assessments and higher risk targeting unless the paradigm changes. An increase in power price can make less-promising resource assessments more

Geothermal Power Generation. http://dx.doi.org/10.1016/B978-0-08-100337-4.00003-6

attractive. However, there is opportunity related to the increasing uncertainty, to the extent that the geothermal industry can extend or develop geophysical technology to better discriminate among the remaining prospects. For example, resource opportunities that are currently more speculative, like deep stratigraphic prospects, may make limited use of MT and gravity surveys to characterize suitable settings before using three-dimensional (3D) seismic reflection to target wells.

Several reviews have summarized geophysical methods that have been routinely applied or tested for geothermal applications. A widely cited 1985 summary of the geothermal geophysics state-of-the-art [82] provides context for early geophysical successes in geothermal exploration but includes misconceptions about geothermal resistivity patterns that commonly confuse investigators unfamiliar with resistivity in a geothermal context. The United Nations University (UNU) in Reykjavik has produced many valuable sets of lecture notes and summaries of geothermal geoscience applications including a 2009 illustrated introduction to geophysics in the context of basalt-hosted geothermal reservoirs of Iceland [30] that updates a 1991 overview [36]. Details for resistivity applications in the same context are outlined in Ref. [35]. The Geothermal Energy Association (GEA) sponsored a 2007 assessment of geophysical applications for nonspecialists with a focus on the western United States [71]. The construction of geothermal resource conceptual models based on resistivity surveys interpreted in the context of geology and geochemistry data is detailed in Ref. [21] for volcano-hosted prospects associated with fumaroles and in Ref. [17] for a sediment-hosted prospect with no thermal fluids at the surface.

In recent years, reports have been commissioned by government and industry groups to summarize general best practices for the integration of geophysics with other geoscience data in support of geothermal resource decisions. The International Geothermal Association (IGA) online geothermal conference database is widely used by geothermal investigators to look for geothermal-specific technical information and analog case histories. The underlying research reports and source data for many publications are held in the United States Department of Energy (DOE) databases summarized in Ref. [83], that also outlines the DOE's general best practice recommendations in 2012 for geophysics and geoscience applications to geothermal development. The 2014 report by IGA Service GmbH [41] is a more comprehensive review of researchers' and consultants' views on geothermal assessment best practices for geothermal developers worldwide. The 2014 U.S. DOE "Geothermal Best Practices Manual" [75] does not focus on specific geophysical methods but on a general conceptual model approach to using geophysics that is likely to be successful in commercial applications worldwide. Consistent with these recently recommended best practices, the following discussion argues that geophysics most effectively illustrates geothermal decision risks when interpreted in the context of geothermal conceptual models consistent with relevant geoscience data, thermodynamic constraints, and uncertainty.

This review outlines issues of integrating geophysics into geothermal resource decisions before introducing a detailed review of the rational for and the components of geothermal resource conceptual models. Because the MT method is the geophysical technique most widely used in a geothermal context, issues and pitfalls in its acquisition and analysis are reviewed in detail, and it is used to illustrate how resource conceptual models can be constrained by integrating geophysical data with supporting geoscience data. Outlines on TEM, gravity, magnetic, seismic monitoring, reflection seismic, and wireline logs are provided to illustrate circumstances in which these methods are given more emphasis. A discussion of issues commonly encountered in managing geophysical applications highlights a key issue for all geothermal geoscientists: developing the experience needed to assess uncertainty and risk in geothermal exploration and development.

3.2   Geophysics in the context of geothermal decision risk assessment

Different styles of geothermal decision making tend to be supported by different types of geophysical analysis. Geothermal decision makers with limited experience tend to be anomaly hunters, targeting wells on data anomalies and favoring the frequentist statistical approaches to reserve assessment like those critiqued earlier [31]. Anomaly hunting (also called data targeting) is based on analogy: ie, if a 5-Ω m resistivity contour outlines some successful wells in one case, the same contour is assumed to define a well target in another case. It is suited to supporting quick, low-value decisions that cannot justify an expensive and time-consuming conceptual analysis. However, among other drawbacks, anomaly hunting does not consider the conceptual validity of an anomaly in the context of a new target. For high-value decisions, recent geothermal best practice has focused on the use of conceptual model approaches for well targeting and resource capacity assessment [41,75]. Geophysical results influence decisions through their effect on conceptual model interpretations that are tested for consistency with respect to all relevant geoscience data and the stringent constraints of the thermodynamics of heat and fluid flow in rocks [17,21].

A conceptual model approach is likely to be ineffective unless it explicitly considers both context and uncertainty. In tests where many experts are each asked to produce a single best interpretation of a potentially decisive geoscience data set that is provided without geological context, the experts' conceptual uncertainty is typically so high that their interpretations have little predictive value [9,23]. Therefore, the practice of requesting geothermal resource assessments based only on a particular geophysical data set is often misleading. The development of a range of integrated geoscience models that are representative of both the conceptual and data uncertainty should also be more representative of prediction reliability and related economic uncertainty. The use of a range of conceptual models has been integrated into widely used geothermal resource decision analyses like the power density approach for estimating resource capacity [22,81].

3.3   Geothermal resource conceptual models

The elements of geothermal resource conceptual models are typically described separately for the natural state and produced state. Except for applications like precision gravity and seismic monitoring that are directed at measuring changes induced by reservoir exploitation, most geophysical methods are directed at constraining the properties of the natural state model and so this is emphasized in this discussion. Fig. 3.1 illustrates the basic elements of natural state geothermal conceptual models. With allowance for differing details of geology and chemistry, these conceptual model elements apply to geothermal reservoirs with sufficient permeability that water moves by buoyant flow, consistent with the great majority of geothermal reservoirs developed for the generation of electricity [32].

The hydrothermodynamic "fluid" elements of the models include the following features.

- The natural state isotherm pattern defines the flow regimen and thermodynamic state; this is the starting point for a numerical simulation of the reservoir. Isotherms are initially constrained by geothermometry, the pressure constraint (and associated boiling point versus depth) from the water table implied by springs, and interpreted permeability pattern. Temperature and pressure logs from wells, if available, are the highest priority data.
- Unless wells are available, the upper fluid boundary condition, particularly the deep water table, is typically inferred from spring chemistry. The deep water table constrains the pressure and the maximum plausible temperature at any depth (boiling point or below).
- Cold meteoric influx zones with the gravity flow of conventional hydrology rather than the buoyant flow of hydrothermal systems are illustrated by the isotherms and can be emphasized using arrows, such as by the white arrow in the model in the right panel of Fig. 3.1.

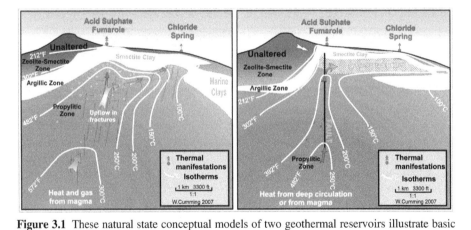

Figure 3.1 These natural state conceptual models of two geothermal reservoirs illustrate basic thermohydrodynamic and geoscience elements of conventional models [17,21]. On the left is a model cross section of a 250−350°C geothermal reservoir with isotherms, alteration zones, and structures. The system outflows to a chloride spring to the right. On the right, the cross section shows a 150−200°C geothermal reservoir with an upflow focused in a single fault and an extensive shallow outflow in a lava breccia with high primary permeability.
Adapted from Cumming W, Mackie R. MT survey for resource assessment and environmental mitigation at the Glass Mountain KGRA. California Energy Commission, GRDA Geothermal Resources Development Account Report. Available at: http://www.energy.ca.gov/2013publications/CEC-500-2013-063/CEC-500-2013-063.pdf; 2007, 119 pp.

- Flow direction arrows clarify implications of isotherms and water table (temperature and pressure). Flow arrows are essential in the very common circumstance where flow is oblique to the cross section [21], in which case it is shown as a circle and dot (an arrow point) if it is coming out of the cross section toward the viewer and a circle with an "x" (an arrow tail) if it is going into the cross section. However, arrows cannot replace isotherms in representing a conceptual model because the isotherms make the constraints of hydrothermodynamics more apparent.
- Conductive heat flow in impermeable rock is illustrated by a linear temperature gradient perpendicular to the isotherms. A high linear temperature gradient implies that a low-permeability barrier (like a clay cap or a fault full of gauge) separates two permeable zones, one with cooler water (like cold meteoric gravity flow above a clay cap) and the second with hotter water (associated with the reservoir). In areas with thin impermeable zones like lake beds with intervening aquifers connected to higher elevation surface water sources, reservoirs can have cold influx zones in the natural state [10]. Thick impermeable zones have a low temperature gradient.
- Boiling zones, steam caps, mixing zones, and zones of gas loss should be graphically illustrated and must be consistent with boiling point versus depth constraints.
- Both of the models in Fig. 3.1 are water dominated (ie, the pressure in the reservoir is hydrostatic). Steam-dominated geothermal reservoirs, in which the pressure is vapor static, are relatively rare but conceptually and economically important. Because steam-dominated reservoirs must be hermetically sealed to prevent influx from the surrounding hydrostatic aquifers at much higher pressure, they lack the outflow zones that are characteristic of water-dominated systems. This has important implications for geophysical interpretation. Over a water-dominated reservoir, a map of the base of the clay cap derived from resistivity data will likely reach its highest elevation at the coolest and probably unprospective end of an outflow, as in the left panel of Fig. 3.1, whereas the clay cap over a steam-dominated system may appear to enclose the reservoir, as is commonly shown in schematics of the clay cap geometry [59]. Simplified geothermal targeting strategies focusing on mapping the apex of the base of the clay cap [2] without considering the likelihood of outflows are higher risk than a conceptual model approach.
- The general fluid properties of the heat source for a geothermal reservoir can usually be inferred from geochemistry, whether it is likely to be associated with recent magmatic activity or deep fluid convection not associated with magmatism. However, specific thermodynamic properties for the heat source and its connection to the producible reservoir are usually omitted because they are poorly constrained. This is consistent with most numerical reservoir simulations where the base is defined by boundary conditions for thermal conductivity, temperature, and fluid mass flow associated with the upflow zone(s). Exceptions include basalt-hosted reservoirs like those in Iceland, Hawaii, and east Africa, where the base of the reservoir is constrained by shallow magma. Deep sedimentary reservoirs like those hosted in the Malm carbonates of Bavaria are conductively heated and a general model of regional heat flow sufficiently constrains the heat source.

The geoscience "rock" elements of the conceptual model are typically customized more than the thermohydrodynamic elements. The understanding of the relationship of rock properties, temperature, stress state, and fracturing to geothermal permeability and fluid flow has rapidly developed in the last decade [25,47,48].

- Below about 2500-m depth, rock properties and stress state favor localized geothermal upflow associated with near-vertical fractures.

- From about 250 to 2500 m in depth, buoyant geothermal upflow and lateral outflow are typically in formation-mediated fracture permeability. The host rock properties are as important to the open-space fracture permeability as are the fault geometry and stress state.
  - In an extensional structural setting in a recent lava-tuff sequence or clean sandstone-shale sequence, permeability is more likely to be distributed throughout a complex network of fractures, like in the model in the left panel of Fig. 3.1.
  - In an extensional structural setting in shales, siltstones, or metamorphic rocks, buoyant flow is likely to be more localized to where the fault and stress geometry are particularly favorable to form open space, such as at fault intersections and terminations, as in the model in the right panel of Fig. 3.1.
- From about 250 to 2500 m in depth, despite the general dominance of fracture permeability in geothermal reservoirs, some formations such as rhyolite lavas and karsted carbonates resist compaction and have sufficient primary permeability to locally dominate thermal flow.
- Shallower than ~ 250 m in depth, thermal flow is commonly hosted in the types of primary formation permeability associated with conventional groundwater aquifers, such as unconsolidated alluvial gravels, lava breccias, and karsted carbonates. If capped by lacustrine clays, such aquifers can host extensive up-dip outflows. Because of the emphasis on the importance of fracture permeability in geothermal reservoirs, less experienced geoscientists may erroneously interpret surface thermal manifestations associated with such formation-hosted outflows as directly emanating from a deeply penetrating fault [52].
- Sediment-hosted, convecting geothermal reservoirs typically include either thick clean sandstones such as those of the Salton Trough in the United States and Mexico [84] or carbonates such as those found in the Menderes Graben of Turkey.
- For most volcano-hosted fields, the details of the magmatic heat source and the deep connection from the heat source to the reservoir are not well constrained. However, in some basalt-hosted geothermal fields, the magma is shallow, more reliably imaged using resistivity and seismic monitoring, and verified by drilling [35]. In these cases, the magma is an important conceptual element constraining the base of the productive reservoir and the isotherm pattern, with the 400°C isotherm usually close to the magma and the 350°C isotherm usually close to the base of the conventional reservoir.
- For geothermal fields heated by large-scale convection in deep fracture systems without evidence of magmatic involvement, the roots of the fractured upflow zone are usually poorly imaged, although their hydraulic properties and chemistry may be generally characterized. Although deep anomalies detected using various geophysical methods have been associated with reservoirs, they have often been inconsistent with the thermodynamic details of the reservoirs.
- At the top of the reservoir, thermal flow is still up-dip but with a significant lateral flow component beneath a dipping cap.
- Faults commonly create flow barriers within fields [63], and many field edges are fault-controlled.
- The low-permeability clay cap is shaded yellow in the models in Fig. 3.1. It is essential to most conventional and all hidden systems, enabling the accumulation of heat in permeable aquifers at relatively shallow depths and preventing catastrophic cooling by surface water influx when the reservoir pressure drops during production [28,32]. In almost all existing commercial geothermal reservoirs, at least part of the cap consists of hydrated smectite or mixed-layer clay (smectite-illite in andesite, smectite-chlorite in basalts). Rocks with a relatively minor amount of smectite tend not to form open space permeability (even when fractured), an essential property of a material that can effectively cap a fractured reservoir [37,48].

- In volcanic rocks, the geothermal system generates its clay cap, whereas in sediments, the impermeable clay is deposited and reservoir flow conforms to that permeability pattern. Because of the conceptual importance of the cap and the exceptionally low resistivity of hydrated smectite clay, resistivity methods dominate geothermal exploration.
- Although almost all commercial geothermal reservoirs have at least a partial smectite clay cap, other types of impermeable cap exist. For example, zones where chlorite and/or illite clay are particularly abundant can act as a cap [25,37]. If the high-temperature part of a geothermal reservoir is exhumed by erosion and exposes high-rank (propylitic) alteration containing abundant chlorite and illite, such rocks are typically highly indurated with low porosity and resist retrograde alteration to hydrated smectite. However, they often have sufficiently low permeability and act as a cap, sometimes marking a high-resistivity edge of a reservoir [33].
- Graphitic-pelitic schist found in Paleozoic metamorphic rocks is typically both low resistivity and low permeability, and so it can act as an easily imaged reservoir cap to permeable formations like karsted dolomites and fractured carbonates [17,45].
- Some types of rock do not hydrothermally alter to smectite clay, like clean sandstones or the trachyte lavas commonly found in east Africa, and so are less likely to host a low-resistivity reservoir cap.

3.4   Geothermal resource models with elements that differ from those in Fig. 3.1

Although the geothermal resource conceptual models in Fig. 3.1 include the main conceptual elements of a majority of the world's developed geothermal fields, as geothermal prospects that best fit such models are explored, the remaining prospects are likely to differ in more significant ways.

3.4.1   Detecting magma in geothermal exploration

The detection of magma using geophysics would seem to be an obvious goal in geothermal exploration, but its feasibility and utility depends on the geological context. At several basalt-hosted geothermal reservoirs, drilling has confirmed that magma at depths shallower than 3000 m has been detected as a low-resistivity body by MT and/or as a transition from brittle to ductile behavior by seismic monitoring [35]. Where related hazards seem manageable, the magma can be integrated into the resource conceptual model, increasing heat availability but decreasing the developable/producible volume. In all of these cases, conventional geochemistry and geology had indicated that both the magma and the overlying neutral resource probably existed; however, the MT and seismic monitoring data verified the depth and location.

In contrast to basalt-hosted reservoirs, detecting andesite magma has proved to be a greater challenge and integrating it into a commercially developable conceptual model is problematic. Active andesite volcanic cones are much more rugged than basaltic shield volcanoes, making acquisition and imaging more challenging. An andesite composite volcano that frequently erupts is very likely to be hot at drillable depths, but its high gas flux is also likely to create a hot magmatic vapor/acid core surrounded by impermeable alteration zones that largely isolate the core from surrounding neutral aquifers [21]. Both the vapor/acid core and the surrounding low-permeability zone should be avoided by wells. Some investigators have proposed that detecting low-resistivity smectite clay alteration draping the flanks of such a volcano would be a reliable indication of a geothermal reservoir target. However, active volcanoes are often mantled by shallow smectite clay alteration caused by gas lost in the repetitive eruptions and/or shallow subboiling sulfate water outflow from summit aquifers. Therefore, unless a surface manifestation is found with geochemistry indicative of the existence of a commercial geothermal reservoir, new targeting strategies and case histories will be needed to make the risks of targeting an andesite-hosted resource with magmatic indications comparable to the risk of targeting a basalt-hosted system with imaged magma.

3.4.2   Deep conductively heated stratigraphic reservoirs

A few developed geothermal reservoirs and many exploration targets may not fit a model dominated by buoyant thermal flow like those in Fig. 3.1. These include deep sedimentary geothermal developments of the Malm carbonate reservoirs in Bavaria [11] and exploration targets in the Pretty Hill sandstone of South Australia. They reach economically interesting temperatures due to regional conductive heating below thick insulating (impermeable) basin sediments. The temperature pattern varies relatively slowly in a manner consistent with the regional temperature gradient. Because temperature and chemistry are reliably predictable, targeting laterally extensive interconnected permeability is the dominant risk issue. In such cases, MT might be used to establish that a suitable thickness of impermeable sediments existed in a basin lacking deep wells. However, reflection seismic data, supported by detailed geology and well logs, are very likely to be the dominant well targeting tools [76]. In these cases, a conceptual model approach is particularly valuable in highlighting the criteria needed to successfully target such zones; for example, these reservoirs can be targeted only in areas where outstanding 3D reflection seismic data can be cost-effectively acquired.

3.4.3   Geothermal prospects with "hidden" or "blind" reservoirs

Although the conceptual models in Fig. 3.1 include thermal manifestations, the conceptual model approach can be adapted to the exploration of "hidden" or "blind" geothermal systems that lack thermal manifestations [13,17]. However, the conceptual model approach does not support the commonly encountered claim that there are likely to be tens of thousands of megawatts electric (MWe) of conventional geothermal capacity "hidden" among already developed fields that had been initially identified based on surface manifestations. The expectation of large cumulative capacity for such fields usually stems from the assumption that geothermal exploration is in a primitive stage of targeting surface manifestations, such as hot springs, in a manner analogous to petroleum explorers targeting oil seeps in the late 1800s. However, this assumption overlooks fundamental physical differences in the conceptual models of petroleum and geothermal fields.

In comparison to petroleum fields that might store their extractable energy for hundreds of millions of years as oil trapped beneath impermeable shale, convecting geothermal fields are dynamic. Even if the convecting water in a permeable geothermal reservoir is perfectly sealed by an impermeable smectite clay cap, the reservoir will lose heat by conduction. If the cap leaks, as seems likely given the fracture permeability that supports all convecting geothermal upflows, the geothermal reservoir loses heat much more quickly by buoyant flow of hot water. Therefore, convecting geothermal reservoirs require continuous replenishment or they will dissipate within a few 1000 years, and leakage to the surface is likely.

Because of the many tens to hundreds of megawatts thermal (MWt) of heat naturally emitted by geothermal reservoirs, ultimately to be radiated at the earth's surface, economically significant convecting geothermal systems are very likely to create surface manifestations, or near-surface manifestations if the water table is deep and the reservoir is far below the boiling point. If near-surface heat loss is mostly conductive, temperatures at depths just below diurnal and seasonal influences are likely to define an underlying convecting resource. If leakage through a generally impermeable cap is significant, hot water with diagnostic geochemistry is likely to be found much shallower than the reservoir. As a result, in the initial exploration of most developed geothermal reservoirs, geophysics has had much lower weight than the geochemistry of hot springs and fumaroles in assessing the probability of the existence of a commercial geothermal resource. (Chapter: Application of geochemistry to resource assessment and geothermal development projects in this volume covers geochemistry in depth.) Where surface geochemistry has not been revealing, geothermal resources have been discovered by shallow drilling to supplement or replace geochemistry information and, importantly, by careful examination of serendipitous data, like temperatures from petroleum or mining bore holes [13].

3.5   Formation properties and geophysical methods

In most cases, the formation properties that are relevant to geothermal exploration are not those directly imaged by surface geophysical surveys (ie, a further inference of another rock property is required to support an interpretation). For example, the MT method is commonly used to image resistivity, which in turn is used to infer the smectite content of the clay that caps almost all commercial geothermal reservoirs. Table 3.1 summarizes geophysical methods commonly used in a geothermal context and the rock and reservoir properties that are typically inferred from them.

A partial list of resistivity methods and their acronyms include:

- Magnetotelluric (MT) - the most widely applied geophysical method in geothermal exploration because it can reliably resolve the bottom of a clay cap at depths well over 700 m, depths that challenge the other methods listed here. However, because MT uses natural signal sources like distant lightning and the sun, it can be overwhelmed by cultural noise (eg, power lines).
- Controlled source audio magnetotelluric (CSAMT) - a variant on MT that uses a generator and transmitter to create a controlled signal instead of relying on natural MT signals. In noisy areas or when natural signals are weak, it can provide images to about 700 m depth, in a typical geothermal context.

Table 3.1 Geophysical survey types most commonly acquired in geothermal exploration, with the rock properties they most directly image, and inferred properties commonly used in geothermal resource assessment

Geophysical survey type	Physical property most directly imaged or inferred	Related properties inferred in geothermal applications
Resistivity methods: MT, TEM, VES, CSAMT, HEM, DC-tomography (E-scan)(Acronyms described below)	Resistivity	Smectite clay content Geometry of base of smectite clay cap Porosity, matrix permeability, pore fluid salinity Phase change (speculative)
Active seismic reflection 2D/3D (and active seismic refraction)	Seismic impedance P-wave velocity Seismic attenuation	Formation interface geometry Faults, structures Alteration, induration Porosity Pore fluid density
Gravity	Bulk rock density contrast	Bulk porosity contrast Inferred structure Intrusions in sediments/tuff
Repeat precision gravity	Pore fluid density change (induced by production/injection)	Steam/water saturation change in pore space due to boiling or condensation. To a lesser degree, pore fluid temperature decrease around injection wells.

Method	Measurement	Application
Seismic (earthquake) monitoring	Event location; Vp, Vs velocity (tomography); Fault failure geometry and stress change; Anisotropy/S-wave splitting	Injectate flow due to cooling stress in >200°C rock deeper than 1000 m; Reservoir compartments; Base of reservoir based on depth of events associated with injection cooling; Magma movement; Cumulative faults and stress state; Phase and bulk compressibility change; Open space foliation orientation, attributed to fracturing
Ground magnetic surveys, aeromagnetic surveys	Magnetic susceptibility; Remnant magnetism in volcanics >1 Ma	Intrusives and lava embedded in sediments and pyroclastics; Demagnetization of mafic/andesite rocks due to sulfate alteration
Self-potential (SP)	Static voltage variation	Shallow fluid flow
Induced polarization (IP)	Chargeability	Disseminated pyrite or clay in generally resistive rocks
Borehole formation logs	Gamma for K, U, and Th; Sonic for Vp (and Vs with added cost); Dual induction for resistivity; Many other potentially useful but more expensive logs	Correlate formations based on K; Porosity trends from Vp, resistivity and nuclear tools; From rock properties, infer susceptibility to fracture permeability
Borehole image logs: (1) acoustic televiewer, (2) microresistivity	Televiewer for borehole breakouts and fracture orientation and aperture; Microresistivity for fracture statistics and formation characterization in blind hole	Stress orientation from breakouts; Fracture properties; Formation characterization when drilling blind

- Time domain electromagnetic (TEM or TDEM) - a method that does not use electrodes and so does not suffer from the "static" distortion that affects resistivity imaging methods that use electrodes. Except in cases where it is feasible to use heavy generators and large loops of wire, the depth of investigation of TEM is <300 m; thus, in a geothermal context, TEM is mostly used to correct MT "static distortion".
- Vertical electrical soundings or Schlumberger soundings (VES) - a method that was used for many decades, now supplanted by methods with easier access requirements, smaller environmental impact, and more sophisticated imaging options.
- Helicopter electromagnetic (HEM) - includes several types of airborne EM systems, with the greatest depth of investigation currently several hundreds of meters. A large area can be quickly surveyed in great detail but survey coverage may be limited for a variety of reasons: it cannot extend over houses, the data are unusable near power lines, and very rugged topography limits the survey because the helicopter must fly at a constant and relatively low ground clearance, usually under 100 m.
- Direct current (DC) tomography - a resistivity approach mainly tested in geothermal settings using the proprietary E-scan system [70]. Its application has been limited by demanding logistics, especially in rugged and/or heavily forested terrain and limited practical resolution below about 1000 m in depth. Like all methods that use electrodes, it suffers from "static" distortion.

3.6 Choosing geophysical methods and designing surveys for geothermal applications

A common argument to justify investment in a particular geophysical method is that the proposed data or analysis costs less than a bad well. This is likely to be true for most particular cases. However, (1) the cumulative cost of all methods routinely proposed would likely be more than a bad well; (2) most are likely to be redundant or irrelevant to the decision; and (3) managing numerous methods would not fit most practical timelines. Based on case history experience, simple tools like decision tables can be used to rank the likely cost and impact on decisions of different types of geophysical data sets and analyses. Although this might disadvantage a promising method that lacks case histories, research organizations and geothermal management have usually given consideration to the potential value of untried technologies. On the other hand, some reasonable skepticism is warranted since most methods said to be untried are merely unreported because of the bias against reporting failure case histories [22].

Geophysical surveys are often specified by analogy to a previous geothermal exploration program that seemed effective, and this is a reasonable starting point. Subsequent review of the geophysical design considers practical issues of cost, access, noise, topography, and safety as well as the expected properties of the targeted geothermal reservoir, especially its likely depth, temperature, and host geology. Therefore, when the survey schedule permits, geophysical designs are based on the results of geology and geochemistry surveys; see chapters "Geology of geothermal resources" and "Application of geochemistry to resource assessment and geothermal development projects," respectively.

The most cost-effective approach to geophysical survey design is usually to place stations closer together in the most prospective area and spaced farther apart in adjacent areas with sufficient flexibility to optimize data quality. Locating proposed measurement stations on a uniform grid is a common but seldom necessary or cost-effective approach. Such grids often place stations in locations where data quality is likely to be poor or unusable, such as gravity stations on cliffs or MT stations under power lines. It is more important that the station spacing be locally adjusted to minimize spatial aliasing (undersampling). For example, MT stations should be closer together near a fumarole on a volcano where the resistivity pattern is likely to vary more rapidly than in a basin adjacent to a volcano. Where access is easy throughout the survey area, it is usually cost-effective to keep a large fraction of the budgeted stations in reserve so that they may be added as the survey progresses and issues like aliasing are revealed, but this is often impractical in areas with difficult access.

For data likely to be interpreted along cross sections like MT or gravity, it is helpful to roughly align stations so that it is easy to produce cross sections oriented to highlight the upflow—outflow elements of the preliminary conceptual model, connecting thermal manifestations. If two-dimensional (2D) modeling is planned, stations will be needed to support profiles perpendicular to the likely strike.

In order to image features at a particular depth, most geophysical surveys have a minimum lateral extent or aperture that must be covered to provide reliable resolution at that depth, also depending on the likely lateral variation in the imaged rock property. If the earth below an MT station is truly one-dimensional (1D), then resistivity can be imaged to 10s or 100s of kilometer. However, in a geothermal context, lateral changes in resistivity typically limit the reliability of a sounding from a single MT station to 1 or 2 km in depth. A wider array of MT stations is required to produce a resistivity image to greater depths using a 3D method. In surveys like precision gravity and seismic monitoring, it is important to anticipate where the signal will likely occur and to arrange surveys to provide an adequate survey aperture for that signal. Thus, a seismic monitoring station would be located near a 2500-m-deep injection well and three more stations would surround it at a distance of 3—4 km. Notwithstanding the limitations implied by station distribution and noise, geophysical images are often extrapolated to the edges of plots, often beyond the extent of reliable resolution, an issue to consider in any geophysical image, especially if station locations are not shown.

A particularly effective approach to optimizing geophysical surveys with respect to access, topography, noise, and geothermal features is to review proposed station locations in a 3D geographic information system (GIS) display including high-resolution geo-rectified satellite images or aerial photographs combined with a digital elevation model (DEM). For many geothermal prospects, suitable imaging is available at minimal cost using Google Earth. Most geoscience field surveys and interpretations benefit from such 3D GIS displays. By importing coordinates for hot springs, alteration, and other important features, survey plans can be optimized. Discerning features like power lines in satellite images typically require a ground sample distance (GSD) of >0.5 m, although 3-m GSD is sometimes usable; there are several commercial satellites providing suitable images. Where suitable images are not available, planning several months in advance to purchase customized satellite images is likely to be cost-effective, more so than lower resolution thermal infrared imaging. In several geothermal concessions undergoing exploration, when 0.5-m GSD satellite images replaced earlier 3 + m Spot and 30 + m Landsat images, new fumaroles and areas of altered ground were discovered that significantly changed exploration assessments and geophysical survey designs.

3.7 Resistivity methods

Resistivity (in Ω m) is the bulk property of a material that indicates its ability to resist the flow of an electric current (in amperes) being induced by an electric field (in V/m). The number of resistivity methods listed in Table 3.1 reflects both evolution of technology and differing cost, depth of investigation and tolerance of noise sources. Any of these methods might be considered to meet specific circumstances but the MT method dominates geothermal applications because its depth of investigation of many 1000s of meters better fits the needs of geothermal targets.

Resistivity in geothermal fields is affected by porosity, pore geometry, temperature, salinity, and water saturation, but the overall pattern of resistivity variation is dominated by the very low resistivity associated with the high-interface conduction of hydrated clays, particularly smectite clay and mixed-layer clays containing hydrated smectite [74]. The transition from low to high resistivity is associated with a transition from smectite to illite clay alteration in andesite and smectite to chlorite alteration in basalt, but the overall pattern is similar. Numerous case histories [2,55] illustrate how resistivity methods image geothermal reservoirs as relatively high-resistivity zones within a low-resistivity halo of low-permeability clays with resistivity typically below 20 Ω m. As a result, resistivity imaging of the smectite clay that caps almost all geothermal reservoirs has been the dominant geophysical approach to targeting geothermal wells and assessing the capacity of geothermal reservoirs in recent decades [21,55].

Although publications from the late 1970s and early 1980s demonstrated the smectite model of geothermal resistivity interpretation based on core and well log studies in geothermal reservoirs of the Salton Trough [26], prior to the publication of numerous case histories worldwide in 2000 [2,3,16,74], resistivity interpretations of geothermal systems often assumed that geothermal reservoirs would be a low- rather than a high-resistivity target [82]. The assumption of a low-resistivity reservoir is based on temperature correlations with resistivity from core studies of clay-free sandstones widely used in the interpretation of well logs in hydrocarbon reservoirs. However, if the entire geological section is considered, including shale and siltstone as well as sandstone, then low-resistivity clay dominates the resistivity pattern in most hydrocarbon as well as most geothermal settings.

Despite abundant evidence from case histories, geothermal well logs, cuttings analyses, and core petrophysics [18,26,29,55,74] supporting the smectite clay model for geothermal resistivity interpretation, confusion about the applicability of this model persists, in part, because the earlier expectation that low-resistivity features would be geothermal reservoirs was so strong that well-regarded investigators concluded that resistivity observations inconsistent with that expectation must be faulty. For example, when a 1981 MT survey detected a higher-resistivity reservoir at the Desert Peak geothermal field [8], the expectation that the reservoir would be relatively low resistivity was so strong that the investigators concluded that the MT method was somehow faulty in this setting, discouraging the use of MT. Basin and Range prospects in the USA for over 20 years and missing an opportunity to establish the prototype for the MT smectite cap interpretation model. Similarly, when VES surveys detected resistive geothermal reservoirs in New Zealand and Indonesia in the 1980s, investigators attributed this to a problem with the VES method, not to a realistic property of the reservoirs. Extensive MT resistivity surveys and cuttings analyses have demonstrated that geothermal reservoirs in New Zealand [10,62] and Indonesia [16,55,69] conform to the smectite model.

Smectite clay zones cap most volcano-hosted geothermal reservoirs because hydrothermal smectite is created from precursor minerals altered by hot gas and fluid buoyantly rising above the reservoir. The clay that caps sedimentary-hosted reservoirs is deposited prior to the emergence of the geothermal system rather than being generated by it, but its effect on interpretation is similar in that the buoyant geothermal flow conforms to the geometry of the base of the impermeable clay.

Smectite is a mineral geothermometer because it becomes unstable as temperatures rise to $>70°C$, and by $\sim250°C$, low-resistivity hydrated smectite has usually completely converted to the more brittle chlorite/illite clay found in neutral geothermal reservoirs. However, because both temperature and permeability affect the conversion of smectite to illite or chlorite [27] the transition can occur over a wide range of temperature. Therefore, although the average reservoir temperature in the resource model on the left in Fig. 3.1 is $100°C$ hotter than the model on the right, isotherms near the top of the permeable reservoir conform to the base of the low-resistivity smectite clay cap in both models. Therefore, although smectite can be an unreliable mineral geothermometer, its ubiquitous role in forming a low-permeability geothermal reservoir cap together with the low resistivity of hydrated smectite make it a very effective geothermal geophysics target.

Smectite reduces permeability, even in small-volume fraction. In particular, the formation of secondary permeability is inhibited in smectite-bearing rocks even when they are fractured [37,47]. Where major fracture zones are exposed at the surface over the shallow parts of a geothermal field, a smectite clay cap typically leaks some hot water and gas locally. The low-permeability smectite cap partitions the geothermal field hydrology into a shallower cooler meteoric zone and a deeper high-temperature zone. If it is very permeable, the meteoric zone is typically high in resistivity. Because rocks with significant smectite content are low in permeability, the smectite alteration cap over a geothermal reservoir accumulates heat by trapping the buoyant thermal upflow. This causes the steep temperature gradient in the clay cap overlying the lower-temperature gradient in the underlying convecting reservoir in Fig. 3.1.

Resistivity maps and especially resistivity cross sections can aid in the construction of a geothermal conceptual model in at least three ways: (1) through the pattern of isotherms implied by geochemistry, wells (if available), and base of the low-resistivity zone; (2) by geometry directly indicative of a prominent structure; and (3) through the intensity of the smectite alteration suggested by resistivity values [17,21]. The natural state temperature pattern is the primary conceptual constraint on bulk permeability in the context of a geothermal reservoir model. Because low resistivity is correlated with low permeability, imaging the base of the low-resistivity zone constrains the minimum depth to the transition from low to high permeability at the top of the reservoir and provides an indirect constraint on temperature. In conjunction with temperature constraints from geochemistry or wells, inferring the temperature and permeability pattern using resistivity data provides sufficient information to target wells on zones with a higher probability of encountering permeable fractures.

Even in cases where the reservoir is dominated by a single fracture, the intersection of the fracture with the clay cap can often be inferred with a lateral resolution less than the depth to the fracture. Highly permeable fractures within a reservoir tend to leak more gas, causing more intense overlying alteration. Sometimes the smectite– illite transition is pushed higher at the apex of the fracture by the higher-temperature upflow, as illustrated by the base of the cap below the fumarole in the models in Fig. 3.1.

3.8 MT surveys

The MT station schematically illustrated in Fig. 3.2 uses a recorder and two sets of grounded electrical field measurement wires about 100 m long, Ex and Ey, and three coil magnetometers, Hx, Hy, and Hz, to measure the natural electric and magnetic fields at the earth's surface caused by electromagnetic waves radiated from the sun and from distant electrical storms [14,77]. Each MT recording system can be carried by a crew of three or four in rugged terrain. After scouting the field site, the crew may move a site up to a few 100 m to locate a more-or-less flat 100×100 m site that avoids sources of electrical noise that would be detected by the electrodes or sources of vibration that would disturb the magnetometers. The 10×120 cm magnetometers are buried in shallow trenches to avoid vibrations, small holes are dug for the electrodes, and the recorder and battery are connected and covered with a tarp. Because the MT signal is commonly best for geothermal applications at night, recordings are run overnight for $12-14$ h, and the system is picked up in the morning and moved to the next site.

Because the electric field in an incident electromagnetic wave induces currents in the earth proportional to the earth's resistivity, resulting in a proportionate reduction of the electric field amplitude, the MT electric field is more directly related to the resistivity at each station than the magnetic field. which The lateral variation of the magnetic field is usually much smaller. varies relatively slowly. The apparent resistivity parameter shown in the center panel of Fig. 3.2 is proportional to the square root of the ratio of the electric to the orthogonal magnetic field. The phase can be thought of as the delay, measured as an angle, between the peaks of the electric and orthogonal magnetic field waves. The phase has a fixed mathematical relationship with apparent

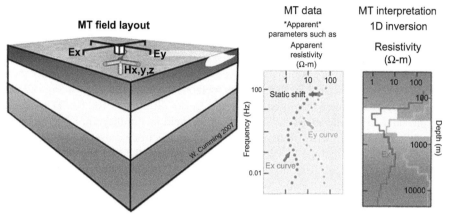

Figure 3.2 MT station layout, data recording, and 1D resistivity models [19].
Adapted from Cumming W, Mackie R. MT survey for resource assessment and environmental mitigation at the Glass Mountain KGRA. California Energy Commission, GRDA Geothermal Resources Development Account Report. Available at: http://www.energy.ca.gov/2013publications/CEC-500-2013-063/CEC-500-2013-063.pdf; 2007, 119 pp.

resistivity so that one can be used to check the reliability of the other [7,57]. An MT sounding is made by computing the apparent resistivity and phase for a range of frequencies, typically <0.01 Hz to >300 Hz for geothermal applications. The electric field of a high-frequency electromagnetic wave dissipates at shallow depths, whereas the field at low frequencies responds to a much thicker and deeper section of the earth. Therefore, by recording MT at a wide range of frequencies, resistivity can be imaged for a wide range of depths. The MT depth of investigation as a function of frequency and average resistivity is roughly approximated by the skin-depth equation.

$$\delta \sim 500(\rho/f)^{1/2} \tag{3.1}$$

where δ is skin depth in m, ρ is resistivity in Ω m, and f is frequency in Hz.

For a typical geothermal clay cap resistivity of 4 Ω m and a frequency of 1 Hz, the skin depth Eq. [3.1] gives a depth of 1000 m, similar to the base of the clay (yellow) zone in the resistivity versus depth models in Fig. 3.2.

Because MT uses natural signals, there is no option to increase the signal strength if cultural noise from power lines or roads disturbs the measurements. Signals from worldwide lightning are usually sufficiently strong to support MT soundings at >1 Hz (shallower than ~ 1 km). However, for frequencies <1 Hz, signal strength depends on the 11-year solar sunspot cycle and day-to-day solar weather. The lowest signal strength relative to noise in geothermal MT soundings is where the solar and lightning sourced signals merge, near 1 Hz. Unfortunately, this roughly corresponds to 1000 m depth, close to the base of the clay cap in a typical geothermal resource. A usually effective approach to mitigating MT noise is to simultaneously record MT at a low-noise location at 10 to 100 km distance, so that this remote station's noise is unlikely to be correlated with the survey station noise. The remote data are used as a reference for the survey data in robust processing software [46]. However, as power transmission expands, in many areas it is becoming increasingly difficult to find suitable remote reference sites that can effectively attenuate noise [73].

Aside from noise, MT is subject to a type of distortion called *static shift*, a potential source of distortion shared by all resistivity methods that use electrodes. In Fig. 3.2, the apparent resistivity data measured for Ex and Ey are different at 100 Hz. They would be the same in a uniform earth. However, in the field layout diagram in left panel of Fig. 3.2, the yellow low-resistivity body near the Ey electrode causes a local distortion in the electric field that the Ex dipole does not detect, so the two apparent resistivity curves calculated from these electric fields, called modes, are separated by a constant offset at all frequencies, called a "static" offset. There are a variety of ways of correcting this, but a geometric average usually works for small distortions. If the static distortion is due to topography, a 3D model including topography can compensate [20,67,80]. In cases where the static distortion is large and caused by local near-surface resistivity changes (eg, surface alteration as in Fig. 3.2), then acquiring supplementary TEM is often the most effective way to mitigate MT static distortion because the TEM method does not use electrodes [5,58].

The MT apparent resistivity curves in Fig. 3.2 show an additional distortion below 0.1 Hz, where the apparent resistivity curves for the Ex and Ey electrodes diverge due to lateral changes in resistivity at a scale larger than the figure. Rectifying this distortion is the objective of 2D and 3D inversions. Fig. 3.3 shows an MT resistivity cross section plotted to show results from three types of inversion [20]. For the 1D plot, a smooth 1D model (called an Occam [14] model) is computed for each MT station, assuming that resistivity varies only with respect to depth below that station, and the 1D resistivity models are stitched together by a contour algorithm to make the cross section. For the 2D plot, resistivity is computed for all elements of a 2D mesh assuming that resistivity only varies in the plane of the cross-sectional mesh and does not vary perpendicular to this profile. The 3D inversion profile is a vertical slice through a 3D mesh of resistivity values varying in all directions. Although 3D MT inversions have been investigated for almost as long as modern smooth 1D and 2D inversions [14,49], realistic geothermal applications of 3D MT inversion have been limited to commercial applications by the largest MT contractors [19] and a few research projects. Over the last decade 3D MT inversion software has become more widely implemented on computer clusters, making computation of realistic models more practical [4,65]. Although 3D inversion has become a standard requirement in geothermal MT contracts, it is not a panacea for MT resistivity imaging in a geothermal context. All current 3D MT inversions require a trade-off between smoothing to minimize the effect of noise in the MT soundings and loss of resolution due to smoothing, a trade-off that can usually be more effectively customized for each MT station in 1D inversion.

As 3D MT inversion research continues, open questions are better resolved but, in the meantime, ambiguity is introduced by, for instance, the ongoing debate in the MT literature regarding whether topographic distortion and static distortion of MT should be addressed by including detailed topography in the 3D inversion [43,67,80] or by

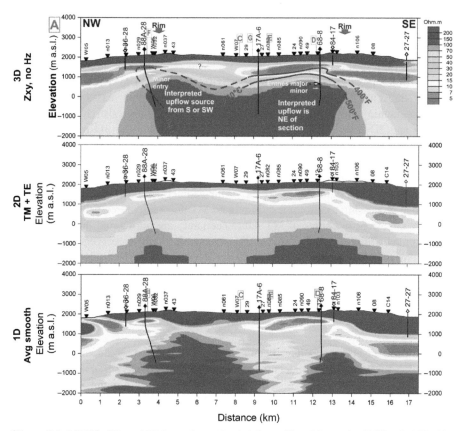

Figure 3.3 MT 1D, 2D, and 3D inversions of resistivity at Glass Mountain, California [19]. The 3D inversion is smoother and more representative of deep resistivity but the overall pattern is resolved by all three inversions to the base of the <7 Ω m (orange-red), low-resistivity clay cap. Some of the conceptual model elements are shown in the top cross section along with ties to wells [20].
Adapted from Cumming W, Mackie R. Resistivity imaging of geothermal resources using 1D, 2D, and 3D MT inversion and TDEM static shift correction illustrated by a Glass Mountain case history. In: Proceedings of the 2010 world geothermal congress, Bali, Indonesia; 2010, 7 pp.

adding 1D TEM constraints to a flat-earth 3D MT inversion [4,5]. Tests of practical resolution and uncertainty are often more tractable using 1D and 2D methods [14], and most MT practitioners who routinely match MT resistivity imaging to geothermal well data emphasize the use of 1D MT inversion approaches for quality assurance of 3D results, for backup interpretations of noisy or sparse MT data, and for resolving shallow resistivity details obscured by 3D inversion smoothing. A great advantage of 1D inversion is that, for each station, it can be truncated at the depth below which it is no longer valid and the cross-sectional plot can be left blank below that depth, a very useful option not available in 2D and 3D inversions (it was not used for the 1D plot in Fig. 3.3 in order to illustrate the deep behavior of the 1D inversion).

To build the conceptual model in the case shown in Fig. 3.3, isotherms were inter-preted from well logs and extrapolated among wells using the 3D MT resistivity pattern, as shown in the upper cross section in Fig. 3.3 and in numerous cross sections and maps [19]. In this prospect, the base of the smectite clay alteration closely corre-sponds to the most rapid increase in resistivity with respect to depth, at the red-to-orange transition in Fig. 3.3. The resistivity interpretation has been validated by petrographic analysis of well cuttings and by a wireline resistivity log that fits the 1D and 2D inversions to the base of the clay cap and the 3D inversion at all depths [19].

Although 3D inversion has been emphasized in geothermal resistivity research in the last decade (eg, Refs. [4,14,19]), suppressing noise and characterizing inversion reliability are arguably higher priorities for the geothermal industry. Improved acqui-sition and processing technologies to mitigate MT noise are prerequisites to effective inversion. Current measures of MT data uncertainty are biased and the measures of inversion reliability such as goodness-of-fit and sensitivity often perversely encourage overfitting of noise. While improvements are developed, the uncertainty of MT ana-lyses conducted in support of high-value decisions could be characterized by completing a quality assurance table that summarizes the reliability of each MT station with respect to its depth of investigation in 1D and 3D MT inversions so that the MT data and inversion reliability can be considered when building resource conceptual models.

Worldwide, MT surveys have been increasingly impacted by the advent of power generation and transmission technology that produces *coherent non−plane-wave noise*, also called *near-field* noise because it distorts MT in a manner analogous to the well-known near-field distortion of CSAMT [85]. Depending on the relative strength of the natural MT signal and the noise signal, such noise can affect MT sta-tions at distances of many tens of km from the noise source. Using a 50- or 60-Hz filter does not attenuate this type of powerline noise. MT stations affected by a coherent noise source often exhibit a characteristic pattern; when the apparent resistivity curves are plotted in an orientation that maximizes differences between the modes (principal axes rotation), one apparent resistivity curve rises at a 45 degree upward trend on a log−log plot, while the phase trends to zero [44]. Because this noise is coherent, the curves look smooth and are often accepted as good data. At distances over 10 km from the source, this noise typically affects MT at frequencies just below 1 Hz [79] so that, in a geothermal context, MT data affected by this noise can often be used to image the base of a clay cap, including in some areas where incoherent noise from local cultural sources makes normal MT difficult to acquire. However, the effect of the noise on the depth of investigation should be modeled [60]. If the coherent noise is not recognized and imaging is attempted deeper than is consistent with the onset of the noise, then it will tend to produce resistivity that is too high, potentially imaging the base of the clay cap shallower than it actually is. This is often accompanied by a deeper low-resistivity feature induced by fading of the coherent noise amplitude at lower fre-quencies, prompting erroneous interpretations of magma or high porosity at great depth. As this type of coherent noise becomes more widespread, mitigating it, or at least recognizing the limitations it implies for MT resistivity imaging will be increasingly important to geothermal applications of MT, particularly for MT surveys conducted to target step-out or injection wells near existing power plants.

3.9   TEM resistivity sounding for correction of MT static distortion

The TEM or TDEM method relies on electromagnetic induction between ungrounded wire loops and the earth to both transmit and detect the signal [15]. A current in a 50- to 500-m-diameter loop is turned on and then abruptly off, inducing "smoke rings" of electrical current that propagate down into the earth. The decaying magnetic field associated with the induced current is detected as a decaying voltage in a smaller detector loop, and this can be used to compute a resistivity model. Because it requires no electrical contact with the earth, TEM is not subject to the static distortion that can affect any method that measures the resistivity of the earth using electrodes.

Because the practical, reliable depth of investigation of TEM is typically <1000 m, TEM is most commonly used in a geothermal context as a supplementary method to constrain MT static distortion [5,16,68]. Model studies have shown that TEM is relatively immune to distortion by small-scale resistivity variations that commonly cause MT static distortion [58]. The most common approach to implementing MT static correction using TEM is as follows: (1) acquire a TEM sounding at an MT station; (2) compute a shallow 1D resistivity model from the TEM; (3) compute from the TEM model synthetic high-frequency MT apparent resistivity and phase curves; and (4) shift the observed MT apparent resistivity curves to fit the MT curve modeled from TEM either as a joint 1D MT-TEM interpretation [19] or as a joint 1D TEM-3D MT inversion [4]. The TEM could be acquired by the MT crew but it is more commonly acquired by a separate TEM crew after the MT is completed. Because a TEM sounding takes only a few minutes to record after the loop is laid out and connected, TEM surveys can be a small fraction of the cost of MT in those areas where it is feasible to use small loops or where access for laying out large loops is easy.

If not suitably designed and processed, TEM data can exacerbate rather than mitigate MT static distortion [20]. A common check is to run MT inversions with and without TEM static corrections or using alternative methods of static correction like polarization analysis relative to topography or 3D topographic modeling. Cross sections of reliably corrected MT stations should be smooth and geologically plausible from station to station. The most common sources of problems with MT static corrections derived from TEM are inverting noisy TEM data [15], relying on TEM distorted by topography [67,80], or using a TEM system too small to penetrate thick resistive rocks at the surface [20].

TEM supplies its own signal rather than relying on natural signals like MT, so the TEM transmitting antenna loop must be large enough and the transmitting electronics powerful enough to provide sufficient signal to penetrate thick surface resistive layers. To provide reliable results for MT static correction, the needed transmitting loop diameter might be 50 m in regions with conductive soils or >300 m in regions with thick resistive surface formations. A heavier transmitter and power source are required for bigger loops. In Iceland, 1200 m of wire can be towed by a snowmobile to lay out a 300-m-diameter loop in a few minutes, sometimes reliably resolving resistivity deeper than 1000 m. However, a TEM station using such a large loop might cost much more than an MT station in rugged areas covered by thick tropical forest. Moreover, because the TEM itself can be distorted in rugged areas [67,80], approaches to MT static correction other than TEM are commonly used in those circumstances [21].

3.10   Awibengkok MT model and validation

The 377-MWe Awibengkok (Salak) geothermal field in Indonesia was initially drilled in 1983 following the acquisition of supporting geology, geochemistry, and MT data [69]. Although the quality of the 1982−1983 MT data was problematic from 0.1 to 10 Hz and the resistivity imaging was done using a relatively primitive 1D Bostick inversion, the resistivity cross section from [33] shown in Fig. 3.4 compares well with similar cross sections produced using new MT and TEM and both 1D and 3D inversions representative of the state-of-the-art [69]. Awibengkok was also among the earliest volcano-hosted geothermal reservoirs to have a geothermal resource conceptual model constructed using the smectite clay cap model to interpret the resistivity.

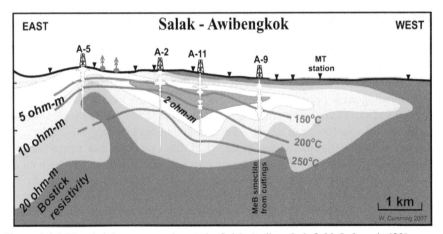

Figure 3.4 MT resistivity cross section at the Salak-Awibengkok field, Indonesia [33]. The color shading illustrates the resistivity pattern, the *red lines* are isotherms and the *yellow plots* in the wells show MeB smectite content. Miscorrelations are mainly due to projecting wells and MT stations to the line of section. Not shown is the deep upflow at just <300°C near well A-9.
Adapted from Gunderson R, Cumming W, Astra D, Harvey C. Analysis of smectite clays in geothermal drill cuttings by the methylene-blue method: for well-site geothermometry and resistivity sounding correlation. In: Proceedings of the 2000 world geothermal congress, Tokyo, Japan; 2000. p. 1175−1181.

The Awibengkok wells and MT have provided the data and materials used in the initial test of the methylene-blue (MeB) method for measuring smectite in order to validate the MT resistivity pattern using well data [33], as shown in Fig. 3.4. Although resistivity well logs have been run through the clay cap and have verified the geothermal resistivity interpretation model based on smectite clay [19,72], running wireline resistivity well logs through the smectite alteration zone is expensive and may put the well at risk since smectite is a swelling clay and a potential drilling hazard. The MeB method, based on a simple blotter test of smectite absorption of MeB dye, provides a low-cost quantitative measurement of smectite content in borehole cuttings that can be conducted at the well-site. It has become the most common method to quantitatively test geothermal MT resistivity interpretations with respect to well data.

3.11   Using MT to build conceptual models and define resource areas and targets

Even in an exploration prospect with no well data, a range of conceptual models representative of uncertainty with the elements shown in Figs. 3.1 and 3.4 can be constructed based only on hydrology, geology, the gas and water chemistry of fumarole and springs, the resistivity pattern, and the constraints of hydrothermodynamics. This process is illustrated in detail for a prospect like that in Fig. 3.1 in Ref. [21], while [17] addresses the more difficult case where there are no surface manifestations in a sediment-hosted prospect.

- The construction of a conceptual model where no well exists is usually most easily started from an initial estimate of the water table. For example, before the Awibengkok wells were drilled (Fig. 3.4), the water table for the exploration was based on a chloride spring off the cross section in Fig. 3.4. In the same way, the water table could be constrained by the chloride spring in the cross sections in Fig. 3.1.
- In Fig. 3.1, the gas geothermometry from the fumarole and cation geothermometry from the chloride hot spring provide two temperature points to plot on the cross section at the depth below the interpreted water table indicated by the boiling point for depth curve. (This is the shallowest depth possible for that temperature and so, unless evidence supports this as a very strongly favored model, other models would locate the same isotherms at greater depth.)
- Starting from these temperatures, isotherm contours can be drawn in a manner that is consistent with the resistivity pattern, with isotherms crowded together in the clay cap where permeability is low and spaced farther apart in the permeable reservoir in a manner consistent with case histories of buoyant flow.
- If the reservoir is (or had been) close to the boiling point at the water table, then the smectite clay cap is likely to extend above the water table, as it does at most fumaroles.
- A buoyant thermal outflow should flow up-dip below a low-resistivity zone corresponding to a smectite cap. The up-dip truncation of an outflow often corresponds to chloride springs. Water >100°C flows up-dip beneath the clay cap, not down-dip as in gravity flow, a common mistake in building conceptual models. For example, geothermal upflows associated with faults in extensional grabens in Turkey and the United States are often capped by impermeable clay-rich lake beds. These intercept the upflow and, because shallow lake beds

typically dip gently down to the center of the graben, they channel the hot water up-dip to the margins of the graben where springs and fumaroles mark the truncation of the lake beds by the basin bounding fault(s).

• In a volcano-hosted reservoir, a tabular low-resistivity zone with a top corresponding to the water table is likely to correspond to a smectite cap over a sub-boiling outflow (or a relict outflow).

• The resistive rock above the low-resistivity clay cap will usually correspond to the unaltered meteoric zone so temperature should be much cooler than 100°C except at thermal features. A high-resistivity zone associated with brittle fractured rock at the surface overlying a thinned cap may correspond to a cold water entry into parts of the reservoir, as is associated with rhyolite domes at Awibengkok [69] and at several New Zealand fields [10].

For both well targeting and capacity assessment, reservoir outlines are needed to define the surficial extent of the resource. A conceptual model approach leads to a different definition of reservoir limits than is used for anomalies. Instead of an assumed boundary associated with a resistivity contour value, the reservoir will be defined based on the productive implications of the conceptual model. In the case of the conceptual model in Fig. 3.1, the median (or P50) [22] reservoir outline might include the >275°C upflow and the tabular >250°C outflow. To estimate a P50 area, these limits would be identified on at least one more cross section. The interpreted isotherms are the most important element that defines the resource area, although these are routinely compared to the data that support the isotherm geometry, especially the surface geochemistry and MT resistivity. Therefore, although the resource area is defined by the resource model, it is common to generate map slices of resistivity or similar plots to help guide the outlines. Of course, well data are the highest priority information, if they exist.

Aside from resistivity cross sections and maps, MT graphics generated in support of the construction of conceptual models typically include maps of the elevation of the base of the low-resistivity zone, because this usually conforms to the top of the reservoir [2], even in cases where secondary non-smectite caps may exist [21]. A map of conductance to a sufficient depth to encompass most of the clay cap will often outline the thickest and most intense clay alteration. To synthesize the results of all of the cross sections and maps, the boundaries of the conceptual elements interpreted in each are transferred to a resource conceptual model map that reconciles them as boundaries for at least one upflow and several reservoir zones [19,21]. Producing such a map is an important objective of the resource assessment.

To characterize uncertainty in decisions based on the resource conceptual model analysis, a range of conceptual models is needed that illustrates the uncertainty in the data and the conceptual assumptions. To define a probability distribution representing Bayesian confidence [42] in the geothermal resource area as part of a resource capacity uncertainty assessment, both optimistic and pessimistic conceptual models are needed [21]. Experience has supported a common initial assumption that geoscientists who practice making assessments using case history data can estimate plausible 10% and 90% levels of confidence, constrained by the reasonable assumption of a lognormal distribution for resource area [61]. A similar range of outlines can guide well targeting, although specific targets can usually be more effectively evaluated using tools like decision tables to develop a self-documenting assessment of the targets. For resources that are expected to be water dominated, the greatest source of uncertainty in the resource outline is usually the proportion of the low-resistivity smectite zone assumed to cap outflow at temperature too low to be economic. Aside from the analysis of the geoscience data, a comparison to case history experience is probably the most effective approach to assessing the resource perimeter uncertainty.

3.12   Deep low-resistivity zones

Low-resistivity zones imaged below a smectite clay cap or in a formation with properties inconsistent with smectite clay, such as in high-rank metamorphic rocks, are commonly interpreted as either in high-fracture permeability saturated with hot saline water or as magma. However, case histories have shown that the former interpretation is usually misleading and the latter interpretation might be cautiously considered for a basalt-hosted system but is otherwise likely to be speculative. The following interpretations of deep low-resistivity zones detected below smectite clay zones are listed roughly in order of frequency that they are observed.

- *Distortion of 3D inversion by incoherent noise*: This is relatively easy to assess by comparing a detailed noise review of each MT station with respect to the location of the deep low-resistivity feature. In many cases, these features are located in gaps between the stations and so are too poorly constrained to be interpreted geologically.
- *Distortion of 3D inversion by coherent noise*: Although coherent noise tends to exaggerate resistivity to unrealistically high values immediately below the clay cap [44], if the natural signal is stronger than the coherent noise below 0.1 Hz, the apparent resistivity will abruptly decline below 0.1 Hz. If this is not edited, a 3D inversion will likely produce a false low-resistivity zone below $\sim 4-10$ km.
- *Graphitic-pelitic schist*: A very useful failure case history [45] indicates that graphitic-pelitic schist is low resistivity and low permeability. Where it overlies a permeable rock, such as dolomite, it may act as a reservoir cap, analogous to a smectite cap.
- *Magma*: MT has probably detected low-resistivity magma beneath several basalt-hosted reservoirs near active volcanoes, such as near Krafla in Iceland. The interpretation of magma using MT can be supported by seismic monitoring surveys.
- *Deep smectite clay*: As is hinted in Fig. 3.1, an outflow can be trapped below a smectite cap and on top of a similarly impermeable smectite cap.
- *Volcanic acid/vapor core zone*: Although low-resistivity minerals like smectite usually convert to chlorite or illite from 100 to 250°C, it is stable to much higher temperature at low pH, making it feasible that an acid vapor-core zone would be low resistivity, but this has not been reliably confirmed.
- *High-temperature clay*: Although resistivity declines as temperature increases from 50 to 250°C in a porous clay-free rock, it increases slightly as temperatures rise from 250 to 350°C. Interface conduction in clay may still lower resistivity at such high temperatures [29], but this effect is minimal in comparison to the reduction of resistivity that is characteristic of hydrated smectite at lower temperatures.

- *Hyper-saline fluid zones near or formerly near the brittle-ductile transition*: Although such rocks are known to occur, because the fluid inclusions are probably not interconnected when near the ductile transition, the resistivity model is still debated.
- *Massive metallic ore bodies*: These do occur but are geologically unlikely to be found associated with a geothermal reservoir and would probably be far more valuable than a geothermal field.
- *High fracture porosity saturated with high-temperature saline fluids*: Because a direct image of a highly fractured permeable zone is exactly the dream of all geothermal exploration managers, it is the most common interpretation made of low-resistivity zones deep below a clay cap. Some investigators have claimed to image such zones connected to the mantle. For a variety of reasons, such interpretations tend to be down-weighted in a conceptual model approach that considers uncertainty. The scale at which a surface method like MT images resistivity at reservoir depth is greater than 100 m. Fluid-filled permeable fractures imaged by a borehole micro-resistivity log are low resistivity at centimeter scale [24], but their cumulative average in a reservoir zone is still higher resistivity than a clay cap. Moreover, interface conduction in clays may be the predominant factor in geothermal reservoir resistivity [29], as well as in the clay cap, consistent with the common observation that permeable zones in geothermal reservoirs are imaged as resistors.

3.13   Gravity methods for exploration and development

Gravity surveys detect density contrasts in the subsurface, not density itself. In exploration applications of gravity, the density differences are mainly due to the different porosity of rock types that are detected as mapped variations in gravity at the surface. In development applications of gravity, repeat surveys are run at intervals of several years and the changes with time in the very precisely measured gravity are mainly due to changes in steam saturation in the pore space of a boiling point reservoir zone. Although directed at very different objectives, both types of gravity survey interpretation have similar types of ambiguity.

Because both types of gravity survey measure variation in a potential field rather than images derived from wave propagation, as is the case for the reflection seismic and MT methods, gravity results are much more ambiguous. A common problem with gravity surveys is aliasing, in which shallow density contrasts that account for a significant part of the observed gravity variation cannot be resolved because the station spacing is too large. This problem can be mitigated by collecting stations closer together. A more serious limitation of exploration gravity surveys in a geothermal context is that, at greater depths, particularly at temperatures >200°C, rocks become more uniformly indurated and density contrasts are much lower. This leads to a common rule of thumb: gravity differences should be initially explained at the depth where expected density contrasts are highest.

Gravity is acquired by occupying stations with a gravity meter. The highly automated Scintrex CG5 gravity meter has become the default for both exploration surveys and development precision surveys. Although a Lacoste & Romberg meter is more delicate, if an operator is available who has a proven record of acquiring reliable data, then its lower cost might be decisive. Very precise absolute gravity meters have an increasingly important role in establishing precision gravity reference stations, but lower cost relative gravity meters like the CG5 will likely meet most precision geothermal needs in the next decade. Gravity requires the accurate measurement of elevation and so a companion differential GPS survey is also run. Repeat precision surveys also require benchmarks that are designed to be stable for decades. Although airborne gravity methods are available, their application to geothermal fields is likely to be limited because land gravity surveys tend to be more cost-effective at the scale of a geothermal prospect. The field procedures and computations required for gravity surveys are well established and, providing that the acquisition contractor has an established reputation for meeting current standards, little quality assurance is usually required by the user of the data (which may not be the case for MT or reflection seismic surveys, for which data reliability depends on the question being asked).

The best known exploration gravity success case in a geothermal setting is at the Salton Sea Geothermal Field, where the gravity more or less outlines the reservoir [84]. This worked so well because the immediately surrounding Salton Trough sediments have a very uniform compaction trend. The lateral variations in density (and gravity) are due to the much higher density of the intrusive rocks and the altered shales (with hydrated smectite converted to dense illite) within the high-temperature reservoir [84]. Similar gravity features are found at the Brawley and Heber fields, also within the Salton Trough.

Exploration gravity works best where the targets of interest have a simple density geometry, like the interface between the low-density impermeable sediments that cap the high-density carbonate-hosted geothermal reservoirs found in extensional grabens in western Turkey and the western United States. These can sometimes be reliably resolved for well targeting at around 200- to 1500-m depth. Although the overall basin geometry at much greater depths can often be characterized based on shallower density contrasts, the more uniform density at greater depth near geothermal reservoirs makes deeper models very ambiguous. A common pitfall is failing to reduce the distance between stations as the interface being mapped becomes shallower; another is interpreting faults between most stations. Despite the pitfalls, gravity is the lowest cost approach to characterizing the geometry of extensional grabens that host many geothermal fields.

Gravity interpretation in volcano-hosted geothermal reservoirs is more emphatically top-down (ie, very shallow density contrasts explain most local gravity variation). At the surface of a volcano, rocks like pumice tuff with a density of 1100 kg/m^3 are commonly found adjacent to lava of density of 2600 kg/m^3, resulting in a lateral density contrast of 1500 kg/m^3. However, within a 250°C geothermal reservoir at 1000-m depth, a similar tuff will have been compacted and hydrothermally altered so that its density may have increased to 2300 kg/m^3, whereas the density of the lava is likely to remain almost unchanged at 2600 kg/m^3, resulting in a much lower lateral contrast of 300 kg/m^3. As a result, gravity variations due to density contrasts at reservoir depth are so much subtler than gravity variations due to surface geology that the deep contrasts cannot be reliably inferred from the gravity data. Moreover, the geological units associated with most of the gravity response can usually be more cost-effectively characterized using satellite imaging and/or geological mapping. On the other hand, joint interpretation of gravity with MT resistivity along profiles can be mutually supportive, with the gravity helping to identify whether surface resistors are unaltered low-density volcanics or unaltered dense lavas. (A common mistake is to assume that all resistive zones are dense.) A potential limitation is that, in volcanic terrain with a rapidly varying admixture of tuffs and lavas at the surface, gravity stations must be located close enough together to avoid aliasing, so that a suitably sampled gravity survey might cost more than MT. Although the difference in the interpretation may be of interest for constraining near-surface hydrology or for planning well casing, it is unlikely to affect resource targets or capacity assessment.

3.13.1 Repeat precision gravity

Repeat precision gravity surveys detect changes in steam/gas saturation in the rock pore space. Therefore, surveys should be designed to have station spacing closer than the depth to the reservoir over parts of a geothermal reservoir that may reach boiling conditions under exploitation. A common misconception is that gravity would increase at injection locations and decrease at production well locations. This can happen but it is incidental. Most boiling occurs near the apex of a reservoir where it is near the boiling point, the typical location for development of a steam cap. Changes in the steam cap account for most gravity changes in a geothermal context, although variations in shallow groundwater aquifer saturation due to changes in rainfall or depletion can complicate interpretations. Cooling related to injection can increase density and gravity but this is usually lower amplitude and confined to the immediate locality of injectors. The locations of the permanent bench marks for the repeat gravity survey should, therefore, be based mainly on the reservoir conceptual model and thermodynamic expectations for where the reservoir water is likely to boil or where a reservoir steam zone is likely to condense under long-term development.

The acquisition of precision gravity data using a relative gravity meter like a Scintrex CG5 requires the use of a reference benchmark. Because the reference may also have changes in absolute gravity, a reference measurement using an absolute gravity meter would reduce this ambiguity, but the cost has inhibited this practice. Depending on the type of gravity meter used, strategies to establish suitable precision vary, but every station should be measured at least twice, preferably three times during each survey. To correct precision gravity surveys for elevation changes that commonly occur in geothermal fields, the elevation of benchmarks must also be measured with precision, usually using differential GPS over large areas or bar code optical levels over smaller survey areas. This survey is also used to monitor field subsidence. Precision gravity surveys in geothermal fields typically achieve a relative precision of $10-20$ µGal.

A simple equation adapted from [1] can aid in planning gravity surveys and assessing shallow groundwater effects after a survey has commenced. The change in gravity due to a change in water saturation (with respect to either steam or air saturation) in a shallow infinite layer is:

$$\Delta g = -2\pi G \Phi (\rho_w - \rho_s)(S_0 - S_1)h \approx -0.0419 \Phi (\rho_w - \rho_s)(S_0 - S_1)h$$

where Δg is gravity change in μGal, Φ is fractional porosity, G is the Universal Gravity Constant, ρ_w is water density in kg/m^3, ρ_s is steam density, $(S_0 - S_1)$ is the change in water saturation fraction, and h is thickness in m.

Given a reservoir porosity of 0.15 and density change of roughly 800 kg/m^3 for a $\sim 250°$C reservoir water,

$$\Delta g \approx -5(S_0 - S_1)h$$

Therefore, if a reservoir simulation model showed that, over the first 2 years of production, the upper zone of the model had boiled so that the percentage of steam (vol.) was now 10% (change in water saturation of 0.1) in a 100-m-thick zone, then the gravity change expected would be <50 μGal, and, if the lateral extent of the zone was comparable to its depth, <25 μGal. This is detectable using repeat precision gravity surveys, so a survey design might be planned to repeat every other year.

Precision gravity data can be interpreted in a variety of ways, aside from simple analytical models used in survey design and quality assurance. The types of 2D models routinely computed for exploration gravity can be used, albeit with very small densities. A repeat gravity survey at the Rotokawa field in New Zealand detected a gravity increase related to the condensation of a shallow steam zone due to shallow injection [40]. Net recharge can be inferred by estimating net mass loss from the reservoir by integrating over the observed gravity [1], although this computation sometimes requires a more extensive survey than one used to fit the gravity using a numerical reservoir simulation. In a gravity simulation, the gravity change is computed for the gravity stations from the fluid density change simulated in a numerical reservoir simulation model. It was this type of computation that supported a revised simulation of the Bulalo field in the Philippines that indicated significantly larger reservoir support than initially modeled, a result with important implications for the eventual expansion of generation from 330 to 458 MW [6].

3.14 Magnetic methods

Magnetic methods measure local variations in the earth's magnetic field that are induced by lateral variations in the magnetic susceptibility and remnant magnetic polarization of rocks. In a geothermal context, magnetite content is the main determinant of susceptibility and, indirectly, of changes in the larger-scale remnant magnetization due to magnetic pole reversals.

There are four main conceptual applications claimed for magnetic surveys: (1) For sediment-hosted geothermal fields, volcanic and intrusive rocks can be detected based on their higher magnetite content [84]. (2) For volcano-hosted geothermal fields, the sensitivity of magnetite to alteration by sulfate water sometimes results in a demagnetized zone where H_2S rising from a high-temperature magmatic reservoir is oxidized by meteoric water [66]. (3) Structural geometry is often inferred based on magnetic patterns, albeit usually with significant ambiguity in geothermal contexts. (4) The depth to the loss of remnant magnetization at the Curie temperature near 570°C has been inferred from spectral analysis, although with uncertainty too high to provide useful constraints in a geothermal context.

Magnetic surveys have some characteristics analogous to gravity surveys in that both measure variations in a potential field that can be tested against hypotheses, but they do not provide the images of the earth from modeling wave propagation, as is the case for the reflection seismic and MT methods. Magnetic and gravity methods share a great success case history at the Salton Sea field. An aeromagnetic survey responded to the relatively high-magnetite content of the intrusive dikes embedded in the sediments, effectively outlining the reservoir in conjunction with gravity data [84].

A commonly attempted application of magnetic surveys in volcano-hosted prospects is the detection of zones demagnetized by sulfate water [66]. However, flying aeromagnetic surveys with the required constant terrain clearance (draped surveys) in rugged volcanic areas can be both costly and dangerous. Moreover, at low magnetic latitudes, such as in Indonesia, the magnetic field inclination is so low that the computations used in the analysis tend to be unstable. Ground magnetic surveys have demanding but more feasible requirements. Because volcanic rocks can induce large variations in magnetic field over distances of a few meters using a detector elevation of <5 m, to avoid aliasing, effective ground magnetic surveys are acquired with a station spacing of <1 m in volcanic areas. In areas with suitable access, magnetic surveys can sometimes be cost-effectively acquired along walked profiles using modern Overhauser or cesium vapor magnetometers. However, ground magnetic surveys for volcano-hosted geothermal prospects are more typically acquired at stations spaced 100s of meters apart, based on practices developed for sedimentary basin surveys, and these data are usually too aliased to be reliably interpreted. Because of these complications and the better imaging capability of resistivity methods, magnetic surveys have had limited impact on the geothermal exploration of volcanic areas.

3.15 Seismic monitoring

For ~30 years, monitoring of generally very small seismic events has been a routine application in geothermal developments that use deep injection into hot rock. It is now an essential tool for defining EGS reservoirs and managing related hazards and nuisance issues have been addressed by a formal standard [51]. An earlier version of the proposed standard has been referenced in regulatory documentation for conventional geothermal reservoirs [50].

Most publications on seismic monitoring of geothermal fields focus on imaging rock properties, commonly tomographic imaging of p- and s-wave seismic velocity. Because of resolution issues, this has seldom been directly used in making resource decisions, although the information is important to the velocity model for observed seismic events and is steadily improving. The most decisive analyses of seismic monitoring data that have affected resource decisions have involved careful correlation of production and injection rates, pressure and temperature, and permeable zones with respect to four-dimensional visualization of the location and timing of the seismic events [63].

A wide range of potential causes of induced seismicity in geothermal reservoirs have been proposed but a limited number fit with geothermal conceptual models.

- Stress related to rock cooling and contraction in response to injection of cold water in hot rock is probability the predominant mechanism in conventional geothermal fields. In some fields there may be a threshold temperature of the rock (>200°C in some cases), a threshold temperature difference between the rock and injectate (>100°C), and a minimum depth to ensure accumulation of stress (generally >1000 m). There are probably at least two subtypes of these cooling-related events: tensile crack formation due to cooling and double-couple seismic events triggered by stress changes. Because these events require cooling, injectate must reach zones where such activity occurs, so a time sequence ploy of event locations might look like a progressively advancing plume.
- Reservoir pressure decline causes compaction-related seismicity, particularly on the top and edges of reservoirs, or where a steam cap forms and increases the rock compressibility. This is a cumulative effect of production and will likely occur as a fuzzy halo of events, unless the events are triggered along a fault by the stress–strain perturbation.
- Local pressure increases around injection wells may trigger seismic events. Although pressure near injectors in permeable geothermal fields seldom increases much, except in high rate stimulation, it may be enough to trigger events. This type of activity is likely to be closely correlated in time with injection pressure but may occur at a distance from the well that is the source of the pressure perturbation, probably along a preferred permeable path.
- Seismic events are commonly triggered throughout geothermal fields by transient pressure changes during a field curtailment for maintenance or due to stress/strain transients caused by distant earthquakes. This is probably related to the perturbed stress state of the reservoir as a whole.
- Pressure increases large enough to hydraulically fracture a formation, similar to those observed during enhanced geothermal system (EGS) well stimulation activities [51], are unlikely to occur in a conventional geothermal setting where rocks are permeable and pressure perturbations are much lower.
- A variety of other causes of seismic events may also have some role, like precipitation of minerals on fractures causing a transition to stick-slip motion, but evidence for them is scanty.

Seismic events that are not induced by the geothermal development will be mixed with the induced events and they may be difficult to differentiate. For example, a non-induced tectonic seismic event may cause a sudden change in field management, like a power plant shut down, that could cause a pressure transient that induces seismic events. Whether they are induced or not, all seismic events near a geothermal field will be relevant to the resource conceptual model.

- Magma movement near or in a geothermal field has been observed in basalt-hosted geothermal fields. It would imply a potential hazard and a likely location for the 400°C isotherm in the conceptual model. The typical conventional production limit is <350°C.
- Seismic events related to regional tectonics are likely to occur independently of the geothermal field development. Seismic monitoring can usefully identify structures in the field but sufficient information to define the overall structure may take many decades or centuries to accumulate.

The physical models for occurrence of seismic events in geothermal fields explain many of the decisive conclusions based on seismic monitoring, such as at the Roto-kawa Geothermal field in New Zealand [63].

- Seismic monitoring of injection fluid movement has improved field management. At the Rotokawa field, injectate was shown to be slowed in its return to the production zone, providing a confirmation of geochemistry and tracer results.
- Major flow pathways and barriers to flow have been detected. At the Rotokawa field, a central field fault has been mapped using the pattern of events from the seismic monitoring. This fault acts as a conceptually important barrier to flow between the injection and production sectors of the field and also hosts significant permeability along the fault.
- The base of the permeable reservoir reached by injection has been characterized using seismic monitoring. At Rotokawa, injection-related events did not extend downward for a significant distance, suggesting that the volume over which the injectate would be cooled is much less than has been indicated at geothermal fields like The Geysers and Awibengkok. The depth of a seismic event is typically less well constrained than its mapped location, often as much as ± 500 m, an issue addressed by investment in improved reservoir seismic velocity models using seismic tomography, sonic logs, and bore hole check shot surveys.
- Exploration applications of seismic monitoring include locating the ductile transition zone and magmatic injection events in basalt-hosted prospects and characterizing volcanic hazards in andesite-hosted prospects.

Recent seismic monitoring research in a geothermal context has focused on EGS commercialization [51], but researchers and geothermal developers continue to collaborate on other geothermal issues. The use of short-term seismic monitoring arrays to characterize permeable structures for targeting exploration wells has remained a research topic since the 1970s. Challenges include the low level of seismic activity that is characteristic of most geothermal prospects until production and deep injection begin. Proposals to use downhole seismometers have merit with respect to data quality, but the cost of an effective array will likely remain cost prohibitive for a decade or more. Widely promoted methods like S-wave splitting tomography will probably remain research topics rather than applications until researchers establish realistic expectations for resolution and uncertainty, including uncertainty with respect to the conceptual interpretation. The development of lower-cost, nontelemetered seismic monitoring systems and the consolidation of the software so that survey costs are not dominated by the installation and basic processing costs to compile seismic event data [39] anticipates the evolution of seismic monitoring into a reservoir monitoring system routinely installed for any geothermal reservoir that uses deep injection into high-temperature rocks.

3.16   Reflection/refraction seismic methods

The reflection seismic method is almost synonymous with geophysics in a petroleum exploration context, historically accounting for the great majority of conventional petroleum exploration geophysics investment. Thus reflection seismic surveys might seem like an ideal approach to locate high-temperature permeable fracture zones.

However, although it is essential to the exploration of deep stratigraphic geothermal reservoirs, this method has had limited success in a geothermal context, especially in volcanic geothermal fields and in other ares where collection of high quality reflection seismic data is difficult and/or very expensive [53].

It is common for geophysical methods to be applied out of context, generally because decision makers lack an appreciation for the conceptual significance of the data. For example, MT resistivity surveys have been acquired in deep sedimentary basins where the very subtle structural and stratigraphic targets are being targeted using high-quality 3D reflection seismic data. The MT resistivity responds to the clay content of the shales and so confirms that the thick insulating impermeable sedimentary section is present, but MT lacks the resolution at several km depth to characterize the detailed stratigraphy or structure of the target. Conversely, many reflection seismic surveys have been acquired to target structures in volcanic geothermal fields with results much less useful than more conventional MT resistivity surveys [56].

The published best practices for targeting geothermal reservoirs in the Malm carbonates of Bavaria focus on high-quality 3D seismic reflection as the essential method required to image the fractures and characterize the karst that may be associated with the fractures [76]. Outstanding 3D reflection seismic data quality is required to conduct these analyses, and fortunately this is characteristic of the area. The importance and uncertainty of a variety of details in the interpretations remain a matter of debate; for example, there is uncertainty regarding the reliability of attributes used to characterize in the characterization of karst in areas with complex structure or where there may be interference from the overlying Purbeck formation. As case histories for the Malm accumulate, a more reliable assessment of the probability of success for analogous targets will be feasible.

At the Travale geothermal field in Italy, 3D reflection seismic imaging of the base of the carbonate reservoir at >280°C has resolved "H marker" reflectors interpreted as the contact metamorphic halo of skarns and hornfels [12]. This identifies the locus of permeability that statistically correlates with ~80% of the production at 2500 to 3500 m in depth. However, individual productive fractures have not been reliably imaged. A deeper "K marker" <5000 m in depth has been interpreted as the >400°C transition to the most recent granite intrusion. In this case, the target is not a structure but a metamorphic zone, implying potentially broader applications of reflection seismic surveys, somewhat comparable to MT, albeit typically at much greater cost.

3.17   Borehole wireline logs

Aside from well integrity logs and reservoir logs such as P-T-S (pressure-temperature-spinner) that are routinely run in a geothermal context, formation well logs directed at measuring rock properties or borehole image logs directed at imaging the borehole rock surface are typically given much less attention in geothermal versus hydrocarbon exploration. In hydrocarbon well completions, formation logs are essential for designing the perforation of cemented well liners to produce hydrocarbon pay zones and avoid water zones. Moreover, the pay zones are evaluated on scales of a few meters, requiring the details provided by logs. In contrast in a geothermal well the perforated liner is hung without cementing, leaving it open to the entire reservoir section.

The calibration of reflection seismic surveys that dominate hydrocarbon exploration requires wireline logs, which explains why such logs are run in wells located in deep stratigraphic geothermal prospects that are targeted using 3D reflection seismic surveys.

Because of changing regional availability, well log costs vary widely by location and time. Several strategies can reduce costs, such as committing to a minimum number of logs with a single contractor and using older logging technology that might address realistic needs. Besides the high cost of the logs themselves, most are run using a special high-temperature telemetry cable that must be rented. Costly rig time may be needed, 8−24 h typically. The lowest cost option for a formation correlation log is the gamma memory tool that uses a simple wireline without telemetry and can be run in casing after the rig has moved off location.

3.17.1 Borehole image logs

There are two main types of borehole imaging logs: (1) borehole televiewers that use a rotating acoustic echo ranger to image the geometry of the borehole and (2) microresistivity imaging logs that use many focused resistivity detectors on paddles pressed against the borehole wall to produce detailed images of resistivity variation [24]. Both can detect the orientation of fractures, although the televiewer usually provides a more reliable assessment of the geometry and aperture of fractures, and a much more reliable assessment of the formation stress state, using analyses of borehole breakouts and induced and natural cracks. On the other hand, microresistivity logs provide constraints on the texture and type of lithology being imaged [34,78], an option that is particularly useful when drilling blind without mud returns. If the fracture and stress state of a reservoir may become a crucial consideration in a large geothermal field, the cumulative information from running televiewer logs in near-vertical wells spaced around the field is likely to add value [25].

3.17.2 Borehole formation logs

Where formation logs are low cost and are unlikely to endanger the well, they should be run. At Pilgrim Hot Springs, Alaska [54], all wells were logged, often through casing, using slick-line gamma memory tools at modest cost, and these proved to be decisive in lithology correlation in a shale-sand sequence. However, at another field, >US$150,000 was spent to mobilize a wireline logging unit to the first well in a field, resulting in almost characterless gamma, resistivity, and sonic logs in relatively uniform andesite with propylitic alteration. A cost-effective option might be to test the likely signal using a memory tool before investing in a full set of logs.

3.17.3 Gamma logs

In a geothermal context, ordinary gamma logs primarily respond to radioactive potassium (with occasional uranium and thorium outliers). Gamma logs tend to be much lower cost than other formation logs, they are quicker to run, and have often proved to be decisive in correlating formations between wells. They are commonly combined at little extra cost with other logs. The lowest cost option for a wireline formation log of a cooled geothermal well (after drilling and circulating) is usually a slick-line gamma memory logging tool.

Geological sections that have the highly variable gamma response typical of most andesite and sand-shale sequences can usually be effectively logged through casing and liners, with a loss of resolution that is usually insignificant in geothermal applications. Lower natural gamma activity, bigger boreholes, more casings, and smaller liners/tools imply slower run times for gamma tools, especially spectral gamma tools.

Gamma spectra logs are often diagnostic of clay in general and that may be important to an assessment but they have not usually been reliably diagnostic of smectite versus mixed-layer chlorite or illite clays in a geothermal context.

3.17.4 Resistivity logs and alternative clay analyses

There are several types of formation resistivity logs run in a geothermal context, most commonly the dual induction log.

There are low-cost alternatives to resistivity logs. For example, MeB smectite measurement in cuttings [33] was initially motivated by the need to validate MT resistivity interpretations without the potentially hazardous procedure of running a resistivity log in smectite-bearing andesite tuff and lahar formations that are notoriously prone to trapping logging tools. The most common clay analysis methods for geothermal applications are cutting analyses using (1) MeB (lower cost and quantitative, but measuring only smectite and some mixed-layer smectite clays), (2) IR reflectance (low cost, nonquantitative, presence/absence), and (3) XRD of glycolated samples (higher cost, semiquantitative, full-spectrum clays) [78].

3.18 SP method

SP is the spontaneous voltage measured between electrodes placed in the ground. Because SP data can be acquired using equipment already available in most university geoscience laboratories, it is commonly found in geothermal prospect files. SP can be caused by shallow ground water flow (streaming potential), mineralized ore bodies, alteration, and stray currents from electric power facilities [38]. SP patterns between surface electrode points and injection wellbores have been measured in order to monitor injection water flow in reservoirs, but this remains a research topic. Most SP patterns detected at the surface in geothermal fields are related to shallow groundwater and subboiling thermal water flow above the clay cap.

Although all geophysical surveys are sometimes misapplied due to insufficient consideration of the conceptual and practical context of the application, SP surveys have been particularly prone to this problem. Rainfall and snowmelt can create large changes in SP over time scales of hours and months, and so the survey must be adjusted to ensure that a reasonable baseline condition for acquisition of the survey is maintained. Quality assurance information includes records of rainfall preceding and during the survey and repeat measurements of suitably located reference stations indicating consistency in the SP measurements. In SP surveys, the electrode spacing is usually determined by the equipment sensitivity, the power line noise level, and the patience of the surveyors. Unfortunately, the spacing needed to avoid aliasing the conceptually significant spatial variations in SP are commonly not considered in the survey design, making the measured SP pattern too ambiguous to interpret.

Well-documented SP surveys conducted with a suitable electrode spacing can address issues like leakage from drilling sumps and flow regimes around hot springs [38], although hazards to surveyors limit options to monitor high-temperature thermal manifestations using SP. A focus on detecting faults has missed more likely applications such as reducing ambiguity in geochemistry interpretations by constraining shallow flow of thermal water near hot springs.

3.19   Geophysics management issues

Strategies for making effective decisions under statistical conditions analogous to those encountered in geothermal exploration and development have been investigated in a wide variety of industries where the value of success justifies reasonable risk-taking but the probability of failure remains high and data are sparse. Realistic assessments of the chance of success of an initial (wildcat) geothermal exploration well are commonly lower than 50% for projects that might be financially viable from a risk-adjusted perspective. With just under 100 geothermal fields worldwide, geothermal case history data are sparse. Any conclusion about geothermal fields cannot be reliably based on a frequentist statistical approach that assumes that a large number (ideally many thousands) of cases have been sampled from a much larger population. To analyze decisions in this context, subjective Bayesian approaches described in popular publications [64] and statistical texts [42] support a decision process in which confidence based on conceptual model assessments can be updated as analysts consider new evidence and additional case histories.

Practical decision making requires that geothermal resource developers make choices regarding what methods they will use. Following the pattern in other businesses [64], geothermal experts who have more experience in assessing geoscience uncertainty and testing their predictions against outcomes tend to estimate higher uncertainty in their predictions and tend to be more skeptical of the incremental improvement provided by new data. For example, experts in a particular type of geophysical data who have limited experience building geothermal conceptual models or comparing predictions against drilling results might conclude that an exploration well has a 90% chance of success based on confidence in their own data. In contrast, an expert who has had more experience with drilling outcomes that test the incremental value of data to geothermal targeting decisions might conclude that the same geophysical data influenced the conceptual models in a manner that increased the chance of success from 40% to 50%, a more realistic expectation and still a cost-effective outcome for a US$150,000 geophysical survey conducted to target a US$7 million well.

Although managers responsible for making geothermal resource decisions should have sufficient experience to be skeptical of an estimate of a 90% chance of success for an exploration well, they are unlikely to be expert in all types of data and analyses that might increase the chance of success from 40% to 50%. A common approach used by managers to test the conclusions of surveys and analyses conducted in support of high-value geothermal resource decisions is to conduct a peer review by outside experts. The peer review can check that the data and analyses are consistent with the conceptual models, that the range of conceptual models considered is representative of the uncertainty in the data and interpretations, and that the assessment process mitigates the cognitive biases most likely to affect the decision.

The dream of most exploration managers is a geophysical method that images permeable fractures from the surface. However, this chapter briefly mentions only one method to accomplish this in a narrow subset of cases: 3D reflection seismic applied to deep stratigraphic targets. Although this type of survey reduces risk of failure to such an extent that it is essential in this context, target risk may still be too high to make this highly subsidized resource economic under current conditions. Yet many geophysical methods are promoted as reliable approaches to image permeable fractures in a wide range of geothermal contexts. A few of these methods appear in some lists of recommended practices. A peer review that considers the application context, the cost, and realistic prospects for reducing risk may reach a different conclusion.

The valuation of geophysical methods is complicated by the need to consider the context and reliability of the case histories that support the valuation. For example, in many cases, a commitment is made to drill a well somewhere in a particular land block before a geophysical survey is completed. The conclusion of the survey may be that all targets accessible from the land block are high risk. Because the drilling commitment forces a choice of a marginal target, in the likely event of a failed well, the geophysical method itself may be considered a failure. Case history claims for data are more likely to be credible if they consider: (1) uncertainty in the geophysical data, (2) uncertainty in the imaging method used to produce models of the subsurface, (3) uncertainty in the conceptual inferences derived from the geophysical images for the resource model properties, and (4) the overall geoscience and management context of the assessment.

Routinely developing a range of conceptual models that are representative of uncertainty can also help mitigate the challenges encountered in implementing a conceptual model approach. Regardless of how competently a conceptual model is constructed, when a well is drilled, at least some predicted model elements will likely be proved wrong, an unappealing of this approach for most geoscientists. Using an anomaly hunting approach is less psychologically challenging because the data that is targeted typically remain valid regardless of the outcome of a well. Moreover, building an integrated conceptual model requires learning about aspects of geothermal systems outside a specialist's area of expertise. Therefore, experts in a particular type of data will usually prefer to target that data or a narrow subset of conceptual elements that their data constrain, that is, focusing only on hot springs (that may be related to an outflow, rather than an upflow) or on faults (that may be boundaries rather than conduits) or on compiling these anomalies without a conceptual context (missing that they may be related to an outflow truncated by a fault). Resource managers can most effectively avoid these pitfalls by supporting a decision risk assessment process that requires the use of a range of conceptual models that realistically illustrate the decision uncertainty.

Increasing interest in Bayesian approaches and the difficulty of training staff to prepare consistent conceptual models have led some consulting groups to promote the assembly of all of a company's prospect data into a Bayesian resource assessment. Many of these amount to elaborate anomaly compilations. In others, the conceptual shape of resource properties may also be considered, but missing the crucial implications of thermodynamics. An analysis from a group nominally expert in statistics but not in constructing geothermal conceptual models might provide results that are easier for an experienced geothermal interpreter to analyze with respect to model properties, but the statisticians are unlikely to provide a reliable "answer."

Although following a well-tested geothermal assessment process can help mitigate the cognitive biases that commonly distort resource predictions [42,61], relevant personal experience in making similar predictions is essential to such a process [64]. However, with <100 geothermal fields developed over the past 50 years, few geothermal exploration staff can have targeted enough wildcat wells to build a balanced perspective on the risks of exploration drilling. Even fewer have completed a retrospective analysis of the effectiveness of different data types and assessment approaches used for geothermal decision making. One remedy for this is to train geothermal management and geoscience staff using simulations of geothermal exploration prediction and decision processes based on case histories of developed geothermal fields.

Acknowledgments

Two anonymous reviewers encouraged a significant expansion of the first draft of this chapter, and Steven Sewell provided a critical final review that clarified many points and corrected a few blunders.

References

[1] Allis R, Hunt T. Analysis of exploitation induced gravity changes at Wairakei Geothermal Field. Geophysics 1986;51:1647−60.
[2] Anderson E, Crosby D, Ussher G. Bulls-eye! - simple resistivity imaging to reliably locate the geothermal reservoir. In: Proceedings of the 2000 world geothermal congress, Tokyo, Japan; 2000. p. 909−14.

[3] Árnason K, Karlsdóttir R, Eysteinsson H, Flóvenz Ó, Guðlaugsson S. The resistivity structure of high-temperature geothermal systems in Iceland. In: Proceedings of the 2000 world geothermal congress, Tohoku-Kyushu, Japan; 2000. p. 923—8.

[4] Árnason K, Eysteinsson H, Hersir G. Joint 1D inversion of TEM and MT data and 3D inversion of MT data in the Hengill area, SW Iceland. Geothermics 2010;39:13—34.

[5] Árnason K. The static shift problem in MT soundings. In: Proceedings of the 2015 world geothermal congress, Melbourne, Australia; 2015. 12 pp.

[6] Atkinson P, Pedersen J. Using precision gravity data in geothermal reservoir engineering modeling studies. In: Proceedings of the 13th workshop on geothermal reservoir engineering. California: Stanford University; 1988. p. 35—40.

[7] Beamish D, Travassos J. The use of the D+ solution in magnetotelluric interpretation. J Appl Geophys 1992;29:1—19.

[8] Benoit WR, Hiner JE, Forest RT. Discovery and geology of the Desert Peak Geothermal Field: a case history. Nev Bureau Mines Geol Bull 1982;97. 82 pp.

[9] Bond C, Gibbs A, Shipton Z, Jones S. What do you think this is? "Conceptual uncertainty" in geoscience interpretation. GSA Today 2007;17:4—10.

[10] Boseley C, Cumming W, Urzúa-Monsalve L, Powell T, Grant M. A resource conceptual model for the Ngatamariki Geothermal Field based on recent exploration well drilling and 3D MT resistivity imaging. In: Proceedings of the 2010 world geothermal congress, Bali, Indonesia; 2010. 7 pp.

[11] Buness H, Von Hartmann H, Rumpel H-M, Beilecke, Musmann P, Schulz R. Seismic exploration of deep hydrogeothermal reservoirs in Germany. In: Proceedings of the 2010 world geothermal congress, Bali, Indonesia; 2010. 5 pp.

[12] Casini M, Ciuffi S, Fiodelisi A, Mazzotti A. Results of a 3D seismic survey at the Travale (Italy) test site. Geothermics 2010;39:4—12.

[13] Casteel J, Trazona R, Melosh G, Niggemann K, Fairbank B. A preliminary conceptual model for the Blue Mountain Geothermal System, Humboldt County, Nevada. In: Proceedings of the 2010 world geothermal congress, Bali, Indonesia; 2010. 6 pp.

[14] Chave A, Jones A. In: Chave A, Jones A, editors. The magnetotelluric method: Theory and practice. Cambridge University Press; 2012. 512 pp.

[15] Christensen A, Auken E, Sørensen K. The transient electromagnetic method. Groundwater Geophys 2006;71:179—225.

[16] Cumming W, Nordquist G, Astra D. Geophysical exploration for geothermal resources, an application for combined MT-TDEM. In: Extended abstracts, 70th annual international meeting, society of exploration geophysics, 19; 2000. p. 1071—4.

[17] Cumming W. Geothermal resource conceptual models using surface exploration data. In: Proceedings, 34th workshop on geothermal reservoir engineering, 2009. California: Stanford University; 2009. 6 pp.

[18] Cumming W, Bruhn D. Preface. The European I-GET project: integrated geophysical exploration technologies for deep geothermal reservoirs. Geothermics 2010;39:1—3.

[19] Cumming W, Mackie R. MT survey for resource assessment and environmental mitigation at the Glass Mountain KGRA. California Energy Commission, GRDA Geothermal Resources Development Account Report. 2007. Available at: http://www.energy.ca.gov/2013publications/CEC-500-2013-063/CEC-500-2013-063.pdf. 119 pp.

[20] Cumming W, Mackie R. Resistivity imaging of geothermal resources using 1D, 2D and 3D MT inversion and TDEM static shift correction illustrated by a Glass Mountain case history. In: Proceedings of the 2010 world geothermal congress, Bali, Indonesia; 2010. 7 pp.

[21] Cumming W. Resource conceptual models of volcano-hosted geothermal reservoirs for exploration well targeting and resource capacity assessment: construction, pitfalls and new challenges. To be presented at the 40th Geothermal Resources Council Annual Meeting, 2016. California: Sacramento; 2016a.

[22] Cumming W. Resource capacity estimation using lognormal power density from producing fields and area from resource conceptual models; advantages, pitfalls and remedies. In: Proceedings, 41st Workshop on Geothermal Reservoir Engineering, 2016. California: Stanford University; 2016b. p. 7.

[23] Dahlberg E. Relative effectiveness of geologists and computers in mapping potential hydrocarbon exploration targets. J Int Assoc Math Geol 1973;7:373–94.

[24] Davatzes N, Hickman S. Interpretation and comparison of electrical and acoustic image logs from a well in the Coso Geothermal Field, California. In: Proceedings, 30th workshop on geothermal reservoir engineering. California: Stanford University; 2005. 11 pp.

[25] Davatzes N, Hickman S. The feedback between stress, faulting, and fluid flow: lessons from the Coso Geothermal Field, CA, USA. In: Proceedings of the 2010 world geothermal congress, Bali, Indonesia; 2010. 12 pp.

[26] Ershaghi I, Dougherty E, Handy L. Formation evaluation in liquid dominated geothermal reservoirs. 1981. Report on US DOE Contract No. AT03–76ET28384, 110 pp.

[27] Essene E, Peacor D. Clay mineral thermometry – a critical perspective. Clays Clay Miner 1995;43:540–53.

[28] Facca G, Tonani F. The self-sealing geothermal field. Bull Volcanol 1967;30(1): 271–3.

[29] Flóvenz O, Spangenberg E, Kulenkampff J, Árnason K, Karlsdóttir R, Huenges E. The role of electrical interface conduction in geothermal exploration. In: Proceedings of the 2005 world geothermal congress, Antalya, Turkey; 2005. 9 pp.

[30] Georgsson L. Geophysical methods used in geothermal exploration. In: Short course IV on exploration for geothermal resources. By UNU-GTP, KenGen and GDC. Lake Naivasha, Kenya; 2009. 16 pp.

[31] Grant M. Resource assessment, a review, with reference to the Australian code. In: Proceedings of the 2015 world geothermal congress, Melbourne, Australia; 2015. 6 pp.

[32] Grant M, Bixley P. Geothermal reservoir engineering. Academic Press; 2011. 378 pp.

[33] Gunderson R, Cumming W, Astra D, Harvey C. Analysis of smectite clays in geothermal drill cuttings by the methylene blue method: for well site geothermometry and resistivity sounding correlation. In: Proceedings of the 2000 world geothermal congress, Tokyo, Japan; 2000. p. 1175–81.

[34] Halwa L, Wallis I, Lozada G. Geological analysis of the volcanic subsurface using borehole resistivity images in the Ngatamariki Geothermal Field, New Zealand. In: Proceedings of the New Zealand geothermal workshop 2013; 2013. 9 pp.

[35] Hersir F, Árnason K. Resistivity methods – MT. In: Short course VI on exploration for geothermal resources. UNU-GTP, GDC and KenGen, Lake Naivasha, Kenya; 2011. 7 pp.

[36] Hersir G, Björnsson A. Geophysical exploration for geothermal resources, principles and applications. Iceland: UNU-GTP; 1991. Report 15, 94 pp.

[37] Hickman S, Davatzes NC. In-situ stress and fracture characterization for planning of an EGS stimulation in the Desert Peak Geothermal Field, NV. In: Proceedings 35th workshop on geothermal reservoir engineering. California: Stanford University; 2010. 11 pp.

[38] Hochstein MP, Mayhew ID, Villanueva RA. Self-potential surveys of the Mokai and Rotokawa high temperature fields. In: Proceedings of the New Zealand geothermal workshop 1990; 1990. p. 87–90.

[39] Hutchings L, Jarpe S, Boyle K, Viegas G, Majer E. Inexpensive, automated micro-earthquake data collection and processing system for rapid, high-resolution reservoir analysis. Geotherm Resour Counc Trans 2011;35:1679−86.

[40] Hunt T, Bowyer D. Reinjection and gravity changes at Rotokawa Geothermal Field, New Zealand. Geothermics 2007;36:421−35.

[41] IGA Service GmbH. Best practice guide for geothermal exploration. 2014. Report written by GeothermEx Inc for International Finance Corporation, edited by Dr. Colin Harvey for IGA Service GmbH, 196 pp.

[42] Jeffrey R. Subjective probability: the real thing. Cambridge University Press; 2004. 142 pp.

[43] Jones A. Three-dimensional galvanic distortion of three-dimensional regional conductivity structures: comments on "Three-dimensional joint inversion for magnetotelluric resistivity and static shift distributions in complex media" by Y. Sasaki and M.A. Meju (2006). J Geophys Res − Solid Earth 2011;116:B12104.

[44] Junge A. Characterization and correction for cultural noise. Surv Geophys 1996;17: 361−91.

[45] Kuyumcu Ö, Destegül-Solaroglu U, Hallinan S, Çolpan B, Turkoglu E, Soyer W. Interpretation of 3D magnetotelluric (MT) surveys: basement conductors of the Menderes Massif, Western Turkey. Geotherm Resour Counc Trans 2011;35:861−6.

[46] Larsen J, Mackie R, Manzella A, Fiordelisi A, Rieven S. Robust smooth MT transfer functions. Geophys J Int 1996;124:801−19.

[47] Lutz SJ, Hickman S, Davatzes NC, Zemach E, Drakos P, Robertson-Tait A. Rock mechanical and petrologic testing in support of well stimulation activities at the Desert Peak Geothermal Field, Nevada. In: Proceedings 34th workshop on geothermal reservoir engineering. California: Stanford University; 2010. 11 pp.

[48] Lutz SL, Zutshi A, Drakos P, Robertson-Tait A, Zemach E. Lithologies, hydrothermal alteration, and rock mechanical properties in Wells 15-12 and BCH-3, Bradys Hot Springs Geothermal Field, Nevada. Geotherm Resour Counc Trans 2011;35:470−6.

[49] Mackie R, Madden T. Three-dimensional magnetotelluric inversion using conjugate gradients. Geophys J Int 1993;15:215−29.

[50] Majer E, Baria R, Stark M. Protocol for induced seismicity associated with enhanced geothermal systems. Report produced in Task D Annex I (9 April 2008), International Energy Agency-Geothermal implementing agreement (incorporating comments by C. Bromley, W. Cumming, A. Jelacic and L. Rybach). 2009. Available at: http://www.iea-gia. org/publications.asp.

[51] Majer E, Nelson J, Robertson-Tait A, Savy J, Wong I. Protocol for addressing induced seismicity associated with enhanced geothermal systems. Report DOE/EE-0662. 2012. Available at: https://www1.eere.energy.gov/geothermal/pdfs/geothermal_seismicity_ protocol_012012.pdf.

[52] McNitt JR. The fault block theory for the migration of geothermal fluid in non-volcanic systems, a model for exploration in extended terrains. Geotherm Resour Counc Bull 1994;232:55−61.

[53] Melosh G, Cumming W, Casteel J, Niggemann K, Fairbank B. Seismic reflection data and conceptual models for geothermal development in Nevada. In: Proceedings of the 2010 world geothermal congress, Bali, Indonesia; 2010. 6 pp.

[54] Miller JK, Prakash A, Daanen R, Haselwimmer C, Whalen M, Benoit D, et al. Geologic model of the geothermal anomaly at Pilgrim Hot Springs, Seward Peninsula, Alaska. In: Proceedings 38th workshop on geothermal reservoir engineering. Stanford, California: Stanford University; 2013. 9 pp.

[55] Muñoz G. Exploring for geothermal resources with electromagnetic methods. Surv Geophys 2014;35(1):101−22.

[56] Nakagome O, Yoshinori I, Uchida T, Horikoshi T. Seismic reflection & VSP in some geothermal fields in Japan − a contribution to a knowledge of fractured reservoir characterization. Geotherm Resour Counc Trans 1996;20:379−85.

[57] Parker R, Booker J. Optimal one-dimensional inversion and bounding of magnetotelluric apparent resistivity and phase measurements. Phys Earth Planet Inter 1996;98: 269−82.

[58] Pellerin L, Hohmann G. Transient electromagnetic inversion: a remedy for magnetotelluric static shifts. Geophysics 1990;55:1242−50.

[59] Pellerin L, Johnston JM, Hohmann GW. A numerical evaluation of electromagnetic methods in geothermal exploration. Geophysics 1996;61:121−30.

[60] Qian W, Pedersen L. Industrial interference magnetotellurics: an example from the Tangshan area, China. Geophysics 1991;56:265−73.

[61] Schuyler J, Newendorp P. Decision analysis for petroleum exploration. 3rd ed. Planning Press; 2013. 588 pp.

[62] Sepulveda S, Glynn-Morris T, Mannington W, Charroy J, Soengkono S, Ussher G. Integrated approach to interpretation of a magnetotelluric study at Wairakei, New Zealand. In: Proceedings 36th workshop on geothermal reservoir engineering. California: Stanford University; 2012. 8 pp.

[63] Sewell S, Cumming W, Winick J, Quinao J, Bardsley C, Wallis I, et al. Interpretation of microseismicity at the Rotokawa Geothermal Field, 2008 to 2012. In: Proceedings of the 2015 world geothermal congress, Melbourne, Australia; 2015. 10 pp.

[64] Silver N. The signal and the noise: why so many predictions fail−but some don't. Penguin Books; 2012. 560 pp.

[65] Siripunavaraporn W, Egbert G, Lenbury Y, Uyeshima M. Three-dimensional magnetotelluric inversion: data-space method. Phys Earth Planet Inter 2005;150:3−14.

[66] Soegkono S, Hochstein M. Application of magnetic method to assess the extent of high temperature geothermal reservoirs. In: Proceedings 20th workshop on geothermal reservoir engineering. California: Stanford University; 1995. 8 pp.

[67] Stark M, Soyer W, Hallinan S, Watts M. Distortion effects on magnetotelluric sounding data investigated by 3D modeling of high-resolution topography. Geotherm Resour Counc Trans 2013;37:521−8.

[68] Sternberg B, Washburne I, Pellerin L. Correction for the static shift in magnetotellurics using transient electromagnetic soundings. Geophysics 1988;53:1459−68.

[69] Stimac J, Nordquist G, Suminar A, Sirad-Azwar L. An overview of the Awibengkok Geothermal System, Indonesia. Geothermics 2008;37:300−31.

[70] Szybinski A, Shore G. Preliminary geological and geophysical characteristics of the Pumpernickel Valley Geothermal System. In: Proceedings 30th workshop on geothermal reservoir engineering. California: Stanford University; 2006. 8 pp.

[71] Taylor M. The state of geothermal technology. Part I: Subsurface technology. Published by the Geothermal Energy Association for the U.S. Department of Energy; 2007. 80 pp.

[72] Uchida T. Three-dimensional magnetotelluric investigation in geothermal fields in Japan and Indonesia. In: Proceedings of the 2005 world geothermal congress, Antalya, Turkey; 2005. 12 pp.

[73] Uchida T, Song Y, Lee T, Mitsuhata Y, Lim S-K, Lee S. Magnetotelluric survey in an extremely noisy environment at the Pohang low-enthalpy geothermal area, Korea. Antalya: World Geothermal Congress; 2005. 11 pp.

[74] Ussher G, Harvey C, Johnstone R, Anderson E. Understanding the resistivities observed in geothermal systems. In: Proceedings of the 2000 world geothermal congress, Tokyo, Japan; 2000. p. 1915−20.

[75] US-DOE. Best practices for risk reduction workshop follow-up manual. July 8, 2014. 40 pp.

[76] Von Hartmann H, Buness H, Krawczyk C, Schuktz R. 3D seismic analysis of a carbonate platform in the Molasse Basin - reef distribution and internal separation with seismic attributes. Tectonophysics 2012;573: 16−25.

[77] Vozoff K. The magnetotelluric method. In: Nabighian M, editor. Electromagnetic methods in applied geophysics, society of exploration geophysicists, vol. 2; 1991. p. 641−711. part B.

[78] Wallis I, McCormack S, Sewell S, Boseley C. Formation assessment in geothermal using wireline tools − application and early results from the Ngatamariki Geothermal Field. In: Proceedings of the New Zealand geothermal workshop 2012; 2012. 8 pp.

[79] Wannamaker PE, Rose PE, Doerner WM, McCulloch J, Nurse K. Magnetotelluric surveying and monitoring at the Coso geothermal area, California, in support of the enhanced geothermal systems concept: survey parameters, initial results. In: Proceedings of the 2005 world geothermal congress, Antalya, Turkey; 2005. 7 pp.

[80] Watts MD, Mackie R, Scholl C, Hallinan S. Limitations of MT static shift corrections using time-domain EM data. In: Extended abstracts, 83rd annual international meeting, society of exploration geophysics, Houston, Texas; 2013. p. 681−4.

[81] Wilmarth M, Stimac J. Power density in geothermal fields. In: Proceedings of the 2015 world geothermal congress, Melbourne, Australia; 2015. 7 pp.

[82] Wright P, Ward S, Ross H, West R. State of the art geophysical exploration for geothermal resources. Geophysics 1985;50:2666−96.

[83] Young C, Reber T, Witherbee K. Hydrothermal exploration best practices and geothermal knowledge exchange on openei. In: Proceedings 36th workshop on geothermal reservoir engineering. California: Stanford University; 2012. 13 pp.

[84] Younker L, Kasameyer P, Tewhey J. Geological, geophysical, and geothermal characteristics of the Salton Sea Geothermal Field, California. J Volcanol Geotherm Res 1982;12: 221−58.

[85] Zonge K, Hughes L. Controlled source audio-frequency magnetotellurics. In: Nabighian M, editor. Electromagnetic methods in applied geophysics, society of exploration geophysicists, vol. 2; 1991. Applications.

Application of geochemistry to resource assessment and geothermal development projects

J.R. Haizlip
Geologica Geothermal Group Inc., San Francisco, CA, United States

4.1 Introduction

The chemistry of geothermal fluids is established by the interaction of water and rock in the reservoir. Since almost all geothermal fluids originate as meteoric water or sea water, the primary variables are the rocks and the physical conditions under which the reactions occur such as temperature, pressure, and time. Therefore, geothermal hydrogeochemistry contains information about the physical conditions that geothermal explorationists want to know about: reservoir rock, reservoir temperature, and sometimes water:rock ratios.

The geochemist can contribute to geothermal development from the earliest stages of exploration to resource assessment, selection of the size and type of power cycle, the design of well fields, power plants and gathering systems, especially separation and injection, environmental mitigation, reservoir management, and power plant and well-field operations and maintenance. During the operational phase of a geothermal project, fluid chemistry is one of the few readily accessible tools available that provides a real-time view of the reservoir response to production and injection and variations in the resource supply to the power plant.

This chapter attempts to expose the reader to some of the methods geochemists apply to collect, evaluate, and interpret hydrogeochemical data from geothermal systems in order to address various issues which arise during geothermal exploration and development. After a very brief introduction to how geothermal fluids acquire their chemistry, this chapter discusses the following subjects in approximate order of a geothermal power project development from exploration to operations:

- Geothermal fluid geochemical data collection: sampling and analysis of the surface manifestations of a geothermal system, well sampling for testing and reservoir monitoring, and geofluid sampling for power plant O&M.
- Predrilling fluid chemistry during geothermal exploration; geochemical processes and interpretation.
- Applying well test geochemistry to resource assessment and development planning.

Geothermal Power Generation. http://dx.doi.org/10.1016/B978-0-08-100337-4.00004-8

- Geochemical aspects of resource design criteria for gathering system and power plant design.
- Geochemical monitoring for reservoir management and supporting power plant operations and maintenance.

This chapter relies heavily on the early work by geochemists who transformed the understanding of geothermal systems by applying basic principles of water–rock interaction to the physical state of the system; see, for example Refs. [4–6,29,31,44,45,77]. Following the exploitation of geothermal resources for power generation, geochemists applied their tools to understand the changes in reservoir conditions and in the resource supply, such as at Wairakei and Ngawha in New Zealand [30,56,57], at Cerro Prieto and Ahuachapán [79,83], Larderello [23,27,81], Icelandic systems [4–6,9], and Mahangdong, Philippines [2]. Powell and Cumming [75] incorporated many of these geochemical graphical and geothermometer tools into a spreadsheet database program. In addition, several geochemical modeling programs have been developed leading to the current geochemical modeling programs most often used: WATCH [4,11,14,59], EQ3/EQ6 [28,86], and Tough-REACT [87].

4.2 Early-phase resource assessment

Geochemistry can make critical contributions to the assessment of resource capacity. Early-stage resource assessments are often based on volumetric [41,42,67,71,84] or power density [85] methods. Both of these methods require an estimate of the geothermal resource temperature and area or volume. The primary contribution of the geochemist to early resource assessment is the reservoir temperature based on geothermometers before drilling allows downhole measurements [39] and reservoir size through the conceptual model, as discussed later. This section discusses liquid and gas geothermometers. The following section discusses geochemical contributions to conceptual models.

4.2.1 Liquid geothermometers and impacting processes

Properly applied geothermometers provide an estimate of subsurface temperatures based on fluid chemistry. Most geothermometers are based on a temperature-sensitive chemical reaction between fluids or between fluids and rock assumed to reach local equilibrium and control the observed fluid chemistry [9]. Others are empirical, focusing on the correlation between fluid chemistry from wells and measured temperature (eg, Ref. [35]). Equilibrium reactions include solubility of individual minerals and temperature-dependent exchange reactions (eg, Refs. [34,45]). Geothermometers can be based on either water-soluble constituents of geothermal brines such as cations, silica, or volatile constituents of geothermal steam such as gases. While virtually all geothermometer reactions are temperature dependent, the best geothermometers are those that are sensitive, that is, that produce a large change in fluid chemistry over the temperature range of interest, typically $100-350°C$.

The appropriate use of liquid-based geothermometers, especially with fluids from surface manifestations, requires that the water must have a component of deep geothermal fluid. Therefore, liquid-based geothermometers cannot be reliably applied to steam, condensate, or steam-heated waters. Some geothermometers are fast reacting, such as silica and K/Mg, and others are slow reacting such as Na/K/Ca, and differences between geothermometers can suggest various processes affecting the fluids closer or farther from the sample point. By relying on ratios, some geothermometers, such as Na/K or Na/Li, are insensitive to mixing or dilution, whereas temperatures from single-component geothermometers such as silica are reduced by mixing with cooler waters.

Many others have discussed geothermometers in general and their applications (eg, Refs. [38,58,61,66,78]). In this section, a few select liquid geothermometers are listed and guidelines for their application are summarized (Table 4.1). The most commonly used liquid geothermometers, silica and Na/K geothermometers, are discussed next.

4.2.1.1 Silica in geothermal fluid, temperatures, and mixing

Silica concentrations in geothermal waters are primarily determined by equilibrium with quartz above $120-180°C$ [34,36,38] or with equilibrium with chalcedony at lower temperatures. Silica geothermometers are dependent on a single constituent; therefore, they can be affected by processes which concentrate silica in geothermal liquid such as boiling or dilute silica such as mixing. A few simple tools have been developed to accommodate the effect of these processes.

For boiling springs, the silica concentration before boiling can be estimated using a silica—enthalpy diagram developed by Fournier (Fig. 4.1 combining figures from Ref. [38]), which uses a graphical technique to compensate for boiling. The quartz geothermometer (maximum steam loss) incorporates the effect of boiling into the temperature equation.

Mixing can produce multiple hot springs with varying silica concentrations and temperatures. Reservoir silica concentrations can be derived by fitting a straight line through the mixed spring chemistry and enthalpy to intersect the quartz solubility line (Fig. 4.1). This model can be applied when there is evidence of mixing such as (1) spring discharge rates are sufficiently high to prevent reequilibration by conductive cooling, (2) conservative elements such as chloride show similar patterns (Section 4.3.3) when plotted against enthalpy, and (3) in well discharge fluids, the measured temperature is much lower than the silica or cation geothermometers, and the silica temperatures are lower than the cation geothermometers.

4.2.1.2 Cation geothermometers and water:rock equilibration

The mixing issue is bypassed when applying the cation geothermometers (Table 4.1) based on ratios such as the Na/K geothermometer. The cation geothermometer depends on the assumption of equilibrium cation exchange between Na and K feldspars. Since in most geothermal systems feldspars alter to clays at lower temperatures,

Table 4.1 Select liquid geothermometers (concentrations in mg/L unless otherwise noted)

Geothermometer	Equation (all constituent concentrations in mg/L)	Application	References
Quartz (no-steam loss)	$t°C = \{1309/(5.19 - \log SiO_2)\} - 273.15$	Liquid <250°C, pH <8–9, no mixing, salinity <2.3%	[33]
Quartz (max steam loss)	$t°C = \{1522/(5.75 - \log SiO_2)\} - 273.15$	Liquid <250°C, pH <8–9, no mixing, salinity <2.3%, boiling at 100°C	[33]
Quartz	$t°C = -42.198 + 0.28831\,SiO_2 - 3.6686 \times 10^{-4}\,SiO_2^2 + 3.1665 \times 10^{-7}\,SiO_2^3 + 77.034 \log SiO_2$	Liquid 20–330°C, salinity <2%	[37]
Chalcedony	$t°C = \{1032/(4.69 - \log SiO_2)\} - 273.15$	$T < 180°C$, pH <8–9, no mixing, salinity <2.3%	[33]
Amorphous silica	$t°C = \{731/(4.52 - \log SiO_2)\} - 273.15$	<150°C	[33,35]
Na/K	$t°C = \{1217/[(\log (Na/K)) + 1.483]\} - 273.1$	Equilibrium with feldspars not clays	[36]
Na/K	$t°C = \{856/[(\log (Na/K)) + 0.857]\} - 273.1$	Equilibrium with feldspars not clays	[78]
Na/K	$t°C = \{933/[(\log (Na/K)) + 0.993]\} - 273.1$	25–250°C, equilibrium with feldspars not clays	[6]
Na/K	$t°C = \{1319/[(\log (Na/K)) + 1.699]\} - 273.15$	250–350°C, equilibrium with feldspars not clays	[6]
Na/K	$t°C = \{1390/[(\log (Na/K)) + 1.75]\} - 273.15$	Equilibrium with feldspars not clays	[45,47]
Na/K	$t°C = \{1178/[(\log (Na/K)) + 1.470]\} - 273.15$	Equilibrium with feldspars not clays	[73]
Na/K	$t°C = \{876.3\ (\pm 26.26)/[(\log (Na/K)) + 0.8775 \pm 0.0508]\} - 273.15$	Statistically derived from well data	[76]
Na/K–Ca	$t°C = \{1647/[\log (Na/K) + \beta\,(\log ((Ca/Na)^{1/2}) + 2.06) + 2.47]\} - 273.15$	$B = 4/3$ for $t < 100°C$ $B = 1/3$ for $t > 100°C$	[35]
Na/Li	$t°C = \{1590/[(\log (Na/Li)) + 0.779]\} - 273.15$	200–350°C, dilute waters	[62]
Na/Li	$t°C = \{1967/[(\log (Na/Li)) + 1.267]\} - 273.15$	60–350°C, sea water source/basalt	[76]
Na/Li	$t°C = \{855/[(\log (Na/Li)) + 1.267]\} - 273.15$		[76]
Li/Mg	$t°C = \{2200/(5.470 - [\log (Li/(Mg)^{1/2})])\} - 273.15$	Applicable at lower temperatures	[63]
K/Mg	$t°C = \{4410/(14 - (\log (K^2/Mg)))\} - 273.15$	Fast-reacting, no acid fluids	[48]

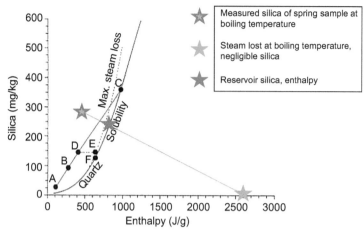

Figure 4.1 Silica enthalpy diagram for estimating reservoir enthalpy from a boiled fluid at the surface. A line fit through silica concentrations from multiple samples of mixed springs (A, B, and D) can be related to a deep reservoir concentration of silica (C) on the quartz solubility line assuming all of the springs represent different levels of mixing between a reservoir fluid in equilibrium with quartz (C) and a cold fluid with negligible silica concentration (~ 0 mg/L) and temperature. (enthalpy ~ 0 kJ/kg) Whereas if boiled, the temperature of boiling can also be evaluated (E) from intersecting the quartz maximum steam loss line at constant silica, and the related unboiled concentration on the quartz solubility line at the same enthalpy (F). In addition, water from a boiled spring (*BLUE STAR*) has lost steam by boiling at surface temperatures (at100°C, enthalpy of steam ~ 2600 J/g) (*GREEN STAR*), the reservoir silica concentration and enthalpy can be estimated from the intersection of a line between measured silica and steam (steam loss line) and the quartz solubility line (*RED STAR*).
Modified from Fournier RO. Water geothermometers applied to geothermal energy. In: D'Amore F, editor. Application of geochemistry in geothermal reservoir development, United Nations Institute for Training and Research, UNITAR/UNDP Center on Small Energy Resources, Rome, Italy; 1991. ISBN:92-1-157178-2. p. 37−69.

the application of this geothermometer requires the assumption of chemical maturity. Giggenbach [47] developed a system to simultaneously evaluate the K-feldspar/Na-feldspar (Na/K) and the K-feldspar/K-mica + chlorite (K^2/Mg) geothermometers using a trilinear diagram (Fig. 4.2) of relative concentrations of K, Na, and Mg. This diagram indicates fluids in equilibrium and partial equilibrium and projected temperatures and "immature" waters to which the geothermometers do not apply.

4.2.2 Noncondensable gas geochemistry and gas geothermometers in geothermal systems

Gas geothermometry is based on equilibrium between gas−gas reactions of reactive gases and between reactive gases and minerals, similar to liquid geothermometers (eg, Refs. [21,27,44]). As with liquid geothermometers, the use and application of the gas geothermometers assume equilibrium between gases, gas and water, or gas

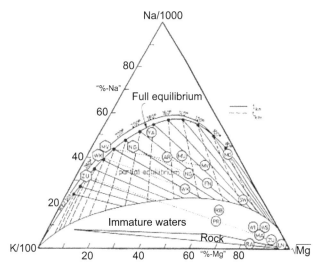

Figure 4.2 Evaluation of Na−K−Mg geothermometer temperatures using the relative proportions of three major cations in a trilinear diagram indicating zones of full and partial equilibrium where temperatures are likely to be reasonable and immature waters to which these geothermometers do not apply. Fluids from wells (hexagons) from several known high-temperature geothermal systems such as Wairakei (WK), Zunil (ZU), Ngawha (NG), plot higher temperatures than related springs (circles) indicating that surface manifestations have cooled, been diluted with groundwater, or do not contain discharge from the deep hot system. Modified from Giggenbach WF. Geothermal solute equilibria. Derivation of Na−K−Mg−Ca geoindicators. Geochim Cosmochim Acta 1998;52:2749−65.

and minerals, except for the empirical gas geothermometer [22]. A select list of gas geothermometers is provided (Table 4.2).

By solving equilibrium reactions for two gas reactions where concentrations are controlled by the temperature-dependent equilibrium constant as well as temperature-dependent vapor−liquid distribution coefficients of each gas (i) ($C_{i,v}/C_{i,l} = B_i$), both reservoir saturation (reservoir-y) and temperature can be resolved with various assumptions [10,25,44]. In vapor-dominated systems the reservoir steam fraction (ie, reservoir-y) is approximately 1, whereas in liquid-dominated systems it approaches 0, and with two-phase systems it lies between 0 and 1. Applied to an exploited geothermal reservoir, estimates of reservoir-y and temperature help identify areas of relative dryness that might be injection targets.

4.3   Contributions to conceptual models

Conceptual models of a geothermal system may be developed before or after exploration drilling. The following section describes some of the key reservoir processes and characteristics that a thorough geochemical evaluation of surface manifestations and exploration wells can constrain or define within a conceptual model.

Table 4.2 Gas Geothermometers Gas. Concentrations mmol/kg steam at 1 atm unless specified as partial pressure, P, or mole fraction, X, Temperature, t in °C

Geothermometer	Equation	Application	References
CO_2–CH_4–H_2–H_2S	$t = 24775/((2 \log CH_4/CO_2 - 6 \log H_2/CO_2 - 3 \log H_2S/CO_2) + 7\log P_{CO_2} + 36.05)$	Empirical geothermometer, where $\log P_{CO_2} = 0.1$atm if < 75vol% $CO_2 = 1$ if > 75vol% $CO_2 = 10 > 75$vol% CO_2 and CH_4 and $H_2S > 2*H_2$	[22]
H_2–CO_2	$t = 190.3 + 55.97(\log H_2 + 1/2\log CO_2) - 0.14(\log H_2 + 1/2\log CO_2)^2$		[72]
H_2S–CO_2	$t = 194.3 + 56.44(\log H_2S + 1/6\log CO_2) + 1.53(\log H_2 + 1/2\log CO_2)^2$		[72]
CO_2	$t = 44.1 + 269.25 \log CO_2 - 76.88 \log CO_2^2 + 9.52 \log CO_2^3$		[7]
H_2S	$\log PH_2S = 6.05 - 3990/T$	T in K, P = partial pressure	[49]
H_2S	$3 \log PH_2S - \log PH_2 = 15.71 - 10141/T$	Pyrite-magnetite T in K, P = partial pressure	[44]
H_2S	$t = 173.2 + 65.04(\log H_2S)$	t $>$ 100°C and Cl $<$ 500 mg/l, part of a series of H_2S geothermometers limited to ranges of t and Cl	[8]
H_2	$t = 212.2 + 36.59(\log H_2)$	t $>$ 100°C and Cl $<$ 500 mg/l, part of a series of H_2 geothermometers limited to ranges of t and Cl	[8]
CO_2/H_2	$\log PCO_2 + 2\log H_2 = 16.298 - 8982/T$	T in K	[58]
CH_4/CO_2	$t = (4625/(10.4 + \log XCH_4/XCO_2)) - 273$	X = mole fraction	[48]

4.3.1 Reservoir temperatures

Geothermometers provide reservoir temperature estimates. Applying multiple geother-mometers (see earlier section) with different kinetics on different phases provides evidence of reservoir processes and temperature distribution.

4.3.2 Water sources and recharge

The geochemical signature of different contributions to reservoir fluids such as groundwater, injection fluid, magmatic influx, or steam condensate is used to identify original source water and recharge during reservoir development. The primary geochemical tools for distinguishing different source waters are the stable isotopes of water, rare or inert gases, and, in some cases, water types.

Reservoir response to production-induced pressure decline usually falls into one of two categories depending on the structure of the reservoir. If the reservoir is sealed when pressure declines, the reservoir fluids will boil, creating two-phase conditions. If the reservoir is not sealed, the boundaries are permeable. Fluid such as cooler groundwater or injectate (recharge) will enter the reservoir in response to declining pressure, flowing along the pressure gradient. While reservoir boiling might increase the pressure decline rate, it increases the enthalpy and excess steam (production of more steam than can be accounted by boiling liquid water at reservoir temperatures) of produced wells. In contrast, recharge can support reser-voir pressure reducing pressure decline but can cause cooling. Therefore, the pres-ence of leaky or sealed boundaries is an important component of a reservoir model and of the reservoir response to production. Before production, the potential for leaky boundaries and real-time recharge are indicated by mixed surface manifesta-tions, regional hydrology, and the degree of equilibrium attained between minerals and water.

4.3.2.1 Distinguishing water sources using stable isotopes

The primary source of water for geothermal systems is local meteoric water. Other sources include volcanic or magmatic waters, connate, and sea water. The main tool used to evaluate the source of water is the stable isotopes of water: oxygen-18 (^{18}O) and deuterium (D), specifically comparing the stable isotopes of geothermal waters to local meteoric waters (Figs. 4.3(a,b) and 4.4). Under two-phase conditions, the heavier stable isotopes, ^{18}O and D, partition into liquid relative to vapor and the lighter isotopes, ^{16}O and H, into vapor. In meteoric water, the partitioning related to these pro-cesses occurs at atmospheric temperature that creates a linear relationship between $^{18}O/^{16}O$ and D/H known as the meteoric water line, shown in the figures. Deviations in the stable isotopic composition of thermal and non-thermal water from this line indi-cate processes such as (1) evaporation or boiling, and precipitation or condensation, (2) temperature and mineral-dependent interaction between water and rock increasing ^{18}O but not D, and (3) mixing with other waters and fluids, which could affect ^{18}O and D [13,18,46,80].

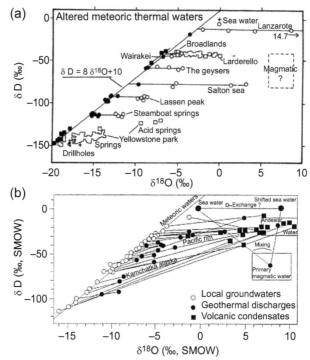

Figure 4.3 Stable isotopes (^{18}O and D) of geothermal, volcanic, and local meteoric waters from different geologic settings. Concentrations are presented as differences, δ, from the isotopic standard for water, standard mean ocean water (SMOW) in parts per thousand, ‰, so δ (‰) $= ((R_x/R_s) - 1) \times 10^3$ where $R_x = {}^{18}O/{}^{16}O$ (or D/H) of the unknown and $R_s = {}^{18}O/{}^{16}O$ (or D/H) of SMOW. The average worldwide meteoric water line is found from $\delta D = 8 \times \delta^{18}O + 10$.

The original discovery of the local meteoric origin of geothermal waters (eg, Refs. [18,80]), was based on similarities in D concentrations between thermal and meteoric waters but enriched ^{18}O in thermal waters relative to meteoric waters. This ^{18}O shift was explained by the high-temperature interaction between oxygen-bearing silicate minerals and water (Fig. 4.3(a); [80]). Shifts from the meteoric water line of both ^{18}O and D can be attributed to mixing with meteoric water from higher elevation, sea water, connate water (metamorphosed sea water), and magmatic and/or andesitic water (Fig. 4.4(b); [46]). Before drilling, the isotopic concentrations of the reservoir rocks are unknown. After drilling, differences in the ^{18}O and D of minerals within and outside of a feed zone can produce quantitative estimate of water throughput based on measureable isotope shifts [13].

Depending on the temperature, boiling shifts both ^{18}O and D composition of the geothermal fluids producing isotopically lighter steam but heavier residual water, while condensation produces heavier liquid and even lighter steam (Fig. 4.4; [58,65]). Therefore, in order to evaluate the source water of geothermal reservoir fluids

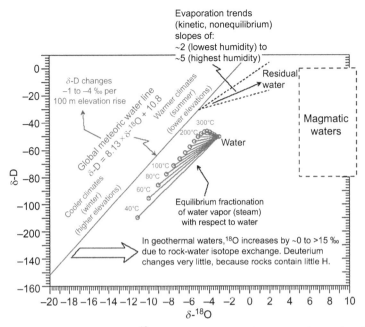

Figure 4.4 Basic processes affecting ^{18}O and D in geothermal water. Units are described in Fig. 4.3. Equilibrium fractionation describes trends in steam from boiling at different temperatures.
Modified from Klein CW. Exploration geochemistry. In: Geothermal Resource Council workshop, September 8–9, 2006. Presentation by GeothermEx, Inc. 47 slides; 2006.

using stable isotopes, isotopic shifts related to subsurface processes such as mixing, boiling or condensation need to be corrected for.

4.3.2.2 Distinguishing water sources using gases

Other indications of source waters involve nonreactive, rare, noble, or inert gases such as argon, Ar, nitrogen N_2, (except where it reacts with H_2 to form NH_3), helium, He, and helium isotopes, helium-3 (^3He) and helium-4 (^4He). Because these gases are not reactive, they are conserved quantities; their relative concentrations can distinguish between the contributions of meteoric, magmatic and crustal water/rock interaction (Table 4.3; [21,48,49,74]).

Trilinear diagrams allow comparison of the relative concentrations of three components: each corner of the triangle represents 100% concentration of the labeled component; the opposite side is 0%; mixtures lie on the inner area, with 33%, 33%, and 33% at the midpoint; end-points can represent multiples of concentrations to accommodate large variations in concentrations. Fig. 4.5 is a trilinear diagram showing the relative concentrations of the three gases, He, Ar, and N_2 in geothermal gases [44,49]. Note that air and air-saturated groundwater (asw in the figure) plot on the line with no He but as mixtures of N_2 and Ar, whereas volcanic gases

Table 4.3 Nitrogen, N_2, argon, Ar, helium, He and He isotopes $^3He/^4He$, and oxygen in noncondensable gas from geothermal fluid sources [21,48,49]

Source fluid	N_2/Ar	He/Ar	$^3He/^4He$	O_2
Meteoric (air-saturated water)	38	<0.001	0.0000014	Negligible
Magmatic	800−2000	0.1	>6−8	Negligible
Crustal	Variable	High	<1	Negligible
Air	84	<0.01	0.0000014	High

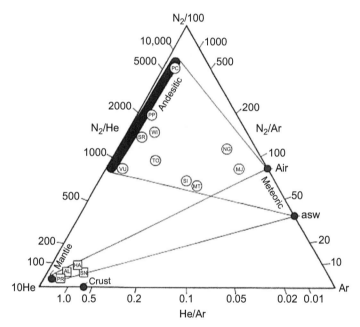

Figure 4.5 Trilinear diagram of He, Ar and N_2 used to distinguish the source of geothermal gases from the relative concentrations of gases. Data from known geothermal wells and surface manifestations (as capital letters) and the composition of crustal rocks (with mantle basalts from divergent plate boundaries or hot spots slightly more He rich), and magmatic input such as volcanic arc andesites (from convergent plate boundaries), and meteoric water (between air and air-saturated groundwater (asw)).

Modified from Giggenbach WF. In: D'Amore F, editor. Application of geochemistry in geothermal reservoir development, United Nations Institute for Training and Research, UNITAR/UNDP Center on Small Energy Resources, Rome, Italy; 1991. ISBN:92-1-157178-2. p. 37−69; Julian RH, Lichti KA, Mroczek E, Mountain B. Heavy metal scaling in a geothermal heat exchanger. In: Proceedings, World Geothermal Congress, 2015, Melbourne, Australia; April 19−25, 2015.

plot on lines with no Ar. Also, gases from fumaroles in the Pacific arc islands of Vanuatu (VU) and White Island (WI) indicate andesitic sources, whereas on Hawaii (HA) the sources are basaltic (ie, mantle and crust), whereas those from the New Zealand geothermal field, Ngawha (NG), are mixed with air-saturated groundwater.

4.3.2.3   Distinguishing geothermal fluids and reservoir processes using water types

The types of geothermal fluids are primarily of water in different phases and different nonwater chemistries: liquid (water or brine, which is typically defined as water with a significant concentration of dissolved solids) and/or vapor (steam with or without gas and insignificant dissolved solids (usually)). Clearly, the presence of steam at the surface suggests boiling, but the geochemistry can also identify boiling by observing the geochemical signature of steam dissolved in water. Fig. 4.6 and Table 4.4 present the chemistry of various types of waters found in geothermal surface manifestations.

Traditional water typing is not often useful in geothermal exploration, but comparative typing can be very useful [3]. Two main tools for comparing geothermal water types, such as Na–Cl versus Ca–HCO_3, are (1) the trilinear diagrams of the primary anions (Cl, HCO_3, and SO_4, Fig. 4.7; [48]) and cations Na, Ca, and K (Fig. 4.2) and (2) the Schoeller diagram, which presents the log of concentrations of numerous constituents. The trilinear diagram can display the relative concentrations of three chemical constituents. The Schoeller diagram presents similar fluids as lines with similar

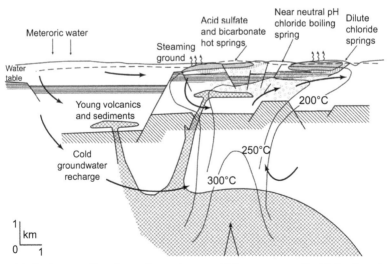

Figure 4.6 Surface manifestations of a moderate to high temperature liquid-dominated geothermal system.
From Chapter 2, this book, adapted by Elders and Moore from Ellis, 1983.

Table 4.4 Types of fluids in geothermal surface manifestations and their origin

Fluid	T,°C	Chemistry	pH	Origin
Water	<35°C	Dilute Ca–Mg bicarbonate (HCO_3^-)	6–7.5	Groundwater
Water	35–95°C	Na–Mg–Cl ± Sulfate (SO_4^{-2}) ± HCO_3^-, higher TDS than groundwater	5–7.5	Mixed thermal brine/groundwater
Water	35–100°C	Na–Cl, lesser HCO_3^- – SO_4^{-2}	5–9	Thermal brine, conductively cooled or boiled
Water	35–95°C	Ca–Mg–HCO_3^-, same as groundwater with possible slight elevation of silica	5–7.5	Moderately conductively heated groundwater
Water	35–95°C	Ca–Mg–SO_4^{-2} ± HCO_3^- + Cl, lower TDS than groundwater, more acid	3–7	Slightly steam heated ground water
Water	>95°C (boiling)	Na–Ca–Cl – HCO_3^- ± SO_4^{-2}, higher TDS than groundwater	5–7.5	Mixed thermal/groundwater
Water	>95°C (boiling)	Na–Ca–Cl – HCO_3^- ± SO_4^{-2}, higher TDS than groundwater	3–7	Slightly steam heated ground water
Steam-pool or vent	>95°C (boiling)	Na–Ca–Cl – HCO_3^- ± SO_4^{-2}, higher TDS than groundwater	<3	Geothermal steam/steam condensate (in pool) from geothermal vapor from vapor reservoir
Steam-vent	>95°C (boiling)	Low TDS, sulfur minerals	<4	Geothermal steam direct discharge of geothermal vapor
Steam-pool or vent	>95°C (boiling)	SO_4^{-2}–HCO_3^- –Low TDS-acid alteration/clays	3–5	Geothermal steam/steam condensate (in pool) from boiling of liquid reservoir
Steam	>95°C (boiling)	Low TDS, minor alteration, HCO_3, lower gas	4–6	Boiling groundwater-heated by steam or brine
Gas	Any	Carbon dioxide (CO_2), hydrogen sulfide (H_2S)		Volcanic gas, possible geothermal
Gas	Any	CO_2 with no sulfide, no steam		non geothermal gas

Groundwater refers to shallow waters, typically (but not always) dilute; thermal brine refers to the deep liquid component of a geothermal system; geothermal vapor is vapor (steam + gas) from deep boiling of thermal brine or reservoir vapor.
Modified from Armansson H. Application of geochemical methods to geothermal exploration. In: Presented at short course II on surface exploration for geothermal resources, organized by UNU-GTP and KenGen, at Lake Naivasha, Kenya; November 2–17, 2007.

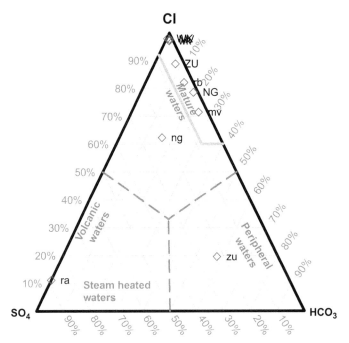

Figure 4.7 Cl—HCO$_3$—SO$_4$ anion trilinear diagram, showing relative proportions of anions from some developed geothermal fields and their surface manifestations: mature fluids are Cl brines, peripheral waters are local ground waters.
Modified from Giggenbach WF. Geothermal mineral equilibria. Geochem Cosmochem Acta 1981;45:393—410; Powell T, Cumming W. Spreadsheets for geothermal water and gas geochemistry. In: Proceedings, 35th Stanford workshop of geothermal reservoir engineering. California: Stanford University; February 2010, SGP-TR-188.

patterns; brines with similar but more dilute chemistry display as parallel lines, whereas different chemistries show as crossed lines.

Predominantly Na—Cl fluids from geothermal wells or springs represent reservoir brine; acidic and relatively dilute Na—SO$_4$ or Ca—SO$_4$ fluids can represent steam or condensed steam; and Ca—CO$_3$ or Ca—Mg—CO$_3$ fluids typically represent ground-water-dominated fluids. The chemistry of steam-heated and volcanic waters are typically related to gas absorption and oxidation. For example, in Fig. 4.6, NG and ng are well waters and clearly mixed spring waters from Ngawha, while the well waters from Zunil, ZU, are Cl-rich brines, and spring waters, zu, are meteoric.

4.3.3 Mixing

In addition to silica mixing diagrams (Fig. 4.1), the simple correlation of the measured temperature (or enthalpy of water at that temperature) of hot springs and the concentration of conservative elements is used to understand mixing and boiling

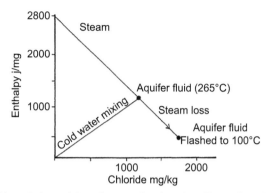

Figure 4.8 Chloride-enthalpy mixing diagram showing the effects of boiling, steam loss, and mixing on a liquid-dominated geothermal fluid with 1200 mg/L chloride (Cl). Steam has high enthalpy and no Cl, cold groundwater with zero enthalpy and no Cl, and reservoir brine has high enthalpy and high chloride. Cl increases and enthalpy decreases with high-temperature boiling and steam loss, and Cl and enthalpy are decreased by mixing with low-temperature groundwater.
From Truesdell AH. Summary of section III geochemical techniques in exploration. In: Section V. U. N. symposium on the development and utilization of geothermal resources, Pisa, 1970. vol. 1; 1976. Geothermics (1970) − special issue 2. p. liii−xixx.

(Fig. 4.8). Conservative constituents of geothermal fluids such as Cl, Br, and usually B are useful tracers of the deep geothermal system as they are nonreactive. Mixing with mostly dilute Ca−Mg−bicarbonate shallow cold groundwaters reduces the Cl concentration and temperature of the thermal water in proportion to cooling. Boiling increases Cl concentration in the remaining water as it cools the fluids in proportion to steam loss.

4.3.4 Upflow and outflow

Upflow zones represent the highest temperatures at the shallowest depth and typically zones of high permeability, often defined by faults. Therefore, such zones can provide good drilling targets, especially for the first well. The geochemistry of these zones is sometimes distinguished by boiling and formation of gas-rich steam. If the gaseous upflow reaches the surface, acidic fluids are produced when the gas-rich steam interacts with shallow groundwater creating acidic, low- (or no) chloride surface manifestations (Fig. 4.6). If there is no subsurface boiling, fluids in upflow zones can be distinguished by higher geothermometer temperatures and/or the least dilution.

Hot and warm springs of outflow zones (Fig. 4.6) are distinguished geochemically by either increasing dilution (mixing with low Cl groundwater, (Fig. 4.7)) or, if cooled conductively, decreasing silica geothermometers, usually without decreasing cation geothermometers. If boiling occurs, fumaroles have decreasing gas concentrations and increasing pH away from the upflow zone [9]. Well fluids in outflow zones can also be distinguished by dilution, lower geothermometer temperatures, "immature" water (Fig. 4.6), and lower gas concentrations.

4.3.5 Reservoir layers

Many geothermal reservoirs have chemically and physically distinct layers which can have a profound effect on resource capacity. Some layers can be characterized by distinct reservoir rocks, mineral alteration, fluid chemistry, and temperature. For example, in Kizildere, Turkey, preproduction chemistry of the deep reservoir indicated higher gas and higher geothermometers than the intermediate reservoir, which had been produced for many years [55], indicating limited circulation between layers in the natural state. In other geothermal areas such as Los Azufres (eg, Refs. [73,83]) and Cerro Prieto (eg, Ref. [83]), in Mexico, layers are distinguished by gas concentrations and enthalpy, and appear to be well connected. With geochemical monitoring, relative changes in and communication between layers with distinct geochemistry can be monitored during reservoir production and injection.

4.3.6 Boiling

Geothermal indications of subsurface boiling include the presence of fumaroles, steam, and excess-steam wells. Gas chemistry of fumaroles and wells allows us to evaluate the presence of high-temperature boiling and reservoir vapor or low-temperature boiling as discussed earlier (Section 4.2.2). The chemistry of any liquid present can indicate whether it is condensate (acidic and dilute) or reservoir liquid (NaCl brine) and, if geothermometers can be used, at what temperatures boiling may have occurred (Section 4.2.1).

Reservoir boiling before or during production indicates a high-temperature system with either low permeability or sealed boundaries. Boiling produces gas-rich steam, which, if separated from the boiled liquid, can move through the reservoir more readily [12] but suggests lower liquid reserves and possibly higher decline rates. In these cases, the geochemist's evaluation of reservoir saturation (reservoir-y) using gas partitioning between liquid and vapor [25] can support reservoir conceptual modeling and estimated resource capacity.

4.4 Geochemical contributions to geothermal power project design

Assuming that the resource assessment indicates that the geothermal system under investigation can support a geothermal power project, there are several aspects of the project design requiring input from the geochemist. In addition to the sustainable enthalpy and mass flow of the geothermal fluid supply to the plant, the project design requires average chemistry of the total fluid at the plant calculated based on a weighted average of each well and, if flashed, the chemistry of each phase after each flash point and when cooled. Based on the chemistry of these phases, issues such as noncondensable gas content (NCG), emissions, scaling, and corrosion can be evaluated.

4.4.1   Noncondensable gas

Geothermal fluids have a wide range of NCG concentrations that can strongly affect power plant and gathering system design and operation. Thus, a good estimate of resource NCG is critical to project design. Because NCG partitions strongly into steam when fluid boils (flashes), the smaller the steam fraction, y, the higher is the gas concentration in the steam at a given concentration of NCG in the total fluid [57]. In multiple-stage flash plants, the NCG that partitions into the liquid during the initial flash is carried over to partition into vapor phase at the next flash level, affecting the gas loading and sizing of gas extraction for each stage. NCG distributes between phases in accordance with the steam fraction, y, distribution coefficients, and gas—water reactions; therefore, the NCG flows through the stages of the flash plant can be estimated using calculations of y, mass flow, and distribution coefficients [23,57] or geochemical modeling programs (eg, Refs. [59,87]).

High levels of NCG can affect the temperature and enthalpy of the steam and brine discharge from the separation station, and the pressure in the turbine after the steam is condensed (flash plant). For example, assume the pressure of the power plant inlet separator is designed for a total pressure of 8 bar, and the gas concentration in the fluid entering the plant is 0.02 kg of NCG/kg of geothermal fluid, the enthalpy is 900 kJ/kg and the salinity is low. Without gas, at the separation pressure of 8 bar,a, the discharge temperature for brine and steam is 170.4°C. Incorporating gas, the temperature drops 4°C, and the brine enthalpy declines from 720.8 kJ/kg to 703.3 kJ/kg.

In addition, high NCG in liquid-dominated reservoirs produces high initial reservoir pressures 2—3 times the water pressure producing high initial flow rates, which can decrease dramatically if NCG decreases with production. Forecasting long-term production in such fields is best done with numerical simulation that incorporates gas [43]. NCG concentrations also affect the pressure distribution and depth of boiling (bubble point) in flowing wells (see Section 4.4.4).

4.4.2   Potential geothermal emissions of concern

Other than the normal impacts of an industrial development in what are often remote areas of natural interest, one key geothermal-specific potential environmental impact is emissions of geothermal fluids, which have the potential to adversely affect the environment and human health and safety. The impact depends on the quantity, concentration, location, and chemistry of the emissions, which depends on the chemistry of the geothermal fluids entering the plant, the partitioning of constituents of concern into steam or liquid (related to B_i) that are then discharged to air, to surface waters, or injection, depending on the project design. Discharge points include cooling towers, well silencers, rock mufflers, brine ponds, etc. While this section is not intended to be an exhaustive assessment of the environmental impacts of geothermal projects, the geochemist should be aware of the importance of potential constituents of environmental concern when providing geofluid chemistry to the power plant engineer (Table 4.5).

Table 4.5 Select emissions according to fluid phase at atmospheric conditions

Vapor phase	Liquid phase
CO$_2$ (climate and air quality) H$_2$S (human health and safety and air quality) Benzene (climate and air quality)	Boron, B Arsenic, As, and mercury, Hg Heavy metals Salt

4.4.3 Scaling

At the design stage of a project, the prediction of the potential and location for scaling is a key aspect of project design. Scaling occurs when geothermal liquids become over-saturated with a given mineral. Steam only produces scale from liquid carryover or corrosion. Scaling from brine can occur as a result of different mechanisms: increased concentrations from boiling and/or boiling-related pH changes and/or cooling. The primary scale minerals of concern are carbonates, particularly in low- to moderate-temperature geothermal systems, as well as amorphous silica and sometimes metal silicates, particularly in higher-temperature systems, and metal silicates and sulfides in saline systems. This section presents a brief discussion of scaling potential in geothermal wells and surface facilities, but there are many more extensive evaluations of scaling and scale mitigations by others. For a more detailed review, refer to Ref. [58] for general scaling, [15,16] for silica, [69] for carbonate, [1,50] for antimony sulfide, [19,29,33,60] for metal silicates, and [20,51] for metal sulfides.

In the early days of geothermal development, scaling was primarily addressed by avoiding conditions that generated scale (suppressing boiling by pumping or keeping temperatures high with high separation pressures, see eg, Refs. [15,40,57]) or by allowing scale and then cleaning the scaled surface (eg, well, pipeline, or heat exchanger). For example, carbonate scale was suppressed in geothermal wells by pumping wells to maintain pressure higher than the gas breakout pressure and using binary technology [69], and silica scale was suppressed by maintaining the plant exit temperature above saturation for amorphous silica [58]. However, these approaches reduced power plant efficiency.

If scaling potential is known before a project is designed, the project design can be modified to prevent scale, if possible, or monitoring and mitigation can be incorporated in the design.

With appropriate fluid chemistry data, the geochemist can predict potential scaling at various conditions in the well production and power conversion process with analytical tools such as mineral saturation or solubility diagrams (eg, Refs. [45,58]) or with computer programs, such as WATCH.

Scaling is indicated if the saturation index for a mineral, SI > 0 where:

$$SI = Q/K$$

Q = activities of products/activities of reactants in the solubility reaction; K = equilibrium constant for the solubility reaction.

SI is typically presented as log (Q/K). Q and K can be calculated or obtained from the output of a geochemical speciation program. If log SI > 0, there is potential for scaling. SI is temperature dependent and assumes equilibrium, which does not always occur in geothermal brines, especially between the power plant and injection well, nor do minerals always precipitate immediately when log SI is at or slightly above 0. Additional discussion of the most common scales and mitigation is presented next.

4.4.4 Carbonate

Carbonate scale most likely occurs in liquid-dominated geothermal developments in producing wells when the carbon dioxide, CO_2, breaks out of the single-phase fluid into the vapor phase according to the following reaction [58]:

$$CaCO_3 + H_2CO_3 \leftrightarrow Ca^{+2} + 2HCO_3^- \text{ or } CaCO_3 + H_2O + CO_2 \leftrightarrow Ca^{+2} + 2HCO_3^-$$

The loss of dissolved CO_2 drives the reaction to the left, precipitating calcite (or carbonates of Mg or Sr) from the liquid phase (Fig. 4.9). Since many geothermal fluids are in equilibrium with calcite in the reservoir liquid, scaling occurs when the gas breaks out of the liquid phase forming a vapor phase during depressurization in a flowing well or plant.

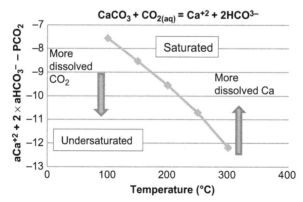

Figure 4.9 Equilibrium for the precipitation of calcium with CO_2 separation. Equilibrium constants: $CO_{2(aq)} = H_2CO_3$.
From Henley RW, Truesdell AH, Barton PB, with a contribution by Whitney JA. Fluid-mineral equilibria in hydrothermal systems. Reviews in economic geology, vol. 1. El Paso (TX): Society of Economic Geologists, The Economic Geology Publishing Company, University of Texas. ISBN:0-9613074-0-4. 267 p.

While carbonate scale can be successfully mitigated and production maintained using downhole scale inhibitors [69], inhibitors need to be injected below the depth of gas breakout. Therefore, designing a carbonate scale mitigation system requires the geochemist not only to forecast the carbonate scale potential, and the amount of inhibitor required, but to predict the depth at which the pressure of gas breakout, P_{BO}, occurs when the well is flowing. P_{BO} is the pressure below which the fluid becomes two phase. P_{BO} is approximately equal to the pressure of saturated water at the feed zone temperature, P_{H_2O}, plus gas pressure, P_{gas}. P_{H_2O} is readily obtained from steam tables at the temperature of the single-phase flowing fluid (reservoir temperature). Assuming the gas is essentially all CO_2, P_{gas} is obtained by multiplying the gas concentration in mole fraction, X_{gas}, by the Henry's law constant, K_H, for CO_2 at the temperature of the single-phase flowing fluid.

$$P_{BO} = P_{H_2O} + X_{gas} \times K_{H\text{-}CO_2}$$

The depth at which P_{BO} occurs in a flowing well depends on the flow rate and WHP. This depth is best modeled using a wellbore simulator or obtained from a dynamic survey but in the latter case is only relevant for the flow rate and wellhead pressure at which the dynamic survey was performed. This depth plus several meters for mixing is the inhibitor tubing set depth.

4.4.5 Silica and silicates

Silica scaling potential is evaluated by estimating amorphous silica SI based on the following:

$$SiO_{2SOL} + 2H_2O = H_4SiO_4$$

Where SiO_{2SOL} is the solid phase of silica (in this case amorphous silica) and H_4SiO_4 is the dissolved form of silica (silicic acid) [15,34]. But in the case of silica, the concentration of dissolved silica is affected by the dissociation of silicic acid:

$$H_4SiO_4 = H_3SiO_4{}^- + H^+$$

As pH increases from neutral or slightly acidic, the dissolved silica or silicic acid, H_4SiO_4, dissociates and the concentration of H_4SiO_4 decreases as it dissociates; therefore, the potential for precipitation of amorphous silica decreases. Thus, the critical factors for amorphous silica precipitation are silica concentration, temperature, and pH [15,16]. These are also key factors for silicate precipitation plus the concentrations of trace constituents of geothermal fluids such as Al and Mg [29].

While SI > 0 indicates the potential for silica to scale, it does not indicate that silica scale will precipitate. Although highly oversaturated fluids will scale, actual precipitation and low levels of saturation depends on kinetics, nucleation, and other factors that are often hard to predict [16]. Therefore, where silica saturation suggests scaling could or might occur, precautions such as scale coupons along with pH and silica measurements should become a routine part of power plant operations and maintenance.

One way to avoid silica scaling is to maintain injection above the amorphous silica saturation temperature. For a reservoir in equilibrium with quartz at 275°C, this might require a maximum allowable temperature after boiling of 180°C [58]. If the temperature of the reservoir is low enough and separation pressure and temperature high enough, there may be no potential for amorphous silica scaling. For example, if reservoir fluids at 195°C in equilibrium with chalcedony are flashed to 110°C, amorphous silica remains soluble until the fluids reach almost 100°C (Fig. 4.10). Alternatively, the pH can be adjusted to increase the solubility of silica [15] by adding acid or acidic gas [64] or manipulated by multi-stage boiling [58]. The sensitivity of predicting the silica scaling potential to fluid chemistry, silica concentrations and pH highlight the importance of good geochemical sampling and analysis for the design phase of the project.

4.4.6 Steam purity

Most turbine manufacturers specify limits to the nonwater constituents of steam or steam purity at the turbine inlet. These typically include constituents of concern for scaling and corrosion such as SiO_2 and Cl at less than 1 mg/kg. Since these constituents are nonvolatile and do not partition into the vapor phase at typical conditions in the gathering system and power plant (excluding superheated steam in vapor-dominated systems), their concentrations in steam at the turbine inlet are related to brine carryover. The design of the steam scrubbing system, including the separator, steam spray wash (spraying water into steam through dispersing nozzles), and steam scrubber, must meet these requirements. Knowing the chemistry of the brine, the geochemist can estimate how efficient the stream scrubbing system has to be to achieve the limits.

Sometimes scaling and corrosion can occur in the turbine despite achieving the turbine manufacturers steam purity limits. Careful monitoring of steam purity allows the power plant engineer to understand and manage these issues.

4.4.7 Corrosion

Other than air-related corrosion, corrosion in geothermal wells and surface facilities is related to the chemistry of the geothermal fluids and the temperature. Most geothermal facilities are designed assuming corrosion related to hydrogen sulfide in steam and air and chloride in brine will be mitigated by appropriate material selection [17]. In the carbon steel parts of the geothermal wells and facilities that carry geothermal fluid, maintaining the pH > 5 may be the primary corrosion mitigation required. Additional corrosion issues [40] requiring estimates of the chemistry of fluids at various stages in the production cycle include:

- Well liners and casings where deep fluid is acidic and/or moderately to highly saline
- Well liners and casings hydrogen embrittlement and sulfide stress cracking related to H_2 and H_2S

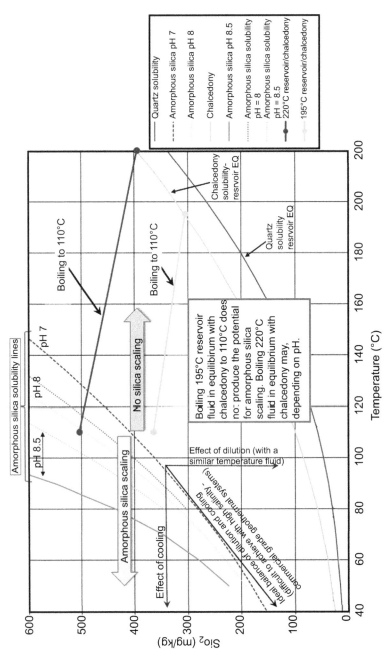

Figure 4.10 Silica fluid concentrations and amorphous silica solubility at various pH values. Reservoir fluids in equilibrium with chalcedony at 220°C (*brown line*) and 195°C (*green line*) after boiling to 110°C cooling and steam loss decrease solubility. Silica scaling is not likely from a 195°C reservoir, but if the reservoir temperature is 220°C, silica may scale, depending on pH.

- At wellhead and shallow depths outside the well casing from oxygen and groundwater water in the presence of H_2S
- Superheated steam in wells and pipelines may contain HCl, which partitions into condensate formed when superheat declines, causing rapid acidic corrosion
- Cooling towers and cold condensate pipelines if CO_2 dissolves without buffering of other gases such as NH_3
- Heat exchangers can experience stress corrosion cracking where O_2 and Cl are present at temperatures $>70°C$
- Electronic equipment in power plants can be destroyed by H_2S

The most severe corrosion conditions are related to high-temperature hypersaline brines such as those found in the Salton Sea and parts of Africa and volatile hydrogen chloride found in some vapor-dominated systems or parts of systems such as The Geysers [26,53,54,82] and Coso in the United States, Larderello [81], and some sites in Japan. While solutions to these corrosion issues have been developed (eg, Ref. [68]), solutions are often site-specific and should be tested at each facility.

4.5 Geochemical tools for geothermal reservoir operation and maintenance

Sustaining geothermal production and injection over the life of a power project requires effective reservoir management. Reservoir management tools for sustaining production and injection capacity of a geothermal reservoir are limited and expensive; therefore, reservoir monitoring to minimize well workovers, drilling or shifting production, and injection between wells is vital.

A thorough reservoir management monitoring program includes period geochemical sampling and analysis of geothermal fluids from each production well and injection well. These data can be used to monitor reservoir cooling, injection or cool groundwater breakthrough, boiling, changes in liquid reserves, and cross-reservoir flow. Sampling and analysis of various geofluid flow streams within the power plant support:

- Managing of gas flows relative to gas extraction capacity
- Scale and corrosion control
- Emissions monitoring and mitigation
- Steam purity

In addition to monitoring the trends in geochemical parameters, geochemical data should be compared with parameters of physical reservoir monitoring such as flow rates, pressures, and temperatures. Simply plotting variations in these parameters with time is an effective project management tool. For example, most reservoirs have two basic responses to production-induced pressure declines: pull in cooler water from the flanks of the system if the edges are permeable, or boil if the edges are not permeable. These changes produce dramatically different geochemical responses. When reservoir cooling or enthalpy decline is observed and both silica and chloride decline, that suggests cooling related to influx of water from outside the geothermal reservoir. If silica declines and chloride increases, then the cooling may be related to injection breakthrough. Wairakei, after 30 years of production, experienced field-wide cooling in response to production that corresponded to declining silica and chloride, geochemical indicators of field-wide dilution by local groundwaters [32,57]. At other fields, such as Cerro Prieto, Ahuachapán [83], and Coso [70], increases in enthalpy of produced fluids in response to production are attributed to boiling of systems with sealed or low permeability boundaries generating vapor-dominated zones, using geochemical observations.

Severe mass flow declines can be caused by inhibitor failure. Monitoring Ca in production well fluids, along with appropriately placed coupons, measures inhibitor effectiveness and allows the reservoir manager to focus mitigation on the inhibitor system.

The key liquid-dominated reservoir conditions that typically change with production and related geochemical and physical monitoring parameters are described in Table 4.6.

In Cerro Prieto, geochemists have been able to track reservoir temperatures, boiling, and recharge using silica and cation geothermometers and chloride concentrations along with total mass flow and enthalpy of production wells. Using geochemical and physical monitoring data [83], differences between measured enthalpies, H_{meas}, enthalpies from geothermometer temperatures, cation H_{NaKCa} and quartz H_{qtz}, can distinguish various reservoir responses to exploitation in plots over time. Here are several examples: (1) when $H_{meas} = H_{NaKCa} = H_{qtz}$, no boiling or mixing has occurred; (2) when $H_{meas} > H_{NaKCa} > H_{qtz}$, near-wellbore boiling has produced excess enthalpy and higher measured enthalpy as well as cooling as indicated by the fast acting quartz geothermometer, but the Na−K−Ca geothermometer still reflects the enthalpy of reservoir liquid; and (3) when $H_{NaKCa} > H_{meas} = H_{qtz}$, mixing with cooler water has taken place and only the slower Na−K−Ca geothermometer retains memory of higher reservoir temperatures.

Table 4.6 Primary parameters for geochemical and physical monitoring

Reservoir process	Geochemical parameters	Physical parameters
Cooling	SiO_2, K/Mg	Temperature or enthalpy
Groundwater breakthrough	Cl^a, Mg, isotopes, Cl/B, SiO_2 G/S, TDS	Temperature, pressure
Injection breakthrough	Cl^a, G/S, TDS, Cl/B, SiO_2	Temperature, pressure
Reservoir boiling	G/S	Enthalpy
Production well scaling	Ca	Mass flow, WHP
Injection well scaling	SiO_2	WHP, injection rate

[a]Cl is a representative conservative constituent fluid and may be supplemented by other conservative parameters.

Vapor-dominated geothermal systems [77] are relatively uncommon among geothermal systems. However, because of their unique characteristics, several have been exploited for many years including The Geysers, USA; Larderello, Italy; Darajat-Kamojang, Indonesia; and Matsukawa, Japan. Geochemical monitoring of vapor-dominated systems is limited to gases, the stable isotopes of water (and gas isotopes), and a few semivolatile constituents at reservoir conditions such as mercury, arsenic, chloride and boron (Table 4.2). Geochemists working in vapor-dominated systems such as Larderello and The Geysers [23−25] developed techniques to use the gas chemistry of produced steam to evaluate reservoir liquid saturation or reservoir-y and temperature using gas geothermometers. Others [52] demonstrated the use of deuterium of water to track injection-derived steam at Larderello and The Geysers before the use of surface water for injection augmentation at The Geysers.

Corrosion in vapor-dominated systems and steam zones of two-phase systems was related to the production of volatile hydrogen chloride (HCl) under high-temperature superheated conditions [26,53,82] by correlating geochemical monitoring with wellhead pressures and temperatures. Geochemical monitoring data from The Geysers has been used to correlate lower gas/steam, superheat, and the potential to produce HCl with injection-derived steam [53].

4.6 Summary

This chapter discusses the application of geochemistry to geothermal exploration, resource assessment, project design, and reservoir monitoring. The subjects are covered in the approximate order of a geothermal power project development. They include:

1. Geothermal fluid geochemical data collection: sampling and analysis of the surface manifestations, wells for initial testing and long-term reservoir monitoring.
2. Geochemical contributions to geothermal exploration before and after drilling the first well.
3. Contributions to resource assessment: using hydrogeochemistry to characterize the resource, understand reservoir processes, and build a conceptual reservoir model.
4. Geochemical aspects of resource design criteria for gathering system and power plant.
5. Geochemical monitoring for reservoir management and supporting power plant operations and maintenance for various types of geothermal systems.

Throughout this chapter, the contribution of geochemistry to geothermal resource exploration, assessment, development and management are emphasized. Because geothermal fluid follows the process flow from the reservoir through the plant and back to the reservoir, the chemistry of geothermal fluids affects the entire geothermal power cycle.

An additional important factor of using fluid geochemistry in geothermal development is the economic value. In comparison to most other geothermal resource tools such as drilling, geophysics, etc., geochemistry is relatively inexpensive. This chapter attempts to provide an introduction to the value of using fluid geochemistry for optimizing all stages of geothermal development.

References

[1] Addison SJ, Brown KL, Hirtz PH, Gallup DL, Winick JA, Siega FL, et al. Brine silica management at mighty river power, New Zealand. In: Proceedings world geothermal congress 2015, Melbourne, Australia; April 19—25, 2015.

[2] Angcoy EC, Arnorsson S. Geochemical modeling of wells discharging excess enthalpy by mechanism of phase segregation in Mahanagdong, Leyte, Philippines. In: Proceedings, 36th Stanford workshop of geothermal reservoir engineering, Stanford University; February 2011. SGP-191.

[3] Armansson H. Application of geochemical methods to geothermal exploration. In: Presented at short course II on surface exploration for geothermal resources, organized by UNU-GTP and KenGen, at Lake Naivasha, Kenya; November 2—17, 2007.

[4] Arnorsson S, Sigurdsson S, Svavarsson H. The chemistry of geothermal waters in Iceland I. Calculation of aqueous speciation from 0-degree-C to 370-degree-C. Geochim Cosmochim Acta 1982;46:1513—32.

[5] Arnorsson S, Gunnlaugsson E, Svavarsson H. The chemistry of geothermal waters in Iceland II. Mineral equilibria and independent variables controlling water compositions. Geochim Cosmochim Acta 1983;47:547—66.

[6] Arnorsson S, Gunnlaugsson E, Svavarsson H. The chemistry of geothermal waters in Iceland III. Chemical geothermometry in geothermal investigations. Geochim Cosmochim Acta 1983;47:567—77.

[7] Arnorsson S, Gunnlaugsson E. New gas geothermometers for geothermal exploration-calibration and application. Geochim Cosmochim Acta 1985;49:1307—25.

[8] Arnorsson S, Gunnlaugsson E. The use of well discharge chemistry to evaluate the characteristics of boiling geothermal reservoirs. In: Proceedings, 5th international symposium on water rock interaction, Reykjavik, Iceland; 1986. p. 17—20.

[9] Arnorsson S. Geochemistry and geothermal resources in Iceland. In: D'Amore F, editor. Application of geochemistry in geothermal reservoir development, United Nations Institute for Training and Research, UNITAR/UNDP Center on Small Energy Resources, Rome, Italy, ISBN 92-1-157178-2. p. 146—90.

[10] Isotope and chemical techniques in geothermal exploration, development and use: sampling, analysis and interpretation. In: Arnorsson S, editor. International Atomic Energy Agency-NAPC Isotope Hydrology Section, ISBN 92-0-101600X. STI/PUB/1086. 351 p.

[11] Arnorsson S, Stefansson A, Bjarnason JO. Fluid-fluid interaction in geothermal systems. Rev Mineral Geochem 2007;65:229—312.

[12] Arnorsson S. Chemical thermodynamics of two-phase geothermal systems. Berlin: ICPWS-XV; September 8—11, 2008.

[13] Blattner. Isotope shift data and the natural evolution of geothermal systems. Chem Geol 1985;49:187—203.

[14] Bjarnason JO. The speciation program WATCH, version 2.1. Reykjavik, Iceland: Orkustofnun; 1994.

[15] Brown KL, McDowell GD, Lichti KA, Bijnen HJC. pH control of silica scaling. In: Proceedings, 5th New Zealand geothermal workshop; 1983. p. 157—62.

[16] Brown K. Thermodynamics and kinetics of silica scaling. In: Proceedings international workshop on mineral scaling, Manila; 2011.

[17] Conover M, Ellis P, Curzon A. Material selection guidelines for geothermal power systems — an overview. In: Casper LA, Pinchback TR, editors. Geothermal scaling and corrosion, ASTM STP 717. American Society for Testing and Materials; 1980. p. 24—40.

[18] Craig H. The isotopic geochemistry of water and carbon in geothermal areas. In: Tongiori E, editor. Nuclear geology on geothermal areas, Spoleto, 1963. Pisa: Consiglio Nazionale delle Ricerche, Laboratorio di Geologia Nucleare; 1963. p. 17–53.

[19] Gallup DL. Iron silicate scale formation and inhibition at the Salton Sea geothermal field. Geothermics 1989;18:97–103.

[20] Gallup DL, Andersen GR, Holligan D. Heavy metal scaling in production well at the Salton Sea geothermal field. Proc Geotherm Resour Counc Trans 1990;14:1583–90.

[21] D'Amore F, Truesdell AH. Gas geothermometry for drillhole fluids from vapor dominated and hot water geothermal fields. In: Proceedings, 6th Stanford workshop on geothermal reservoir engineering. Stanford University; December 1980.

[22] D'Amore F, Panichi C. Evaluation of deep temperature of hydrothermal systems by a new gas-geothermometer. Geochim Cosmochim Acta 1980;44:549–56.

[23] D'Amore F, Celati C, Calore C. Fluid geochemistry applications in reservoir engineering (vapor-dominated systems). In: Proceedings 8th Stanford workshop on geothermal reservoir engineering. Stanford University; December 1982.

[24] D'Amore F, Celati R. Methodology for calculating steam quality in geothermal reservoirs. Geothermics 1983;12:129–40.

[25] D'Amore F, Truesdell AH. Calculations of geothermal reservoir temperatures and steam fractions from gas compositions. In: Proceedings Geothermal Resources Council annual meeting, vol. 9; 1985.

[26] D'Amore F, Truesdell AH, Haizlip JR. Production of HCl by mineral reactions in high-temperature geothermal systems. In: Proceedings 15th Stanford workshop on geothermal reservoir engineering; 1990.

[27] Applications of geochemistry in geothermal reservoir development. In: D'Amore F, editor. Application of geochemistry in geothermal reservoir development, United Nations Institute for Training and Research, UNITAR/UNDP Center on Small Energy Resources, Rome, Italy, ISBN 92-1-157178-2.

[28] Daveler SA, Bourcier B. EQ3NR — input file user's guide LLNL (Version EQ3NR R114). July 7, 1989. www.LLNL/pubs.

[29] Ellis AJ. Quantitative interpretation of chemical characteristics of hydrothermal systems. Geothermics 1970;2(special issue no. 2):516–28.

[30] Ellis AJ, Mahon WAJ. The chemistry of the Broadlands geothermal area, New Zealand. Am J Sci 1972;272:48–68.

[31] Ellis AJ, Mahon WAJ. Chemistry and geothermal systems. New York: Academic Press; 1977. 392 p.

[32] Fahlquist L, Janik C. Procedures for collecting and analyzing gas samples from geothermal systems. U S Geol Surv Open File Rep 1992:92–211.

[33] Fournier RO. Silica in thermal waters: laboratory and field investigations. In: Proceedings, international symposium on hydrogeochemistry and biogeochemistry, Tokyo, 1970. Hydrogeochemistry, vol. 1. Washington (DC): Clark; 1973. p. 122–39.

[34] Fournier RO, Truesdell AH. An empirical Na-K-Ca geothermometer for natural waters. Geochim Cosmochim Acta 1973;37:1255–75.

[35] Fournier RO, Rowe JJ. The solubility of amorphous silica in water at high temperatures and high pressures. Am Mineral 1977;62:1052–6.

[36] Fournier RO. Application of water geochemistry to geothermal exploration and reservoir engineering. In: Ryback L, Muffler LJP, editors. Geothermal systems: principles and case histories. New York: Wiley; 1981. p. 109–43 [Chapter 4].

[37] Fournier RO, Potter RW. A revised and expanded silica (quartz) geothermometer. Geotherm Resour Counc Bull 1982;11:3–9.

[38] Fournier RO. Water geothermometers applied to geothermal energy. In: D'Amore F, editor. Application of geochemistry in geothermal reservoir development, United Nations Institute for Training and Research, UNITAR/UNDP Center on Small Energy Resources, Rome, Italy, ISBN 92-1-157178-2. p. 37−69.

[39] Fridriksson T, Armannsson H. "Application of geochemistry in geothermal resource assessments", presented at short course on geothermal development in Central America − resource assessment and environmental management, organized by UNU-GTP and LaGeo, in San Salvador, El Salvador. November 25−December 1, 2007.

[40] Fridriksson T, Thorhallsson S. Geothermal utilization: scaling and corrosion. Iceland Geosurvey; 2006. power point presentation.

[41] Garg SK, Combs J. A reexamination of USGS volumetric "Heat in place" method. In: Proceedings 36th Stanford workshop of geothermal reservoir engineering, Stanford, California, January 31−February 2, 2011; 2011. SGP-TR-191.

[42] Garg SK, Combs J. A reformulation of USGS volumetric "heat in place" resource estimation method. Geothermics 2015;55:150−8.

[43] Garg SK, et al. Reservoir simulation and Kizildere geothermal field. In: Proceedings World Geothermal Congress; 2015.

[44] Giggenbach WF. Geothermal gas equilibria. Geochem Cosmochem Acta 1980;44:2021−32.

[45] Giggenbach WF. Geothermal mineral equilibria. Geochem Cosmochem Acta 1981;45: 393−410.

[46] Giggenbach WF, Stewart MK. Processes controlling the isotopic composition of steam and water discharges from steam vents and steam-heated pools in geothermal areas. Geothermics 1982;11:71−80.

[47] Giggenbach WF. Geothermal solute equilibria. Derivation of Na-K-Mg-Ca geoindicators. Geochim Cosmochim Acta 1988;52:2749−65.

[48] Giggenbach WF. In: D'Amore F, editor. Application of geochemistry in geothermal reservoir development, United Nations Institute for Training and Research, UNITAR/ UNDP Center on Small Energy Resources, Rome, Italy, ISBN 92-1-157178-2. p. 37−69.

[49] Giggenbach WF. The origin and evolution of fluids in magmatic-hydrothermal systems. In: Barnes HL, editor. Geochemistry of hydrothermal ore deposits. 3rd ed. NY: John Wiley & Sons; June 1997.

[50] Gill J, Rodman D, Muller L. Inhibition of antimony sulfide (Stibnite) scale in geothermal Fields. In: Proceedings annual Geothermal Resources Council meeting; 2013.

[51] Hardardottir V. Metal-rich scales in the Reykjanes geothermal system, SW Iceland: sulfide minerals in a seawater-dominated hydrothermal environment [Ph.D. thesis]. University of Ottawa; 2011.

[52] Haizlip JR. Stable isotopic composition of steam from wells in the Northwest Geysers, Sonoma County, California. Geotherm Resour Counc Trans 1985;9(Pt 1):311−6.

[53] Haizlip JR, Truesdell AH. Hydrogen chloride in superheated steam and chloride in deep brine at The Geysers geothermal field, California. In: Proceedings, 13th workshop on geothermal reservoir engineering, Stanford, California, January 19−21, 1988; 1988.

[54] Haizlip JR, Truesdell AH, Bloomfield K, Driscoll AJ. Changes in plant inlet gas chemistry with reservoir condition, location, and time over 15 years of production at The Geysers, CA, U.S.A.. In: Proceedings of the World Geothermal Congress, vol. 3; 1995. p. 1939−44.

[55] Haizlip JR, Tut-Halkidir FS. High noncondensible gas liquid dominated geothermal reservoir Kizildere, Turkey. Geotherm Resour Counc Trans 2011;35.

[56] Hedenquist JW, Stewart MK. Natural steam-heated waters in the Broadlands−Ohaaki geothermal system, New Zealand: their chemistry, distribution and corrosive nature. In: Proceedings Geothermal Resources Council annual meeting, vol. 9; 1985.

[57] Henley RW. Applied chemistry in the exploration and development of New Zealand geothermal systems. N Z J Technol 1985;1:207−21.

[58] Henley RW, Truesdell AH, Barton PB, with a contribution by Whitney JA. Fluid-mineral equilibria in hydrothermal systems. Reviews in economic geology, vol. 1. El Paso (TX): Society of Economic Geologists, The Economic Geology Publishing Company, University of Texas; 1984, ISBN 0-9613074-0-4. 267 p.

[59] Iceland Chemistry Group. Presents the chemical speciation program WATCH. 2010. Upgrade Version 2.4. http://www.geothermal.is/software.

[60] Julian RH, Lichti KA, Mroczek E, Mountain B. Heavy metal scaling in a geothermal heat exchanger. In: Proceedings, World Geothermal Congress, 2015, Melbourne, Australia; April 19−25, 2015.

[61] Karingithi CW. Chemical geothermometers for geothermal exploration. In: Presented at short course IV on exploration for geothermal resources, organized by UNU-GTP, KenGen and GDC, at Lake Naivasha, Kenya, November 1−22, 2009; 2000.

[62] Kharaka YK, Lico MS, Law LM. Chemical geothermometers applied to formation waters, Gulf of Mexico and California basins. Am Assoc Pet Geol Bull 1982;66:588.

[63] Kharaka YK, Mariner RH. Chemical geothermometers and their application to formation waters from sedimentary basins. In: Naeser ND, McCollon TH, editors. Thermal history of sedimentary basins. NY: Springer-Verlag; 1989. p. 99−117.

[64] Klein CW. Management of fluid injection in geothermal wells to avoid silica scaling at low levels of silica oversaturation. Proc Geotherm Resour Counc Trans October 1995;19.

[65] Klein CW. Exploration geochemistry. In: Geothermal Resource Council workshop, September 8−9, 2006. Presentation by GeothermEx, Inc. 47 slides; 2006.

[66] Marini L. Geochemical techniques for the exploration and exploitation of geothermal energy. In: Geochemical and Geophysical Methodologies in Geothermal Exploration. Italy: University of Genova; 2000.

[67] Muffler LPJ, editor. Assessment of geothermal resources of the United States− 1978, vol. 790. U.S. Geologica Survey Circular; 1978. 170 p.

[68] Meeker KA, Haizlip JR. Factors controlling pH and optimum corrosion mitigation in chloride-bearing geothermal steam at The Geysers. Geotherm Resour Counc Trans 1990;14.

[69] Michels D. Calcium carbonite deposition: contrasts in mechanisms among boilers, de-salinators, and geothermals. In: Presented to American Institute of Chemical Engineers, San Francisco; November 27, 1984.

[70] Monastero FC. An overview of industry-military cooperation in the development of power operations at the coso geothermal field in Southern California. GRC Bull September/October 2002:188−94.

[71] Nathenson M, Muffler LJP. ,Geothermal resources in hydrothermal convection systems and conduction-dominated areas. In: White DE, Williams DL, editors. Assessment of geothermal resources of the United States, vol. 726. U.S. Geological Survey Circular; 1975. 160 p.

[72] Nehring N, D'Amore F. Gas chemistry and thermometry of the Cerro Prieto geothermal field. Geothermics 1984;13:75−90.

[73] Nieva D, Nieva R. Developments in geothermal energy in Mexico. Part 12. "A cationic geothermometer for prospecting of geothermal resources". Heat Recovery Syst CHP 1987; 7:243−58.

[74] Powell T. A review of exploration gas geothermometry. In: Proceedings 25th Stanford geothermal engineering workshop, January, 2000, Stanford, CA; 2000. SGP-TR-165.

[75] Powell T, Cumming W. Spreadsheets for geothermal water and gas geochemistry. In: Proceedings, 35th Stanford workshop of geothermal reservoir engineering. California: Stanford University; February 2010. SGP-TR-188.

[76] Santoyo, Diaz-Gonzales. A new improved proposal of the Na/K geothermometer to estimate deep equilibrium temperatures and their uncertainties in geothermal systems. In: Proceedings World Geothermal Congress 2010, Bali, Indonesia; April 25−29, 2010.

[77] Truesdell AH, White DE. Production of superheated steam from vapor dominated geothermal reservoirs. Geothermics 1973;2(3−4):154−72.

[78] Truesdell AH. Summary of section III geochemical techniques in exploration. In: Section V. U. N. symposium on the development and utilization of geothermal resources, Pisa, 1970, vol. 1; 1976. Geothermics (1970) − special issue 2. p. 1iii−xixx.

[79] Truesdell AH, Manon A, Jimenez ME, Sanchez AA, Fausto LJJ. Geochemical evidence of drawdown in the Cerro Prieto geothermal Field. Geothermics 1979;8:257−65.

[80] Truesdell AH, Hulston JR. Isotopic evidence on environments of geothermal systems. In: Fritz P, Fontes JCh, editors. Handbook of environmental isotope geochemistryvol. 1. The Terrestrial Environment; 1980. p. 179−226 [chapter 8].

[81] Truesdell AH, D'Amore F, Haizlip JR. The rise and fall of chloride in larderello steam. In: Proceedings, 14th workshop on geothermal reservoir engineering, Stanford, California; January 24−26, 1989.

[82] Truesdell AH, Haizlip JR, Armannsson H, D'Amore F. Origin and transport of chloride in superheated geothermal steam. Geothermics 1989;18(1/2):295−304.

[83] Truesdell AH, Lippmann MJ, Quijano JL, D'Amore F. Chemical and physical indicators of reservoir processes in exploited high-temperature, liquid-dominated geothermal fields. In: Proceedings 1995 World Geothermal Congress, Florence, Italy; 1995.

[84] Williams CF. Development of revised techniques for assessing geothermal resources. In: 29th workshop on geothermal reservoir engineering Stanford University; 2004. p. 26−8.

[85] Wilmarth M, Stimac J. Power density in geothermal fields. In: Proceedings World Geothermal Congress 2015, Melbourne, Australia; April 19−25, 2015.

[86] Wolery TJ, Daveler SA. EQ6 a computer program for reaction path modeling of aqueous geochemical systems − user's guide and documentation. LLNL; March 30, 1989. www LLNL/pubs.

[87] Xu T, Sonnenthal E, Spycher N, Pruess K. TOUCHREACT: a new code of the TOUGH family of nonisothermal multiphase reactive geochemical transport in variably saturated geologic media. In: Proceedings of the TOUGH symposium 2003, Lawrence Berkeley National Laboratory, Berkeley, California; May 12−14, 2003.

Geothermal well drilling

L.E. Capuano, Jr.
Capuano Engineering Co., Santa Rosa, CA, United States

5.1 Introduction

Geothermal well drilling is very similar to oil and gas drilling. In principle, they are the same—a rotary drilling rig turning the bit to the right. Formations are penetrated and the drilled cuttings are removed from the bottom of the hole and carried to the surface by a drilling medium (ie, mud, water, aerated liquids, or compressed air). However, geothermal well drilling has many different variables than oil and gas drilling, and it is these variables that will be discussed in this chapter.

For many years, the geothermal industry has estimated the cost of development to be between US$4,500,000 and US$5,500,000 per installed megawatt. Approximately 50% of this geothermal development cost is expended in the process of drilling and completing the wells associated with production and reinjection of the resource. In this chapter, the design of a cost-effective drilling program for geothermal wells is discussed. Prior to determining an accurate well cost estimate, the complete drilling program must be developed. A drilling program pulls together all necessary programs into one complete package. These individual programs include the basic well plan, the casing and cementing program, the bit program, the mud program, the directional program, the blow-out preventer system and wellhead design, the rig selection, as well as all associated procedures and processes needed for an efficiently drilled well.

To understand what is included in the well cost estimate, one must understand the various costs that contribute to the total well cost. The total well cost can be divided into two major classes of costs as follows:

1. Nonrecurring costs
2. Daily operating costs

Nonrecurring costs are those costs that are fixed and occur only once during the drilling of the well. These costs are easy to calculate at the start of the well and include:

1. Mobilization of the drilling rig
2. Demobilization of the drilling rig after the job is completed; these costs are spelled out in the drilling contract
3. Casing costs (length of casing run times cost of casing per foot plus transportation to location)
4. Cementing costs of casing to be run. Cementing costs for lost circulation or sidetrack plugging represent more of a daily operation cost.
5. Wellhead and control valves that will be left on the well after the well is completed
6. Drilling site, sump, and road construction

Geothermal Power Generation. http://dx.doi.org/10.1016/B978-0-08-100337-4.00005-X

Daily operating costs are those operating costs that accumulate and occur daily or less frequently. These costs are as follows:

1. Daily rig rate and crew cost including crew subsistence and travel
2. Bit and stabilization costs
3. Fuel costs
4. Mud and mud cleaning costs including mud chemicals and additives as well as engineering services
5. Directional drilling cost, tools, and engineers
6. Engineering and on-site supervision
7. Insurance
8. Drilled cutting handling and disposal
9. Electric logging and mud logging

In many cases, the nonrecurring costs can be easily calculated and are relatively low in proportion to the accumulated daily operating costs. The daily operating costs are a function of the number of days spent on location drilling and completing the well. To get the best cost per foot for drilling a well, one must get the most hole drilled, in meters or feet per rig day. Therefore, it is very important to have the right drilling program with all associated procedures and schedules before the well is spudded. More time that is spent in planning and developing these procedures will pay off in the actual drilling operation and help with selecting the proper rig for the project. These planned programs are common in all drilling operations, in both oil and gas as well as geothermal.

To develop a geothermal drilling program, one must understand the differences between oil and gas well drilling and geothermal well drilling. A major difference between a geothermal field and an oil and gas field involves the type of resource being produced as well as the environment in which each resource exists. Geothermal fluids often occur in fractures and fissures of hard volcanic formations rather than the pore space of sedimentary formations. As a consequence, geothermal drilling operations deviate from common methods known in the oil and gas drilling industry.

The physical and geological differences between oil/gas and geothermal drilling are discussed first.

Rock formations: As mentioned earlier, geothermal fluids (in a liquid or gaseous phase) occur in pathways such as fractures and fissures in hard igneous and metamorphic rocks rather than porous sedimentary formations targeted for the production of oil and gas. Hence, the main target in geothermal drilling is those pathways. Typical oil and gas exploration tools do not provide the same results in geothermal exploration.

High temperature: The "normal" geothermal gradient encountered around the world is $1-2°F$ per 100 feet of depth, which leads to temperatures of $80-160°F$ at a depth of about 8000 ft. Some very deep oil wells can experience a bottom-hole temperature of 300°F or more at 15,000 ft. To be able to produce electricity from steam or hot water, a temperature of about 300°F is needed when the geothermal fluid (or geofluid) reaches the surface. Hence, above-average subsurface temperatures are required for an economically reasonable geothermal project. Geothermal gradients of $5-10°F$ per 100 ft are to be expected in geothermal fields. This often results in bottom-hole temperatures between 500 and 700°F at a depth much shallower than 7000−8000 ft. Normal oil and gas drilling tools have temperature limitations.

Reservoir pressures: Many of the geothermal reservoirs occur at subhydrostatic conditions. This implies that drilling with mud may not be feasible and might cause irreversible formation damage if this gel mud is lost to productive fractures, thereby reducing or destroying permeability.

Large well and casing design: Since the production rate of the geofluid is related to the well diameter, the largest possible casing diameter is planned for the production casing. This implies that the first casing must have a very large diameter in order to finish with a reasonably sized casing at the reservoir depth since the well casings step down in size as the well deepens; see Fig. 5.1.

Well completion program: Geothermal wells are completed barefoot (ie, normally open hole completions) and in many cases with a string of slotted casing hung across the production interval.

A good understanding of the differences between the two types of drilling as well as an understanding of what makes up a drilling program and final cost estimate is essential in successfully planning and carrying out an effective operation at the well site.

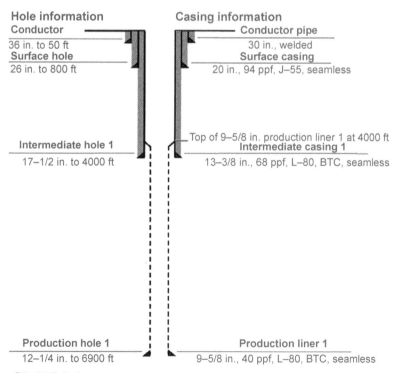

Figure 5.1 Well design.
Original diagram created by Capuano Engineering Co.

5.2 Getting started

Prior to setting up a drilling program or to providing drill cost estimates for a geothermal well, some relevant factors influencing the drilling program and the final cost of the drilling operation need to be determined. The following are a series of questions that need to be answered before starting the design process.

- Well location: Depending on where the well will be located, different regulatory agencies have jurisdiction over the well and different regulations apply. Specific regulations for depths, casing, and drilling methods as well as blow-out preventer requirements need to be considered separately.
- Elevation: Not only the accessibility of a site depends on the elevation above sea level but also the operational reliability of the engines used on the rig. Their functionality decreases with decreasing amounts of oxygen caused by increasing elevation. As a rule of thumb, an engine loses 3% of its operational efficiency for every 1000-ft increase in elevation.
- Accessibility: This point determines whether additional roads need to be constructed in order to set up the rig and hence play an important role for the cost estimates.
- Water availability: A water truck can provide about 100 barrels of water per load at a cost of approximately US$800 per day. The number of loads a water truck can haul will depend on the distance to the water source. A rig's daily water needs are approximately 700–800 barrels. Water quality is not critical other than it should be pH-neutral.
- Well type: The casing size and the well depth depend on the type of well planned. A monitoring well usually has a smaller diameter than an injection or production well. Injection wells might reach deeper than production wells to inject the returning water deeper into the geothermal system for reheating and returning to production.
- Flowing conditions: Depending on whether the well will flow to the surface (artesian flow) determines the necessity of a pump to be placed in the well below the water level. In case a pump is needed, the casing's internal diameter needs to be as large as possible to accommodate the appropriate downhole pump. These pumps can cost as much as US$200,000.
- Depth: The depth of the well is dependent on the geological situation. The more accurate the geological prognosis, the better is the information on where the production casing needs to be set and cemented into the reservoir formation at or near reservoir temperature.
- Temperatures: The local geothermal gradient may differ greatly depending on the geological setting of the site. Temperatures encountered during drilling as well as at reservoir depth determine the economic viability as well as the workability of the geothermal field. The higher the temperature at depth, the more energy can theoretically be produced at the surface. Higher temperatures also have a significant effect on the longevity of the drilling equipment and the casing.
- Resource type: The type of the resource (vapor or liquid dominated) is of great importance for planning the completion of the well (ie, is a pump needed or should the well have open flow potential).
- Sumps: Fluids from the well testing and water produced during drilling may need a sump on site. In many areas of the world disposal sumps are permitted; however, in areas that do not allow a sump, a cuttings cleaning program with a method of cuttings disposal is necessary. If a sump is permitted, then the type of acceptable sump liner (vinyl or clay) must be determined.

Once all these questions are thoroughly answered, the drilling program can be developed. This chapter will work through the drilling program components in the order that they are needed.

The initial step in the process is to develop is a well profile. The size and depths of the casing to be set can be determined based on the casing requirements of the regulatory agencies involved and the lithology expected, as well as the expected temperature profile of the subsurface environment. Designing of the casing program starts from the bottom of the hole and works its way upward. First, the preferred size of hole across the producing interval of the well needs to be determined, for instance, an $8\frac{1}{2}$ -in. open hole; all sizes are diameters unless otherwise stated. To have an $8\frac{1}{2}$ -in. open hole in the reservoir, the production casing must be set into the top of the reservoir and it must have an internal diameter to accommodate the $8\frac{1}{2}$ -in. bit, ie, at least a $9\frac{5}{8}$-in. casing (weight, grade, and thread connections can be determined when actually designing the casing string).

To cement a $9\frac{5}{8}$-in. casing, at least a $12\frac{1}{4}$ -in. open hole is required. This then must be filled with cement providing a complete bond between the casing and the wellbore. To drill a $12\frac{1}{4}$ -in. hole, there needs to be a casing large enough to accommodate this $12\frac{1}{4}$ -in. bit; therefore, at least a $13\frac{3}{8}$-in. surface casing (weight, grade, and thread connections can be determined later in the actual casing design) is necessary. In order to cement a $13\frac{3}{8}$-in. casing, a $17\frac{1}{2}$ -in. hole is needed and this would require a 20-in. casing set as the anchor point, at $300-400$ ft depth. Last, conductor pipe that is set prior to the rig moving to the location would have to be large enough to allow a 26-in. hole to be drilled for the 20-in. casing.

Once the size of the holes to be drilled and the casings that will be set in these holes are determined, the depth to which these holes and casings should be drilled and set, respectively, must be defined. Again, the process starts at the bottom, where the $9\frac{5}{8}$-in. casing setting depth should be dictated by formation and temperature. The $13\frac{3}{8}$-in. casing setting depth must be deep enough to allow the $9\frac{5}{8}$-in. casing to go as deep as necessary to case off any problem or poorly consolidated formations or any colder water entries. A visualization of a typical well design used in the geothermal industry is given in Fig. 5.1.

Each program and its component procedures and schedules to complete the well program and the cost estimate are designed separately. A helpful methodology is to establish a step-by-step drilling procedure of the well by defining each step in detail as they should occur in the course of drilling the well. Additionally, estimates of the penetration rate for each section of the drilled hole are carried out (ie, the 26-in. section, the $17\frac{1}{2}$ -in. section, the $12\frac{1}{4}$ -in. section, and the $8\frac{1}{2}$ -in. section). Based on this penetration rate, the amount of hole drilled per day, including connection time, trip time for new bits and directional survey time can be estimated. This will assist in developing a drilling curve, that is, depth versus time, which will be essential in determining the final well cost estimate.

5.3 Casing design

The casing is a major contributor to the final well cost, and thus it is crucial to select the appropriate casing grades, sizes, and lengths according to the given geological setting and the regulatory requirements. The casing not only prevents the well from collapsing but also prevents unintentional exchange between the surrounding formations and the well (eg, cold water mixing). To design the casing of a well, the influence of the following parameters on the selection of the casing needs to be clarified:

- Type and temperature of the resource (liquid or vapor or both)
- Temperature of the rock
- Chemistry of the resource (corrosion potential)
- Depth of the reservoir
- Necessity of a pump (downhole or surface)
- Regulations
- Is directional drilling necessary to reach the anticipated reservoir?

To choose the casing matching the parameters determined above, several casing properties need to be accounted for: grade (tensile strength, burst strength, collapse resistance) and the weight, given as lb-m per foot of the casing string.

5.3.1 Grade

To identify the casing grade required, API (American Petroleum Institute) steel-grade tables can be used (see Table 5.1). The API developed standards to characterize casings in a normalized manner. Extensive API tables list the strength properties, the physical dimensions, and the quality control test procedures for casings with outside diameters (OD) between $4\frac{1}{2}$ and 20 in. In a geothermal environment, hydrogen sulfide

Table 5.1 API steel grades

API grade	Yield stress, psi		Minimum ultimate tensile strength, psi	Minimum elongation, %
	Minimum	Maximum		
H-40	40,000	80,000	60,000	29.5
J-55	55,000	80,000	75,000	24.0
K-55	55,000	80,000	95,000	19.5
N-80	80,000	110,000	100,000	18.5
L-80	80,000	95,000	95,000	19.5
C-90	90,000	105,000	100,000	18.5
C-95	95,000	110,000	105,000	18.5
T-95	95,000	110,000	105,000	18.0
P-110	110,000	140,000	125,000	15.0
Q-125	125,000	150,000	135,000	18.0

Applied Drilling Engineering by Bourgoyne, Chenevert, Milheim, and Young, SPE Textbook Series, vol. 2, p. 302.

(H_2S)-induced stress corrosion cracking also needs to be taken into consideration when choosing the casing grade. Higher-strength steel casings are prone to H_2S-induced stress cracking, which limits the use of the available casing grades.

The fabrication of casings is carried out using a seamless process, an electrical resistance or electric-flash welding technique [1]. The finished goods from the manufacturer are API controlled (in case of API casings) and are allowed some minor deviations from the specified standard. For example, the pipe wall thickness of a casing is permitted a tolerance of 12.5% of its *nominal wall thickness* for the API-specified wall thicknesses. The inner diameter (ID) of a pipe is allowed a certain *drift*, which is limited by a standard-sized mandrel.

The formulas given in the sections below were taken from [1] and can be referenced there for more details.

5.3.2 Tension

The tensional force required to cause permanent damage on a pipe (F_{ten}) works against the force exerted by the pipe walls itself (F_2). The most commonly used safety factor for tension is 1.75.

$$F_2 = \sigma_{yield} A_s \qquad [5.1]$$

where σ_{yield} is the minimum yield strength in psi, and A_s is the cross-sectional area of the casing given as

$$A_s = \frac{\pi}{4}\left[(OD)^2 - (ID)^2\right], \text{ where the units of } A_s \text{ are in}^2 \qquad [5.2]$$

$$T_{ten} = \frac{\pi}{4}\sigma_{yield}\left[(OD)^2 - (ID)^2\right] \qquad [5.3]$$

Loading is calculated by multiplying the weight per foot of the casing by its length. The greatest tensional load occurs in the top joint of casing and it is calculated as if the casing were hanging in air.

5.3.3 Burst

The burst pressure is the minimum pressure needed to cause irreversible damage on the pipe by bursting. It is calculated considering zero external pressure and no axial loading. In that case, the force F_1 required to burst a pipe from the inside is counteracted only by the strength of the pipe walls (F_2).

$$F_1 = P_{br}L\frac{d}{2}\,d\theta \qquad [5.4]$$

where P_{br} is the internal pressure acting on the internal surface area along the pipe; L stands for the length of the pipe; $d/2$ represents the radius; and $d\theta$ is the angle delineating the area $(0°-360°)$ affected by the internal pressure.

$$F_2 = \sigma_s tL \frac{d\theta}{2} \qquad [5.5]$$

where σ_s is the strength of the steel pipe acting over the steel area tL (thickness of the pipe t times its length L).

Totaling the forces for static conditions $(F_1 - 2F_2 = 0)$ and solving for the burst pressure will give

$$P_{br} = \frac{2\sigma_s t}{d} \qquad [5.6]$$

To account for the worst-case scenario by the allowable deviation in wall thickness mentioned above (87.5% of nominal wall thickness), this equation should be adjusted to

$$P_{br} = 0.875 \frac{2\sigma_{yield} t}{d_n} \qquad [5.7]$$

where d_n stands for the nominal wall thickness.

The most common burst safety factors is 1.25, calculated in the bottom joint with hydrostatic pressure as well as surface wellhead pressure (blow-out preventer or BOPE pressure limit) with zero pressure on the outside of the casing.

5.3.4 Collapse

To compute the collapse pressure, more complex calculations are necessary that are not elaborated in this chapter. For a more detailed discussion on collapse pressure, see Bourgoyne et al. [1]. Collapse can occur in a pipe when the internal pressure of the pipe is zero (no fluid inside) and the external pressure exceeds the casing's wall strength. Collapse is not as much determined by the casing grade but depends more on the wall thickness and its geometry [1].

The most common safety factor for collapse is 1.15, calculated at the bottom of the casing string, that is, load with hydrostatic pressure externally and zero pressure internally. Collapse is the lowest design factor and is usually the first design factor considered.

5.3.5 Weight

The weight of a casing in pounds per foot is listed in the API tables for each casing grade. The weight determines the depth to which this casing type can be set and the rig capacity needed to handle and place the casings.

5.4 Mud program

The purpose of the mud, also called drilling fluid, is to remove cuttings from the bottom of the borehole and transport them to the surface. Besides that, the drilling fluid has lubricating characteristics for the drill bit and cools it at the same time. Moreover, the stability of the borehole is enhanced during the drilling process by the hydrostatic pressure provided by the drilling fluid against the borehole walls.

To assemble the proper drilling fluid for a geothermal well, the mud engineer has to consider the following factors:

- Lithology, including the thickness, strength, permeability, and pore pressure of the formations
- Water quality and availability on site to account for possible reactions of the water with additives and necessary cleaning processes
- High temperatures encountered downhole, which may alter the mud's properties
- H_2S and CO_2 existence, which dissolve in water and alter the mud pH

Since the downhole situation varies with depth, the mud's properties need to be monitored closely so that the program may be adjusted according to the circumstances given along the borehole.

Key properties that need to be checked frequently as the drilling process continues are:

- The **density** of the mud is controlled by adding inert solids. These may be formed by the rock crushing action of the drill bit or if a higher density is necessary by adding API barite with a specific gravity of 4.2 g/mL [1]. The density of the fluid needs to be adjusted according to the downhole formation pressure. In geothermal wells, the pressures are usually low and the density of the drilling mud should be kept as low as possible to prevent it from entering into the surrounding rock.
- The **viscosity** of the drilling fluid needs to be high enough to be able to transport the cuttings to the surface to keep the hole clean, to prevent lost circulation and differential sticking and other problems. The primary recommended viscosifier for geothermal drilling is API grade bentonite (sodium montmorillonite). For a while, synthetic polymers have been added to the drilling fluid as viscosifiers since they provide instantaneous viscosity increase and encapsulate cuttings making the separation process easier. Unfortunately, these synthetic polymers often lose their advantageous properties within a short time under elevated temperatures [6].
- The **alkalinity** of the drilling fluid controls the possible effect of contaminants such as H_2S and CO_2 as well as corrosion rates but also steers reactions of additives (lignite and polymers) with the mud. The addition of caustic soda (NaOH) or caustic potash (KOH) regulates the pH of the drilling fluid [6]. It is recommended to keep the pH near 10.5.
- At high temperatures ($>350°F$), the **solids** in the drilling fluid absorb high amounts of the free water and consequently increase the fluid's viscosity and cause gelation tendencies. Therefore, a smooth and effective separation process at the surface is essential [6].
- In the past, lignite has served as the most common **filtrate reducer** in geothermal drilling. In recent years, project specific needs for filtrate control became standard practice and polymer filtrate reducers (resistant to high-temperature alterations) are becoming more common [6].
- Especially during directional drilling, when the drill bit is in contact with the borehole wall, proper **lubrication** and **cooling** of the drill bit is crucial for the advancing of the drill head. Only time-proven lubricants such as graphite or TORKease that function in a high-temperature environment should be applied in such cases [6].

Once the mud engineer has identified an appropriate mud mixture for the geothermal drilling project, which usually consists of a freshwater-based mud with active solids (viscosifiers) and inert solids for density control, the mud needs to be circulated to the drill bit where it serves its purpose as a lubricating and cooling agent as well as the medium to transport cuttings to the surface. In the geothermal industry, mud usually can be used as the drilling fluid in the uppermost section of the borehole but often needs to be changed to air or an aerated fluid as the production interval is reached to avoid clogging the fractures that allow fluid movement.

Due to subhydrostatic pressures in the resource reservoir, it may be necessary to drill with straight compressed air or an aerated water or mud mixture. Compressed air requirements will depend on the size and the depth of the hole. Aerated systems are a mixture of air and liquid phase. The mud pumps will start to pump mud or water and then air will be injected into the system. To get the proper ratio of air and liquid close observation is needed at the mud pits. If the level in the pits rises, that indicates too much air is being pumped because the well is producing reservoir water. If the level in the pit drops, this indicates too much liquid is being pumped as the formation is taking liquid. The ratio can be adjusted until the mud pit level stays constant.

5.5   Directional program

A directional drilling program might be necessary if the target horizon is not accessible from a location directly above it. This could be due to topographical obstacles (lakes or mountains) or legal barriers (eg, protected land). The advantage of directional drilling includes intersecting a liquid-bearing fracture at a more beneficial angle compared to a vertical intersection. Moreover, directional drilling allows having multiple wells originating at the same surface location and deflecting into different directions (angles) as they go deeper. This enables tapping one resource from different positions (angles) or to explore further into the underground. Multiple wells on a given drill pad also reduce the total costs of drill site construction since only one access road is needed, the rig is skidded within a short time and distance, only one disposal pit is needed, steam gathering pipe work costs are lowered, and overall supply costs are reduced [4].

Planning a well that markedly deviates from vertical to reach its target reservoir is a complex process. After determining the above-mentioned reservoir and casing depth in the first step, the geometry of the well needs to be established. S-shaped or J-shaped wells are mostly applied in the geothermal industry and examples are displayed in Fig. 5.2 [4].

The most important factor to consider for a directional drilling program in the geothermal industry is the selection of proper material that can withstand the elevated temperatures and the angle with which the well deviates from vertical (building angle). A mud motor equipped with a measure while drilling (MWD) assembly only lasts a short time at temperatures above 302−347°F (150−175°C). Hence, mud coolers are

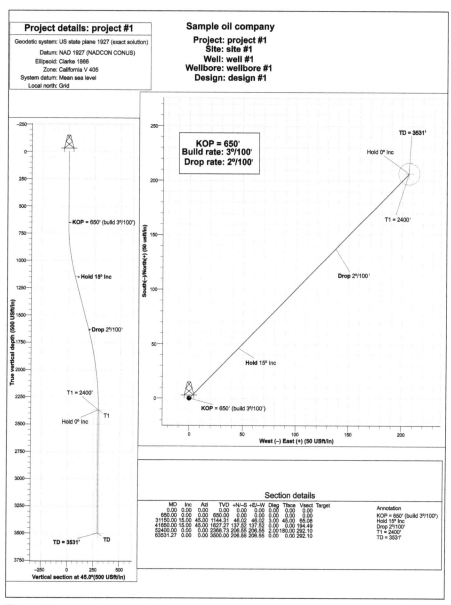

Figure 5.2 Directional drilling plan.
Example diagram provided by Scientific Drilling, Bakersfield, CA.

needed for drilling motors. Simple placement of stabilizers in the bottom-hole assembly (BHA) can achieve changes in the well deviation, ie, reduce or increase deviation. However, stabilization placement cannot affect direction; directional changes or adjustment can be accomplished by the use of mud motors.

5.6   Wellhead design and blow-out preventer systems

The wellhead system is the permanently installed top of the production casing that will remain on the well after the drilling has been completed. It consists of a series of weld-on casing-heads, flanges, and valves that control the flow during the production phase of the well. In geothermal wells, these wellheads and master control valves are usually installed on the top of the production casing prior to drilling out the cement from the bottom of the casing. The master valves are then opened and a system of blow-out preventer valves installed above the master control valve. Drilling is then conducted through the opened master valve. After the well is completed or the drilling is finished, the drill pipe and drilling tools are pulled out of the hole and the master valve closed, blow-out preventers removed, and the rig demobilized off location.

The casing-head will vary with the type of resource being developed:

In *dry steam—dominated* resource areas, the casing-head consists of a full bore (ie, the same bore as the production casing) slip-on, weld-on, thick-walled tubing with a flange on top for the installation of the master valve and blow-out preventer system. The casing-head is also equipped with two smaller flanged side outlets for the installation of choke lines and fill-up lines. All flanges in the wellhead and master valve system are the same size and dimensions with the same size and pattern of bolt holes. The master valve and blow-out preventers are stacked up and then bolted together. Between each set of flanges a steel gasket is placed in a matching set of ring grooves. When being bolting together, the gasket is compressed to complete the seal.

If the resource is a *liquid-dominated* resource, then the casing-head consists of an expansion spool on top of a full bore wellhead. The casing-head is installed on the surface casing and the expansion spool is then installed on top of the casing-head after the production casing is run and cemented. The purpose of the expansion spool is to absorb the elongation of the production casing internally, which is caused by large temperature changes during production. When the wells are shut in, the production casing will drop back down to its original length. To maintain an adequate seal during production, this expansion needs to be absorbed internally by the expansion spool. Expansion spools allow for approximately 18 in. of internal expansion. The master control valve is installed above the expansion spool followed by the blow-out preventer components that are stacked up and bolted on.

The blow-out preventer system consists of a variety of well control valves that are hydraulically controlled. These types of control valves may be ram type, bag type, or slab gate (a slab gate is a hydraulically controlled valve that consist of a plate of steel across the valve opening to open and shut the valve). A *ram type* preventer is made up of two hydraulically controlled rams that sit on opposite sides of the wellbore enclosed in a ram body. These rams can be fitted with a gasketed hole in the center that can close and hold pressure around the drill pipe. The hole is usually the same size as the drill pipe being used; the device will be used when the pipe is in the hole and the well starts to flow. The rams can also be fitted without a hole in the center. In that case, these rams, called complete shutoff rams (CSO rams), are closed when the drill pipe is out of the hole. A *bag-type* preventer is also known as an annular preventer and consists of a rubber element in a cylindrical body. A piston in this device will compress the rubber and seal off around any size of pipe in the hole.

All blow-out preventers have the ability to close in a matter of seconds. All closing elements are controlled by a device known as an accumulator. The accumulator maintains the hydraulic system to close these valves instantly. The accumulator is hydraulically operated with a nitrogen gas charge backup in case the hydraulic system fails. In some cases, the regulatory agency may require a third backup system. The closing stations are usually maintained in remote locations from the accumulator itself. One closing unit is located on the rig floor right behind the driller's station. Another closing station is located farther away from the rig in case the rig has been impacted and abandoned and the unit needs to be closed as the rig crew moves away from potential problems.

The various types of blow-out preventer equipment or stacks (BOPE) are listed next; see Fig. 5.3:

1. A Class I BOPE system. A minimum assembly consisting of any device installed at the surface that is capable of complete closure of the wellbore with drill pipe out of the hole. This device must be closed whenever the well is unattended.
2. A Class II BOPE system. A minimum system of annular and/or ram type preventers capable of providing complete closure of the wellbore and closure around the pipe in use.

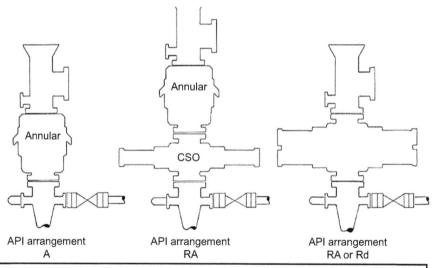

| API arrangement A | API arrangement RA | API arrangement RA or Rd |

API BOP-stack arrangement legend
R = Single ram preventer w/1 set of rams, either blank or for pipe.
Rd = Double ram preventer w/2 sets of rams, positioned according to operator's choice.
A = Annular blowout preventer
S = Drilling spool w/side outlet connections for choke and kill lines.

Figure 5.3 Blow-out prevent systems.
Drawing from Blowout Prevention in California, Equipment Selection and Testing. Publication by California Department of Conservation Division of Oil, Gas and Geothermal Resources, p. 5.

3. A Class III BOPE system. A minimum system consisting of annular preventer, which is capable of closing around any size or shape of pipe, a blind or CSO ram type preventer, and pipe ram-type preventers capable of closure around a specific size pipe, that is, drill pipe, drill collar, or casing. Rams are changed as pipe sizes changed.
4. A Class IV BOPE system. A minimum system of annular preventer, blind (CSO) ram type preventer, and two or more pipe ram type preventers capable of closure around all pipe, exclusive of drill collars and smaller pipe. The blind rams must be above at least one set of pipe rams.
5. A Class V BOPE system. This class only applies in the case of a subsea BOPE stack and will not be discussed here.

During air or aerated drilling, the operation will require the use of a diverter on top of the blow-out preventer stack. The diverter will allow drilling under a flowing situation. The well will discharge the air and cuttings under pressure and the diverter will direct this discharge away from the rig. The drilling can continue even if the well is producing geothermal fluids until enough resource has been encountered to determine that the well is commercial.

All blow-out preventer equipment should be pressure rated to give a clear safety margin over the expected pressure to be encountered in the well. All components of the blow-out preventer stack should have a known height and outside dimensions so as to be able to calculate the maximum height of the final stack. This height is used to determine the substructure height and clearance to fit the stack under the drilling rig floor.

5.7   Cementing program

The most important aspect of a good geothermal well completion is a good casing cement job. In the oil and gas industry, the purpose of cement is for isolation. However, in the geothermal industry, the purpose for cementing the entire string is to limit the elongation of the casing as a result of the temperature change from the time of placement to the production phase when high-temperature geofluids are carried from the reservoir to the surface. In addition, it assists in preventing thermal fatigue due to thermal expansion and contraction.

The difference between oil and gas completions and geothermal completion is that in oil and gas completions the casing is cemented across the only zones of potential production and the cement volume is usually calculated to cement across the zone of interest plus several 100–1000 ft above and below the zone. After the productive zone is cemented for isolation, the well is then completed with packers and tubing. Perforations penetrate the casing, the cement, and then the production zone. The well is then completed with packers inside the casing isolating the perforated section. Afterward, tubing is run through the packers to carry the oil or gas out the perforations into the tubing and to the surface. In this manner, several zones can be produced simultaneously.

In this section, the cementing formula or recipe to provide high-temperature resistant cement is discussed. This is a cement mixture that will withstand the high stress of thermal cycling that these geothermal wells are likely to experience. In this cycle, the cement will experience shut-in well conditions, which will be the rock temperature at depth to high temperature during flowing conditions. This cycle could involve a temperature variation of as much as 300°F. During this cycle, the casing will want to elongate due to the very large increase in temperature when production begins. Hence, the cement must have sufficient bonding to control the elongation.

Casing cement is a nonrecurring cost incurred in drilling and completing the well. It is a nonrecurring cost because the volume of cement to fill all the annular space between the casing strings and the borehole can be calculated with an acceptable amount of excess for enlarged hole or hole washouts. This calculated amount is usually brought to location in advance to reduce rig waiting time for cement arrival; thus, it is a nonrecurring well cost.

In geothermal wells, the cement in the annulus is carried up the annulus from total depth of the casing to the top of the casing. It is important to bring the cement entirely to the surface. Cement volumes are easy to calculate, and the estimated cost of cement are calculated in dollars per cubic foot or barrel of mixed cement. Excess calculations range from 100% in the open hole section of the well to 30% in the cased hole sections. When determining the amount of cement needed in each section, it is better to have too much compared to running short, thus leaving uncemented voids in the annulus. If loss of circulation (LOC) occurs during cementing, then excess cement costs may be encountered in carrying out cementing top jobs (ie, placing cement into the top of the annular space using tubing to finish securing the casing to the borehole). In the cementing procedure section of this chapter, various methods of placing the cement effectively in the annulus are discussed.

Once the volume of cement needed is known, designing a cement blend or recipe to appropriately perform the best job is essential. Designing the cement slurry starts with the expected temperature that the cement will be exposed to during the placement as well as the geofluid temperature to be experienced during production. The basic cement slurry consists of API Class G cement as the base. Additives are then blended into the dry cement to control various properties of the cement slurry. Controlling fluid loss of the cement is important to ensure that the free water in the cement is tied up and will not be lost to the formation as the cement sets in the hole. Small amounts of bentonite gel are added to control the free water. Besides, since it is important to control the pumping qualities of the cement, friction reducers are added to give the slurry better pumping factors. Silica flour is added to enhance the compressive strength of the cement, in many cases as much as 40% silica flour is used. This makes it very durable; the cement will last a long time at high temperatures while the silica flour will prevent strength retrogression of the hardened cement. Other additives may be used for retarding the cement setting time. Setting time can be adjusted to give more pumping time as well as accelerating the setting for short strings of casing. The type of retarders used to extend the setting time depends on the bottom-hole temperature to which the cement is exposed.

Another aspect of wellbore cementing is selecting the appropriate density of the cement to be used. Geothermal wells have highly stressed formations, and in many cases the fracture gradients for these formations are very sensitive to hydrostatic pressure changes. Therefore, it is crucial to have an effective cement at a relatively low hydrostatic pressure gradient. Mixed Standard Class G cement will have a weight of approximately 16.5 pounds per gallon (PPG), which is twice the weight of water. If the well was drilled using water and then a highly dense cement was introduced, circulation losses could be created by opening up fractures or creating new fractures.

When designing a cement slurry, the pore pressure and the fracture gradient of the formation have to be considered to assign the appropriate density of the cement. It is of outmost importance to use the lowest density of the cement slurry without jeopardizing the quality of the cement. In the early days of geothermal cementing, the industry would make a 1:1 mix Portland cement with perlite (perlite is an inert organic material that is mined and finely ground and is extremely lightweight). This mixture would reduce the density of the cement slurry from 16.5 PPG to approximately 11.5 PPG. Perlite proved inappropriate over time in that it broke down and created permeability in the cement. The industry then attempted to reduce the density with the addition of ceramic spheres. This could reduce the density to below 10.0 PPG. The next progression in developing lightweight cement slurries was with the practice of injecting the slurry with air or gas bubbles. Nitrogen injection into the cement slurries also proved very effective. With the proper proportion of nitrogen to cement, the mixture slurry could be reduced to a density less than water. This cement could float on water and proved very effective and durable.

Once the cement blend has been specified, the cement must be tested at the estimated wellbore conditions. When testing the cement it is very important that the cement is mixed using the same water that will be used at the drill site. Testing of these slurries is carried out in cement lab under control conditions. The slurry is mixed to the appropriate density and then all relevant properties will be determined, including setting time and temperature, fluid loss and pumping qualities. During this mixing, the quantity of water required to mix the cement slurry to the specified density is determined since this is a calculation needed in the field when the actual cement job takes place. In addition, the yield of the cement per sack (ie, one sack of cement will yield × cubic feet of slurry) is calculated. This determines how many sacks of cement need to be transported to the well site. Cured cubes of the hardened cement will then be compression tested in the laboratory for compressive strength over time.

5.8 Cement placement

Just as important as the cement mixture is the method of cement placement into the wellbore. There are many placement methods available in the cementing industry. The various types of placement will be discussed in the following section. Methods of placement are as follows:

1. Conventional cementing
2. Inner string cementing
3. Liner cementing
4. Squeeze cementing and plug cementing
5. Multistage cementing
6. Reverse cementing

5.8.1 Conventional cementing

This is the most common method of cementing casing; see Fig. 5.4. This method involves the running of the casing into the hole with one or two float valves (two check valves separated by one or two joints of casing). Fig. 5.4(a) shows the start or cementing. These valves allow one-way flow through the valve and prevent backflow (U-tubing) into the casing. This is very important because the cement is denser than the placement fluid. The process is to insert a bottom rubber plug into the top of the casing, which will be pumped down ahead of the cement. This plug is a full-diameter wiper plug that wipes the casing as it is pumped down. The calculated volume of cement is then pumped following the bottom plug (the volume is the amount to fill the annular space plus excess). Fig. 5.4(b) is during the cementing job. Once the bottom plug engages the top float valve, the pressure will rupture the center bladder of the plug and the cement is then pumped through the bottom wiper plug. When all the calculated volume of cement is pumped, a top rubber plug is inserted into the casing. The plug is made of solid rubber and it separates the cement from the displacement water or mud. It also wipes the casing as it is pumped down. Fig. 5.4(c) shows casing and cement at the end of the cement job. Once the top plug encounters the bottom plug all the cement is in place. Pressure is released above the placement fluid and no flow back should be encountered if the float valves hold. After the cement has set, installation of the casing head can be performed.

The disadvantage to this method is that you get only one chance to fill the annulus with cement and if the hole is washed out or the volume is short to fill

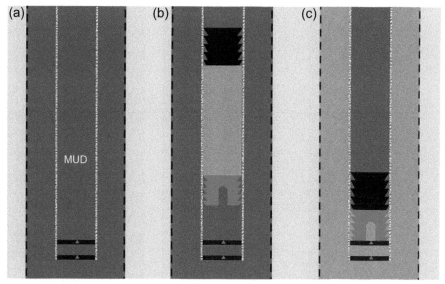

Figure 5.4 Conventional cementing. (a) At the start of cementing, (b) during cementing, and (c) at the end of cementing.
Diagram created and provided by Resource Cementing, Rio Vista, CA.

the annulus, the cement might not reach to the surface inside the annular space. Top cement jobs would be required to fill the annular space properly.

5.8.2   Inner string cementing

This method of cementing is very similar to the conventional method described earlier, in that the casing is run into the hole with two float valves on the bottom separated by one or two joints of casing, see Fig. 5.5. Fig. 5.5(a) shows the start of the cement job. The main difference is that the top float valve has a stab-in adaptor above the float valve. The drill pipe is then run into the casing with a stab-in stinger on the bottom. The stab-in stinger will then stab into the stab-in adaptor and the casing is then circulated with mud or water to establish good circulation rate through the drill pipe. Fig. 5.5(b) shows circulating and cementing through the drill pipe. Cement is mixed and pumped down the drill pipe and up the annular space to fill the annulus from bottom to top with cement. When using this method, the amount of cement needed to fill the annulus is calculated and pumped until the cement is coming back to the surface in the annulus. At that point, the mixing of cement can be halted and the drill pipe can be removed from the stab-in adaptor and the cement in the drill pipe may be dropped in the drill pipe on top of the float valve. This method of cementing provides full control of the cementing operation. If the hole is in the gauge this operation may save cement and, if the hole is washed out, additional cement can be mixed until the entire annulus is full. Fig. 5.5(c) shows the cement job completed.

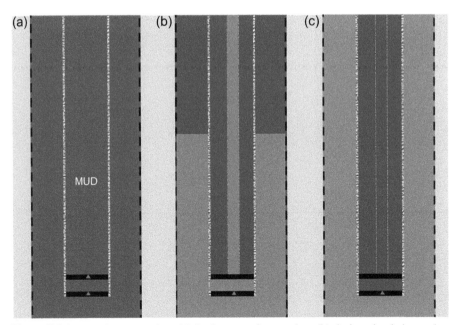

Figure 5.5 Inner string cementing. (a) At the start of cementing, (b) during circulating and cementing, and (c) at the end of cementing.
Diagram created and provided by Resource Cementing, Rio Vista, CA.

5.8.3   Liner cementing

The term "liner" means that this casing does not come back to the surface but rather is hung using a liner hanger from the bottom of the prior run casing, see Fig. 5.6. These liners are usually hung approximately 200 ft up inside the prior cemented-in casing. Fig. 5.6(a) shows the liner is hung and circulation started. The liner is run on a drill pipe which screws into the liner hanger; the liner hanger is screwed into the top of the liner and is then lowered into the hole on this drill pipe. The liner is equipped with float valves located on the bottom of the liner separated by one or two joints of liner. Once the liner is in place the liner is hung by rotating the drill pipe, which allows casing slips on the liner hanger to engage the outer casing in which it is hung. The liner hanger is equipped with a drillable bushing through which the setting tool is run. On the end of the setting tool are two rubber wiper plugs. Both wiper plugs are open full bore, and the bottom plug has a smaller flow-through opening than the top plug. The calculated volume is then mixed and pumped down the drill pipe. A dart type wiper plug is lowered ahead of this cement which will separate the mud from the cement as well as wipe the drill pipe as it is pumped down. This dart will engage the bottom wiper plug on the liner setting tool and shear off the plug. Now the cement is pumped down and the bottom casing wiper plug performs the same task as in the inner string cementing job. Fig. 5.6(b) shows the casing and cementing during cementing job. After all the cement is pumped the other larger diameter dart is dropped which separates the cement from the displacement fluid and is then pumped down wiping the drill pipe as it is pumped. This dart will then engage the top casing wiper plug and shear off. This wiper will then be pumped down until the top plug encounters the bottom plug on top of the top liner float valve. Cement will fill the annular space and re-enter the annular area between the drill pipe and the upper casing, filling completely the area between the liner and the casing. Fig. 5.6(c) shows completion of the cementing job. Once the cement is in place the liner hanger setting tools are then disengaged from the liner and removed from the well, excess cement will be deposited on top of the liner hanger; all cement and liner hanger internal items are drillable.

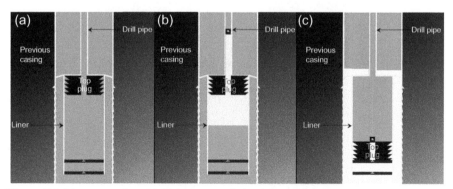

Figure 5.6 Liner cementing. (a) Liner hung and circulating, (b) during cement job, and (c) completion of cementing.
Diagram created and provided by Resource Cementing, Rio Vista, CA.

5.8.4 Squeeze cementing and plug cementing

These methods of cementing are not involved with the practice of cementing in a casing or liner string, but rather a repair operation, or an abandonment procedure or sidetrack operation, see Fig. 5.7 (packer is set above the hole and cement is squeezed below the packer and out of the hole). This method calls for a calculated amount of cement to be mixed and then placed in the wellbore at specific locations. Squeeze cement is an amount of cement mixed and deposited in the casing across an area in the casing that has holes or leaks. The bottom of the casing is plugged and a packer is set above the area of the holes or the squeeze pressure can be achieved by closing the pipe rams on the drill pipe and holding this pressure to squeeze cement into the holes or leaks. The cement is then squeezed into the holes or the leaky areas. This is an effort the fill the holes with cement to stop the leakage.

Plug cementing is mixing cement and setting it at a specified location in the casing to abandon the well as per the regulatory requirements. It is also possible to set a plug in cased hole or open hole for the purpose of sidetracking the well to an alternative location. These plugs are set as balanced plugs (ie, placed in the hole to be balanced inside and outside the drill pipe that emplaced it), so that it would be safe to remove the drill pipe without contaminating the cement plug.

5.8.5 Multistage cementing

This method of cementing is used in circumstances where the formation will not hold the increased hydrostatic loads during cementing, which is the case in many geothermal areas. This is accomplished by running the casing into the hole with a ported sub in the casing string (see Fig. 5.8). This ported sub is located just above the leak formations.

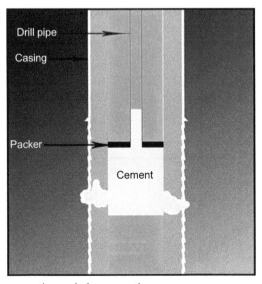

Figure 5.7 Squeeze cementing and plug cementing.
Diagram created and provided by Resource Cementing, Rio Vista, CA.

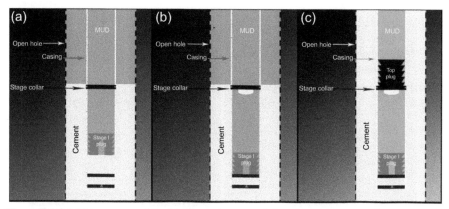

Figure 5.8 Multistage cementing. (a) Cementing stage 1, bottom zone cemented up to port, (b) port opened and casing circulated from port up, and (c) stage 2 cemented from port up and then port closed with plug.
Diagram created and provided by Resource Cementing, Rio Vista, CA.

The ported sub will have a sliding sleeve covering the hole. The sleeve can be shifted either with hydrostatic pressure or mechanically by passing a dart or wiper plug through it. After cementing through it the sleeve can be closed. The principle here is to mix enough cement to fill the bottom annular space between the bottom of the casing up to the sliding sleeve. Fig. 5.8(a) shows the cementing stage 1, bottom zone. Cement is mixed and plugs are used. Once this bottom section of cement is in place, then the sleeve is opened and the hole circulated from the sleeve to the surface while the bottom section of cement hardens. Fig. 5.8(b) shows circulation through port. Once the bottom sets, the volume of cement needed to fill the top portion of the annular space from the ported collar to the surface is mixed. The cement is pumped in with a wiper plug that will close the sleeve. Fig. 5.8(c) shows completion of cementing top stage.

The problem with this form of cementing is that of a potential leak in the casing string at the ported collar. Over a period of time at high temperature, this sleeve will start leaking and repairs will have to be done during the life of the well.

5.8.6 Reverse cementing

Reverse cementing is just as the name implies—the opposite of conventional cementing. Cement is pumped into the annular space and up into the casing, which is equipped with a check valve that will allow flow in (ie, upward) but not out preventing any U-tubing of the cement (see Fig. 5.9). This method requires that only a guide shoe be run on the bottom of the casing without a float or check valve. The cement is then mixed and pumped down the annulus and returns up inside the drill pipe while the pressure is monitored. Cement volume is calculated and pumped based on caliper logs. In this manner, the hydrostatic load on weak formations can be kept low. Cement can be designed with much lower amounts of retarders as the cement will be exposed to a high temperature for less time. In conventional cementing the cement is exposed to the bottom-hole temperature and then has to remain in a liquid state until it returns to

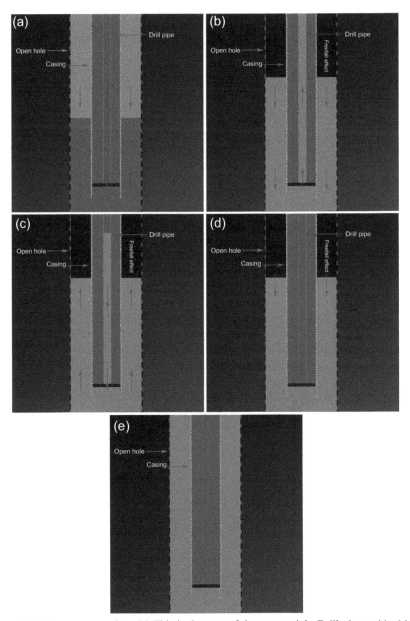

Figure 5.9 Reverse cementing. (a) This is the start of the cement job. Drill pipe stabbed into float collar to unset float on bottom of casing and cement mixed and pumped down the annulus and the mud returns up through the drill pipe to the surface. (b) This is during the job with cement returning up the drill pipe to the surface. (c) This is a point in the job when good cement has reached the surface through the drill pipe at which point the tail cement can be pumped down the drill pipe and around the casing shoe. (d) This is after tail cement has be pumped and is in place around casing shoe and up 200' or so. The drill pipe is pulled out of the casing and the float closes preventing flow back. (e) This is after completion of the job. Tail cement is in place and top cement job performed on the annulus to bring cement from the prior top of cement to the surface in the annulus.

the surface. In reverse cementing, the cement is exposed to bottom-hole temperature only once, when it reaches the bottom. When cement is in place, the top of the casing is shut in to prevent the cement and mud from U-tubing into the casing. To know when the cement is in place, radioactive tracers are placed in the lead cement and then monitored via wireline until it reaches the surface.

5.9 Hydraulic and bit program

As mentioned before, one purpose of the drilling fluid is to keep the borehole stable while drilling. This is possible because the fluid exerts a continuous hydrostatic pressure against the borehole wall as it is circulated through the drill string and up the annular space. The drilling fluid is pumped from the mud tank through the surface equipment down to the rotating drill bit where its hydraulic pressure penetrates the rock, fractures it, removes the cuttings, and brings it back to the surface where the cuttings are separated from the mud. To optimize the circulation of the mud and keep it as effective as possible on the borehole bottom, the mud hydraulics need to be calculated properly.

The following parameters are considered to keep the fluid pressure up:

- Flow rate of the drilling fluid in gallons per minute (GPM)
- Flow area through which the drilling fluid is pumped, eg, pipes and hoses (in^2)
- Length of the circulation loop, ie, total distance over which the mud flows (ft)
- Fluid properties; most importantly the density of the mud in pounds per gallon (PPG) and its viscosity in centipoise (cP)

Pressures generated by the mud pump on the surface are lost throughout the system as the mud circulates from the mud tank to the drill bit and back to the mud tank. The sum of all pressure losses in the system determines the total pump pressure.

$$\Delta P_{system} = \Delta P_{surface\ equipment} + \Delta P_{pipe} + \Delta P_{collar} + \Delta P_{downhole\ tools}$$
$$+ \Delta P_{bit} + \Delta P_{annulus}$$

[5.8]

To optimize the pressure at the bit, where the mud performs its primary job, two methods are used:

1. *The maximum jet impact force method*: The borehole bottom is best cleaned when the *force* exerted by the nozzles is at maximum. This is the case when 48% of the pressure drop occurs at the bit and the rest (52%) occurs elsewhere in the system.
2. *The maximum hydraulic horsepower method*: The borehole bottom is best cleaned when the *hydraulic horsepower* at the bit is at maximum. This is the case when 65% of the pressure drop in the system occurs at the bit and the remaining 35% occur elsewhere in the system.

To determine which method should be applied, the final anticipated borehole depth and the downhole conditions (lithology, permeability, pressure, temperature) need to be considered. Usually, the first method is applied in the upper section of the borehole, where high volumes of cuttings and large-diameter boreholes need high flow rates to effectively remove cuttings and clean the drill bit. The second method is more advantageous in the deeper parts of the borehole, where diameters are smaller, penetration rates are lower, and higher exit velocities are needed to transport the cuttings back to the surface [3].

The pressure at the bit multiplied by the flow rate is a measure of the power available for cleaning the borehole bottom [3]. To optimize the use of the available pressure and channel it to increase the bit performance, adjustments on the nozzles have to be considered. The effectiveness of the nozzles depends on the flow rate of the mud, as well as the nozzle diameter, its distance to the rock on the bottom, the arrangement of nozzles to each other, and the pressure with which the mud is released. Depending on the specific conditions, these parameters can be adjusted to prevent the bit from losing its mechanical effectiveness on the rock.

5.9.1   Bits

As mentioned earlier, the drill bit is responsible for mechanically penetrating and crushing the rock underneath it. A major difference in geothermal drilling from oil and gas drilling is the medium in which the reservoir is located. Igneous or metamorphic formations—common media for geothermal reservoirs—are hard, abrasive, and fractured and deform in a brittle manner as opposed to plastic deformation in weaker formations. This and the high-temperature environment encountered at depth impose great strains on the drill bit and the lubricant, decreasing its downhole performance.

To select the proper drill bit, the costs and the wearing of the bit need to be evaluated. The costs for a bit are US$20,000–25,000 for a roller cone bit and US$150,000–250,000 for a drag bit. The bit's wearing (dullness) depends on the compressive strength of the rock, the chemical composition of the brine (corrosion), and the proper application of the bit downhole (rpm, alignment, weight on bit, etc.). Improper selection of a bit results in lower penetration rates, fast wearing of the teeth, and frequent bit changes, which result in higher drilling costs overall.

The drilling bits commonly used in the geothermal industry are described next.

5.9.1.1   Roller cone bits

Roller cone bits assembled with three cones (tri-cones) are mostly used in the geothermal industry (see Fig. 5.10(a)). Three rotating cones, assembled with either inserted or milled teeth, crush and grind the rock underneath it. The hardness (grade) of the teeth and the quality of the bearings and seals determine how long a bit lasts in a high-temperature, high-pressure, hard, and abrasive environment.

Roller cone bits are usually preferred for geothermal drilling because they perform better in hard rocks and are more economic than drag bits. Also, drag bits (no moving parts and relays on scrapping action) need good fluid circulation to perform; this keeps the drag bit cool or else cracking may occur due to the extreme heat created from friction, which is not the case in many geothermal areas, as many geothermal have vast loss circulation zones. In addition, drag bits cannot be used for the large-diameter holes needed for the surface segment of the well (26 in.). Drag bits are hardly ever used in

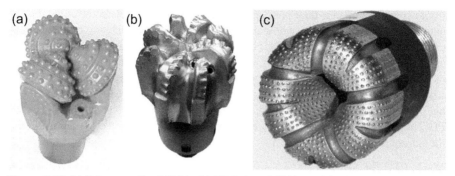

Figure 5.10 (a) Tri-cone roller drill bit, (b) PDC drag drill bit, and (c) Impregnated drill bit. Photograph obtained from the Internet.

geothermal drilling and are mostly used in water well drilling in upper zone soft formations. Off-the-shelf bits are usually applied, but special bits could be manufactured if necessary.

An important factor influencing the performance of the drill bit is the lubricant's (drilling mud) ability to effectively cool the drill bit while drilling and removing the cuttings from underneath the bit. Severe damage can result on the cones when the mud loses its properties due to high bottom-hole temperatures. In such a case, the drill bit overheats caused by the friction of the bit's rotating motion and the additional formation temperature at depth. The cones can crack or even fall off entirely in such a case. These and other kinds of damage, such as chipped or broken-off teeth, indicate inappropriate use or selection of bits (misalignment, hardness, etc.) and/or unexpected rough conditions given downhole.

To determine whether the selected bit was an appropriate choice, a dull grading system for roller bits was developed. The International Association of Drilling Contractors (IADC) determined eight parameters to be graded to examine the bit application. Depending on the results, the bit hardness can be adjusted as well as the weight on the bit or the rotation speed.

1. Inner cutting structure
2. Outer cutting structure
3. Dull characteristics
4. Location
5. Bearing seals
6. Gauge
7. Other dull characteristics
8. Reason pulled

5.9.1.2 Drag bits

Due to their wedging and shearing motion, drag bits tend to be more efficient than roller cone bits, see Fig. 5.10(b). In addition, they have no moving parts downhole, which increases longevity of the bit tremendously. What has kept the geothermal industry from using them often is their price, particularly in larger diameters, the circulation requirements, and unsatisfactory downhole performance. A special drag bit, using polycrystalline diamond compact (PDC) cutters, which use synthetic diamond cutters, are applied under special circumstances, when the price—wear relationship justifies. Technical studies were carried out to improve performance and reduce costs of PDC bits when drilling in hard rock typical of geothermal drilling [7]. The results were scattered but promising, and more recent studies carried out by Raymond et al. [5] suggest that PDC drag bits have the potential to be more widely applied in geothermal drilling in the future.

5.9.1.3   Impregnated diamond bits

Self-sharpening, diamond-impregnated drill bits that consist of synthetic grit-sized diamonds in a tungsten carbide matrix are ideal for drilling in hard, abrasive formations (see Fig. 5.10(c)). As the bit wears out, a new layer of diamonds emerges from the slightly softer matrix. However, such bits are very expensive, and the matrix tends to get very hot when air or aerated mud is used instead of a "regular" mud and can break. This, in addition to higher downhole temperatures, prevents the geothermal industry from applying this type of drill bit.

5.10   Drilling curve

Proposed well lithology will need to be identified to estimate the penetration rate for each section of the hole. Once this penetration rate is determined, a drilling curve can be established, that is, a chart displaying drilled depth versus days on location (see Fig. 5.11). In this curve, the total time required to drill each section is estimated, including drilling time, survey time, trip time, as well as rig service time. Each flat spot on the curve represents the time when no drilling is done, but logs are run, casing and cementing are carried out, blow-out preventers are installed, and drilling assemblies are changed to drill the next section. Additional time is added to the end of the drilling to provide time to perform well testing and completion time at the end of the well. Further estimates are needed for the initial mobilization time and rig-up time to drill as well as for the rig-down and demobilization. Once all times have been estimated, the drilling curve can be completed and the most important part of this exercise, namely, the total number of days required to drill and complete the well can be determined. This total number of days will be used in the final well cost estimate. Fig. 5.11 gives an example of how such a drilling curve might look.

5.11   Mud logging

A mud logging unit is installed on the rig when geologic information must be retrieved on a timely basis. If the formations drilled are not well known or if a specific geologic

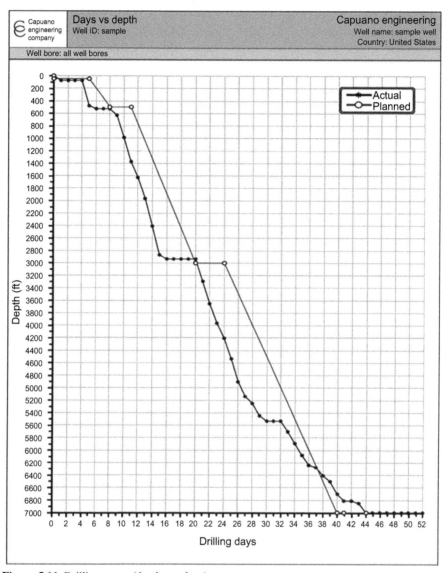

Figure 5.11 Drilling curve (depth vs. days).
Original example created by Capuano Engineering Co.

horizon is targeted, mineralogical data from cuttings brought to the surface with the mud might be essential. For example, secondary hydrothermal mineralization is a key indicator to locate a viable horizon as a geothermal reservoir. Changes in the mineralogical composition of the formations can require adjustments in the mud program.

The main task of a mud logging unit is to monitor and collect all necessary information to ensure an uninterrupted drilling process. This includes the measurement of gases, mud flow rates (in and out), temperatures, drill rates, depth, and pressures.

Gases monitored include hydrogen sulfide (H_2S) and carbon dioxide (CO_2). H_2S is poisonous, very corrosive when dissolved in water, and flammable as a gas. When dissolved in water (of the mud) it lowers the mud pH increasing corrosion tendencies. As a gas it is transported to the surface when air is applied as the drilling fluid. Carbon dioxide is soluble in water and can form carbonic acid in the drilling fluid which reduces the mud pH and makes it more corrosive.

It is important to detect the gases in a timely manner using a mud-gas trap during mud drilling or an air-gas trap during air drilling, so appropriate abatement measures can be taken.

The depth and the drill rate of the drill rig are also monitored by the mud logger. A sudden increase in the drill rate might help indicate fracture zones, which are of great interest in the geothermal drilling process [2].

5.12   Drilling rig selection and special considerations

One of the last items to complete in the drilling program is the recommendation of the components for the most appropriate drilling rig to successfully drill and complete the well at the best price. Drilling rigs are not normally owned and operated by the field developer or operator, so the developer will engage a rotary drilling rig contractor for this service. There are various types of drilling contracts that exist between the contractor and the developer. These contracts are as follows:

- **Day work contract**: The developer leases the rig and crews at a price per day or per service. The contractor will furnish the rig with all necessary equipment requested by the developer with crews to operate the rig on a 24/7 basis. In this contract, the risk is all on the developer. This can be the most economically priced contract if the developer and their drilling personal have a good understanding of the area being drilled.
- **Turn-key contract**: The developer will request that the contractor furnishes him a well with a specific casing program and total depth for a fixed total price. This contract is usually the most expensive for the developer because the risk is all on the contractor. The total cost can be 30–40% higher than a day work contract.
- **Footage contract**: Here the contractor will provide the developer a well at a cost per foot drilled. Again, this is a higher priced contract than the day work contract in that the risk is on the contractor.
- **Hybrid contract**: This is usually a blend between a turn-key and a day work contract.

Once the drilling program has been completed, its components will be used to specify the type and size of the drilling rig that is required for this drilling project. Big is not always best; a proper-sized rig will perform better at a more economical price. The casing program will provide the maximum loads needed to be lifted or lowered into the well, which determines the derrick rating to be used. The size of holes to be drilled will determine both the pump sizes and capacities needed to clean the hole during drilling, and the bit sizes in turn will dictate the size of opening in the rotary table needed. The required blow-out preventers and the height of the blow-out preventer stack will determine the required clearance under the rig, thereby fixing the structure height requirements. These are the most important components to be considered.

Based on the findings determined in the drilling program, recommendations can be made for the following:

1. Derrick capacity and height; depth rating
2. Substructure height
3. Pump sizes and output requirements
4. Rotary table size and opening
5. Drill pipe and drill collars
6. Power source, depending on whether the rig is a mechanical or an electrical one
7. Mud surface storage
8. Blow-out preventer equipment
9. Forklift and man-lifts

When provided the requested information in response to a request for proposal (RFP) on the rig, the drilling contractor will also provide information on the rig such as:

1. Fuel consumption
2. Number of loads into which the rig breaks down for transport
3. Complete inventory of equipment and tools provided with the rig
4. Crew listing with experience
5. Work history of the rig
6. Geothermal experience
7. Down-time history of the rig

The financial response to the RFP from drilling contractors to provide this drilling service will include the following pricing quotes:

- Day work rate: The cost for a 24-h operating day as well as crew size.
- Crew subsistence cost: Usually housing and travel time costs.
- Complete mobilization cost: Rig, manpower, trucks, and crane to move the rig in, rig it up and be prepared to drill. The total cost will include everything up to the actual spud in of the well.
- Complete demobilization cost: The complete cost to rig down, clean pits, and move rig off location to a specific site.
- Stand-by rates: The costs for a 24-h period for the rig to stand by and wait, with and without crews.
- Force majeure: A rate that the rig will charge in case of uncontrolled acts of God, such as snow storms, rain storms, floods, etc.
- Down-time clause. A rig is usually allotted a certain number of repair hours per month for routine rig maintenance. A standard clause allows 24 h of repair time per month with no more than four in any one occurrence.
- Special considerations: All tubular goods should be inspected prior to start-up of the operations and again upon completion of the drilling. Repairs to these tubular goods will be the responsibility of the operator. Blow-out preventer elements will be inspected before and after

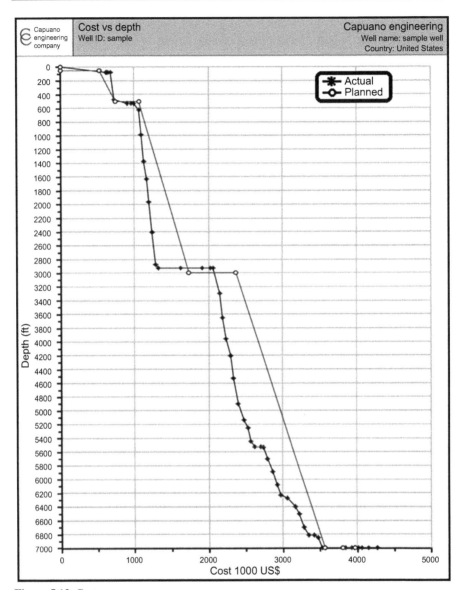

Figure 5.12 Cost curve.
Original example created by Capuano Engineering Co.

the drilling. Drill collars will be inspected after a determined number of hours in use. Certain corrosive drilling fluids will be allowed, but additional rig cleaning after drilling will be charged to the operator. If air drilling is needed, special consideration will be given to the air string of the drill pipe or the operator can provide their own drill pipe for drilling this section of the well.

CAPUANO ENGINEERING COMPANY

WELL COST ESTIMATE

NAME OF OPERATOR: **GRC Drilling Workshop**
FIELD NAME: Reno, NV
Well Name: Production Well
Estimated Number of Days: 61

SUMMARY OF ESTIMATED COSTS

EQUIPMENT RENTAL AND SERVICES	$ 4,049,327
MATERIALS, CONSUMABLES AND RELATED SERVICES	$ 950,450
TOTAL DRILLING COST	**$ 4,999,777**

Code	COST CATEGORIES	Total Cost
	EQUIPMENT RENTAL AND SERVICES	**$ 4,049,327**
0100	RIG MOBILIZATION and DEMOBILIZATION	100,000
0200	CONTRACT DRILLING RIG	1,973,500
0300	PLANNING, ENGINEERING AND PROJECT MANAGEMENT	225,300
0400	DRILLING FLUIDS AND SOLIDS CONTROL	121,527
0500	DIRECTIONAL DRILLING SERVICES	266,275
0600	CEMENT and SERVICES	779,400
0700	AIR DRILLING SERVICES	33,000
0800	GEOLOGIC EVALUATION AND RESERVOIR ENGINEERING	142,000
0900	DRILLING TOOLS RENTAL AND REPAIR	159,700
1000	WELL CONTROL EQUIPMENT RENTAL AND SERVICES	85,250
1100	RIG SITE LOGISTICS	36,875
1200	ROAD AND LOCATION CONSTRUCTION	105,500
1300	TRUCKING AND TRANSPORTATION	21,000
1400	COMPLETION SERVICES	-
1500	FISHING TOOLS AND SERVICES	-
	MATERIALS, CONSUMABLES AND RELATED SERVICES	**$ 950,450**
1600	BITS	245,000
1700	CASING, TUBING AND SERVICES	426,000
1800	CASING ACCESSORIES AND SERVICES	78,000
1900	WELLHEAD EQUIPMENT AND SERVICES	66,500
2000	TURNKEY CONTINGENCY AND PROJECT PROFIT	134,950

Figure 5.13 Cost estimation form.
Original diagram created by Capuano Engineering Co.

5.13   Cost estimate

The cost estimate is the last item to be developed in the drilling program and all the components of the program will be used to accurately complete this component. The drill curve (ie, depth vs. days) is an important factor and will determine the number of days needed to drill and completed the well. The total number of days indicated in the curve is used to complete the well cost estimate, found from the contractor day rate multiplied by the total number of days.

The project manager will take the various components of the program and send them out to third party suppliers for bidding on the equipment, goods, and services. These supplies include mud engineering and additives, directional services, fuel, casing, cementing, all rentals not supplied by the drilling contractor, site construction costs, and many other components of the drilling operation. All of these actual prices will be used to compile the final cost estimate.

A cost curve (ie, a plot of depth vs. cost) will be created to represent the estimated daily expenditures to monitor overall drilling progress as well as cost control of the project (see Fig. 5.12).

The on-site rig supervisor will be responsible to complete and provide a daily drilling progress report as well as compile all the cost accumulated in that 24-h period, as well as the total accumulated cost of the well to date as it is being drilled.

A typical well cost summary form is attached Fig. 5.13, which shows the principal line items for determining a well cost estimate.

Acknowledgments

I would like to take the opportunity thanks the many people who assisted me in preparing this chapter on geothermal drilling.

Our intern, Linda Wyman, was a great assistance in writing, rewriting, and editing this work.

My son, Louis E. Capuano, III, coworker and fellow drilling engineer, provided material and background information.

My good friend Jerry Hamblin reviewed the material and provided valuable editing.

Ron DiPippo is a good friend who requested that I write this chapter. He assisted in providing many valuable comments, suggestions, and edits.

References

[1] Bourgoyne AT, Millheim KK, Chenevert ME, Young FS. Applied drilling engineering. SPE textbook series, vol. 2; 1986.
[2] Cahill B, Palmer R. An introduction to mudlogging. Geothermal Resource Council Bulleting; January 1992.
[3] Finger J, Blankenship D. Handbook of best practices for geothermal drilling. Albuquerque, New Mexico and Livermore, California: Sandia National Laboratories; 2010.

[4] Hole H. Directional drilling of geothermal wells. Dubrovnik, Croatia: Petroleum Engineering Summer School; June 08, 2008. Workshop #26 June 9—13.

[5] Raymond D, Knudsen S, Blankenship D, Bjornstad S, Barbour J, Schen A. PDC bits outperform conventional bit in geothermal drilling project, vol. 36. GRC Transactions; 2012.

[6] Tuttle JD. Drilling fluids for the geothermal industry — recent innovations, vol. 29. GRC Transactions; 2005.

[7] Wise JL, Roberts T, Schen A, Matthews O, Pritchard WA, Mensa-Wilmot G, et al. Hard-rock drilling performance of advanced drag bits, vol. 28. GRC Transactions; 2004.

Characterization, evaluation, and interpretation of well data

R.N. Horne
Stanford University, Stanford, CA, United States

6.1   Upward convective flow in reservoirs

In the context of understanding wellbore measurements, it is important to point out that one of the most prominent characteristics of geothermal reservoirs is that they are almost always dynamic—the fluids within them are in continuous motion, albeit at very slow velocities. Unlike oil and gas reservoirs, geothermal reservoirs tend to be unconfined, and water flows through them driven by buoyancy forces and regional hydraulic gradients. A conceptual model is illustrated in Fig. 6.1. A high-temperature heat source (usually of volcanic origin) heats water that saturates the pore space of the reservoir, causing it to reduce in density and become buoyant relative to the cooler fluid that surrounds it. This buoyancy causes the water to rise, bringing it closer to the surface and in many cases producing discharge features such as hot springs, steaming ground, and geysers. Upwardly migrating water is replaced by cooler water that flows in from the sides to replace it, producing a convective circulation of water through the reservoir.

The magnitude of the convective circulation can be appreciated by calculating the forces. At a depth of around 3 km, we might expect a "normal" geothermal gradient to result in water temperature around 120°C, which would have a density of 943 kg/m^3. If water is heated to 250°C by an intrusion, then the density would reduce to 800 kg/m^3, resulting in a buoyancy-driven dynamic pressure gradient that is about 15% above hydrostatic. This rather considerable gradient causes the water to flow upward in the reservoir.

The consequence of this excess gradient is that the pressure in a wellbore, which would normally be hydrostatic, differs from the pressure in the reservoir. Hence, the wellbore and reservoir are not in static equilibrium, and fluid is likely to flow between them. A common example is illustrated in Fig. 6.2, which shows a well with two feed zones; due to the fractured nature of igneous rocks in which most geothermal reservoirs are found, most geothermal wells are productive only from a limited number of discrete fractures or feed zones. Because the well pressure gradient and reservoir pressure gradients are different, the reservoir pressure exceeds the well pressure at the lower feed zone—hence, fluid will flow into the well at that zone. Similarly, the well pressure exceeds the reservoir pressure at the upper feed zone—hence, the fluid will be injected into the reservoir at that zone. Thus, reservoir fluid enters the well at the lower feed zone, flows upward, and leaves the well at the upper feed zone, even when the wellhead valve is shut. As a result, a log of temperature as a function of depth

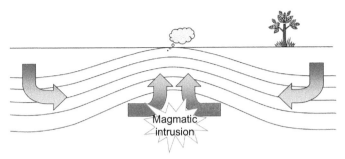

Figure 6.1 Conceptual illustration of the dynamic state of a geothermal reservoir, due to buoyancy-driven convection.

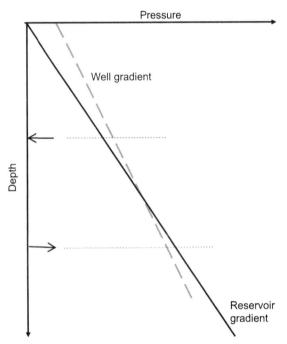

Figure 6.2 Pressure gradient in the well (hydrostatic) differs from that in the reservoir (greater than hydrostatic). In a well with two feed zones, fluid will enter the well at the lower feed zone, flow upwards and leave the well at the upper feed zone.

will measure only the temperature of the fluid passing the logging tool; this is just the temperature of the fluid entering at the lower feed zone. Importantly, this apparently "isothermal" section does not reflect the temperature—depth profile in the reservoir, and is indicative only of the reservoir temperature at a particular depth (the lower feed zone in this specific case). Fig. 6.3 shows an example from the Wairakei geothermal reservoir in New Zealand.

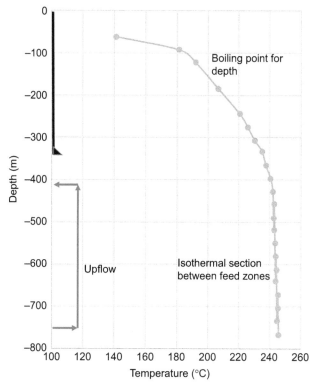

Figure 6.3 Static temperature log from well WK24 in New Zealand. Even though the well is shut at the wellhead, fluid enters from the feed zone at 760 m depth, flows upwards, and exits from the feed zone at 410-m depth.

These internal flows within the wellbore are a complication during analysis of the pressure and temperature data from the wells.

6.2 Pressure and temperature profile analysis

Having emphasized the importance of understanding the internal flows that may occur in geothermal wells, even when they are shut-in, we can now consider the interpretation of pressure and temperature profiles, as measured with wireline logging tools.

It is clear that it is necessary to identify the number and location of the main feed zones in the well, in order to be able to interpret the pressure and temperature profiles.

6.2.1 Feed zone locations

The most common method used to locate permeable zones in a well is an injection test. This is sometimes called a "pump" test. In an injection test, cool surface-temperature

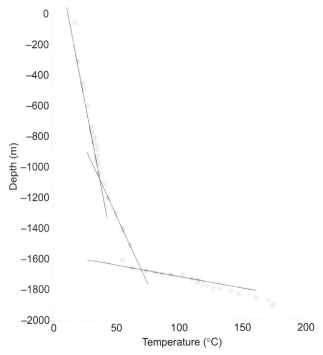

Figure 6.4 Example of a temperature profile during an injection test. Feed zones are at 1080 and 1680 m in depth.

water is pumped down the well, and a temperature log is run simultaneously. An example of the observed behavior is shown in Fig. 6.4. The temperature recorded by the instrument is that of the injected water down to the top of the first feed zone into which the water is injected, although the fluid rises slowly in temperature due to heat transfer from the hotter surroundings (see Section 6.6). After reaching the top feed zone, some of the fluid exits the wellbore into the formation, so the recorded well temperature rises more quickly below that depth because the fluid velocity is reduced and the heat transfer is greater. If more than one feed zone is present, there will be other gradient changes in the profile. The profile shown in Fig. 6.4 suggests feed zones at 1080 and 1680 m depth.

There are several compelling advantages to an injection test. First, the instruments are run in a cool flow stream and are thus less likely to be damaged by high temperatures. Second, due to the pressure of the injection, there is unlikely to be production within the well itself, although this does sometimes happen, and the interpretation has less ambiguity. Third, the injection test provides other useful information concerning the injectivity of the well (see Section 6.3).

After injection is stopped, well temperatures are measured at intervals as the well warms up. The temperatures at the feed zones are usually found to rise slower than

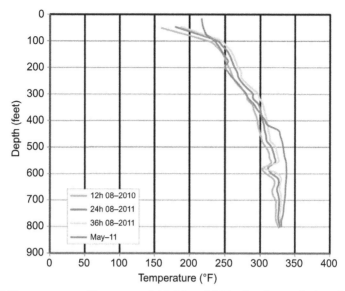

Figure 6.5 Temperature profiles measured 12, 24, and 36 h after the conclusion of drilling, as well as a profile measured 9 months later.
Reproduced from Kolker, A., Stelling, P., Cumming, W., Rohrs, D., 2012. Exploration of the Akutan geothermal resource area. In: Proceedings, Thirty-Seventh Workshop on Geothermal Reservoir Engineering, Stanford University, Stanford, California, January 30—February 1, 2012, SGP-TR-194.

the rest of the profile due to the quantity of cool water they accepted. Fig. 6.5 shows an example of this behavior in well TG-2 in Akutan, Alaska (from Kolkar et al., 2012). The major permeable zone is at 580-feet depth, identified by a cooled "notch" in the temperature profiles that disappeared as the well heated up. It should be noted that these measurements were taken after the drilling circulation stopped—hence, this was not a formal injection test, but injection had still taken place during the drilling process.

The disparity between well pressure and reservoir pressure can have some surprising consequences as the well pressure profile moves in response to well operations that cause the wellhead pressure to change. Fig. 6.6 illustrates the way that the well profile shifts in response to production and injection—the result of injection, which moves the profile to higher pressure, is that injection tends to be greater into the upper feed zone. Contrarily, production is greatest from the lower feed zone.

6.2.2 Estimating reservoir pressure and temperature

Having determined the approximate locations of the main feed zones, the interpreter can then estimate the reservoir pressure and temperature. This is most easily seen in the context of an example.

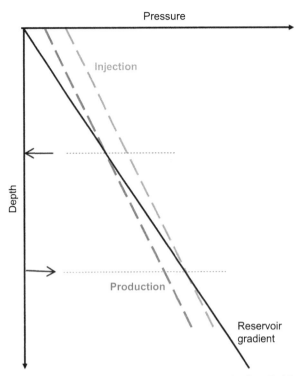

Figure 6.6 Pressure profiles in the well during injection or production. Raising the wellhead pressure during injection causes injection into the upper feed zone only. Reducing the wellhead pressure during production causes production from the lower feed zone only. Compare with Fig. 6.2 (shut-in well).

Fig. 6.7 shows the pressure and temperature profiles in a static, liquid-dominated well. The well is shut-in at the wellhead and stands with a liquid level at around 150-m depth. The well is cased to 950-m depth and has a slotted liner to around 1400-m depth. Although the well is deeper than 1400 m, the wireline tool cannot measure the pressure and temperature below the bottom of the slotted liner, because the liner is run with a "casing spear" at the bottom, which prevents the tool from passing out through the end of the liner.

There is a typical isothermal section, suggesting upflow in the well from one feed zone to another. The isothermal section actually continues up into the cased section of the well—this cannot be an upflow because the casing prevents any water from leaving (or entering) the well at shallower depths. This apparently isothermal section within the casing is not unusual, and is usually attributed to bubbles of steam rising in the casing and heating the fluid. So the interpretation of the temperature profile is that there are two feed zones, at around 1380- and 950-m depths; fluid enters the well at 1380 m and exits at 950 m. The measured temperature in the isothermal section is 229°C, which is the temperature of the fluid entering at 1380 m.

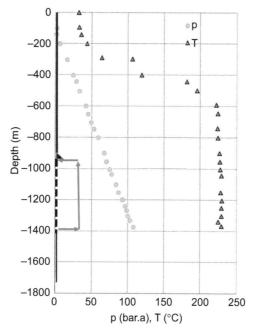

Figure 6.7 Pressure and temperature profiles in a static well with two feed zones. The feed zones are at around 1380 and 950 m in depth. The isothermal section above the casing shoe (950 m) is due to the rise of steam bubbles.

Based on this understanding of the feed zone locations and the internal flow in the well, we can conclude:

1. The reservoir temperature is 229°C at 1380-m depth. Water at this temperature enters the well at this depth and flows upward, causing the isothermal measurement up to the point where the water exits the well at 950-m depth and reenters the formation.
2. The reservoir pressure is 83.5 bar,a at 1115-m depth, halfway between the two feed zones. Because the well pressure must equilibrate with the reservoir pressure and the pressure drop due to injection and production at the feed zones, the depth at which the reservoir pressure and well pressure are the same is midway between the feed zones. The common interpretation that this point is half way is based on an assumption that the productivity and injectivity of the feed zones are the same.

It is important to note that the interpretation of the pressure and temperature profile provides only a single estimate of reservoir pressure at a specific depth, and a single estimate of reservoir temperature at a (different) specific depth. The detailed temperature−depth profile in the well is not indicative of the temperature−depth profile in the reservoir, except in the particular (and unusual) case where the fluid in the well is genuinely static, which can occur, for example, if the well has no feed zones, if there is only one feed zone, or if the well is completely cased.

6.2.3 Plotting reservoir pressure and temperature with depth

Given that only one pressure point and one temperature point are estimated in each well, it requires the assembly of such points from many wells in order to construct the actual profile of reservoir pressure and reservoir temperature as a function of depth. This is done commonly during the development stage of a project. Two examples are shown here.

Fig. 6.8 shows the profile of reservoir pressure as a function of depth in the Tongonan Reservoir in the Philippines, plotted by Whittome and Smith (1979) using the methods described here. Interestingly, the indicated pressure gradient in the reservoir was reported by Whittome and Smith (1979) to be about 10% in excess of hydrostatic, as has been discussed in Section 6.1.

Fig. 6.9 shows the assembly of the reservoir temperature profile at Los Azufres, Mexico, as determined by García-Gutiérrez (2009). Each point is estimated from an individual well (with the well number indicated within the circle) using the methods described here. In comparison to pressure, there is less consistency in the implied temperature−depth values, due to the wells lying in different regions of the reservoir.

6.2.4 Temperature transient analysis

As explained in Section 6.2.1, it is common to conduct a series of pressure−temperature logs at a succession of time intervals after injection (or drilling) stops. This succession of times depends on how quickly the well warms up but typically might be something like 12 h, 24 h, 3 days, 7 days, 15 days, or 30 days. Due to the reduction of temperature due to the cooler injection (or circulation during drilling),

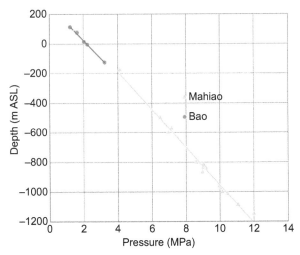

Figure 6.8 Interpreted pressure profile in the Tongonan geothermal reservoir, Philippines. Derived from Whittome, A.J., Smith, E.W., November 1979. A model of the Tongonan geothermal field. In: Proceedings 1st New Zealand Geothermal Workshop. University of Auckland, pp. 141−147.

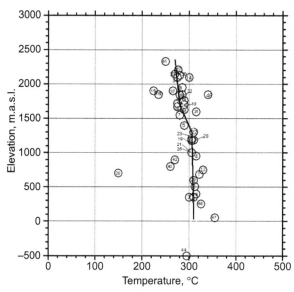

Figure 6.9 Reservoir temperature at the Los Azufres geothermal field, Mexico, in its initial state.
Reproduced from García-Gutiérrez, A., 2009. Initial thermal state of the Los Azufres geothermal reservoir. GRC Trans. 33. aggarcia@iie.org.mx.

it usually takes days or weeks before the well temperatures stabilize (see Fig. 6.5 for example). There are important situations in which the drilling engineer would like to have a more precise estimate of the reservoir temperature, for example, to decide whether to terminate drilling and begin the completion process. In such cases, it would not be reasonable to leave the rig idle while the well warms up. Hence, there is a need for a method to extrapolate the early buildup of temperature, to estimate its likely value at later times. For this, we can use temperature buildup analysis.

In pressure transient testing (described later in Section 6.4), it is common to analyze *pressure* buildup tests using the Horner plot, p vs $\log(t_p + \Delta t)/\Delta t$, where t_p is the producing time and Δt is the shut-in time. Horner analysis is sometimes also applied to temperature buildup analysis, by plotting T vs $\log(t_c + \Delta t)/\Delta t$, where t_c is the circulation time. However it is important to note that the $\log(t)$ solution used to derive the Horner method for pressure does not actually apply to temperature. This is because the pressure solution is governed by a pressure-gradient boundary condition (ie, specified flow rate) whereas the temperature solution is governed by a temperature boundary condition (ie, specified circulation temperature). Hence, Horner analysis of temperature buildups usually yields incorrect results. A better approach to temperature buildup analysis was developed by Brennand (1984), who proposed a plot of temperature T vs $1/(\Delta t + \alpha t_c)$, where α is a parameter proposed by Brennand to have a value 0.785 based on temperature analysis at Tongonan in the Philippines. Such a plot is expected to follow a straight line behavior; extrapolating to infinite shut-in time Δt corresponds to extrapolating $1/(\Delta t + \alpha t_c)$ to 0.

Table 6.1 **Example temperature buildup data**

t (days)	T (°C)	Brennand $1/(\Delta t + 0.785t_c)$	Horner $(t_c + \Delta t)/\Delta t$
0.1	104	0.2079	61
3	147	0.1297	3
7	165	0.08540	1.8571
14	180	0.05345	1.4286
40	198	0.02237	1.15

As an example, a series of temperature buildup measurements is shown in Table 6.1, after a circulation of 6 days. The data are plotted in Brennand and Horner plots in Fig. 6.10. The two plots demonstrate how much easier it is to extrapolate the Brennand plot than the Horner plot. The Brennand plot extrapolates to an intersect of 208.3°C;

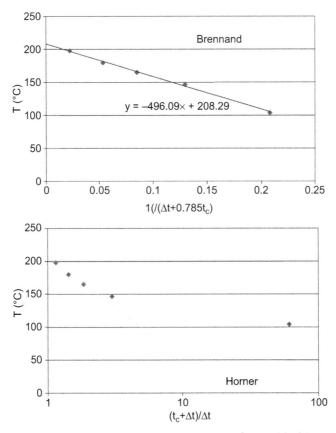

Figure 6.10 Example of temperature buildup plots. The data from Table 6.1 are plotted on a Brennand plot (left) and Horner plot (right).

however, the Horner plot is not only nonlinear but would only extrapolate to a value close to 208°C after the last few points (which took 40 days to measure). The Brennand plot extrapolates linearly to a similar intersect even using only the earliest few points.

6.3 Injection testing

As described earlier in Section 6.2.1, an injection test is a common way to determine feed zone locations in the well. However, injection tests have several other useful functions as well. Pumping cold water into the well immediately after completion has been found to enhance well productivity due to thermal stimulation of the formation. In addition, measurement of downhole pressure and surface flow rate during the injection allows for the calculation of the injectivity, II (kg/s)/bar, which is defined as the injection rate $\dot{m}$ (kg/s) divided by the pressure differential at the feed zone (the difference between the well pressure during injection p_{well} and the original reservoir pressure $p_{reservoir}$):

$$II = \frac{\dot{m}}{p_{well} - p_{reservoir}} \tag{6.1}$$

A common objective in geothermal well testing is to determine the ultimate productivity prior to completion, so injectivity testing is an important kind of well test. Fig. 6.11 shows the relationship between injectivity and ultimate productivity observed in a producing liquid-dominated geothermal field.

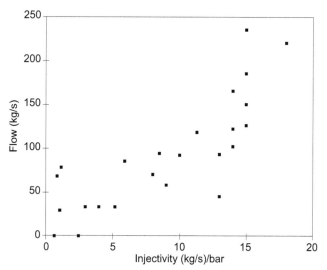

Figure 6.11 Correlation between injectivity and ultimate production from the well. Each data point is from a different well in the field.

As explained earlier, it is common that feed zones higher in the well are most receptive to injection, and feed zones lower in the well are most likely to be the producing ones (see Fig. 6.6). Hence, it is not necessarily valid to assume that a well with a high injectivity will automatically have a high productivity, because injection may be into one zone and production may be from a different zone. If the different zones had different permeability, then the relationship between flow rate and pressure drop would differ between them. Nonetheless, Fig. 6.11, which shows real field measurements, does reveal a useful degree of correspondence between injectivity and ultimate production. The general trend is upwards to the right, although the reader should note that two wells had finite (but small) injectivity but did not produce and two different wells had very small injectivity yet produced effectively.

6.4 Discharge tests

Another common well test in newly completed wells is the discharge test. A discharge test not only confirms the ability of the well to flow but also is used to determine the characteristics of that flow. In general, reducing the wellhead pressure will increase the output of the well (up to a limit), so the design of the turbine inlet pressure at the power plant is linked to the well characteristics. Lower wellhead pressure results in greater flow from the well but lower output of the turbine. Selection of the most appropriate wellhead pressure (and turbine inlet pressure) is an optimization problem that depends on the results of the discharge test.

During the discharge test, the well is flowed at a range of wellhead pressures, by opening the wellhead valve fully and then closing it incrementally. As the well is closed, the flow will reduce and ultimately will stop when the well reaches the *maximum discharge pressure*. An example is shown in Fig. 6.12.

In this example, the maximum discharge pressure is around 16 bar,a, at which point the well would cease flowing. Note that once the well stops flowing, the wellhead

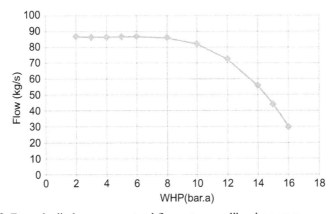

Figure 6.12 Example discharge curve, total flow rate vs wellhead pressure.

pressure would not remain at 16 bar,a but would reduce further as the fluid column in the wellbore collapsed and the liquid level settled farther down in the casing.

This curve is useful in the design of the development but also reveals the characteristics of the well and the reservoir. The well in Fig. 6.12 is an example of a "wellbore-limited" flow. The flat region of flow at smaller wellhead pressures shows that lowering the wellhead pressure below about 6 bar,a does not produce additional flow—the well is effectively choked within the casings. Another scenario is the flow may be "reservoir limited" in which case the flow rate continues to increase as the wellhead pressure reduces. Recognizing that the flow is wellbore limited suggests that the development team could usefully consider drilling and completing wells with larger-diameter casings.

Finally, it is important during a discharge test to evaluate the flowing enthalpy. Flowing enthalpy is very informative about reservoir conditions and is an important consideration in power plant design. Typically, both enthalpy and flow rate are measured using a separator with an orifice plate to measure the steam flow rate and a weir to measure the liquid flow rate. A simpler configuration uses a lip-pressure measurement at a horizontal atmospheric discharge pipe (together with the liquid flow measurement at the weir). This is known as the James lip-pressure method.

6.5 Pressure transient tests

Geothermal wells are frequently like oil or gas wells from the point of view of pressure transient testing, and therefore their buildup and drawdown tests are often interpreted using the same software used for oil or gas well tests. However, geothermal wells more often show the effect of fractures. Pressure transient tests in geothermal wells can be complicated by the difficulty of keeping the pressure tools in the well at high temperature for a sufficient time, so geothermal well tests tend to be short in duration. On the other hand, geothermal wells are usually very highly productive and are less prone to wellbore storage effects, which means that the tests do not need to be so long.

6.5.1 Well-test interpretation

Geothermal well tests can often be treated using standard interpretation techniques, although it may be necessary to take account of the effects of two-phase fluid properties. This section will describe the extra considerations associated with the interpretation of two-phase tests.

6.5.1.1 Compressibility

Grant and Sorey (1979) showed that the compressibility of two-phase mixtures of steam and water in porous rock can be written as

$$c_t = \frac{(1-\phi)\rho_r c_r + \phi S_w \rho_w c_w}{\phi \rho_s h_{fg} \frac{\partial p_{sat}}{\partial T_{sat}}} \qquad [6.2]$$

where c_r and c_w are the heat capacities of rock and water, respectively; ρ_r, ρ_w, and ρ_s are the densities of rock, water, and steam, respectively; h_{fg} is the enthalpy change due to boiling; p_{sat} and T_{sat} are the saturation pressure and temperature; and ϕ is the porosity.

For most pressures and temperatures of geothermal interest, the compressibility can be estimated more easily from the approximation:

$$\phi c_t = \overline{\rho c}\, 0.0192\, p^{-1.66} \qquad\qquad [6.3]$$

where compressibility c_t is in bar^{-1}, mean volumetric heat capacity $\overline{\rho c}$ is in kJ/m^3-°C, and pressure p is in bar,a.

Many geothermal wells also contain noncondensable gas, usually CO_2. For cases in which the reservoir fluid contains significant gas, an additional contribution may be included in the compressibility computation to take account of the change of volume as the gas comes in or out of solution. The effect of the gas is generally to *reduce* the compressibility.

6.5.1.2   Viscosity

Based on a mass balance and Darcy's law, we can write that the effective kinematic viscosity v_t is

$$\frac{1}{v_t} = \frac{k_{rw}}{v_w} + \frac{k_{rs}}{v_s} \qquad\qquad [6.4]$$

where k_{rw} and k_{rs} are the relative permeabilities to water and steam, respectively. From a heat balance, we can determine the total density and the ratio of the relative permeabilities:

$$\frac{1}{\rho_t} = \frac{1}{h_{fg}}\left[\frac{h_t - h_w}{\rho_s} + \frac{h_s - h_t}{\rho_w}\right] \qquad\qquad [6.5]$$

$$\frac{k_{rw}}{k_{rs}} = \frac{v_w}{v_s}\left[\frac{h_s - h_t}{h_t - h_w}\right] = \frac{v_w}{v_s}\frac{1 - x}{x} \qquad\qquad [6.6]$$

where x is the mass fraction of steam or steam quality.

Based on the estimates of ρ_t and v_t, we can determine the dynamic viscosity $\mu_t = \rho_t v_t$. Often, it is more convenient to work directly with mass flow rate $\dot{m}$ instead of volumetric flow rate $\dot{V}$, in which case it may be easier to work with kinematic viscosity v directly. Many of the equations of well-test analysis include the group of terms $\dot{V}B/\mu$, which is the volumetric flow rate under reservoir conditions divided by the dynamic viscosity. This group of terms can be conveniently replaced in many geothermal situations, using the equivalent mass flow rates:

$$\frac{\dot{V}B}{\mu} = \frac{\dot{m}}{\rho^2 v} \qquad\qquad [6.7]$$

The relative permeabilities of steam and water have been measured but have long been assumed to follow a linear relationship in which the sum of k_{rw} and k_{rs} is always unity. In reality, the steam−water relative permeability curves area expected to more closely follow the Brooks−Corey relations. However, it is often more convenient (and offers only a small inaccuracy) to use the linear form. This linear relationship simplifies the preceding equations, hence

$$\nu_t = x\nu_s + (1 - x)\nu_w \qquad\qquad\qquad\qquad [6.8]$$

6.5.2 Example

A geothermal well undergoes 118 h of production at 78 t/h, then is shut in for a buildup test. The producing enthalpy of the well was 1600 kJ/kg and the reservoir feed temperature was 260°C at a pressure of 47 bar,a.

At 260°C and 47 bar,a, the properties of water and steam are as shown in Table 6.2 (from steam tables).

Hence, we can find the steam mass fraction and the total mixture viscosity and density:

$$x = \frac{h - h_w}{h_s - h_w} = \frac{1600 - 1134.9}{2796.4 - 1134.9} = 0.28$$

$$\nu_t = x\nu_s + (1 - x)\nu_w = 0.28 \times 15.5 + (1 - 0.28) \times 1.34 = 16.46 \times 10^{-7}\ \text{m}^2/\text{s}$$

$$\frac{1}{\rho_t} = \frac{1}{h_{fg}}\left[\frac{h_t - h_w}{\rho_s} + \frac{h_s - h_t}{\rho_w}\right]$$

$$= \frac{1}{2796.4 - 1134.9}\left[\frac{1600 - 1134.9}{23.74} + \frac{2796.4 - 1600}{783.9}\right]$$

Hence, $\rho_t = 78.68$ kg/m^3, and $\mu_t = 1.295 \times 10^{-4}$ Pa s (=0.1295 cp).

The compressibility can be estimated from a typical value of $\overline{\rho c}$ of 2500 kJ/ (m^3°C), as:

$$\phi c_t = \overline{\rho c}\ 0.0192\ p^{-1.66} = 2500 \times 0.0192 \times 47^{-1.66} = 0.0805\ \text{bar}^{-1}$$

Table 6.2 **Water and steam properties at 260°C and 47 bar,a**

	Water	Steam
ν (m^2/s)	1.34×10^{-7}	15.5×10^{-7}
ρ (kg/m^3)	783.9	23.74
h (kJ/kg)	1134.9	2796.4

Table 6.3 **Pressure measurements during well buildup**

Δt (h)	Δp_{ws} (bar)	Δt (h)	Δp_{ws} (bar)
0.0024	0.174	0.1219	3.39
0.0049	0.348	0.1708	3.65
0.0073	0.695	0.2442	4.00
0.0098	1.13	0.3667	4.26
0.0122	1.30	0.6111	5.13
0.0171	1.57	0.8556	6.35
0.022	1.74	1.2194	7.48
0.0244	1.91	1.5861	8.17
0.0292	2.00	1.8361	8.43
0.0367	2.09	2.4417	9.22
0.0414	2.17	3.4167	10.2
0.0489	2.43	3.8611	10.4
0.0586	2.61	6.3056	11.8
0.0733	2.78	8.3056	12.6
0.0856	2.96	10.9722	13.2
0.0975	3.13		

The pressure history is as shown in Table 6.3.

Looking at the diagnostic plot, Fig. 6.13, we can see that there is sufficient data to expect a straight line on the Horner plot, Fig. 6.14. The slope m of the Horner straight line is (in consistent units, eg, $\dot{m}/\rho$ in m^3/s, μ in Pa s, k in m^2, and h in m):

$$m = 0.1832 \frac{(\dot{m}/\rho)\mu}{kh}$$

From Fig. 6.13, the slope of the line is 7.09 bar/cycle; hence

$$kh = 0.1832 \frac{(\dot{m}/\rho)\mu}{m} = 0.1832 \frac{((78000/3600)/78.68) \times 1.295 \times 10^{-4}}{7.09 \times 10^5}$$
$$= 9.2 \times 10^{-12} \text{ m}^3 = 9.2 \text{ d} \cdot \text{m}$$

These data can also be analyzed as a fractured well; however, the estimated permeability is about the same. This is to be expected because a fractured well acts like a well with negative skin effect. Many geothermal wells exhibit this behavior.

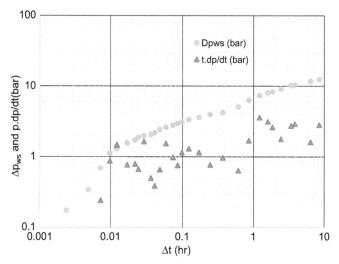

Figure 6.13 Diagnostic plot (Δp and $t \cdot dp/dt$ vs Δt) for example buildup test. The circles represent Δp and the triangles $t \cdot dp/dt$.

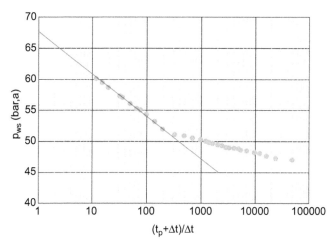

Figure 6.14 Horner plot (p vs ($t_p + \Delta t$)/Δt) for example buildup test, showing the inferred Horner straight line which extrapolates to a value of 67.5 kPa at a Horner time of 1.

The Horner plot can be extrapolated to a Horner time of 1, which corresponds to infinite shut-in time (Δt). This very useful value, known as the Horner-extrapolated pressure, is an estimate of the current value of the average reservoir pressure. In Fig. 6.14, the estimate of average reservoir pressure is 67.5 bar,a at a Horner time of 1.

6.6   Wellbore heat loss

Wellbore temperatures may be measured directly or estimated by calculation. The accurate estimation of downhole conditions in geothermal wells is of great importance,
because the wellbore environment may be particularly hostile to downhole instruments. High temperatures and corrosive chemicals can cause failure of the instrument
or the wireline, and thus measurements may be reliable for a short duration only. There
are also some parameters (eg, enthalpy, steam quality, brine concentration) that cannot
be measured downhole at this time. This section addresses the problem of evaluating
heat transmission through the flowing well as a means of calculating downhole temperatures and enthalpies from measured wellhead information. The analysis is separated into two parts: one for single-phase flow (either water or steam) and the other
for two-phase flow in the well.

6.6.1   Single-phase flow

The calculation of heat transfer during injection of a single-phase liquid or gas into a
well was presented by Ramey (1962, 1964). The results of Ramey (1962) were
expanded by Horne and Shinohara (1979), who presented single-phase flow heat transmission equations for both production and injection.

  Although it is not difficult to derive the equations for production from those for
injection, it is usually helpful to state each individually.

  The single-phase flow analysis is based on the determination of the well fluid temperature T as a function of depth in terms of the reservoir temperature, which also is a
function of depth. Ramey's study made use of the substantially differing time scales
between heat transfer in the wellbore due to fluid movement and heat transfer to the
earth due to transient conduction. Thus, the depth and time dependence could be
decoupled, and solutions for temperature in terms of depth were obtained, except
that some "constants" of the equation are slowly varying functions of time.

6.6.1.1   Production

The equation for the evaluation of the temperature in a producing geothermal well is

$$T = (T_{bh} - ay) + aA\left(1 - e^{-y/A}\right) + (T_0 - T_{bh})e^{-y/A} \qquad [6.9]$$

where y is the distance upward from the bottom of the well (assuming the feed zone is
at the bottom), T_0 is the inflowing fluid temperature T_{bh} is the downhole reservoir
temperature (usually equal to T_0), and a is the geothermal gradient (the increase of
formation temperature with increase in depth). The parameter A is a "diffusion depth"
and is a slowly varying function of time, t, given by

$$A(t) = \frac{\dot{m}cf(t)}{2\pi K} \qquad [6.10]$$

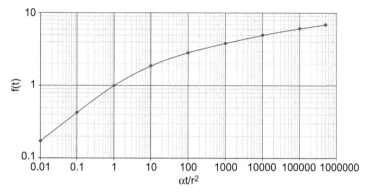

Figure 6.15 Function $f(t)$, calculated based on Ramey (1962, 1964).

where $\dot{m}$ is the mass flow rate, c is the specific heat of the fluid (assumed constant), K is the thermal conductivity of the formation, and $f(t)$ is a dimensionless time function representing the transient heat transfer to the formation. The function $f(t)$ may be found from Fig. 6.15 or alternatively may be approximated for flowing times greater than 30 days or so by the equation:

$$f(t) = -\ln\frac{r}{2\sqrt{\alpha t}} - 0.290 \qquad [6.11]$$

where α is the thermal diffusivity of the formation and r is the radius of the casing.
The total heat loss rate, q, for a well of depth H may be found from

$$q = \dot{m}c(T_0 - T_1) \qquad [6.12]$$

where T_1 is T from Eq. [6.12] with $y = H$. Alternatively, q may be determined directly from

$$q = a\dot{m}c\left(H + A\left[e^{-H/A} - 1\right]\right) \qquad [6.13]$$

These equations have been presented such that a, A, y, $\dot{m}$, and q are all positive numbers for the parameters as defined (upward flow, heat loss, etc.). Attempting to use these equations for injected flow by reversing the sign of these parameters will usually lead to error because the solution of the differential equation is different for injected flow (see the next section).

Implicit in the derivation of Eqs. [6.12]–[6.16] is that the geothermal gradient is constant; that is, temperature increases linearly with depth. If this is not the case, then the formation temperature–depth profile may be broken up into a number of linear segments and the equations applied successively to each segment.

6.6.1.2 Injection

The temperature in a well into which single-phase liquid or gas is injected is given by

$$T_1 = (T_{surf} + az) - aA + (T_{inj} - T_{surf} + aA)e^{-z/A} \tag{6.14}$$

where A is defined in Eq. [6.13], z is the distance downward from the top of the well, T_{surf} is the surface temperature of the earth, and T_{inj} is the temperature of the injected fluid. As before, a, A, and z are all still positive with the equation as written. The total heat loss rate q from a well of depth H can be determined from

$$q = \dot{m}c(T_{inj} - T_2) \tag{6.15}$$

where T_2 is T from Eq. [6.6], with $z = H$. Alternatively, q may be obtained directly from

$$q = -\dot{m}c\left[aH - (T_{inj} + aA - T_{surf})\left(1 - e^{-H/A}\right)\right] \tag{6.16}$$

6.6.1.3 Example of single-phase calculations

In an injection test, 1.26 kg/s of 60°C water is pumped into a 914-m-deep well. The downhole reservoir temperature is 204°C, and the surface temperature is 21°C. The object of the experiment is to evaluate the effect of injecting cold water into a hot reservoir, so it is necessary to determine the temperature of the water when it reaches the permeable depth. After a time of 30 days, it is expected that the system will be relatively stable. Well radius is 0.1 m, and the following parameters of the formation have been evaluated:

$$K = 2 \text{ W}/(\text{m°C}) \quad \alpha = 10^{-6}\text{m}^2/\text{s}$$

Evaluating $f(t)$ from Eq. [6.14] (or from Fig. 6.15),

$$f(t) = -\ln\frac{0.1}{2\sqrt{10^{-6} \times 30 \times 24 \times 3600}} - 0.290 = 3.182$$

The geothermal gradient, a, is $(204-21)/914$, which yields 0.2°C/m. The specific heat capacity, c, of water is 4100 J/kg-°C. Thus, from Eq. [6.13]:

$$A(t) = \frac{1.26 \times 4100 \times 3.182}{2\pi \times 2} = 1308 \text{ m}$$

Then the downhole fluid temperature will be, from Eq. [6.6]:

$$T = 204 - 1308 \times 0.2 + [60 - 21 + 1308 \times 0.2]e^{-914/1308} = 91.9°C$$

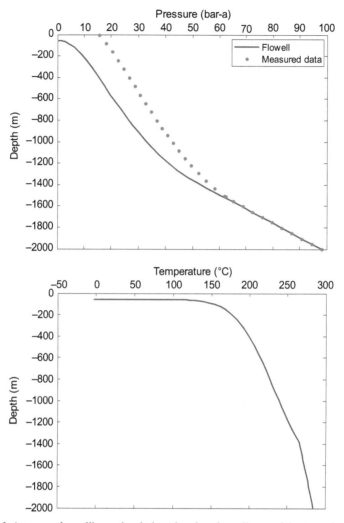

Figure 6.16 An example wellbore simulation showing the collapse of the two-phase column due to condensation of steam from heat loss. Flow cannot reach the surface. The measured data show the well flowing condition after the well had been heated and was flowing normally.

This is the solution to the problem. Should we have needed it, the total heat loss could have been calculated from Eq. [6.15] or [6.16]. Using Eq. [6.15]:

$$q = 1.26 \times 4100 \times (60 - 91.9) = -164,800 \text{ W} \quad \text{or} \quad -164.8 \text{ kW}$$

Thus, there is a heat *gain* by the injectate at a rate of 164.8 kW.

6.6.2   Two-phase flow

The vertical flow of a steam/water mixture in the wellbore is complicated by boiling and a change of phase during transport. The problem of calculating the temperatures in the well is replaced by that of calculating the heat loss because the temperature is restricted to remain on the saturation line for phase equilibrium. The saturation temperature depends on the pressure which is itself difficult to determine for two-phase flow.

Nowadays, full analyses of two-phase flow in wellbores are all computer based and thus do not provide a handy method without the necessary computer software.

One example of the importance of estimating heat loss in two-phase flow arises when newly drilled wells in elevated topography have liquid levels that lie well below the wellhead. Wells depend on boiling within the casing to "gas lift" the water out of the well—percolating like a coffee pot. Without boiling, the well is not able to flow. Such wells may be stimulated by injecting air or nitrogen at the wellhead to depress the liquid level back into the formation, heating all of the liquid. Releasing the wellhead pressure causes the hot liquid to rush back up the wellbore, boiling as it goes, hence lifting the bubbly mixture to the surface. In this situation, wellbore heat loss can be a significant problem, because $f(t)$ and A will be small numbers and Eq. [6.16] shows that the heat loss will be very large. Such a large heat loss can be sufficient to completely condense all the steam that has been produced in the wellbore, collapsing the gas lift and causing the flow to subside before reaching the surface. Therefore, the well cannot be started.

An example is shown in Fig. 6.16, which shows simulated pressure and temperature profiles in a well in which heat loss causes a collapse of the two-phase mixture before the fluid can reach the wellhead. The simulation used a wellbore simulator FloWell, by Gudmundsdottir and Jonsson (2015).

Such wells have been stimulated successfully by successively depressing and releasing the liquid level that causes the wellbore to be heated up or, in extreme cases, by injecting steam or hot water (from nearby wells) to heat the casing.

6.7   Summary

Geothermal wells are tested on completion to determine their properties, to provide information to improve understanding of the reservoir, and to aid in designing the field development strategy. Although the sequence of tests varies depending on the development team, as well as on the type of the reservoir, a common portfolio of tests could include:

1. An injection test on completion, with the determination of injectivity (Section 6.3), and a measurement of the downhole temperature profile to infer feed zone locations (Section 6.2.1).
2. A pressure transient test (pressure fall-off) after injection stops (Section 6.5).
3. A succession of pressure and temperature profiles measured after the injection is concluded to determine the reservoir pressure and temperature (Section 6.2.2).
4. A discharge test to determine the discharge characteristics of the well (Section 6.4).
5. A pressure transient test (pressure buildup) after discharge stops (Section 6.5).

During drilling, temperature buildup analysis (Section 6.2.4) may be used to obtain a quick estimate of the reservoir temperature at the current depth. Before or during discharge, estimation of wellbore heat losses (Section 6.6) may be used to evaluate production problems and to estimate true reservoir enthalpy.

Well measurements do not end after the well has been completed. Most developers conduct an ongoing campaign of well measurement to monitor and understand changes in the reservoir as it undergoes production. For example, it is common to perform discharge tests at regular intervals (eg, once per year) to evaluate reservoir depletion and to re-optimize the production system. Pressure buildup tests allow for the estimation of the current average reservoir pressure, as well as provide a view of reservoir depletion.

References

Brennand, A.W., 1984. A new method for the analysis of static formation temperature tests. In: Proceedings of the 6th New Zealand Geothermal Workshop, pp. 45–47.

García-Gutiérrez, A., 2009. Initial thermal state of the Los Azufres geothermal reservoir. GRC Trans. 33. aggarcia@iie.org.mx.

Grant, M.A., Sorey, M.L., 1979. The compressibility and hydraulic diffusivity of a water-steam flow. Water Resour. Res. 15, 684–686.

Gudmundsdottir, H., Jonsson, M.T., April 2015. The wellbore simulator FloWell — model enhancement and verification. In: Proceedings of the World Geothermal Congress, Melbourne, Australia.

Horne, R.N., Shinohara, K., January 1979. Wellbore heat loss in production and injection wells. J. Pet. Technol. 116–118.

Kolker, A., Stelling, P., Cumming, W., Rohrs, D., 2012. Exploration of the Akutan geothermal resource area. In: Proceedings, Thirty-Seventh Workshop on Geothermal Reservoir Engineering, Stanford University, Stanford, California, January 30–February 1, 2012. SGP-TR-194.

Ramey Jr., H.J., April 1962. Wellbore heat transmission. J. Pet. Technol. 427–435. Trans., AIME, 225.

Ramey Jr., H.J., November 1964. How to calculate heat transmission in hot fluid injection. Pet. Eng. 110–120.

Whittome, A.J., Smith, E.W., November 1979. A model of the Tongonan geothermal field. In: Proceedings 1st New Zealand Geothermal Workshop. University of Auckland, pp. 141–147.

Reservoir modeling and simulation for geothermal resource characterization and evaluation

7

M.J. O'Sullivan, J.P. O'Sullivan
University of Auckland, Auckland, New Zealand

7.1 Review of resource estimation methods

7.1.1 Introduction

The two most common simple methods for resource estimation are the power density method and the stored heat method. The power density method was introduced by Grant [58], who produced a plot of MW_e/km^2 versus temperature for several geothermal projects (replotted as Fig. 7.1). There is a lot of scatter in the data about the best-fit curve.

The main problem with Grant's analysis is that he does not define how the reservoir area should be calculated. This issue was addressed in a recent review of the power density method by Wilmarth and Stimac [155], who estimated the reservoir area as a merged 500 m buffer around all current production well tracks. This has the merit of rigorously defining the reservoir area but obviously depends on the drilling strategy, land access, and infrastructure as well as the actual size of the reservoir.

Willmarth and Stimac used data from 53 producing geothermal fields, which they divided into four categories: main sequence, rifts, arcs, and mature. For the "main sequence" and "rifts" systems, they found an approximately linear trend with a slope of $\sim 2\ MW/km^2$ per $10°C$.

For the other two categories ("arcs" and "mature"), the data fall well below this line.

At best, the power density method provides a rough first approximation for resource estimation, and to improve its accuracy, data should be used from existing fields most like the field for which the size of the resource is to be estimated.

Neither of the studies mentioned above resolves the issue of how to calculate the reservoir area: Grant does not discuss the matter, and the method suggested by Wilmarth and Stimac is not appropriate when only a few wells have been drilled.

The stored heat method, a method based on exergy, and a fourth alternative of using numerical simulation for geothermal resource estimation are discussed in later sections.

Geothermal Power Generation. http://dx.doi.org/10.1016/B978-0-08-100337-4.00007-3

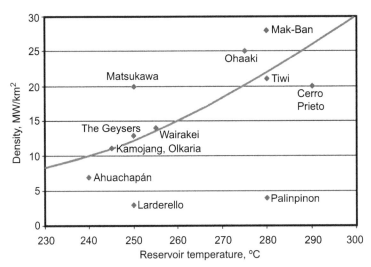

Figure 7.1 Power density versus temperature.
After Grant, MA. Resource assessment, a review, with reference to the Australian code.
In: Proc., WGC 2015, Melbourne, Australia, 2015.

7.1.2 Stored heat calculation

If a geothermal reservoir is subdivided into number of subregions of volume V_i with an average temperature of T_i, $i = 1, 2, ..., M$, then the thermal energy stored in the reservoir (called both "thermal energy in place" and "stored heat" in the literature) can be calculated as:

$$U = \sum_{i=1,M} \rho_t c_t [T_i - T_{\text{ref}}] V_i \tag{7.1}$$

The subscript "ref" refers to some cut-off, or base, energy level and $\rho_t c_t$ is the effective, or total, heat capacity of the saturated rock given by:

$$\rho_t c_t = (1 - \phi)\rho_r c_r + \phi[\rho_l c_l S_l + \rho_v c_v S_v] \tag{7.2}$$

Often a single average temperature is used for the whole reservoir, that is, $M = 1$.

The basic assumption in using Eq. [7.1] to calculate U is that all the thermal energy in the reservoir can be extracted uniformly so that at the end of the life of the project the whole reservoir is at a temperature of T_{ref}. In reality, this can never be achieved, so to calculate U_{rec}, the recoverable energy, U is multiplied by a recovery factor R_f:

$$U_{\text{rec}} = R_f U \tag{7.3}$$

Then, to calculate the power capacity (MW_e) of the project, U_{rec} is multiplied by a conversion efficiency η_c and divided by the planned lifetime L of the project and a load factor F.

$$MW_e = \frac{\eta_c R_f U}{LF} \qquad [7.4]$$

There is obviously considerable uncertainty in the parameters in Eq. [7.4], particularly the recovery factor, and it has become common to apply a Monte Carlo approach to the calculation of U_{rec}, with various assumptions made about the range and statistics of the parameters [53,126].

In the first application of the stored heat method by the USGS [148], a recovery factor of 25% was used and the value of T_{ref} was 59°F (15°C), a typical ambient temperature.

Recently there have been several reviews of the stored heat method [53,59,125,127,153,154,162]. The most recent review by Grant [59] analyzes data from the other reports and concludes from past experience that a mean value for the recovery factor is 11% (range 3–17%).

One exception to this conclusion were the experiences from the Philippines and Iceland reported by Sarmiento and Bjornsson [127], who found that a recovery factor of 25% had given good results for estimates compared with actual performance. However, as pointed out by Grant [59], these stored heat estimates had used a T_{ref} of 180°C, corresponding to the minimum temperature at which wells could supply steam to flash plants, and if the lower USGS value was used, the recovery factor would have to be adjusted down to $\sim 10\%$.

There has recently been further discussion on what value of T_{ref} should be used and what improvements to the stored heat method could be made [53,162]. However, in their discussion of T_{ref}, these authors introduced issues such as the type of plant to be used, which meant they were really discussing the product of recovery factor multiplied by the conversion efficiency ($\eta_c R_f$) rather than the recovery factor alone.

If the stored heat method is to be used, then it is probably best to keep separate the consideration of the two factors, say by adopting a simple and standard approach of using $T_{ref} = 15°C$ and $R_f = 10\%$, and then putting extra effort into the calculation of the conversion efficiency, η_c [33], taking account of the type of plant being used and the type of resource.

7.1.3 Australian standard

In the early 2000s, Australia developed an interest in geothermal energy and because of their strong background in the minerals and oil and gas industries decided a formal *Geothermal Reporting Code* was required [3,4]. Around this same period, there was a worldwide increase in interest in geothermal energy and in some places, where the use of geothermal energy was well established, privatization of geothermal projects had recently occurred (eg, New Zealand, Philippines). In this environment, the Australian code attracted considerable international interest and the Canadians set up a similar

Table 7.1 Default values of recovery factor recommended by the Australian code

Type of reservoir	Mean value of R_f	Range of R_f
Fracture dominated reservoirs (including EGS)	14%	8–20% with a uniform probability
Sedimentary reservoirs or porous volcanic-hosted reservoirs, of "moderate" porosity (<7% on average)	17.5%	10–25% with a uniform probability
Sedimentary or porous volcanic-hosted reservoir of exceptionally high average porosity (>7%)	2.5 times the porosity	Up to a maximum of 50%

code [17]. Recently, the International Geothermal Association (IGA) established a working group to produce a global classification scheme for geothermal energy consistent with the UNFC-2009 framework [139].

In the "Lexicon" attached to the Australian Code, several methods of resource assessment are discussed, but the two main methods recommended are stored heat and computer modeling.

The recommendations given by the Australian Code on recovery factors are shown in Table 7.1.

The mean values given in Table 7.1 were attributed to the USGS [153]. However, as pointed out by Grant [59], based on past experience, the values in Table 1.1 are too large. As mentioned above, the authors agree with Grant that an appropriate default value is $R_f = 10\%$.

The Lexicon to the Australian Code discusses the use of Monte Carlo simulation techniques for providing probabilistic assessment of resources and reserves, and several studies have used this approach [53,126]. Grant [59] points out in his discussion of Monte Carlo simulations for various geothermal fields that the rigor implied by the use of such a sophisticated method is illusory if overly restrictive estimates of parameter ranges or inappropriate statistical distributions are used.

7.1.4 Exergy

The Lexicon also includes this statement: "In most cases estimates based on a defined conversion efficiency are based on the first law of thermodynamics: that is to say a simple difference in energy. An alternative approach is to take the second law into account and define an efficiency based on 'exergy.' The latter has been adopted in the new USGS national geothermal resource inventory (Williams et al. [153]). While this is a good approach from a theoretical perspective, it is not as readily understandable by a nontechnical audience and will give factors which are significantly numerically larger than an 'efficiency' based on energy only, which could make it difficult to compare different projects, so it is strongly recommended this approach should not be adopted."

This recommendation is unfortunate as exergy, E, is the best thermodynamic measure of available work [32,80]. In [153], exergy is defined by:

$$E = m_{\text{WH}}[h_{\text{WH}} - h_0 - T_0(s_{\text{WH}} - s_0)] \qquad [7.5]$$

Here h and s are enthalpy and entropy, respectively, the subscripts "WH" and "0" refer to wellhead and reference conditions (eg, 15°C) and m_{WH} is the extractable mass. The size of the power plant, MW_e, for a given project lifetime (eg, 25–30 years) can then be calculated by multiplying the rate of exergy use, $\dot{E}$, by a utilization efficiency, η_u:

$$\text{MW}_\text{e} = \eta_\text{u}\dot{E} = \dot{m}_{\text{WH}}[h_{\text{WH}} - h_0 - T_0(s_{\text{WH}} - s_0)] \qquad [7.6]$$

In Eq. [7.6], $\dot{m}_{\text{WH}}$ is the production rate (kg/s) for the project. Williams et al. [153] give a plot of η_u as a function of temperature for several existing power plants, showing a value of 0.4 for temperatures >175°C. Thus, the use of Eq. [7.6] offers much more certainty in terms of the thermodynamics than the stored heat or energy in place approach of Eq. [7.4].

There remains uncertainty in how to calculate $\dot{m}_{\text{WH}}$, but extrapolating the output from a few wells to estimate it and then using Eq. [7.6] should give a better result than Eq. [7.4]. Better still may be the use of computer modeling, discussed in the next section.

7.1.5 Computer modeling

As noted by Grant [59], the Australian Code Lexicon is somewhat negative about the early use of computer modeling for resource assessment. It states: "At the exploration stage, when limited data are available, numerical simulation is unlikely to give a more realistic estimate of long term capacity than simpler volumetric methods."

Experience shows it is useful to carry out computer modeling as early as possible as it provides a useful check on the conceptual models and may help to guide the exploration and monitoring program (ie, to help decide which data are most critical for understanding the system behavior).

An early-stage computer model will include all the data available for the stored heat calculation, that is, downhole or geothermometer temperatures, but it will include much more in terms of consistency with the conceptual model and behavior that conforms to the laws of conservation of mass and energy. The authors agree with Grant that computer modeling should be preferred to a stored heat estimate.

Sanyal and Sarmiento [126] stated that the stored heat method should be preferred over numerical simulation because it could provide the probabilistic assessment required for booking of reserves under the categories of proved, probable, and possible. However, with the recent increase in inexpensive computer power, particularly parallel computational resources, a Monte Carlo approach to simulate future scenarios is now possible [27,93] and should soon become routine.

In the next section, a recommended approach to modeling is discussed: natural state, history matching, and future scenario simulations [108]. At the early stage, only a natural state simulation can be carried out, and the amount of data for quantitative calibration is limited. However, qualitative calibration can be carried out, making sure that the numerical model is consistent with the conceptual model or else, as is often the case, discovering and fixing problems with the conceptual model. Obviously, the quality of the model will improve as the amount of data increases, particularly when there are a few years of production history to allow calibration of a production history model.

For continuity and to extract the most value from modeling, it should continue throughout the life of a geothermal project as a key tool for planning and managing the project. Because computer modeling is a specialty skill, it has often been used only on a sporadic basis (eg, when some key decision on project expansion has to be made), but more value can be gained from computer modeling by carrying it out as a continuing process.

7.2 Computer modeling methodology

7.2.1 Basic equations

The governing equations can be written integral form, expressing conservation of mass:

$$\frac{d}{dt}\int_V A_m dV = -\int_S \mathbf{n}\cdot\mathbf{F}_m dS + \int_V q_m dV \qquad [7.7]$$

and conservation of energy:

$$\frac{d}{dt}\int_V A_e dV = -\int_S \mathbf{n}\cdot\mathbf{F}_e dS + \int_V q_e dV \qquad [7.8]$$

Here V is an arbitrary control volume, with surface area S and outward pointing unit normal vector $\mathbf{n}$. In Eqs. [7.7] and [7.8], A_m and A_e are the amounts of mass and energy per unit volume of reservoir, vectors $\mathbf{F}_m$ and $\mathbf{F}_e$ are the mass and energy fluxes, and q_m and q_e are the amount per unit volume of mass or energy produced or injected by wells.

If the reservoir contains water with significant quantities of dissolved chemicals, then an additional conservation equation for the mass of each chemical can be added in the form:

$$\frac{d}{dt}\int_V A_c dV = -\int_S \mathbf{n}\cdot\mathbf{F}_c dS + \int_V q_c dV \qquad [7.9]$$

For the general case of two-phase flow, the amounts per unit volume of mass or energy are calculated using:

$$A_m = \phi[\rho_l S_l + \rho_v S_v] \tag{7.10}$$

and

$$A_e = (1 - \phi)\rho_r c_r T + \phi[\rho_l u_l S_l + \rho_v u_v S_v] \tag{7.11}$$

Here the subscript l refers to the liquid phase (water), v to the vapor phases (steam), and r to the properties of the rock matrix. In Eq. [7.11], ϕ is porosity, ρ is density, and S is saturation (volume fraction). Additionally, in Eq. [7.11], u is specific internal energy, c_r is specific heat of rock and T is temperature.

The fluxes are defined using a two-phase version of Darcy's law. For the flux of water

$$\mathbf{F}_{ml} = -\frac{kk_{rl}}{\nu_l}[\nabla p_l - \rho_l \mathbf{g}] \tag{7.12}$$

and for the flux of steam:

$$\mathbf{F}_{mv} = -\frac{kk_{rv}}{\nu_v}[\nabla p_v - \rho_v \mathbf{g}] \tag{7.13}$$

Here $\mathbf{g}$ is the gravity vector, k is the permeability of the rock, k_{rl} and k_{rv} are the relative permeabilities for water and steam, respectively, and ν_l and ν_v are the kinematic viscosities.

The total flows of mass, energy, and chemical are given by:

$$\mathbf{F}_m = \mathbf{F}_{ml} + \mathbf{F}_{mv} \tag{7.14}$$

$$\mathbf{F}_e = h_l\mathbf{F}_{ml} + h_v\mathbf{F}_{mv} - \kappa\nabla T \tag{7.15}$$

$$\mathbf{F}_c = x_l\mathbf{F}_{ml} + x_v\mathbf{F}_{mv} \tag{7.16}$$

In Eq. [7.15], h_l and h_v are enthalpies of water and steam and κ is thermal conductivity (of the saturated rock matrix). In Eq. [7.16], x_l and x_v are the mass fractions of the chemical in the liquid phase and vapor phase, respectively. Optionally, Eq. [7.16] can be modified to include mass transport by diffusion and sometimes hydrodynamic dispersion has also been included [99,100].

7.2.2   Numerical techniques

Several simulators have been developed for solving the above equations numerically. The most widely used of these is the TOUGH2 code [118]. The numerical technique used in TOUGH2 is described in the manual as the "integral finite difference" approach, but it is the same as is more commonly called the "finite volume method." It has the advantage of conserving mass and energy in the discrete form.

For a computational block or element, and its neighbors, the discrete versions of Eqs. [7.7] and [7.8] are:

$$V_i\left(A_{\mathrm{m}i}^{n+1} - A_{\mathrm{m}i}^n\right) = -\sum_j a_{ij} F_{\mathrm{m}ij}^{n+1} \Delta t_{n+1} + Q_{\mathrm{m}i}^{n+1} \Delta t_{n+1} \qquad [7.17]$$

and

$$V_i\left(A_{\mathrm{e}i}^{n+1} - A_{\mathrm{e}i}^n\right) = -\sum_j a_{ij} F_{\mathrm{e}ij}^{n+1} \Delta t_{n+1} + Q_{\mathrm{e}i}^{n+1} \Delta t_{n+1} \qquad [7.18]$$

Here the subscript "i" refers to the ith computational block while the double subscript "ij" refers to a quantity evaluated at the interface between block i and block j. The volume of block i is V_i and the area of the ij interface is a_{ij}. The superscript "n" is the time-step counter, while the size of the current time-step is Δt_{n+1}.

The time-stepping used in Eqs. [7.17] and [7.18] is the simple implicit Euler method. Experience has shown that the significant nonlinearity introduced by boiling and recondensing requires the use of a fully implicit method.

The discrete version of Darcy's law is given by:

$$F_{\mathrm{m}lij}^{n+1} = -\left(\frac{kk_{\mathrm{rl}}}{\nu_1}\right)_{ij}^{n+1} \left[\frac{p_{1j}^{n+1} - p_{1i}^{n+1}}{d_{ij}} - \rho_{1ij}^{n+1} g_{ij}\right] \qquad [7.19]$$

and

$$F_{\mathrm{m}vij}^{n+1} = -\left(\frac{kk_{\mathrm{rv}}}{\nu_{\mathrm{v}}}\right)_{ij}^{n+1} \left[\frac{p_{vj}^{n+1} - p_{vi}^{n+1}}{d_{ij}} - \rho_{vij}^{n+1} g_{ij}\right] \qquad [7.20]$$

Here d_{ij} is the distance between block centers.

The key to a successful numerical scheme is the way the interface quantities are evaluated. A straightforward average is applied to calculating the densities in the gravity term, and a harmonic mean is usually used for calculating the rock permeability k_{ij}. However, for the relative permeabilities and kinematic viscosities, upstream weighting must be used.

The total flows are calculated using:

$$F_{\mathrm{m}ij}^{n+1} = F_{\mathrm{m}lij}^{n+1} + F_{\mathrm{m}vij}^{n+1} \qquad [7.21]$$

and

$$F_{eij}^{n+1} = h_{lij}^{n+1} F_{mlij}^{n+1} + h_{vij}^{n+1} F_{mvij}^{n+1} - \kappa_{ij}^{n+1} \left[\frac{T_j^{n+1} - T_i^{n+1}}{d_{ij}} \right] \qquad [7.22]$$

In Eq. [7.22], harmonic weighting is used to calculate the thermal conductivity at the interface, whereas upstream weighting is used for evaluating enthalpies.

The implicit Euler method for time-stepping and upstream weighting of mobilities (ie, relative permeability divided by kinematic viscosity) and enthalpies have both been implemented in well-known geothermal simulation codes with two-phase capability, such as TOUGH2 [118], TETRAD [140], FEHM [164], and STAR [116].

7.2.3 Equations of state

One of the recent advances in geothermal reservoir simulation is the extension of the capabilities of the equation of state (EOS) modules. For example, Kissling [75] and Croucher and O'Sullivan [24] have developed an EOS for water, using the updated IAPWS-97 thermodynamic formulation [142], that can handle supercritical conditions. The second of these has been incorporated into the iTOUGH2 framework [85].

Several versions of a water–carbon dioxide EOS have been developed, but these have been aimed at conditions appropriate for carbon sequestration studies [119,120] and do not cover a high enough temperature range for geothermal conditions.

An EOS that could handle water–sodium chloride–noncondensable gas mixtures was first set up for TOUGH2 by Battestelli et al. [7]. EOS for the simpler water–sodium chloride mixture that can handle high temperatures, pressures, and chloride content have been investigated by several researchers [40,76,77,82,146] but are not available as open-source code or with the standard version of TOUGH2.

Borgia et al. [10] introduced a new EOS for TOUGH2, called ECO2H, that can handle temperatures and pressures in the range of $10 \leq T \leq 243°C$ and $P \leq 67.6$ MPa, and salinity from zero up to full halite saturation. However, this is not adequate for conditions at depth in gassy geothermal systems like Ohaaki and Rotokawa (New Zealand) where temperatures exceed 300°C.

7.3 Computer modeling process

7.3.1 Conceptual modeling and computer modeling

The very first stage of computer modeling of a geothermal system is conceptual modeling. This is the process of gathering and synthesizing all the data, geosciences, and reservoir engineering to create an understanding of the key features of the particular system. In practical terms, the conceptual model will consist of a brief commentary and a few sketches of the system, say, a plan view and two or three vertical sections, showing features such as approximate geology, faults, resistivity, isotherms, surface features, deep upflows, and chemistry.

With the improvement of software tools for model development and data visualization, it is now becoming more common to use computer modeling to test and modify conceptual models [106]. This means developing a computer model at a much earlier stage of field development than has traditionally been the case. It should not be expected that such early-stage models will give highly accurate predictions of future scenarios for the system, although their predictions should be better than those from simple stored heat analysis. However, through sensitivity studies [93] and data-worth analysis [143], they can be used to help guide exploration and monitoring, and they can provide a rigorous way of testing and improving conceptual models.

7.3.2 *Interaction with geological and other geoscience models*

Recently there has been an increase in the use of software such as LEAPFROG [1,78] for developing computer-based geological models of geothermal systems [92,98]. Such software can be used to create and visualize a three-dimensional (3D) geological model, including faults, and can include and overlay other data such as magnetotelluric (MT) or temperatures. Thus, this type of software can be used as a tool for creating and visualizing 3D conceptual models. A demonstration LEAPFROG model [1], based on synthetic data, is shown in Fig. 7.2.

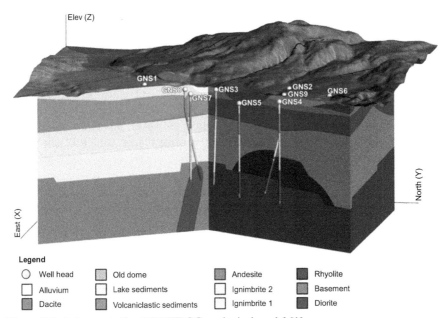

Figure 7.2 A demonstration LEAPFROG geological model [1].
Original provided by S. Alcaraz, GNS Science, Wairakei, New Zealand.

The geological modeling software can also be interfaced directly with reservoir modeling software. For example, a LEAPFROG geological model can be used to assign rock types to a TOUGH2 model [98,111,112] and can display results from a TOUGH2 model on top of a geological model.

7.3.3 Model design

The next step after conceptual modeling is model design. At this stage, the size and shape of the model and the level and type of discretization must be decided. It is usual to make the areal extent of models large enough so that the cold or warm recharge area surrounding the hot reservoir has similar dimensions to the reservoir; for example, if the reservoir is 4 km across then a zone of 4 km will be added on each side. Similarly, the models should be at least 500 m deeper than the deepest well, giving total depths of 3–4 km. Such models are not large enough to include the total depth of a large-scale convective geothermal system requiring the input of some hot water at the base of the model (see Section 7.3.4).

The type of grid system used in geothermal models is variable. Some simulators such as TOUGH2 allow the use of a general irregular unstructured grid, but this level of generality is seldom used for two reasons: (1) Some simulators such as TETRAD [140] and older versions of the PETRASIM [138] implementation of TOUGH2 [118] require the use of structured grids, and (2) mesh generation, model construction, and visualization of model results for complex grids are all difficult tasks.

A standard rectangular grid is obviously easy to set up, but if mesh refinement is used, with smaller blocks near the production and injection wells, then blocks with a large aspect ratio will occur at the outer part of the model. Some modelers have used mesh refinement with one large block joining two or more smaller blocks [41,122,150], or else have used some triangular blocks to link a coarse and fine mesh [43,106]. Both of these techniques are satisfactory provided care is taken to condition the grid so that lines joining block centers are approximately orthogonal to block boundaries [26]. A utility within the PyTOUGH library [25] is available for optimizing two-dimensional (2D) grids, or a layer of a 3D grid. An optimized grid for a model of Rotorua (New Zealand) [122] is shown in Fig. 7.3.

Some modelers have used a Voronoi grid to achieve mesh refinement [74], which is an attractive option because it guarantees an orthogonal grid; see Fig. 7.4. Most of these model grids were generated with AMESH [63], which, while guaranteeing orthogonality, may create grids that are poorly conditioned because the sides of a block are very different in size. This problem was recognized by Sieger et al. [133], who described a method for improving these undesirable grids.

Many geothermal models have been based on a simple layer structure with constant-thickness, horizontal layers. Most of the models developed at the University of Auckland are of this type, although in some cases the top few layers have some blocks missing and the very top layer has variable thickness, so that a water table or ground surface with a varying elevation can be accurately represented in the model.

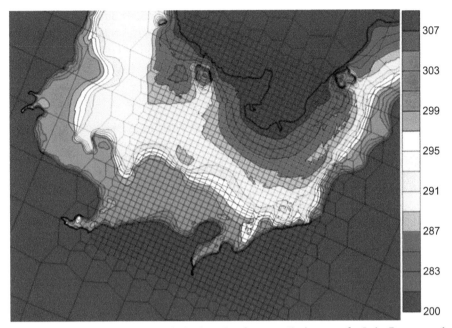

Figure 7.3 Model grid showing optimized mesh refinement. Bathymetry for Lake Rotorua and near-shore land contours are shown.

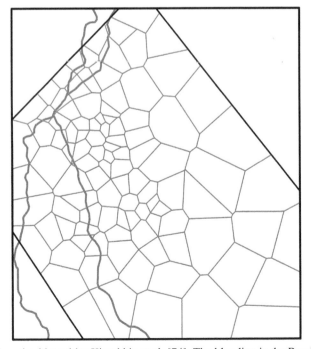

Figure 7.4 Voroni grid used by Kiryukhin et al. [74]. The blue line is the Pauzhetsky river.

PETRASIM supports more general 3D models with layers of varying elevation and thickness, and there is a range of general mesh generation software (see TOUGH2 Website) that can be used for generating model grids.

7.3.4 Boundary conditions

The allocation of boundary conditions is one of the most important aspects of model design. There is always a trade-off between the importance of the boundary conditions and the size of the model. On the one hand, if the results from the model are very sensitive to the choice of boundary conditions, then the model should be made larger, but on the other hand, increasing the size of the model will increase the computational time. The types of boundary conditions used are somewhat different for the sides, top, and base of the model.

Sides. The two options for the side boundary conditions for a geothermal model are:

- No flow of mass or heat.
- Specified pressures and temperatures (usually corresponding to one-dimensional [1D] hydrostatic conditions and a constant vertical heat flux).

Often, the no-flow boundary conditions are used for natural state models; pressure-dependent recharge is added for production history and future scenario simulations [160] to allow flow into or out of the model if the pressure changes near the boundary. However, if the boundary recharge is too important, probably the model should be increased in size.

Some models have included injection or extraction of fluid at the model boundaries, but this practice should be avoided because for internal consistency the permeability structure of the model should determine the heat and mass flows; using injection and production at the side boundaries to achieve a desired temperature distribution may give results that are not compatible with the model geology.

Top. There are two basic approaches to the top boundary conditions of a geothermal model:

(i) Use the location of the water table as the top of the model.
(ii) Use ground surface as the top of the model.

In case (i), the pressure is set at atmospheric conditions (assuming little pressure change through the vadose zone) and the temperature is set at an estimated value (possibly varying across the model). With this type of boundary condition, water and/or steam can flow out of the top of the model, whereas only water can flow in.

In case (ii), the vadose zone is included in the model and an air–water equation of state (EOS) must be used. Now atmospheric pressure and temperature are fixed at the top of the model, but in addition, some infiltration of rainfall must be included [14,86,109].

The trouble with using the case (ii) approach is that an air–water model will run much more slowly than an all-water model, and it is probably only worthwhile if some surface or near-surface behavior is of interest. This is the case for Wairakei [109,160] as the effect of production on the surface features (steaming ground and hot springs) is of interest, as is subsidence, which is related to shallow pressure changes. Most modeling studies have concentrated on the deep reservoir behavior and have used the case (i) approach.

For fields where the CO_2 content is important, a CO_2–water EOS has been used; where the case (ii) approach has been used [21,151], the atmosphere and the vadose zone contains CO_2 rather than air.

Some geothermal systems extend under a water body such as a lake or the ocean, and for them the models have used a "dry" atmosphere boundary condition for blocks on dry land and a "wet" atmosphere boundary condition for the blocks under water [112,122,160].

Base. The usual deep boundary conditions are a background heat flow (say, 65–80 mW/m^2) over most of the base of the model. Sometimes the heat flux is increased near the upflow zone. Then, the hot upflow zone is represented either by the injection of very hot water or by a hot plate at a constant high temperature and pressure. The problem with the use of a hot plate boundary condition is that it provides an infinite source of hot water and for future scenario simulations, after a time, a spurious quasi-steady state may be set up where the deep recharge from the hot plate matches the specified net production rate (ie, the difference between production and injection), assuming lateral recharge and recharge from above are small. When this occurs, the model will predict that the required steam flow for the project will last forever. This problem can be partly avoided or at least delayed by extending the model deeper so that the hot plate is further separated from the production zone. Because of this problem, a mass flux boundary condition should be preferred rather than a hot plate boundary condition.

In some cases, recharge boundary conditions at the base have been used to allow some extra deep recharge to be induced by a pressure drop in the production zone. However, care must be taken with the choice of the recharge coefficient to ensure the balance between recharge from the base, the sides, and the top of the model is physically realistic. If the deep recharge coefficient is too large the same problem as for the hot plate will occur (ie, a spurious endless supply of deep hot recharge).

It would be preferable to be able to avoid issues with the base boundary conditions by making the model large enough and deep enough (say 6–8 km) to include the whole convective system down to the brittle–ductile transition zone, and recent developments with a supercritical version of TOUGH2 [24,75] make this theoretically possible. However, there are practical difficulties with this approach as this style of model runs slowly [106].

In the case of vapor dominated systems, it is necessary to use hot plate base boundary conditions. It can be proved [89] that a mass flux boundary condition will always produce a relatively wet two-phase zone from which wells will not produce dry steam. The nonphysical behavior of the model can be delayed beyond the target simulation time (say 25–30 years) by taking the base of the model deep enough and making the permeability low enough in the deep layers of the model.

7.3.5   Natural state, production history, and future scenarios

One of the major differences between an oil and gas reservoir and a hot geothermal reservoir is that there is fluid movement in the latter due to large-scale convection driven by heat flow from depth. As the hot convective plume rises, its progress is influenced by the geological structure of the system. Thus, the temperature distribution in a convective geothermal system is controlled by the geology, and consequently the temperature distribution can be used to infer the geological structure. This is the basis of natural state modeling of a geothermal system.

In calibrating a natural state model, the permeabilities and the location of the deep upflow are adjusted until a good match is achieved between the results predicted by the model and the data for the downhole temperatures and the locations and strength of the surface features (eg, hot springs, steaming ground, large shallow temperature gradients). For each step of the calibration process, the model is run until a stable steady state is reached, requiring a simulation time of, say, $\sim 10^6$ years. Each natural state simulation may take a few hours to run, and the calibration process may take weeks of a modeling effort to obtain a good match.

Once a good natural state model is obtained, further calibration against production history data (if available) should be made. This involves adjusting porosities and some local permeability so that the model results match the observed changes over time of production enthalpies and downhole pressures. The measured injection and production rates are used as input for the model. At each stage, the natural state model should be rerun to check that it is still matching the data.

Finally, future scenarios are run to test future production and injection options. For these simulations, the production and injection rates for existing wells or new make-up wells are not known, and it is usual to assume that the wells operate on deliverability, with their production rate declining as the reservoir pressure declines. Some simulators, such as AUTOUGH2 [159], allow the automatic addition of make-up wells, from a preselected list, in order to maintain a target total steam or mass flow.

It is usual to run future scenario simulations for periods of 25–50 years into the future. The accuracy of the model predictions is likely to decrease with time, and recently some modelers have been adding uncertainty estimates to their predictions [27,93]. It is also worthwhile to extend the period for future scenario simulations to check that the long-term behavior of the model is physically credible. It is likely that most geothermal reservoirs will ultimately fail by the intrusion of cold water laterally, from above, or from the reinjection wells into the hot reservoir. The model should be able to simulate this long-term process.

In some cases, a geothermal project may fail due to excessive pressure drawdown in some wells, but if the wells are operating on deliverability, the model should be able to reproduce this process.

7.3.6   Dual-porosity versus single-porosity models

For geothermal systems where there is likely to be significant interaction between production and injection or where local boiling near the feed zones for some wells causes them to produce excess enthalpy fluid (ie, enthalpy values greater than saturated liquid values at the reservoir temperature, say, values from 1500 to 2500 kJ/kg), then a dual-porosity model is likely to give more accurate results than a single-porosity model for production history simulations [5]. The results from single- and dual-porosity

Table 7.2 Parameters used in dual-porosity models

Parameter	Typical values used	Authors' suggested starting values
Volume of fractures V_f	1−10%	2%
Fracture porosity	0.01−0.90	0.9
Matrix permeability	$0.0001−1.0 \times 10^{-15}\,\mathrm{m}^2$	$0.1 \times 10^{-15}\,\mathrm{m}^2$
Matrix porosity	0.004−0.20	0.01
Number of matrix blocks	One block: $V_m = 100 - V_f$	Two blocks: $V_{m1} = 20$, $V_{m2} = 78$
	Two blocks: $V_{m1} = 20$, $V_{m2} = 80 - V_f$	
Fracture spacing	15−300 m	100 m

natural state models should be identical, but in some cases the method for calculating the effective thermal conductivity can lead to very minor differences in the natural state temperature distribution.

In a dual-porosity model, high-permeability, low-volume fractures are embedded in a low-permeability matrix. In TOUGH2, this dual continuum is represented by the MINC system [117] with each block of the single-porosity model partitioned into a fracture block and one or more matrix blocks. The main difficulty in setting up a dual-porosity model is in deciding on the model parameters, particularly the volume percentage of fractures, fracture spacing, fracture porosity, number of matrix blocks, and permeability of the matrix. Unfortunately, there is little guidance from the literature; there have not been many published modeling studies on dual-porosity models, and some of those that have appeared have not included details of the parameters used. Austria [6] carried out a survey of the literature augmented with some private communications and found the values listed in Table 7.2. The starting values recently used by the authors in setting up new dual-porosity models are listed in the third column of Table 7.2. Calibration has often resulted in the fracture porosity being reduced, probably an indication that V_f should be decreased.

7.3.7 Model calibration

Model calibration used to be mainly carried out manually using many adjustments of the parameters to improve the model (both natural state and history matching), testing both the skill and the patience of the modeler.

Besides the challenges of this process, the modeler is often faced with challenges in processing the data required for calibration. For example, the downhole temperature profiles may be affected by wellbore processes such as interzonal flows or a shallow bleed of steam, and interpretation is required to obtain a reservoir temperature profile for model calibration. Often, the data on surface features are not quantitative, and the mass and heat flows from individual or groups of thermal manifestations are not known. Similarly accurate estimates of the total mass and heat flows from the surface of the system may not be available.

Difficulty is frequently experienced in matching the absolute values of pressure versus time in production history matching, although a good match to pressure drop is obtained. This problem appears to arise because of a difference in elevation between the block in the model where the results are extracted and the elevation where it is suggested the data were measured.

Recently, there has been increased use of inverse modeling for the calibration of geothermal models using software such as iTOUGH2 [60,150] or PEST [43,103]. Both iTOUGH2 and PEST use gradient-based methods, such as the Levenberg-Marquadt method to solve the nonlinear optimization problem posed by choosing the best parameter values for a geothermal model. What makes the geothermal problem difficult for inverse modeling is the length of time required for one forward run and the number of model parameters to choose. Theoretically, there could be a permeability and a porosity value assigned to each block in the model, but this is usually reduced to values for a smaller number of rock types assigned to the model blocks. Pilot points [39] instead of rock types have been used in some cases [102].

A statistical sampling approach to model calibration through the Markov Chain Monte Carlo method (MCMC) has also been used [27]. The calibration problem requires the location of the minimum of an objective function that measures the fit of the model results to the data (eg, the sum of squares of errors). Unfortunately, the objective function may have many local minima and the global minimum may reside in a part of the parameter space that the modeler considers undesirable. As explained by Cui et al. [27], the MCMC approach has some theoretical advantage over gradient-based methods in dealing with these two problems as it can find a region of parameter space where all models are good. The main difficulty with the MCMC approach is the length of time required for a forward run. The test of the method [27] on a 3D geothermal model required 3 weeks of computer time for 10,000 forward runs. However, by using parallel computing, this process could be sped up.

In general, the use of inverse modeling for calibration of geothermal models does not simplify the process. Instead of facing the challenges associated with selecting permeabilities, porosities, and boundary conditions, the modeler is faced with new challenges such as assigning weighting factors for regularization, controlling the minimization process effectively, and dealing with the issues of local minima. Also, the model structure must be defined (using either rock types or pilot points) regardless of whether manual calibration or inverse modeling is used. These factors mean that inverse modeling should not be seen as a "black-box" tool capable of automatically calibrating a geothermal model and relieving the modeler of the task. Instead, inverse modeling provides a framework for supporting modelers to rigorously and methodically calibrate geothermal models while also generating useful statistical information such as parameter sensitivities.

General experience with inverse modeling is that it is very useful for improving a model that has been partly manually calibrated; however, if a model is poor at matching the data, inverse modeling may not help.

7.4 Recent modeling experiences

7.4.1 Introduction

A literature survey found 102 articles on reservoir modeling studies postdating the survey by O'Sullivan et al. [108]. There are many more articles on generic modeling studies and reservoir physics studies involving numerical modeling.

Geothermal reservoir modeling appears to be now considered a commercially sensitive topic, and there have been many more modeling studies than those in the open-file literature. Also, because of concerns about commercial sensitivity, many of the published studies do not give complete details of the models used.

The continuing increase in inexpensive computing power has resulted in an increase in complexity of geothermal models. In the early 2000s, most models contained <5000 blocks, but now models with 50,000−100,000 blocks are common. This is still an order of magnitude smaller than is common for models of oil and gas reservoirs.

One of the recent advances in numerical modeling is coupling of reservoir and wellbore simulators [49,110], and in some cases, modeling of the surface equipment has also been included [41,64].

7.4.2 Dual porosity

There seems to be particular reluctance to report details of the parameters used in dual-porosity models. Of the nine studies examined that used a dual-porosity model, only three gave most or all of the dual-porosity parameters [5,16,66]. For the remaining studies, some gave essentially no information beyond the fact that a dual-porosity model has been used [22,64,104], while the others gave limited information [41,67].

A dual-porosity model should be preferred when returns of cold water from injection wells to production wells are important [22] or when tracer tests are being used for model calibration [113]. In some geothermal reservoirs, there is a small volume of highly fractured rock surrounding the feed zone of some wells, and this leads to local boiling and production of excess enthalpy fluid. This is a challenging phenomenon to match in a geothermal model, and some modelers have found dual-porosity models work best [5,66]. If a large part of the reservoir is boiling, such as at Wairakei [160], a single-porosity model is able to match the performance of the high enthalpy wells very well; however, if boiling is more local and variable, then a dual-porosity model will work better.

7.4.3 Boundary conditions

Most geothermal models use the case (i) boundary conditions (discussed above) with the water table as the top of the model, but recently there has been an increase in the number of models including the vadose zone [21,122,160]. Interesting models that include the vadose zone near a surface boundary that changes with time are the models of the Lihir geothermal system, which has a goldmine being excavated in the shallow zone [105,151]. The model in [105] made extensive use of Python scripting tools [25] to adjust the model grid and top boundary conditions as the pit was excavated.

There are a few models that have used "hot plate" bottom boundary conditions [18,64]. As mentioned above, these boundary conditions are undesirable as they may lead to overly optimistic predictions, and heat and mass flow boundary conditions should be preferred. Several of the published studies did not make clear what base boundary conditions were used.

A few modelers have had to use large injection or extraction of mass at the side boundaries of their models to achieve a match to the measured temperatures [93]. This practice should be avoided as it is an indication that either the permeability distribution in the model is not compatible with the temperature distribution or the model is not large enough.

7.4.4 Modeling EGS

The idea of an enhanced or engineered geothermal system (EGS) is old, dating back to the early 1970s work at Fenton Hill (New Mexico, USA) [13], but interest in EGS has increased since the MIT study of its potential [136].

The idea of an EGS is simple in principle:

(i) Create a zone of artificial permeability in rock at high temperatures by drilling a deep well and pumping in cold water to fracture the rock.
(ii) Drill a second well to intersect the fractured zone.
(iii) Pump cold water down one well and produce hot water from the second well.

A good EGS requires a large cloud of uniform fractures to provide the ideal underground heat exchange system. To date, this ideal has not been achieved, and for various reasons none of the past EGS projects have been commercially successful.

The two best-documented EGS projects are Soultz-sous-Forêts (France) [56] and Cooper Basin (Australia) [19,65]. For both of them, the poststimulation (TH) behavior has been modeled using TOUGH2 [83,134].

Recently, there has been a considerable amount of research on modeling the THM problem of the development of the fracture cloud at Soultz-sous-Forêts, Cooper Basin and elsewhere. The first difficulty in such modeling studies is deciding on the mechanism for the stimulation of fractures. McClure and Horne [88] identified four mechanisms: pure opening mode (POM), pure shear stimulation (PSS), primary fracturing with shear stimulation leak-off (PFSSL), and mixed-mechanism stimulation (MMS); they suggested that field data are consistent with the MMS concept. Dempsey et al. [29] investigated the role of thermoelastic and poroelastic stress changes in stimulation, using data from the Desert Peak (Nevada, USA) EGS project to test their models.

Several different numerical methods have been used for the hydromechanical (HM) part of the stimulation modeling, including finite elements [81] and discrete fracture networks [36,87]. There are also some hybrid methods combining finite elements and DFN [79]. Probably the simplest approach is to solve the TH problem using the finite volume method and the M problem using finite elements, such as using

TOUGH2 and FLAC3D [124], TOUGH2 and ABAQUS [114], or FEHM [70,71]. This approach requires the use of some constitutive laws for representing shear failure or development of rock damage and how this affects permeability and porosity.

In order to evaluate the different approaches, a code comparison study has been set up [152]. It involves seven benchmark problems and has 12 participant groups from the United States. Some results for Benchmark Problem 2 were reported [57]. There is wide variation in the results, more so for the temperatures and flows than for the mechanical deformations.

There is an interesting variant of EGS under investigation in Japan, namely, looking at the deep brittle—ductile transition zone [2].

7.4.5 TOUGHREACT, scaling, and other chemical processes

Chemical reactions were included as an option in TOUGH2 with the development of CHEM-TOUGH2 [149] and TOUGHREACT [156,158].

Some of the studies with CHEM-TOUGH2 have been generic, such as evolution of permeability in granite [135], deposition of silica around an injection well [95], and formation of cap-rocks in geothermal systems [129]. Other studies have investigated particular geothermal fields, such as deposition of quartz at Wairakei [94] and using chemistry to constrain a model of Kakkonda (Japan) [128].

Kiryukhin et al. [73] used data from nine geothermal fields in Japan and Russia (Kamchatka) on the distribution of secondary minerals to carry out a THC model calibration study. They used a simple geochemical system including quartz, K-feldspar, Na-feldspar, cristobalite, and a $Cl-Na-CO_2$ enriched fluid. They showed that precipitation dominates over dissolution with a reduction of porosity and permeability resulting from K-feldspar and quartz deposition. This result was confirmed by the observed abundance of K-feldspar in the production zone of several of the geothermal fields they considered.

TOUGHREACT was used for studies of water—rock interaction and the dissolution and deposition of minerals at Yellowstone [34] and at Dixie Valley (Nevada, USA) [145]. It has also been used in several studies of THC problems associated with injection in geothermal fields [90,157].

7.5 Current developments and future directions

7.5.1 Model calibration

In a hot convective geothermal system, the geological structure controls the rising convective plume, and thus there is a correlation between the subsurface temperature distribution and the permeability structure. This makes the calibration of a model of the natural state of a geothermal system a very useful process and one that is not available for a groundwater reservoir model or oil and gas reservoir models.

Previously, most natural state models have been calibrated manually. This is a lengthy process requiring skill, experience, and good physical intuition from the modeler. If, for example, a zone in the model is too cold, the following steps may be taken:

(i) Find the flow direction for the zone and find the surrounding temperature distribution.
(ii) If there is a cold flow toward the cold zone, decrease the permeability to lower this cold flow.
(iii) Alternatively, if there is a hot flow toward the cold zone, increase the permeability to raise the hot flow.

A similar algorithm can be specified for a zone of the model that is too hot. This is a difficult iterative process as the local 3D temperature, flow, and permeability structure have to be thought about together. Also, adjustments in one part of the model may degrade the match to data in other parts.

For production history matching, measured data for pressure versus time and production enthalpy versus time may be available. For reservoirs that boil during production, with wells that produce excess enthalpy, the production enthalpy provides very good calibration data as it is very sensitive to changes in porosity and permeability near the feed-zone [107].

Calibration of the natural state model provides constraints on the large-scale permeability structure, whereas production history matching determines more local permeability values and determines porosity values.

As mentioned, the use of inverse modeling with iTOUGH2 [44] or PEST [37] has been increasing, with several recent modeling studies incorporating inverse modeling. Although inverse modeling has been effective in improving geothermal models, many problems have been encountered in its use, some of which are discussed next.

7.5.1.1 Long run-times

As of 2015, a large complex geothermal model may take 4–5 h for a natural state simulation and a similar time for a production history run. The most obvious solution to this problem is parallelization and inverse modeling, which requires many forward runs can be very effectively parallelized. Both PEST and iTOUGH2 offer parallelized simulation as an option through Beo-PEST [130] and iTOUGH2-PVM [46], respectively.

One method for dealing with long run-times is to use a hierarchy of coarse-to-fine models, with results from the coarse model being used to update parameters in the fine model. This approach was used by Cui et al. [27] in using an adaptive delayed acceptance Metropolis–Hastings algorithm to speed up the MCMC method.

7.5.1.2 Runs fail to finish

A natural state model that is running quickly up to a large time step and a stable steady state may start behaving badly if the model parameters are changed; that is, it may take a long computational time to reach a steady state or in some cases may go into an irregular oscillatory state. Some of the issues with a slow approach to a steady state appear to be related to the iterative conjugate gradient solvers used in TOUGH2 not reaching a sufficiently accurate solution. It may be that a more accurate linear solution could be achieved by more sophisticated preconditioning.

If a steady state is not reached during inverse modeling of a natural state model, it has the potential to cause the whole process to fail. PEST and iTOUGH2 deal with this problem by assigning a large value to the objective function (the measure of fit of the model results to the measured data) if the model fails to finish.

7.5.1.3 Large number of parameters

Large complex geothermal models typically contain 50,000—100,000 blocks and up to three permeabilities plus a porosity for every block. Thus, there are potentially a very large number of unknown parameters. In the most common rock-type approach, a relatively small number of rock types are defined, usually corresponding to the different geological formations, and then these are assigned to all the blocks in the model. In the less common pilot point approach, rock parameters are assigned to a number of spatially distributed pilot points (not necessarily at nodal points), and some form of interpolation scheme is used to assign parameters to all the blocks in the model.

Some inverse modeling has used the SVD-assist (SVDA) option in PEST to reduce the number of model parameters. This method combines two important regularization methodologies, ie, truncated singular value decomposition and Tikhonov regularization [37]. In an inverse modeling study of Mt. Apo (Philippines), the use of SVDA reduced the number of model parameters from 936 to 200 in one case [5].

In some modeling studies, the upflows of heat and mass at the base of the model have been selected as parameters for PEST or iTOUGH2 to estimate [69]. In some cases, the upflows at individual blocks have been used, but more commonly a smaller number of upflow zones have been used with the model parameter being a multiplier for all upflows within the zone.

7.5.1.4 Parameter refinement

In some cases, the best-fit inverse model still does not match the data as well as is desired and refinement of the model parameters should be undertaken. Heuristic methods have been used [103] where the top two or three rock types are identified in terms of their influence on the objective function and then they are broken down into two or more sub-rock types based on some simple criteria, such as inside or outside the reservoir or shallow or deep. An application of this technique to inverse modeling of a geothermal field in the Philippines increased the number of rock types from 17 to 57 while reducing the objective function by a factor of 2.5 [103].

7.5.1.5 Excess enthalpy

As mentioned, production enthalpy data from wells that produce at excess enthalpy are very useful for model calibration. However, it is difficult to use the data for inverse modeling in a straightforward fashion because of the very nonlinear dependence of production enthalpy on permeability and porosity. If the porosity and permeability near a particular well are too high, then the reservoir will not boil and the well will produce very hot water. For some modest changes in permeability and porosity, this behavior will not change and the corresponding derivatives of the production enthalpy residuals (difference between model values and field data) will be zero. In a manual calibration process, a modeler can recognize that still greater changes in permeability and porosity are required to achieve boiling, but an inverse modeling code cannot directly recognize this problem. Adding extra terms to the objective function, minimizing the difference between the block pressure and the boiling pressure, works well for matching excess enthalpy wells [5].

7.5.2   Uncertainty predictions and data worth analysis

As reservoir models have become established tools for predicting the future behavior of geothermal systems, an increasing effort has been made to calculate rigorously the uncertainty in model predictions. By their nature, most reservoir models are deterministic giving a single solution for a scenario based on a single set of model parameters. In reality, there are many sources of errors and uncertainty in both the model parameters and the process for generating predictions. Quantifying and analyzing uncertainty are fundamental parts of using numerical models to support decision-making [38].

The main sources of errors and uncertainty in reservoir models can be grouped into four general categories. The categories often overlap and should not be thought of as strict groupings but rather as an aid to identifying and understanding different sources of error.

7.5.2.1   Systematic errors and inconsistencies

These are errors in the model description of the real system. Examples include errors in the locations of geological structures; limitations in the physical and chemical process included oversimplified models of well behavior. These errors are generally the most difficult to identify and to correct and unfortunately usually have the greatest impact on model predictions [45]. This problem was addressed by Wellmann et al. [147], who considered uncertainties in the geological data.

7.5.2.2   Uncertainty in model parameters

All model parameters have some associated uncertainty. This can come from measurement errors as described below or from an inability to measure the parameter directly. Examples include the amount of deep upflow entering a reservoir, rock permeabilities and porosities, and fluid chemistry.

7.5.2.3   Discretization errors

Numerical models require discretization of both time and physical space. The discretization process can introduce errors such as numerical dispersion. It also causes errors and uncertainty as a result of the aggregation of spatially distributed real-world processes into block-average quantities [9].

7.5.2.4   Measurement errors

All measurements include errors and uncertainty, which affect not only model parameters but also calibration data. The errors introduced can be systematic such as those caused by biased equipment or they can be random as caused by operator mistakes.

Having identified sources of uncertainty in models, there are many different methods for quantifying their effect on model predictions. The topic is a challenging field of research in its own right and, as recently as 2011, the U.S. Department of Energy *Workshop on Mathematics for the Analysis, Simulation, and Optimization of Complex Systems* stated that the "computational demands associated with the analysis of the creation and propagation of uncertainties in a complex multiphysics, multiscale subsurface system severely affect all existing approaches to uncertainty analysis" [35]. However, much progress has been made in developing techniques for both groundwater flow models and models used in the petroleum industry; good summaries have been presented by several authors [84,101,132].

While there are significant differences between techniques, they all share some common characteristics [91]. In all of the techniques, the subsurface is parameterized and the parameters are estimated. There are many different methods of parameterization available including homogeneous regions, pilot points, and geostatistics. The parameter space is then sampled and scenarios are simulated for each sample. This produces statistics for the model predictions that can be used to determine the prediction uncertainty. The high computational cost of this process arises from the fact that subsurface models have many parameters, and therefore many samples are required to adequately describe the parameter space.

Most approaches fall into one of two main categories: nonlinear regression methods and Bayesian methods. In general, the nonlinear regression methods are less computationally expensive than the Bayesian methods. However, the Bayesian methods can provide more accurate statistics for highly nonlinear problems with local minima [84]. The general details of the two approaches are given next.

- Nonlinear regression methods
  These methods provide uncertainty quantification by developing regression confidence intervals (RCIs) for the predictive uncertainty resulting from parametric uncertainty. To generate the RCIs, optimal model parameters must first be estimated and the estimation uncertainties calculated using the least-squares method. The uncertainties in parameter estimates can be transformed into uncertainties in the predictions by using either linear or nonlinear theory to produce the RCIs and then combining the effects of other model errors [132]. Many examples of using this approach in subsurface modeling are based on either of the two software packages PEST [37] and iTOUGH2 [44], and detailed descriptions of the mathematical background and practical applications can be found in both users' guides.
- Bayesian methods
  In these methods, Bayesian theory is used to generate credible intervals (BCIs) from model parameters and predictions, both of which are considered to be random variables [132]. The predictive uncertainty can then be calculated from the BCIs by incorporating the effects of other model errors [51]. The methods within this general category reflect different

approaches for reducing the computational cost of generating the probability density functions of the parameters and predictions by sampling sufficiently from each of them. Common methods include MCMC, randomized maximum likelihood, Ensemble Kalman Filters, proxy/surrogate models, and generalized likelihood uncertainty estimation (GLUE). Details of these methods and others can be found in review articles [48,101,132] and the references therein.

There are few examples of uncertainty quantification applied to geothermal reservoir models. This is largely because of the additional complexity inherent in geothermal systems where high temperature and complex chemistry lead to highly nonlinear forward models that are computationally expensive to solve. However, progress is being made by adopting techniques from groundwater modeling, the petroleum industry, carbon sequestration, and nuclear waste storage (eg, [30,47,68]).

Some recent projects are:

- Cui et al. [27] carried out a formal uncertainty analysis of downhole temperature predictions from a reservoir model using a modified version of the Metropolis Hastings algorithm for MCMC sampling. The results were promising but the study did not consider any transient predictions from the reservoir model.
- Vogt et al. [141] estimated the permeability fields of the reservoir at the EGS at Soultz-sous-Forêts, using an Ensemble Kalman Filter. Long-term predictions of the performance of the system were made including an uncertainty analysis. However, the EGS system has a relatively low temperature, and many of the challenges presented by quantifying uncertainty in a hot, natural system were not addressed.
- Moon et al. [93] used a Monte Carlo method to propagate uncertainties through forecasting scenarios for a large-scale reservoir model of the Ngatamariki geothermal field in New Zealand. The parameter values and their distribution were identified using iTOUGH2. Ten parameters were selected for the uncertainty quantification, and 89 samples were used for the Monte Carlo simulation. The results are promising, but the sampling method was not described and no formal explanation of how well the samples represent the parameter space is given.

7.5.2.5 Data worth analysis

Data worth analysis is a field of research related to uncertainty quantification that has been used extensively in groundwater modeling [97] and more recently in TOUGH2 simulations [143]. It is based on the premise that the worth of data increases in proportion to its ability to reduce the uncertainty of key model predictions [28]. The methodology can be used to compare the relative worth of different data of the same type (eg, two different downhole temperature sets) and also to compare the worth of different data types (eg, pressure transients vs. flowing enthalpies). Details of the theory and the method are presented by Dausman et al. [28] who show that if both the model parameters and the noise associated with historical data display Gaussian variability, then equations can be formulated which describe the constraining effect of these data on the potential model parameter values.

Both PEST and iTOUGH2 have utilities designed specifically for carrying out data-worth analysis that also help the user to interpret the results of the process. Useful examples and exercises are presented by Doherty [37] and Finsterle [44].

7.5.3 Simulator improvements

For most of the post-2000 modeling studies reviewed, TOUGH2 was the simulator used. In four cases TETRAD [140] was used [12,16,41,67]. Two studies used the STARS "Advanced Process and Thermal Reservoir Simulator" [23], which was originally developed for modeling thermal problems in the oil and gas industry but can be applied to geothermal reservoirs [31,161]. The similarly named STAR simulator [116] was used in four studies [54,55,96,104]. The HYDROTHERM simulator [72] was used in three studies [8,50,131]. A handful of other simulators were used in studies of individual fields. All of the simulators appear to have similar functionality.

For studies of EGS and hot water systems where boiling does not take place, some simulators that can only handle single-phase hot water have been used effectively, such as SHEMAT [11,20,121] and FEFLOW [62,123].

A few new geothermal simulators are currently being developed, based on the same methodology as TOUGH2 and the other established codes, but using more modern programming languages and more integrated parallelization. Some examples of such codes are listed below:

  (i) TOUGH+, a version of TOUGH2.2 rewritten in modern FORTRAN [163]
 (ii) OpenFoam [137]
(iii) OpenGeoSys (http://www.opengeosys.org/) [144]
 (iv) Geothermal Super Models project [15].
  (v) DuMu^x (http://www.dumux.org/) [42]
 (vi) PFLOTRAN (http://www.pflotran.org/) [52]
(vii) STOMP (http://stomp.pnnl.gov/)
(viii) FALCON [115] built on the MOOSE framework [61]
 (ix) OOMPFS [49]

It remains to be seen if any of these codes can achieve the wide use enjoyed by TOUGH2.

References

[1] Alcaraz SA, Chambefort I, Pearson R, Cantwell A. An integrated approach to 3-D modelling to better understand geothermal reservoirs. In: Proc., WGC 2015, Melbourne, Australia; 2015.

[2] Asanuma H, Tsuchiya N, Muraoka H, Ito H. Japan beyond-brittle project: development of EGS beyond brittle-ductile transition. In: Proc., WGC 2015, Melbourne, Australia; 2015.

[3] Australian Geothermal Reporting Code Committee (AGRCC). Australian code for reporting of exploration results, geothermal resources and geothermal reserves: the geothermal reporting code second edition (2010). Australian Geothermal Energy Group AGEG and the Australian Geothermal Energy Association; 2010a. 34 pp.

[4] Australian Geothermal Reporting Code Committee (AGRCC). Geothermal lexicon for resources and reserves definition and reporting second edition (2010). Australian Geothermal Energy Group AGEG and the Australian Geothermal Energy Association; 2010b. 90 pp.

[5] Austria JJC, O'Sullivan MJ. Dual porosity models of a two-phase geothermal reservoir. In: Proc., WGC 2015, Melbourne, Australia; 2015.

[6] Austria JJC. Dual porosity numerical models of geothermal reservoirs [Ph.D. thesis]. Auckland, NZ: University of Auckland; 2015.

[7] Battestelli A, Calore C, Pruess K. The simulator TOUGH2/EWASG for modelling geothermal reservoirs with brines and non-condensible gas. Geothermics 1997;26(4): 437−64.

[8] Bellani S, Gherardi F. Thermal modeling of an area west of the Mt. Amiata geothermal field, Italy. GRC Trans 2009;33:431−5.

[9] Blasone RS, Vrugt JA, Madsen H, Rosbjerg D, Robinson BA, Zyvoloski GA. Generalized likelihood uncertainty estimation (GLUE) using adaptive Markov chain Monte Carlo sampling. Adv Water Resour 2008;31(4):630−48.

[10] Borgia A, Pruess K, Kneafsey TJ, Oldenburg CM, Pan L. Numerical simulation of salt precipitation in the fractures of a CO_2-enhanced geothermal system. Geothermics 2012; 44:13−22.

[11] Borozdina O, Ratouis T, Ungemach P, Antics M. Thermochemical modelling of cooled brine injection into low enthalpy sedimentary reservoirs. GRC Trans 2012;36:151−7.

[12] Bowyer D, Holt R. Case study: development of a numerical model by a multi-disciplinary approach, Rotokawa geothermal field, New Zealand. In: Proc., WGC 2010, Bali, Indonesia; 2010.

[13] Brown DW. Hot dry rock geothermal energy: important lessons from Fenton Hill. In: Proc., 34th workshop on geothermal reservoir engineering, Stanford University, Stanford (CA, USA); 2009.

[14] Burnell JG, Kissling WM, Gordon DA. Rotorua geothermal field: modelling and monitoring. In: Proc., 29th NZ geothermal workshop, Auckland, NZ; 2007.

[15] Burnell J, O'Sullivan MJ, O'Sullivan JP, Kissling W, Croucher A, Pogacnik J, et al. Geothermal supermodels: the next generation of integrated geophysical, chemical and flow simulation modelling tools. In: Proc., WGC 2015, Melbourne, Australia; 2015.

[16] Butler SJ, Sanyal SK, Henneberger RC, Klein CW, Gutiérrez PH, de León VJS. Numerical modeling of the Cerro Prieto geothermal field, Mexico. In: Proc., WGC 2000, Kyushu − Tohoku, Japan; 2000.

[17] Canadian Geothermal Energy Association (CANGEA). Canadian geothermal code for public reporting. Reporting of exploration results, geothermal resources and geothermal reserves, 2010 edition. Canadian Geothermal Code Committee; 2010. 34 pp.

[18] Cei M, Barelli A, Casini M, Romagnoli P, Bertani R, Fiordelisi A. Numerical model of the travale geothermal field (Italy) in the framework of the i-GET European project. GRC Trans 2009;33:1041−5.

[19] Chen D, Wyborn D. Habanero field tests in the Cooper Basin, Australia: a proof-of-concept for EGS. GRC Trans 2009;33:159−64.

[20] Clauser C. Numerical simulation of reactive flow in hot aquifers—SHEMAT and processing SHEMAT. Heidelberg: Springer; 2003.

[21] Clearwater EK, O'Sullivan MJ, Mannington WI, Newson JA, Brockbank K. Recent advances in modelling the Ohaaki geothermal field. In: Proc., 36th New Zealand geothermal workshop, Auckland, NZ; 2014.

[22] Clearwater J, Burnell J, Azwar L. Modelling the Ngatamariki geothermal system. In: Proc., 33rd New Zealand geothermal workshop, Auckland, NZ; 2011.

[23] Computer Modeling Group Ltd.. Advanced process and thermal reservoir simulator, version 2010. Calgary, Canada: Computer Modeling Group, Ltd.; 2010.

[24] Croucher AE, O'Sullivan MJ. Application of the computer code TOUGH2 to the simulation of supercritical conditions in geothermal systems. Geothermics 2008;37: 622−34.

[25] Croucher AE. PyTOUGH: a Python scripting library for automating TOUGH2 simulations. In: Proc., 33rd New Zealand geothermal workshop, University of Auckland, Auckland, NZ; 2011.

[26] Croucher AE, O'Sullivan MJ. Approaches to local grid refinement in TOUGH2 models. In: Proc., 35th New Zealand geothermal workshop, Rotorua, NZ; 2013.

[27] Cui T, Fox C, O'Sullivan MJ. Bayesian calibration of a large − scale geothermal reservoir model by a new adaptive delayed acceptance Metropolis Hastings algorithm. Water Resour Res 2011;47(10). W10521, 26 pp.

[28] Dausman AM, Doherty J, Langevin CD, Sukop MC. Quantifying data worth toward reducing predictive uncertainty. Groundwater 2010;48(5):729−40.

[29] Dempsey D, Kelkar S, Davatzes N, Hickman S, Moos D, Zemach E. Evaluating the roles of thermoelastic and poroelastic stress changes in the Desert Peak EGS stimulation. In: Proc., 39th workshop on geothermal reservoir engineering Stanford University, Stanford (CA, USA); 2014.

[30] Dempsey D, O'Malley D, Pawar R. Reducing uncertainty associated with CO_2 injection and brine production in heterogeneous formations. Int J Greenh Gas Control 2015;37: 24−37.

[31] Deo M, Roehner R, Allis R, Moore J. Reservoir modeling of geothermal energy production from stratigraphic reservoirs in the Great Basin. In: Proc., 38th workshop on geothermal reservoir engineering, Stanford University, Stanford (CA, USA); 2013.

[32] DiPippo R. Geothermal power plants-principles, applications, and case studies. Oxford: Elsevier; 2005. 450 pp.

[33] DiPippo R. Geothermal power plants: evolution and performance assessments. Geothermics 2015;53:291−307.

[34] Dobson PF, Salah S, Spycher N, Sonnenthal EL. Simulation of water−rock interaction in the Yellowstone geothermal system using TOUGHREACT. Geothermics 2004;33: 493−502.

[35] DOE. A multifaceted mathematical approach for complex systems. 2011. Retrieved from US DOE Report on Workshop: http://science.energy.gov/∼/media/ascr/pdf/program-documents/docs/Multifaceted_Mathematical_Approach_for_Complex_Systems.pdf.

[36] Doe T, McLaren R, Dershowitz W. Discrete fracture network simulations of enhanced geothermal systems. In: Proc., 39th workshop on geothermal reservoir engineering Stanford University, Stanford (CA, USA); 2014.

[37] Doherty J. PEST: model-independent parameter estimation, user manual. 5th ed. Corinda, Australia: Watermark Numerical Computing; 2010. http://www.sspa.com/pest.

[38] Doherty JE, Hunt RJ, Tonkin MJ. Approaches to highly parameterized inversion: a guide to using PEST for model-parameter and predictive-uncertainty analysis. 2010. U.S. Geological Survey. Scientific Investigations Report 2010−5211, 71 pp. Available at: http://pubs.usgs.gov/sir/2010/5211.

[39] Doherty JE, Fienen MN, Hunt RJ. Approaches to highly parameterized inversion: pilot-point theory, guidelines, and research directions. 2010. U.S. Geological Survey. Scientific Investigations Report 2010−5168, 44 pp.

[40] Driesner T, Heinrich CA. The system H_2O−NaCl. Part I: correlation formulae for phase relations in temperature−pressure−composition space from 0 to 1000°C, 0 to 5000 bar, and 0 to 1 X_{NaCl}. Geochim Cosmochim Acta 2007;71:4880−901.

[41] Enedy SL, Butler SJ. Numerical reservoir-modeling of forty years of injectate recovery at the Geysers geothermal field, California, USA. GRC Trans 2010;34:1221−7.

[42] Faigle B, Helmig R, Aavatsmark I, Flemisch B. Efficient multiphysics modelling with adaptive grid-refinement using an MPFA method. Comput Geosci 2014;18: 625−36.

[43] Feather BM, Malate RCM. Numerical modeling of the Mita geothermal field, Cerro Blanco, Guatemala. In: Proc., 38th workshop on geothermal reservoir engineering, Stanford University, Stanford (CA, USA); 2013.

[44] Finsterle S, Pruess K. Automatic calibration of geothermal reservoir models through parallel computing on a workstation cluster. In: Proc., 24th workshop on geothermal reservoir engineering, Stanford University, Stanford (CA, USA); 1999.

[45] Finsterle S. iTOUGH2 User's Guide, Report LBNL-40040, Earth Sciences Division Lawrence Berkeley National Laboratory. Berkeley (CA 94720, USA): University of California; 2007. 137 pp.

[46] Finsterle S. Parallelization of iTOUGH2 using PVM, Report LBNL-42261, Earth Sciences Division, Lawrence Berkeley National Laboratory. Berkeley (CA 94720, USA): University of California; 2010. 37 pp.

[47] Finsterle S. Enhancements to the TOUGH2 simulator implemented in iTOUGH2, Report LBNL-TBD, Earth Sciences Division, Lawrence Berkeley National Laboratory, Berkeley (CA 94720, USA). 2015.

[48] Floris FJT, Bush MD, Cuypers M, Roggero F, Syversveen AR. Methods for quantifying the uncertainty of production forecasts: a comparative study. Pet Geosci 2001;7(S): S87−96.

[49] Franz P. OOMPFS − a new software package for geothermal reservoir simulation. In: Proc., WGC 2015, Melbourne, Australia; 2015.

[50] Fujimitsu F, Kanou R. Numerical modelling of the hydrothermal system in Unzen Volcano, Japan. In: Proc., 25th New Zealand geothermal workshop, Auckland, NZ; 2003.

[51] Gallagher M, Doherty J. Parameter estimation and uncertainty analysis for a watershed model. Environ Model Softw 2007;22(7):1000−20.

[52] Gardner WP, Hammond GE, Lichtner PC. High performance simulation of environmental tracers in heterogeneous domains. Ground Water 2015;53:71−80.

[53] Garg SK, Combs J. A reformulation of USGS volumetric "heat in place" resource estimation method. Geothermics 2015;55:150−8.

[54] Garg SK, Haizlip J, Bloomfield KK, Kindap A, Haklidir FST, Guney A. A numerical model of the Kizildere geothermal field, Turkey. In: Proc., WGC 2015, Melbourne, Australia; 2015.

[55] Garg SK, Goranson C, Johnson S, Casteel J. Reservoir testing and modeling of the Patua geothermal field, Nevada, USA. In: Proc., WGC 2015, Melbourne, Australia; 2015.

[56] Genter A, Fritsch D, Cuenot N, Baumgartner J, Graff J-J. Overview of the current activities of the European EGS Soultz project: from exploration to electricity production. In: Proc., 34th workshop on geothermal reservoir engineering, Stanford University, Stanford (CA, USA); 2009.

[57] Ghassemi A, Kelkar S, McClure M. Influence of fracture shearing on fluid flow and thermal behavior of an EGS reservoir − geothermal code comparison study. In: Proc., 40th workshop on geothermal reservoir engineering, Stanford University, Stanford (CA, USA); 2015.

[58] Grant MA. Geothermal resource proving criteria. In: Proc., WGC 2000, Kyushu − Tohoku, Japan; 2000.

[59] Grant MA. Resource assessment, a review, with reference to the Australian Code. In: Proc., WGC 2015, Melbourne, Australia; 2015.

[60] Gunnarsson G, Arnaldsson A, Oddsdóttir AL. Model simulations of the geothermal fields in the Hengill Area, South-Western Iceland. In: Proc., WGC 2010. Bali, Indonesia; 2010.

[61] Guo L, Huang H, Gaston D, Permann C, Andrs D, Redden G, et al. A parallel, fully coupled, fully implicit solution to reactive transport in porous media using preconditioned Jacobian-free Newton-Krylov. Adv Water Resour 2013;53:101−8.

[62] Harada A, Itoi R, Yoseph B. Three dimensional numerical model of aquifer for springs and hot springs in Karang volcanic area, West Java, Indonesia. In: Proc., 35th New Zealand geothermal workshop, Rotorua, NZ; 2013.

[63] Haukwa C. AMESH, a mesh creating program for the integral finite difference method: a user's Manual, Report LBNL-45284, Earth Sciences Division, Lawrence Berkeley National Laboratory. Berkeley (CA 94720, USA): University of California; 1998. 53 pp.

[64] Hernandez D, Clearwater J, Burnell J, Franz P, Azwar L, Marsh A. Update on the modeling of the Rotokawa geothermal system: 2010−2014. In: Proc., WGC 2015, Melbourne, Australia; 2015.

[65] Hogarth RA, Bour D. Flow performance of the Habanero EGS closed loop. In: Proc., WGC 2015, Melbourne, Australia; 2015.

[66] Itoi R, Kumamotoa Y, Tanakaa T, Takayamab J. History matching simulation of the Ogiri geothermal field, Japan. In: Proc., WGC 2010, Bali, Indonesia; 2010.

[67] Jaimes-Maldonado JG, Velasco RA, Pham M, Henneberger R. Update report and expansion strategy for Los Azufres geothermal field. In: Proc., WGC 2005, Antalya, Turkey; 2005.

[68] Jordan AB, Stauffer PH, Zyvoloski GA, Person MA, MacCarthy JK, Anderson DN. Uncertainty in prediction of radionuclide gas migration from underground nuclear explosions. Vadose Zone J 2014;13(10).

[69] Kaya E, O'Sullivan MJ. Three-dimensional model of the deep geothermal resources in part of the Taupo Reporoa Basin. In: Proc., New Zealand geothermal workshop, Auckland, NZ; 2011.

[70] Kelkar S, Lewis K, Hickman S, Davatzes NC, Moos D, Zyvoloski G. Modeling coupled thermal-hydrological-mechanical processes during shear stimulation of an EGS well. In: Proc., 37th workshop on geothermal reservoir engineering, Stanford University, Stanford (CA, USA); 2012.

[71] Kelkar S, Dempsey D, Zyvoloski G, Pogacnik J. Investigation of mesh sensitivity in coupled thermal-hydrological-mechanical models: examples from Desert Peak, Nevada, USA and Ngatamariki, New Zealand. In: Proc., 35th New Zealand geothermal workshop, Rotorua, NZ; 2013.

[72] Kipp Jr KL, Hsieh PA, Charlton SR. Guide to the revised ground-water flow and heat transport simulator: HYDROTHERM − version 3. 2008. U.S. Geological Survey, Techniques and Methods 6−A25, 160 pp.

[73] Kiryukhin AV, Xu T, Pruess K, Slovtsov I. Modeling of thermo-hydrodynamic-chemical processes: some applications to active hydrothermal systems. In: Proc., 27th workshop on geothermal reservoir engineering, Stanford University, Stanford (CA, USA); 2002.

[74] Kiryukhin AV, Asaulova NP, Finsterle S. Inverse modeling and forecasting for the exploitation of the Pauzhetsky geothermal field, Kamchatka, Russia. Geothermics 2008; 37:540−62.

[75] Kissling WM. Extending MULKOM to super-critical temperatures and pressures. In: Proc., WGC 1995, Florence, Italy; 1995. p. 1687−90.

[76] Kissling WM. Transport of three-phase hyper-saline brines in porous media: examples. Transp Porous Media 2005a;60:141−57.

[77] Kissling WM. Transport of three-phase hyper-saline brines in porous media: theory and code implementation. Transp Porous Media 2005b;61:25−44.

[78] Knight RH, Lane RG, Ross HJ, Abraham APG, Cowan J. Implicit ore delineation. In: Proc., exploration 07: fifth decennial international conference on Mineral exploration, Toronto, Canada; 2007. p. 1165−9.

[79] Kohl T, Megel T. Predictive modeling of reservoir response to hydraulic stimulations at the European EGS site Soultz-sous-Forets. Int J Rock Mech Min Sci 2007;44:1118−31.

[80] Lee KC. Classification of geothermal resources by exergy. Geothermics 2001;30: 431−42.

[81] Lee SH, Ghassemi A. Three-dimensional thermo-poro-mechanical modeling of reservoir stimulation and induced microseismicity in geothermal reservoir. In: Proc., 36th workshop on geothermal reservoir engineering, Stanford University, Stanford (CA, USA); 2011.

[82] Lewis KC, Lowell RP. Numerical modeling of two-phase flow in the NaCl-H$_2$O system: introduction of a numerical method and benchmarking. J Geophys Res 2009;114: B05202. http://dx.doi.org/10.1029/2008JB006029.

[83] Llanos EM, Zarrouk SJ, Hogarth RA. Numerical model of the Habanero geothermal reservoir, Australia. Geothermics 2015;53:308−19.

[84] Lu D, Ye M, Hill MC, Poeter EP, Curtis GP. A computer program for uncertainty analysis integrating regression and Bayesian methods. Environ Model Softw 2014;60:45−56.

[85] Magnusdottir L, Finsterle S. An iTOUGH2 equation-of-state module for modeling supercritical conditions in geothermal reservoirs. Geothermics 2015;57:8−17. http://dx.doi.org/10.1016/j.geothermics.2015.05.003.

[86] Mannington WI, O'Sullivan MJ, Bullivant DP. An air/water model of the Wairakei−Tauhara geothermal system. In: Proc., WGC 2000, Kyushu − Tohoku, Japan; 2000. p. 2713−8.

[87] McClure MW, Horne RN. Discrete fracture network modeling of hydraulic stimulation: coupling flow and geomechanics, Springer briefs in earth sciences. Heidelberg, Germany: Springer International Publishing; 2013. http://dx.doi.org/10.1007/978-3-319-00383-2.

[88] McClure MW, Horne RN. An investigation of stimulation mechanisms in enhanced geothermal systems. Int J Rock Mech Min Sci 2014;72:242−60.

[89] McGuinness MJ, Blakeley MR, Pruess K, O'Sullivan MJ. Geothermal heat pipe stability: solution selection by upstreaming and boundary conditions. Transp Porous Media 1993; 11:71−100.

[90] McLin KS, Kovac KM, Moore JN, Kaspereit D, Berard B, Xu T, et al. Modeling the geochemical effects of injection at Salton Sea geothermal field, California: comparison with field observations. GRC Trans 2006;30:507−11.

[91] McVay DA, Lee WJ, Alvarado MG. Calibration improves uncertainty quantification in production forecasting. Pet Geosci 2005;11(3):195−202.

[92] Milicich SD, van Dam MA, Rosenberg MD, Rae AJ, Greg Bignall G. "Earth research" 3-dimensional geological modelling of geothermal systems in New Zealand − a new visualisation tool. In: Proc., WGC 2010, Bali, Indonesia; 2010.

[93] Moon H, Clearwater J, Franz P, Wallis I, Azwar L. Sensitivity analysis, parameter estimation and uncertainty propagation in a numerical model of the Ngatamariki Geothermal Field, New Zealand. In: Proc., 39th workshop on geothermal reservoir engineering, Stanford University, Stanford (CA, USA); 2014.

[94] Mroczek EK, White SP, Mountain BW. Precipitation rates of quartz from Wairakei (New Zealand) geothermal field brine at temperatures of between 200°C and 250°C and calculation of aquifer surface areas. In: Proc., 22nd New Zealand geothermal workshop, Auckland, NZ; 2000.

[95] Mroczek EK, White SP, Graham D. Estimating the quantity of silica deposited in the reservoir around an injection well. In: Proc., 24th New Zealand geothermal workshop, Auckland, NZ; 2002. p. 187−90.

[96] Nakanishi S, Pritchett JW, Tosha T. Changes in ground surface geophysical signals induced by geothermal exploitation − computational studies based on a numerical reservoir model for the Oguni geothermal field, Japan. GRC Trans 2001;25:657−64.

[97] Neuman SP, Xue L, Ye M, Lu D. Bayesian analysis of data-worth considering model and parameter uncertainties. Adv Water Resour 2012;36:75−85.

[98] Newson J, Mannington W, Sepulveda F, Lane R, Pascoe R, Clearwater E, et al. Application of 3D modelling and visualization software to reservoir simulation: LEAPFROG GEOTHERMAL and TOUGH2. In: Proc., 37th workshop on geothermal reservoir engineering, Stanford University, Stanford (CA, USA); 2012.

[99] Oldenburg CM, Pruess K. A two-dimensional dispersion module for the TOUGH2 simulator, Report LBL-32505, Earth Sciences Division, Lawrence Berkeley National Laboratory. Berkeley (CA 94720, USA): University of California; 1993. 66 pp.

[100] Oldenburg CM, Pruess K. EOS7R: radionuclide transport for TOUGH2. Report LBL-34868, Earth Sciences Division, Lawrence Berkeley National Laboratory. Berkeley (CA 94720, USA): University of California; 1995. 75 pp.

[101] Oliver DS, Chen Y. Recent progress on reservoir history matching: a review. Comput Geosci 2011;15(1):185−221.

[102] Omagbon JB. Automated calibration of geothermal models [ME thesis]. Department of Engineering Science, University of Auckland; 2011. 128 pp.

[103] Omagbon JB, O'Sullivan MJ. Use of an heuristic method and PEST for calibration of geothermal models. In: Proc., 33rd New Zealand geothermal workshop, University of Auckland, Auckland, NZ; 2011.

[104] Osada K, Hanano M, Sato K, Kajiwara T, Arai F, Watanabe M, et al. Numerical simulation study of the Mori geothermal field, Japan. In: Proc., WGC 2010, Bali, Indonesia; 2010.

[105] O'Sullivan JP, Croucher A, O'Sullivan MJ, Stevens L, Esberto M. Modelling the evolution of a mine pit in a geothermal field at Lihir Island, Papua New Guinea. In: Proc., 33rd New Zealand geothermal workshop, auckland, NZ; 2011.

[106] O'Sullivan JP, Kipyego E, Croucher AE, Ofwona C, O'Sullivan MJ. A supercritical model of the Menengai geothermal system. In: Proc., WGC 2015, Melbourne, Australia; 2015.

[107] O'Sullivan MJ. Modelling of enthalpy transients for geothermal wells. In: Proc., 9th New Zealand geothermal workshop, University of Auckland, Auckland, NZ; 1987. p. 121−5.

[108] O'Sullivan MJ, Pruess K, Lippmann MJ. State of the art of geothermal reservoir simulation. Geothermics 2001;30:395−429.

[109] O'Sullivan MJ, Angus Yeh A, Mannington WI. A history of numerical modelling of the Wairakei geothermal field. Geothermics 2009;38:155−68.

[110] Pan L, Oldenburg CM. T2Well − an integrated wellbore-reservoir simulator. Comput Geosci 2014;65:46−55.

[111] Pearson SCP, Alcaraz SA, White PA, Tschritter C. Improved visualisation of reservoir simulations: geological and fluid flow modelling of the Tauranga low-enthalpy geothermal system, New Zealand. GRC Trans 2012;36:1293−7.

[112] Pearson SCP, Alcaraz SA, Barber J. Modelling the effects of direct use at the Tauranga low-temperature geothermal system. In: Proc., 35th New Zealand geothermal workshop, Rotorua, NZ; 2013.

[113] Pham M, Klein C, Ponte C, Cabeças R, Martins R, Rangel G. Production/injection optimization using numerical modeling at Ribeira Grande, São Miguel, Azores, Portugal. In: Proc., WGC 2010, Bali, Indonesia; 2010.

[114] Pogacnik J, O'Sullivan JP, O'Sullivan MJ. Linking TOUGH2 and ABAQUS to model permeability enhancement using a damage mechanics approach. In: Proc., WGC 2015. Melbourne, Australia; 2015.

[115] Podgorney R, Lu C, Huang H. Thermo-hydro-mechanical modeling of working fluid injection and thermal energy extraction in EGS fractures and rock matrix. In: Proc., 37th workshop on geothermal reservoir engineering, Stanford University, Stanford (CA, USA); 2012.

[116] Pritchett JW. STAR: a geothermal reservoir simulation system. In: Proc., WGC 1995, Florence, Italy; 1995. p. 2959−63.

[117] Pruess K, Narasimhan TN. A practical method for modeling fluid and heat flow in fractured porous media. Soc Petroleum Eng J 1985;25(1):14−26.

[118] Pruess K, Oldenburg C, Moridis G. TOUGH2 User's guide version 2.0, Report LBNL-43134, Earth Sciences Division, Lawrence Berkeley National Laboratory. Berkeley (CA 94720, USA): University of California; 1999. 210 pp.

[119] Pruess K. ECO2N: a TOUGH2 fluid property module for mixtures of water, NaCl, and CO_2, Report LBNL-57952, Earth Sciences Division Lawrence Berkeley National Laboratory. Berkeley (CA 94720, USA): University of California; 2005. 74 pp.

[120] Pruess K. ECO2M: a TOUGH2 fluid property module for mixtures of water, NaCl, and CO_2, including super- and sub-critical conditions, and phase change between liquid and gaseous CO_2, Report LBNL-4590E, Earth Sciences Division Lawrence Berkeley National Laboratory. Berkeley (CA 94720, USA): University of California; 2011. 94 pp.

[121] Rath V, Wolf A, Bücker M. Joint three-dimensional inversion of coupled groundwater flow and heat transfer based on automatic differentiation: sensitivity calculation, verification, and synthetic examples. Geophys J Int 2006;167:453−66.

[122] Ratouis TMP, O'Sullivan MJ, O'Sullivan JP. An updated numerical model of Rotorua geothermal field. In: Proc., WGC 2015, Melbourne, Australia; 2015.

[123] Rühaak W, Pei L, Bartels J, Heldmann C-D, Homuth S, Sass I. Thermo-hydro-mechanical-chemical coupled modeling of geothermal doublet systems in limestones. In: Proc., WGC 2015, Melbourne, Australia; 2015.

[124] Rutqvist J, Dobson PF, Garcia J, Hartline C, Oldenburg CM, Vasco DW, et al. Pre-stimulation coupled THM modeling related to the Northwest Geysers EGS demonstration project. In: Proc., 38th workshop on geothermal reservoir engineering, Stanford University, Stanford (CA, USA); 2013.

[125] Sanyal SK, Butler SJ. An analysis of power generation prospects from enhanced geothermal systems. GRC Trans 2005;29:131−7.

[126] Sanyal SK, Sarmiento Z. Booking geothermal energy reserves. GRC Trans 2005;29:467−74.

[127] Sarmiento ZF, Björnsson G. Reliability of early modeling studies for high-temperature reservoirs in Iceland and the Philippines. In: Proc., 32nd workshop on geothermal reservoir engineering, Stanford University, Stanford (CA, USA); 2007.

[128] Sato M, White SP, Osato K, Sato T, Okabe T, Doi N, et al. Modeling of chemistry and rock alteration at a deep-seated geothermal field. In: Proc., 25th workshop on geothermal reservoir engineering, Stanford University, Stanford (CA, USA); 2000.

[129] Sato T, Sato M, Ueda A, Kato K, Kissling WM, White SP. Model study of formation of the cap rocks for geothermal system using CHEMTOUGH2. In: Proc., WGC 2005, Antalya, Turkey; 2005.

[130] Schreüder W. Running BeoPEST. Principia Mathematica. 2009. Document can be found via the following website address: http://www.prinmath.com.

[131] Setyawan A, Ehara S, Fujimitsu Y, Saibi H. Assessment of geothermal potential at Ungaran Volcano, Indonesia deduced from numerical analysis. In: Proc., 34th workshop on geothermal reservoir engineering, Stanford University, Stanford (CA, USA); 2009.

[132] Shi X, Ye M, Finsterle S, Wu J. Comparing nonlinear regression and Markov chain Monte Carlo methods for assessment of prediction uncertainty in vadose zone modeling. Vadose Zone J 2012;11(4).

[133] Sieger D, Alliez P, Botsch M. Optimizing voronoi diagrams for polygonal finite element computations. In: Proc., 19th international meshing Roundtable, IMR 2010, Chattanooga, Tennessee, USA. Springer; 2010, ISBN 978-3-642-15413-3.

[134] Siffert D, Haffen S, Garcia MH, Geraud Y. Phenomenological study of temperature gradient anomalies in the Buntsandstein formation, above the Soultz geothermal reservoir, using TOUGH2 simulations. In: Proc., 38th workshop on geothermal reservoir engineering, Stanford University, Stanford (CA, USA); 2013.

[135] Sutopo, White SP, Arihara N. Modelling the evolution of granite permeability at high temperature. In: Proc., WGC 2000, Kyushu − Tohoku, Japan; 2000.

[136] Tester J, editor. The future of geothermal energy: impact of enhanced geothermal systems (EGS) on the United States in the 21st century. Massachusetts Institute of Technology; 2006. http://mitei.mit.edu/system/files/geothermal-energy-full.pdf.

[137] Thorvaldsson L, Palsson H. Modeling hydrothermal systems with OpenFoam. In: Proc., WGC 2015, Melbourne, Australia; 2015.

[138] Thunderhead Engineering. PetraSim user manual. 2015. https://www.rockware.com/product/documentation.php?id=148.

[139] UN. United Nations framework classification for fossil energy and Mineral reserves and resources 2009, ECE Energy Series No. 39. New York and Geneva: United Nations; 2010.

[140] Vinsome KW, Shook M. Multi-purpose simulation. J Petroleum Sci Eng 1993;9:29−38.

[141] Vogt C, Marquart G, Kosack C, Wolf A, Clauser C. Estimating the permeability distribution and its uncertainty at the EGS demonstration reservoir Soultz−sous−Forêts using the ensemble Kalman filter. Water Resour Res 2012;48(8).

[142] Wagner W, Cooper JR, Dittman A, Kijima J, Kretzschmar H-J, Kruse A, et al. The IAPWS industrial formulation 1997 for the thermodynamic properties of water and steam. ASME J Eng Gas Turbines Power 2000;122:150−82.

[143] Wainwright HM, Finsterle S. Global sensitivity and data-worth analysis in iTOUGH2, user's guide, Report LBNL-TBD, Earth Sciences Division, Lawrence Berkeley National Laboratory. Berkeley (CA 94720, USA): University of California; 2015.

[144] Wang W, Fischer T, Zehner B, Böttcher N, Görke U-J, Kolditz O. A parallel finite element method for two-phase flow processes in porous media: OpenGeoSys with PETSc. Environ Earth Sci 2015;73(5):2269−85.

[145] Wanner C, Peiffer L, Sonnenthal E, Spycher N, Iovenitti J, Kennedy BM. Reactive transport modeling of the Dixie valley geothermal area: insights on flow and geothermometry. Geothermics 2014;51:130−41.

[146] Weis P, Driesner T, Coumou D, Geiger S. Hydrothermal, multiphase convection of H_2O-NaCl fluids from ambient to magmatic temperatures: a new numerical scheme and benchmarks for code comparison. Geofluids 2014;14:347−71.

[147] Wellmann JF, Finsterle S, Croucher AE. Integrating structural geological data into the inverse modelling framework of iTOUGH2. Comput Geosci 2014;65:95−109.

[148] White DE, Williams DL. Assessment of geothermal resources of the United States-1975. 1975. U.S. Geological Survey Circular 726, 155 pp.

[149] White SP. Multiphase non-isothermal transport of systems of reacting chemicals. Water Resour Res 1995;31:1761−72.

[150] White SP, Okabe T, Sato T, Sato M, Shiga T, Takahashi Y. Modelling the deep geothermal system of the Uenotai reservoir. In: Proc., WGC 2005, Antalya, Turkey; 2005.

[151] White SP, Burnell J, Melaku M, Johnstone R. The Lihir open pit goldmine revisited. In: Proc., 28th NZ geothermal workshop, University of Auckland, Auckland, NZ.; 2006.

[152] White MD, Phillips BR. Code comparison study fosters confidence in the numerical simulation of enhanced geothermal systems. In: Proc., 40th workshop on geothermal reservoir engineering, Stanford University, Stanford (CA, USA); 2013.

[153] Williams CF, Reed MJ, Mariner RH. A review of methods applied by the U.S. Geological Survey in the assessment of identified geothermal resources. 2008. U.S. Geological Survey Open-File Report 2008−1296, 27 pp.

[154] Williams C. Evaluating the volume method in the assessment of identified geothermal resources. GRC Trans 2014;38:967−74.

[155] Wilmarth M, Stimac J. Worldwide power density review. In: Proc., 39th workshop on geothermal reservoir engineering, Stanford University, Stanford (CA, USA); 2014.

[156] Xu T, Pruess K. Coupled modeling of non-isothermal multi-phase flow, solute transport and reactive chemistry in porous and fractured media: 1. Model development and validation. Report LBNL-42050, Earth Sciences Division Lawrence Berkeley National Laboratory. Berkeley (CA 94720, USA): University of California; 1998. 39 pp.

[157] Xu T, Sonnenthal E, Spycher N, Pruess K. TOUGHREACT − a simulation program for non-isothermal multiphase reactive geochemical transport in variably saturated geologic media: applications to geothermal injectivity and CO_2 geological sequestration. Comput Geosci 2006;32:145−65.

[158] Xu T, Spycher N, Sonnenthal E, Zhang G, Zheng L, Pruess K. TOUGHREACT version 2.0: a simulator for subsurface reactive transport under non-isothermal multiphase flow conditions. Comput Geosci 2011;37:763−74.

[159] Yeh A, Croucher AE, O'Sullivan MJ. Recent developments in the AUTOUGH2 simulator. In: Proc., TOUGH Symposium 2012, Berkeley, CA, USA; 2012.

[160] Yeh A, O'Sullivan MJ, Newson JA, Mannington WI. An update on numerical modelling of the Wairakei-Tauhara geothermal system. In: Proc., 36th New Zealand geothermal workshop, Auckland, NZ; 2014.

[161] Yeltekin K, Parlaktuna M, Akin S. Modeling of Kizildere geothermal reservoir, Turkey. In: Proc., 27th workshop on geothermal reservoir engineering, Stanford University, Stanford (CA, USA); 2002.

[162] Zarrouk SJ, Simiyu F. A review of geothermal resource estimation methodology. In: Proc., 35th New Zealand geothermal workshop, Rotorua, NZ; 2013.

[163] Zhang K, Moridis G, Pruess K. TOUGH+CO_2: a multiphase fluid-flow simulator for CO_2 geologic sequestration in saline aquifers. Comput Geosci 2011;37:714−23.

[164] Zyvoloski GA. FEHM: a control volume finite element code for simulating subsurface multi-phase multi-fluid heat and mass transfer. Document: LAUR-07-3359. Los Alamos (NM, USA): Los Alamos National Laboratory; 2007. 47 pp.

Part Two

Energy conversion systems

Overview of geothermal energy conversion systems: reservoir-wells-piping-plant-reinjection

R. DiPippo
University of Massachusetts Dartmouth (Emeritus), N. Dartmouth, MA, United States

8.1 Introduction

At the Las Pailas geothermal power plant in Costa Rica (see Fig. 8.1), it takes roughly 12 minutes for the geofluid to flow from the production wells through the separators and heat exchangers and return to the reservoir via the reinjection wells. In that time, a total of about 400,000 kg of geofluid passes from the production to reinjection wells, and the power plant turns out about 7325 kilowatt-hours (kWh) of electricity at an average power of 35,400 kW. At a price of, say, $0.10/kWh, the value of that electricity would amount to $3660 per hour of operation. If the plant runs 85% of the time over the course of 1 year, that means over $27 million in revenues accrue to the plant owner, which happens to be the Costa Rican Electricity Authority.

Now, if the reinjection wells are properly sited and the field is sensibly managed, the geofluid will continue to circulate through the reservoir on some finite but imprecisely known time scale and return to the production wells ready for another trip through the power plant. This, combined with natural means of recharge to the reservoir such as rainfall, makes for a renewable and sustainable operation that can in principle produce electricity as long as the equipment does not fail.

In the sense just described, geothermal energy takes its place alongside the other major renewable sources of energy — solar, wind, and hydro — with a very important distinction: geothermal energy performs regardless of time of the day, weather conditions, droughts, etc. It is always there, except for periods of maintenance, which are planned and infrequent.

The objective of this chapter is to provide a road map for the ensuing chapters that will deal specifically with the details of geothermal energy conversion into electricity. Here we will describe the arc of the life of the geofluid and explain the processes needed, starting from the reservoir and ending with the disposal of the geofluid after the energy has been extracted to make electricity. The presentation will be primarily qualitative and descriptive, leaving the more quantitative aspects to the later chapters, but will include very simple diagrams and selected photographs to illustrate the processes and equipment under consideration.

Geothermal Power Generation. http://dx.doi.org/10.1016/B978-0-08-100337-4.00008-5

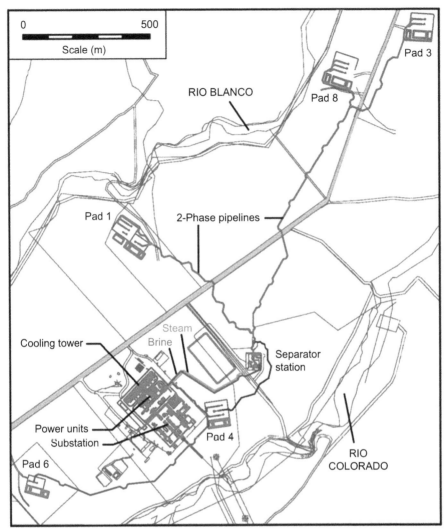

Figure 8.1 Field layout at Las Pailas geothermal plant, Guanacaste Province, Costa Rica. Production wells are at pads 1, 3, and 8; reinjection wells are at pads 4 and 6. The power plant is a binary type that uses separated steam and brine as heating media for isopentane; the condensers are served by a water cooling tower [1].

8.2   It begins with the reservoir

A dictionary tells us that a *reservoir* is "a place where anything is collected and stored, usually in large quantities." Water reservoirs may be naturally formed, such as ponds, or purpose-built such as by construction of dams. In the case of a geothermal reservoir, what exactly is collected and stored? In theory, it is energy, but in practice, it is hot water or steam. The underground geothermal reservoir consists of a large volume of high-temperature, fractured rock that holds hot geofluid—liquid, steam, or a mixture of the two—within its open spaces. Usually the spaces are extremely narrow but extend over a vast volume providing for a great amount of storage.

Where does the geofluid come from? There are two main sources of the geofluid, usually liquid composed mainly of water but with any number of dissolved elements. These are (1) condensed magmatic gases and (2) meteoric water. These two are normally found together in varying ratios depending on the site-specific nature of the system. The age of the fluid in the reservoir can vary over a very wide spectrum, with meteoric water being the youngest. The age can be accurately determined using geochemical techniques that assess the amount of different chemical components found in the fluid.

Not every geothermal reservoir will be valuable as a commercial asset. What are the essential features that make for a commercial hydrothermal reservoir? Five specific features must be present.

1. There must be a *source of heat*. This allows the rock formation to reach elevated temperatures relative to background. Cooling magma bodies, called plutons, embedded within the shallow crust of the earth are typical heat sources. They are found close to the boundaries of the massive plates that form the earth's crust, and at localized "hot spots" within the body of plates. These plutons may be the remnants of volcanic activity that occurred long ago, hundreds of thousands or even millions of years ago, or may be the forerunners of future volcanic activity.
2. There must be a *permeable volume of rock*. This is the body of the reservoir that will host the fluid needed to drive the power plant.
3. There must be a *supply of liquid* in the permeable formation. The liquid serves as the agent for energy transfer to sweep the thermal energy from the hot rocks and convey it through the formation via convection cells.
4. There must be an *impermeable cap* to seal the permeable formation from the surface of the earth. Unless this is present, the geofluid will dissipate by rising to the surface through fissures or faults where it will issue forth as steam vents, hot springs, geysers, boiling pools, and the like. It may happen that systems starting out without an impermeable cap actually develop one over a long period of time. This self-sealing process occurs due to precipitation of chemical compounds such as silica and calcite, leaving the hot fluid contained within the reservoir volume.
5. There must be a *recharge mechanism* for the reservoir. If the reservoir holds a finite amount of fluid that is isolated from its surroundings, perhaps having been implanted eons ago, then production will reduce the amount of fluid and eventually deplete the reservoir. Recharge can be achieved by natural processes, such as rainfall or snow melt that percolate through the ground, into the fractured formation, and eventually finds its way into the hot fractured rock. It can also be achieved to some degree by reinjecting the produced geofluid back to the formation after it is used in a power plant. This technique is not without risk because the fluid being returned is much lower in temperature than the fluid in the formation. Unless the injection wells are properly sited relative to the production wells, it is possible for the reinjected liquid to cool the formation and to mix with and cool the fluid being produced. This kind of "short-circuit" can quench a reservoir if allowed to continue.

Actual geothermal reservoirs may not have all five characteristics as just described. The cap rock may not be a perfect seal, the heat source may have a marginal temperature, the permeability may be minimal, natural recharge may be insufficient, and the formation may be deficient in fluid. To the extent that there is a serious short-fall in any one of the five essentials, the system may not be commercial. For example, without an adequate supply of liquid in place, the formation becomes a mass of hot fractured rock with little or no means available to effectively extract the thermal energy in place. Since such systems occur in many places, efforts have been directed to introduce liquid into such formations in an attempt to create a hydrothermal resource where none previously existed. Research programs such as hot dry rock (HDR), hot fractured rock (HFR), engineered geothermal systems (EGS), and enhanced geothermal systems (EGS) have been some of the experimental approaches aimed at this kind of development.

Another example is a system that is hot and saturated with fluid but lacks sufficient permeability. In this case the fluid cannot flow easily through the formation and con-vection cells are either nonexistent or weak. This is not uncommon and will be dis-cussed in Section 8.3.

Assuming it has been confirmed that a true hydrothermal system exists at some location, one needs to know how much energy it contains. Equally important is how long it will last under commercial exploitation. These are questions for the geo-scientists who will conduct the surveys that will result in a qualitative and, to some degree, quantitative characterization of the system. The methods used will yield esti-mates to these questions but will be burdened with uncertainty until deep wells probe the system for a better understanding of the complex rock formations lying thousands of meters beneath the surface (see Fig. 8.2). The problem is that deep wells are

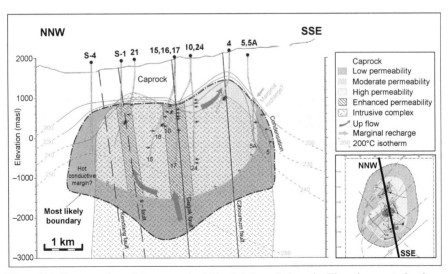

Figure 8.2 Cross section of the reservoir at Darajat, Java, Indonesia. There is a central volume with high permeability surrounded by zones of lower permeability; the highest permeability is associated with two main faults that act as the main geofluid conduits. After Ref. [2].

expensive, and one would like to know first if the system has sufficient potential to warrant the expenditure.

8.3   Getting the energy out of the reservoir

After the geoscientists have concluded their work and issued a report, usually called a "prefeasibility report," a decision is reached on whether to continue or discontinue the project. A positive finding leads to the drilling of wells. Unlike fossil fuels such as coal, oil, or natural gas, which can be gathered by mining or drilling and carted or piped away for use at distant power plants, geothermal energy must be used close to the source. Hot water and steam lose much of their stored energy when transported over long distances via pipelines. This is particularly important if the intended use is for electricity generation.

Many drilling programs begin with a series of relatively shallow, small-diameter wells aimed at determining the three-dimensional temperature pattern in the formation. These are called "temperature gradient holes." Once the primary zone of hot, upwelling fluid is inferred, a deep, large-diameter "discovery well" is drilled into the heart of the presumed productive zone. Since the future of the prospect is often riding on the success of the first deep well, the target should be the place with the highest probability of success.

A successful well will produce a steady, high-volume flow of high-temperature geofluid. The hotter the fluid, the greater potential it has for power generation. Tests are carried out on the produced fluid to determine various properties of the fluid and the formation. Since geofluid normally contains dissolved solids, some of which may be unsuitable to mix with ground waters, a means must be provided to contain the geofluids, especially the liquids, during well tests. This can be a stumbling block if there is only one well in the field. Lacking a well to take injection, large holding ponds are constructed to prevent contamination of the surroundings.

Given the inconvenience of coping with produced fluids, one may ask: Why does the fluid have to be withdrawn from the reservoir? In other words: Can the geothermal energy be used in situ? In fact, some have tried to do exactly that but without success. Once a well is drilled, some form of heat exchanger (picture a coiled tube) may be constructed and lowered down the well. Cold fluid is pumped into one end at the surface, and hotter fluid is recovered at the surface at the other end after the fluid has circulated through the piping in the well.

There is in fact nothing wrong with this idea as long as the objective is to provide space heating for buildings or other so-called direct heat uses. But the generation of electricity is much more demanding. Electricity-generating plants generate *power*. Power is energy per unit time. A power plant generally converts thermal energy to electrical energy, and the rate of conversion is critical. It is physically extremely difficult to transfer heat continuously and steadily from hot rocks in a well at a sufficiently high rate to yield meaningful amounts of electrical power. The basic difficulty lies in the mode of heat transfer: conduction versus convection. The latter is what happens in a geothermal well that extracts hot fluid from the formation, whereas the former is what happens in the case of a downhole heat exchanger. Unless the geofluid that presumably resides in the fractured rock can be induced to flow through the formation and the heat exchanger, the rate of energy transfer will be too slow to yield much power.

Inconvenient though it may be, one needs to plan on an array of wells, both producers and injectors, in order to extract large amounts of energy from the reservoir at a high enough rate to yield electrical power suitable for delivery to an electrical grid.

A few more questions need to be addressed about wells:

How deep can wells be drilled? Obviously, the wells must be designed to reach the most productive layers or fractures in the reservoir. This depth varies widely from field to field, and even in different sectors of the same field. It is routine nowadays to drill vertically as deep as 2000 m, and up to 4000−5000 m for very deep production zones using specialized drill rigs.

Can wells be drilled horizontally? This is routinely done in oil and gas fields, but is rarely attempted in geothermal fields. The typical sedimentary formations in oil and gas fields are much easier to drill than the hard, fractured and hot rocks in geothermal fields. Deviated wells are normally drilled in many geothermal fields, particularly those in rugged terrain, to minimize the number of drilling pads needing to be constructed. Wells can be drilled with a horizontal offset at depth of more than 1000 m, enabling several wells to be drilled from a single pad without interfering with each other (see Fig. 8.3). Deviated wells are best deployed at geothermal fields marked by steeply dipping faults that act as fluid conduits. If there are several fluid-carrying, nearly vertical faults, then a sharply deviated well approaching horizontality could be a strong producer by tapping several faults at once.

How can we improve the performance of a poor well? Suppose a well shows good thermal characteristics but the flow rate is less than desired. A technique called "stimulation" may be employed to enhance the permeability. The technique is similar to hydraulic fracturing or "hydrofracking" in the oil and gas industry. By injecting

Figure 8.3 Wellpad at The Geysers, California, showing four wellheads. Wells are drilled directionally, allowing compact wellpads without interference at depth.
Photograph courtesy of Calpine Corporation.

water into the well under pressure, small fractures may be enlarged or extended, forming a larger network of interconnected porosity, thereby allowing higher flow rates of fluid from the formation into the well.

Lastly, How many wells are needed to support a commercial-sized power plant? If we accept that 50 MW constitutes a commercial-size power plant, since the average power capacity of a typical geothermal well is roughly $5-10$ MW (many are less than 5 MW, many are higher than 10 MW, and a few even reach $40-50$ MW in the early stages of exploitation), then it will take about $5-10$ production wells. Assuming that reinjection must be used to help recharge the reservoir and avoid environmental problems, and recognizing that poorer wells are likely to be used for reinjection, it might require $7-12$ reinjection wells. Thus, the 50 MW plant would likely need from 12 to 22 wells in total to begin operating. In the course of a drilling campaign to create this many wells, it is inevitable that some will be failures. These might be suitable as monitoring wells.

8.4   Connecting the wells to the power station

A piping system gathers and delivers the geofluid from the wells, which may be scattered widely across the field, to a central point where the power station is constructed. The system includes many elements, such as valves of all sizes and types, straight pipes, elbows, bends, expansion loops, interconnecting pipes, pipe supports, steam traps, drains, separators, flash vessels, steam headers, liquid tanks, moisture removers, emergency holding ponds, and vent stations.

At the world's largest liquid-dominated geothermal power facility, the 570-MW Cerro Prieto plant in Mexico, the gathering system has 140 km of high-pressure and low-pressure steam pipes connecting 165 production wells to four separate power stations. In general, the site terrain plays an important role in the layout and design of the piping system since gravity increases the pressure in down-comers and causes pressure reductions in risers.

There are many different options for conveying the geofluid from wells to power plant (see Fig. 8.4). For example, many plants use two-phase, liquid−vapor pipelines from each well to a separator station located near the power house (see Fig. 8.5). This works very well when the wellheads are situated higher than the power house. The opposite alternative has separators at each well with steam lines running to the power house and liquid pipelines to reinjection wells. By using satellite separator stations in the well field, various combinations are possible. However, only one will be optimal for the given field.

The designer of the piping system has to balance thermodynamics and economics. For example, pressure losses must be kept low to preserve the power potential of the geofluid. By using large-diameter pipes, pressure losses are much lower than using smaller-diameter pipes. However, large pipes are more expensive than small ones. And for the same mass flow, the velocity of the fluid increases as the diameter is reduced, leading to more severe erosion potential, which in turn requires heavier-gauge, more-expensive piping. All of these factors and more weigh in on finding the optimal gathering system.

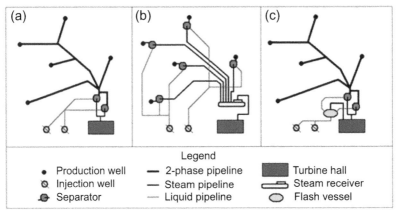

Figure 8.4 Three examples of geofluid gathering systems. (a) Two-phase, liquid–vapor pipelines connect each well to the power house where separators yield steam for the turbines and liquid for reinjection. Schematic shown would be for a single-flash plant. (b) Individual separators at each well yield steam, which is sent to the turbine hall via individual steam pipelines, and liquid, which is sent to reinjection wells. This is also for a single-flash plant. (c) Similar to Case (a) except the liquid from the separators is flashed to yield low-pressure steam for a dual-pressure turbine. This is a double-flash plant. Many other possibilities exist. After Ref. [3].

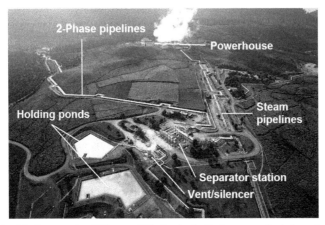

Figure 8.5 Aerial photograph of Wayang Windu single-flash power plant, Java, Indonesia. The two-phase pipelines carry geofluid from the several production wells from the area at the top left of the image and beyond; reinjection wells are located off-image at the bottom and the upper right [3].
Original photograph from Indonesian Internet Forum: "Apa itu Geotermal?" [4].

For some resources with challenging geofluids, elaborate facilities are necessary to process the geofluid and purify the steam so that it can be used in a turbine. For example, geofluids found at the Salton Sea geothermal area in southern California, USA, are very hot ($\sim 350°C$) and highly contaminated with dissolved solids ($\sim 250,000$ ppm), to such an extent that the fluids defied use for decades. Two techniques have evolved to cope with these highly corrosive fluids: (1) flash-crystallizer/reactor-clarifier technology and (2) pH modification. The chemical processing and treatment facilities actually dwarf the power generating equipment but have proved to be effective and reliable, allowing over 350 MW of power production from several plants in the Salton Sea area.

8.5   Central power station

As with gathering systems, many different design options are available for the actual energy conversion equipment. If the resource is one of the rare dry steam systems, the options reduce to a single choice: a steam turbine. Once the steam is purged of any impurities such as rock dust and any liquid that may have formed in the piping, the steam is simply admitted to a turbine that rotates and drives a generator, thereby producing electricity. This is very similar to what occurs in any other central power station except that the steam is natural and not made by boiling pure water with heat obtained from burning coal, oil or gas, or from the heat of radioactive decay of uranium or some other nuclide.

If the resource is liquid dominated, the most common type of geothermal resource, the power plant may be a flash-steam unit wherein the steam, after separation from the two-phase geofluid mixture, is deployed in a turbine, as in the dry steam case. Up to three stages of flash/separation are now being used in plants around the world; the efficiency increases with the number of flash/separation processes. Or a binary plant may be used wherein the hot geofluid is the heating medium to boil a secondary working fluid, typically an organic substance such as a refrigerant or hydrocarbon that has a low-boiling point temperature (see Fig. 8.6). In this way, the geofluid never comes in contact with the turbine but merely yields some of its thermal energy to provide the heat necessary to power the plant.

Binary plants are the best option for lower-temperature geofluids, say 150°C or less, whereas flash plants are the best option for high-temperature geofluids. Binary plants have the added advantage of being extremely benign as regards environmental impact, assuming reinjection is used to dispose of the cooled geofluid, as is essentially standard practice nowadays.

The exact characteristics of the geofluid and the local conditions will determine which of the many conversion system options is best for any given prospect. Once that choice is made, engineers will establish the specifications for the plant to optimize the electrical output subject to a suite of constraints, which include site conditions, environmental regulations, financial/economic aspects, labor costs, anticipated plant

Figure 8.6 Aerial photograph of Ngatamariki binary power plant, North Island, New Zealand. Air-cooled condensers (ACC) dominate the layout of this four-unit, 100MW (gross), 82 MW (net) steam-brine power station. The heat exchangers of two of the units can be seen next to the ACC; the other two are on the opposite side. The separator is seen just above the ACC next to the emergency holding pond.
Photograph courtesy of Ormat Technologies, Inc.

lifetime, etc. Once the plant is built, it must be maintained to perform at peak levels throughout its lifetime. Periodic shut-downs are planned to inspect various pieces of equipment and repair or replace worn or fouled items.

Maintenance of the wells is also vital to keeping the electrons and revenues flowing. Since it is unlikely that all the wells that initially feed the plant will last the entire lifetime of the electromechanical equipment, replacement wells will need to be drilled. Indeed, the gathering system will likely undergo changes over the 30-plus years that a plant will operate. This sets geothermal plants apart from conventional plants in which a continuous supply of fuel is available and where only the electromechanical equipment need be monitored and maintained.

Defining and monitoring appropriate measures of plant and well performance are essential for long-term success. This is relatively easy to do today with electronic monitoring and control systems that can record and display a myriad of parameters essentially continuously in real time. Furthermore, since geothermal plants are very reliable, it is normal to monitor several power units from a central control room that can be far from the units themselves, even in a different country.

Ultimately, one would hope to achieve sustainable operation wherein the resource potential is used at a rate compatible with the recharge to provide a profitable venture for the long term. The oldest field in existence, Larderello in Italy, has been generating electricity for 110 years. The first field developed in the United States, The Geysers, has been in operation for 55 years. Over the years many of the power units have been upgraded with newer, more-efficient equipment to achieve this remarkable longevity.

8.6 Geofluid disposal

With worldwide environmental awareness, nearly all countries have regulations regarding the disposal of potentially hazardous fluids and other materials. This was not always the case and some early geothermal power plants simply disposed of the waste geothermal liquid by dumping it into rivers and streams, and discharged noncondensable gases directly to the atmosphere. In many cases, those early plants were rather isolated from population centers so the impact on humans was not significant. If rivers were not handy, then holding ponds were created out of the natural landscape. If the ground was porous, the liquid percolated back into the soil. However, since waste liquids are usually supersaturated with dissolved substances such as silica, precipitation rapidly settles in the ponds, rendering the surface impermeable, and creating ever larger lagoons. The only natural process for removing geofluid from these ponds is slow evaporation. In this way, the famous Blue Lagoon was created next to the Svartsengi power station in southwestern Iceland, which is now a popular tourist bathing attraction (see Fig. 8.7).

With tight constraints on fluid disposal, it has become nearly universal practice to reinject waste geofluids and to treat gases before discharge if they can adversely impact human health. Given the potential difficulty in locating reinjection wells so that they will not interfere with producing wells, together with a requirement for full reinjection, the power that can be developed from a given field can be restrained by a lack of sufficient reinjection capability, regardless of the capability of production wells.

In light of this, reinjection wells may be placed into one of the two categories: (1) disposal wells or (2) recharge wells. In the first case, wells are sited on the periphery

Figure 8.7 Blue Lagoon at Svartsengi, Iceland. Silica-laden waste geofluid from the geothermal power plant (background) serves as a bathing attraction for local residents and tourists. Public domain photograph.

of the field, downstream of the main production zone. While they may contribute to the maintenance of pressure in the reservoir, most likely the fluid will move away from the reservoir and never be seen again. In the second case, wells are judiciously sited within the production zone and are intended to replace some of the fluid being withdrawn via the production wells. The latter approach is clearly a more risky venture owing to the potential for cooling of the production area and is usually only attempted after a solid understanding of the reservoir is acquired from some years of operating experience.

A mathematical model of the reservoir is essential to planning the development of a resource. Such a model allows computer simulations of various scenarios to determine the response of the reservoir to different modes of exploitation. The information needed to construct a useful model is accrued bit by bit over several years and continues throughout the life of the plant to guide decisions regarding replacement wells.

8.7   Conclusions and a look ahead

The next six chapters will delve into the details of all the energy conversion options available to the designers of geothermal power plants. First, the basic principles of thermodynamics, fluid mechanics, and heat transfer are presented and applied to the major components of the power plants. Then, flash-steam and dry-steam plants that may be used at medium- to high-temperature hydrothermal resources are described. One chapter is devoted to total-flow systems that have been under development for nearly 50 years and now seem to be on the verge of becoming commercial. Owing to their inherent simplicity, they hold promise providing they can demonstrate competitive energy conversion efficiency.

One chapter covers binary plants in depth and includes many design variations that make binary plants flexible to mate with the wide variety of geothermal resources encountered. The final chapter in this part of the book explores hybrid power systems wherein two or more different sources of energy or types of power systems are combined in a single plant in a synergistic manner to achieve efficiencies greater than could be achieved using the individual entities separately in their own power plants. Some of the systems combine geothermal plants of the flash-steam and binary types using solely geothermal energy as input, while others combine geothermal with solar photovoltaic, concentrating solar power, biomass, and even fossil fuels.

From Part Two of this volume, the reader will gain an understanding of the wide variety of designs available to convert geothermal energy into electricity. Those with scientific or technical backgrounds will also add a working knowledge of the intricacies of these plants and be able to select those designs that match up well with geothermal resources over the wide spectrum that are found around the world. Finally, they should be able to put together preliminary systems and assess their theoretical performance to compare competing designs against each other.

References

[1] DiPippo R, Moya P. Las Pailas geothermal binary power plant, Rincón de la Vieja, Costa Rica: performance assessment of plant and alternatives. Geothermics 2013;48:1−15.
[2] Rejeki S, Rohrs D, et al. Geologic conceptual model update of the Darajat geothermal field, Indonesia. Proceedings of the World geothermal congress 2010, Bali, Indonesia, 25−29 April 2010.
[3] DiPippo R. Geothermal power plants: principles, applications, case studies and environmental impact. 4th ed. Oxford, England: Butterworth-Heinemann: Elsevier; 2016.
[4] Indonesian Internet Forum: "Apa itu Geotermal?" ["What is geothermal?"], http://www.kaskus.co.id/thread/5315f6a7bdcb17c90b8b460a/apa-itu-geotermal/.

Elements of thermodynamics, fluid mechanics, and heat transfer applied to geothermal energy conversion systems*

9

R. DiPippo
University of Massachusetts Dartmouth (Emeritus), N. Dartmouth, MA, United States

9.1 Introduction

This chapter provides the essential elements of thermodynamics, fluid mechanics, and heat transfer to set the stage for the concepts presented in Chapters 10—14: "Flash steam geothermal energy conversion systems: single-, double-, and triple-flash and combined-cycle plants," "Direct steam geothermal energy conversion systems: dry steam and superheated steam plants," "Total flow and other systems involving two-phase expansion," "Binary geothermal energy conversion systems: basic Rankine, dual—pressure, and dual—fluid cycles," and "Combined and Hybrid Geothermal Power Systems." Readers with a scientific or technical background will be able to use these working equations to assess the performance of existing geothermal power plants and to create preliminary designs for new plants. The complete specification of any geothermal plant involves a multidisciplinary team of professionals, not solely energy conversion specialists. Nor can this brief chapter replace a thorough examination of these complex topics, which requires dedicated study over several semesters in an engineering curriculum plus extensive work experience.

9.2 Definitions and terminology

The major elements of a geothermal power plant are the wells, piping system, geofluid processing equipment (eg, separators and flashers), turbines, condensers, heat exchangers, cooling towers, and pumps. Each of these may constitute a *system* from the point of view of the application of the laws of physics embodied in thermodynamics, fluid mechanics, or heat transfer. A *system* is a well-defined region of space containing an element or elements of interest.

---

* Parts of this chapter are drawn from "Geothermal Power Plants: Principles, Applications, Case Studies and Environmental Impact, Fourth Edition, 2016, Appendix D: Elements of Thermodynamics" by the author, used with permission of the publisher Elsevier.

Systems may be classified as either *closed* or *open*. A closed system is one for which no *mass* crosses the system boundary, whereas mass does cross the boundary of an open system. A special case arises if a closed system has no energy interactions with its surroundings; such a system is called *isolated*.

Any system that is *perfectly thermally insulated* from its surroundings is called *adiabatic*. Well-insulated steam or hot water pipes are often modeled as adiabatic, that is, having no heat loss.

A system is described by specifying values for its *properties*, physical characteristics that can be measured or calculated. Properties can be: *extensive*—dependent on the size or extent of the system, such as mass, volume, energy; *intensive*—independent of the size or extent of the system, such as temperature, pressure; or *specific*—an extensive property per unit mass, such as specific volume. When the array of properties has a fixed set of values, the system is said to exist in a certain *state*.

When a system operates such that the properties within the boundaries do not change with time, the operation is called *steady*. The properties may vary from point to point inside the system but at any point in space the properties are invariant. When the system properties vary with time, the operation is called *unsteady* or *transient*. Geothermal plants undergo transient behavior during startup and shutdown but normally can be approximated as being steady for normal operations.

Properties are found from and related to each other through *equations of state (EOS)*. Many analytic forms of such equations exist and form the basis for correlations that are incorporated into electronic representations useful for spreadsheet calculations. Some of the more useful EOS are the van der Waals, Beattie-Bridgeman, or Benedict-Webb-Rubin equations. Software packages are available that give accurate properties for a large number of substances and their mixtures. The National Institute of Standards and Technology (NIST) offers one called the *NIST Standard Reference Database 23*, commonly known as *REFPROP* [1]. A free limited version is available for students.

When a system operates from some initial state to a final state, at least one, but usually more, of its properties undergoes changes. This is called a *process*. If the process is ideal, the system remains in equilibrium during the process. Such a process is called *reversible*. Since this is physically impossible because there needs to be a disturbance of the equilibrium to cause a change of state, reversible processes are not realizable in practice, but serve as the ideal limit for real processes, which are called *irreversible*.

This terminology is important when applying the laws of thermodynamics because it matters greatly if the system under study is open or closed, adiabatic or not, operating steadily or not, and if the process may be idealized as reversible or not.

9.3   First law of thermodynamics for closed systems

The most general form of the first law for a closed system is

$$Q_{1,2} - W_{1,2} = E_2 - E_1 \qquad\qquad\qquad [9.1]$$

where $Q_{1,2}$ is the heat transfer, $W_{1,2}$ is the work transfer, and $E_2 - E_1$ is the change in the energy of system during the process. This form applies to all closed thermodynamic systems, regardless of the type of process. Note that heat transfer is defined as positive when heat enters the system and work is defined as positive when it is delivered from the system to its surroundings.

If the process is assumed to be ideally *reversible*, then the following differential form may be used:

$$\delta Q - \delta W = dE \qquad [9.2]$$

where the terms are defined as follows: δQ, heat exchanged during an infinitesimal step in the process; mathematically, this is an *imperfect differential* that integrates to $Q_{1,2}$ from the following equation:

$$Q_{1,2} = \int_1^2 \delta Q \qquad [9.3]$$

δW = work exchanged during an infinitesimal step in the process; another *imperfect differential* that integrates to $W_{1,2}$ using

$$W_{1,2} = \int_1^2 \delta W \qquad [9.4]$$

dE, change in total system energy during an infinitesimal step in the process; this is a *perfect differential* that integrates to $E_2 - E_1$ using

$$E_2 - E_1 = \int_1^2 dE \qquad [9.5]$$

It is important to note the difference between the heat and work terms and the energy terms. Energy, being a perfect differential, is a property of the system, whereas heat and work are imperfect differentials and are not properties of the system. Heat and work are manifestations of energy transfer between the system and its surroundings. They have physical meaning only when a heat transfer or work transfer process is under way, not before or after the process. Thus, it is as meaningless to ascribe a certain amount of heat to a system as it is to ascribe a certain amount of work to a system. Both arise only from processes; neither heat nor work *per se* can be stored in a system. Energy is what is stored in a system and its transfer is called "heat" when the energy transfer is driven by a temperature difference and "work" when it is caused by a pressure difference or an unbalanced force.

This has implications for assessments of the potential of geothermal resources where often the term "heat in place" is seen. Strictly speaking from a thermodynamic point of view, this is meaningless; energy is in place as manifested by the high temperature of the rock formation and the fluid that may be contained therein. A process involving a temperature difference is needed before the thermal energy can be recovered, that is, a heat transfer process.

For normal geothermal power plant operations that involve the flow of fluids into and out of various components of the facility, it is rare to encounter a truly closed system. The closest one might be considered to be the complete loop of the working fluid in a binary power plant (see Chapter 13: Binary Geothermal Energy Conversion Systems: Basic Rankine, Dual Pressure and Dual-Fluid Cycles), but even there the loop is not perfectly closed as there will always be small intentional discharges or unintentional, unwanted leaks from the system. Nearly always the components that compose geothermal plants are open systems.

9.4 First law of thermodynamics for open steady systems

The first law for open systems undergoing a *steady process* is

$$\dot{Q} - \dot{W} = -\sum_{i=1}^{n} \dot{m}_i \left(h_i + 0.5\, \mathcal{V}_i^2 + g z_i \right) \tag{9.6}$$

Each term is defined as follows: $\dot{Q}$, rate of heat transfer (thermal power) between the system and its surroundings ($+$ when heat enters the system, $-$ when it leaves); $\dot{W}$, rate of work transfer (mechanical or electrical power) between the system and the surroundings ($+$ when work is delivered to the surroundings from the system, $-$ when it enters the system); i, index that accounts for all inlets and outlets of the system; n, total number of inlets and outlets; $\dot{m}_i$, mass flow rate crossing each inlet or outlet ($+$ for inlets, $-$ for outlets); h_i, specific enthalpy of the fluid at each inlet or outlet; $\mathcal{V}_i$, velocity of the fluid at each inlet or outlet; z_i, elevation of each inlet or outlet relative to an arbitrary datum; and g, local gravitational acceleration, 9.81 m/s^2.

Sometimes Eq. [9.6] is written as

$$\dot{Q} - \dot{W} = \sum_{j=1}^{p} \dot{m}_j \left(h_j + 0.5\, \mathcal{V}_j^2 + g z_j \right) - \sum_{k=1}^{q} \dot{m}_k \left(h_k + 0.5\, \mathcal{V}_k^2 + g z_k \right) \tag{9.7}$$

where the first sum is taken over all the outlets, $j = 1, p$ and the second sum is taken over all the inlets, $k = 1, q$. In this form, the mass flow rate is always a positive number.

The *principle of conservation of mass* in steady state requires that

$$\sum_{i=1}^{n} \dot{m}_i = 0 \tag{9.8}$$

$$\text{or} \sum_{j=1}^{p} \dot{m}_j = \sum_{k=1}^{q} \dot{m}_k \qquad [9.9]$$

depending on whether one uses Eq. [9.6] or Eq. [9.7] for the first law.

9.5 First law of thermodynamics for open unsteady systems

The first law for open systems undergoing an *unsteady process* is mathematically more complex than for steady operation. Rather than an algebraic working equation, a differential equation must first be integrated before numerical values can be substituted. Since the power plants discussed in this volume will generally be assumed to be operating steadily, this form of the first law is not needed for our current purposes but is included here for the sake of completeness. The most general form is

$$\delta Q - \delta W - P dV = d(mu) - \sum_{i=1}^{n} dm_{\xi i}\left(h_\xi + 0.5\,\mathscr{V}_\xi^2 + gz_\xi\right)_i. \qquad [9.10]$$

The new terms in the equation are defined as follows: PdV, expansion or compression work exchanged during an infinitesimal step in the process via a reversible deformation of system boundary, negligible for rigid systems; $d(mu)$, change in system internal energy during an infinitesimal step in the process; mathematically this is a *perfect differential* that integrates to $m_2u_2 - m_1u_1$; dm, increment of mass entering (positive) or leaving (negative) the system during an infinitesimal step in the process; also a *perfect differential*; and ξ, subscript denoting properties of streams entering or leaving the system.

For the unsteady case, the *principle of conservation of mass* requires

$$\sum_{i=1}^{n} \dot{m}_i = \frac{dm}{dt} \qquad [9.11]$$

That is, the net influx or outflow of mass across the boundary equals the time rate of change of mass within the system.

9.6 Second law of thermodynamics for closed systems

The second law appears in many forms, some verbal and some mathematical. Often it is expressed as a negative statement or a mathematical inequality because the nature of the second law is to place restrictions on what can be accomplished during energy conversion processes. Here we summarize some of the common expressions of the second law.

9.6.1 Clausius statement

It is impossible for heat to flow *spontaneously* from a body of lower temperature to one of higher temperature. If this were not true, it would be possible for refrigerators to operate without any motive energy being supplied.

9.6.2 Kelvin–Planck statement

It is impossible to operate a cycle such that the only effects are the transfer of heat from a *single source* and the delivery of an equal amount of work from the cycle to the surroundings. If this were not true, it would be possible to operate power plants without rejecting any heat to the surroundings, and the plants would have a thermal conversion efficiency of 100%.

9.6.3 Clausius inequality

Clausius also gave us a mathematical expression of the second law, namely:

$$\oint \frac{\delta Q}{T} \leq 0 \qquad\qquad [9.12]$$

which represents a *closed line integral* or a sum taken completely around the cycle of operations. The equality sign is used when all the heat transfer takes place *reversibly*, that is, when there is no temperature difference across the heat transfer surface, whereas the inequality applies to *irreversible* heat transfer across a finite temperature difference.

The Clausius inequality may also be expressed in the following manner:
It is impossible for any system to operate on a *cycle* such that

$$\oint \frac{\delta Q}{T} > 0 \qquad\qquad [9.13]$$

9.6.4 Existence of entropy

The second law implies the existence of a property called entropy, S, which is found from the equation:

$$dS \equiv \frac{\delta Q^{\text{rev}}}{T}. \qquad\qquad [9.14]$$

The term dS is a mathematical *perfect differential*, and Eq. [9.14] may be integrated to obtain:

$$S_2 - S_1 = \int_1^2 \frac{\delta Q^{\text{rev}}}{T}. \qquad\qquad [9.15]$$

This is significant in light of the fact that δQ^{rev} by itself is an *imperfect differential*. The quantity $1/T$ is an *integrating factor* for heat, creating a quantity that is a *perfect differential*. To carry out the integration of Eq. [9.15], it is necessary to substitute for the reversible heat transfer using Eq. [9.2].

Since entropy is a *thermodynamic property* and a *mathematical potential*, we are only interested in *differences* in entropy, not the actual value of entropy. Furthermore, once the entropy difference has been found between a given pair of states by using Eq. [9.15], the result will be *the same no matter what type of process connects the two states, even if the process is irreversible*. This latter point is often not understood by beginners to the subject who mistakenly believe that the entropy change between two fixed states will differ if one process is irreversible and one is ideally reversible. Only the initial and final states determine the entropy change, not the process between them. Nevertheless, the *calculation* of the entropy change must be done using a theoretical (and usually convenient) *reversible process*, as shown explicitly in Eq. [9.15].

9.6.5 Principle of entropy increase (PEI)

This statement of the second law is intrinsically tied to the concept of an irreversible process. The PEI may be stated as follows:

When an *adiabatic* system undergoes an *irreversible* process from an initial state 1 to a final state 2, the entropy of the system can only *increase*, that is,

$$S_2 - S_1 > 0 \qquad\qquad\qquad\qquad [9.16]$$

Since all real processes are irreversible and since the selection of the system boundary is arbitrary, some people have extended the boundary to include the entire universe and have concluded that the entropy of the universe is constantly increasing. This implies that the universe is bounded by an adiabatic wall, which has yet to be verified.

The PEI may be put into an alternative, negative form in keeping with the general format of the second law, namely:

When an *adiabatic* system undergoes an *irreversible* process from an initial state 1 to a final state 2, it is impossible for the entropy of the system to *decrease*.

If an *adiabatic* system were to undergo an ideally *reversible* process, then the entropy of the system would remain *constant*, that is, $S_2 = S_1$.

9.7 Second law of thermodynamics for open systems

The most general form of the second law that applies to open systems is

$$\dot{\theta}_{\text{p}} = \frac{dS}{d\tau} - \sum_{i=1}^{n} \dot{m}_i s_i - \int_{\tau_1}^{\tau_2} \frac{1}{T} \frac{dQ}{d\tau}. \qquad\qquad [9.17]$$

Each new term is defined as follows: $\dot{\theta}_p$, rate of entropy production for the system caused by irreversibilities; τ, time; s_i, specific entropy of the fluid at each inlet or outlet, S_i/m_i; T, absolute temperature (in K or °R) associated with the heat transfer Q.

The term on the left-hand side vanishes for reversible processes, while the first term on the right-hand side vanishes for steady operation, and the last term on the right vanishes for adiabatic systems.

9.8 Exergy and exergy destruction

The term "exergy" was coined by Z. Rant in 1956 [2], but J.W. Gibbs developed the concept of technical available work in 1873 [3]. Rant effectively created a simple one-word descriptor for Gibbs' concept.

The importance of exergy is that it is not a conserved quantity. In any actual process, energy is conserved according to the first law, but some exergy is always destroyed owing to irreversibilities caused, for example, by friction, heat transfer across a finite temperature difference, heat loss, or mixing. The value of the exergy, or the maximum thermodynamically possible work output that can be obtained from the given system, is greater in proportion to the disequilibrium existing between the system and its surroundings. The exergy ascribed to a particular state of a system depends on the state of the surroundings of the system. When a system is caused to change its state such that it eventually comes into equilibrium with its surroundings, there is no further potential for work output and the system then has zero exergy. Furthermore, if the process followed by the system were completely ideal, that is, reversible, with no losses whatsoever, then the amount of work obtained would be the maximum possible value and that is what is called the *exergy* of the initial state.

Here we will consider only thermal and mechanical equilibrium, leaving aside chemical equilibrium since such effects are usually not important in geothermal power applications. Using Eq. [9.17] for the case of steady flow and the ideal case of a reversible process, and combining the resulting equation with the first law for open, steady systems, it is easy to show that for the case of one inlet and one outlet the following equation represents the maximum thermodynamic work:

$$\dot{W}_{max} = \dot{m}\{h(P_1, T_1) - h(P_0, T_0) - T_0[s(P_1, T_1) - s(P_0, T_0)]\} \qquad [9.18]$$

The properties inside the simple parentheses are the independent variables used to find the enthalpy h and entropy s. The subscript 1 refers to the initial state of the system and the subscript 0 refers to the surroundings, also called the "dead state" since the system may be considered thermodynamically "dead" when it is in equilibrium with the surroundings. Using Rant's terminology, Eq. [9.18] may be rewritten as the exergy:

$$\dot{E} = \dot{m}\{h_1 - h_0 - T_0[s_1 - s_0]\} \qquad [9.19]$$

where a shorthand notation has been adopted. This equation gives the "exergetic power" whereas the following equation gives the specific exergy, that is, the exergy per unit mass:

$$e = h_1 - h_0 - T_0(s_1 - s_0) \tag{9.20}$$

Unlike energy, exergy is not a true system property. The value of a property depends only on other properties of the *system*; the value of the exergy depends not only on other system properties but also on the properties of the surroundings. Thus, there are no "exergy tables" because one would need to define a particular surrounding as a "standard surrounding" to make the numbers meaningful, but they would only apply for that particular surrounding. Geothermal plants operate in a variety of surroundings, say, from the hot and humid tropical climate in Costa Rica to the frigid, snow-bound region on Russia's far-eastern Kamchatka peninsula. Theoretically, if geofluids with the same thermodynamic properties happen to be found in these two areas, the latter would have significantly greater exergy.

When heat transfer occurs between two systems, there is exergy associated with the heat. The equation for it follows from the Carnot efficiency for the maximum thermal efficiency of an ideal cycle operating between two systems at different temperatures:

$$\eta_{\max} = 1 - \frac{T_L}{T_H} \tag{9.21}$$

In this equation, T_L is the temperature of cooler system and T_H is the temperature of the hotter system, both in kelvins, or absolute degrees. Since the maximum work that can be obtained from a cycle receiving a quantity of heat Q is the product of that heat and the Carnot efficiency, with the important condition that the cooler system be the dead state, we arrive at the expression for the exergy of the heat:

$$E_Q = Q\left[1 - \frac{T_0}{T_H}\right] \tag{9.22}$$

When work transfer takes place, the exergy associated with the work is the value of the work itself for obvious reasons.

Consider any open system operating steadily, with any number of inlets and outlets where mass enters and leaves the system, while heat and work transfers are taking place: an exergy analysis may be performed to assess losses. The method simply calculates all the incoming exergy due to heat, work, or mass transfers and a similar calculation for all the outgoing terms. Since the first sum will always exceed the latter due to irreversibilities, the difference will be loss in exergy. If the system receives one thermal input via heat transfer and produces one mechanical output via a work transfer, then

$$\dot{E}_{\text{loss}} \equiv \Delta\dot{E} = \sum \dot{E}_{\text{in}} - \sum \dot{E}_{\text{out}} = \dot{E}_Q + \sum_{i=1}^{n} \dot{m}_i e_i - \dot{E}_W - \sum_{j=1}^{k} \dot{m}_j e_j \tag{9.23}$$

One may define the exergetic efficiency for any system is a straightforward way as simply the ratio of all the output exergies to all the input exergies. This is sometimes called a "brute-force" definition. Clearly this ratio must be less than 1. Alternatively, using definitions based on the intended purpose of the system is termed a "functional" method. An example of the latter in the case of a steam turbine would be the ratio of the electricity generated (ie, the exergy of the output) to the change in exergy of the steam as it passes through the turbine. In other words, the functional efficiency shows the fraction of the exergy given up by the steam that was turned into electricity.

9.9   Thermodynamic state diagrams

It is helpful to view processes in thermodynamic state diagrams. These two-dimensional diagrams provide a "road map" for visualizing, understanding, and analyzing the changes that occur when systems undergo various changes of state. Processes may be characterized by whatever property remains constant during the process. Thus, a process at constant temperature is called "isothermal," one at constant pressure is called "isobaric," etc.

The framework for the diagram is an appropriate projection of the working fluid's three-dimensional surface of equilibrium states, assuming it is a simple pure substance. In the case of geothermal power plants, the most useful choices for the independent properties, that is, the coordinates of the two independent axes of the diagram, are (1) temperature–entropy, T–s, and (2) pressure–enthalpy, P–h. The former is handy for flash-steam plants, whereas the latter is particularly helpful for binary plants.

Fig. 9.1 shows a skeleton schematic T–s diagram for water over the spectrum of possible phases, while Fig. 9.2 gives the liquid–vapor and superheated portions of the phase space to scale.

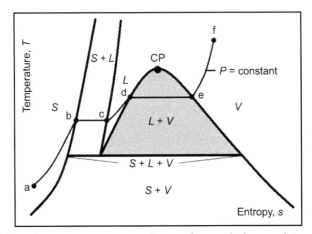

Figure 9.1 Schematic temperature–entropy diagram for a typical pure substance, like water. Key: *S*, solid; *L*, liquid; *V*, vapor; *CP*, critical point; *a-b-c-d-e-f*, typical subcritical isobar, *shaded area* represents two-phase, liquid–vapor boiling states.

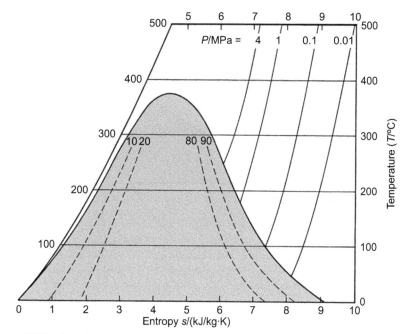

Figure 9.2 Portion of temperature−entropy diagram for water, showing the liquid−vapor and superheated steam regions, to scale. *Dashed lines* are constant quality (%).

Fig. 9.3 shows a schematic skeleton log(pressure)−enthalpy (log P−h) diagram for all possible phases; Fig. 9.4 gives the liquid, liquid−vapor, and superheated vapor regions for propane to scale.

9.10   Bernoulli equation

The energy equation, based on the first law of thermodynamics, is often called the Bernoulli equation when used in fluid mechanics:

$$P_j + 0.5\rho \, \mathcal{V}_j^2 + \rho g z_j = P_k + 0.5\rho \, \mathcal{V}_k^2 + \rho g z_k \qquad [9.24]$$

where the equation is applied along a section of pipe between points j and k. This follows easily from Eq. [9.7] if the pipe is assumed adiabatic and does not include any work-producing or -consuming devices (turbines or pumps). The enthalpy h has been replaced using its definition

$$h \equiv u + Pv = u + P/\rho \qquad [9.25]$$

together with the assumptions that the fluid has constant density ρ and friction is negligible. Thus, the internal energy u remains constant and Eq. [9.24] results.

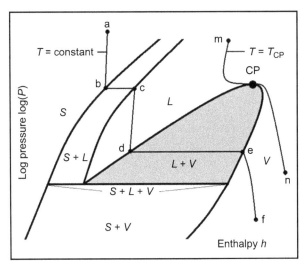

Figure 9.3 Schematic log (pressure)−enthalpy diagram for a typical pure substance. Key: *S*, solid; *L*, liquid; *V*, vapor; *CP*, critical point; *a-b-c-d-e-f*, typical subcritical isotherm; *m-n*, critical isotherm.

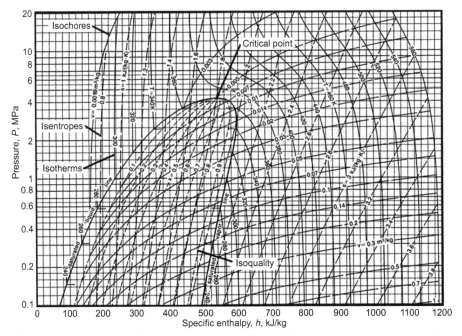

Figure 9.4 Portion of semilog pressure−enthalpy diagram for propane, showing to scale the liquid, liquid−vapor, superheated and supercritical regions. Propane is a possible working fluid for binary cycles.

A special case involves a horizontal pipe where there is no change in the elevation z; then a simplified Bernoulli equation takes the form

$$P_j + 0.5\rho \, \mathscr{V}_j^2 = P_k + 0.5\rho \, \mathscr{V}_k^2 \qquad [9.26]$$

or

$$P_j - P_k = 0.5\rho\left(\mathscr{V}_k^2 - \mathscr{V}_j^2 \right) \qquad [9.27]$$

This indicates that a fluid flowing through a pipe subject to the stated assumptions, will travel faster where the pressure is lower, and vice versa.

A convenient form of Eq. [9.24] expresses the terms in units of length. This is achieved by dividing the equation by ρg. The sum of the three terms is called the total head $\mathscr{H}$, which remains constant:

$$\mathscr{H} \equiv z + \frac{P}{\rho g} + 0.5\frac{\mathscr{V}^2}{g} = \text{constant} \qquad [9.28]$$

9.11   Pressure loss calculations

A major item of interest is the pressure change along a pipe. The pressure of geothermal steam or a working fluid should be maintained as high as feasible to preserve its ability to drive a turbine. Further, the turbine has a certain design inlet pressure that must be met by the motive fluid. For a geothermal steam plant, the change in pressure from the wellhead to the turbine must yield that value of pressure to assure that the turbine will operate at its design point. This creates a situation that may be a challenge for the gathering system designer.

Considering that there are three possible flow situations in a geothermal pipe—liquid only, steam only, and two-phase liquid–steam flows—and each will have a distinctive pressure loss, it is necessary to know which type of flow is present at all points along the pipe and what equations are needed to accurately predict the change in pressure. In general, the pressure change consists of three terms—accelerational, frictional, and hydrostatic:

$$\Delta P_{\text{Total}} = \Delta P_{\text{Accel}} + \Delta P_{\text{Frict}} + \Delta P_{\text{Hydrostatic}} \qquad [9.29]$$

The first term comes into play only when the fluid changes phase, as when a liquid flashes into steam, and the last term is necessary only when the pipe changes its elevation while carrying liquid. The density of steam or other vapors is very small, meaning that any pressure change due to elevation change is negligible. Thus, the frictional pressure loss constitutes the main effect in most cases, but the other terms need to be accounted for when appropriate.

The pressure loss due to friction between the fluid and the interior pipe wall may be estimated from the following working equations.

9.11.1 Liquid-only flows

$$\Delta P_{\text{Frict, Liq}} = f \left[\frac{L}{D} \right] \left[\frac{\rho \mathscr{V}^2}{2} \right]$$ [9.30]

where the friction factor f (dimensionless) is found from

$$f = \left\{ 2 \log_{10} \left[\frac{\varepsilon}{3.7D} + \frac{5.74}{\text{Re}^{0.9}} \right] \right\}^{-2}$$ [9.31]

The Reynolds number Re (dimensionless) is defined as

$$\text{Re} \equiv \frac{4}{\pi} \frac{\dot{m}}{\mu D}$$ [9.32]

The ε in Eq. [9.31] is the absolute roughness of the inside surface of the pipe. In S.I. units, the pressure loss from Eq. [9.30] will be in units of pascals, Pa.

When the piping system contains fittings such as elbows, valves, etc., one may use an equivalent length of straight pipe to account for the pressure loss in the fitting, provided that the fluid is single phase and incompressible, as is the case here. For example, a flanged 8-in diameter regular 90° elbow is equivalent to 3.7 m of straight 8-in diameter pipe, whereas an 8-in diameter regular 45° bend is equivalent to 2.3 m of straight 8-in diameter pipe. See Ref. [4] for detailed tables for fittings of various sizes.

9.11.2 Steam-only flows

$$\Delta P_{\text{Frict, Steam}} = 1.8472 \times 10^{-5} \frac{L \dot{m}^{1.85}}{\rho D^{4.97}}$$ [9.33]

The units of the terms in Eqs. [9.30]–[9.32] may be any compatible units. However, Eq. [9.33] is not dimensionally homogeneous, and the equation is valid only when the following units are used: $\{L\} = \{D\} = $ m, $\{\dot{m}\} = $ kg/s, $\{\rho\} = $ kg/m^3. Then the pressure loss will have units of kilopascals, kPa.

Whereas the frictional pressure loss is well known for liquid-only and steam-only flows, the situation is not as straightforward when both phases are present. See Refs. [5,6] for a thorough discourse on the subject. The reason is that there are many possible spatial configurations of the two phases within the pipe, and this strongly influences the pressure loss. Because this situation occurs in boilers of all kinds, much research has been carried out on this subject. As approximations, the following sets of equations may be used for two-phase, liquid–vapor flows of water and steam. One set applies to low-to-medium quality, x (steam mass fraction at any pipe section) and one applies to high-quality flows.

9.11.3 Steam–liquid water flow, low-to-medium steam quality

$$\Delta P = \frac{8}{\pi^2} \frac{fL_{eq}}{D^5} \dot{m}^2 x^2 v_S \Phi^2 \qquad \qquad [9.34]$$

The friction factor f may be found from Eq. [9.31] using the Reynolds number based on the mass flow rate and viscosity of the steam phase. The factor Φ is found from

$$\Phi = 1.676 X^{0.133} \qquad \qquad [9.35]$$

where X (dimensionless) is given by

$$X = \left\{ \left[\frac{1-x}{x} \right]^{1.8} \left[\frac{v_L}{v_S} \right] \left[\frac{\mu_L}{\mu_S} \right]^{0.2} \right\}^{0.5} \qquad \qquad [9.36]$$

9.11.4 Steam–liquid water flow, high steam quality

The above Eqs. [9.34] and [9.36] may be used for this case but with a different equation for the factor Φ:

$$\Phi = \exp\left(1.5195 X^{0.38}\right) \qquad \qquad [9.37]$$

Since the steam quality, x, is subject to change along the pipe, Eqs. [9.34]–[9.37] should be applied successively over small lengths of pipe to track changes in the flow properties from the starting point to the end point. However, even using step-by-step calculations over small increments of length will only produce estimates of the two-phase pressure drop, subject to an uncertainty of 30–40% at best. For this reason, field tests are sometimes carried out to measure the real two-phase pressure losses using the actual geofluids prior to designing the entire gathering system.

9.12 Principles of heat transfer applied to geothermal power plants

Heat transfer is the primary purpose of some components of geothermal plants, while for others it represents a loss. Among the former are preheaters, evaporators, recuperators, and superheaters in binary power plants, plus condensers and cooling towers in all types of geothermal plant. Among the latter are steam pipelines, pumps, and turbines. Through the use of insulation, heat losses can be kept to very small values. The main heat transfer components will be the focus of this section.

Eqs. [9.6]−[9.9] and [9.23] will be applied to design and assess the performance of heat exchangers. Heat exchangers may be classified as follows:

- Shell-and-tube (no mixing)
  - Preheaters
  - Evaporators
  - Superheaters
  - Recuperators
  - Condensers
- Plate-and-frame (no mixing)
  - Preheaters
  - Condensers
- Direct contact (fluids mix)
  - Low-level condensers
  - Barometric condensers
  - Water cooling towers

The analysis of *shell-and-tube heat exchangers* is basically the same regardless of the application and will be used here to illustrate how the equations are applied. Fig. 9.5 depicts some basic designs in simple schematics; designs A and B are called

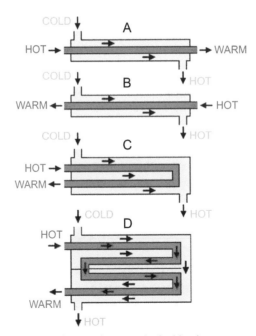

Figure 9.5 Basic shell-and-tube heat exchangers: A, double-pipe, cocurrent flow, one tube pass, one shell pass; B, double-pipe, countercurrent flow, one tube pass, one shell pass; C, shell-and-tube, two tube pass, one shell pass; D, shell-and-tube, four tube pass, two shell pass.

double-pipe and are the simplest form of a shell-and-tube design. It is assumed that the objective of each design is to heat the cold fluid using the hot fluid. In geothermal binary plants, the incoming hot fluid is usually hot brine, b, obtained by pumping from a well and the cold fluid is the cycle working fluid, wf, usually an organic compound.

First law analysis: The basic principle of energy conservation states that the heat delivered by the brine as it cools equals the heat absorbed by the working fluid plus any heat loss through the walls of the heat exchanger. To a first approximation, the heat loss may be neglected when the system is properly insulated. Thus,

$$|\dot{Q}_b| = \dot{m}_b(h_{b1} - h_{b2}) = |\dot{Q}_{wf}| = \dot{m}_{wf}(h_{wf2} - h_{wf1}) \qquad [9.38]$$

Subscripts 1 refer to inlet states and subscripts 2 refer to outlet states, making each term in Eq. [9.38] positive. Eq. [9.38] can be applied to any of the designs shown in Fig. 9.5. Normally in a geothermal binary plant there are two heat exchangers in series, a preheater, PH, and an evaporator, EV. It is convenient to view the heat transfer process in a temperature–heat transfer diagram. Fig. 9.6 is such a diagram for a PH-EV set of heat exchangers.

Second law analysis: Heat exchangers may be assessed using the exergy concept. The exergy given up by the hot fluid equals the exergy picked up by the cold fluid plus the exergy destroyed in the process; ie,

$$\Delta\dot{E}_{HOT} = \Delta\dot{E}_{COLD} + \dot{E}_{LOST} \qquad [9.39]$$

where

$$\Delta\dot{E}_{HOT} = \dot{m}_b[h_{b1} - h_{b2} - T_0(s_{b1} - s_{b2})] \qquad [9.40]$$

$$\Delta\dot{E}_{COLD} = \dot{m}_{wf}[h_{wf2} - h_{wf1} - T_0(s_{wf2} - s_{wf1})] \qquad [9.41]$$

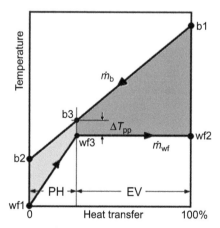

Figure 9.6 Temperature-heat transfer diagram for preheater–evaporator combination. The point of closest approach between the brine cooling line and the working fluid heating line is called the pinch-point, *pp*; typically this temperature difference is about 5°C.

The second law or utilization efficiency may be calculated using either the brute-force, BF, or the functional, FUN, method.

Brute-force utilization efficiency:

$$\eta_{U,\ BF} = \frac{\dot{m}_{wf2}e_{wf2} + \dot{m}_{b2}e_{b2}}{\dot{m}_{wf1}e_{wf1} + \dot{m}_{b1}e_{b1}} \tag{9.42}$$

Functional utilization efficiency:

$$\eta_{U,\ FUN} = \frac{\dot{m}_{wf}(e_{wf2} - e_{wf1})}{\dot{m}_{b}(e_{b1} - e_{b2})} \tag{9.43}$$

Eq. [9.42] is straightforward; Eq. [9.43] gives the fraction of the exergy given up by the incoming brine that is transferred to the working fluid.

Sizing heat exchangers: Once the required amount of heat transfer is determined from thermodynamic system analysis, the heat exchangers must be suitably sized to deliver the correct amount of thermal energy. This means providing the right amount of surface area through which the heat transfer takes place. The basic equation is

$$\dot{Q} = U \times A \times \text{LMTD} \tag{9.44}$$

where U is the overall heat transfer coefficient, A is the total heat transfer area, and LMTD is the log-mean temperature difference that is defined as follows:

$$\text{LMTD} \equiv \frac{\Delta T_{\text{HOT end}} - \Delta T_{\text{COLD end}}}{\ln[\Delta T_{\text{HOT end}}/\Delta T_{\text{COLD end}}]} \tag{9.45}$$

Eq. [9.45] is strictly applicable to a very simple double-pipe heat exchanger. For example, for the simple case B in Fig. 9.5, the LMTD may be found from

$$\text{LMTD} = \frac{(T_{b1} - T_{wf2}) - (T_{b2} - T_{wf1})}{\ln[(T_{b1} - T_{wf2})/(T_{b2} - T_{wf1})]} \tag{9.46}$$

For more complicated geometries, Eq. [9.44] is modified to include a correction factor F for the LMTD:

$$\dot{Q} = U \times A \times \text{LMTD} \times F(R, P, \text{HXer type}) \tag{9.47}$$

The factor F depends on two dimensionless parameters, R and P, and the configuration of the heat exchanger:

$$R \equiv \frac{T_a - T_b}{t_b - t_a} \tag{9.48}$$

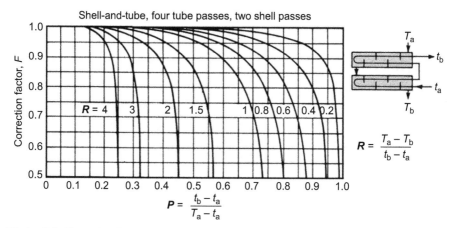

Figure 9.7 Correction factor F for a shell-and-tube heat exchanger with four tube passes and two shell passes.
After Chemical Engineering Calculations. LMTD correction factor charts, http://checalc.com/solved/LMTD_Chart.html [accessed 12.01.15].

and

$$P \equiv \frac{t_b - t_a}{T_a - t_a} \qquad\qquad [9.49]$$

Temperatures denoted by T refer to the shell-side fluid; temperatures denoted by t refer to the tube-side fluid; subscripts a refer to inlets and subscripts b refer to outlets. Fig. 9.7 is an example of a chart that gives F for case D in Fig. 9.5, namely, four tube passes and two shell passes. Charts and correlations for other configurations can be found in the literature, including a spreadsheet add-in [7,8].

The overall heat transfer coefficient, U, in Eqs. [9.44] and [9.47] is very important and can only be determined by experimentation. Tables are available that give estimates for U for a variety of fluids and situations, but the range of values is often very large. Table 9.1 gives some examples that may be germane to certain kinds of geothermal power plants.

9.13 Example analyses for elements of geothermal power plants

The preceding equations will here be applied to some of the common elements of geothermal power plants. Although the specific objectives will vary depending on the particular element, use will be made of the first law, second law, and thermal and utilization efficiencies, as appropriate. REFPROP will be used to obtain all required fluid properties.

Table 9.1 Overall heat transfer coefficients [9−11]

Hot fluid	Cold fluid	$U/(\text{W/m}^2\,^{\circ}\text{C})$
Heat exchangers (general)		
Forced liquid	Forced liquid	900−2500
Forced liquid	Condensing vapor	1000−4000
Boiling liquid	Forced liquid	300−1000
Boiling liquid	Condensing vapor	1500−6000
Forced gas	Forced gas	10−30
Forced gas	Condensing vapor	10−50
Heaters		
Steam	Water	1500−4000
Steam	Organic fluids	500−1000
Steam	Gases	30−300
Evaporators		
Steam	Aqueous solutions	1000−1500
Steam	Light organics	900−1200
Steam	Heavy organics	600−900
Coolers		
Organic fluids	Water	250−750
Organic fluids	Brine	150−500
Water	Brine	600−1200
Gases	Brine	15−250
Condensers		
Aqueous vapors	Water	1000−1500
Organic vapors	Water	700−1000
Hydrocarbons	Water	400−550
Vapors with noncondensable gases	Water	500−700
Condensing steam	Air	300−450
Light organics	Air	300−700
Heavy organics	Air	50−150
Gases	Air	50−300
Condensing hydrocarbons	Air	300−600
Ammonia	Water	800−1400
Refrigerants	Water	300−1000

9.13.1 Production well

A production well taps a hot water aquifer at a temperature of 220°C and a pressure of 2.8 MPa. The geofluid is assumed to be pure water for the analysis. The wellhead pressure is maintained at 0.45 MPa by the wellhead valve. The total mass flow rate is 60 kg/s.

Find: (a) the wellhead quality, (b) wellhead temperature, (c) mass flow rates of steam and liquid, and (d) exergy loss from the reservoir to the wellhead if the dead state is at 25°C and 0.1 MPa.

The well flow is assumed to be adiabatic and isenthalpic; that is, the enthalpy remains constant up the well: $h_2 = h_1$. REFPROP yields the properties of the geofluid and the results are given in Table 9.2; **bold** numbers are known (given) and *italic* numbers come directly from REFPROP. For (a), the wellhead quality is 0.1512; for (b) the wellhead temperature is 147.9°C (ie, the saturation temperature for 0.45 MPa); for (c) the steam flow rate is 9.072 kg/s (ie, 0.1512 × 60 kg/s) and 50.928 kg/s for the liquid.

The specific exergy is calculated from Eq. [9.20]. The reservoir value is 197.89 kJ/kg; the wellhead value is 178.51 kJ/kg. Thus, the specific exergy loss (d) is 19.38 kJ/kg. This is about 10% of the initial exergy, so the well may be viewed as being 90% efficient from a second law standpoint. However, since the well is assumed to flow spontaneously, some of the exergy "loss" is actually used to lift the geofluid from the reservoir to the wellhead, so the efficiency is in fact somewhat higher when this is accounted for.

Table 9.2 Results of production well analysis

State	Location	Temp. (°C)	Press. (MPa)	Quality	Entropy (kJ/kg K)	Enthalpy (kJ/kg)	Exergy (kJ/kg)
1	Reservoir	**220**	**2.8**	n.a.	*2.5168*	*943.71*	197.89
2	Wellhead	*147.9*	**0.45**	*0.1512*	*2.5818*	*943.71*	178.51
0	Dead state	**25**	**0.1**	n.a.	*0.3672*	*104.92*	**0**

9.13.2 Separator

Building on the previous example, suppose a cyclone separator receives the full flow from that well. It will produce two streams, liquid (hot water) and vapor (steam), at the wellhead pressure, assuming the separator is close to the wellhead and neglecting pressures losses between the two.

Find: the exergy loss in the separation process. The results are given in Table 9.3; note that the inlet temperature, pressure, and quality carryover from the previous example and are shown in **bold** (given).

The total inlet exergy flow is 10,710.37 kW (ie, 178.51 kJ/kg × 60 kg/s); the outlet steam flow carries 6385.06 kW of exergy (703.82 kJ/kg × 9.072 kg/s); and the outlet

Table 9.3 **Results of cyclone separator analysis**

State	Location	Temp. (°C)	Press. (MPa)	Quality	Entropy (kJ/kg K)	Enthalpy (kJ/kg)	Exergy (kJ/kg)
2	Two-phase inlet	147.9	0.45	0.1512	2.5818	943.71	178.51
3	Steam outlet	147.9	0.45	1.0	6.8560	2743.4	703.82
4	Liquid outlet	147.9	0.45	0.0	1.8205	623.14	84.93

liquid flow carries 4325.32 kW (84.93 kJ/kg $\times$ 50.928 kg/s). The outlet exergy flow rate is the sum of the exergy in the steam and the liquid, namely, 10,710.37 kW. Since there is no dissipation taking place in the separator according to our assumptions, the two phases are merely partitioned, and the inlet and outlet exergy values are equal.

9.13.3 Flash vessel

A flash vessel is used in a double-flash plant. The inlet is a saturated liquid at a given pressure and the two outlets are steam and liquid at a lower pressure. The liquid from the previous example is used as input; the outlet pressure is 0.15 MPa.

Find: (a) the temperature of the outlet steam and liquid, (b) mass flow rate of steam and liquid, and (c) exergy lost in the process. The results are shown in Table 9.4. Note that the enthalpy is constant for the flash process, just as it was in the well-flow process.

Table 9.4 **Results of flash vessel analysis**

State	Location	Temp. (°C)	Press. (MPa)	Quality	Entropy (kJ/kg K)	Enthalpy (kJ/kg)	Exergy (kJ/kg)
4	Liquid inlet	147.9	0.45	0.0	1.8205	623.14	84.92
4'	Two-phase flash	111.4	0.15	0.070087	1.8394	623.14	79.284
5	Steam outlet	111.4	0.15	1.0	7.2230	2693.1	544.12
6	Liquid outlet	111.4	0.15	0.0	1.4337	467.13	44.233

For (a) the outlet temperature is 111.4°C, for (b) the steam mass flow rate is 3.569 kg/s (ie, 50.928 kg/s × 0.070087), and the liquid flow rate is 47.359 kg/s. For (c) the inlet exergy associated with the high-pressure liquid is the same as previously found, namely, 4325.47 kW, and the sum of the steam and liquid exergy leaving is 4037.10 kW, giving a loss of 288.37 kW or 6.7% of the inlet exergy. Thus, the flasher may be seen as 93.3% efficient from an exergy standpoint.

9.13.4 Pipelines

Next consider the two streams of fluid leaving the separator in Section 9.13.2. The steam travels to the steam header (SH) at the power house, while the liquid travels to a reinjection well. Each pipe will be analyzed subject to certain design specifications to determine the pressure change from the separator to the end points.

Steam pipeline: A 16-in diameter, well-insulated, Sch. 30 steel pipe carries the steam. There is a total of 500 m of straight pipe with four 45° bends and two gate valves, one each at the separator and SH ends. There are no significant changes in piping elevation.

Using the equivalent length method for steam, each 45° bend is roughly equivalent to a length of pipe equal to the length of the bend measured along its centerline, namely, about 0.48 m; the two fully open gate valves are equivalent to their actual length, about 1.0 m each. Thus, the equivalent length of all pipes and fittings is $1 + 500 + 4 \times 0.48 + 1 = 503.9$ m.

Only the frictional pressure drop will be important. Thus, Eq. [9.33] applies here:

$$\Delta P_{\text{Frict, Steam}} = 1.8472 \times 10^{-5} \frac{L \dot{m}^{1.85}}{\rho D^{4.97}} \qquad [9.33]$$

in which $L = L_{\text{eq}} = 503.9$ m, $\dot{m} = 9.072$ kg/s, $\rho = 2.4161$ kg/m^3, and $D = 0.38733$ m, resulting in a pressure drop of 25.4 kPa. Since the initial pressure is 450 kPa, this is only a 5.6% pressure loss, and is less than the usual allowable limit of 10%. Also, the steam velocity turns out to be about 32 m/s, right in the range of 30–40 m/s recommended for steam flow.

The exergy analysis takes into account the pressure loss, assuming the steam remains saturated. This means there will be a small drop in temperature while the entropy increases (due to friction effects) and the enthalpy decreases; see Table 9.5. Thus,

Table 9.5 **Results of steam pipeline analysis**

State	Location	Temp. (°C)	Press. (MPa)	Quality	Entropy (kJ/kg K)	Enthalpy (kJ/kg)	Exergy (kJ/kg)
3	Steam inlet	147.9	0.4500	1.0	6.8560	2743.39	703.82
7	Steam outlet	145.8	0.4246	1.0	6.8755	2740.80	695.43

the exergy drops by 8.39 kJ/kg or by 1.2%. So the steam pipeline is 98.8% efficient based on preservation of exergy.

Liquid pipeline: An 8-in diameter, well-insulated, Sch. 40 steel pipe carries the hot water. There is a total of 1500 m of straight pipe with four 90° elbows, 10 45° bends, and two gate valves, one each at the separator and injection well. There is a drop in elevation of 25 m between the separator and the injection well.

Using the equivalent length method for liquid, each 90° elbow is roughly equivalent to 6 m of piping; each 45° bend and gate valve are equivalent to 3- and 1.4-m equivalent length, respectively. Thus, the equivalent length of all the pipes and fittings is

$$1.4 + 1500 + 4 \times 6 + 10 \times 3 + 1.4 = 1556.8 \text{ m}$$

Owing to the drop in elevation, there will be an increase in the pressure from gravity effects; this will offset some of the frictional pressure loss. Thus, the pressure at the end of the hot water pipeline will be found from

$$P_8 = P_4 - f\left[\frac{L}{D}\right]\left[\frac{\rho \mathcal{V}^2}{2}\right] + \rho g \Delta \mathcal{V} \qquad [9.50]$$

where Eqs. [9.31] and [9.32] will also be needed. The following values are known: $L = L_{eq} = 1556.8$ m, $\dot{m} = 50.928$ kg/s, $\rho = 918.96$ kg/m^3, $D = 0.2027$ m, $\Delta \mathcal{V} = 25$ m, $\mu = 185.25 \times 10^{-6}$ Pa s, $\varepsilon = 45 \times 10^{-6}$ m, and $g = 9.81$ m/s^2.

The velocity of the liquid is 1.72 m/s, within the usual range of 1.5–2.0 m/s for liquid flow in pipes. The Reynolds number from Eq. [9.32] turns out to be 1,356,212, indicating turbulent flow. The friction factor f from Eq. [9.31] is 0.01479. Thus, Eq. [9.50] can be solved:

$$P_8 = P_4 - 0.01479\left[\frac{1556.8}{0.2027}\right]\left[\frac{918.96 \times 1.718^2}{2 \times 10^{-6}}\right] + \frac{9.18.96 \times 9.81 \times 25}{10^{-6}}$$

$$P_8 = 0.450 - 0.154 + 0.225 = 0.521 \text{ MPa}$$

As it happens, the 25-m drop in elevation more than offsets the frictional pressure drop and the hot water arrives at the injection well at a higher pressure than it had leaving the separator, namely, 0.521 MPa. Note that the liquid leaving the separator is saturated and should immediately be directed downward into the reinjection pipeline to avoid flashing in the pipe.

The exergy analysis takes into account the change in pressure as with the steam pipeline, but the drop in elevation means there is another term in the exergy equation. The change in potential energy was neglected when Eq. [9.7] was combined with Eq. [9.17] to get Eq. [9.20], but here it needs to be taken into account. Thus, Eq. [9.20] is modified to:

$$e = h_1 + g(z_1 - z_0) - h_0 - T_0(s_1 - s_0) = h_1 + gz_1 - h_0 - T_0(s_1 - s_0)$$
$$[9.51]$$

Table 9.6 **Results of liquid pipeline analysis**

State	Location	Temp. (°C)	Press. (MPa)	Quality	Entropy (kJ/kg K)	Enthalpy (kJ/kg)	Exergy (kJ/kg)
4	Liquid inlet	147.9	0.45	0.0	1.8205	623.14	84.9
8	Liquid outlet	147.9	0.521	Subcooled	1.8203	623.17	85.24

where the reference elevation z_0 has been set to zero, in this case, at the elevation of the injection well; thus, z_1 is the difference in elevation. The results are shown in Table 9.6. The hot water has slightly more exergy at the end because of the potential energy being converted to a higher pressure and higher exergy.

9.13.5 Turbine

Two cases will be examined: a steam turbine typical of a flash-steam or dry-steam power plant and an organic-fluid turbine typical of a binary power plant.

Steam turbine: The turbine receives steam at the same conditions as the steam leaving the steam pipeline in Section 9.13.4. The steam is discharged from the turbine at a pressure of 12.5 kPa. The isentropic efficiency for the turbine is assumed to be 0.82; this is the ratio of the actual output to the ideal output if the expansion process were isentropic (reversible and adiabatic). The turbine specific output is equal to the enthalpy drop, $h_7 - h_9$, and comes to 439.2 kJ/kg. Multiplying by the mass flow of steam, 9.072 kg/s, gives the power output, 3.984 MW. About 10 such wells would be needed to support a commercial power plant. The state-point data are given in Table 9.7.

The steam exergy is reduced by 528.07 kJ/kg, while the turbine delivers 439.2 kJ/kg of work output. Thus, the turbine may be seen as 83.2% efficient using a functional definition of utilization efficiency.

Table 9.7 **Results of steam turbine analysis**

State	Location	Temp. (°C)	Press. (MPa)	Quality	Entropy (kJ/kg K)	Enthalpy (kJ/kg)	Exergy (kJ/kg)
7	Steam inlet	145.8	0.4246	1.0	6.8755	2740.80	695.43
9s	Ideal outlet	50.24	0.0125	0.8377	6.8755	2205.20	159.83
9	Steam outlet	50.24	0.0125	0.8782	7.1736	2301.61	167.36

Table 9.8 **Results of isopentane turbine analysis**

State	Location	Temp. (°C)	Press. (MPa)	Quality	Entropy (kJ/kg K)	Enthalpy (kJ/kg)	Exergy (kJ/kg)
7′	Vapor inlet	150	1.868	1.0	1.3669	526.64	119.13
9s′	Ideal outlet	77.47	0.152	Superheated	1.3669	431.59	
9′	Vapor outlet	83.82	0.152	Superheated	1.4018	443.95	26.036
0	Dead state	25	0.1		−0.02152	−6.4489	0

Organic-fluid turbine: The turbine receives 60 kg/s of isopentane as a saturated vapor at 150°C. The discharge pressure is 0.152 MPa. Since isopentane is a so-called "dry fluid," the turbine expansion process stays in the superheated region; this means the turbine has a higher efficiency than one with a wet expansion process. The isentropic efficiency is assumed to be 87%. Table 9.8 shows the state-point data and results.

The specific output is 82.69 kJ/kg and the power output is 4.96 MW. The isopentane gives up 93.094 kJ/kg of exergy in driving the turbine to yield 82.69 kJ/kg of output. Thus, from a functional utilization efficiency standpoint, the turbine is 88.8% efficient.

9.13.6 Condenser

All permanent geothermal power plants have a condenser; only temporary back-pressure steam units discharge the turbine exhaust to the atmosphere. In binary plants the condenser returns the organic working fluid from the turbine exhaust to a saturated or subcooled state, closing the cycle of operations. In flash-steam and dry-steam plants, the condenser allows the turbine to operate over a larger pressure ratio to produce about twice as much power compared to simply discharging the steam to the atmosphere. Here two types of condenser will be examined: *a direct-contact condenser* in which the steam is condensed by mixing with cold water, a process involving both heat and mass transfer, and a *surface-type condenser* in which there is no mixing and heat transfer occurs across metal tubes or plates.

Direct-contact condenser: The condenser receives 115 kg/s of steam and water with a quality of 0.8782 from the turbine at a pressure of 0.0125 MPa. It mixes with 2700 kg/s of cooling water (CW) at 27°C. The condensate leaves the condenser as a subcooled liquid to be pumped to the water cooling tower. Table 9.9 summarizes the data where state 9 was determined in a previous example (see Table 9.7).

Table 9.9 **Results of direct-contact condenser analysis**

State	Location	Temp. (°C)	Press. (MPa)	Quality	Entropy (kJ/kg K)	Enthalpy (kJ/kg)	Exergy (kJ/kg)
9	Two-phase inlet	50.24	0.0125	0.8782	7.1736	2301.61	167.36
10	CW inlet	27	0.1	Subcooled	0.39515	113.28	0.02792
11	Liquid outlet	48.4	0.0125	Subcooled	0.68314	202.68	3.5625

The cooling water approaches the condenser at essentially ambient pressure but is sprayed into the condenser which is maintained at a vacuum. Thus, the CW pressure during mixing is the same as for the condensing steam. An exergy analysis shows that the condenser destroys much of the exergy that enters. The rate of exergy entering is 19.322 MW whereas 10.028 MW leave. A brute-force utilization efficiency would make the condenser efficiency only 51.9%.

Surface-type, shell-and-tube condenser: The exhaust from the steam turbine in Section 9.13.5 will be used here as in the previous example. This time the heat of condensation is transferred to cooling water (CW) passing through tubes inside the condenser shell where the steam flows (see Fig. 9.5 but with the hot and cold fluids switched). The CW enters the condenser tubes at 30°C and the terminal temperature difference is 5°C. Since the steam enters as a liquid−vapor mixture and leaves as a saturated liquid (assumed), its temperature stays constant at 50.24°C. The CW leaves at 45.24°C (Table 9.10).

The required mass flow of CW is found from Eq. [9.7] where only the enthalpy terms are used:

$$\dot{m}_9(h_9 - h_{12}) = \dot{m}_{CW}(h_B - h_A) \tag{9.52}$$

Table 9.10 **Results of shell-and-tube condenser analysis**

State	Location	Temp. (°C)	Press. (MPa)	Quality	Entropy (kJ/kg K)	Enthalpy (kJ/kg)	Exergy (kJ/kg)
9	Two-phase inlet	50.24	0.0125	0.8782	7.1736	2301.61	167.36
12	Liquid outlet	50.24	0.0125	0.0	0.70692	210.35	4.14113
A	CW inlet	30	0.11	Subcooled	0.43672	125.83	0.18335
B	CW outlet	45.24	0.11	Subcooled	0.64172	189.53	2.75772

Thus, the mass flow rate of the CW is 3791.7 kg/s. The total exergy of the inflowing streams is 19.942 MW, while the total outflowing exergy is 10.933 MW, resulting in a destruction of 9.009 MW of exergy. Alternatively, if the condenser is viewed as an exergy transfer device, the steam loses 18.770 MW of exergy while the CW picks up 9.762 MW; the exergy transfer is 52.0% efficient. In reality, no use is made of the exergy acquired by the CW since it merely dissipated to the atmosphere in a water cooling tower. In that sense, no credit should be given to the exergy leaving the condenser with the CW.

Eq. [9.22] may be used as an alternative approach to the utilization efficiency of the condenser. The heat released as the steam condenses could theoretically be used to generate power in a cycle, the best one being a Carnot cycle. The thermal power given up by the condensing steam can be found by using the left-hand side of Eq. [9.52] and comes to 241.530 MWt. Using the dead state as the best possible lower temperature for the Carnot cycle would yield 18.851 MWe. This is the exergy associated with the condensation heat transfer. However, any real power plant that would try to use this waste heat would have a much lower efficiency than a Carnot cycle and may not be cost-effective.

9.13.7 Preheater and evaporator

The final example pertains to the main heat exchangers (HXers) in binary power plants. The preheater (PH) raises the temperature of the organic working fluid to the boiling point, followed by the evaporator (EV) that boils the fluid to produce saturated vapor for use in a turbine. This example will treat the pair as arranged in series with the heating provided by a hot geofluid that cools continuously as it passes through the HXers; see Fig. 9.8. The working fluid is isopentane, i-C_5H_{12}; it enters the PH at 50°C (state 13) and leaves the EV at 130°C as a saturated vapor (state 7'). The geofluid enters at 160°C and 1.0 MPa (state C); it leaves the EV at 135°C (state D), that is, the pinch-point temperature difference is 5°C. The geofluid mass flow rate is 350 kg/s.

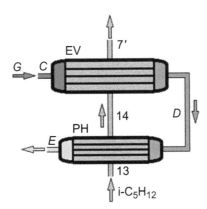

Figure 9.8 Preheater—evaporator schematic.

Table 9.11 Results of preheater–evaporator analysis

State	Location	Temp. (°C)	Press. (MPa)	Quality	Entropy (kJ/kg K)	Enthalpy (kJ/kg)	Exergy (kJ/kg)
C	Geofluid in	160	1.0	Subcooled	1.9421	675.70	101.22
D	Pinch-point	135	1.0	Subcooled	1.6865	568.20	69.932
E	Geofluid out	111.8	1.0	Subcooled	1.4406	469.47	44.515
13	i-C_5H_{12} in	50	1.3125	Subcooled	0.16314	52.918	4.3105
14	Pinch-point	130	1.3125	0.0	0.75303	267.17	42.687
7'	i-C_5H_{12} out	130	1.3125	1.0	1.3316	500.44	103.46

Find: (a) the isopentane flow rate, (b) geofluid outlet temperature, (c) heat transfer and (d) exergy efficiency of the HXers. Pressure losses will be neglected in this analysis. Fig. 9.6 provides the basis for the analysis.

(a) The mass flow rate of isopentane is found from the first law for the evaporator: 161.29 kg/s. (b) The geofluid outlet temperature is found from the enthalpy calculated by applying the first law to the preheater: 111.8°C. Note: the outlet temperature might be found using the slope of the cooling line (see Fig. 9.6) but the line is not exactly straight owing to variable specific heat for the geofluid. Such an approach yields a very good approximation (112.0°C) but not the actual temperature. (c) The heat transfer in the preheater and evaporator are found from the first law: 34.556 MWt and 37.624 MWt, respectively. (d) The exergy given up by the geofluid in the evaporator and preheater are 10.951 MW and 8.896 MW, respectively. The exergy gained by the isopentane in the same HXers are 9.801 MW and 6.190 MW, respectively. Thus, the utilization efficiencies for these HXers based on the transfer of exergy are 89.5% and 69.6%, respectively. Taken as a unit, the PH-EV combination has an overall utilization efficiency of 80.6%. The EV is more effective in transferring exergy because the LMTD is smaller than for the PH, 14.0°C versus 22.6°C. The smaller the LMTD, the better aligned are the heating and cooling curves, and the lower the irreversibility associated with the heat transfer process. The state-point data are summarized in Table 9.11.

Note that the HXers are 100% efficient in the transfer of heat from the hot fluid to the cold one, assuming perfect insulation, based on the first law. The second law accounts for the irreversibility in the transfer of heat across a temperature difference and leads to the efficiency values calculated here.

9.14 Conclusions

The basic principles presented in this chapter apply to all the geothermal energy conversion systems described in the remaining chapters of this part of the book, regardless of how complex the system may be. The working equations based on the first and second laws of thermodynamics, the energy equation of fluid mechanics, and the fundamental laws of heat transfer, together with a database for the properties of substances, such as REFPROP among others, form a package that allows both assessment of existing plants as well as design tools to develop plants to match various geothermal resource characteristics.

Sources of further information

Bejan A. Advanced engineering thermodynamics, 2nd ed. Wiley-Interscience; 1997.

DiPippo R. Geothermal energy as a source of electricity: a worldwide survey of the design and operation of geothermal power plants. US. Department of Energy, DOE/RA/28320−1, US. Government Printing Office. (Washington, D.C.); January 1980.

Armstead HCH. Geothermal energy. 2nd ed. Spon E & FN. (London, England); 1983.

Geothermal energy systems: exploration, development, and utilization. In: Huenges E, editor. (Weinheim, Germany): Wiley-VCH; 2010.

DiPippo R. Principles, applications, case studies and environmental impact. In: Geothermal power plants. 4th ed. (Oxford, England): Butterworth-Heinemann, Elsevier; 2016.

Bejan A. Heat transfer. New York: John Wiley & Sons; 1993.

Sourcebook on the production of electricity from geothermal energy. Kestin J, editor-in-chief; DiPippo R, Khalifa HE, Ryley DJ, editors. US. Department of Energy, DOE/RA/28320−2, US. Government Printing Office. (Washington, D.C.); August 1980.

References

[1] National Institute of Standards and Technology (NIST). Reference fluid thermodynamic and transport properties. U.S. Dept. of Commerce; http://www.nist.gov/srd/nist23.cfm.

[2] Rant Z. Exergie ein neues Wort für 'technische Arbeitsfähigkeit', (Exergy, a new word for 'technical ability to do work'). Forsch Ingenieurwes 1956;22:36.

[3] Gibbs JW. The scientific papers of J. Willard Gibbs, vol. I − Thermodynamics. New York: Dover Publications, Inc.; 1961. Reprint of original publication by Longmans, Green and Company, 1906.

[4] The Engineering Tool Box. Resistance and equivalent length of fittings, http://www.engineeringtoolbox.com/resistance-equivalent-length-d_192.html.

[5] Wolverine Tube, Inc. Two-phase pressure drops. In: Engineering data book III; 2006 [Chapter 13], http://www.wlv.com/wp-content/uploads/2014/06/databook3/data/db3ch13.pdf.

[6] Hewitt, G.F., "Pressure drop, two-phase flow," Thermopedia: http://www.thermopedia.com/content/37/?tid=104&sn=1302.

[7] Oko COC, Diemuodeke EO, Katsina MB. Spreadsheet add-in for heat exchanger logarithmic mean temperature difference correction factors. Int J Comput Appl (0975 − 8887) April 2012;44(5).

[8] Chemical Engineering Calculations. LMTD correction factor charts, http://checalc.com/solved/LMTD_Chart.html [accessed 12.01.15].

[9] The Engineering Tool Box. Overall heat transfer coefficient, http://www.engineering toolbox.com/overall-heat-transfer-coefficient-d_434.html.

[10] Engineering Page. Typical overall heat transfer coefficients (U-values), http://www. engineeringpage.com/technology/thermal/transfer.html.

[11] Wiley Online Library. Introduction to thermo-fluids systems design. Appendix C — Heat Exchanger Design. August 30, 2012. http://onlinelibrary.wiley.com/doi/10.1002/9781118 403198.app3/pdf.

Flash steam geothermal energy conversion systems: single-, double-, and triple-flash and combined-cycle plants

10

W. Harvey, K. Wallace
POWER Engineers, Hailey, Idaho, United States

10.1 Flash steam cycles

Chapters 7 and 8 described the thermodynamic fundamentals, including governing equations, of flash steam plants using steam turbines. As some of the equipment used in a steam plant will also be covered in other chapters, the current chapter will primarily focus on the specific challenges and innovations that are applicable to harnessing a two-phase, high-enthalpy resource, versus higher enthalpy dry steam or lower enthalpy single phase resources.

It should be noted that although "flash plant" is widely used in reference to cycles that employ a steam turbine, in fact there are binary cycles that also flash the incoming geofluid into separate steam and water phases, as described in Chapter 12. Thus, some of the techniques for cycle design, mitigation of scale, and use of byproducts described here may apply to either steam turbine or binary cycles employing a flash. For the purposes of this chapter, following general use, a reference to "flash plant" implies a cycle that relies primarily on a steam turbine to generate the majority of its power.

10.1.1 Single flash

The mainstay of the geothermal industry for higher-temperature resources has historically been, and continues to be in many locations, the single-flash plant. A typical flash cycle is shown in Fig. 10.1. Geofluid is led from the wells to a high-pressure (HP) separator, which produces separated steam and brine. The separated steam is led through a demister to remove fine carryover droplets, before entering the steam turbine. The single-pressure steam turbine would be similar to those used for dry steam plants, covered in this chapter. The steam turbine exhausts to heat rejection equipment, consisting of the condenser, circulating water or hotwell pumps, and cooling tower. Additional flashes using the liquid from the HP separator will be discussed later; for the single-flash plant, the liquid from the first separator would simply be injected back to the reservoir.

Geothermal Power Generation. http://dx.doi.org/10.1016/B978-0-08-100337-4.00010-3

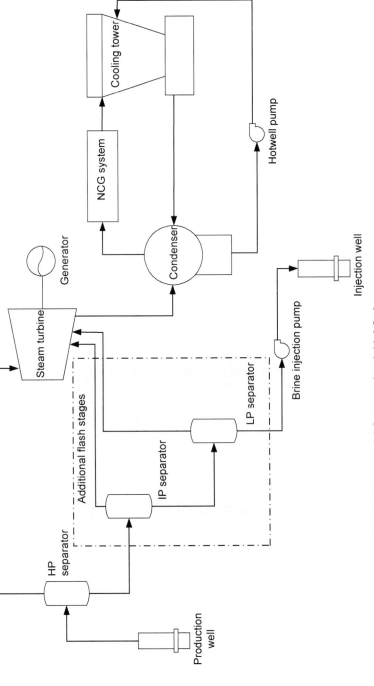

Figure 10.1 Flash plant schematic, showing potential for second and third flash stages.

The separator pressure is selected depending on the characteristics of the field, the wells' ability to deliver fluid at a selected wellhead pressure, and a natural performance tension that exists between a desire for higher-pressure steam at the inlet to the turbine (tending toward higher separator pressure) and a desire for higher mass flow rate of steam to the turbine (tending toward a lower separator pressure). DiPippo [8] describes the process of selecting this flash pressure for maximum power output, considering the well performance curves and plant characteristics.

Single-flash plants generally carry out separation in the 6–20 bar,a (absolute) range. As a result, the brine to be injected back to the reservoir is already at an elevated pressure, reducing the head requirements of brine injection pumps compared with multiple-flash plants with lower separator pressures.

The resultant concentration of solids in the brine due to flashing and the temperature of the brine dictate the propensity for precipitation of solids. Because the resultant injection temperature of the separated brine is also high for single-flash plants, this cycle can have a large margin of safety from the onset of scaling in the brine.

One of the key strengths of the single-flash plant − its high separator pressure and brine temperature − is also a limitation; it may forego an opportunity to generate more power. Exergy analysis, presented in Chapter 9, is a useful way to illustrate the potential for power generation from various streams. Consider an example where 500 kg/s of geofluid, considered as pure water, is available at an enthalpy of 1200 kJ/kg. If separated at 8 bar,a, this would produce 117 kg/s of steam and 383 kg/s of liquid ("brine"). Table 10.1 shows the resultant exergy flows, assuming the dead state is evaluated at 20°C and 1.013 bar,a.

An exergetic efficiency of a single-flash plant, based on the original exergy of the geofluid prior to the flash, may be around 30–35%; thus, a single-flash plant may be expected to generate around 50–55+ MW of net electrical power with this resource. It can be seen that the exiting stream of brine to be injected carries with it significant exergy. If some fraction of this can be harnessed in a useful manner, the cycle can generate more power, if the total geofluid flow remains the same. Alternatively, if a fixed output is desired, increasing the power extracted per kg/s of fluid may allow the geofluid flow to be reduced, lowering the cost of wells and the gathering system.

A fundamental tradeoff for flash plants is the attempt to increase the resource utilization efficiency − generally by increasing the number of flashes and cycle complexity − weighed against the investment of drilling additional wells. At locations

Table 10.1 Single-flash illustrative exergy flows

	Mass flow (kg/s)	Temperature (°C)	Quality or enthalpy	Total exergy (MW)
Geofluid from reservoir	500	Enthalpy of 1200 kJ/kg		162
Steam from separator	117	170	1	95
Brine from separator	383	170	0	48

where the resources are sizable and energetic and drilling costs are low, the single-flash plant provides a simple, reliable, and low-cost power plant solution and may be the most economical from a total project cost perspective. It can be observed that through 2012, all the flash plant (nonbinary) geothermal projects in Indonesia, Kenya, and Costa Rica — major global producers — were single flash. However, where drilling costs are high, a more efficient and costly design for the surface facilities may result in better overall project economics, slower resource decline due to reduced geofluid extraction rate, and mitigation of the drilling risks. The John Featherstone triple-flash plant in the Salton Sea region, an innovative and complex project with aspects to be discussed later, is an example of increasing the number of flashes to improve the resource utilization.

10.1.2   Double and triple flash

Double- and triple-flash plants are a logical extension of the single-flash process. Double- and triple-flash plans were installed at Wairakei as far back as the late 1950s/early 1960s, so the basic concept is well established. Reviewing Fig. 10.1, we see the double- or triple-flash configurations that can produce some combination of high-, intermediate-, and low-pressure (HP/IP/LP) steam. The flash pressures and temperatures would normally be distributed between the equivalent resource temperature and the condenser temperature following the philosophy of an "equal temperature split"; this evaluation process is described in DiPippo [8]. However, practical constraints, such as achievable wellhead delivery pressures, and the desire to keep the LP steam pressure above atmospheric pressure to avoid air in-leakage or piping design complications when operating at a vacuum, generally modify these idealized temperature or pressure spacings. While thermodynamically an infinite number of flashes would result in the highest power, the physical and economic constraints, and the diminishing returns of increasing the number of flashes, limit current plants to no more than three flashes.

Table 10.2 compares single- and double-flash performance parameters for a set of assumed resource conditions of 500 kg/s total geofluid flow (pure H_2O) and 1200 kJ/kg enthalpy, corresponding approximately to an equivalent single-phase resource temperature of 273°C. These values assume realistic turbine and generator efficiencies. The HP separator pressure has been kept the same for both cases for ease of illustration, although this may not be an optimum thermodynamically.

Table 10.2 illustrates that the additional flash stage can theoretically add significant power, in this example, 22% more. While this increase in efficiency and potential output is highly resource and cycle specific, and often limited by the resource chemistry (to be discussed later), a more conservative assessment would be to estimate that a double flash can add around 10—15% net output compared with single flash, and a triple flash might add another 5—7%, with flash pressures redistributed for each case.

The incremental increase in capital cost for a greater number of flashes lies primarily in the design of the separator station(s) and the steam turbine. The separator station must be configured with multiple stages of separators and demisters to handle each

Table 10.2 **Sample comparison of single- and double-flash cycle parameters**

	Units	Single Flash	Double Flash
HP separator pressure	bar,a	8	8
HP steam flow	kg/s	117	117
HP brine flow	kg/s	383	383
LP separator pressure	bar,a	—	2
LP steam flow	kg/s	—	38
HP turbine output	MW	60	60
LP turbine output	MW	—	13.5
Net output	MW	56.5	69.0
Exergetic efficiency	%	35%	43%

stage of flash. The turbine design then must accommodate admission of steam at multiple pressures. While multiple-pressure turbines are long-standing common practice at fossil thermal plants, they are more complex in geothermal applications, due to the relatively lower pressures of the steam and the resultant need for larger-diameter piping. Generally, steam at each pressure level is admitted to the turbine via two pipelines ("right hand" and "left hand"), each equipped with main stop and governor valves. A triple-pressure turbine thus may have six inlet pipelines and 12 of these large hydraulically actuated control valves, adding cost and requiring additional space. Fig. 10.2 shows the triple-pressure turbine at Kizildere II, operated by Zorlu Energy

Figure 10.2 Triple-pressure turbine at Kizildere. In the foreground are the left-hand inlet steam lines: IP (top), HP (middle), and LP (bottom), running right to left to the turbine. The right-hand inlet steam lines run closer to the powerhouse wall in the background. The single-flow HP turbine has a single exhaust duct (green) at the top of the turbine. The double-flow IP/LP turbine to its left has two larger exhaust ducts.

in Turkey. In the foreground are the left-hand inlet steam lines: IP (top), HP (middle), and LP (bottom).

Following the turbine, the configuration of a multiple-flash power cycle would be similar to that for a single-flash or dry steam plant. The turbine exhausts to a condenser, which is supplied with circulating water from a cooling tower. Noncondensable gases from the condenser are removed, generally using a multistage configuration of steam ejectors, and for hybrid systems, last-stage vacuum pumps. An array of circulating and cooling water pumps drive the water flows between the condenser and cooling tower and provide cooling water to other components such as lubricating oil and generator air coolers.

10.1.3   Silica scaling

Chapter 16 provides a fuller and more chemistry-focused discussion regarding amorphous silica and silicate deposition in geothermal fluid processing systems. In this section, we focus on the effects of injection temperature and concentration constraints on the performance of a flash plant. Using the same hypothetical cycle parameters as shown in Table 10.2, we can scrutinize the conditions of the brine to be injected. Table 10.3 shows how the additional flashes further concentrate the constituents in the brine to be injected, as well as reduce the injection temperature. Table 10.3 assumes a representative concentration of 563 ppm silica in the geofluid, estimated from the resource temperature.

Through each flash, the concentration of silica in the brine is increased and the temperature is reduced. Due to the lowering temperature, the solubility of constituents such as amorphous silica (SiO_2) is reduced, hence the increasing number of flashes brings the brine closer (or past) the point of precipitation. The single-flash plant results

Table 10.3 **Sample comparison of single- and double-flash cycle parameters regarding silica**

	Units	Single flash	Double flash
Net output	MW	56.5	69.0
HP flash pressure	bar,a	8	8
LP flash pressure	bar,a	–	2
Silica concentration in injectate	ppm	735	815
Injection temperature	°C	170	120
Equilibrium amorphous silica concentration at the injection temperature	ppm	744	458
Ratio of silica concentration to the equilibrium concentration (silica saturation index)		0.99	1.78

in hotter, less-concentrated brine, and if the flash pressure is selected appropriately, problems with injection scaling may not be encountered, as the injectate is still at a concentration lower than the equilibrium concentration. The deeper double-flash increases the concentration of silica to over 1.7 times the equilibrium concentration for the conditions shown in Table 10.3.

Assessing and manipulating the potential for solids precipitation are key design focuses in the multiple-flash project design. Overcoming these injection issues is not a challenge specific to flash plants; binary projects may also be motivated to operate at lower injection temperatures, in order to harvest the maximum energy possible from the geofluid. Thus, several strategies have evolved to control scaling in injectate in order to achieve higher plant output. More detailed descriptions of the chemistry and methods can be found in Chapter 16.

10.1.3.1   Control of solubility

The simplest and most direct method of avoiding precipitation in the injectate is to select the cycle parameters such that the fluid is returned to the reservoir at a temperature above that for which precipitation would occur. The single-flash cycle in Table 10.3 can be seen to have some margin from precipitation of solids in the injectate. However, the revenue stream from the missed opportunity of harnessing an additional 10+ MW from the double flash could equate to around $8 million per year, if evaluated at $100/MWh. Solutions that require a modest investment in capital and/or operations and maintenance (O&M) costs thus might be attractive.

10.1.3.2   Control of precipitation kinetics

It has been shown that dosing injectate with chemicals, such as acid to reduce the pH ("pH-mod"), can greatly slow the precipitation kinetics [15]. If, after dosing, the separated brine can be led through the gathering system, down the injection wells, and be distributed into the reservoir away from the wellbore before solids come out of solution and can begin reheating such that the scaling potential is reduced, this can be an effective strategy that allows greater harvesting of energy from the geofluid.

The pH-mod and other scale inhibitor chemical techniques have been used in locations such as the Salton Sea (California), New Zealand, and Turkey. These systems incur a lower capital cost than the crystallizer-reactor-clarifier (CRC) process to be discussed next but may require "up-alloying" of components that handle or are exposed to acid chemicals. They also require a continuous supply of chemicals and corresponding operating cost.

10.1.3.3   Solids separation

Another technique to avoid precipitation in the injection system is to remove solids at the surface. The CRC process is shown in Fig. 10.3. In this process, geofluid for multiple-flash plants is flashed in the initial and subsequent separators (crystallizers). Constituents in the brine are encouraged to precipitate by adding seed particles that serve as nucleation sites in the later-stage crystallizers instead of their depositing on

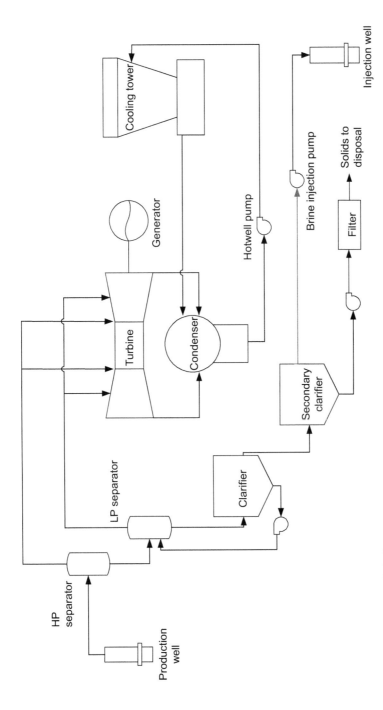

Figure 10.3 Schematic of CRC process

Adapted from Newell DG, Messer PH, Whitescarver OD, Messer PH. Salton Sea Unit 3 47.5 MWe geothermal power plant. Geothermal Resources Council Bulletin; May 1989.

vessel and piping walls. The solids in the supersaturated brine then are removed from the clarifier using filters and the purified injectate is returned to the reservoir. The John Featherstone triple-flash plant in California, commissioned in 2012, uses this technique. A high concentration of minerals in the geofluid, coupled with the triple-flash design that was chosen to maximize the extraction of energy from the geofluid, leads to a high concentration of constituents in the brine at a low brine temperature.

The CRC process requires a larger capital investment than the pH-mod process and produces a solid waste stream that must be disposed of or harvested if there are useful byproducts. Featherstone et al. [13] provide a comprehensive review of this technology and contrast it with the pH-mod technique.

Designers tune cycle parameters such as flash pressures and injection temperature, in conjunction with exploring tools available to predict and avoid injection scaling. These tools continually attract research and development efforts due to the significant potential benefits through greater revenues or the avoidance of higher and uncertain drilling costs.

10.2 Mixed and combined cycles

The usual cycle choice for a high enthalpy resource producing two-phase fluid would be a plant using a single- or multiple-pressure steam turbine. As noted earlier, there are also examples of binary plants that harness flashed steam resources; binary technology will be covered in more detail in Chapter 12. Chapter 14 discusses mixed and combined cycles along with hybrid systems.

At certain sites, creative combinations of steam turbine and binary technologies can result in improved performance and economics compared with either alone. This section will therefore discuss "mixed" and "combined" cycles, using terminology following that used by Bombarda and Macchi [4].

10.2.1 Mixed cycles

A mixed cycle uses a steam turbine plus binary technology in a manner where the two units can operate independently of one another. The simplest example would be that of a single-flash plant with a steam turbine harnessing the flashed steam, and a binary plant harnessing the brine, as shown in Fig. 10.4.

The potential of the mixed cycle can be illustrated using data from the sample project presented in Tables 10.2 and 10.3. Assume there is a strategy for preventing solids precipitation from the brine from the single-flash plant at $170°C$, such that additional energy can be extracted from the brine down to $150°C$. The thermal energy extracted from the brine's temperature difference might equate to 32 MW_{th}. If converted to electricity at a first law efficiency of 12.5%, this might generate an additional 4 gross MW_e, or a 7% boost to the cycle output.

Although the mixed cycle may be purpose-built as an original single plant, it is common to encounter these as retrofits to an existing flash plant. The advantage of

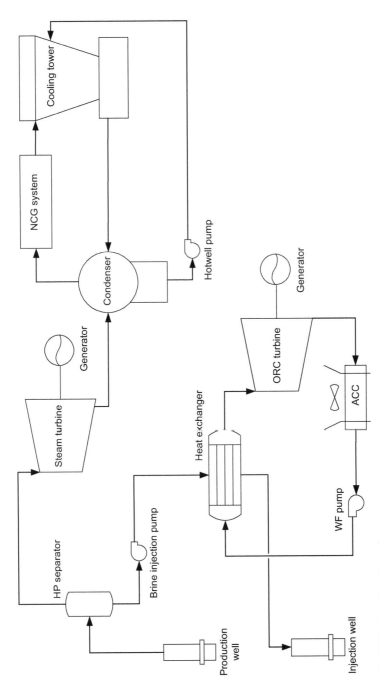

Figure 10.4 Mixed-cycle schematic.

Figure 10.5 Beowawe bottoming binary plant. The flash plant's cooling tower is in the background to the left; the newer cooling tower for the bottoming plant is to the right. The binary heat exchangers are in the foreground; the binary turbine and a packaged electrical building sit behind the exchangers.

a retrofit is that the operation of the flash plant over a period of time prior to the binary addition allows the operators to characterize the chemistry of the brine injectate more precisely. A more accurate determination of the appropriate injection temperature reduces the likelihood that scaling will occur and helps optimize the potential output of the binary "bottoming" plant.

Although there are evident advantages in assessing the extra potential at single-flash plants due to the higher brine temperature, even double-flash plants can have margin in the brine temperature suitable for a bottoming binary unit. A 16.7-MW$_g$ (gross) double-flash plant was commissioned at Beowawe (Nevada) in 1985. In 2011, a 2.5-MW$_g$ TAS/Barber Nichols binary unit, shown in Fig. 10.5, was installed. This reduced the brine injection temperature from 96 to around 68°C [3].

Purpose-built or retrofit bottoming binary units combined with flash plants to form a mixed cycle can offer increased output, without additional extraction of geofluid from the resource, and can leverage existing infrastructure. The effect of colder injection temperatures must be evaluated by the resource specialists to ensure that unduly accelerated cooling of the resource does not take place. Depending on the availability of makeup water, the binary plant may be air or water cooled.

10.2.2 Combined cycles

In contrast to the mixed cycles, which may operate independently of one another, a combined cycle couples the two technologies directly together. Many configurations are possible; in the following sections, we discuss air- and water-cooled options.

10.2.2.1 Air-cooled combined cycles

A typical air-cooled combined cycle is shown in Fig. 10.6. Geofluid is flashed to steam that passes through a backpressure steam turbine and exhausts to a binary vaporizer. Brine from the separator can be led to preheaters or, alternatively, to a separate

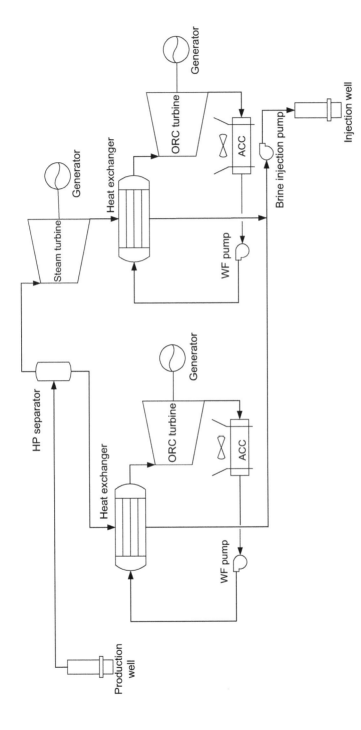

Figure 10.6 Air-cooled combined cycle configuration.
Adapted from Menzies AJ, Brown P, Searle J, King R. The Mokai geothermal power plan, New Zealand: analysis of performance during first year of operation. GRC Trans 2001;25.

bottoming binary cycle. The balance of the binary cycles would be similar to standard binary plants. The combined cycle may have several advantages over all-flash or all-binary cycles. For large units, such as the 3 × 110−MW Sarulla project currently under development in Indonesia, steam topping turbines may result in lower capital costs. The combined cycle can use air cooling; while air cooling for a flash plant is not impossible, it has certain limitations that have resulted in it not yet being widely used in the geothermal industry. For the cycle shown in Fig. 10.6, the use of air cooling results in virtually all the geofluid (other than a small amount vented with noncondensable gases [NCGs] from the vaporizer) being returned to the reservoir. At some locations, such as certain fields in New Zealand, subsidence can be an issue, such that net withdrawal of fluids from the reservoir is restricted by regulations. Air-cooled plants may also be desirable for enhanced geothermal systems (EGS) resources (see Chapters 1 and 3), if there will be a limited supply of water to circulate to the reservoir as the geofluid.

The 60-MW Mokai I project is an example of an air-cooled combined cycle. The steam turbine is 31 MW, with 4 × 4.6−MW Ormat "steam-binary" units powered from the exhaust steam. Additional energy is harvested from the flashed brine by 2 × 4.9−MW "brine binary" units [25].

10.2.2.2 Water-cooled combined cycles

A water-cooled combined cycle configuration is shown in Fig. 10.7. The HP steam from the first flash separator drives a backpressure steam turbine and associated binary plant. However, the brine from the separator is further flashed and used in an LP steam turbine. The combined condensate from the HP and LP cycles is sufficient to allow evaporative cooling.

These cycles have been applied to high enthalpy projects in Turkey, where the geofluid contains a high percentage of NCGs. A conventional flash plant using a condensing turbine would require large investments in capital cost and parasitic load for the NCG removal equipment. The HP portion of the cycle, consisting of a backpressure steam turbine with a bottoming binary, does not require such equipment, and the LP portion requires only a modest NCG removal system for the condenser, since most all the NCGs are carried into the HP steam with the first flash. The deeper flashes provide sufficient makeup water to allow the complete cycle to be water cooled.

Evaporative cooling can offer lower capital costs and operations closer to the design point for more of the year compared with air-cooled units. Since deviations from the design point result in reduced turbine and other component efficiencies, a water-cooled unit can usually generate some 3−5% more annual energy output compared with a comparable air-cooled unit of the same design point net capacity, depending on the climate. Moya and DiPippo [27] noted that a comparison of air-cooled versus water-cooled units for the Miravalles Unit 5 binary plant revealed that the air-cooled condenser would cost more than three times as much, weigh more than two and a half times as much, cover about three times as much surface area, and consume about three times more fan power than a water-cooled tower. There are thus powerful motivations for seeking out water-cooled solutions.

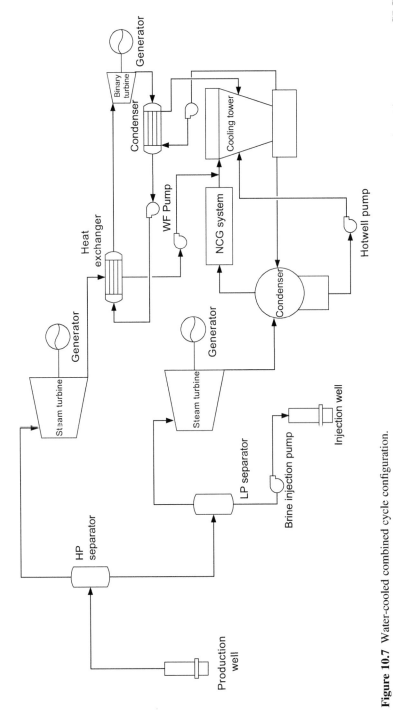

Figure 10.7 Water-cooled combined cycle configuration.
Adapted from Dunford T, Lewis B, Wallace K, Harvey W. Combined cycle strategies for high enthalpy, high noncondensable gas resources. GRC Trans 2014;38.

Combined cycle configurations, of which only several of the possible combinations have been presented, thus offer additional degrees of freedom, that is, flexibility between steam and binary equipment economics, performance advantages, reduced sensitivity to geofluid conditions and composition, and a range of cooling options. For larger units that merit the additional investment in cycle complexity, these cycles can provide unique solutions for site-specific challenges.

10.3 Cogeneration and coproduction from flashed brines

Unlike plants harnessing fluid in a single phase, such as dry steam or low-temperature binary units, a flash plant will have multiple product streams of condensate, brine, and noncondensable gases. Developers have shown commendable ingenuity in attempting to harness aspects of these different streams for additional benefit and revenue. Albertsson and Jonsson [1] outline the paradigm of a "resource park" concept for geothermal installations, which integrates a variety of uses, supports the sustainable development of the society, bridges technical and social cultures, and considers activities on a time scale of centuries. Flash plants, such as the Svartsengi resource park in Iceland, can contribute to these objectives with a variety of useful products. We discuss three basic resources that can be harvested from flash plants: thermal energy, fluids, and solids.

10.3.1 Thermal energy

Although power generation is the focus of this book, it would be a disservice to not at least briefly highlight the significant contribution to project revenue and the impact on the cycle design that features to incorporate district heating or serve other thermal loads can introduce. A case study that illustrates these points is the design of the Nesjavellir flash plant in Iceland. Fig. 10.8 shows a cycle diagram of the first phase of the project, nominally 60-MW gross output. The single pressure turbine operates at an elevated exhaust pressure of 0.2 bar,a. Makeup water from an external source is preheated as it flows through the tubes of the surface condenser while condensing the turbine exhaust steam on the shell side. The water is heated further using separated brine and then deaerated. The hot water is then sent to the district heating system for the city of Reykjavik. The turbine and condenser are designed with flexibility in the steam flow and condenser pressure/temperature to accommodate variations in annual heating demands. Energy from both the steam turbine exhaust and the brine from the single-flash separator is harnessed to provide 128 MW of thermal output (hot water) at the design point, via a pipeline 27 km long. During colder conditions, the geofluid flow rate is increased from 326 to 373 kg/s, and the turbine exhaust pressure is increased from 0.2 to 0.35 bar,a, resulting in an increase of thermal power from 128 to 227 MW [2].

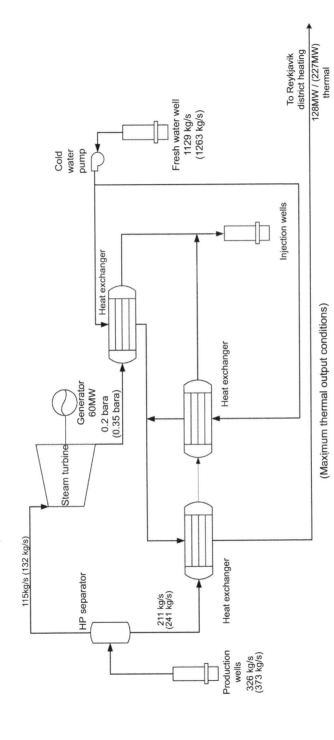

Figure 10.8 Cycle diagram of the Nesjavellir plant.
Adapted from Ballzus C, Frimannson H, Gunnarsson GI, Hrolfsson I. The geothermal power plant at Nesjavellir, Iceland. In: Proceedings of the world geothermal congress; 2000 May 28–June 10; Kyushu–Tohoku, Japan; 2000.

Table 10.4 **Nesjavellir revenues from production of electricity and hot water**

	Electricity 2013	Hot water 2013	Electricity 2012	Hot water 2012
Revenue (M$)	8.1	2.7	8.7	1.9
% of total revenues	75%	25%	82%	18%

Adapted from Reykjavik Energy. 2013 Annual Report; 2013

Nesjavellir was expanded to 120-MW electrical power and 290-MW thermal power in 2005. The plant revenues give an indication of the value of the thermal power. Table 10.4 shows excerpts from Reykjavik Energy's 2013 financial statement, showing revenues from the electricity and hot water generation at Nesjavellir. It can be seen that the hot water contributed 18–25% of total revenue over the 2-year period.

The contributions that a cogeneration plant can make are naturally highly dependent on the surroundings. A flash plant in Iceland such as Hellisheidi or Nesjavellir relatively close to population centers clearly is more advantageously sited to contribute to district heating systems compared with projects in the tropics. However, an increasing number of projects throughout Europe, regardless of power plant technology, are being configured to include a cogeneration mode. Buescher [5] presents data from several binary projects in Germany that produce both thermal and electrical output, such as Bruchsal (5.5 MW$_{th}$, 0.55 MW$_e$), Landau (5 MW$_{th}$, 3.6 MW$_e$), and Sauerlach (4 MW$_{th}$, 5 MW$_e$). Buescher states that in 2015, 27 of the 33 European countries are expected to be equipped with geothermal district heating systems. Geothermal district heating is also making inroads in China, with an installed capacity of 2946 MW$_{th}$ in 2015, a 2.8-times increase over 2009 [23]. Flash plants advantageously sited to serve thermal energy users should consider this potential.

10.3.2 Fluids

Using the resource park perspective, any product stream can be considered for potential revenue, and several types of fluids can be harnessed from flash plants.

Noncondensable gases in geothermal steam are extracted from the condenser and discharged to the atmosphere. These are predominantly carbon dioxide (CO_2) but may contain other constituents such as hydrogen (H_2), hydrogen sulfide (H_2S), or methane (CH_4).

Dunstall and Foster [10] report on a project in New Zealand that uses geothermal energy for heating of a greenhouse, and uses CO_2 to enrich the atmosphere by several 100 ppm to increase growth rates. They describe strategies to balance the temperature, humidity, and CO_2 concentrations in a synchronized manner to maximize the growth rates of the plants. Noxious gases such as H_2S must be filtered out to tolerable levels.

Greenhouses in Kenya and Iceland also use geothermal CO_2 to enhance greenhouse growth of various species of flowers and vegetables. Thermal energy and CO_2 may also be used to accelerate the growth of algae, which can be harvested for fuel or other useful byproducts [31].

At locations in Turkey and Iceland, geothermal CO_2 is harvested for industrial uses such as in bottling of carbonated beverages. The George Oleh Renewable Methanol Plant, operated by Carbon Recycling, Inc. (CRI), captures CO_2 and H_2 from the exhaust of the Svartsengi geothermal plant in Iceland, combining those gases with hydrogen from electrolysis to synthesize methanol. The methanol is blended with gasoline and used for vehicle fuel. The plant started operations in 2012 and expanded in 2014 to a production capacity of 4 million liter of methanol per year [6].

10.3.3 Solids

Solids, as in the case of the silica scaling mentioned earlier, often are regarded as a nuisance that the cycle designer must accommodate. In some cases however, valuable minerals can be carried by the geothermal brine. Table 10.5 shows typical geofluid compositions from two fields in southern California.

A pilot plant for zinc recovery from plants at the Salton Sea was operated by CalEnergy, starting in the late 1990s but was shut down in 2004 [17]. As lithium batteries find increasing applications worldwide, lithium extraction is drawing greater attention. In 2011, Simbol conducted research near the John Featherstone plant in the Salton Sea region, on the recovery of lithium from geothermal brine, but scaled back operations in 2015. Despite literature that describes the potential for mineral recovery and ongoing research, it appears that deployment on a commercial scale is still a work in progress.

While there are many flash plants that harness the multiple potential benefits, the resource park at Svartsengi serves as a particular exemplar. Providing electrical power, district heating to nearby communities, thermal waters and spa products at the nearby Blue Lagoon, and gases for fuel synthesis, research, and development actively continues there on other potential outputs. It would be hoped that many other flash plants would follow their leadership to convert more "waste streams" to "value streams."

Table 10.5 Selected species content (in ppm) in geofluid

Species	Salton Sea	Brawley
Lithium (Li)	194	219
Manganese (Mn)	0.03–3.9	0.1
Zinc (Zn)	0.1–15.9	12.8

Adapted from Gallup DL. Geochemistry of geothermal fluids and well scales, and potential for mineral recovery. Ore Geol Rev 1998;12:225−36.

10.4 Equipment research and development

In this section, we discuss some of the ongoing and future research and development thrusts related to flash plants. These advances are primarily focused on improvements in efficiency and economics, although environmental aspects such as water usage and abating gaseous emissions are also priorities. Advances in binary technology that may be applicable to combined cycles are covered in Chapter 12.

10.4.1 Gathering systems and separator stations

The gathering system conveys the geofluid from the production wells to the power plant and to the injection wells or other means of disposal. In contrast to the single-phase geofluids used for dry steam or low-temperature binary units, the design of gathering systems for flash plants must include an appropriate strategy for transfer and separation of the two-phase geofluid into steam (including noncondensable gases) and liquid ("brine"). In this section, we discuss strategies and innovations in the gathering systems for flash plants.

10.4.1.1 Separator station design options

The configuration of the gathering system can take two basic approaches. Fig. 10.9 shows the initial separation taking place at the wellheads for a double-flash plant.

The HP steam and brine are conveyed separately to the plant. Measurements of the separated steam and brine mass flow rates from individual wells can be made accurately, in contrast to the imprecision in measuring two-phase flow from the wells into the separators. These measurements provide better data for ongoing reservoir modeling calibration and production adjustments. The single-phase fluids can also be directed to flow more easily uphill to the plant, if the terrain demands, without the risks of slugging that might be induced for two-phase piping over the wide range of conditions from warmup/startup to full load operation. Fig. 10.10 shows a wellhead separator at the Olkaria field.

The HP brine would be pumped or gravity flowed to the plant area, where for a double-flash plant, LP separator(s) would produce lower pressure steam, as well as brine for reinjection. Due to the higher specific volume of the lower-pressure steam and the need for large-diameter piping to the turbine to keep pressure drops reasonable, the lower-pressure separators for double- or triple-flash plants would be located closer to the turbine.

Wellhead separation results in a large number of separators, each with their attendant pressure, temperature, flow, and level instrumentation and, where required, brine transfer pumps and other accessories. This results in a large quantity of equipment in the field, as well as separate steam and brine pipelines.

In contrast, the centralized separator station, shown in Fig. 10.11, relies on conveyance of two-phase fluids from the wells to a common station closer to the turbine. The reduced number of vessels and greater concentration of other devices and construction

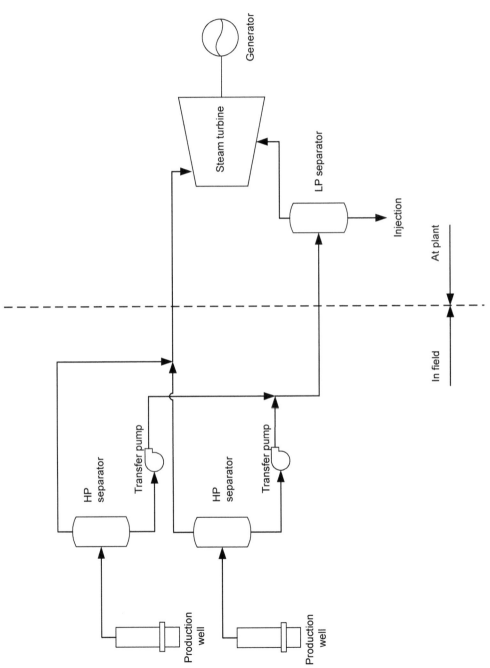

Figure 10.9 Wellhead separation configuration.

Figure 10.10 Wellhead separator at Olkaria. Steam exiting the vessel from the bottom passes through and exits from the top of a ball check valve assembly (to the lower right of the separator) that prevents liquid carryover. The brine line is seen exiting from the side of the separator and turning to the foreground.

work at a central location leads to lower capital costs. The feasibility of centralized separation depends on the terrain.

Fig. 10.12 shows a double-flash centralized separator station at Germencik I (Turkey). The two vertical HP separators to the right receive two-phase geofluid from the production wells. The horizontal vessels to the left are the LP separators.

Predictions of the appropriate flow regimens and pressure drops in two-phase piping across the spectrum of horizontal, vertical, and sloped piping have and continue to be a subject of ongoing research, and more-detailed descriptions of these challenges can be found in other works (eg, [8]; see also Chapter 9). Although improvements in the correlations for two phase flow regimes and pressure drops have led to more centralized separator station designs being implemented in newer plants, wellhead separation or compromises (clusters of satellite stations in the field where necessary) may still be appropriate where there are site-specific challenges.

10.4.1.2 Separator and demister design options

The design of vessels to separate the liquid and vapor phases falls into several basic categories: vertical cyclone and horizontal. Zarrouk and Purnanto [32] provide a detailed review of design aspects for several of these types, with key points summarized here.

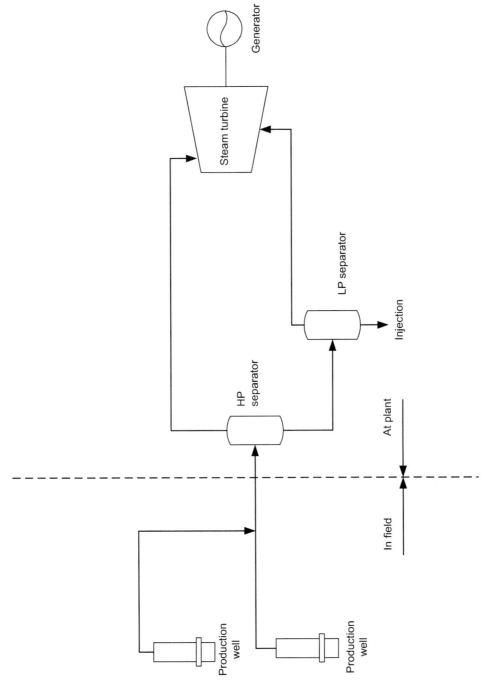

Figure 10.11 Centralized separation configuration.

Figure 10.12 Centralized separator station at Germencik I. Vertical HP and LP demisters can be seen behind the rightmost LP separator. The roof of the powerhouse can be seen behind the LP separators, and the cooling tower behind the HP separators. In the foreground is the brine pond, which can store excess brine during startup or operating excursions.

Although they are similar in appearance and physical operation, there are several kinds of separation vessels in a flash plant, each with distinct purposes. The first stage of separation of the two-phase geofluid into steam and liquid phases takes place in the separator proper. Since the separated steam may contain some liquid droplets due to less than perfect removal in the first stage, a second set of vessels is often added to "demist," "scrub," or "polish" the steam to achieve higher values of steam quality, generally 99.9% or better. This scrubbing may be done in conjunction with steam washing (covered in Chapter 11) to further reduce the impurities in the steam to the turbine. This final stage of moisture removal is carried out in vessels that may bear names such as "demisters," "scrubbers," "humidity separators," "mist eliminators," or something similar.

When two-phase fluid enters a vertical cyclone separator, such as that shown in Fig. 10.10, the largest liquid droplets fall by gravity to the bottom and exit the vessel as brine. Steam and noncondensable gases that may still entrain smaller droplets of brine rotate around the vessel (creating a "cyclone" pattern) as they approach an outlet pipe aligned vertically in the center of the vessel. This cyclone creates centrifugal forces that whirl the droplets to the vessel wall. These droplets coalesce on the wall and drain to the bottom of the vessel. The steam and noncondensable gases that flow to the pipe in the center of the vessel are extracted from a nozzle at the top or bottom of the vessel.

Lazalde-Crabtree [19] provided detailed sizing equations for vertical cyclone separators with a bottom outlet. Due to the reliance on centrifugal forces to achieve the separation, the vertical cyclone separator outlet steam quality is dependent on the steam velocity at the separator inlet and within the upflow zone in the separator. At low steam velocities, the centrifugal forces are low, resulting in less efficient separation and lower exit steam qualities. As the velocities increase, the outlet steam quality increases until it reaches a threshold where the steam entrains droplets within it, and

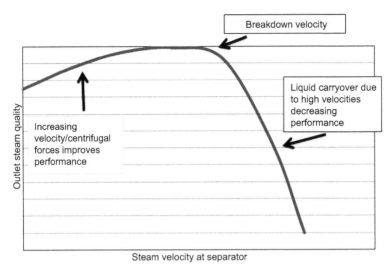

Figure 10.13 Separator outlet quality as a function of separator inlet velocity.

liquid may creep up the vessel and pipe walls despite gravity. As a result, the outlet steam quality decreases rapidly after a certain "breakdown" velocity. Fig. 10.13 shows this qualitatively; a fuller comparison of modeled versus field data showing this breakdown tendency can be found in Zarrouk and Purnanto [32]. Thus, the design of vertical cyclone separators must review normal and off-design operation to verify the separator operates in an appropriate range.

Horizontal separators, such as the LP separators in Fig. 10.13, are designed to slow the steam as it travels horizontally, such that there is sufficient residence time for liquid droplets to fall and collect in a liquid level under the influence of gravity and be removed from the bottom of the vessel as the steam is removed from the top. The steam outlet quality of a horizontal separator is thus a function of residence time (volumetric flow rate, area, and length), size of the droplets, and their drag coefficients. While horizontal separators consume a greater plan area, they are not as tall as vertical ones, making it easier to house them inside buildings and access the level instrumentation. Perhaps partly as a result, horizontal separators appear to be more prevalent in Icelandic and Russian designs for colder climates.

Steam demisters generally rely on refinements of these schemes to achieve higher outlet steam qualities. Vertical cyclone demisters are often fitted with secondary means to collect water deposited on the walls of the inner tube that carries the steam out. The larger liquid droplets are then sometimes "recycled" into the main body of the vessel to help coalesce and separate out smaller droplets. Another demister style uses chevron-type or mesh pad internals, which rely on steam passing through a set of angled vanes or pads of wire. Liquid droplets, with larger momentum than the steam, are collected on the vanes or pads and fall to the bottom of the vessel.

In summary, there are a range of separator and demister designs, and there may be several appropriate choices for a particular application. For separators or demisters with complex internals or small passages, care must be taken to ensure the design is suitably corrosion, vibration, and scale resistant.

10.4.1.3   Reboilers

Separators are used to segregate the geofluid into gaseous and liquid phases; however, the gaseous phase contains both steam and NCG. A flash plant equipped with a condenser requires considerable investment in NCG removal equipment and parasitic loads. A reboiler is an upstream treatment option that can be used to reduce the content of NCGs in the steam. Reboilers may be direct-contact or surface heat exchangers, and operate on principles similar to a deaerator in a fossil plant. The mixture of steam and NCGs is admitted to the reboiler. The steam is condensed by liquid pumped to the vessel and the NCGs are vented. The hot condensate is subsequently flashed to produce relatively NCG-free steam. The remaining liquid from the flash is pumped back to the reboiler to condense the incoming steam. Fig. 10.14 shows reboilers in some direct-contact or indirect-contact (surface exchanger) configurations; other configurations are possible.

The reboiler incurs additional capital cost, a pressure drop, and a loss of exergy due to the heat exchange across the temperature difference and the flash process. These costs must be weighed against the reduction in size of the NCG removal system. Although reboilers have not yet seen wide application in the geothermal industry with only a selected few installations at locations such as the Salton Sea, they remain another option for designers dealing with high-NCG resources.

10.4.1.4   Vent stations and turbine bypass

During steady state, geofluid flows from the wells to the separators, steam flows from the separators to the turbine, and brine flows from the separators to injection wells. In the event of a plant trip or a reduction in load, the valves admitting steam to the turbine close, and the pressure in the gathering system rises. To control this transient condition, pressure control valves are actuated upstream of the turbine, discharging excess steam to the atmosphere through a silencer or rock muffler. While vent stations are most commonly used, an alternative is to bypass the steam around the steam turbine to the condenser, if the cooling system is still operating. These bypasses may be useful in locations where the discharge of steam or NCGs (particularly H_2S) to the atmosphere is restricted.

While vent stations or turbine bypasses relieve the short-term transient, for longer periods of load reduction it would be preferable to reduce the total flow from the wells, rather than continuing to extract geofluid for little benefit. However, changing flows from the wells may cause degradation to the reservoir, depending on its characteristics.

The baseload capability of a geothermal plant is a strong comparative advantage compared with other renewable energy sources such as wind or solar, and the design of power plants to more effectively load follow is becoming of increasing importance with the larger quantity of such intermittent sources on the grid. Load following strategies are also important for islands or small grids where a geothermal plant may be a

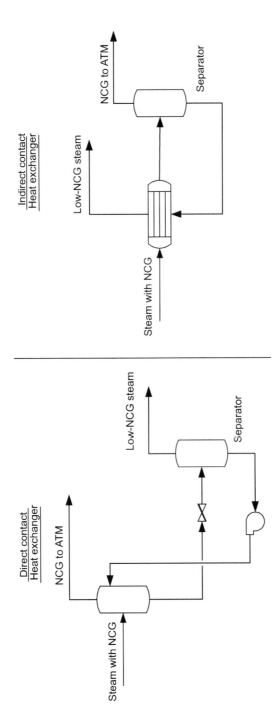

Figure 10.14 Direct- and indirect-contact reboilers. Adapted from Duthie RG, Nawaz M. Comparison of direct contact and kettle reboilers to reduce noncondensables in geothermal steam. GRC Trans 1989;13.

major contributor to the total load, and where significant diurnal load variations are expected. Due to reservoir, plant, and local environmental conditions, the design of a geothermal plant to accommodate changes in load, while best preserving its geofluid, can be a site-specific challenge.

10.4.1.5 Brine injection pumps

A final component in the design of a flash plant gathering system is the strategy for reinjection of the brine into the reservoir. From the viewpoint of the separation station and flash plant designer, the appropriate and timely selection of the brine reinjection pumps can be a challenge, especially for new projects where the injection characteristics may not yet be well known. Reinjection wells are often the last to be drilled and tested, and thus the reinjection pressure for the required flow rates may not be known until late in the project. Depending on the flow rate and injection pressure required, these pumps may consume significant parasitic power, and may have long lead times. After commissioning, the reinjection characteristics or strategies may change as the reservoir matures. Thus creative solutions for the brine injection pumps such as considerable margin on head, variable frequency drives, multiple pumps each carrying some fraction of the total flow, or other design approaches with flexibility may be used.

10.4.2 Steam turbines

The steam turbine for a geothermal flash plant is a major driver in the cost and schedule of the project. The cost of the turbine, generator, and associated auxiliaries might represent around 15—20% of the total power plant cost. Research into improving efficiency, increasing size, and reducing delivery time all can improve the project economics.

10.4.2.1 Condensing steam turbines

One critical parameter impacting turbine performance is the velocity of steam exiting the last-stage blades. For the same exiting volumetric flow rate, a turbine with a smaller exhaust area will have a higher exhaust velocity, and hence higher kinetic energy losses ("leaving losses"). This leaving loss effectively raises the turbine exhaust enthalpy, resulting in lower enthalpy drop and work output per unit mass of steam passing through the turbine. Fig. 10.15 shows an indicative set of curves of exhaust loss as a function of annulus velocity and turbine configurations for a rotational speed of 3600 rpm.

Higher mass flow rates (in the pursuit of larger units) and lower condenser pressures (to extract more work from each unit of mass passing through the turbine) combine to drive volumetric flow rates through turbines higher. To cope with this, designers can increase the number of flow paths by using double-flow configurations or multiple turbines. Another approach can be to increase last stage-blade lengths to provide greater annular area in order to keep leaving losses within reason.

Bottom exhaust double flow turbine configurations have been historically popular for large units, such as the many 50- to 55-MW units installed at the Geysers and

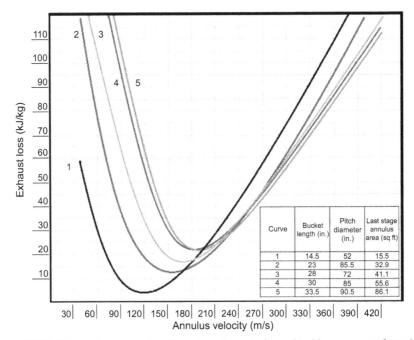

Curve	Bucket length (in.)	Pitch diameter (in.)	Last stage annulus area (sq ft)
1	14.5	52	15.5
2	23	85.5	32.9
3	28	72	41.1
4	30	85	55.6
5	33.5	90.5	86.1

Figure 10.15 Exhaust losses as a function of annulus velocity and turbine geometry for turbine at 3600 rpm.
Adapted from Cotton KC. Evaluating and improving steam turbine performance. New York: Cotton Fact; 1998.

throughout Indonesia. The bottom exhaust steam turbine provides a low pressure drop between the turbine and condenser, but requires an elevated turbine deck and pedestal, results in a taller powerhouse, and takes longer to construct. Smaller units historically were top exhaust, with the turbine and condenser mounted at the same elevation. This allows a shorter construction period and easier access for operations and maintenance, but introduces a larger pressure drop between the turbine and condenser, reducing performance.

Longer last-stage blades designs made possible through improvements in metallurgy and analysis allow single-flow units to reach over 50 MW, such as the 55 MW MHPS turbine installed at Los Azufres Unit 17 (Mexico). As larger single-flow machines are developed, axial exhaust configurations become more appealing. Fig. 10.16 shows these three configurations.

The axial exhaust configuration has several advantages over either the bottom or top exhaust machines. These include:

- Lower powerhouse and smaller turbine pedestal, compared with the bottom exhaust
- Easier turbine maintenance than the top exhaust, as it is not necessary to remove the balance bellows to remove the turbine upper casing
- Lower exhaust pressure drop compared with the longer crossover duct with two 90 degree bends for the top exhaust

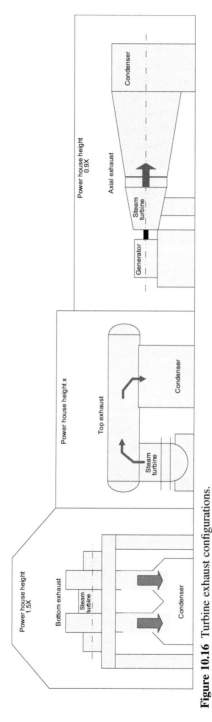

Figure 10.16 Turbine exhaust configurations.
Adapted from MHI. A new geothermal steam turbine with a single cylinder axial exhaust design (outline of the Hellisheidi geothermal turbine in Iceland). Tech Rev December 2007;44(4).

Conversely, the axial exhaust configuration has several limitations that must be addressed in the design. The maximum size of the unit using current blade technology is constrained to around 50–55 MW, due to the single-flow design. If larger units are desired, they generally will be double flow, bottom exhaust. Water induction relates to the risk that a rising water level in the condenser would come up to the level of the spinning last stage blades, causing damage to the turbine. A trip of the pumps removing water from the condenser in fossil or geothermal plants results in a turbine trip. For surface condensers, the rate of water level rise following a trip of the condensate pumps is not significant since the source of the condensate (incoming steam) is cut off. However, if a direct-contact condenser is used, circulating water can continue to flow into the condenser after a turbine trip because it is drawn into the condenser by vacuum. Vacuum breakers are generally provided, but the distance and hence margin against water induction are less for axial exhaust than for bottom or top exhaust designs. To provide additional margin, the speed of vacuum breaking may be increased, additional water shutoff valves may be added, a surface condenser may be used, and/or the condenser may be set at a lower elevation, increasing the cost of civil works and introducing some pressure drop.

Due to the advantages and regardless of these limitations, the advantages of axial exhaust machines have led to them being adopted in several recent flash plant projects such as Hellisheidi (Iceland, multiple units, each 45 MW), Los Azufres Unit 17 (Mexico, 55 MW), Maibarara (Philippines, 20 MW), and Alasehir (Turkey, 30 MW). It is anticipated that axial exhaust units will continue to be a competitive option for developers, occupying a middle ground between the bottom and top exhaust configurations, and may reach higher MW levels with advances in blade technology.

10.4.2.2 Wellhead and backpressure turbines

Projects such as Nga Awa Purua at 140 MW (New Zealand) or the 2×70–MW units at Olkaria IV (Kenya) currently represent the larger capacity ends of the geothermal flash plant development spectrum. Large projects may take more than 7–10 years to develop, considering the phases of exploration, drilling, plant design, construction, and commissioning. The burden of many years of investment without cash flow can be an insurmountable obstacle for private developers.

One strategy to improve return on investment is to install small wellhead generators. These are typically in the 2- to 5-MW size range, and may harness the flow from one or several wells on a common wellpad. The gathering system is modest, as the units are located at the wellpad. The wellhead unit can be more closely tailored to the deliverability curve of the well, unlike larger units where wells with many different characteristics must be grouped to deliver steam at a common pressure to a large unit. Wellhead units thus may have particular appeal for weaker or geographically isolated wells. Fig. 10.17 shows the 2.4-MW Eburru power plant in Kenya. This project uses a condensing steam turbine, direct-contact condenser, and fiberglass cooling tower.

Wellhead units thus can serve as the vanguards in a new field. Placed online soon after the wells are drilled, they can provide an early source of revenue and serve as a

Figure 10.17 Eburru wellhead power plant [24]. The turbine and generator are located deeper within the enclosure; the condenser is in the foreground. The hotwell pump is to the left of the condenser, and the vacuum pump to the right. The cooling tower is not shown in this view.

long-term flow test for the wells. Small units can often be installed more rapidly than large units, especially if surplus or reconditioned units can be used. If later a larger, more efficient centralized plant is installed, the wellhead units can be moved to a new site.

Backpressure units that exhaust to atmosphere rather than to a condenser and cooling tower generate only about half the energy that would be produced from a condensing unit, but require very little infrastructure and are more portable, allowing them to "hop" between wells more easily. Wellhead units could also use binary technology; if air-cooled condensers are used, no cooling water or associated treatment systems are required.

This phased development approach has been used in many locations. One example is at San Jacinto in Nicaragua. Initially two 5-MW backpressure units, which had formerly been intended for use in El Salvador, started operation in 2005. These units were operated through 2013, at which time the steam was transferred to larger and more efficient $2 \times 38.5-$MW condensing units. The backpressure turbines were subsequently removed and transferred to another project in Mexico, where they are providing yet more generation in advance of the installation of a larger condensing unit.

Similar "phased development" approaches using small wellhead units have been used at large fields such as Los Azufres in Mexico and Olkaria in Kenya. Studies (eg, [21]) indicate that the internal rate of return (IRR) of a project may be improved by several percentage points with earlier wellhead generation. The increased confidence in well deliverability characteristics that allows better tuning of the subsequent large condensing plant design may be another appealing consideration.

Small wellhead units may have several drawbacks. While backpressure units have been successfully "hopped" at several locations, there is less history on the time and degree of effort required to move condensing units, with their associated cooling towers and large equipment foundations. The smaller turbines are often equipped with gearboxes and have lower cycle efficiencies than the larger flash plants. There are no extensive O&M cost data on maintaining multiple small units versus a single large unit; some economies of scale might be expected in favor of the large unit. In order to take full advantage of the earlier potential completion of a wellhead unit, transmission lines and substations must also be ready early. There are thus several logistical issues that have led to staged asset developments using wellhead units not being universally adopted. As more projects are planned, the supply chain for small units matures, and more capital and O&M cost data are gathered, it is anticipated that these might be more widely employed.

10.4.2.3   Biphase turbines

One thrust of turbine research for flash plants that has been under way for decades has been the development of a prime mover capable of accepting a two-phase mixture of water and steam. This would allow better harnessing of the total geofluid flow without the need for separation. This topic is discussed in more detail in Chapter 11, Total-Flow Systems. To date, few of these machines have entered commercial service at flash plants.

10.4.2.4   Supercritical cycles

Most flash plants harness resources with a downhole single-phase equivalent temperature around 180–300°C. There are certain locations where extremely high pressure and temperature resources can produce highly superheated or supercritical fluid in the 400–600°C range [16]. Harnessing this high-temperature, HP (potentially in the 100–200 bar range) fluid would present new cycle, equipment, and material challenges compared with conventional technology and might rely on either steam turbine or binary technology for power generation. The Iceland Deep Drilling Project (IDDP) is one venture that has drilled several high-temperature wells around Iceland. It remains to be seen how the challenging resource conditions, with HP, high temperature, and corrosive constituents such as chlorides, will be best addressed to produce power economically; this remains a research project.

10.4.3   Condensers

The previous section has focused on the "hot end" of the cycle — primarily advances in turbine technology. Of equal importance to the efficiency of the cycle is the "cold end" — the performance of the condenser and cooling water systems.

There are two basic types of condensers used in geothermal service: direct-contact and surface. Fossil plants, where maintaining the purity of boiler water is essential, use surface condensers. For the cycle equipped with a surface condenser, circulating water is pumped from the cooling tower, by circulating water pumps, through the tubes of the condenser. Warm water returns to the cooling tower. The turbine exhaust steam on the shell side condenses and forms condensate, which is pumped to the cooling tower via condensate pumps.

With a geothermal flash plant there is no true "boiler" that requires high water purity; thus, the turbine exhaust steam and cooling water may be mixed in a direct-contact condenser. In the direct-contact condenser, turbine exhaust steam combines with circulating water drawn from the cooling tower into the condenser by virtue of the latter's vacuum. The resultant condensed mixture is pumped from the condenser to the cooling tower by hotwell pumps.

There are many tradeoffs in the selection of direct-contact versus surface condensers. The advantages of the direct-contact condenser are more efficient heat exchange due to the direct mixing of the steam and circulating water. Direct-contact condensers are also generally lower cost and less prone to fouling, as they have no tubes. Surface condensers can be configured to have lower parasitic loads, since the circulating water pumps that must overcome the condenser tube side pressure drop may require lower head than the hotwell pumps, which must draw suction from the low direct-contact condenser pressure. Surface condensers also provide a higher purity stream of water as condensate, unmixed with the circulating water that may contain water treatment chemicals and oxygen. Any fraction of the condensate not used as cooling tower makeup can then be used more effectively for purposes such as steam wash or irrigation. Some noxious NCGs, such as H_2S, can be absorbed into the circulating water in a direct-contact condenser. These NCGs can subsequently be released to the atmosphere as the circulating water is sprayed into the cooling tower. At locations where H_2S emissions are strictly regulated, this secondary emissions path may be difficult to treat to achieve the target levels of H_2S reduction. In contrast, the surface condenser produces a much smaller stream of condensate that may contain H_2S, making it easier to abate before it reaches the cooling tower.

The direct-contact condenser requires a careful manipulation of the elevations between the cooling tower and condenser in order to ensure adequate flow of water to the condenser, drawn in by condenser vacuum and the elevation difference, but guarding against an excessive elevation difference that would increase the parasitic loads of the hotwell pumps. As mentioned in the axial exhaust turbine discussion, water induction can also be a concern for the direct-contact condenser. The surface condenser allows greater flexibility in plant layout and relative elevations and provides greater margin from water induction, making it well suited for axial exhaust configurations.

Both the direct-contact and surface condensers will likely continue to see application in the future, depending on the site-specific conditions and required features. It is not the intent of this work to identify a clear preferred solution; the appropriate application of the condenser type must be evaluated on a case-by-case basis. As the majority of flash plants installed use direct-contact condensers, we discuss here innovations that have improved performance for that style. Refer to Chapter 11 for more discussions of surface condensers, which are more prevalent at dry steam plants such as in The Geysers.

10.4.3.1 Direct-contact condenser improvements

The condenser has several purposes. One is to cool and condense the exhaust steam from the turbine in a "main" section. The second is to cool the existing NCGs prior to exiting the condenser, to reduce their volume and moisture to lower the work of the NCG removal system; this process takes place in a "gas cooler" section, which may be a subsection of the main shell of the condenser or be in a separate vessel. Geothermal direct-contact condenser designs at flash plants dating from the 1970s to the presently used sprays or trays to mix the cooling water, exhaust steam, and NCGs. The condensers were generally in a "spray−spray" configuration, with sprays used in both the main and gas cooler sections, or a "spray−tray" configuration, with sprays for the main section and trays for the gas cooler.

Table 10.6 shows the impact of condenser performance on the sizing of major components for a sample single-flash plant as earlier outlined in Table10.2. For the purpose of this comparison, it is assumed the gross power and condenser pressure are kept constant, and the benefits of a more efficient design are seen in the sizing of the balance of plant components. Subcooling, defined as the difference between the bulk condensing temperature and the temperature of the exiting condensate, is a measure of condenser efficiency. The greater the subcooling, the more cooling water is required, and, consequently, the less efficient is the condenser. The gas cooler approach is the temperature difference between the NCGs exiting the condenser and the cooling water temperature. A lower gas cooler approach means reduced temperature and vapor load to the NCG system. Improved condenser designs that reduce subcooling and gas cooler approach result in smaller hotwell pumps and cooling

Table 10.6 **Direct-contact condenser performance impacts on balance of plant**

Parameter	Units	Conventional condenser	Improved condenser
Condenser pressure	bar,a	0.09	0.09
Subcooling	°C	4	2
Gas cooler approach	°C	5	2
Cooling water flow	m^3/h	14,408	12,590 (−13%)
Total flow of NCGs and water vapor to ejectors	kg/h	5945	5485 (−8%)
Hotwell pump power	kW	1262	1107 (−13%)
Cooling tower fan power	kW	1011	854 (−16%)
First-stage ejector steam consumption	kg/h	11,388	10,252 (−10%)

towers and reduced parasitic loads. Thus, the evaluation of condenser improvements must consider the broader plant impact.

Recent innovations in direct-contact condenser designs include the Advanced Direct Contact Condenser (ADCC), originally developed by the National Renewable Energy Laboratory (NREL) for use in Ocean Thermal Energy Conversion (OTEC) projects but which has been applied effectively to geothermal projects. The technology is licensed by SPX Heat Transfer LLC. The ADCC relies on structured packing, similar to counterflow cooling tower film fill, to allow film-wise heat transfer between the various phases in the condenser. Refer to Chapter 11 for a schematic of the ADCC internal arrangement, where the packing is used in both the main and gas cooler sections. The flow of steam and cooling water is cocurrent in the main section, and the flow of NCGs and water is countercurrent in the gas cooling section. Fig. 10.18 shows an ADCC main condenser installed at the San Jacinto single-flash plant in Nicaragua. For this project, the intercondenser and aftercondenser of the NCG system are also ADCCs and are mounted on the main condenser shell to reduce interconnecting piping and structural steel.

A limitation of the ADCC is that the steam and cooling water flow paths must be cocurrent in the main section; this means the design is better suited for a bottom or top exhaust turbine configuration than axial exhaust. ADCC condenser technology has been increasingly applied over the past decade, with units in The Geysers (retrofit), Mexico, Indonesia, Nicaragua, Kenya, and Turkey. Their economic competitiveness is a function of the condenser capital cost as weighed against the cost and parasitic power impact on the balance of the plant: hotwell pumps, cooling tower, NCG system, and associated piping and valves.

Figure 10.18 ADCC at San Jacinto, Nicaragua. The two-stage ejector NCG system, exhausting to their associated rectangular intercondenser and aftercondenser, is mounted on the near side of the main condenser.

10.4.4   Cooling system considerations for flash plants

Chapter 14, Waste Heat Rejection Methods, covers cooling systems in more detail. In this section, we briefly discuss the evolution of specific cooling tower technology applied to flash plants and the potential for dry cooling.

10.4.4.1   Water cooled

The state of the art for wet cooling towers for flash plants has steadily evolved over the past decades. Throughout the 1980 and 1990s, wood and concrete towers with splash bar fill were widely used, and some still find their way into recent projects. Cooling towers using wood structures may be economical but require care to keep the wood damp during outages to prevent splintering and can be prone to fires, such as the wildfire that destroyed the Unit 5/6 tower at The Geysers in 2013. Concrete towers are robust but require extensive formworks and work above grade. Depending on local market conditions, these may be a costly option and present construction duration and safety challenges.

Towers built with fiberglass-reinforced polyester (FRP) are becoming increasingly popular. These offer lower shipping weight and installation labor compared with wood or concrete and better fire and environmental resistance than wood. They can be configured with either the traditional crossflow splash bar fill, which offers good fouling resistance, or more efficient counterflow film fill versions, which can reduce capital costs and parasitic fan power loads, although the water chemistry must be monitored more carefully to prevent fouling. Film fills come in a variety of configurations, some with more tortuous paths for better heat transfer between the water and air that are more efficient but more prone to fouling, and low clog configurations with performance and fouling potential that lies between splash bar and high efficiency film fill.

The lighter structures of FRP provide less mass, thus the structures and mechanical equipment (fan, motor, and gearbox) must be designed and monitored to prevent undue vibration. FRP tower price is sensitive to the price of oil, which is used as a feedstock in its production, and so the competitiveness of any of the three tower structure types is often worth revisiting for any particular project due to market conditions.

10.4.4.2   Air cooled

Air cooling is becoming a more common option for conventional fossil combined-cycle plants, especially in arid areas where there is little makeup water available for wet cooling. Geothermal flash plants usually produce sufficient water from the condensation of steam to allow wet cooling. Typically around 75% of the condensed steam from a flash plant is evaporated in the cooling tower, and around 25% may be available for makeup and blowdown from the cooling tower to keep the concentration of solids in the tower at an acceptable level. A plant consuming 1000 t/h of geofluid that undergoes a 20% flash producing 200 t/h steam thus might evaporate 150 t/h of water and return 800 t/h of brine and 50 t/h of cooling tower blowdown to injection or disposal.

During the hottest periods of summer days, the net water balance may turn negative, and the operating plant would see a slow reduction in cooling tower water level. However, the large volume of the water in the cooling tower basin often permits this depletion during the day to be replenished by more condensate production at night, so that an external source of makeup water is usually not required for flash plants.

An air-cooled unit would have more geofluid consumption per unit of net power output due to the less efficient cycle design but would return a higher percentage of the geofluid consumed to the reservoir, theoretically 100%. Since binary plants must use surface condensers, they are well suited to being air cooled. Cooling configurations for binary plants are covered in more detail in Chapter 11.

Some coal-fired or combined cycle fossil plants use air-cooled condensers (ACCs) that are similar to those used in binary geothermal projects but in horizontal or A-frame configurations. For a geothermal flash plant, however, there are several challenges in attempting to use air cooling with this configuration. These include [22]:

- The difficulty of removing large quantities of noncondensable gases present in the geothermal steam. There are only trace amounts of NCGs present in fossil or binary plant working fluids.
- The potential for deposition of solid impurities from the geothermal steam, including products such as oxidized sulfur from H_2S, on the tube side of the ACCs, which would be difficult to clean. The purer working fluid in a conventional fossil or binary plant does not present this difficulty to the same degree.
- In place of an ACC, one could use a direct-contact or surface condenser cooled by water that is subsequently cooled by water to air heat exchangers. These mitigate the NCG removal and solid deposition issues, but the intermediate loop introduces higher capital costs and parasitic loads.

As a result, the performance and cost impact of supplying flash plants with air cooling is a considerable burden unless the valuation of the increased water flow rate to the reservoir is sufficiently high to merit the investment. The Mutnovsky flash plant in Russia is a unique example of a flash plant using an A-frame type ACC. The ACC circumvents issues with freezing in cooling towers that would have been an issue for this cold and remote location [29]. Another approach to capturing the performance and economic benefits of steam turbine technology for high-enthalpy resources that require air cooling is to use a combined cycle, such as the Mokai project in New Zealand discussed earlier.

10.4.4.3 Hybrid cooling

Complete air cooling for flash plants may be a challenge, but several technologies exist to reduce the quantity of water evaporated from wet cooling towers. Lindahl and Jameson [20] describe a set of configurations, which may differ by vendors, but generally rely on a set of air-cooled heat exchangers that initially remove energy from the warm circulating water sent to the cooling tower. After passing through the dry cooling section, the water is further cooled in a conventional wet cooling section in the tower. These configurations can also mitigate the visual plume from the tower, especially on colder days. Chapter 15 discusses hybrid cooling in more detail. To date, flash plants have not used hybrid cooling towers; however, as the value of water increases, these may see broader application.

10.4.5   Noncondensable gas removal systems

The performance of a flash plant is contingent on maintaining a low condenser pressure. The condenser pressure is dictated by two factors, either of which may be limiting:

- The condenser performance, which is a function of the thermal duty (amount of steam that must be condensed) and the cooling water flow rate and temperature
- The NCG removal system capacity

A common misperception is that the gas removal system "sets" the condenser pressure. If the NCG system is of sufficient capacity, the condenser pressure is controlled by the condenser characteristics, and no significant further reduction in condenser pressure results if the NCG system has surplus capacity. The condenser pressure would rise or fall naturally as the cooling water temperature or steam flow vary. It is in situations where the NCG system has insufficient capacity that improvements in it can improve the condenser pressure.

NCG systems are in principle similar to those used for conventional steam turbine Rankine cycle plants, which must remove air from the condenser that enters the cycle due to in-leakage. The difference for a geothermal flash plant is that the steam itself contains larger quantities of NCGs that originate from the geothermal fluids and may be composed of mixtures of gases such as CO_2, H_2S, H_2, CH_4, and other constituents. The quantity depends on the resource characteristics and can vary widely, perhaps from 0.5% to 30% by mass of the total HP steam plus NCG content. The NCG content and composition may also change over time as the reservoir evolves. As one example of variability that can be experienced, Kotaka et al. [18] reported on the NCG system at the Kawerau double-flash plant in New Zealand. The design basis was originally 2.3% by mass NCG content in the HP steam, but during commissioning the level was as high as 3.2−3.5%, later dropping to less than 2.8%. These variations were accommodated with a flexible NCG system with several trains of different capacities.

Flash plants with direct-contact condensers also have air liberated from the cooling water. Thus, NCG systems at flash plants must deal with much higher and variable volumes of NCGs to be removed, and often more corrosive mixtures of gases than would be encountered in conventional fossil plants. Configuring the NCG system for efficiency, reliability, and flexibility thus is a major task for the flash plant designer.

10.4.5.1   Ejectors

Ejectors are the most common NCG system component. These use motive steam to draw NCGs and water vapor from the condenser and compress it to higher pressures. As a typical condenser pressure might be around 0.1 bar,a, and a typical ejector compression ratio may be 3:1 to 4:1, it generally takes two to three stages of compression to exhaust to atmospheric pressure, about 1 bar,a. Multiple stages of ejectors usually are configured with intercondensers that cool the NCGs and condense steam in order to reduce the work of the subsequent stages.

Ejectors have the advantage of being inexpensive and rugged, with no moving parts. However, the motive steam they consume can represent a significant reduction in generation compared with what could otherwise be provided if passed to the turbine.

10.4.5.2   Liquid ring vacuum pumps

Liquid ring vacuum pumps operate by drawing NCGs into a rotating vane, eccentrically mounted in a liquid ring. NCGs are drawn into the vane sections of larger area and gradually compressed, exiting via ports placed where the vane area is smaller. These are generally motor driven and have much higher efficiencies than ejectors; one can thus with an investment of parasitic electric power save on the motive steam consumption, which can instead be used in the turbine. Due to limitations as a result of vaporization of the water used for the liquid ring, these generally cannot operate at suction pressures as low as the condenser pressure, and thus are more often used as the final element in a multiple (two- or three-) stage hybrid system. Fig. 10.19 shows the hybrid NCG system at the Germencik double-flash plant.

10.4.5.3   Turbocompressors

Turbocompressors are another option, analogous to LRVPs but generally higher speed, more costly, and higher efficiency. Turbocompressors are found in many plants in Larderello (Italy) and others influenced by those designs, such as Cerro Prieto, Kızıldere I, and Miravalles Units 1 and 2. These may be steam turbine or electric-motor driven, often through a gearbox, and may operate at variable speeds to

Figure 10.19 Hybrid NCG system at Germencik. The last-stage liquid ring vacuum pumps are in the foreground; ejectors and associated condensers are housed in the steel structure behind.

accommodate changes in NCG content. While they offer promising motive steam savings over an ejector installation, the higher capital and O&M costs can be daunting for the project economics.

10.4.5.4 Other NCG removal strategies

Other strategies to deal with high-NCG concentrations at flash resources include the use of reboilers, combined cycles, or binary cycles. The decision of which type of NCG system to use depends on factors such as the value of steam, O&M considerations, anticipated variability of NCG content in the future, and capital cost constraints. Each has applications in certain settings, and systems may also be retrofitted and reconfigured with different types later in plant life as the reservoir characteristics evolve. While some general aspects of NCG systems are covered here, additional insight into system design is covered in Chapter 11.

10.4.6 Environmental systems – H$_2$S abatement

Geothermal steam often contains H$_2$S, which is generally vented to atmosphere. H$_2$S is a noxious gas that can lead to respiratory problems and environmental corrosion even at minor (several ppm) levels of concentration. While there are many nations that do not have emissions limitations on H$_2$S, several jurisdictions, such as California in the United States, Italy, and Japan, have emissions limits that often require flash plants to install abatement systems for the vented NCGs. There is a proliferation of techniques, and more nations are considering limiting these emissions. Some of the H$_2$S abatement techniques used include:

1. Incinerators or regenerative thermal oxidation, when the NCGs contain sufficient flammable constituents such as H$_2$S, H$_2$, or CH$_4$. The burner/scrubbers at The Geysers, and the AMIS systems at Larderello (covered in Chapter 19), are examples of these processes.
2. Liquid redox methods, which convert H$_2$S to elemental sulfur, which is disposed of as a solid. These methods would include the Stretford systems at The Geysers and LO-CAT at Coso in California.
3. BIOX, which uses biocides in the cooling water to convert H$_2$S to water-soluble sulfates. This technique is used at the John Featherstone plant in the Salton Sea.
4. Separation/injection where H$_2$S is removed from the NCGs and reinjected to the reservoir. Experiments are under way at the Hellisheidi plant in Iceland for this technique.

Abatement equipment can be a major capital and operating cost burden. Farison et al. [12] provide a comprehensive comparison of operating incinerator and Stretford systems and outline key O&M improvements based on that extensive experience.

10.5 Summary

Several decades ago flash plants were commonly configured as single-flash systems with bottom exhausting turbines, exemplified by the many 50- to 60-MW plants throughout Indonesia and the Philippines installed in the 1980 and 1990s. More recent trends for flash plants include combined and binary cycles applied to flash resources, axial exhaust turbines being applied up to larger MW ratings, double-flow bottom exhaust turbines reaching up to the 70- to 130-MW range in a single casing, a greater use of hybrid and flexible NCG systems, and the introduction of H_2S abatement systems. Improvements in scaling prediction and control have led to improved plant efficiencies made possible through the application of multiple flashes and bottoming units better tuned to the geofluid chemistry. Flash plants are increasingly being viewed as a "resource park" providing cogeneration or coproduction harnessing gases, brine, and potentially valuable solids. The plethora of available cycle configurations, variations in resource chemistry, and other site-specific factors make flash plants well suited to custom designs that can maximize the economic output of the project.

References

[1] Albertsson A, Jonsson J. The Svartsengi resource park. In: Proceedings of the world geothermal congress; 2010 April 25−29; Bali, Indonesia; 2010.
[2] Ballzus C, Frimannson H, Gunnarsson GI, Hrolfsson I. The geothermal power plant at Nesjavellir, Iceland. In: Proceedings of the world geothermal congress; 2000 May 28− June 10; Kyushu-Tohoku, Japan; 2000.
[3] Benoit D. The long-term performance of Nevada geothermal projects utilizing flash plant technology. GRC Trans 2014;38.
[4] Bombarda P, Macchi E. Optimum cycles for geothermal power plants. In: Proceedings of the world geothermal congress; 2000 May 28−June 10; Kyushu-Tohoku, Japan; 2000.
[5] Buescher E. Green district heating − actual developments of deep geothermal energy in Germany. In: Proceedings of the world geothermal congress; 2015 April 19−25; Melbourne, Australia; 2015.
[6] Carbon Recycling International. Expansion of Carbon Recycling International environmentally friendly fuel plant celebrated. 2015. http://www.carbonrecycling.is/index.php? option=com_content&view=article&id=72%3Aexpansion-of-carbon-recycling-international-environmentally-friendly-fuel-plant-celebrated&catid=2&Itemid=6&lang=en [accessed June 2015].
[7] Cotton KC. Evaluating and improving steam turbine performance. New York: Cotton Fact; 1998.
[8] DiPippo R. Geothermal power plants: principles, applications, case studies, and environmental impact. 3rd ed. Oxford, UK: Butterworth-Heinemann; 2012.
[9] Dunford T, Lewis B, Wallace K, Harvey W. Combined cycle strategies for high enthalpy, high noncondensable gas resources. GRC Trans 2014;38.
[10] Dunstall MG, Foster B. Geothermal greenhouses at Kawerau. Geo-Heat Cent Q Bull 1998; 18(1).
[11] Duthie RG, Nawaz M. Comparison of direct contact and kettle reboilers to reduce noncondensables in geothermal steam. GRC Trans 1989;13.
[12] Farison J, Benn B, Berndt B. Geysers power plant H_2S abatement update, vol. 34. GRC Transactions; 2010.
[13] Featherstone J, Butler S, Bonham E. Comparison of crystallizer reactor clarifier and pH mod process technologies used at the Salton Sea geothermal field. In: Proceedings of the world geothermal Congress; 1995 May 18−31; Florence, Italy; 1995.

[14] Gallup DL. Geochemistry of geothermal fluids and well scales, and potential for mineral recovery. Ore Geol Rev 1998;12:225−36.

[15] Gallup DL. Brine pH modification scale control technology. 2. A review. GRC Trans 2011;35.

[16] Hjartarson S, Saevarsdottir G, Ingason K, Palsson B, Harvey W, Palsson H. Utilization of the chloride bearing, superheated steam from IDDP-1. Geothermics 2014;49:83−9.

[17] Klein K, Gaines L. Reducing foreign lithium dependence through co-production of lithium from geothermal brine. GRC Trans 2011;35.

[18] Kotaka H, Shingai H, Gray T. Gas extraction system in Kawerau geothermal power plant. In: Proceedings of the world geothermal Congress; 2010 April 25−29; Bali, Indonesia; 2010.

[19] Lazalde-Crabtree H. Design approach of steam-water separators and steam dryers for geothermal applications. Geothermal Resources Council Bulletin; September 1984.

[20] Lindahl P, Jameson R. Plume abatement and water conservation with the wet/dry cooling tower. Cooling Tower Institute 1993 annual meeting. 1993. Technical Paper TP93-01.

[21] Long M, Raman R, Harvey W. Staged asset development - commercial and technical advantages of using a wellhead generation unit. In: Proceedings of the 4th African rift geothermal conference, 2012, 21−23 November, Nairobi, Kenya; 2012.

[22] Louw R, Wallace K, Harvey W. Air cooling options for flash plants. GRC Trans 2015;35.

[23] Lund JW, Boyd TL. Direct utilization of geothermal energy 2015 worldwide review. In: Proceedings of the world geothermal congress; 2015 April 19−25; Melbourne, Australia; 2015.

[24] Mendive DL, Green LH. Wellhead geothermal power plant at Eburru, Kenya. GRC Trans 2012;36.

[25] Menzies AJ, Brown P, Searle J, King R. The Mokai geothermal power plan, New Zealand: analysis of performance during first year of operation. GRC Trans 2001;25.

[26] MHI. A new geothermal steam turbine with a single cylinder axial exhaust design (outline of the Hellisheidi geothermal turbine in Iceland). Tech Rev December 2007;44(4).

[27] Moya P, DiPippo R. Miravalles unit 5: planning and design. GRC Trans 2006;30.

[28] Newell DG, Whitescarver OD, Messer PII. Salton Sea Unit 3 47.5 MWe geothermal power plant. Geothermal Resources Council Bulletin; May 1989.

[29] Povarov OA, Nikolskiy AI. Experience of creation and operation of geothermal power plants in cold climate conditions. In: Proceedings of the world geothermal Congress; 2005 April 24−29; Antalya, Turkey; 2005.

[30] Reykjavik Energy. 2013 Annual Report. 2013.

[31] Suryata I, Svavarsson HG, Einarsson S, Brynjolfsdottir A, Maliga G. Geothermal CO_2 bio-mitigation techniques by utilizing microalgae at the Blue Lagoon, Iceland. In: Proceedings of the world geothermal congress; 2010 April 25−29; Bali, Indonesia; 2010.

[32] Zarrouk SJ, Purnanto MH. Geothermal steam-water separators: design overview. Geothermics 2015;53:236−54.

Direct steam geothermal energy conversion systems: dry steam and superheated steam plants

11

K. Phair
Global Power Solutions, Golden, CO, United States

11.1 Introduction

11.1.1 Early applications

Development of geothermal energy for producing electric power essentially began at steam-dominated resources at Larderello in Italy and at The Geysers in the United States. Fumaroles and other surface manifestations brought geothermal resources to the attention of developers in these regions. For more information on the development at Larderello, see chapter "Larderello: 100 years of geothermal power plant evolution in Italy." For more information on the development at The Geysers see chapter "Fifty-five years of commercial power generation at the Geysers geothermal field, California: the lessons learned."

Not only were dry steam plants some of the first geothermal power plants, but the installed cost of a dry steam power plant has traditionally been the lowest of any type of geothermal power plant. Unfortunately, dry steam geothermal resources are very rare in nature.

11.1.2 Unit size

As for any geothermal power plant, it is prudent to select the size for a new unit using the best available resource productivity information that is available. For the first unit on a newly developed resource, the available resource data are typically limited to well flow test data. Review of recent multiunit developments on dry steam resources shows a progression in unit size from small (~ 10 MW) for the first unit trending to larger (up to ~ 80 MW) for later units. Dry steam units on exceptionally productive resources with high-energy density in Indonesia have been even larger at over 120 MW.

Economics of scale play an important role when determining optimum unit size. In general terms, with sufficient available resource production, the most cost-efficient dry steam geothermal power plant would use the largest available double-flow steam turbine that is suitable for geothermal service. However, resource production capability must be considered in the selection of unit size.

With most recent geothermal power plants being developed to supply electrical power under a Power Purchase Agreement (PPA), the capability of the resource should

Geothermal Power Generation. http://dx.doi.org/10.1016/B978-0-08-100337-4.00011-5

be known before PPA negotiations are completed. The optimum PPA would allow sales of power at or near the capability of the resource.

11.1.3   Power cycle

The power cycle for geothermal steam power plants is comparatively simple. In general, the principles and techniques as applied to traditional thermal power plants apply equally to geothermal steam plants. However, geothermal projects demand consideration of detailed differences in the process fluids, equipment applications, and even procurement methods to achieve the greatest economic benefit from a geothermal resource.

The potential economic benefit of a project will be strongly enhanced by detailed optimizations to extract the greatest value from a geothermal resource. Geothermal fluids vary between resources and frequently between wells in a given resource. From this reality it follows that each geothermal power plant must be engineered for the resource on which it operates. Further, operating experience has shown that geothermal resource production can and does change over time. This adds to the engineering challenge to incorporate off-design capability and flexibility to the power cycle at minimum additional cost.

11.1.4   Steam quality

Steam quality is crucial to reliable long-term operation of a geothermal steam plant. The challenges of preventing scale formation and/or erosion are best managed upstream of the power cycle equipment. Where traditional power cycles use superheated steam to improve cycle efficiency, geothermal steam power cycles typically use saturated steam to protect against degradation from scale, corrosion, and erosion.

11.1.5   Steam systems

Geothermal steam supply systems do more than transport the steam from the wells to the turbine. Equipment in the steam system processes the produced geothermal steam to deliver clean steam of sufficient quality to the power plant for long-term reliable operation.

11.1.6   Turbines

Geothermal steam turbines generally follow the design practices of the low pressure turbines at traditional thermal power plants. Recent enhancements in the design and construction of geothermal steam turbines have improved efficiency, reliability, and durability. In the 1980s, turbine overhauls were required on a biannual schedule. New materials and construction features now offer production runs of up to 10 years or longer between overhauls.

11.1.7   Condensers

Worldwide, the majority of geothermal steam plants use direct contact condensers. Where control of hydrogen sulfide (H_2S) emissions is required, surface condensers have been the economic choice. Recent developments in direct contact condenser technology have produced a packed condenser that has operated well in a number of installations.

11.1.8   Gas removal systems

A geothermal steam plant in The Geysers was one of the first geothermal power plants to use a hybrid gas removal system. Hybrid gas removal systems using a first-stage steam jet ejector and a second-stage liquid ring vacuum pump have since become the preferred configuration for gas removal systems at new plants.

11.1.9   Cooling systems

On paper, the cooling system at a geothermal steam plant is similar to the cooling system at a traditional thermal power plant. However, the cooling system has a much greater influence on the economic performance of a geothermal steam plant than a traditional thermal power plant. Physical differences in equipment design and materials of construction are required to achieve reliable service in geothermal applications.

11.1.10   Plant auxiliaries

The engineering and design of auxiliary systems and equipment for a geothermal steam plant are similar to the same systems and equipment at a traditional thermal power plant. However, the fluids and environmental conditions at a geothermal steam plant typically require corrosion resistant materials and additional protective equipment and measures to perform reliably in geothermal service.

11.1.11   Engineering materials

In the 1970s, active testing of materials at existing and new geothermal plant sites contributed a baseline of knowledge for materials selection. These studies revealed that some materials that worked successfully at one resource suffered unacceptable corrosion at other resources. In the intervening years, certain materials have become near standard for new geothermal steam plants. However, the variations in geothermal fluids and the surrounding environmental conditions at geothermal steam plant sites will continue to require plant designers to actively participate in the selection of suitable materials for a new geothermal steam plant.

11.2   Power cycle

11.2.1   Overview

The power cycle for generating electricity from steam-dominated resources is comparatively simple, consisting of a turbine-generator, condenser, and waste heat rejection system. In principle, these elements are similar to the components used in traditional thermal power plants. However, the simplicity of the power cycle and the unique nature of the geothermal fluids have led to refinements in the power cycle to improve both overall plant efficiency and economic performance with high reliability and availability.

Both direct contact and surface condensers are used in geothermal power plants operating on vapor-dominated resources. Typical power cycles with direct contact and surface condensers are shown in Figs. 11.1 and 11.2. From a thermodynamic perspective, the direct contact condenser offers slightly better performance than a surface condenser. However, operational costs in the form of auxiliary power requirements for the hotwell pumps and increased chemical costs for emissions control, when required, have resulted in using surface condensers in some geothermal steam power plants.

Geothermal steam typically contains a small amount of noncondensable gases that must be managed to maintain efficient operation of the thermal power cycle. Whereas a typical thermal power plant will have a small air removal system to extract and

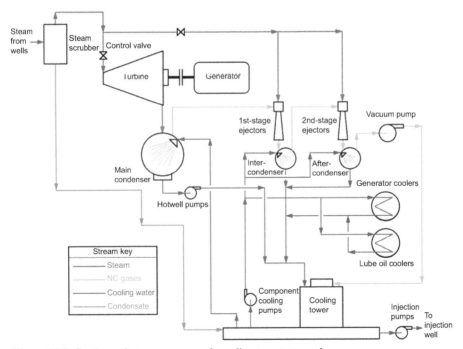

Figure 11.1 Geothermal steam power cycle — direct contact condenser.

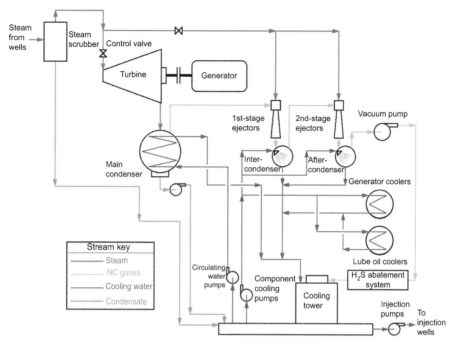

Figure 11.2 Geothermal steam power cycle — surface condenser.

compress air in-leakage from the condenser, geothermal steam power plants require a significantly larger system to remove the air in-leakage plus the noncondensable gases that entered the condenser with the geothermal steam. Gas removal systems at geothermal steam power plants use steam jet ejectors or ejectors with mechanical compressors in a hybrid system to remove the noncondensable gases and air from the condenser.

Most geothermal steam power plants use evaporative cooling systems with mechanical draft cooling towers. As with the other power cycle components, cooling towers and other cooling system components require specific differences from traditional power plant cooling system components to provide reliable service in geothermal applications.

11.2.2 Optimization of turbine inlet conditions

Production of geothermal steam is typically represented by a deliverability curve that presents steam flow as a function of wellhead pressure. Deliverability curves are resource specific and vary from resource to resource and even well to well. Wellhead deliverability curves should be combined and adjusted to reflect gathering system losses to establish the deliverability curve at the power plant. Further, deliverability curves vary over time. Selection of the power cycle inlet parameters must consider the specific resource deliverability and any forecasts of deliverability in future years.

11.2.3   Optimization of turbine exhaust pressure and cooling system

Selection of the optimum turbine exhaust conditions is a multifaceted effort that will have far reaching effects on the long-term economic performance of a geothermal steam power plant. Inputs to this development include turbine performance and costs, condenser and cooling tower performance and costs, annual ambient conditions, project owner's economic factors, and the terms of the PPA.

11.2.4   Optimization around the concentration of noncondensable gases

As the concentration of noncondensable gases in the geothermal steam increases, the optimum configuration and design of the condenser and noncondensable gas removal system changes. Additionally, the optimum turbine exhaust pressure may rise.

11.3   Steam quality

11.3.1   Geothermal steam

Steam produced from vapor-dominated geothermal resources is a mixture of vapor, liquid, gas, and solid constituents that varies from resource to resource and frequently from well to well within a resource. The design of power plant systems and equipment should be appropriate for the steam analysis for the specific resource.

11.3.2   Vapor

At the wellhead, produced steam may be superheated, saturated, or wet steam. In all cases, the steam is processed to remove undesirable chemicals, compounds, liquids, and solids before delivery to the power plant.

11.3.3   Liquid and solid constituents

To minimize erosion of turbine steam path components and damage to sealing and close-fitting running surfaces, liquid and solid contaminants in the steam are removed from the steam by centrifugal and inertial separators installed upstream of the turbine admission valves.

11.3.4   Noncondensable gases

Noncondensable gases produced with the geothermal steam are expanded in the steam turbine to produce power. The gases are removed from the process at the condenser.

11.3.5   Chemical constituents

Early in the development of a project, produced steam is analyzed for chemical composition. The analysis is then examined for chemicals and compounds that can contribute to scale formation, corrosion, or must be managed or treated to meet emissions regulations. For additional information on scale and corrosion control see chapter "Silica scale control in geothermal plants - historical perspective and current technology."

11.4   Steam systems

Although geothermal steam is produced at several wells distributed across a field, the steam is typically delivered to the power plant boundary at a single location. To provide clean, high-quality steam to the steam turbine, dry steam geothermal power plants include steam pretreatment systems and equipment to remove chemicals, compounds, liquids, and solids that could result in scale formation, corrosion, or erosion damage to the steam turbine and other power plant equipment.

11.4.1   Steam pretreatment

11.4.1.1   Scrubbers

Steam pretreatment is typically a multistep process that begins at the wellhead. An inertial separator frequently called a "rock catcher" is located near the wellhead. As the name implies these devices serve to remove large solids from the flow stream. A piping network delivers the steam from all the production wells to the power plant site where a cyclone separator or "scrubber" serves to remove liquid droplets and small solid particles from the steam.

11.4.1.2   Steam desuperheating and washing

At some dry steam geothermal resources, the produced steam is superheated. It was found that the superheated steam transported compounds to the steam turbine that contributed to scale formation on turbine steam path components as the steam expanded and cooled in the steam turbine. To effectively remove the scaling compounds upstream of the turbine, the steam was desuperheated by injecting clean water sprays into the steam system upstream of the scrubber. The scaling compounds concentrate in the liquid droplets. When the steam enters the scrubber, the liquid droplets and the scaling compounds are removed from the steam flow as scrubber drain flow. Steam exits the scrubber at saturated conditions. The desired effect of the desuperheating is purification of the steam.

Detailed design of the desuperheating/steam washing system needs to account for the composition of the compounds to be removed, the analysis of the spray water, and the residence time necessary for the scaling compounds to dissolve into the spray water. The steam must remain wet when it enters the scrubber to effectively remove the scaling compounds. The design of the scrubber should include efficient collection and drainage of liquid entering the separator. Scrubber design efficiencies typically exceed 99%.

11.4.1.3 Inertial demisters

A steam strainer or demister is typically installed immediately upstream from the turbine stop and control valves. This inertial separator serves to remove any droplets or solid particles that have managed to pass through the upstream separators and scrubbers.

11.4.2 Steam piping

11.4.2.1 Piping material

Dry geothermal steam is usually not very corrosive. However, geothermal steam condensate is much more corrosive. Steam piping at dry steam geothermal power plants is typically carbon steel in accordance with ASME/ASTM A53 Gr B or A106 Gr B. Piping wall thickness selection typically includes a corrosion allowance of 0.125 in. (3 mm). Experience at a dry steam geothermal power plant in California indicated that an alternative corrosion resistant material should be selected where the steam pH is ≤ 5.

11.4.2.2 Arrangement considerations

Because geothermal steam condensate is corrosive to the steam piping, it is prudent to apply best industry practices to locate condensate collection pots in the steam piping. Condensate collection piping at dry steam geothermal power plants is typically Type 316L stainless steel. Condensate piping operating at less than 225°F (107°C) can be fiberglass reinforced plastic (FRP).

Dry geothermal steam systems can, and have, carried unanticipated amounts of solid or liquid contaminants. It has been beneficial to use piping tees for simple horizontal or upward changes of direction for in-plant piping. The unused run connection on the tee is capped with a weld neck flange and blind flange set that serves as a maintenance cleanout location for any accumulated debris.

Thermal growth and piping stress are typically managed with piping expansion loops rather than metallic expansion joints to minimize the risk of corrosion or scale-related system failures of thermal expansion control devices.

11.4.2.3 Condensate management

The corrosive nature of steam condensate requires detailed attention to the condensate collection and drainage system design to effectively and reliably remove condensate from the steam piping. Drain pots should be located immediately upstream of any upward direction change and upstream of any valve station. For geothermal dry steam applications, drain pot diameter is typically about half the steam header diameter. Because the temperature of the geothermal steam and condensate are very close, effective control of the drain flow is achieved by controlling the liquid level in the drain pot.

Drain pot length should provide approximately 2.5 ft (0.76 m) of measurable liquid depth. It is important to design the drain pot and liquid drainer for the highest load that is typically encountered during plant or system startup and warmup. However, the drainage system should not be so large as to not operate on a reasonable frequency during normal operation. Liquid drainers have been found to be effective and reliable in draining condensate from geothermal steam headers.

11.4.2.4   Flow measurement devices

Flow measurement of clean saturated steam provides flow data with significant uncertainty. Flow measurement of saturated geothermal steam may face increased uncertainty from scale formation at or near differential pressure measurement ports in the flow measurement device. Erosion may also alter flow measurement device performance contributing to additional uncertainty.

To minimize measurement uncertainty, great care should be applied to the piping design and configuration both upstream and downstream of the flow-measuring element. Piping arrangement guidelines from the flow element manufacturer should be taken as the absolute minimum, with further guidance from the ASME Performance Test Codes.

Any flow measurement device that operates using differential pressure measurements will be subject to inaccuracies in geothermal service due to scale formation or erosion of the flow measurement device. A venturi was selected for SMUDGEO #1 as it offered the highest accuracy for the measurement of steam flow. Accuracy was very important as the venturi served as the billing meter for the purchase of steam. Although the venturi worked satisfactorily for a few years, it is no longer in service as the differential pressure ports in the venturi suffered from plugging and scale-induced uncertainties.

While not approved by the ASME Performance Test Codes, insertion type flow elements have attained wider use in geothermal dry steam flow measurement applications. Insertion type flow elements can be removed from the flow path with the system in operation permitting inspection, cleaning, and recalibration of the complete flow element assembly prior to reinsertion into the system. At the time of reinsertion, the geothermal portion of the measurement uncertainty has been essentially eliminated.

11.5   Turbine-generators

11.5.1   Steam turbines in geothermal service

11.5.1.1   Turbine size and configuration

Geothermal steam turbines have ranged in size from less than 10 MW upward to 120 MW. The smaller units use single-flow turbines, the mid-range units have double-flow turbines, and the larger capacities are either two- or four-flow machines.

Single-flow turbines offer the advantage of a low-profile installation with either a top or axial exhaust. Double- and four-flow machines are typically mounted on an elevated pedestal with downward exhaust into the condenser that is installed directly below the turbine.

Advances in turbine technology have raised the capacity of a single-flow turbine to more than 40 MW. Single-flow machines remain useful for applications at small or newly developing geothermal resources. For larger capacity units, double-flow machines typically offer better economics than a four-flow machine of equal capacity. The improved economics are driven by reduced equipment cost, reduced construction cost, and reduced maintenance cost for the double-flow machine.

11.5.1.2   Last-stage blades

Durability concerns and preferences for use of last-stage blades (LSB) with a history of reliable service in geothermal applications have constrained the choices for geothermal steam turbines. Whereas single-flow turbines for small projects use shorter LSBs, a majority of geothermal steam turbines in 60 Hz service use LSBs between 17 and 26 in. (432 and 660 mm). The constraint on LSB length resulted in the later units at The Geysers having four-flow turbines.

Project economics and 50 Hz service at the Darajat Stage II project in Indonesia provided an opportunity to technically and economically justify the use of longer LSBs. Similitude relationships were applied between 25 in. (635 mm) blades with proven reliability in geothermal service at 60 Hz, and longer 30 in. (762 mm) blades operating at 50 Hz. The Darajat Stage II project has been operating reliably since 2000. Other geothermal projects are now operating with up to 31 in. (787 mm) LSB.

For many years, the preferred capacity for a new geothermal power plant or unit addition remained constant at 55 MW. Improvements in turbine design, including longer LSBs and known economies of scale, have contributed to an increase in the preferred size for a new large capacity dry steam geothermal power plant or unit addition to approximately 80 MW. However, optimum turbine capacity for a new project may be limited by either total capability of the geothermal resource or the energy density of the geothermal resource.

11.5.2   Steam path considerations

11.5.2.1   Blade profile and shape

Through computational fluid dynamic (CFD) analysis, geothermal turbine manufacturers have developed improved blade profiles that enhance turbine efficiency. Additionally, the CFD analysis has contributed to changes in the radial shape of the blades to reduce secondary flow losses. These techniques were applied to the steam turbine for Darajat Stage II, which contributed to the plant being one of the most efficient geothermal power plants in the world.

11.5.2.2 Integral shrouds

Improvements in forging techniques have allowed turbine manufacturers to provide rotating blades with integral shrouds. Integral shroud blades (ISB) provide benefits in steam path efficiency, stress reduction, blade vibration mitigation, and interstage sealing, improving both efficiency and reliability.

11.5.2.3 Moisture removal

All the major geothermal steam turbine suppliers use interstage moisture removal devices for both improved thermal efficiency and reduction of erosion potential. Techniques differ between the turbine manufacturers, but the goal is the same. Mitsubishi uses hollow nozzles in the stationary blading to assist in moisture removal as shown in Fig. 11.3. Toshiba locates all rotating blade seal fin strips on the blade shrouds to provide a free path for moisture to exit the steam path as shown in Fig. 11.4.

11.5.3 Design features for geothermal applications

11.5.3.1 Design for scale prevention/mitigation

To achieve long-term reliability and efficiency the risk of scale formation in the steam supply system and turbine steam path must be minimized. Early in the development of a new project, the geothermal steam chemistry should be analyzed for potential scale-forming compounds. Because pressure, temperature, and concentrations change through the power cycle, the potential for scale formation may increase as the steam expands through the turbine.

Effective mitigation of scaling risk has been achieved by washing the geothermal steam with geothermal steam condensate or another clean water source. Steam washing is discussed in Section 11.4.1.

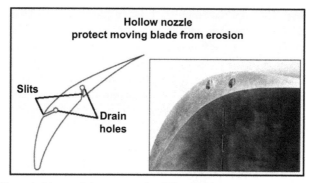

Figure 11.3 Steam turbine moisture removal — Mitsubishi.
Development of Large Capacity Single Cylinder Geothermal Turbine paper presented at the 1998 GRC Annual Meeting.

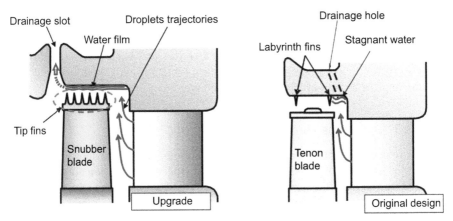

Figure 11.4 Steam turbine moisture removal – Toshiba.
Turbine Upgrade for Geysers Power Plant paper presented at the 2011 GRC Annual Meeting.

Geothermal steam power plants have been typically operated as base load units. As such, the turbine stop and control valves are held in near-constant positions for long periods. Scale can form on valve stems and discs that can hinder valve operation or prevent full valve closure. Turbine stop and control valves at geothermal steam plants are subjected to frequent stem-free testing to confirm operability of the valves. The stem-free testing permits partial or full stroking of individual valves with the turbine remaining in service.

11.5.3.2   Moisture limit in turbine exhaust steam

Geothermal steam turbines operate almost entirely in the wet region of the Mollier Chart steam dome. Because the turbine inlet conditions are constrained by the resource deliverability, efforts to achieve greater generation from the available resource will seek to extend the turbine expansion line to a lower exhaust pressure. As the exhaust pressure becomes lower, the moisture content of the steam in the later stages of the turbine increases. The moisture droplets bring negative effects of reducing steam path efficiency and increasing erosion of the laterstage blade surfaces. Some geothermal steam turbines have longer than standard erosion protection inserts, typically Stellite strips brazed on the leading edge of the later-stage blades to mitigate the erosion potential of the wet steam. Turbine manufacturers have typically established limits for allowable moisture content in the turbine. Dry steam geothermal plants at The Geysers in California and in Indonesia that were designed for high efficiency have operated successfully with up to 14% moisture in the turbine exhaust steam.

11.5.3.3   Moisture removals and stage drain management

Moisture removal flow paths are frequently subject to scale formation; therefore it is beneficial, wherever possible, to have the moisture removal lines with the flow control orifices routed outside the turbine casing with sight flow indicators to provide visual confirmation the drain flow path remains open during operation. In many cases the internal structure of the turbine will prevent routing of the drains outside the turbine casing. At a minimum, these turbines should be provided with replaceable stainless steel orifices accessible from inside the turbine casing.

11.5.3.4 Material enhancements for wear parts

Recent advancements in thermal spray coatings have been applied to geothermal steam turbines to reduce corrosion and wear on critical surfaces of turbine rotors. Specifically, the gland seal areas on geothermal steam turbines experience high corrosion rates resulting from the mixture of air and geothermal steam within the seal labyrinth. Thermal spray of a cobalt-based alloy on the gland seal labyrinth on the turbine rotor has been shown to reduce or eliminate corrosion damage to the gland seal surface. An alternative to thermal spray is weld overlay of the gland seal area with a high nickel alloy.

11.5.4 Materials selection for turbine internals

11.5.4.1 Rotors

Turbine manufacturers have developed special alloys for geothermal steam turbine rotors. Until recently, Cr-Mo-V steel has been the preferred alloy for turbine rotors. For projects with elevated risks of stress corrosion cracking, some manufacturers offer rotor forgings of 12 Cr steel. For a new project, a detailed steam analysis should be presented to the turbine manufacturer for confirmation of the preferred alloy for the turbine rotor.

11.5.4.2 Rotating blades

Recent geothermal steam turbines have been produced with rotating blades manufactured from a variety of materials. Material choices typically include 12 Cr steel, 17-4 PH precipitation hardened stainless steel, and titanium. The turbine manufacturers will evaluate each stage for blade stress, corrosion, and erosion potential based on the detailed fluid conditions and composition at each stage. It is not uncommon for different materials to be selected for different stages in a geothermal steam turbine.

11.5.4.3 Stationary blades

Early geothermal steam turbines used stationary blades and diaphragm rings fabricated from carbon steel. In geothermal wet steam service these carbon steel components suffered significant corrosion damage. The damage not only reduced turbine efficiency but also contributed to extended overhaul outages as the corrosion extended the time required to disassemble the turbine. More recent machines have used materials and configurations with greater corrosion and erosion resistance for the stationary blades and diaphragm rings.

11.5.4.4 Corrosion and erosion protection

Joint faces in the carbon steel turbine casing are subject to corrosion and erosion. These problematic areas have been improved with stainless steel inserts to prevent corrosion and subsequent erosion of the joint surface.

11.5.5 Generators

11.5.5.1 Cooling

The larger units (>50 MW) at The Geysers used hydrogen cooling for the generators. Hydrogen cooling is used in a majority of large power plants to provide improved generator efficiency. In geothermal applications, hydrogen cooling (as opposed to using an open ventilated system with ambient air for cooling) also protected the generator internals from the potentially corrosive atmosphere that typically exists around geothermal power plants. These benefits, however, came with higher initial and operating costs of the hydrogen cooled generator, hydrogen storage and supply system, and the required fire protection systems.

Improved project economics were achieved in the 1990s with totally enclosed water to air cooled (TEWAC) generators. TEWAC generators circulate ambient air, rather than hydrogen, through the generator internals to remove waste heat. As in a hydrogen cooled generator, the waste heat is transferred to a cooling water system in tubular heat exchangers installed in the generator casing.

11.5.5.2 Corrosion protection

Air-cooled generators are typically provided with a particulate filter for the air inlet to the generator. For geothermal applications, in consideration of the corrosive atmosphere that typically exists around geothermal power plants, an additional filter is required to remove hydrogen sulfide from the air that enters the generator. The air filter media is typically either activated carbon or potassium permanganate.

The tube sides of the generator air coolers are typically fabricated from Type 316L stainless steel to protect against corrosion from the cooling water.

11.5.6 Designing for efficiency and reliability

Seeking improvements in thermal performance and efficiency must be balanced with equipment capability to provide reliable service. Much has been achieved over the past 30 years. In the early 1980s, turbine overhauls were planned for every 2 years. Through diligent shared development between the plant operator at The Geysers and Toshiba (the turbine OEM), equipment enhancements have been developed and proven, allowing extended run periods of at least 10 years with unit availabilities greater than 95%.

11.6 Condensers

11.6.1 Water-cooled direct contact condensers

Worldwide, most geothermal steam power plants use water-cooled direct contact condensers. Direct contact condensers eliminate concern for heat exchange surface cleanliness, fouling, and corrosion that can occur in geothermal service; therefore they are preferred for dry steam geothermal power plants.

Barometric and spray/tray direct contact condensers have been used in geothermal power plants. However, the spray/tray type has become preferred as it does not require the additional excavation or building height required for a barometric direct contact condenser.

More recently, a new type of direct contact condenser has been developed and proven in service. The Ecolaire® Advanced Direct Contact Condenser (ADCC™)[1] from SPX Heat Transfer LLC uses licensed technology that uses both spray nozzles and a fill pack to create heat transfer and condensation in the vapor space above the fill pack and in the turbulent film flow through the fill pack. External and cutaway views of the ADCC are presented in Figs. 11.5 and 11.6.

The ADCC has been shown to provide notable performance benefits at reduced installed cost as compared to traditional direct contact condensers. ADCC performance benefits include:

- Lower condenser pressure
- Reduced cooling water flow
- Reduced auxiliary power requirements for the condenser gas removal system
- Low susceptibility to fouling and/or corrosion

The ADCC contributions to reduced installed costs include:

- Reduced height
- Reduced foundation costs

ADCC units have been in successful operation in dry steam geothermal service at The Geysers Units 5, 6, and 11 in California and Kamojang IV in Indonesia, as well as other flash steam geothermal power plants.

The installations at The Geysers Units 5, 6, and 11 were completed as retrofit replacements of spray/tray direct contact condensers. Unit performance tests at Unit 11 were completed both before and after the installation of the ADCC. Test results report a 2% increase in gross generation from reduced condenser pressure and more than 4% increase in net generation from reduced auxiliary power requirements for the condenser gas removal system.

Both economics and performance capability contributed to selecting the ADCC for the initial design of the 60 MW Kamojang IV power plant in Indonesia. As with a majority of recently constructed geothermal projects, the design was driven

---

[1] Ecolaire® and ADCC™ are trademarks of SPX Heat Transfer LLC, Tulsa, Oklahoma.

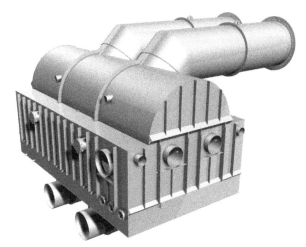

Figure 11.5 Advanced direct contact condenser − external view.
ADCC Geothermal Condenser cut sheet (IN-HT2-14.pdf).

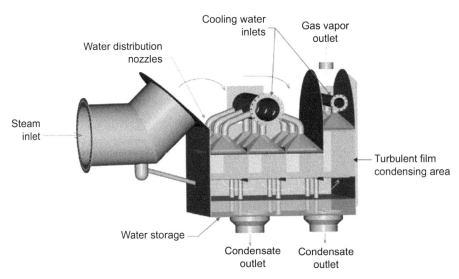

Figure 11.6 Advanced direct contact condenser − cutaway view.
ADCC Geothermal Condenser cut sheet (IN-HT2-14.pdf).

by a requirement for low installed cost. The performance of the ADCC at Kamojang
IV met expectations from plant startup in 2006 and continues to operate successfully.

11.6.2 *Water-cooled surface condensers*

Increasing production and generation at The Geysers in the mid-1970s led to regulatory requirements to control hydrogen sulfide emissions. Initial control techniques for

the existing direct contact condenser units required significant chemical additions to the circulating water that was pumped from the condenser hotwell to the cooling tower. Studies showed that abatement costs would be substantially reduced by using surface condensers rather than direct contact condensers in the power cycle. While it was impractical to retrofit surface condensers to the existing units, all new units built in The Geysers after 1980 use surface condensers. The primary differences between the power cycles with direct contact condensers and surface condensers can be seen by comparing Figs. 11.1 and 11.2.

Note that the condensate flow path in Fig. 11.2 directs the condensate to the cooling tower basin. As the steam condenses, a portion of the hydrogen sulfide that entered the condenser with the geothermal steam dissolves into the condensate. Routing the condensate directly to the cooling tower would allow the hydrogen sulfide to be stripped from the condensate and be released to the atmosphere. By directing the condensate flow to the cooling tower basin at a location remote from the circulating water pump intake structure, sufficient residence time is provided for the hydrogen sulfide to be oxidized by the chemicals and compounds in the cooling water so that when the condensate and cooling water mixture is delivered to the cooling tower, there is little or no hydrogen sulfide remaining in the mixture that could be stripped and vented to the atmosphere.

Surface condensing technology is well developed and understood for traditional thermal power cycle applications. Adaptation of proven surface condenser technology for geothermal applications required design and material changes to address:

• Corrosion mitigation
• Scale management
• Collection and removal of accumulated noncondensable gases

The existing direct contact condensers fabricated from Type 304L stainless steel had shown good resistance to corrosion. This material selection was carried forward for the shells of the surface condensers.

Initially, Type 304 stainless steel was also selected for the condenser tubes. However, these tubes were subject to pitting corrosion under low-flow situations or as a result of surface scale formation. Surface condensers installed in The Geysers from the mid-1980s used Type 316 stainless steel owing to its greater resistance to pitting corrosion at a relatively small increase in cost. Because future resistance to corrosion to the potentially changing environment in geothermal service cannot be reliably predicted, titanium tubes were selected for three units in The Geysers. These condensers, installed between 1983 and 1989, have performed well without loss of reliability from tube corrosion.

Surface condenser design for traditional thermal power cycles typically includes a cleanliness factor to account for accumulation of heat transfer inhibiting film or scale on the cooling water side of the tubes. However, since impure geothermal steam is condensing on the tubes, the shell side of the tubes in geothermal condensing service is subject to the formation of heat transfer inhibiting scale. Allowing for film or scale on both sides of the condensing tube wall resulted in selection of condenser cleanliness factors between 0.7 and 0.8.

11.6.3   Multipressure condensers

The generation capability of a geothermal steam plant with a multiflow steam turbine and a surface condenser can be increased by arranging the surface condenser as separate shells for each turbine exhaust flow and delivering the cooling water to the shells in series. The series water flow results in progressively higher operating pressures in each condenser shell as the cooling water temperature rises. The average operating pressure for a multipressure condenser will be lower than the operating pressure in an equivalent single pressure condenser resulting in higher total available enthalpy drop for the turbine expansion. The SMUDGEO #1 dry steam geothermal plant in The Geysers was one of the first geothermal power plants to apply this technology to increase both net generation and plant efficiency.

11.6.4   Noncondensable gas cooling zones and external coolers

The gas cooling section of the condenser must cool the noncondensable gases to near the cooling water inlet temperature with minimum resistance to gas flow through the gas cooling section. While all geothermal condensers operate with integral gas cooling sections, some plants use a secondary, external noncondensable gas cooler. External gas coolers, sometimes called precoolers or precondensers, serve to further cool the noncondensable gases to reduce the vapors of saturation that would otherwise have to be compressed by the gas removal system.

Precoolers provide value over the operating life of a project. Because precoolers are not essential for operable condensing and gas removal systems, projects driven by low installed cost typically do not use precoolers.

11.6.5   Air-cooled condensers

Recent trends in traditional thermal power plants have included an increasing number of air-cooled condenser installations. The selection of air cooling over evaporative cooling is typically driven by a lack of water supply for makeup to an evaporative cooling system.

Air-cooled condensers offer environmental and operational benefits including:

- No need for makeup water for evaporative cooling
- No visible plumes from an evaporative cooling tower
- Retention of geothermal fluid inventory

The benefits are countered by:

- Increased equipment cost
- Larger plant area requirements
- Reduced generation capacity
- Greater seasonal fluctuation of generation

Air-cooled condensers have not been selected for use in current dry steam geothermal power plants. Future use of air-cooled condensers as part of a hybrid wet-dry cooling system may offer better economic performance than air cooling for a geothermal steam plant.

11.6.6   Designing for efficiency and reliability

Achieving both efficiency and reliability in the design and operation of condensers for geothermal steam plants begins with a balanced design of the entire cooling system. A system designed for very high theoretical efficiency may hinder reliable performance. Similarly, a design focused strictly on reliability may lead to inefficient use of the geothermal resource.

Power industry norms for cooling tower range and approach, condenser subcooling, and tube velocities usually serve well in geothermal service, but each should be examined for suitability for the specific project and conditions of operation.

Selecting suitable and cost-efficient materials for geothermal steam condensers is critical to achieving long-term reliability. In recent years, Type 316L stainless steel has become the preferred material for geothermal steam condenser shells and internals. For shell and tube condensers, Type 316 stainless steel tubes generally provide reliable performance. However, application of tubing materials such as duplex stainless steel or titanium in specific applications can offer improved reliability and resulting economic benefits over the life of a project.

As for traditional thermal power plants, maintaining the efficiency and reliability of a steam condenser and the associated cooling water system requires active maintenance of the cooling water chemistry. Geothermal steam plants typically use condensed geothermal steam as makeup water for the cooling cycle. While this practice reduces the need for surface water makeup streams, it adds and concentrates chemicals and compounds that are carried in the geothermal steam condensate in the cooling water, increasing the risk of scale formation and corrosion. Variations in the geothermal fluids of different resources require cooling water chemistry to be specifically engineered for each project.

For additional information on geothermal power plant condensers, see chapter "Waste heat rejection methods in geothermal power generation."

11.7   Gas removal systems

11.7.1   Function

Dry steam produced from geothermal wells typically contains a small percentage of noncondensable gas. The gas, which is mostly carbon dioxide but also contains small amounts of hydrogen sulfide and other compounds, is transported to and through the steam turbine with the geothermal steam. On entering the condenser, the expanded, low-pressure steam is condensed into liquid. To maximize generation, the condenser operates at a vacuum. To maintain the operating vacuum, the noncondensable gas must be extracted from the condenser. Refer to the process flow diagrams presented as Figs. 11.1 and 11.2 for a diagrammatic representation of the gas removal systems. These process flow diagrams show three-stage hybrid gas removal systems.

Development of early geothermal power plants operating on dry steam resources used a two-stage steam jet ejector system to extract the noncondensable gases from the condenser and compress the gas for release to the atmosphere. Steam jet ejectors offered simplicity, reliability, and low cost. Furthermore, the economic factors for these early projects did not assign a fuel cost to the motive steam consumed by the ejectors. Beginning in the early 1980s, changes in steam supply contracts and the recognition that geothermal resources were not limitless led to significant efficiency improvements in the design and operation of condenser gas removal systems. The advances developed for condenser gas removal systems in the 1980s contributed to the current de facto standard two-stage hybrid gas removal system specified for most new geothermal dry steam and flash steam power plants.

Condensers at traditional thermal power plants include an air removal system to extract and compress the small amount of air that enters the vacuum space in the condenser through turbine shaft seals. The amount of air in-leakage is very small, requiring only a small air removal system that is typically provided with the condenser. Early geothermal steam plants followed the traditional thermal power plant approach of including the gas removal system with the condenser procurement. The drive to improve the efficiency of new geothermal steam plants in the 1980s led to the recognition that the goals of maximum efficiency and minimum cost were best met by separating the procurement of the condenser and gas removal system into separate packages.

11.7.2   Steam jet ejectors

Steam jet ejectors are mass flow machines that are ideally suited for extracting and compressing noncondensable gas from a condenser operating at high vacuum. Compared with other mechanical compressors, steam jet ejectors offer the benefits of no moving parts and low cost. However, the energy efficiency of steam jet ejectors is significantly less than a mechanical compressor. Efficient and reliable operation of steam jet ejectors requires delivery of motive steam within a relatively narrow range of pressure. Geothermal steam plants typically include a pressure control valve to provide motive steam to the ejectors at a constant pressure that is slightly less than the main steam header pressure.

To reduce the load on subsequent stages of compression, ejector condensers are typically located immediately downstream of steam jet ejectors. The high fluid velocities in steam jet ejectors create high noise levels that are typically mitigated by choosing an appropriate location for the ejectors and by using thermal/acoustic insulation.

Steam jet ejectors at geothermal steam plants are typically fabricated from Type 316L stainless steel to resist corrosion from the noncondensable gas and steam mixture. Because operating conditions may change during the operating life of the geothermal power plant, the motive steam nozzles for steam jet ejectors should be replaceable.

11.7.3 Ejector condensers

Ejector condensers improve system efficiency by condensing the ejector motive steam and a portion of the saturation vapors to reduce the compression load on the flowing stages in the gas removal system. As for the ejectors, ejector condensers at geothermal steam plants are typically fabricated from Type 316L stainless steel.

Ejector condensers are either direct contact or surface condensers typically in the same form as the main condenser. Shell and tube ejector condensers should be straight tube units, single or double pass with removable heads. Access to the tube sheet is necessary to permit effective and efficient cleaning of the tubes.

Ejector condensers should be designed for 100% operation of all ejector elements. Cooling water flow is typically provided at the design flow rate for all operating conditions. This approach not only simplifies system controls but, in the case of shell and tube condensers, maintains cooling water velocity to protect against local crevice corrosion of the stainless steel tubes.

Condensate drains from ejector condensers should be routed to either the main condenser or the cooling tower basin. Drains to the main condenser should enter the main condenser shell at a relatively high elevation to assist in separating any dissolved NCG from the drain flow within the condenser. This consideration is particularly important for projects requiring treatment of noncondensable gases for the removal of hydrogen sulfide.

11.7.4 Liquid ring vacuum pumps

Liquid ring vacuum pumps are low-speed, positive-displacement, volume flow compressors that use a rotating ring of water to seal the compression elements during operation. Liquid ring vacuum pumps are significantly more efficient than steam jet ejectors, but as volume flow machines they are not suitable for the first stage of a condenser gas removal system.

Liquid ring vacuum pumps in geothermal gas removal service typically operate below synchronous speed. Where the required pump speed cannot be achieved with a multipole motor, a speed reducer must be used. Gear reducers are the preferred means to achieve the required pump speed. Belt drives have proved to be unreliable in geothermal service due to corrosion of the sheaves from atmospheric contaminants.

The water supply for liquid ring makeup should be carefully selected for best performance. Water quality should be evaluated to minimize corrosion. Temperature should be checked as high temperature will reduce the flow capacity of the vacuum pump.

11.7.5 Axial or radial compressors

Axial and radial flow compressors are high-speed volume flow machines. Axial and radial flow compressors offer excellent efficiency. However, these machines are expensive compared to liquid ring vacuum pumps and they are sensitive to operating conditions. Axial or radial flow compressors offer economic solutions for geothermal steam plants with high noncondensable gas mass flows. However, the geothermal fluids should be closely evaluated and treated if necessary for scaling and erosion potential.

The full range of expected operating conditions should be within the surge and choke limits of the compressor. Using a variable frequency drive may increase the operational flexibility and range of the compressor. Slight scale buildup on the compressor rotating elements can result in high vibration alarms and trips.

11.7.6   Hybrid systems

In the early 1980s, driven by economic considerations, the SMUDGEO #1 dry steam plant in The Geysers was one of the first geothermal plants to use a hybrid gas removal system. The system consisted of two stages of steam jet ejectors with ejector condensers and a third stage of liquid ring vacuum pumps. This system greatly reduced the consumption of motive steam, providing a significant contribution to making SMUDGEO #1 one of the most efficient geothermal power plants in the world.

Axial compressors were chosen for the second (final) stage of the gas removal system at a 2 × 65 MW dry steam geothermal power plant constructed in The Geysers in the late 1980s. Early operation of the compressors was not trouble free. And, unfortunately, major problems with the quality of the produced geothermal steam led to early termination of plant operation before the operational issues with the compressors were resolved.

Hybrid gas removal systems for current steam geothermal power plants use steam jet ejectors with an ejector condenser for the first stage and a final stage of mechanical compression, typically a liquid ring vacuum pump driven by an electric motor with a gear speed reducer. Hybrid gas removal systems provide greater efficiency, flexibility, and range than an all-ejector system.

11.7.7   Design for efficiency and reliability

11.7.7.1   Number of stages

The number of stages of compression is primarily driven by the first-stage suction pressure which is less than the operating condenser pressure. Because typical design conditions for geothermal steam plants are near the transition between requiring two stages of compression and three stages of compression, selecting a three-stage system frequently offers greater energy efficiency. The benefit of the improved efficiency must be balanced against the increased cost of the additional stage of compression.

11.7.7.2   Stage equipment selection

Due to the very high specific volume of the noncondensable gas and vapor stream that exits the condenser, steam jet ejectors have been universally applied for the first stage of compression in geothermal steam plant gas removal systems. Ejector stages are typically composed of multiple ejector elements of different sizes to provide a wide range of efficient operation from much less than the design noncondensable gas flow to a selected capability higher than the design noncondensable gas flow.

Three-stage systems typically use steam jet ejectors and an ejector condenser for the second stage in the same configuration as the first stage. The final stage is typically a liquid ring vacuum pump.

11.7.7.3 Installed spare capacity for rotating equipment

Usual power plant design guidelines for reliability call for installed spare capacity for all rotating equipment required for normal operation. To provide the desired reliability and minimize project costs, a steam jet ejector with an ejector condenser has frequently been installed in parallel with the final stage liquid ring vacuum pump.

11.7.7.4 Instrumentation

Because operating conditions at geothermal steam power plants may change over time, extensive system instrumentation should be installed to permit continuous observation and evaluation of system and equipment performance. Of particular importance are local instruments or test taps for pressures and temperatures at the inlet and outlet of the main condenser, steam jet ejectors, ejector condensers, vacuum pumps, and compressors.

11.7.7.5 Motive steam pressure

Steam jet ejectors are sensitive to motive steam pressure. Ejector operation may become unstable if the motive steam pressure falls approximately 10% below the design motive steam pressure. Because geothermal steam header pressure may vary over time, recent geothermal steam plants use a pressure control valve to reduce the steam header pressure approximately 10 psi (69 kPa) for use in steam jet ejectors. The pressure reduction provides a stable steam supply pressure to the steam jet ejectors.

11.8 Cooling systems

To date, geothermal steam plants have used wet, evaporative cooling systems to transfer waste heat to the atmosphere. For traditional thermal power plants, there are now alternative dry and hybrid wet-dry cooling systems available to meet a wide variety of requirements for specific projects.

11.8.1 Evaporative cooling

Most geothermal steam plants use crossflow mechanical draft cooling towers with splash fill. With the introduction of counterflow cooling towers with high efficiency film fill to the power industry, the technology was applied to geothermal power plants. While initial results were promising, extended operation with geothermal fluids resulted in significant fouling of the film fill. The fouling led to plugging and, in extreme cases, collapse of the fill pack. These problems lead to industry-wide selection of wider spaced fill packs frequently described as low-clog film fill. Counterflow cooling towers with low-clog film fill have provided reliable and economic service at geothermal steam plants.

In recent years, FRP has largely replaced wood as the primary structural material for cooling towers in geothermal service. FRP provides a variety of benefits over wood for cooling tower structures. FRP is available with fire retardant ratings that can significantly reduce the cost and complication of the plant fire protection systems. Environmental concerns with compounds used for pressure treatment of wood for preservation are eliminated.

11.8.2   Dry and hybrid cooling

A new dry steam geothermal power project is evaluating both dry cooling and hybrid wet-dry cooling as a means to manage geothermal fluid inventory over the life of the project. A hybrid cooling system offers most of the benefits of both evaporative and dry cooling but costs more than either evaporative or dry cooling. A hybrid cooling system uses both evaporative and dry cooling circuits in parallel. The dry cooling system operates under most operating conditions. Under high-temperature ambient conditions, the evaporative system serves to increase the generation capacity.

11.8.3   Cooling systems and project optimization

Cooling system design for a geothermal steam plant strongly affects both performance and cost of the completed project. The most successful geothermal steam plant designs were subjected to an extensive optimization of the entire cooling system to achieve the greatest economic benefit from the project. The following parameters will influence the optimum cooling system design for a geothermal steam plant:

* Value of saleable power
* Owner's economic factors for the project
* Selection of design ambient conditions
* Value of auxiliary power
* Incremental cost of condenser duty
* Incremental cost of cooling tower duty
* Incremental cost of cooling water pumping power
* Incremental cost of NCG compression

11.8.4   Cooling system piping

FRP piping has provided excellent service in geothermal cooling systems. FRP is not subject to corrosion from geothermal fluids. The smooth inner surface minimizes scale formation and reduces piping friction loss. It is, however, important that FRP piping be properly designed and installed. Design considerations should include consideration of exposure to vacuum conditions during operational transients.

11.9 Plant auxiliaries

A geothermal steam plant includes auxiliary services and systems that are similar in function to systems at a traditional thermal power plant. However, certain adaptations are necessary to achieve reliable operation in geothermal service. The adaptations are required to mitigate the corrosion potential of the geothermal fluids as well as the atmosphere surrounding a geothermal power plant.

11.9.1 Auxiliary cooling

As can be seen in Figs. 11.1 and 11.2, geothermal steam plants typically draw auxiliary cooling water from the cooling tower. This water contains various elements and compounds from the geothermal fluids that enter the system as makeup, so the auxiliary cooling water system must be protected from corrosion from the cooling water. Many geothermal steam plants installed in the 1970 and 1980s used components fabricated from Type 304 stainless steel. Although this material has provided reliable service for many years, examination of certain components has shown increasing corrosion damage. Experience at geothermal steam plants has found that the increased corrosion resistance of Type 316L stainless steel provides sufficiently greater corrosion resistance at a small increase in cost. In recent years, Type 316L stainless steel has been the preferred material for auxiliary cooling systems at geothermal steam plants.

11.9.2 Compressed air

The ambient air surrounding a geothermal power plant is typically contaminated with varying amounts of hydrogen sulfide. To prevent corrosion to the internals of plant pneumatic control devices and service air users, purification equipment is installed at the inlet to the compressed air system. Chemical filters with activated carbon or potassium permanganate media have been used successfully to remove corrosive compounds from ambient air at geothermal steam plants.

11.9.3 Fire protection

Much of the approved or listed fire protection equipment available today is of brass or bronze construction. These materials are subject to rapid corrosion when exposed to geothermal fluids or the contaminants in the atmosphere surrounding geothermal power plants. To maintain code compliance and meet insurer requirements, listed or approved components should be specified with chrome plating and be protected from the geothermal fluids and atmosphere to the greatest extent possible. Fire water supply should be from a clean water source. Emergency connections to a geothermal fluid source can be included in the design as an extreme emergency water source.

11.9.4   Design features for geothermal applications

The atmosphere surrounding a geothermal power plant is corrosive to many traditional metallic materials used in power plant construction. Of special concern are electrical and electronic components. To achieve reliability of the electrical systems at geothermal steam plants, electrical equipment must be protected from the corrosive atmosphere. To the greatest extent possible sensitive electrical equipment should be installed in a controlled air space. The controlled air space must be pressurized with purified air and conditioned to maintain a suitable temperature. Geothermal steam plants have successfully used filter media such as potassium permanganate and activated carbon to remove hydrogen sulfide and other contaminants from makeup air.

Electrical equipment that cannot be located in a controlled air space should be protected by installation in a suitably rated enclosure such as NEMA 4X, IEC IP55, or IP65. Further protection can be provided by consumable filtration packets placed inside the enclosure.

11.10   Engineering materials

11.10.1   Power plant materials in geothermal service

Engineering materials used in traditional thermal power plants may or may not survive in geothermal service. Geothermal fluids contain numerous corrosive elements and compounds that are not present in the power cycle fluids of traditional thermal power plants. In many cases, materials used in traditional thermal power plants will experience rapid corrosion if used in geothermal service.

General guidelines based on industry-wide experience can serve as initial materials selection recommendations. However, these recommendations may not be applicable to a specific geothermal resource. Therefore, the actual fluid chemistry, temperature, and pressure should be considered when selecting materials for a specific geothermal service.

11.10.2   Steam service

Carbon steel has provided reliable service for geothermal steam systems. Dry geothermal steam causes progressive general corrosion of carbon steel. However, the corrosion rate is reasonably predictable, allowing the application of a suitable corrosion allowance to the original design pressure boundary wall thickness to manage the risk of corrosion. Locations such as piping bends are also subject to erosion-corrosion that may accelerate the loss of pressure boundary wall thickness. Historically, a 0.125 in. (3 mm) corrosion allowance has been applied to new geothermal steam systems.

Periodic inspection of wall thickness at certain locations in the steam system should be performed to monitor the remaining wall thickness for safety.

Stainless steel is not recommended for geothermal steam service because the corrosive environment, operating temperature and pressure, and the resulting tensile stress in the piping constitute conditions that support stress corrosion cracking that could lead to catastrophic failure of the system pressure boundary.

11.10.3   Condensate service

Geothermal steam condensate is much more corrosive than dry geothermal steam. Because condensate piping operates at a temperature below the threshold for stress corrosion cracking, austenitic stainless steel typically provides reliable service. The corrosion resistance of austenitic stainless steel is largely the result of a hard oxide layer on the surface. Liquid velocities should not exceed approximately 12 ft/s (3.6 m/s) to minimize the potential for erosion that would remove the oxide layer.

Early geothermal steam plants used Type 304 stainless steel for condensate and other corrosive services. The material has performed well, but new plants have upgraded to Type 316 stainless steel. Type 316 stainless steel provides a significant increase in corrosion resistance at a small increase in cost.

To minimize the risk of sensitizing the base material during welding, most austenitic stainless steel used in geothermal steam plants is specified as low carbon or L grade.

11.10.4   Noncondensable gas service

Austenitic stainless steel has served well in geothermal noncondensable gas service. As with condensate service, current geothermal steam plants typically use Type 316L stainless steel for noncondensable gas service. Some projects have transitioned to FRP piping after the final stage of gas compression with excellent reliability.

11.10.5   Cooling water service

Large-diameter cooling water piping for most recent geothermal steam plants has been FRP. FRP is inert to corrosion from geothermal fluids and provides a smooth interior surface to minimize piping friction loss. For cooling water systems, FRP piping should be designed to accept full vacuum without failure.

11.10.6   Auxiliary services

Auxiliary cooling water systems should use austenitic stainless steel and/or FRP piping.

Plant compressed air systems should be fabricated from a corrosion resistant material. Many new plants have used thin wall (Schedule 5S or 10S) austenitic stainless steel piping for compressed air systems.

11.10.7   Electrical equipment

Copper is subject to corrosion in the presence of even small amounts of hydrogen sulfide. As such, electrical equipment must be protected from the corrosive atmosphere. Small-gauge control cables are typically tinned copper wire. Large-gauge copper power cables have not been tinned. Junction boxes rated at IEC IP55, IP65, or NEMA 4X have provided adequate protection for electrical components located outside controlled airspaces.

Silver and cadmium components should be avoided.

11.11   Summary

Over the past 30 years, geothermal steam plant efficiency, reliability, flexibility, and durability have been improved through continued development and innovation. These advancements have come through the joint efforts of plant owners, operators, equipment OEMs, engineers, and technicians.

With greater understanding of initial and long-term resource production parameters, plant power cycles and systems can be optimized to provide greater economic benefit from a geothermal resource. Major plant equipment including steam turbines, condensers, gas removal system, and cooling towers have benefited from geothermal-specific development with innovative solutions that contribute to higher efficiency and reliability. Geothermal steam plants now operate at availability factors in excess of 95%. Technology has been developed to extend operating periods between overhauls from 2 to 10 years or more.

And yet there are additional opportunities to improve the capability of geothermal steam plants. Application of dry or wet dry hybrid cooling can enhance project operation and stewardship of the geothermal resource and the natural environment. Retrofitting of current technology in existing plants can provide efficiency, operability, and durability results similar to new equipment. And as grid management becomes more important, geothermal steam plants can contribute operational flexibility to provide necessary grid support and ancillary services.

Relevant literature

[1] Saito, S., Suzuki, T., Ishiguro, J., Suzuki, T., 1998. Development of large capacity single-cylinder geothermal turbine. Geothermal Resources Council Transactions 22.
[2] Maedomari, J., Avery, J., 2011. Turbine upgrade for geysers geothermal power plant. Geothermal Resources Council Transactions 35.
[3] Henderson, J., Bahning, T., 1997. Geysers advanced direct contact condenser results. Geothermal Resources Council Transactions 21.
[4] Hodgson, S.F., California Department of Conservation, Division of Oil, Gas, and Geothermal Resources, 2010. A Geysers album, five eras of geothermal history. Publication No. TR49, 2nd ed.

[5] Sison-Lebrilla, E., Phair, K., 2010. The life and times of SMUDGEO: a historical perspective. Geothermal Resources Council Transactions 34.

[6] Phair, K., 1992. Design, construction and operational experience with modular equipment at the Bear Canyon power plant. Monograph on The Geysers Geothermal Field. Geothermal Resources Council Special Report 17.

Total flow and other systems involving two-phase expansion

12

I.K. Smith
City University London, London, United Kingdom

12.1 Total flow

12.1.1 Introduction

The majority of geothermal resources are located at a drilled well bottom, some distance below ground level, as hot pressurized liquids that emerge as self-flowing, two-phase mixtures of water and steam at the wellhead. Power is then obtained from these brines by separating the liquid and vapor components and passing the dry steam through a turbine, in a single-flash steam plant, as shown in Fig. 12.1.

The process can be improved by flashing the separated water to a lower, intermediate pressure, separating the additional steam thus formed, and admitting it to an intermediate stage of the expansion process in the turbine to boost the total power output, in a double-flash steam plant, as shown in Fig. 12.2.

Clearly, there is a considerable loss of available energy (or exergy), due to the separation of the liquid. The power output of the system could be increased if it were possible to expand the entire mixture as wet vapor in a single expander, as shown in Fig. 12.3, in what is known as a total flow system.

The efficiency of a two-phase expander that is required to produce the same power output as that from a flash-steam system with an assumed turbine efficiency of 82% and a condensing temperature of 45°C, is shown, for a single-flash plant, in Fig. 12.4 and, for a double-flash plant, in Fig. 12.5, over a range of initial temperatures and dryness fractions in both cases.

In the majority of cases in practice, the geofluid enters the expander with a dryness fraction in the range of 5—15%. As can be seen from these figures, a total flow expander needs only to achieve an adiabatic efficiency of the order of 50% to produce the same power output as a single-flash steam system and ~60% in the case of a double-flash system. Apart from its simplicity, there is therefore a strong incentive for the use of a total flow system as the preferred method of power recovery from emerging brines if two-phase expansion can be achieved with efficiencies significantly greater than these values.

Unfortunately, there are major obstacles to the efficient expansion of two-phase geofluids in conventional turbines. These are mainly due to erosive effects of the liquid droplets on the rotor blades and the inability of the entrained droplets to respond to the induced aerodynamic forces that should lead them to flow smoothly through the blade

Geothermal Power Generation. http://dx.doi.org/10.1016/B978-0-08-100337-4.00012-7

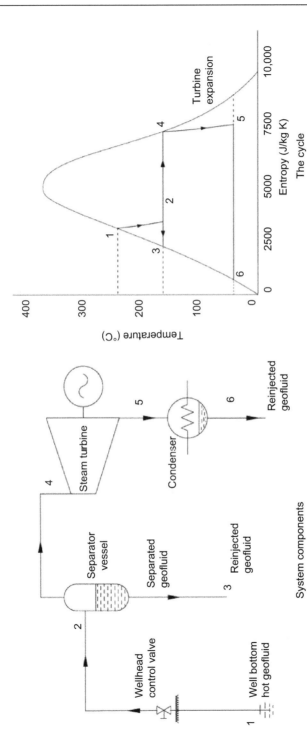

Figure 12.1 Single-flash system.

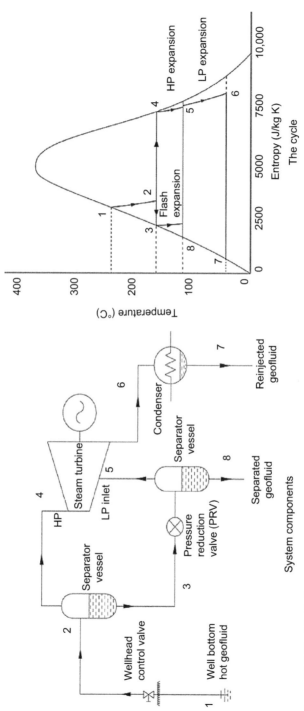

Figure 12.2 Double-flash system.

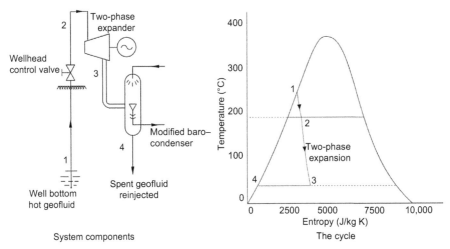

System components   The cycle

Figure 12.3 Total flow system.
Adapted from US DOE Handbook on Geothermal Energy 1982.

passages. Hence, the droplets impact on the blade surfaces, and this leads to blade erosion and low expander efficiencies.

Both of these difficulties can be overcome by the use of positive displacement machines instead of turbines. These operate with much lower internal fluid velocities, thereby minimizing erosive effects, while the work output results from pressure–volume changes within the machine, so that fluid dynamic effects play a relatively small part in the power production process. However, the low fluid velocities lead

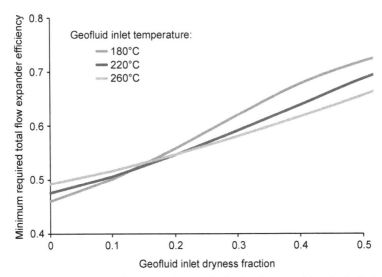

Figure 12.4 Efficiency of total flow expander required to break even with a single-flash system assuming condensation at 45°C.

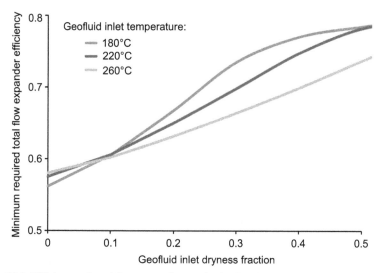

Figure 12.5 Efficiency of total flow expander required to break even with a double-flash system assuming condensation at 45°C.

to the need for much larger machines than can be produced by current manufacturing methods, if power is to be developed on a large scale, and such machines cannot operate efficiently over the huge volume changes required to fully expand the wet vapor. These are indicated in Fig. 12.6 for a range of operating conditions.

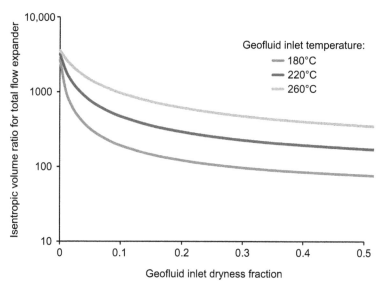

Figure 12.6 Volume ratio required for an isentropic total flow expander assuming condensation at 45°C.

Smith I, Stosic N, Kovacevic A. Power recovery from low grade heat by means of screw expanders. Elsevier; 2014. ISBN:978-1-78242-189-4.

Despite these difficulties, there have been repeated efforts over the past 40 years to produce two-phase expanders that would overcome them. The following summarizes the most significant work carried out to date.

12.1.2 Previous work

The concept of maximizing power recovery by two-phase expansion is not new and can be traced back to Ruths [1], who, in 1924, proposed to increase the total work output from a steam accumulator by expanding the liquid water contained in it, in preference to the separated steam. However, the mechanism required to expand the two-phase fluid was described only vaguely, with a reciprocating machine shown in his diagrams. Prior to that, Baumann [2] had reported on problems associated with the expansion of wet steam in large steam turbines. Subsequently, numerous studies have been carried out on two-phase expansion of water–steam mixtures, many of which are referred to by Ryley [3]. Austin and Higgins [4] defined the total flow concept in 1973, followed by a comparative study of its relative advantages over flash steam or binary plant by Austin and Lundberg [5] in 1974, but the radial inflow turbines that they proposed for the two-phase expansion process do not appear to have been tried in geothermal applications. Radial inflow turbines for use as two-phase process expanders of organic fluids in the chemical industry were proposed by Swearingen [6], whose company, Rotoflow, installed a number of such units with a claimed adiabatic efficiency of 67% in one case. Lund [7] mentioned a 500-kW two-phase turbine having been produced in Russia, at an undisclosed earlier date, but no details of its performance were given. Extensive experimental and analytical studies were also carried out on both axial flow and radial outflow turbines at the Lawrence Livermore and Jet Propulsion Laboratories.

The first practical demonstration of total flow geothermal power recovery appears to have been inspired by Sprankle's patent [8], obtained in 1973, to pass the entire geofluid through a screw expander. This was followed by laboratory test work by Weiss et al. [9–11] and further tests at various geothermal sites. To date, no commercially successful total flow system is known to have been installed and in operation, but the following summarizes most of what has already been done and possible future developments.

12.1.3 Screw expanders

As already mentioned, wet steam screw expanders were tested in the field at a number of sites, as described by McKay [12], Steidel et al. [13], Gonzalez Rubio and Illescas [14], and Carey [15]. The machines used were modified screw compressors of the oil-free type and the results were generally disappointing in that, although Sprankle had claimed adiabatic efficiencies of 75% to be possible from such machines, the majority of test results published indicated efficiencies only of the order of 50%. In the case of Gonzalez Rubio and Illescas [14] the tests were carried out at Cerro Prieto, Mexico, with flow directly from a geothermal well, over a wide range of inlet and exit pressures and a maximum efficiency of 68% was reported, but the operating conditions at which this efficiency was obtained were not given. Unfortunately, during a 2-month endurance test, there were eight failures.

The experimental work was accompanied by analytical work by Margolis [16], Ng et al. [17], Steidel et al. [18], and DiPippo [19]. However, the analyses were mainly of a parametric character and not concerned with the detailed physical processes within the machine.

Taniguchi et al. [20] carried out a series of tests of two-phase flow in a small screw expander using Refrigerant-12 as the working fluid and accompanied this with detailed modeling of the thermodynamic and fluid flow processes within the machine, from which they obtained good agreement between the measured and predicted performance. An important finding obtained from this was that the delay time for the initiation of flashing within the machine could be taken as zero for analytical purposes (ie, the performance of the machine was not affected by any delay in the onset of flashing within the expander).

Smith et al. [21] modeled the expansion process in screw machines, based on previous work by Stosic and Hanjalic [22,23] on screw compressors, which was generally similar to the approach of Taniguchi et al. [20]. Predictions obtained with their model were compared, not only with their own test results, carried out on a closed cycle system, using R113 as the working fluid, but also with the results of Taniguchi, with R12, and those of Steidel et al. [13], with water–steam mixtures. Fair agreement was obtained between predicted and measured performance in all cases and, from this work, a much improved understanding of both the advantages and the limitations of twin screw machines was obtained and, indeed, of other machines of the positive displacement type, such as single screw, scroll, and reciprocating expanders. This is given in greater detail by Smith et al. [24]. Resulting from their work, screw expanders designed to operate on wet steam are now commercially available, designed primarily for enhancing the performance of industrial steam plant [25]. Based on operating data obtained from these machines, a better understanding of the characteristics of screw expanders, and their possible use in total flow systems, can be summarized as follows:

Positive displacement machines affect pressure changes by admitting a fixed mass of fluid into a working chamber where it is confined and then compressed or expanded and from which it is finally discharged. Such machines operate on a sequence of intermittent processes. Thus, reciprocating machines take in fluid either once or twice per cycle, depending on whether the pistons are single or double acting, respectively, whereas rotary screw expanders usually induce four or five separate charges of fluid per revolution. Such intermittent operation is relatively slow, although it is much faster in rotary machines than in reciprocators. Hence these machines are comparatively large. They are therefore better suited for smaller mass flow rates and power inputs and outputs. A number of types of machine operate on this principle such as reciprocating, vane, scroll, and rotary piston machines, as shown in Fig. 12.7.

Because of its pure rotary motion and its ability to run at higher rotational speeds than the others illustrated, the twin screw is the only one that can be considered for the power outputs required in large scale geothermal applications. Twin screw machines are widely used in industry as compressors and, unlike turbo-machines, can be adapted to act as expanders with little change to the mechanical design, but

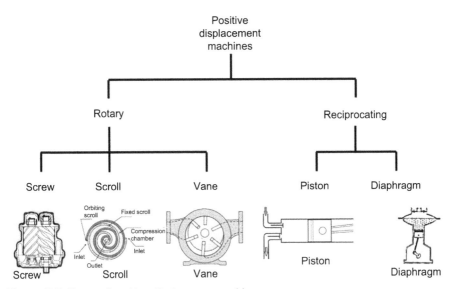

Figure 12.7 Types of positive displacement machines.
Smith I, Stosic N, Kovacevic A. Power recovery from low grade heat by means of screw expanders. Elsevier; 2014. ISBN:978-1-78242-189-4.

important details need to be modified in order to optimize their performance. An assembled view of a typical twin screw machine is shown in Fig. 12.8, in this case, with axial admission to the external high pressure port and a radial external low-pressure discharge port.

Internally the passages are more complex, as shown in an exploded view in Fig. 12.9. From there, it can be seen that admission to the rotors at the high pressure

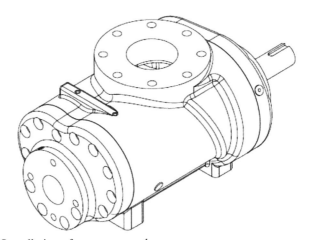

Figure 12.8 Overall view of a screw expander.
Smith I, Stosic N, Kovacevic A. Power recovery from low grade heat by means of screw expanders. Elsevier; 2014. ISBN:978-1-78242-189-4.

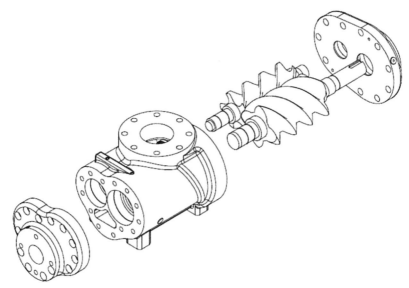

Figure 12.9 Exploded view of a screw expander.
Smith I, Stosic N, Kovacevic A. Power recovery from low grade heat by means of screw expanders. Elsevier; 2014. ISBN:978-1-78242-189-4.

end is mainly axial, with a small radial component of area, while discharge begins in the axial direction but is mainly radial, the inlet and outlet flow being from or into a surrounding plenum chamber, respectively, with the latter leading to the external discharge port.

Its principle of operation is best understood by examination of the succession of diagrams in Fig. 12.10.

As can be seen, this is based on volumetric changes in three dimensions rather than two. The machine consists essentially of a pair of meshing helical lobed rotors, which rotate within their surrounding casing that totally encloses them. The two meshing rotors effectively form a pair of helical gear wheels with their lobes acting as teeth. These are normally described as the male or main rotor and the female or gate rotor, respectively. The space between any two successive lobes of each rotor and its surrounding casing forms a separate working chamber. Starting at the high pressure end, as rotation proceeds, the volume of each chamber increases from zero to a value determined by the size of the inlet port. During this period of rotation, fluid flows into this space in the filling process. As rotation continues beyond this point, the working chamber is cut off from its connection with the inlet port and the line of contact between the rotors advances with a consequent increase in its volume, thus causing the trapped fluid to expand and its pressure to decrease.

The volume reaches a maximum when the entire length between the lobes is unobstructed by meshing contact with the other rotor. At approximately this point, the working chamber becomes exposed to the low-pressure discharge port, at the opposite end of the rotors, and the trapped fluid begins to flow out of it. As can be seen in the final

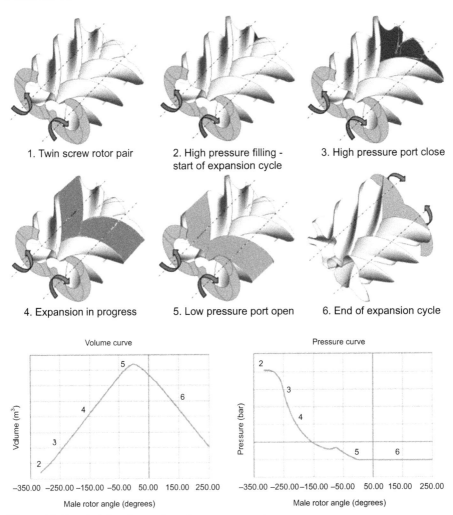

Figure 12.10 Principle of operation of a screw expander.
Smith I, Stosic N, Kovacevic A. Power recovery from low grade heat by means of screw expanders. Elsevier; 2014. ISBN:978-1-78242-189-4.

diagram in Fig. 12.10, viewed from the underside of the rotors, the continuing rotation leads to the line of contact between the lobes receding, thus decreasing the volume of the working chamber and expelling the fluid at approximately constant pressure, until the volume between the adjacent lobes is zero and all the fluid is discharged. This entire process is completed in 720° or two revolutions of the male rotor. The idealized processes, thus described, are illustrated on pressure—volume diagrams in Fig. 12.11(a) and (b) when acting as a compressor or as an expander, respectively.

Although the diagrams are superficially identical, there is an important difference between them. Regardless of the built-in volume ratio of the machine, for given

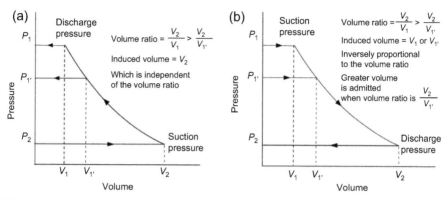

Figure 12.11 (a) Ideal compressor processes. (b) Ideal expander processes.

suction conditions, a compressor of a specified size will take in the same volume of gas or vapor. However, in the case of an expander, the volume inflow decreases as the built-in volume ratio is increased by decreasing the size of the high-pressure inlet port.

It should also be noted that the expansion process in a positive displacement machine is not governed by its built-in volume ratio but rather by the external conditions governing the overall process. Thus, on the one hand, if the condensing or discharge pressure is low, underexpansion will take place in the machine and expansion will continue outside the working chamber after the discharge port is exposed. On the other hand, if the discharge pressure is too high, there will be overexpansion within the working chamber and backflow into it as the discharge port is exposed, leading to a catastrophic loss of power and efficiency. These possibilities, as well as that of ideal expansion, are shown in Fig. 12.12.

When gases are compressed or expanded over normal operating pressures, the assumption of constant pressure flow through the high-pressure port of a compressor or expander, as shown in Fig. 12.12, is approximately correct. However, when liquids or dense two-phase mixtures are expanded, account must be taken of the significant acceleration of the fluid as it passes, intermittently, through the inlet port into each

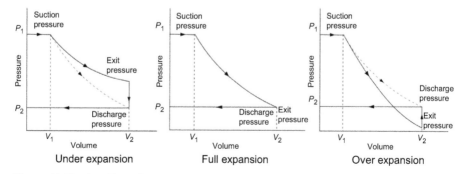

Figure 12.12 The effect of varying the built-in volume ratio of a positive displacement expander.

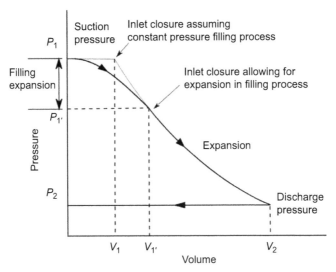

Figure 12.13 Effect of intake port expansion on overall volumetric expansion ratio.

working chamber section. The pressure drop associated with this acceleration, which is proportional to ρV^2, becomes significant because of the high-density ρ, which is orders of magnitude greater than that of normally compressed gases and the high local velocity V. Hence, the expansion process begins at admission of the fluid and not after closure of the high-pressure port. Accordingly, the overall volumetric expansion ratio of the working fluid passing through the machine can be significantly greater than the machine built-in volume ratio, as shown in Fig. 12.13.

Geometric considerations limit the built-in ratio of a screw expander to somewhere between 4:1 and 5:1. This can be raised by increasing the number of rotor lobes, but more lobes decrease the space between them. Increasing the built-in volume ratio reduces the mass flow admitted and hence the recoverable power from a given size of machine.

The best operating conditions for a screw expander are determined by a balance between internal friction losses, which are proportional to the rotor tip speed squared, leakage losses, which are more or less independent of the speed, and the built-in-volume ratio. However, leakage losses become less significant as the mass flow increases as a result of increased rotational speed or decreased built-in volume ratio. It follows that if the built-in volume ratio is too high, the mass flow admitted will be reduced to the point where leakage losses will start to dominate. Then, not only will the efficiency of the expander decrease but also the power output from a given size machine will be diminished owing to the reduced mass flow admitted. In practice, built-in volume ratios of the order of 4:1 represent an upper limit for screw expanders.

When all these factors are taken into account in a detailed model of the expansion process, it can be shown that the ideal expansion process for two-phase flow in a screw expander is achieved with a limited amount of underexpansion, as shown in Fig. 12.14.

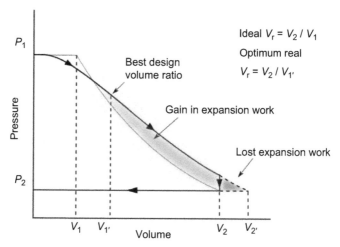

Figure 12.14 Tradeoff between gain and loss of expansion work.

This is because reduction of the built-in volume ratio permits a greater mass flow of fluid to pass through the machine and hence reduces the effects of leakage, the ideal point being reached where the reduction in efficiency due to underexpansion just balances the gain in efficiency due to reduced leakage. Moreover, at this point, the power output is greater than that achieved with full expansion and the size of machine required for a given mass flow is reduced, thereby lowering the expander cost. However, even with a limiting built-in volume ratio of 4:1, because expansion begins in the filling process, overall expansion of the working fluid within the machine can be as high as 30:1, while maintaining adiabatic efficiencies of the order of 70% or more.

A further limitation to the design of screw machines is that the rotor bearings are subject to very high loads, while their dimensions are governed by the center line distance between the main and gate rotors. Because of this, the maximum pressure difference that can be considered for screw machines with an acceptable bearing life is ~ 20 bar.

Since the expansion work is equal to $\int V\, dP$, where V is the volume and P is the pressure, it follows that the greater the *pressure drop* within the machine, the larger will be the power developed. However, if the *pressure ratio* across the machine is too large, then the working fluid will not be fully expanded within it and the efficiency will be low.

When all these factors are taken into account, it can be shown that a screw expander is best suited to working with as high a pressure difference as is possible, up to a limit of approximately 20 bar, to maximize the power output, and with a low-pressure ratio to maximize its efficiency.

Finally, since screw expanders operate with low fluid velocities compared with turbines, they are relatively large. The largest rotor diameter normally manufactured is ~ 850 mm, above which manufacturing costs become very high. Such rotors are rare and very expensive and the upper limit of sizes in regular production is of the order of 500 mm. This limits power outputs from a single machine to <3 MW.

As shown in Fig. 12.6, the required volume ratios for full expansion of geofluids, down to condensing conditions, are at least 100:1, while, as previously explained, even with pure liquid admission, the volumetric expansion ratio within a screw expander is limited to ~30:1. Consequently, the geofluid left the early experimental machines substantially under expanded, as shown by Smith et al. [21], and this is why they operated with disappointingly low efficiencies. Consideration should therefore be given to the possibility of screw expanders cascaded in series, so that the low-pressure machine, or machines, admits the total outflow from the high-pressure expander, as shown in Fig. 12.15.

The more critical problem of cascading is that while it increases the efficiency of the expansion process, it limits the attainable power output, partly because it reduces the pressure drop over each stage but, more significantly, because the mass flow is limited by the size of the first-stage expander. Thus, if the entire expansion were done in a single expander of the same size as the second stage, it could be made to swallow a far greater mass flow and produce greater power output than if it were preceded by a smaller first stage. However, the working fluid would leave it significantly under expanded and its efficiency would be reduced. This is exemplified by the ability of the 440-mm machine tested at Cerro Prieto to develop power outputs of up to 930 kW, though not at high efficiencies. It follows that the operating conditions under which the cited maximum efficiency of 68% was obtained [14] must have been associated with discharge pressures well above atmospheric.

In principle, it is possible for a cascaded total flow system using two screw expanders in series to expand geofluids in a total flow system with efficiencies of >70%. However, the total power output from a unit with a 510-mm rotor diameter unit in the second stage would be only of the order of 500 kW. It can be concluded from cost estimates derived from existing commercial wet steam screw expanders, and screw compressors, of this size, that the installed cost of such a system would be in excess of US$3000 per KWe.

Despite this, recently Yu et al. [26] reported on the use of cascaded screw expanders in geothermal power recovery systems in China. The system described

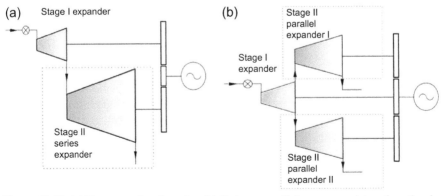

Figure 12.15 (a) Two expanders in series. (b) Series — parallel arrangement to reduce the size of the low-pressure—stage machines.

is a total flow arrangement in which a single screw expander was used for the high-pressure stage and two larger machines operate in parallel for the second stage. The sizes of these machines are not stated specifically, but it would appear from their description that these may be of 800- and 1000-mm male rotor diameters, respectively, and are reported as the largest ever made in China. Machines of this size are rarely made, although there are anecdotal reports of 1200-mm screw compressors used in Russian gas pipelines. The design conditions quoted are for admission of wet steam with a dryness fraction of $\sim$6% at 133.5°C and condensing at 59°C. The estimated isentropic efficiencies of the expanders are 75% with a total power output of $\sim$6 MW from four of these sets in parallel (ie, 1.5 MWe from each set of three expanders), implying that the power output from each of these huge machines is only 500 kWe. This figure could probably be significantly increased if the pressure difference across these machines was larger, bearing in mind that there is a total pressure drop of <3 bar across the two stages.

12.1.4   Screw expander–turbine combinations

12.1.4.1   Single-flash systems

In geothermal steam plant where downhole pumping is used to boost the well flow rate and/or prevent flashing in the well, which may lead to precipitation, or clogging, the brine has to be flashed to a significantly lower temperature and pressure at the surface in order to optimize the power recoverable from the separated steam, as shown in Figs. 12.16 and 12.17.

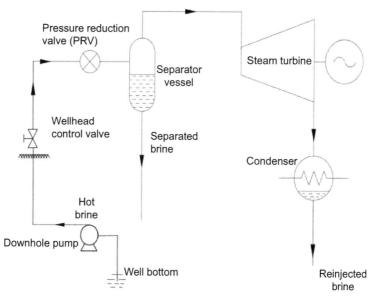

Figure 12.16 Single-flash steam system with downhole pumping.

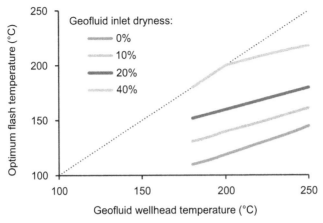

Figure 12.17 Graph of flash temperatures for maximum power recovery from pumped geothermal wells using a flash steam system.

In such cases, significant gains in power output are possible by replacing the required pressure reduction valve (PRV) either wholly or partially, by a screw expander, as shown in Fig. 12.18, for a single-flash system.

Depending on the temperature and dryness of the brine immediately prior to the PRV, it can be shown from data obtained from existing commercially available wet

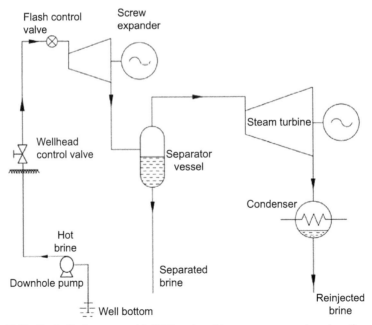

Figure 12.18 Single-flash system with PRV replaced by a screw expander when the well is pumped.

steam screw expanders that all or the major part of the expansion can be carried out in a single machine, within the normal range of manufactured rotor sizes, leading to gains in output and efficiency from 8% to 20% in systems of up to 10 MW output. From the commercial point of view, such an installation can be highly cost-effective, though the best return on capital is not necessarily achieved at the highest attainable gain in power output.

12.1.4.2 Double-flash systems

Should a double-flash system be used for power recovery from a self-flowing resource, then there is the possibility of replacing the PRV, used to enhance the steam flow in the lower pressure turbine stages, by a two-phase expander, as shown in Fig. 12.19.

Here again, it can be shown from data obtained from commercially available wet steam screw expanders that all or the major part of the expansion can be carried out in a single machine, within the normal range of manufactured rotor sizes. In this case, though, since only the separated liquid flows through the two-phase expander, the power recoverable from this expansion is a smaller fraction of the total system power output than in the single-flash case. Accordingly, the percentage improvement gained by the use of screw expanders is rather less than if the total flow were expanded, but, it is possible, in principle, for cost-effective power recovery from a single unit in systems with a total power output of up to 25 MW, using commercially available machines.

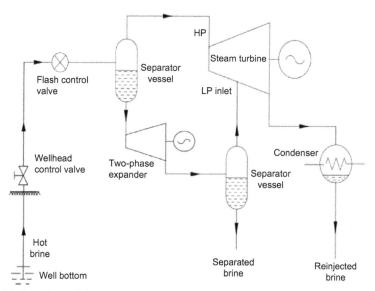

Figure 12.19 Enhanced double-flash system with second-stage PRV replaced by a screw expander.

12.1.4.3   Parallel screw expander and turbine

An ingenious combination of a screw expander to operate in parallel with a conventional steam turbine is reported [27] as shown in Fig. 12.20.

In addition to the conventional inlet port, at the high-pressure end of the rotors, and exit port, at the low-pressure end, additional exit ports are included at the high-pressure end face of the rotors to which the rotor spaces are exposed before rotation is sufficient for expansion to be complete. Partially expanded vapor is drawn from these additional ports. Since the motion of the rotors imparts a velocity to the fluid trapped within them, presumably reversal of the direction of flow caused by the opening of these ports leads to phase separation with the water continuing in the same direction through the passages between the rotor lobes and the partially expanded steam reversing its direction to emerge from the screw expander and entering a conventional turbine. The residual water then expands further as the screws continue to rotate while the separated steam expands in parallel in the turbine. The additional steam formed in the screw expander is again separated out to enter the same turbine in a second stage, where it further expands together with the steam previously drawn off. No record has been found of such a system being built or tested.

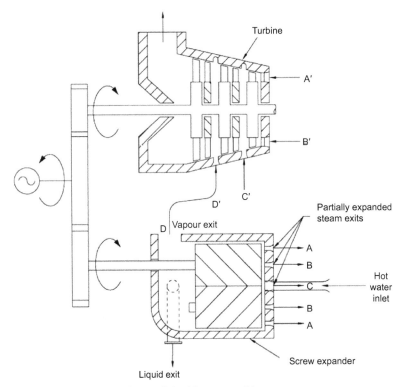

Figure 12.20 Screw expander in parallel with steam turbine.
Adapted from Japanese Patent JPS53-134,139,22, Mitsubishi Ltd., November 1978.

12.1.5 Turbines

12.1.5.1 The biphase turbine

An early attempt to overcome the problems associated with the efficient expansion of liquid—vapor mixtures through the turbine rotor was the biphase turbine, described in principle by Studhalter [28], in which the separator vessel of a normal flash steam plant was replaced by a high-efficiency two-phase nozzle followed by a rotary separator. There, the liquid water total head was recovered to produce additional power in a Pelton-type turbine rotating on the same axis as the separator, as shown in Fig. 12.21, while the separated steam was expanded in a conventional turbine.

The biphase rotary separator turbine was more satisfactory than a single-flash system since its purpose was to augment the power of flash steam plant rather than to replace it. However, the overall adiabatic efficiency of the separator unit was only of the order of 35%. Although it was successfully demonstrated at Roosevelt Hot Springs (Utah), and a 9-MW plant was installed at the Desert Peak (Nevada) geothermal plant [29,30]; there are no further reports of its development.

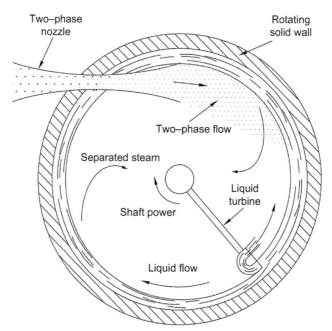

Figure 12.21 Principle of recovery of liquid kinetic energy in biphase rotary separator turbine. Modified from Studhalter WR. Rotary separator turbine enhances geothermal power recovery. Modern Power Systems; September 1982. p. 63, Fig. 2.

12.1.5.2   Axial flow turbines

Extensive research and development programs were carried out on two-phase nozzles by Alger [31] and Comfort [32], at the Lawrence Livermore National Laboratory, and by Elliott [33] at the Jet Propulsion Laboratory. This led to the development of two-phase nozzles with adiabatic efficiencies of up to 82% with water–steam mixtures and up to 90% with organic working fluids. From these studies, it was predicted that turbine efficiencies of >70% were possible, provided droplets of <5 μm could be produced in the nozzle. These and other analytical studies and experimental studies are referred to by Ryley [3], who was unable to reconcile many of the analyses quoted with the test results, but he concluded that the dominant requirement was to obtain smaller droplets. There are great difficulties in obtaining very small droplets and, even if they are formed, there is the further problem of their agglomerating into larger droplets in their passage through the expander. On the basis of his extensive test work, and despite the high nozzle efficiencies obtained in his tests, Elliott's [33] best prediction for the adiabatic efficiency of two-phase axial flow turbines was only 55%. More recent experimental work on wet steam nozzles by Vahaji et al. [34] appears to have been less successful than that of Elliott; they quote maximum nozzle efficiencies of only 65%, although this was for water–steam mixtures at more adverse conditions with maximum temperatures of <100°C.

Welch and Boyle [35] describe a variant of an axial flow impulse turbine described as a variable phase turbine; they claim two-phase nozzle efficiencies of 90–97% and rotor efficiencies of 78–85% with no erosion effects on the blades provided they are made of titanium. However, no test results are given of such a unit in a total flow installation.

12.1.5.3   Radial inflow turbines

Apart from the apparently untried recommendations of Austin and Lundberg [5] and Swearingen's use of such machines in chemical plant [6], no further work has been located on the development of radial inflow turbines for total flow applications. It would seem that more work in this area may be justified, given Swearingen's design of the rotor blade shape to maintain the flow through them parallel to the direction of the net vector force, thereby minimizing the effect of liquid impact on the rotor, and his claim of an isentropic efficiency of 67% using an unspecified working fluid.

12.1.5.4   Radial outflow (reaction) turbines

A radial outflow turbine of the Hero type, where the expansion process occurs in the rotor, and only reaction forces apply, appears to offer several advantages over axial or radial inflow turbines. This was first investigated by Comfort [36] and subsequently by House [37]. The basic layout of such a machine, as tried, is shown in Fig. 12.22. House's test results were disappointing with the best turbine efficiency recorded being only 33%.

Fabris [38] attributed the main faults of House's configuration to the high nozzle inlet pressure resulting from the centrifugal acceleration of the working fluid moving outward from the center of the rotor to its periphery, thereby subcooling the fluid and delaying the onset of flashing, while the very short nozzles would be inadequate for full efficient expansion at their exit. On this basis, he proposed a nozzle configuration as shown in Fig. 12.23 to minimize the lateral forces on the accelerating and expanding

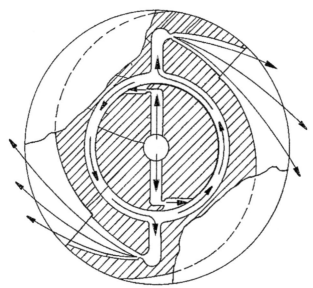

Figure 12.22 Radial outflow reaction turbine by House.
Adapted from US Patent 4,332,520, June 1, 1982, Fig. 1.

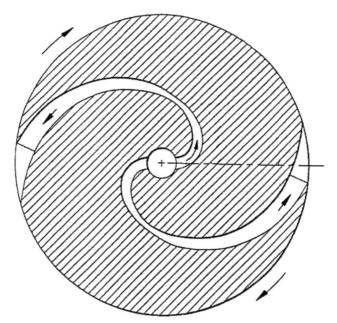

Figure 12.23 Radial outflow reaction turbine by Fabris.
Adapted from US Patent Application 07/601,911, Fig. 3.

fluid and to maximize the length of the nozzle in which expansion took place. Subsequent [39] efficiencies of up to 50% were obtained from tests carried out on a turbine with this configuration. It was claimed that these could be further improved if additional stages were included to recover power from the fluid leaving the reaction stage.

12.1.6 Conclusion

The early concept of the use of screw expanders, or indeed of any positive displacement device as an alternative to flash steam systems has been shown to be impractical with a single-stage machine and uneconomic with a two-stage expansion system. However, there are prospects for further improving the power output from flash steam systems by replacing the pressure reduction valve by a single-stage screw expander in a pumped single-flash system or for expansion to intermediate pressure in a double-flash system.

The remaining alternative of expanding through a two-phase turbine has not yet been developed successfully, due to difficulties in producing a two-phase working fluid with widely dispersed droplets of sufficiently small diameter. However, should these difficulties be overcome, and two-phase expanders with adiabatic expansion efficiencies of more than 70% become possible for large scale plant, then the economic case for including total-flow expanders in geothermal power generation would be greatly enhanced.

12.2 Alternative systems for power recovery based on two-phase expansion

12.2.1 Engineered geothermal systems with water as the power plant working fluid

Artificially created geothermal systems are derived from drilling two wells in areas where the temperature gradient with depth is high, fracturing the rock between the wells, and then pumping water through the system so that heat is recovered from the hot fractured rock and then transferred to a binary system power plant at the surface. These are called enhanced (or engineered) geothermal systems (EGS) and are discussed in chapter "Enhanced geothermal systems: review and status of research & development." The cooled water is then reinjected to form a continuous closed loop. In this case, the artificially created hot brine has to be maintained at a high pressure throughout the circuit to keep the gaps open in the fractured rock while minimizing the power required to circulate it. In such cases, binary power plant systems are required with heat transferred from the recirculated geofluid to a closed- cycle power plant system. Due to the very high cost of producing such reservoirs, it is important to maximize the power recoverable from them.

In some regions, attainable downhole temperatures in excess of 280°C have been recorded [40]. At these high temperatures, traditional organic Rankine cycle systems are less attractive, due to poor matching between the geofluid and the working fluid in the primary heat exchanger, and thermal stability problems associated with many of the commonly used working fluids. Under these circumstances, a closed-cycle double-flash steam plant is a viable option, as shown in Fig. 12.24.

Detailed studies carried out for such a system [41] have shown that the power output of a 25-MW system can be raised by 10% by substituting a 510-mm screw expander having an adiabatic efficiency of 65% in place of the first-stage PRV, as shown in Fig. 12.25, thus yielding a cost-effective improvement. The power output could be increased to 29 MW by also replacing the intermediate pressure PRV by two such screw expanders in parallel. However, estimates based on current machine prices showed that this further improvement would not be cost-effective.

Fig. 12.26 shows a 250-kW screw expander gen-set designed and assembled at City University, London, for installation in place of a PRV in a 1-MW high-temperature EGS pilot plant.

12.2.2 Binary systems using organic working fluids

As already described, screw expanders can expand two-phase fluids with good efficiency over volume ratios of up to $\sim 30{:}1$ (effective), when the working fluid enters the machine in the pure liquid or very dense vapor phase. In the case of binary cycle systems, using organic working fluids, critical temperatures are much lower than that of water, while in most cases, the critical pressure is in the region of only 25−40 bar. A consequence of this, as explained by Smith et al. [21,24] is that organic fluid vapors have much higher pressures and, hence, are far denser than steam at normal condensing pressures and lower vapor pressures than steam at higher temperatures. This implies that volume ratios for the expansion of the working fluid from saturated liquid down to normal condensing temperatures can be orders of magnitude less than that of water. In addition, organic liquids are $\sim 30{-}50\%$ denser than water. This implies that there will be a greater expansion in the filling process at the inlet to the expander. As a consequence, a single-screw expander can operate with good efficiency over a large temperature range. This can be used to good effect in binary plant to include two-phase expansion processes in the cycle, thereby improving both the heat recovery and the overall plant efficiency. The following two systems can therefore be considered, as viable with currently available machinery.

12.2.2.1 Binary system for medium enthalpy sources

For pumped wells, producing saturated liquid brines at the wellhead with temperatures in the 180−200°C, a closed-cycle system using neo-pentane or pentane as the working fluid with a first-stage screw expander, followed by a separator and a second-stage turbine [42], as shown in Fig. 12.27, can produce more net power than either a subcritical or supercritical organic Rankine cycle (ORC) system. This is due to both better matching in the primary heat exchanger and less feed pump work in the case of supercritical systems, since this compound system works at significantly lower pressure differences.

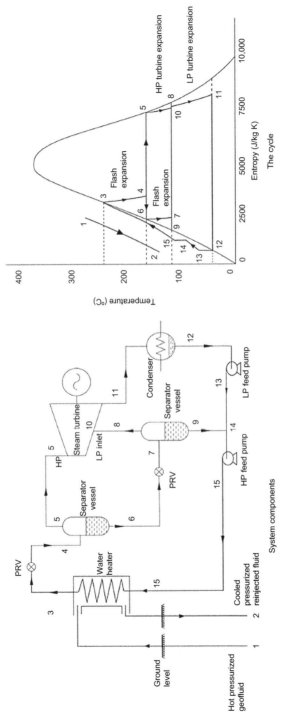

Figure 12.24 Closed-cycle double-flash steam system for high-temperature EGS resources.

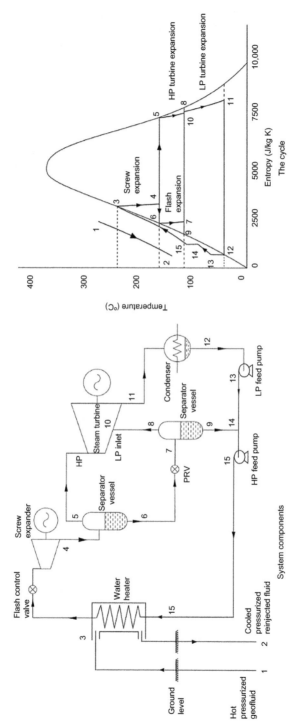

Figure 12.25 Enhanced closed-cycle double-flash steam system for high temperature EGS resources.

Figure 12.26 A 250-kW pressurized hot water screw expander gen-set for EGS application. The author is on the left; Prof. Dan Wright, CEO of Heliex Power Ltd., is on the right. Photo courtesy of Geodynamics Limited, Milton, Australia.

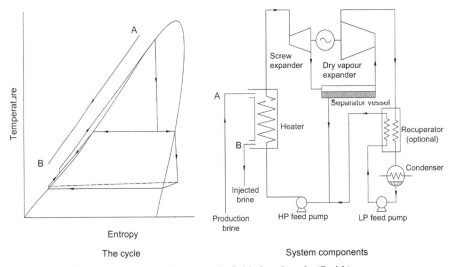

Figure 12.27 Combined screw-turbine organic fluid closed-cycle (Smith) system.

12.2.2.2 Trilateral flash cycle and wet organic Rankine cycle systems

For geofluids with wellhead conditions of saturated liquid at temperatures of 90−110°C, a possible alternative to an ORC system is shown in Fig. 12.28, also using an organic working fluid. This has been called a trilateral flash cycle (TFC) system by

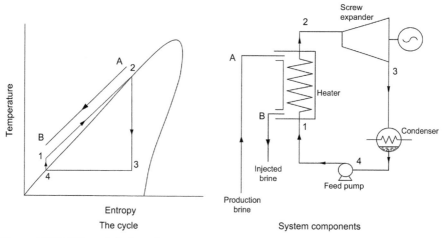

Figure 12.28 Trilateral flash cycle system.

Smith et al. [21,43,44], and by heating the working fluid in the liquid phase only, it is, in principle, possible to maximize the system efficiency. However, the TFC requires a much greater feed pump power than the conventional dry vapor ORC. This is due, mainly, to the lower energy input in heating the working fluid with no evaporation and hence the need for a much greater working fluid mass flow rate per unit of heat received, partly because higher working fluid temperatures and pressures are attainable if there is no evaporation.

More detailed studies have shown that, in practice, when using a screw expander, higher overall efficiencies may be possible if the working fluid is pressurized to a lower pressure and partially evaporated with an inlet dryness fraction of the order of 5−10%, as shown in Fig. 12.29.

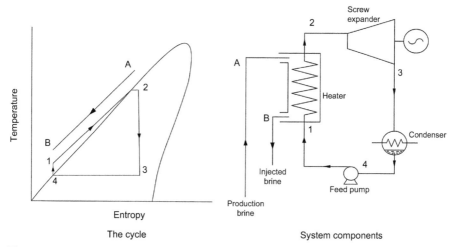

Figure 12.29 Wet organic Rankine cycle system.

The volume ratio of the expansion process is thereby greatly reduced, permitting a single screw expander to operate with adiabatic efficiencies of the order of 75%, while the working fluid mass flow rate required is significantly reduced. Overall, this reduces the feed pump power requirement, thereby resulting in a greater net power output from the system then would be obtained from a TFC system at these lower temperatures. It has been estimated that using screw expanders of $\sim$ 500-mm rotor diameter, gross outputs of up to $\sim$ 4 MWe are possible from a single machine. From a given resource in this temperature range, this system could produce up to twice the gross output of an equivalent ORC system expanding dry vapor. In fact, though, because the feed pump power required at these temperatures is still very large, a wet ORC (WORC) system of this type would produce a gain in net output of only 30—50% over an ORC system expanding dry vapor. It should be noted, though, that in both the WORC and TFC systems, this large increase in power is derived from much greater heat extraction from the geofluid and its correspondingly lower injection temperature. This is associated with much larger heat exchangers than are needed for conventional ORC systems. Careful attention to costing, including that of developing the resource, is therefore necessary in order to determine which system is the most suitable.

12.2.3 Conclusion

The difficulties associated with matching the falling temperature of the geofluid with the rising temperature of the working fluid in a binary cycle plant primary heat exchanger can be much reduced by including a two-phase expansion process in the cycle to enable heating of the working fluid to be carried out entirely in the liquid phase. This enables more heat to be recovered and higher maximum working fluid temperatures to be attained in the binary cycle. In some cases, a single screw expander can be used for the two-phase expansion process to produce larger power outputs than would be possible from dry vapor expansion in turbines.

The increased power output is associated with significantly larger heat exchangers, both due to the recovery and rejection of more heat and the much reduced mean temperature difference in the primary heat exchanger. Careful costing is therefore required to determine whether the increased power output is obtained cost effectively. This is most likely to be the case where drilling and site development costs are high.

It should be noted that all the binary systems described here are based on the use of screw expanders to their best advantage. If a two-phase turbine can be developed with adiabatic efficiencies of 75% or more, then such a turbine-driven TFC system would supplant those described in the binary systems described here.

References

[1] Ruths J. Method and means of discharging heat storage chambers containing hot liquid used in steam power and heating plants. UK patent 217,952. 1924.
[2] Baumann K. Some recent developments in large steam turbine practice. J Inst Elect Eng 1921;59:565.

[3] Ryley DJ. Critical appraisal of some aspects of the analysis of the wet steam nozzle as used in total flow machines. Geothermics 1984;14(2−3):437−47.

[4] Austin Al, Higgins GH, Howard JH. The total flow concept for recovery of energy from geothermal hot brine deposits. UCRL-51366 Lawrence Livermore Laboatory; April 03, 1973.

[5] Austin AL, Lundberg AW. Electric power generation from geothermal hot water deposits. Mech Eng 1975;97(12):18−25.

[6] Swearingen JS. Flashing liquid runs turboexpander. Oil Gas J 1976;74(27).

[7] Lund JW. The basics of geothermal power conversion. In: Presentation at the Conference on Renewables, "Day of Geothermal Power", Bonn; 1984.

[8] Sprankle RS. Electrical power generating systems. US patent 3,751,673. August 1973.

[9] Weiss H, Shaw, G. Geothermal two-phase test facility. In: Second UN Symp on Dev and Use of Geothermal Resources, San Francisco, Calif., USA, 20−29 May, 1975.

[10] Weiss H, Steidel RF, Lundberg A. Performance test of a Lysholm engine. UCRL-51861 Lawrence Livermore Laboratory; July 3, 1975.

[11] Steidel RF, Weiss H, Flower JE. Characteristics of the Lysholm engine as tested for geothermal applications in the Imperial Valley. Trans ASME J Eng Power 1982;104: 231−40.

[12] McKay R. Helical screw expander evaluation project. JPL Publication; March 1982. p. 82−5.

[13] Steidel RF, Pankow D, Berger RE. Performance characteristics of the Lysholm engine as tested for geothermal power applications. In: Proc Intersoc Energy Convers Eng; 1981.

[14] Gonzalez Rubio JL, Illescas F. Test of a total flow helical rotor screw expander at Cerro Prieto, Mexico. Trans Geotherm Resour Counc 1981;5:425−7.

[15] Carey B. Total flow power generation from geothermal resources using a helical screw expander. In: Proc. 5th New Zealand Geothermal Workshop; 1983.

[16] Margolis DL. Analytical modelling of helical screw turbines for performance prediction. Trans ASME J Eng Power July 1978;100:482−7.

[17] Ng KC, Bong TY, Lim TB. A thermodynamic model for the performance analysis of screw expander performance. Heat Recov Syst CHP 1990;10(2):119−33.

[18] Steidel RF, Pankow DH, Brown KA. The empirical modeling of a Lysholm screw expander. In: Proc 18th Intersoc Energy Conversion Eng Conf; 1983. p. 286−93.

[19] DiPippo R. Effect of expansion-ratio limitations on positive displacement total flow geothermal power systems. Trans Geotherm Resour Counc 1982;6:343−6.

[20] Taniguchi H, Kudo K, Giedt WH, Park I, Kumazawa S. Analytical and experimental investigation of two-phase flow screw expanders for power generation. Trans ASME J Eng Gas Turbines Power 1988;110:628−35.

[21] Smith IK, Stosic N, Aldis CA. Development of the trilateral flash cycle system. Part 3: the design of high efficiency two-phase screw expanders. Proc IMechE, Part A 1996;210(A2): 75−93.

[22] Stosic N, Hanjalic K. The development and optimization of screw engine rotor pairs on the basis of computer modeling. In: 12th Int Conf on Compressor Engineering. Purdue University; July 1994.

[23] Hanjalic K, Stosic N. Application of mathematical modeling for screw engines to the optimization of lobe profiles. In: VDI Tagung "Schraubmaschinen '94", Dortmund, Germany; October, 1994.

[24] Smith I, Stosic N, Kovacevic A. Power recovery from low grade heat by means of screw expanders. Elsevier; 2014, ISBN 978-1-78242-189-4.

[25] www.Heliexpower.com.

[26] Yu Y-f, Hu D, Wu F-z. Applications of the screw expander in geothermal power generation in China. In: Proc World Geothermal Congress 2015, Melbourne, Australia; April 19–25, 2015.

[27] Masashi F. Hot water prime mover. Japanese Patent JPS53–134,139. November 22, 1978.

[28] Studhalter WR. Rotary separator turbine enhances geothermal power recovery. Modern Power Systems; September 1982.

[29] Butz JR. Case study of two-phase flow at Roosevelt Hot Springs, Utah KGRA. 1980.

[30] Cerini DJ, Diddle CP, Gonser WC. Project development desert peak 9 MW power plant. Trans Geotherm Resour Counc August 1984;8.

[31] Alger T. The performance of two-phase nozzles for total flow geothermal impulse turbines. 2nd UN Symposium on Development and Use of Geothermal Resources, San Francisco, Calif., USA, May 20–29, 1975.

[32] Comfort W. The design and evaluation of a two-phase turbine for low quality steam-water mixtures. Lawrence Livermore Laboratory; 1977. UCRL 52281.

[33] Elliott DG. Theory and tests of two-phase turbines. JPL Publication; 1982. p. 81–105.

[34] Vahaji S, Akbarzadeh A, Date A, Cheung SCP, Tu JY. The efficiency of a two-phase nozzle as a motion force for power generation from low temperature resources. WIT Trans Eng Sci 2014;83:179–89.

[35] Welch P, Boyle P. New turbines to enable efficient geothermal power plants. Geotherm Resour Counc Annu Meet 2009;33:672–9.

[36] Comfort WJ. Applicability of the hero turbine for energy conversion from low quality two-phase, inlet fluids. ASME Trans 1978:45–53.

[37] House PA. Performance tests of the radial outflow reaction turbine for geothermal applications. Lawrence Livermore Laboratory Report UCID-17902. 1978.

[38] Fabris G. Two-phase reaction turbine with no separation gradual flashing nozzles. US Patent Application US523 6349. 1993.

[39] Fabris G. Two phase turbine for cogeneration, geothermal, solar and other applications. FAS Engineering; 2006.

[40] www.geodynamics.com.au.

[41] Smith IK. Geodynamics Ltd, Milton, Australia. Private Communication. 2008.

[42] Smith IK, Stosic N, Kovacevic A. An improved system for power recovery from higher enthalpy liquid dominated fields. Trans Geotherm Resour Counc 2004;28:561–5.

[43] Smith IK. Development of the trilateral flash cycle system. Part 1: fundamental considerations. J Power Energy Proc IMechE 1993;207:179–94.

[44] Smith IK, Stosic N, Kovacevic A. Screw expanders increase output and decrease the cost of geothermal binary power plant systems. In: GRC annual meeting, Reno, Nevada; September 2005.

Bibliography

[1] Austin AL, Lundberg AW. Comparison of methods for electric power generation from geothermal hot water deposits. ASME74-WA/Ener-10. 1974.

[2] Lorensen LE. Materials screening program for the LLL geothermal project. ASME NW Sect; 1974. p. 105–13.

[3] Cerini DJ. Field evaluation of rotary phase-separator/expander engine. EPRI (WS-78-97). 1978. p. 49–52.

[4] Kestin J, DiPippo R, Khalifa HE, Ryley DJ. Sourcebook on the production of electricity from geothermal energy. US DOE/RA/4051-1. March 1980.

[5] DiPippo R. Geothermal energy as a source of energy. US DOE/RA/28320-1. January 1980.

[6] Holt B, Sims AV, Campbell RG. Assessment of advanced geothermal energy concepts. EPRI Adv Power Syst Div Rep 1981. p. 5A.24−31.

[7] Cerini DJ. Rotary separator turbine for wellhead power production. EPRI Adv Power Syst Div Rep 1984:5.33−4.

[8] Bauer EH. Evaluation of the performance gains available by using total flow expanders in geothermal power systems. In: Proc Intersoc Energy Conversion Engineering Conf; 1985. p. 697−705.

[9] Demuth OJ. Preliminary assessment of the velocity pump reaction turbine as a geothermal total-flow expander. In: Proc of the Intersociety Energy Conversion Engineering Conf; 1985. p. 706−11.

[10] Akagawa K, Fujii T, Ohta J, Takagi S. Cycle performance of total flow turbine systems (1st report, utilization of saturated hot water). Trans Jpn Soc Mech Eng Part B 1986;52(480): 3052−8.

[11] Akagawa K, Fujii T, Ohta J, Takagi S. Cycle performance of total flow turbine systems (2nd report, utilization of saturated wet steam). Trans Jpn Soc Mech Eng Part B 1988; 54(502):1509−15.

[12] Hudson RB. Technical and economic overview of geothermal atmospheric exhaust and condensing turbines, binary cycle and biphase plant. Geothermics 1988;17(1):51−74.

[13] Elliott DG, Weinberg E. Acceleration of liquids in two-phase nozzles. Jet Propulsion Laboratory Technical Report 32-987. 1988.

[14] Chujun G, Caren L. Thermodynamic principles of total flow power production from hot-water geothermal resources. Trans Geotherm Resour Counc 1988;12:389−95.

[15] Hu L, Wang Z, Fan W, Pang F, Lu C. Organic total flow system for geothermal energy and waste heat conversion. In: Proc of the Intersociety Energy Conversion Engineering Conf, vol. 5; 1989. p. 2161−5.

[16] Sekioka M. Analysis on direct uses of geothermal energy in Japan with respect to inlet temperature and flow rate. Trans Geotherm Resour Counc 1989;13:63−8.

[17] Bunch TK, Kornhauser AA. Efficiency of a flashing flow nozzle. In: Proc of the Intersociety Energy Conversion Engineering Conference, vol. 3; 1996. p. 1610−5.

[18] Barbier E. Geothermal energy technology and current status: an overview. Renew Sustain Energy Rev 2002;6(1−2):3−65.

[19] DiPippo R. Ideal thermal efficiency for geothermal binary plants. Geothermics 2007;36: 276−85.

[20] Hunt RD. Evaluation of the effect of gravity on the total-flow geothermal cycle. Trans Geotherm Resour Counc 2010;34(2):960−5.

[21] Pierce M. Improving binary cycle efficiency by eliminating parasitic loads. Trans Geotherm Resour Counc 2011;35(2):1331−6.

[22] DiPippo R. Geothermal power plants, principles, applications and case studies. 3rd ed. Elsevier Science; 2012.

[23] Hecht-Méndez J, De Paly M, Beck M, Bayer P. Optimization of energy extraction for vertical closed-loop geothermal systems considering groundwater flow. Energy Convers Manag 2013;66:1−10.

Binary geothermal energy conversion systems: basic Rankine, dual–pressure, and dual–fluid cycles

13

G. Mines
Idaho National Laboratory (Retired), Idaho Falls, ID, United States

13.1 Introduction

While geothermal energy can be used directly, its conversion to electrical power and the subsequent transmission to customers remove the constraint that the geothermal energy and its direct use be co-located. The conversion of geothermal energy to electrical power is predominately accomplished using the "dry" steam, flash-steam, and binary power cycles. In each of these power cycles, a pressurized vapor is expanded in a turbine coupled to an electrical generator. Typically, the fluid exiting the turbine is condensed, with the associated latent heat being rejected to the ambient (backpressure steam turbines are an exception). Previous chapters have dealt with geothermal steam plants; here, the focus is on the binary plants.

In the binary cycle, heat from the geothermal fluid is transferred to a secondary working fluid circulated in a closed loop. This heat is used to vaporize the working fluid that is subsequently expanded in a turbine coupled to an electrical generator. There is evidence that the first geothermal binary plant operated in Italy on the island of Ischia in the 1940s, around the time of World War II; see chapter "Larderello: 100 years of geothermal power plant evolution in Italy," Section 19.9.1 in this volume. Later, a larger binary plant was brought into service at Paratunka in Russia's Kamachatka peninsula in the 1960s, and it is considered the first commercial binary geothermal plant. Today, the binary plants make up ∼47% of the number of geothermal units operating worldwide and provide ∼14% of the geothermal capacity [1]. The relatively low capacity for binary plants is indicative of these plants using lower-temperature, lower-enthalpy resources resulting in smaller power stations.

The selection of a conversion system will depend on a number of factors, though likely is driven by economic considerations. Typically, the higher-temperature resources will use the flash-steam power cycle, while the binary power cycle is used with the lower-temperature resources. The binary cycle has the advantage of being able to extract more energy from the geothermal fluid. This advantage is reduced when the geothermal fluid requires a minimum temperature constraint be imposed on the geothermal fluid exiting the plant to prevent mineral precipitation in surface equipment and injection wells. Silica is common to geothermal fluids and is frequently

Geothermal Power Generation. http://dx.doi.org/10.1016/B978-0-08-100337-4.00013-9

the basis for the temperature constraint. Because the solubility of most minerals increases with temperature, as energy is extracted from geothermal fluid, its temperature and solubility decrease, and the potential for mineral precipitation increases. With higher temperature resources, the minimum temperature required to prevent precipitation is higher, and the potential for the binary cycle to extract more energy for the geothermal fluid is reduced.

The commercial availability of geothermal production pumps capable of operating at higher temperatures is another factor that limits the use of the binary cycle at these temperatures. While the binary cycle can operate using steam as the heat input, it would be more beneficial (in terms of power production) to expand that steam directly is a steam turbine. Typically, the binary plants are used at temperatures less than 200°C, because [1] their potential performance advantage at higher temperatures is diminished by temperature constraints to prevent mineral precipitation and [2] temperature and setting depth limitations of current downhole production pumps.

13.2   Binary power cycle

In the binary cycle, heat is transferred from the geothermal fluid to a secondary working fluid. During this heat transfer process, the pressurized working fluid is vaporized. The working fluid leaving the geothermal heat exchangers is subsequently expanded through a turbine producing work, or electrical power. The lower-pressure working fluid leaving the turbine is condensed and pumped back to the geothermal heat exchangers, completing the closed cycle. This process is depicted in Fig. 13.1, with the heat rejection accomplished with an air-cooled condenser. The heat transfer processes for this cycle are shown in Fig. 13.2.

In Fig. 13.2, the working fluid is shown leaving the vaporizer with a small level of superheat. The inlet vapor is superheated to ensure that the vapor is "dry" and that it remains in the vapor phase throughout the turbine expansion. The level of superheat that is added depends on the working fluid selected. Some fluids tend to superheat on expansion (isobutane, R245fa, isopentane), while others tend to desuperheat and begin to condense on expansion (propane, ammonia, water, R134a). Those fluids that tend to desuperheat will require more superheat before entering the turbine. If a significant amount of superheat were required, an additional geothermal heat exchanger would be included in the basic cycle shown in Fig. 13.1, located between the vaporizer and turbine, and using geothermal fluid directly from the wellhead. The geothermal fluid leaving this superheater would then be directed to the vaporizer. When a minimal amount of superheat is required, the superheating is accomplished in the vaporizer as depicted in Fig. 13.1.

The basic cycle processes are also depicted in Fig. 13.3, which is a temperature–entropy plot of the cycle conditions shown in Fig. 13.2. The working fluid used in this depiction of the cycle superheats when expanded.

The heat exchangers used in the binary power plants are typically shell and tube exchangers with the geothermal fluid on the tube size of the exchanger and the working fluid on the shell side. This configuration facilitates mechanical cleaning of the heat exchanger if fouling occurs due to the exposure to the geothermal fluid. Plate heat

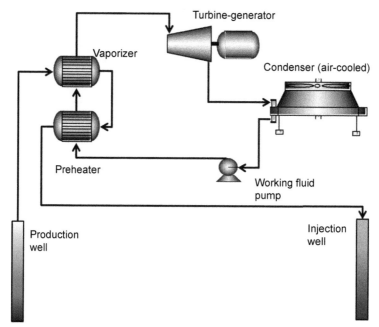

Figure 13.1 Basic binary cycle with air-cooled condenser.

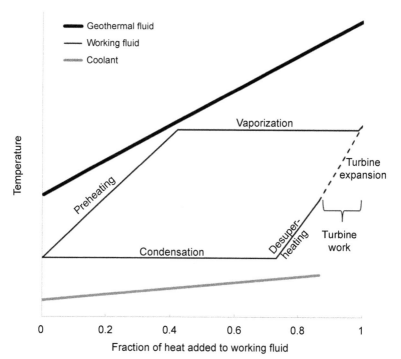

Figure 13.2 Heat transfer processes in basic binary cycle.

Figure 13.3 Temperature—entropy plot of basic binary cycle.

exchangers have been considered as an alternative to the shell and tube design for the geothermal heat exchangers, particularly the liquid preheaters, because of their compact size and performance. Generally turbines or turbo-expanders are used as the expansion devices used to drive the electrical generators. These rotary devices are either axial or radial inflow turbines with varying degrees of reaction. Screw or helical expanders are also used in a limited number of applications. These types of expanders are typically smaller, although a 1-MWe screw expander is used at the Lightning Dock facility in New Mexico, USA [2].

The heat rejection in the binary cycle is accomplished in either water-cooled or air-cooled condensers. With the water-cooled condensers, heat is rejected evaporatively to the ambient, while with the air-cooled condensers, the heat rejected via the sensible heating of air. Because evaporative heat rejection consumes water, it is necessary to have an adequate, reliable supply of water for the water-cooled plants. If there is not an adequate supply of makeup water, air-cooled condensers are used.

13.3   Binary cycle performance

13.3.1   Performance metrics

The different metrics used to define the performance of binary cycles include thermal efficiency, second law efficiency, and specific output (also referred to as brine effectiveness or brine utilization). The specific output is the mechanical work, or net plant electrical output, divided by the geothermal flow rate.

13.3.2 Thermal efficiency

The thermal efficiency is commonly used as an indicator of plant performance, whether it is a fossil, nuclear or a binary geothermal plant. It is the ratio of the mechanical power delivered $(\dot{W}_m)$ to the heat input to the power cycle $(\dot{Q}_{in})$.

$$\text{Thermal efficiency} = \frac{\dot{W}_m}{\dot{Q}_{in}} \qquad [13.1]$$

The mechanical work $(\dot{W}_m)$ is the net plant power, or the generator output less the parasitic power requirements to operate the power plant. These include pumping and fan power requirements within the plant, exclusive of any geothermal fluid pumping for either production or injection. The geothermal pumping power is excluded because it is external to the cycle and site specific; thus, their inclusion is not indicative of the power cycle performance. The plant parasitic requirements are included as they are indicative of performance. The heat input to the cycle (Q_{in}) is defined as the heat extracted from the geothermal fluid or

$$\dot{Q}_{in} = \dot{m}_{gf} \times (h_{gf-in} - h_{gf-out}) \qquad [13.2]$$

where $\dot{m}_{gf}$ is the geothermal flow rate, and h_{gf-in} and h_{gf-out} are the enthalpy of the geothermal fluid entering and leaving the power cycle. Typical thermal efficiencies for binary plants will be in the $8-12\%$ range.

13.3.3 Second law efficiency

The principles presented in this section apply equally well to all types of geothermal plants but will be applied here to binary plants. The second law efficiency is an indicator of how well the power cycle converts the geothermal fluid's exergy, or available energy, to work. The exergy term represents the maximum work (power) that could be done in bringing the geothermal fluid into equilibrium with the surroundings using processes that are completely reversible. The surroundings are generally taken to be the ambient conditions to which heat is rejected. Once the geothermal fluid and ambient conditions are specified, the fluid's exergy is effectively a property of the fluid. If either the geothermal fluid or ambient conditions change, so, too, does the geothermal fluid exergy.

The specific exergy, or exergy per unit mass, is defined as

$$e = (h_{in} - h_0) - T_0(s_{in} - s_0) \qquad [13.3]$$

where h_{in} and s_{in} are the enthalpy and entropy of the geothermal fluid entering the cycle or plant, and h_o and s_o are the enthalpy and entropy of the fluid at the surroundings (ambient conditions), respectively. T_o is the surroundings or ambient temperature, appropriate for the chosen heat rejection method.

No power cycle is composed of completely reversible processes. Each process and component deviate from the ideal; the degree to which it deviates is the irreversibility associated with that component or process. The irreversibility term for each of the cycle components can be expressed as

$$\text{Irreversibility} = \sum_{\text{all fluids}} -\dot{m}_{\text{fluid}}(T_o)(s_{in} - s_{exit}) \qquad [13.4]$$

where $\dot{m}_{\text{fluid}}$ is the flow rate of a given fluid. Reducing the irreversibility for a component reduces the exergy loss for the cycle and results in more power production. For a pump, fan or turbine, this relationship applies with only a single fluid in the summation. The relationship can be used to determine the irreversibility associated with frictional losses in piping systems as well. The irreversibility of the individual components and processes dissipate the exergy of the geothermal fluid entering the cycle, as does the exergy of each of the fluid stream exiting the plant. This dissipation of the inlet exergy establishes the net power for the cycle

$$\text{Power}_{\text{net}} = \sum_{\text{all fluids}} \dot{m}_{\text{fluid}}(e_{\text{fluid–in}}) - \sum_{\text{all processes}} \text{Irreversibility}_{\text{process}}$$
$$- \sum_{\text{all fluids}} \dot{m}_{\text{fluid}}(e_{\text{fluid–exit}}) \qquad [13.5]$$

The second law efficiency indicates the rate at which the exergy is converted to useful work or power and is defined as

$$\eta_{\text{II}} = \frac{\dot{W}_{\text{m}}}{\dot{m}_{\text{gf}}(e)} \qquad [13.6]$$

where $\dot{W}_{\text{m}}$ is the net mechanical work (power) as used in the definition of thermal efficiency, $\dot{m}_{\text{gf}}$ is the geothermal flow rate, and e is the specific exergy. The specific output term is the mechanical work, or power, per unit mass, so the second law efficiency can be determined as the ratio of the specific output to the specific exergy. If the ambient and resource conditions are constant, the specific output provides the same indication of cycle performance as the second law efficiency.

Increasing the second law efficiency and net power produced is accomplished by reducing some combination of the irreversibility associated with one or more processes or decreasing the exergy of one or more of the fluid streams leaving the plant. The irreversibility associated fan, pump, or expander is the loss associated with the inefficiency of the device. The selection of the type of binary cycle and the working fluid has some impact on the irreversible exergy losses for these components, though that impact is secondary to the impact that will occur when reducing the irreversibility associated with the heat exchanger processes and reducing the exergy of the fluid streams leaving the plant or power cycle. These reductions are accomplished through a combination of the selection of the cycle configuration, working fluid, and process conditions, though invariably reducing the irreversibility of a heat exchanger process will increase the costs of the components impacted.

13.3.4 Efficiency metrics versus power

Because of the cost of supplying and disposing of the geothermal fluid used in the plant, there is incentive to use that fluid more efficiently. Plants designed to minimize total project costs on the basis of $/kW should also have lower generation costs, or $/kWh.

The efficiency metrics discussed are functions of the plant design at a given set of resource and ambient conditions. The elements of a plant design that affect these efficiencies include the type of binary cycle and working fluid used, component efficiencies, heat exchanger approach temperatures, and the turbine inlet and exhaust conditions. A given plant design is based on a combination of these elements for the specified resource and ambient conditions, with that design having a unique net power output and corresponding efficiencies.

The two efficiency metrics discussed are ratios of the net plant power to either an energy or an exergy term. Because a plant and its equipment are designed for specified resource and ambient conditions, the exergy term is fixed and independent of the design. Consequently, the net power varies directly with the second law efficiency [Eq. [13.6]].

The heat extracted from the geothermal fluid [Eq. [13.2]] is not necessarily fixed and is dependent on the plant design. As a result, plants can yield the same thermal efficiency but have significantly different levels of net power production from the same geothermal inlet and conditions, as depicted in Fig. 13.4. In this figure, the power outputs for different plant designs are shown as a function of the thermal efficiencies resulting from those designs. The points shown reflect different combinations of the parameters that can be varied in the design of an air-cooled binary plant with fixed geothermal fluid (125°C and 296 kg/s) and ambient (10°C) design conditions.

This figure illustrates the lack of a direct relationship between power production and thermal efficiency; designs that yield a certain thermal efficiency can have significant variation in the power that is produced.

Because of the direct relationship between the second law efficiency and net plant output, this efficiency metric is used in the subsequent discussion of the merits of the binary power cycles.

13.3.5 Limitations on performance

Mineral solubility nearly always increases with the temperature of the geothermal fluid, calcium carbonate being the most important exception for geofluids. This can result in a minimum temperature constraint being placed on the operation of the binary

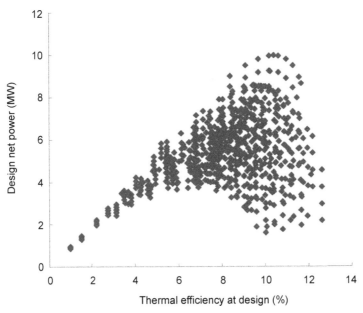

Figure 13.4 Relationship between thermal efficiency and net plant output.

plant to prevent mineral precipitation both on the surfaces exposed to the geothermal fluid within the plant and in the injection piping and wells. One temperature constraint commonly imposed is to prevent the precipitation of amorphous silica, which goes into solution as quartz in the reservoir [3,4]. The temperature at which amorphous silica begins to precipitate can be estimated based on the temperature of the geothermal resource [5]. A constraint may also be placed on the temperature leaving the plant to meet the requirements of a downstream usage of the geothermal fluid.

With a constraint in place, the exergy of the geothermal fluid leaving the plant is fixed (for the resource and ambient conditions), and the improvements to cycle performance must be accomplished by a reduction of a combination of component irreversibility and the exergy of the coolant leaving the plant.

13.4   Types of binary cycles

13.4.1   Subcritical boiling cycles

In the basic binary cycle shown in Fig. 13.1, the working fluid is vaporized at a single pressure. Although this basic cycle is widely used in current commercial facilities, there are continuing efforts to improve the performance of the cycle and lower generation costs either by reducing the flow rate that must be supplied from the well field for a fixed power or by producing more power from a constant geothermal flow. As discussed, unless efficiencies of rotating equipment (fan, pump, and turbine) are improved, achieving higher plant performance will be accomplished by reducing the exergy losses in the heat exchange processes. The cycles developed to achieve performance improvements have focused on reducing the exergy losses (irreversibility) associated with the transfer of heat between the geothermal fluid and the working fluid and by decreasing the exergy of the geothermal fluid leaving the plant.

In the development of the binary cycle technology, one of the first attempts to reduce the losses in this heat exchange process was to vaporize the working fluid at two separate pressures, with the working fluid vapor being directed either to two separate turbines or to a multistage, dual-pressure turbine with two inlets. There are variations to how this dual boiling cycle can be configured; one possible scheme is shown in Fig. 13.5. As shown, the working fluid flow leaving the preheater is split, with a portion being vaporized and sent to the low-pressure turbine, and the remainder being pumped up to the pressure of the high-pressure vaporizer.

The heat transfer processes for the dual boiling cycle as depicted earlier are shown in Fig. 13.6.

The area between the curves for the cooling of the geothermal fluid and the heating of the working fluid is indicative of the irreversibility (exergy loss) in the heat exchange process (see Section 9.12 of chapter "Elements of thermodynamics, fluid mechanics and heat transfer applied to geothermal energy conversion systems"). The differences in the areas between these curves as depicted in Figs. 13.2 and 13.6 illustrates how the addition of the second boiling stage reduces the irreversibility with this heat exchange process and increases cycle performance.

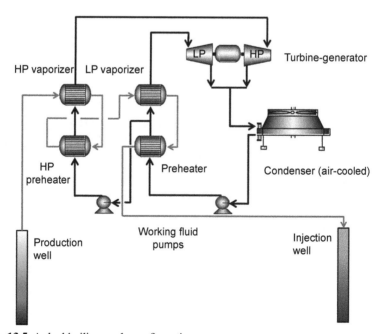

Figure 13.5 A dual boiling cycle configuration.

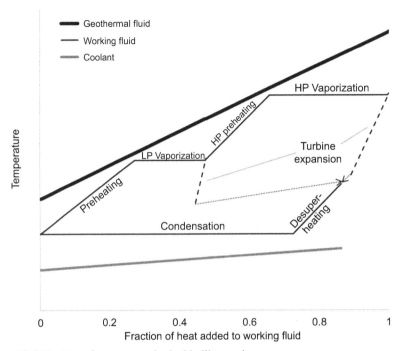

Figure 13.6 Heat transfer processes in dual boiling cycle.

The first use of the dual boiling cycle in an operating facility was in the U.S. Department of Energy (DOE) plant built on the Raft River resource in south central Idaho, USA. That plant operated with a 140°C resource, generating ~ 5 MW$_e$ by using isobutane as a working fluid in a cycle configuration similar to that shown in Figs. 13.5 and 13.6. Since then, the concept of using dual boiling cycles to increase second law efficiency has been used elsewhere, including the current plant that is installed at the same resource.

One iteration on this cycle that was considered to provide further performance improvement was to add a third level of boiling [6]. By boiling a working fluid at a third pressure (or temperature), it would be possible to further reduce the exergy loss in the transfer of heat between the working fluid and geothermal fluid.

13.4.2   Trilateral cycle

Another cycle configuration that addresses the irreversibility of the heat addition process is the "trilateral" cycle [7]. In terms of the equipment configuration, this cycle is similar to the basic binary cycle depicted in Fig. 13.1, with the exception that there is no vaporizer. A heated working fluid liquid leaving the geothermal heat exchanger (preheater) is expanded directly to the condensing pressure. The cycle is depicted in the temperature–entropy plot in Fig. 13.7. The shape of this cycle on a T–s plot

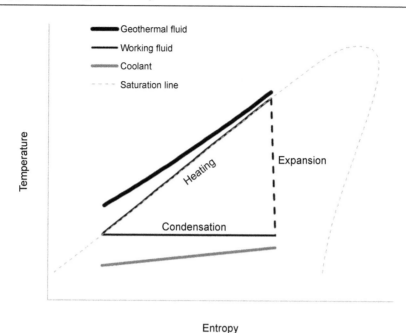

Figure 13.7 Trilateral cycle processes.

can be made to approximate the triangle cycle that DiPippo indicates would approach the thermodynamic ideal for a sensible heat source [3].

As shown in this figure, the expansion of the heated liquid into the two-phase region is accomplished in an expander having an isentropic efficiency of 85%. This is what one might expect with the turbo-expanders used in binary cycles where a pressurized vapor is being expanded. It is unlikely that a turbo-expander used for vapor expansions would achieve this level of efficiency for expansions of liquid into the two-phase region. In the 1990s, tests were conducted with supercritical binary cycles where turbine expansions were allowed to enter the two-phase region [8]. In this testing, the resulting degradation in turbine efficiency was related to the degree to which the expansion process entered the two-phase region. At the extreme condition tested, an isobutane working fluid was expanded from inlet conditions of 130.5°C and 41.4 bar(a) to an exhaust pressure of 4 bar(a) in an axial flow, impulse turbine. The isentropic efficiency of the turbine decreased by ~20% relative to what it attained when expanding a dry vapor. Proponents of the trilateral cycle recognize this issue and propose the use of other types of expanders that are better suited for these expansions [9,10]. This is discussed in detail in chapter "Total flow and other systems involving two-phase expansion" of this volume.

If expansion efficiencies are attained similar to those achieved in conventional binary cycle, this cycle increases power output because of the lower irreversibility associated with the heat addition. If they are not, the turbine irreversibility will increase and negate some portion of the benefit derived from reducing the exergy loss in the heat exchanger. This cycle is also impacted more by the imposition of any constraint on the temperature of the geothermal fluid leaving the plant, suggesting it may have more applicability with lower-temperature resources.

13.4.3 Supercritical cycle

Efforts to provide further improvements beyond those achieved with the multiple boiling cycles lead to the consideration of the supercritical cycle where the working fluid vaporizes at a pressure above the critical pressure of the fluid. A supercritical cycle configuration is effectively the same as the basic binary cycle depicted in Fig. 13.1, with the exception that during the heat addition processes, the working fluid pressure is above its critical pressure. The heat addition process for this cycle is shown in Fig. 13.8, while Fig. 13.9 depicts the cycle process on a temperature–entropy diagram. Like the trilateral cycle, the shape of this cycle on the T–s plot approaches that of DiPippo's triangle cycle.

Because heat is added to the working fluid while at a pressure greater than the critical pressure, there is no discrete phase change. The working fluid density continues to decrease as heat is added, and the fluid leaves the vaporizer as a high-pressure vapor (above the fluid critical pressure). As depicted in Figs. 13.8 and 13.9, this vapor is then expanded thru a turbine, and condensed at a temperature (or pressure) typical of those occurring in subcritical plants.

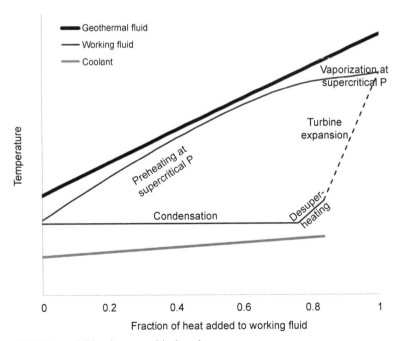

Figure 13.8 Heat addition in supercritical cycle.

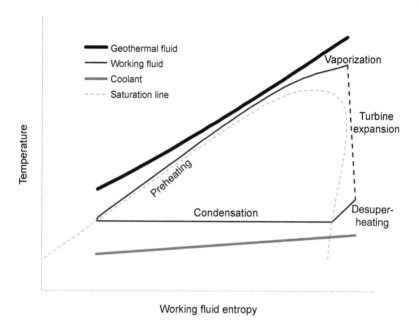

Figure 13.9 Supercritical cycle processes on temperature–entropy diagram.

As shown in Fig. 13.9, the vapor entering the turbine typically has an entropy greater than the maximum saturated entropy that would occur between the turbine inlet and exhaust pressures, assuring that the turbine expansion does not enter the two-phase region. The fluid used to generate this figure (isobutane) has a retrograde curve and tends to superheat when expanded isentropically. Some fluids (propane and R134a) tend to desuperheat with expansion. With these fluids, the temperature of the working fluid entering the turbine is based on the saturated vapor entropy at the turbine exhaust pressure.

The heat addition process for the supercritical cycle relative to those for the basic binary cycle and the dual boiling cycle are shown in Fig. 13.10. This figure is based on the same scenarios used in Fig. 13.6. The supercritical cycle depicted in Fig. 13.10 uses an isobutane working fluid. With the supercritical cycle, the area between the geothermal fluid cooling and working fluid heating curves is reduced even more than achieved with the dual boiling cycle, further reducing the irreversibility (exergy loss) associated with the heat transfer process. Because the fluid in the supercritical cycle does not undergo an isothermal phase change during vaporization, it has the potential to minimize the temperature difference (area difference) in the heat exchanger component.

Although multiple studies have shown the performance benefits of the supercritical cycle, there has been minimal application of the cycle in operating facilities. The U.S. DOE funded work on the cycle both in a prototype test facility and in a demonstration plant near Heber, California, in the 1980s. That work did not identify any particular issues with the cycle; however, there was no commercial application of the cycle until

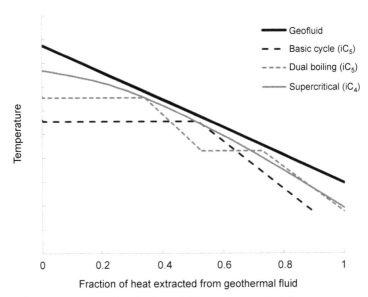

Figure 13.10 Temperature profiles for heat addition process with different power cycles (maximum heat extraction limited by geothermal fluid temperature constraint).

recently in plants at Neal Hot Springs in Oregon and San Emideo in Nevada. Both these plants use the R134a working fluid, and they have similar resource temperatures ($\sim 140°C$). Other binary plants have been designed with turbine inlet pressures just below the critical pressure, including the binary facilities at Mammoth Lakes. During initial operation, one of the facilities at Mammoth briefly operated at supercritical turbine inlet pressures with no issues. The Mammoth facilities use an isobutane working fluid and have a resource temperature of $\sim 165°C$.

13.5 Selection of working fluid

The selection of the working fluid for a binary power cycle is critical to cycle performance. The fluid providing superior performance will be dependent on the geothermal conditions, the type of binary cycle being used, and any operating or design constraints placed on the plant. Other factors impacting the selection include the fluid flammability, toxicity, chemical aggressiveness, potential hazards to the environment, and cost.

These additional factors are potential deterrents to the fluid selection and one or more apply to nearly all candidate fluids. With the exception of cost, there are numerous compilations of these different factors. An example is DiPippo's summary of this information for common working fluids for binary plants [3].

With the exception of fluids that are expensive or that are restricted or banned because environmental hazards imposed, engineering design controls can mitigate the remaining factors. These factors are addressed by the engineering codes and standards used in the design, fabrication, and construction of a plant and its equipment. Agencies having regulatory authority over the construction and operation of these facilities will require that these codes and standards be applied. These agencies serve as the first points of contact in determining requirements imposed on binary plants built within their jurisdiction. Those requirements may be deciding factors in the fluid selection.

While single-component working fluids are dominated in operating binary plants, working fluid mixtures are an option with potential to improve performance. Mixtures that have been proposed and used include ammonia—water and blends of common hydrocarbons or refrigerants. In the following discussion on working fluids, it is assumed that the common hydrocarbon working fluids are single component fluids. In practicality, they are not. In most, the major constituent exceeds 90%, with the remainder being primarily the adjacent hydrocarbons; for example, propane, butane, pentane, and isopentane are adjacent to isobutane and are the primary impurities in commercially available isobutane. For additional cost, the amount of impurities can be minimized, although, generally, the blends that are sold will perform similarly to that predicted for a pure fluid.

13.5.1 Single-component working fluids

The single-component working fluids are used in nearly all operating binary facilities. These fluids are primarily a hydrocarbon or engineered refrigerant having a normal boiling point temperature that is at or below ambient temperatures. Though ammonia has been considered as a possible alternative [11,12], aside from the mixtures in the Kalina cycle facilities, it has not been used in a commercial geothermal binary power cycle.

Those fluids having potential application in binary cycles can be initially screened using their normal boiling point temperature. Fig. 13.11 was generated from the refrigerants in the National Institute of Standards and Technology (US Department of Commerce) Reference Fluid Thermodynamic and Transport Properties (REFPROP) database.

Refrigerants with potential application in binary cycles generally have normal boiling point temperatures between −50°C and 50°C, as bracketed in the figure by the dashed lines. In total there are 66 fluids meeting this criteria. Though both of these limits are somewhat arbitrary, there is some basis for their selection. At the lower limit, the critical temperature is ∼ 100°C. If the critical temperatures are lower, the power cycles used will likely be supercritical with high operating pressures, including the potential for condensing pressures to exceed the fluid critical pressure. Operation at these pressures would likely be cost prohibitive for the lower temperature resources more likely to use these working fluids. As the resource temperature increases, it is probable that the fluids used will have increasing normal boiling point temperatures. If a subcritical boiling cycle is being used, it is likely the fluid will have a higher normal boiling point temperature, while a fluid using a supercritical cycle will have a lower normal boiling point and, more importantly, a lower critical temperature.

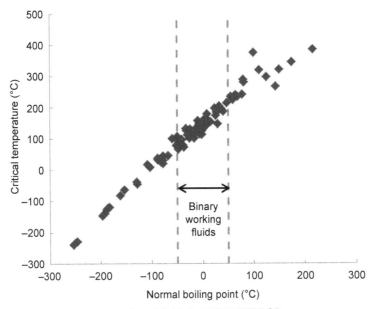

Figure 13.11 Temperature properties of fluids in NIST REFPROP.

At the higher temperature limit in the figure, it is probable that the working fluid would be condensing at pressures less than 1 atm or roughly 1 bar(a). With subatmospheric condensation, there is an increased probability of air leakage into the working fluid system. Air is noncondensable and will adversely impact performance, as well as pose a potential safety risk if the working fluid is flammable. If operation at subatmospheric conditions in the working fluid system is to be avoided, it will influence the selection of the working fluid. In Fig. 13.12, fluid condensing pressures are shown when condensation occurs at 0 and 40°C; fluids in the shaded region will be unacceptable. If this constraint is imposed, it may also affect the selection of the heat rejection system used for plants in colder climates where air-cooled plants are more readily able to operate at lower condensing temperatures.

The candidate hydrocarbon working fluids depicted in Figs. 13.11 and 13.12 are either a flammable liquid (ie, isopentane and pentane) or a liquefied petroleum gas (ie, isobutane, n-butane, and propane), with isopentane and isobutane most commonly used in commercial plants. These fluids are flammable and require engineered fire protection systems to mitigate the associated hazard. They are also relatively inexpensive with low toxicity and environmental risk.

The engineered refrigerants most commonly considered for geothermal binary plants are R134a and R245fa. The supercritical facilities at Neal Hot Springs in Oregon and San Emideo in Nevada use the R134a working fluid. It is also used in the binary bottoming units installed at the Dixie Valley and Beowawe flash plants in Nevada, as well as a bottoming unit at Coso in California. These engineered refrigerants have the

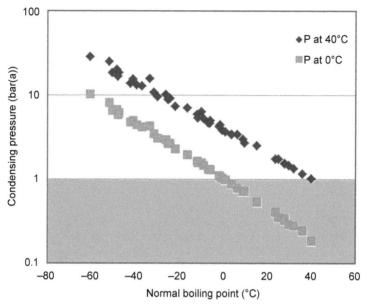

Figure 13.12 Condensing pressures of candidate working fluids at 10 and 40°C.

advantage of not being flammable. However, relative to the hydrocarbons, they have higher Global Warming Potentials [13] by at least two orders of magnitude and their cost will be significantly higher than alternative hydrocarbon fluids. Some literature indicates [14] that R134a will begin to decompose at temperatures above 200°C, but this is likely not an issue because there are other working fluids that would provide superior performance or combination of cost and performance at this temperature.

Both DiPippo and Milora provide approaches and relative rankings for the impact of the working fluid selection on the turbine size and cost. These are first approximations based on the premise that all fluids are expanded from the same inlet temperature to the same condensing temperature. Based on the assumptions made, ammonia would have the smallest turbine exhaust area. While this is the case for the assumptions made, some caution should be used. Ammonia would have a sonic velocity at the exit conditions that is two to three times the sonic velocity of the other fluids. At an equivalent exit velocity, its advantage relative to propane and R134a would be negated. In addition, ammonia goes through a higher isentropic enthalpy change in the turbine. As a consequence, the optimal tip speed of an ammonia turbine rotor (proportional to the square root of the enthalpy change) is higher. This would require the ammonia turbine rotor either to have a larger diameter or to have a higher rotational speed that would necessitate a speed-reducing gear box to mate it to the generator. As the authors indicate, the methods used provide a first approximation. And while the turbine is a major contributor to the cost of the binary plant, there is cost associated with supply and disposing of the geothermal fluid. The selection of the working fluid should consider all these factors.

13.5.2 Mixed working fluids

With single-component working fluids, binary cycle performance improvements have resulted from efforts to reduce the irreversibility in the heat exchange process. This has been accomplished by matching the heating curve of the working fluid to the cooling line of the geothermal fluid. The multiple boiling, trilateral, and supercritical cycles all accomplish this to varying degrees. The use of mixed working fluids allows this approach to be extended to the heat rejection process. When undergoing isobaric phase changes, nonazeotropic or zeotropic, mixtures are nonisothermal, with the difference between the bubble and dew point temperatures (called the "glide") being a function of the formulation of mixture components and concentration.

The characteristics of an ideal working fluid [15] are shown in Fig. 13.13 for a geothermal resource that does not have a constrained temperature leaving the plant. With this ideal fluid, the temperature difference between the fluids in the heat addition process would remain constant throughout the heat exchange process. Similarly, the temperature difference between the coolant and working fluid would be constant throughout the condensation process.

If the geothermal outlet temperature were constrained, then the temperature difference between the working fluid and geothermal fluid would not be constant as depicted here but would increase as the geothermal fluid is cooled.

Perhaps the most recognized cycle that uses a working fluid mixture is the *Kalina cycle*, which makes extensive use of recuperative heating in a cycle that uses an ammonia–water working fluid mixture. The negative impact of an exit brine temperature constraint on performance can be mitigated (but not totally avoided) by the use of recuperative heating, where the fluid exhausting the turbine preheats the working

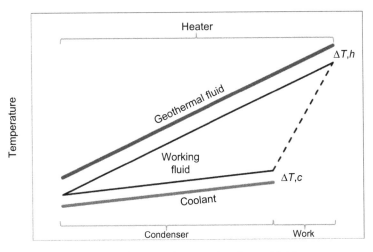

Figure 13.13 Binary cycle with ideal working fluid.

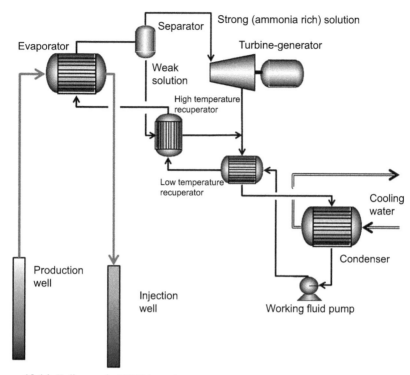

Figure 13.14 Kalina cycle KCS34 version.

fluid prior to its entering the geothermal heat exchangers. There are several variations of the Kalina cycle, one of which is shown schematically in Fig. 13.14.

In the KCS34 version of the cycle depicted in Fig. 13.14, the boiling of the working fluid in the evaporator is incomplete. The two-phase fluid leaving the evaporator is separated, with the ammonia-rich vapor going to the turbine and the liquid having a higher concentration of water going to preheat the working fluid entering the evaporator. After leaving the high-temperature recuperator, this weak solution combines with the vapor leaving the turbine. That stream then performs the initial heating of the working fluid discharging from the circulation pump. Because the vapor enters the turbine saturated, it will be two phase exiting the turbine and will remain two phase until it enters the condenser and the condensation process is completed. The preheating of the working fluid in the low-temperature recuperator can be accomplished because of the significant difference between the dew point and bubble point temperatures of the working fluid mixture. With the particular compositions of ammonia and water used, the cycle can be designed such that the geothermal fluid is used primarily to vaporize the working fluid, while the internal recuperation provides most, if not all, of the preheating of the working fluid.

An example of the heat exchange processes in the Kalina cycle is shown in Fig. 13.15. These conditions are representative of those found at the Kalina cycle plant

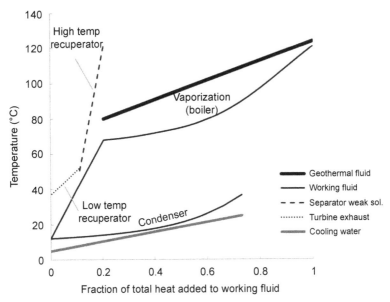

Figure 13.15 Heat addition in Kalina cycle plant at Husavik.

at Husavik, Iceland. In this facility, nearly all working fluid preheating is accomplished with the recuperative heat exchangers using the weak liquid solution leaving the boiler/separator (Fig. 13.14) and the turbine exhaust stream. The fluid (vapor and liquid) exhausting the turbine is two phase and remains so when combined with this weak solution. It partially condenses in the low-temperature recuperator, with final condensation in a water cooled condenser. The nonisothermal behavior shown of the exhaust in this exchanger and in the condenser is characteristic of this particular mixture, namely, 82% (wt.) ammonia in the condenser.

The advantages of using mixed working fluids as a means of increasing cycle performance was recognized prior to the introduction of the Kalina cycle. The U.S. DOE funded the development of binary cycles using mixed working fluids in the 1970s and 1980s [8]. Funded efforts included testing on a prototype scale and application in the Heber Geothermal Binary Demonstration plant located in the Imperial Valley of California.

This early work primarily focused on the use of mixed hydrocarbon working fluids in supercritical cycles, where their impact on reducing the irreversibility associated with the heat addition process is similar to that achieved in this cycle with single-component working fluids. The performance advantages of the mixed fluids beyond those achieved with the supercritical vaporization were associated with the nonisothermal behavior of the fluids during condensation. Nonisothermal condensing behavior is shown in Fig. 13.16 for the condensation of both a mixture of isobutane and hexane, and pure isobutane. For both fluids, the minimum internal approach temperature is 9°C. These condensing curves show that the use of mixtures reduces the

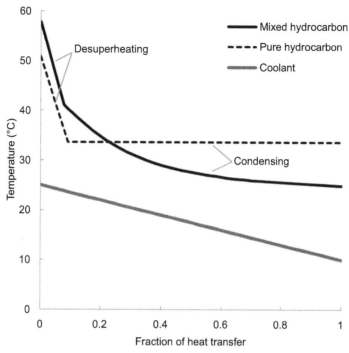

Figure 13.16 Condenser heat rejection profile with pure fluid (isobutane) and mixed working fluid (90% [wt.] isobutane, 10% [wt.] hexane).

area between the curves for coolant and working fluid; this area is indicative of the irreversibility (exergy loss) associated with this heat rejection process.

While supercritical vaporization of mixtures provides high levels of performance, it also avoided the potential issues associated with the boiling of the mixed fluids. To achieve the nonisothermal boiling, it is necessary to keep the liquid and vapor phases in equilibrium throughout the boiling process. During a phase change (boiling or condensation) at a subcritical pressure, the concentration of the mixture constituents in the liquid and vapor phase will vary at any point between the bubble point and dew point temperatures. In boiling, the first vapor that forms (bubble point) has higher concentrations of the lighter, more volatile constituent. Conversely, the last liquid to vaporize (dew point) has a higher concentration of the heavier constituent. At a given temperature in the boiling process, there is a unique composition for each phase that allows the nonisothermal vaporization to proceed. Keeping the phase concentrations in equilibrium produces the desired "integral" vaporization process.

An example of that process is shown in Fig. 13.17 for a mixture of 70% isobutane (i-C_4H_{10}), 30% hexane (C_6H_{14}) that is being vaporized at 20 bar(a). At this pressure, pure isobutane fluid would boil at a temperature of $\sim 100°C$, while pure hexane would boil at $\sim 207°C$. At this pressure a 70–30% iC_4–C_6 mixture would begin to boil at the bubble point temperature of $\sim 121°C$ (point A on the figure), and with the mixture

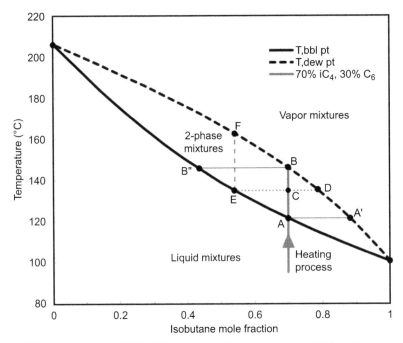

Figure 13.17 Temperatures of 70–30% isobutane–hexane mixture at 20 bar(a).

being completely vaporized at the dew point temperature of ~146°C (point B). The composition of the first vapor at the bubble point (point A′) would have ~88% isobutane, while the last liquid to vaporize (point B″) would have an isobutane concentration of ~43%.

To obtain the desired "integral" phase change at any temperature in the boiler, the compositions of the liquid and vapor phase are defined by the compositions at the dew and bubble point curves. For example, at a temperature of 135°C (point C), the equilibrium vapor concentration would have ~79% isobutane (point D), while the equilibrium liquid concentration of isobutane would be ~54% (point E). If at point C the vapor phase were separated from the liquid phase, the liquid would begin to boil at the new composition (point E), and to achieve complete vaporization, it would be necessary to heat this new mixture (54% iC$_4$, 46% C$_6$) to ~163°C (point F), assuming that the phases remain in equilibrium during this vaporization process. If the vapor phase were continually stripped away or separated from the liquid phase, it would be necessary to heat the mixture to the saturation temperature of pure hexane (~207°C) to achieve complete vaporization.

It was postulated that this phase separation would occur in the kettle boiler design commonly used in the subcritical, boiling binary cycles. This concern contributed to the early selection of the supercritical cycles when using the mixed working fluids.

Likewise, this concern also applies to the condensation process. As during boiling, throughout the condensation process, concentrations of mixture components differ

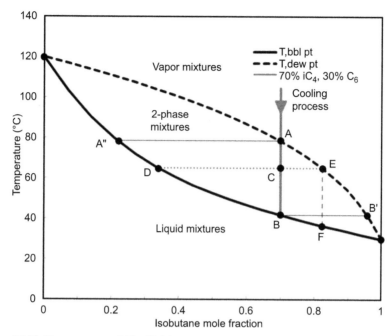

Figure 13.18 Temperatures of 70−30% isobutane−hexane mixture at 4 bar(a).

between the vapor and liquid phases. Achieving the desired condensation behavior of the mixture requires that these concentrations remain in equilibrium. An example of the condensation process is shown in Fig. 13.18 for the same isobutane−hexane mixture shown in Fig. 13.17.

Ideally, the condensation process would occur between points A and B depicted in Fig. 13.18. The first liquid to condense at the dew point would have a concentration given by point A″, while the last vapor to condense would have an isobutane concentration given by point B′. At a condensing temperature of 65°C (point C), the phases would be in equilibrium if the vapor phase had the isobutane concentration corresponding to point E, and the liquid, a concentration corresponding to point D. As with the boiling process, if the liquid and vapor phases were separated, the vapor would begin to condense at point E and would have to be taken to point F to entirely condense (assuming no additional separation of the liquid and vapor phases). This would mean that the final condensing temperature would be ∼5°C lower to achieve total condensation. If the coolant temperature and flow rate were fixed, the condenser pressure would have to increase to completely condense the working fluid. As a first approximation for this example, the pressure would have to increase from 4 bar(a) to 4.6 bar(a), assuming that the phase separation occurred at the same 65°C. The consequence of this increased condensing pressure is a higher turbine exhaust pressure and less power output. If the vapor were continuously separated from the liquid phase, the working fluid would have to be cooled to the saturation temperature of pure isobutane at 4 bar(a). Again, with fixed coolant inlet temperature and flow rate, the condenser pressure would have to increase, adversely impacting power output.

While the supercritical vaporization alleviated the concern as to whether the integral vaporization of mixtures could be achieved, there was concern as to whether condensers could be designed to achieve this integral condensation process. At the time this work was performed, the condensers considered (and used) for binary plants were water cooled, where convention would have the condensation occurring on the shell side of the heat exchanger. There was concern that this condenser design would result in the undesired phase separation that would negate, at least in part, the benefit from using the mixed working fluid. The U.S. DOE funded work on the use of mixtures that included their testing in a prototype facility where condensation of the working fluid occurred inside the condenser tubes. This testing indicated the phases remained in equilibrium during the condensation process, and the desired "integral" condensation was attained with the condenser design tested [16,17].

In binary cycles, a typical mixture has a higher concentration of the more volatile constituent (ie, having a lower normal boiling point temperature). As the concentration of the less volatile constituent is increased, the difference between the bubble and dew point temperatures increases until reaching a maximum at an intermediate concentration, after which it decreases until they are equal again when the fluid is composed entirely of the less volatile constituent, as shown in Figs. 13.17 and 13.18. The magnitude of this temperature difference is also shown in Fig. 13.19 for four different mixture combinations. The bubble and dew point differences in this figure are each

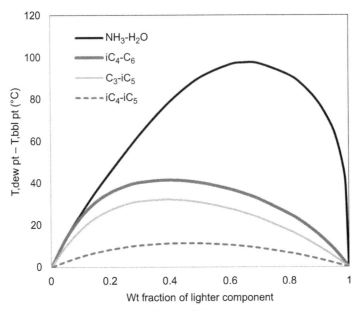

Figure 13.19 Effect of mixture composition on difference between bubble and dew point temperatures (generated using NIST REFPROP).

based on a fixed pressure that corresponds to saturation at 30°C for each of the lighter constituents [ammonia (NH_3), isobutane (i-C_4H_{10}), and propane (C_3H_8)]. This temperature is representative of a possible condensing temperature.

This figure illustrates how mixtures might be selected for the different cycles discussed. As shown, the ammonia—water mixtures can have considerably larger temperature differences between the bubble point and dew point. This makes this fluid combination better suited for a binary cycle where the geothermal energy provides the heat to vaporize the working fluid and internal recuperation provides the heat to preheat the working fluid. In supercritical cycles, mixture constituents and concentrations are selected to match this temperature difference more closely to the temperature change of the coolant used in condenser. As shown, the hydrocarbon fluids would provide greater flexibility in matching the coolant temperature rise.

Two of the hydrocarbon mixtures shown in Fig. 13.19 are the fluids tested at the Heat Cycle Research Facility [16,17]. The mixtures tested were nonadjacent hydrocarbons (propane—isopentane and isobutane—hexane). The third hydrocarbon working fluid mixture combination shown (the adjacent hydrocarbon mixture of isobutane—isopentane) was used at Heber Binary Demonstration Plant. The use of the nonadjacent hydrocarbons produces a larger difference between the bubble and dew point temperatures and some increased flexibility in designing and optimizing the supercritical cycle.

Other mixture combinations have been considered, including the use of mixtures of halocarbons as working fluids [18]. The results suggested these mixtures of engineered refrigerants would provide levels of performance comparable to those achieved with hydrocarbons. However, the refrigerants considered in the study (R-114, R-22) have been discontinued owing to their detrimental effects on the environment. This work suggested that tertiary mixtures of these fluids might provide some additional performance improvement, whereas similar studies with the paraffin-series hydrocarbons did not show a potential benefit when using tertiary mixtures.

13.6 Cycle performance comparison

As previously discussed, the irreversibility associated with the individual components or processes and the exergy in the fluid streams leaving the power cycle adversely impact the power that can be produced. The relationship that shows the dissipation of the geothermal fluid exergy [Eq. [13.5]] is repeated here.

$$\text{Power}_{net} = \sum_{\text{all fluids}} \dot{m}_{\text{fluid}}\left(e_{\text{fluid–in}}\right) - \sum_{\text{all processes}} \text{Irreversibility}_{\text{process}}$$
$$- \sum_{\text{all fluids}} \dot{m}_{\text{fluid}}\left(\dot{e}e_{\text{fluid–exit}}\right)$$

This relationship can be used to compare the different power cycles and consider where potential improvements to the cycles can be made.

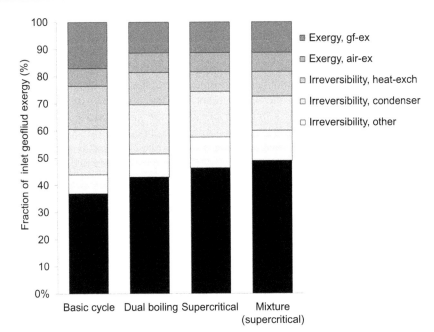

Figure 13.20 Exergy dissipation for different binary cycle concepts. The height of the black columns gives the second law efficiency for the cycle.

In Fig. 13.20, the dissipation of the geothermal fluid exergy is shown for selected binary cycle concepts that have been discussed (not shown are the trilateral and Kalina cycles). This figure was generated assuming an air-cooled plant with a 175°C resource with a 10°C sink and a 15°C temperature rise for the coolant (air). In the analysis, the minimum internal temperature approaches for the heat addition and heat rejection processes were the same for each cycle. For the assumptions made, the turbine inlet conditions were found that produced the maximum net power for a fixed geothermal flow rate, where net power is the difference between the generator output and the parasitic power required for pumping and fans. Although the analysis is relatively simple, it is informative in that it allows one to assess where the cycle deviates from the ideal and how each of the different cycle concepts affects the dissipation of exergy. The basic cycle and the dual boiling cycle shown in this figure are based on an isopentane working fluid. The supercritical cycle uses isobutane, and the supercritical mixture cycle uses a mixture of isobutane and hexane (96% isobutane by mass). These working fluids are not optimal for the individual cycles; they were selected as representative of the resource conditions selected. None of the cycles was evaluated using recuperation to offset the effects of the minimum temperature constraint imposed on the geothermal fluid.

Because of a minimum temperature constraint is imposed on the geothermal fluid leaving the cycle, the exergy of the geothermal fluid leaving the plant (E_{gf-ex}) is the same for all cycles, except the basic cycle. For the conditions considered, the temperature of the geothermal fluid leaving the optimal basic cycle was above this constraint ($\sim 60°C$). Relative to the basic cycle, the irreversible loss associated with the geothermal heat exchangers (I_{hx}) is lower for the dual boiling cycle, and even lower for the supercritical cycle (both with and without mixed working fluids). The irreversibility in the condenser is $\sim 25\%$ lower in the supercritical cycle with the mixed working fluid than it is in those cycles using a single-component working fluid. The other irreversibilities (I_{other}) shown are those associated with friction and the inefficiency of the pump, fan, and turbine. Because the supercritical cycles operate at high pressures, they have higher irreversibilities associated with the pump. In addition, because they produce more power, they have higher irreversibility associated with the turbine.

The net effect is that the net output for the dual boiling cycle is $\sim 16\%$ higher than the basic cycle, while the supercritical cycle and supercritical cycle with a mixture produce $\sim 26\%$ and 33% more power, respectively. Again, the cycles shown in Fig. 13.17 are not optimized, but these results are illustrative of how each cycle configuration impacts the second law conversion efficiency and net power generation from a given geothermal flow rate.

These conditions were also used to assess the performance of the trilateral cycle using an isopentane working fluid. If the cycle expander were able to achieve the isentropic efficiency of 85% assumed for the turbines used in the cycles shown in Fig. 13.20, the second law efficiency of the trilateral cycle would be slightly higher than that of the supercritical cycle with the isobutane working fluid. However, if the trilateral cycle expander efficiency is adversely impacted by the two-phase expansion, its performance would decline. If the trilateral expander efficiency were 75%, then the second law efficiency would be less than that for the dual boiling cycle.

The performance of the Kalina cycle-system 12 has been compared with that of the supercritical cycle with mixed hydrocarbons [19]. This analysis was done for a resource temperature of 182°C, with outlet temperature constraints of 71°C and 77°C. The Kalina cycle had a specific net output per mass flow of geothermal fluid ranging from 19.4 to 19.9 Wh/kg; the supercritical cycle with an isobutane—heptane mixture had a specific output of 20.6—20.9 Wh/kg. The analysis was done with a common set of assumptions for equipment efficiencies, minimum approach temperatures, and cooling water inlet conditions. Interestingly, the differences in performance decreased when the temperature constraint increased. This analysis was done for one of the early version of the Kalina cycle where the ammonia—water mixture is completely vaporized in the geothermal heat exchangers. Bliem subsequently examined how closely these cycles were approaching practical limits of performance [15]. His analysis indicated that for typical rotating equipment efficiencies and minimum internal approach temperatures, cycles using the mixed working fluids (both the Kalina cycle and the supercritical cycle) were approaching the performance of a cycle with the idealized working fluid with similar equipment constraints.

13.7   Design considerations

To achieve the performance benefits identified for the binary cycle configurations discussed, it is critical that the flow path in heat transfer processes be countercurrent when the working fluid temperature is increasing or decreasing as heat is added or rejected. This applies to all preheating, supercritical vaporization, and desuperheating processes, as well as all heat exchangers in cycles using mixed working fluids. If flow paths deviate from countercurrent in these heat exchangers, component irreversibility will increase and performance will decrease. The lack of a countercurrent flow path is not a performance deterrent when a single component fluid is being condensed or vaporized, provided there is minimal or no desuperheating or preheating occurring.

13.7.1   Geothermal heat exchangers

In addition to providing the countercurrent flow path when required, the heat exchangers designed for duty with geothermal fluids must account for potential fouling of surfaces exposed to the geothermal fluids. This fouling can occur in the form of mineral precipitation as the geothermal fluid temperature (and pressure) decreases in the heat exchanger. It can also occur in the form of corrosion products on the heat exchanger surfaces. Both of these fouling mechanisms will be a function of the chemistry of the geothermal fluid, while that due to corrosion will depend on the materials of construction. If the geothermal fluid chemistry increases the probability that fouling will occur with cooling, it is common to impose a minimum temperature constraint to prevent that fouling. Whereas a constraint to minimize precipitation of amorphous silica is most common, there are other chemical species (eg, antimony sulfide) that may be present and have similar potentials for mineral precipitation with decreasing temperature. While hydrocarbon and engineered refrigerants are chemically benign, it would be prudent to account for some fouling on surfaces exposed to these fluids when designing the heat exchanger.

If the rate at which geothermal fouling occurs is high, it will be necessary periodically to isolate and clean the heat exchangers. If this is anticipated, it will impact the selection and design of the exchangers. To accommodate the cleaning, a shell and tube exchanger is typically selected with the geothermal fluid inside the tubes. "U-tube" bundles would not be utilized. In some instances, plate heat exchangers may be used in binary plant applications because of their performance and compact size. While proponents of their use include their ease of cleaning as a benefit for their application, there are plants that have replaced the plate exchangers with shell and tube exchangers because they were not conducive to repeated disassembly and reassembly for cleaning.

As discussed previously, the heat exchanger designs for mixed working fluids must be able to provide countercurrent flow paths and achieve integral phase changes to realize the potential performance benefits. Both criteria were achieved in the designs used for the hydrocarbon mixture testing at the prototype Heat Cycle Research Facility [16,17]. The geothermal heat exchangers used in this prototype facility, as well as those used in the supercritical binary plant at Neal Hot Springs, use a single-pass shell

(TEMA E shell). These heat exchangers require internal tube supports to prevent both tube sag and flow-induced vibration [20]. Baffles providing this support direct flow across the tube bundle, producing a deviation from the desired countercurrent flow path. The type of baffling used and the spacing of that baffling are options considered in the design of the heat exchanger. Reducing the baffle spacing increases the velocity and the flow path length, both of which increase the pressure drop of the shell side fluid. It also increases the number of times that the flow is directed across the bundle and allows the effective flow path to approach the ideal countercurrent. The design of the heat exchanger will consider these various factors to produce an exchanger that best meets the specifications provided [21]. Because they operate at higher pressures, the heat exchangers used in these cycles tend to have smaller shell diameters, with longer tubes to attain the required surface area. (For a given operating pressure, the required thickness of the shell wall increases with the shell diameter.) Using smaller-diameter shells decreases the number of tubes, while increasing the tube velocity and associated pressure drop. The increased length and reduced diameter also tend to increase the shell side pressure drop, though those effects are mitigated by the higher densities due to the supercritical pressures.

When a temperature limit is placed on the geothermal fluid leaving the plant, heat recuperation can increase power output resulting in improvements in both cycle thermal efficiencies. While heat recuperation will mitigate a portion, but not all, of the adverse impact of imposing the temperature limit on power production, some portion of that benefit may be offset by the additional pressure drop imposed on the working fluid system. That pressure drop increases the turbine exhaust pressure, which in turn lowers power output. While it is possible to design heat exchangers to minimize this pressure drop, doing so will increase the size and cost of the recuperator. When considering the benefits of recuperation in a power cycle, it is important to account for these pressure drops, in terms of their impact on both performance and cost.

13.7.2   Heat rejection

The availability of an adequate supply of water suitable for makeup to an evaporative heat rejection system will determine whether an evaporative heat rejection system can be used. If water is not available, heat will be rejected by sensibly heating air in the condenser. When water is available, it is probable that an evaporative heat rejection system, with a water-cooled condenser will be used. An evaporative heat rejection system effectively rejects heat to the lower ambient "wet" bulb temperature. This lower sink temperature increases the exergy of the geothermal fluid entering the plant and will result in more power production for an equivalent second law conversion efficiency. The evaporative heat rejection system with cooling tower, condenser, and cooling water circulation pumps will typically have a capital cost equivalent to or less than the air-cooled condensers. Because makeup water will have to be provided and the cooling water treated, the evaporative rejection system will likely have higher operating costs.

The choice of the heat rejection system can impact the selection of the working fluid, specifically whether a mixed working fluid will be used. Achieving the benefits associated with these fluids requires that both the flow paths be countercurrent and the "integral" condensation process be achieved. This is easier to accomplish in designs where condensation occurs inside a tube rather than on the shell side of a condenser. Though condensation would occur inside the tubes in an air-cooled condenser, the flow of air is typically across the tube bundle and not countercurrent to the working fluid flow. One can approach the desired countercurrent flow path needed by using multiple tube passes. This is difficult to accomplish without increasing the working fluid pressure drop through the condenser. Use of multirow passes also increases the probability that more tube rows would be required than typically used. Additional tube rows will either result in increased fan power, or a lower airflow rate, which would increase condensing pressure. If a tube pass has multiple tube rows, there is increased potential for vapor and liquid phase separation and for deviation from the desired "integral" condensation process. Along with the higher tube pressure drop, any deviation from the integral condensation process would increase the turbine exhaust pressure and decrease power output. While air-cooled condensers provide the potential to realize some performance benefit from using mixtures, that benefit will not be as large as suggested by using the assumption of countercurrent flow [22].

Regardless of the type of condenser used, the use of mixed working fluids will impact sizing of a condenser. Condenser tests at the Heat Cycle Research Facility with propane–isopentane mixtures indicated that the overall heat transfer coefficient decreased as the concentration of isopentane in the mixture increased. These tests were conducted with the condenser at three orientations: vertical, 60° off the horizontal and 10° off the horizontal. The vertical orientation was used for the initial testing because it allows gravity to assist in keeping the liquid and vapor phases in equilibrium. The other orientations were tested to determine whether they produced any deviation from the required integral condensation [17].

In considering the use of mixed working fluids in air-cooled condensers, the vertical condenser performance at the Heat Cycle Research Facility was modeled using the Aspen Exchanger Design and Rating (EDR) software to determine whether the software could depict the trends observed during the earlier testing [22]. The overall heat transfer coefficients determined from the test data, and those predicted by the EDR software are shown in Fig. 13.21. In this figure, the orientations given are for the condenser tube relative to the horizontal. The EDR software models performance only for vertical and horizontal tube orientations. The predicted performance for the vertical condenser orientation is included only to show that the observed trends in condenser performance with changing composition would be expected.

13.7.3 Operation at "off-design" conditions

Binary plants are designed to operate at a specific set of geothermal and ambient conditions. Given that ambient conditions are continually varying, the plant will operate a significant portion of the time at conditions other than "design." As a first approximation, the effect of this off-design operation on plant output can be estimated as the product of the second law efficiency and the exergy of the geothermal fluid at those ambient and resource conditions. While this approximation is indicative of the effect

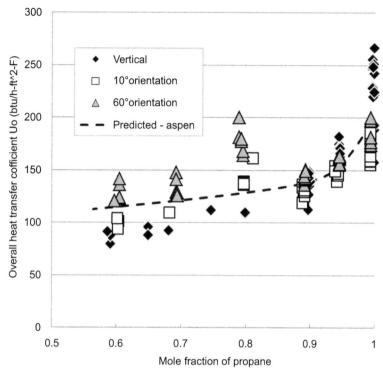

Figure 13.21 Effect of mixture concentration on overall condenser performance (from INL/EXT-13-30178).

that nondesign resource or ambient conditions will have on output, it tends to be optimistic, as shown in Fig. 13.22. This figure shows a typical summer ambient temperature profile and the corresponding variation in power and in the exergy relative to the power and geothermal fluid exergy at the conditions for which the plant is designed. Design conditions for this particular scenario are a geothermal fluid temperature of 150°C (300°F) and an ambient temperature of 10°C (50°F). Power is the predicted output for an air-cooled binary plant, where the simulation used matches the design and output of an operating facility. The second law efficiency tends to be near optimal close to the conditions for which it is designed. When resource or ambient conditions change, it is probable that this conversion efficiency will be adversely impacted. In this figure, the deviation from unity for both the exergy ratio and power ratio is indicative of the decrease that occurs in the second law efficiency with an increasing ambient temperature.

If the resource temperature or flow rate increases or if the ambient temperature decreases, the plant will produce more output; however, the conversion of the geothermal fluid exergy to power will be lower (ie, lower second law efficiency).

Some have implied that the adverse impact of higher ambient temperatures on the power generation from air-cooled plants is that the condensers are somehow less

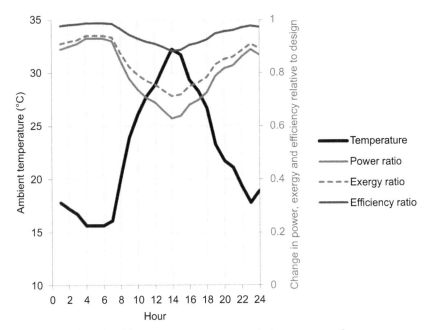

Figure 13.22 Effect of ambient temperature on power during a summer day.

efficient. The decrease in power is primarily a result of the decrease in the geothermal fluid available energy, and to a lesser extent to the increase in the turbine irreversibility associated with the efficiency decrease due to the higher exhaust pressures.

Despite the impact that the ambient temperature has on power output, these air-cooled plants are commonly designed for a mean or average annual air temperature. With this design ambient, their average power output over a year will be close to their design value. An air-cooled plant can be designed for a higher ambient temperature; however, because the plant output is a direct function of the geofluid exergy, a 20-MW plant designed for 25°C is going to cost more and require more geofluid flow than a 20-MW plant designed for 10°C. If the second law conversion efficiency remains constant, the difference in plant size and flow rate required would be in proportion to the geofluid available energy at the two ambient temperatures. For a 175°C resource, the exergy with 10°C ambient is 22% higher than with a 25°C ambient. This represents the increase in plant size and flow rate that would be needed for an equivalent power output at the high ambient temperature, assuming constant second law utilization efficiency.

When an evaporative heat rejection system is used, the response to changes in the ambient conditions differs. The wet bulb temperature is representative of the heat sink for these plants, and these temperatures tend not to vary as significantly throughout a summer day because the relative humidity tends to decrease with increasing air temperature. Consequently, the power generation does not go through the same magnitude of diurnal changes in power as occurs in an air-cooled plant. Plants with evaporative cooling are limited in colder climates when the temperatures approach or fall below freezing. To prevent icing damage, operation of the cooling tower will be curtailed, typically by reducing airflow. When this occurs, the plant is unable to take advantage of the higher exergy associated with the lower sink temperature. As a consequence the output from a water-cooled plant is likely not to experience the variation in output that will occur with an air-cooled plant. While air-cooled plants do not have this operational constraint, operation may be curtailed at lower ambient temperatures either to prevent the working fluid system from going to subatmospheric pressures at the condenser or to keep generation below the maximum generator power rating.

13.8 Economic considerations

Reducing heat exchanger irreversibility produces the potential performance benefits discussed above and shown in Fig. 13.20 for the different types of binary cycles. These reductions are realized by reducing the average temperature difference between the fluids in the heat transfer processes. The consequences of doing this are increased heat exchanger surface area and higher capital costs for equipment. As a first approximation, the change in the heat exchanger size and cost can be estimated as the change in the product UA, where U is the overall heat transfer coefficient, and A is the heat transfer surface area. The UA product is the ratio of the total heat transfer to the log-mean temperature difference ($LMTD$) for the heat transfer process. The changes in UA and the net power produced relative to the basic binary cycle is shown in Fig. 13.23 for the dual boiling, supercritical and mixture (supercritical) cycles depicted in Fig. 13.20.

The ratios of the UA for the geothermal heat exchangers (UA,hx) are all greater than one, consistent with the reduced irreversibility for this heat exchanger in these three cycles. The supercritical cycle using the mixed working fluids also has a higher UA for the condenser (UA,cond), which is also consistent with the reduced irreversibility associated with that heat transfer process using mixtures.

It is not just the heat exchanger cost that is impacted by the choice of cycle. The pumping power is significantly higher for the two supercritical cycles, which increases the size and cost of the turbine-generator beyond the increased power shown since the power to run the pump comes off the generator bus. To assess how the cycle selection could impact generation cost, a simple approach was developed [23] that examined iterative change in cost and power. To assess these impacts, reference binary plant costs were used for the two lower-temperature resources in the Next Generation Geothermal Power Plants report [24]. Those costs were assigned to the basic cycle plant, and then relative costs were determined for the relative changes in the heat exchanger sizing. Those changes are summarized in Table 13.1.

All of the cycles showing improved performance relative to the basic cycle had relative increases in cost. When looking at the ratio of cost to power, the dual boiling cycle has a lower cost per unit power than the basic binary cycle, while the two supercritical cycles have higher costs. This ratio would be indicative of the relative cost in terms of $/kW.

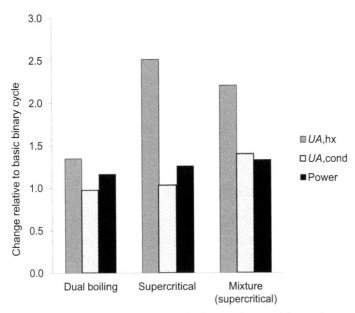

Figure 13.23 Effect of type of binary cycle on the heat exchanger sizing and power output.

Table 13.1 **Initial estimate of power cycle selection on cost**

Component	EPRI cost	Basic binary	Dual boiling	Supercritical	Supercritical mixtures
Geothermal HXers	0.041	0.041	0.055	0.103	0.091
Turbine-generator	0.126	0.126	0.146	0.177	0.180
Working fluid pump	0.009	0.009	0.009	0.058	0.047
Heat rejection	0.155	0.155	0.152	0.161	0.217
Total	**0.331**	**0.331**	**0.361**	**0.499**	**0.536**
Cost relative to basic cycle	N/A	1	1.093	1.507	1.619
Power relative to basic cycle	N/A	1	1.166	1.260	1.327
Relative cost to power	N/A	1	0.938	1.196	1.220

Based on plant cost alone, one would not consider either the supercritical cycle or the supercritical cycle using mixtures. However, the capital cost of the plant is only one cost contributor to the project. There are costs associated with the development of the well field, operation of the facility, including the pumping costs for the geothermal fluid.

Continuing with this simplistic approach, one can estimate the point at which the additional cost for a performance improvement offsets the beneficial impacts the improvement has on the other costs. Simply, this is a ratio of total cost to power that equals unity. If plant and field costs are equivalent, then if 20% more power is produced, the total cost can increase by 20% also. If the geothermal flow is fixed and it is assumed that field costs are constant, the plant cost can increase by 40% before it is less cost-effective to use a 20% more efficient plant. Fig. 13.24 shows this "breakeven" cost increase for different scenarios of plant to wellfield cost. As the plant and field cost ratios vary, so does the breakeven cost. Even with scenarios where the field cost is small relative to the plant, there is benefit from increasing performance and power output, as long as the increased plant cost is below this threshold.

Again, this is a very simplistic approach to show why and when the binary cycles with improved performance can provide a benefit. The cost and performance analyses provided for the binary cycles discussed have not been optimized. This discussion is provided for reference in describing how these power cycles can impact performance as well as cost. When selecting a particular cycle for a project, comparisons of different options should be done on an equivalent basis, with common assumptions. If cycles are evaluated in this manner with each optimized to a defined performance criterion, it is probable that the cycle that more efficiently uses the geothermal flow (higher second law efficiency) will have a higher plant capital cost. As shown, this is not necessarily a deterrent to using that cycle, once one includes all project costs.

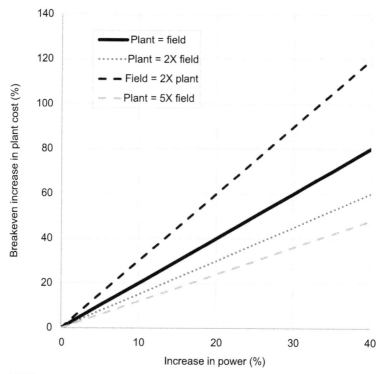

Figure 13.24 Breakeven costs associated with increased plant performance.

References

[1] Bertani R. Geothermal power generation in the world 2010−2014 update report. In: Proceedings World Geothermal Congress. Melbourne Australia. s.n.; 2015.

[2] Biederman TR, Braz J. Geothermal ORC systems using large screw expanders. In: West Lafayette IN: International Compressor Engineering Conference; 2014.

[3] DiPippo R. Geothermal power plants. Oxford UK: Elsevier; 2012.

[4] Fournier RO, Rowe J. The solubility of amorphous silica in water at high temperatures and high pressures, vol. 62. Chantilly VA: American Mineralogist; 1977.

[5] Gunnarsson I, Arnorsson S. Amorphous silica solubility and the thermodynamic properties of H_4SiO_4 in the range of 0 to 350 C at P_{sat}. Geochim Cosmochim Acta s.l. 2000;64(13).

[6] Demuth OJ. Analysis of binary thermodynamic cycles for a moderately low-temperature geothermal resource. Idaho Falls, ID: Idaho National Engineering Laboratory; 1979.

[7] Smith I. Trilateral flash cycle system: a high efficiency power plant for liquid resources. In: Proceedings of the World Geothermal Congress. Auckland, New Zealand: International Geothermal Association; 1995.

[8] Mines GL, Nix G, Bruton C, Bucher T, Kukacka L, Premuzic E. Energy conversion, 1976−2006. A history of geothermal energy research and development in the United States. s.l.: U.S. Department of Energy, Geothermal Technologies Program; September 2010.

[9] Smith IK, Stosic N, Kovacevic A. Screw expanders increase output and decrease the cost of geothermal binary power plant systems. In: Transactions of Geothermal Resources Council Annual Meeting. Reno NV: Geothermal Resources Council; 2005.

[10] Boyle P, Hays L, Kaupert K, Welch P. Performance of variable phase cycle in geothermal and waste heat recovery applications. In: Transactions of Geothermal Resources Council Annual Meeting. Las Vegas: Geothermal Resources Council; 2013.

[11] Milora SL, Tester J. Geothermal energy as a source of electrical power, thermodynamic and economic design criteria. Cambridge, MA, USA: MIT Press; 1977.

[12] Forsha MD, Nichols K. Factors affecting the capital costs of binary power plants. In: Transactions of Geothermal Resources Council Annual Meeting. Sparks, NV, USA: Geothermal Resources Council; 1991.

[13] EPA. Ozone layer protection − Alternatives/SNAP. US Environmental Protection Agency. [Online] http://www.epa.gov/ozone/snap/subsgwps.html, Intergovernmental Panel on Climate Change, Fourth Assessment Report; 2007.

[14] Greenhut A, Tester J, DiPippo R, Field R, Love C, Nichols K, et al. Solar-geothermal hybrid cycle analysis for low enthalpy solar and geothermal resources. In: Bali Indonesia: Proceedings World Geothermal Congress; 2010.

[15] Bliem CJ, Mines G. Advanced binary geothermal power plants limits of performance. Idaho Falls, ID: Idaho National Engineering Laboratory; 1991.

[16] Demuth OJ, Bliem C, Mines G, Swank W. Supercritical binary geothermal cycle experiments with mixed-hydrocarbon working fluids and a vertical, in-tube counterflow condenser. EGG-EP-7076. s.l.: Idaho National Engineering Laboratory; 1985.

[17] Bliem CJ, Mines G. Supercritical binary geothermal cycle experiments with mixed working fluids and a near-horizontal in-tube condenser. Idaho Falls, ID: Idaho National Engineering Laboratory; 1989. EGG-EP-8800.

[18] Bliem CJ. Zeotropic mixtures of halocarbons as working fluids in binary geothermal power plants. In: Philadelphia PA: Proceedings of 22nd Intersociety Energy Conversion Engineering Conference; 1987.

[19] Aspects of Kalina technology applied to geothermal power production. Idaho Falls, ID: Idaho National Engineering Laboratory; 1989.

[20] Mukherjee R. Effectively design shell-and-tube heat exchangers. Chem Eng Prog February 1998;94.

[21] Wolverine Tube, Inc.. Chapter 2, Section 2.5. Preliminary design of shell and tube heat exchangers. In: Heat transfer databook; June 2014 [Online], http://www.wlv.com/wp-content/uploads/2014/06/databook/ch2_5.pdf.

[22] Mines GL, Wendt D. Air-cooled condensers for next generation geothermal power plants: final report. Idaho Falls ID: Idaho National Laboratory; 2013.

[23] Demuth OJ, Whitbeck J. Advanced concept value analysis for geothermal power plants. Idaho Falls, ID: Idaho National Engineering Laboratory; 1982.

[24] Brugman J, Hattar M, Nichols K, Esaki Y. Next generation geothermal power plants. Palo Alto, CA: Electric Power Research Institute; 1995.

Combined and hybrid geothermal power systems

14

R. DiPippo
University of Massachusetts Dartmouth (Emeritus), N. Dartmouth, MA, United States

14.1 Introduction and definitions

This chapter discusses integrated power plant systems that consist of either different types of geothermal energy conversion systems or some type of geothermal plant combined with a plant using at least one other different source of energy. The former are called "combined systems," and the latter are called "hybrid systems." In all cases the objective in combining different kinds of plant is to achieve synergy, meaning that the integrated system is capable of superior performance compared to separate individual plants. This may mean higher utilization or thermal efficiency, more net power, or a more economic outcome. Geothermal combined systems may consist of different types of flash-steam units and/or binary plants in an integrated combination that achieves advantages and benefits not realizable in separate units.

A hybrid geothermal–natural gas power plant for The Geysers geothermal field in California, USA, was proposed and considered in 1984 [1]. Among the first attempts to conceptualize and analyze hybrid geothermal power plants was carried out at Brown University in the late 1970s and early 1980s [2–11]. A commercial proposal that stemmed from one of the Brown studies concerned a geothermal-coal power plant at Roosevelt Hot Springs, Utah, that would have supplied electricity to the city of Burbank, California; it went through a thorough engineering verification before being shelved in favor of a conventional coal plant [12], despite indications of significant advantages. The equations presented here draw upon the studies by the Brown research group. Another concept was put forth by Hiriart and Gutierrez in 1995 and involved the use of solar superheating of separated steam at Cerro Prieto, Mexico [13]. These and several other more recent projects will be described in this chapter.

Since there are many options available to measure the performance of an integrated plant, equations are developed to provide the basis to assess the synergistic characteristics of these designs. First, basic thermodynamic principles are applied to this situation (Section 14.2), followed by theoretical integrated designs involving various flash-steam units (Section 14.3) and ones that combine flash-steam units with binary units (Section 14.4). Next geothermal and fossil energy resources are integrated to form hybrid plants of differing designs (Section 14.5), followed by plants that combine geothermal and solar energy resources (Section 14.6). Nuclear, hydroelectric and wind power plants do not lend themselves to hybrid designs with geothermal plants. As appropriate throughout the chapter, examples of plants either in operation or ones that have been subjected to extensive research or feasibility studies will be described in some detail.

Geothermal Power Generation. http://dx.doi.org/10.1016/B978-0-08-100337-4.00014-0

14.2 General thermodynamic considerations

Fig. 14.1 depicts the general scheme of a power system involving three different sources of primary energy. Geothermal, fossil and solar energy are selected for this illustration. The stand-alone, state-of-the-art (SOTA) geothermal, fossil, and solar plants are denoted as G, F, and S, respectively. Each plant receives input energy, produces work output and discharges waste energy, as shown. In the figure and equations that follow, asterisks are used to emphasize that the three plants represent the best that can be achieved by each individual type of plant given current technology. The hybrid plant, H, receives the same energy inputs as the three separate plants and produces work while rejecting some waste energy.

The energy terms may be heat transfer in cases where this is the appropriate form of input, but more generally they should be the exergy delivered to and rejected from each plant. This allows various types of energy to be placed on the same thermodynamic footing. Thus, using exergy allows for a valid comparison among heat obtained from burning fossil fuels, hot pressurized geofluids used either directly as in flash plants or indirectly in binary plants, and radiant solar energy used either directly as in photovoltaics (PV) or indirectly as in concentrating solar thermal plants (CSP). Here it is assumed that the analyses of all these individual systems is well known and may be carried out based on established techniques.

To perform the assessment of the worth of the hybrid system there are several metrics available; see Eqs. [14.1]–[14.4]. First, an overall assessment is found by calculating the overall hybrid figure of merit, F_H, by comparing the output of the hybrid plant to the sum of the outputs of the individual SOTA plants, ie,

$$F_H = \frac{W_H}{W_G^* + W_F^* + W_S^*}.$$
                                                                                         [14.1]

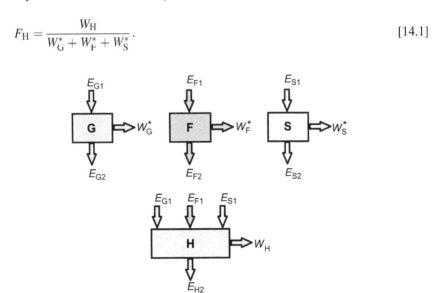

Figure 14.1 Schematic depiction of individual geothermal, fossil, solar, and hybrid plants.

Obviously for the hybrid plant to hold a thermodynamic advantage over the separate plants it must be true that $F_H > 1$.

It is also of interest to find figures of merit for the individual sources of energy. Thus, geothermal, fossil, and solar figures of merit for the hybrid plant may be defined as follows:

$$F_G = \frac{W_H - \left(W_F^* + W_S^* \right)}{W_G^*} \qquad [14.2]$$

$$F_F = \frac{W_H - \left(W_S^* + W_G^* \right)}{W_F^*} \qquad [14.3]$$

$$F_S = \frac{W_H - \left(W_G^* + W_F^* \right)}{W_S^*} \qquad [14.4]$$

Defined this way, each of these three equations attributes to each energy source, respectively, all the synergistic power gain from the hybrid plant by debiting the hybrid output by the maximum that could be obtained from the other two sources using SOTA power plants.

There are other useful measures of the efficacy of hybrid plants, but they are beyond the scope of this chapter; the reader may consult earlier sources [4–8].

The SOTA outputs may be estimated using a utilization efficiency, η_u, for each type of energy input, where the efficiency gives the fraction of the supplied "fuel" exergy that can be converted into useful output. Thus, one arrives at Eqs. [14.5]–[14.7]:

$$W_G^* = \eta_{uG} E_{G1} \qquad [14.5]$$

$$W_F^* = \eta_{uF} E_{F1} \qquad [14.6]$$

$$W_S^* = \eta_{uS} E_{S1} \qquad [14.7]$$

By using SOTA individual plants, a high bar is set for hybrid designs. These may be established with regard to the best available plants of the various types of energy under consideration. As examples, SOTA plants might be a dual-reheat, supercritical-pressure coal-fired plant with seven feedwater heaters, or a natural-gas—fired aeroderivative gas turbine, or a triple-flash geothermal steam plant, or a dual-pressure hot water geothermal binary plant, or a parabolic trough or power tower CSP plant, or a high-efficiency PV plant.

Representative values for the utilization efficiencies depend on the nature and characteristics of the "fuel" and the plant design. For example, a SOTA geothermal binary plant may exhibit η_u values ranging from 10% to 16% for resource temperatures ranging from about 120°C to 165°C, whereas flash-steam plants may have values from 35% to 55% for plants of single-, double-, or triple-flash design.

Coal plants can have values from 35% to 45% depending on steam pressure, number of reheats, and the number of feedwater heaters; simple gas-fired turbine plants may approach values up to 40%, while CSP plants may be limited to about 30%. PV solar plants generally can convert about 10–15% of incident solar energy to electrical output. The highest efficiency among fossil plants is achieved using gas turbine and steam turbine combined cycles (GTSTCC) where 60% conversion efficiency is achievable. However, these plants are themselves combined systems and might not be appropriate as a comparison for assessing geothermal hybrid systems.

There are considerations besides thermodynamics that bear on the suitability of a hybrid plant. For example, the environmental impacts must not be worse than individual plants and preferably better. A reasonable method must be present to absorb the waste energy from the plant. Permits must be obtained for not just one type of energy use, but for each one in the hybrid plant. Even a highly efficient power plant will not be built unless the economics are favorable, permits are obtainable, and funding can be arranged.

Central to the matter of economics is the relative proximity of the different energy sources. Fossil fuels can be mined or obtained by drilling and then moved to the location of the power plant, whereas geothermal "fuels" must be used very near their source. Solar energy is available anywhere, albeit in different strengths and for different time durations. The geothermal energy component therefore determines where the hybrid plant must be built. Typically, solar energy will be available at the site in some form, but a fossil fuel may be problematic. However, except in the case of mine-mouth coal plants, the fuel will need to be shipped or piped somewhere for exploitation in a power plant. Certainly natural gas is almost always sent by pipeline to power stations. So, a site-specific economic analysis is critical to determining the efficacy of using the fossil fuel in a conventional plant or shipping it to a geothermal site instead.

Thus, thermodynamics, the focus of this chapter, provides a necessary, but not sufficient, condition for the superiority and acceptance of any hybrid or combined power plant.

14.3   Combined single- and double-flash systems

Geothermal field development often occurs in discrete stages, starting with small wellhead units and progressing to simple single-flash plants and then possibly to combined single- and double-flash plants. Fig. 14.2 shows a combined single- and double-flash power station composed of two single-flash units and one double-flash unit. The synergy is obvious in this case. Since all of the waste brine from the first two units is used as the input to the third unit, the combined plant will generate more power than the two single-flash units. If the units were operated as stand-alone units, the geofluid needed for the double-flash unit would have to be obtained from new wells drilled on purpose for that unit. That extra geofluid would

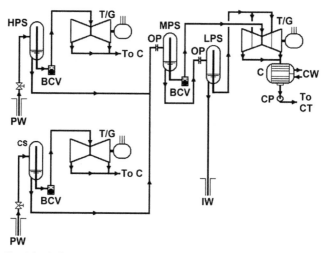

Figure 14.2 Combined single- and double-flash plant; see Nomenclature.

mean more exergy would need to be extracted from the reservoir for the same power that is produced by the combined system.

An example of this type of plant used to exist at the Cerro Prieto I power station in Baja California, Mexico. Originally four single-flash units of 37.5 MW each were constructed. The waste brine was disposed off into a large evaporation pond. As soon as confidence was established in the longevity of the resource, a fifth unit was constructed to put all the waste brine to use in the manner shown in Fig. 14.2. Two lower pressure steam flows were created by flash and separation processes to drive a dual-pressure, dual-admission turbine that added 30 MW more power without the need to drill more wells. Thus, a 20% amplification of power output and utilization efficiency was achieved.

A similar result was obtained at Ahuachapán in El Salvador when a third unit, a double-flash unit, was added to two existing single-flash units. The original power capacity of 60 MW (2 × 30 MW) was increased to 95 MW with the 35 MW low-pressure unit, a 58% amplification, that was achieved through flashing the brine from wellhead separators and using other low-pressure wells that could not be used with the original units.

14.4 Combined flash and binary systems

Many fields begin their operating lives with a single-flash unit until confidence in the field justifies additional units. One common way to extend the output of the initial unit is to add a bottoming binary unit that capitalizes on the waste geofluit that typically is reinjected from the first unit. This option creates a combined flash-binary plant and is an alternative to the combined flash plant described in Section 14.3. As in that case, the

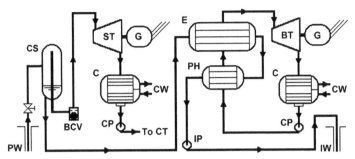

Figure 14.3 Combined flash-binary power plant; see Nomenclature.

synergy is obvious as additional power can be generated from the same amount of geo-fluid being extracted from the reservoir. A flow diagram for such a system in shown in Fig. 14.3.

There are numerous examples of this arrangement. For example, Unit 5 at Miravalles in Costa Rica, where a pair of binary cycles capitalize on the roughly 900 kg/s of 165°C separated brine coming from three single-flash units, adding some 15 MW to the 137 MW from the flash units, and raising the utilization efficiency of the whole operation by about 13% with no additional wells [14]. Similar designs are found at Brady's Hot Springs, Beowawe, Steamboat Hills and Dixie Valley, all flash plants with bottoming binary cycles in the state of Nevada, USA.

An alternative design to the combined flash-binary system shown in Fig. 14.3 is the integrated flash-binary plant shown in Fig. 14.4. This type of plant is designed as a unit

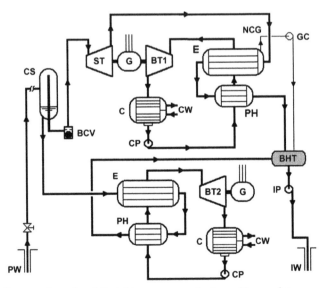

Figure 14.4 Integrated two-level flash-binary power plant; see Nomenclature.

and installed as a single plant. It is particularly appropriate for high-temperature resources in areas of environmental sensitivity where emissions of any kind are to be avoided. One distinguishing feature of this plant is that all the geofluid extracted from the reservoir in reinjected. This includes even the noncondensable gases (NCG) that are released from the pressurized liquid brine as the geofluid passes through the plant equipment. In this design the NCGs are captured from the steam at the evaporator of the upper binary cycle, compressed, delivered to the waste brine holding tank where they go back into solution, and are finally reinjected along with the waste brine.

Plants of this general design are in operation at Puna, Hawaii, USA; Amatitlán, Guatemala; Rotokawa and Ngatamariki, New Zealand; and elsewhere.

14.5 Geothermal-fossil hybrid systems

Geothermal and fossil energy resources lend themselves to hybrid configurations, presuming the resources are co-located or the transportation costs to bring the fossil fuel to the geothermal site are acceptable. Many different design options are available depending on the nature of the fossil fuels.

14.5.1 Fossil-fueled superheated geothermal steam plants

Using fossil fuels to enhance geothermal resources was proposed as long ago as 1924 by Caufourier [2]. His idea was to pass saturated steam obtained by several stages of flashing hot water from natural springs through a furnace fired with coal to obtain superheated steam, which then drove a multistage steam turbine to make electricity; see Fig. 14.5. It is not known if this idea was ever brought to fruition, but a simple assessment showed that it would have had about 65% utilization efficiency [15].

A similar proposal was made to superheat the geothermal steam at the dry steam power plant at The Geysers in 1984 by using a natural gas supply that was handy to

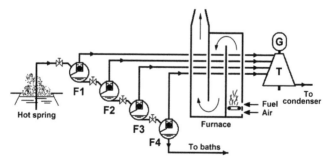

Figure 14.5 Multistage geothermal steam plant with fossil superheating [2,15]; see Nomenclature.

the site [16]. There were two schemes; the first would operate with gas-fired heaters, while the other would involve constructing a gas turbine unit with the exhaust heat supplying the geothermal steam superheating, much like the now common combined steam and gas turbine plants. Neither proposal was ever built even though the utilization of both the fossil and geothermal "fuels" would have been greatly improved.

In general, any geothermal resource may be enhanced through superheating. Figure 14.6 shows such a system built on a geothermal double-flash plant; by eliminating the flash vessel (FV) and the low-pressure turbine (LPT), the system becomes a single-flash with superheat.

A parametric study of the plant in Fig. 14.6 [5] showed that for geofluid inlet temperatures from 175°C to 250°C and 540°C superheat temperature, this system has an overall figure of merit F_H ranging from 1.05 to 1.075, a geothermal F_G ranging from 1.06 to 1.10, and a fossil-fuel F_F ranging from 1.20 to 1.34.

A study conducted for this chapter considered an optimized double-flash plant receiving geofluid from a 245°C reservoir and having a condenser operating at 52°C with a fossil-fired superheater added along with a recuperator, as shown in Fig. 14.6. Steam and water properties were obtained from NIST Refprop software [17] that was embedded into an Excel spreadsheet written to analyze the plant. The stand-alone double-flash plant produced 111.2 kW per kg/s of geofluid from the production well. Assuming a dead-state at 30°C, this gives a utilization efficiency of 44.9%. The superheater raises the geosteam temperature from 283.5°C leaving the recuperator to 540°C, considered to be the metallurgical limit. The hybrid plant produces 151.5 kW per kg/s, an increase of 36.2%. The input exergy of the heat added in the superheater is found by assuming it could be used in an ideal power cycle between the average temperature of the combustion gases and the dead-state temperature. Thus, the total exergy input to the hybrid plant comes to 288.8 kW per kg/s, giving a utilization efficiency of 52.5%. The processes are shown in a scale temperature-entropy diagram; see Fig. 14.7. Pressure losses in piping and heat exchangers were ignored, but wet turbine efficiencies were calculated from the Baumann rule to account for performance degradation caused by moisture [15].

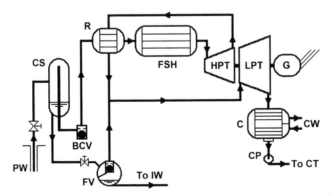

Figure 14.6 Double-flash geothermal plant with fossil superheating; see Nomenclature.

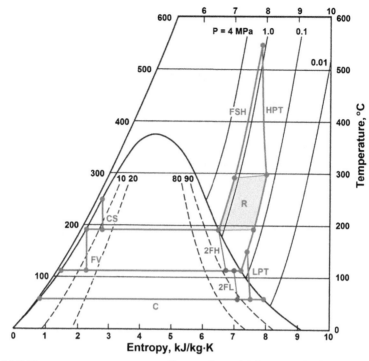

Figure 14.7 Temperature-entropy process diagram (to scale) for example cited. Process labels correspond to Fig. 14.6; turbine processes 2FH and 2FL are for the basic double-flash plant, both of which are replaced by the HPT and LPT of the hybrid system; see Nomenclature.

14.5.2 Coal-fired plants with geothermal-heated feedwater

Whereas the previous system was basically a geothermal plant with an assist from the fossil fuel, the system described in this section is the opposite, that is, a fairly conventional fossil fuel, typically coal-fired, central power station with a geothermal assist. Fig. 14.8 shows a simplified schematic of a typical arrangement.

The geothermal fluid provides some of the heating of the feedwater, eliminating some or all of the low-pressure feedwater heaters normally placed between the condensate pump (CP) and the deaerator (DA) that require steam from the low-pressure turbine (LPT) to effect the heating. This allows the use of lower temperature geofluids that could not be used effectively in a geothermal power plant, while allowing more power to be generated by the fossil plant owing to higher steam mass flow through the lower pressure stages of the turbine.

A systematic parametric study [4] of a geothermal preheat hybrid system built on a standard subcritical fossil-fueled central station found that over a geofluid temperature range from $150°C$ to $250°C$ the overall F_H ranged from about 1.02 to 1.05, the fossil F_F ranged from about 1.02 to 1.06, and the geothermal F_G ranged from 1.50 to 1.44. The same study using a supercritical fossil plant showed nearly identical F_H and F_F as

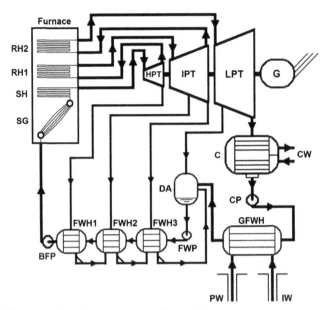

Figure 14.8 Geothermal preheat system; see Nomenclature.

the subcritical case, but the F_G ranged from 1.71 to 1.55 over the same temperature range.

An engineering proposal for a 750-MW hybrid coal-geothermal power plant of the geothermal-preheat type was put forth by the city of Burbank, CA, in the late 1970s. Several assessments were conducted [18] for a plant to be sited at Roosevelt Hot Springs, UT, and the conclusions were as follows:

- A well-designed hybrid plant could generate electricity at a lower cost than either a conventional SOTA coal plant or a SOTA geothermal plant.
- The geothermal energy would contribute more than 20% of the energy input to the hybrid plant.
- The hybrid plant would use the geothermal energy far more efficiently than any conceivable all-geothermal plant. For high-grade geothermal resources, the hybrid plant would have a 20% higher utilization efficiency; for low-grade resources, the improvement would be 150–200%.

A further study by the Ralph M. Parsons engineering firm verified the technical and economic feasibility of the hybrid power plant [12]. Despite the advantages held by the hybrid design, a conventional coal-fired plant, the 2×950 MW Intermountain Power Plant, was built instead.

14.5.3   Compound hybrid geothermal-fossil plants

It is possible to combine the two previous hybrid systems to form a compound hybrid plant, as shown in Fig. 14.9. The plant uses a single-flash geothermal unit that derives

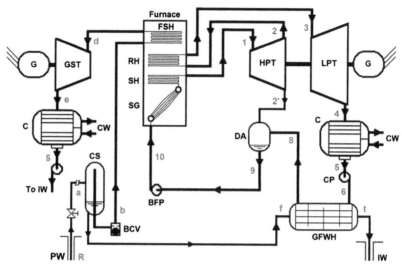

Figure 14.9 Compound fossil-geothermal hybrid plant. Note: State-points are keyed to Fig. 14.10; see Nomenclature.

superheat from the furnace of a conventional fossil-fired central plant while providing a portion of the feedwater heating to the fossil plant by means of the hot separated geothermal brine. A double-flash geothermal plant may also be used in a similar manner but with more complexity [6,9].

With reference to Fig. 14.9, there are several adjustable parameters that define the operating pattern of such a plant. The mass fraction of steam entering the cyclone separator (CS) is fixed by the reservoir geofluid condition and the wellhead pressure, the latter being a degree of freedom in the design. Likewise, the steam temperature entering both fossil turbines, HPT and LPT, can be specified along with their pressures. The fraction of steam extracted from the HPT to feed the deaerator (DA) may be set so as to produce a saturated liquid entering the boiler feed pump (BFP). Both condensers may be assumed to operate at the same pressure (vacuum) since both are supplied with cooling water from a cooling tower. The geothermal feedwater heater (GFWH) can be designed for a certain terminal temperature difference, TTD, and this will allow the determination of the geothermal−to−pure water mass flow ratio. The three turbines, including the geothermal steam turbine (GST), will operate essentially as dry expanders and may be characterized by typical isentropic efficiencies.

A study was performed for this chapter using the following assumptions:

Geofluid production mass flow rate = 100 kg/s
Geothermal wellhead temperature = 200°C
Geothermal quality (dryness fraction) at inlet to CS = 0.20
HPT and LPT inlet temperatures = 540°C
HPT and LPT inlet pressures = 16.5 and 4.0 MPa, respectively
GST inlet temperature = 510°C
Condensing pressure = 0.0056 MPa

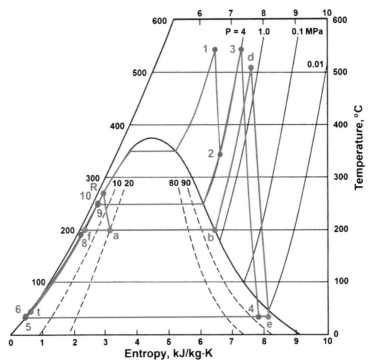

Figure 14.10 Temperature-entropy process diagram for hybrid plant shown in Fig. 14.9. Notes: State 5 applies to both pure water and geothermal steam condensate; states 5 and 6 and states 9 and 10 are indistinguishable on this diagram.

GFWH TTD = 10°C
HPT isentropic efficiency = 0.82
LPT and GST isentropic efficiency = 0.85
CP and BFP isentropic efficiency = 0.80
Pressure and heat losses were ignored.

Table 14.1 shows the properties at the state points; Table 14.2 gives the main results. The efficiency of the plant may be measured in several ways. The overall thermal efficiency based on the total heat added from the fossil fuel is 53.0%. Looking only at the pure steam side of the plant, the thermal efficiency is 48.0%, obtained by subtracting the output of the GST from the total net power and dividing by only the heat input to the pure water and steam from the fossil fuel. The synergy created by combining the simple fossil-fuel plant with the single-flash geothermal plant with the geothermal brine heater (BH) and the fossil-fired geothermal steam superheater allows the plant to gain nearly five percentage points of efficiency, which is about a 10% improvement.

A full parametric analysis was carried out in DiPippo et al. [6] for a system defined in Fig. 14.9 for three values of wellhead temperature, four values of wellhead quality (or dryness fraction), and including 10% pressure losses in all piping and heat

Table 14.1 **State-point properties for compound hybrid plant shown in Fig. 14.9**

State	T C	P MPa	s kJ/kgK	h kJ/kg	x	$\dot{m}$ kg/s
1	540.00	16.5	6.4298	3406.45	SH	91.79
2	341.95	4	6.5516	3073.08	SH	91.79
2′	341.95	4	6.5516	3073.08		11.30
3	540.00	4.000	7.2075	3537.47	SH	80.49
4	34.91	0.0056	7.8532	2410.27	0.9363	80.49
5	34.91	0.0056	0.5039	146.24	0.0000	80.49
6	35.25	4	0.5071	151.26	CL	80.49
8	190.00	4	2.2315	808.69	CL	80.49
9	250.35	4	2.7968	1087.49	0.0000	91.79
10	254.36	16.5	2.8041	1106.95	CL	91.79
Res	269.37	5.45	2.9708	1240.22	0.0000	**100**
a	200.00	1.55	3.1505	1240.22	0.2000	**100**
b	200.00	1.55	6.4302	2792.01	1.0000	20
d	510	1.555	7.5828	3495.00	SH	20
e	34.91	0.0056	8.1515	2502.15	0.9743	20
f	200.00	1.55	2.3305	852.27	0.0000	80
t	45.25	1.555	0.6412	190.83	CL	80

Table 14.2 **Performance of compound hybrid plant shown in Fig. 14.9**

Heat input to steam generator (SG)	211,069 kWt
Heat input to reheater (RH)	37,378 kWt
Heat input to geosteam superheater (FSH)	14,060 kWt
Total heat input	**262,507 kWt**
Power output from HPT	30,600 kWe
Power output from LPT	90,727 kWe
Power output from GST	19,857 kWe
Total turbine power output	**141,183 kWe**
CP power required	404 kWe
BFP power required	1786 kWe
Total pump power required	**2190 kWe**
Net plant power output	**138,993 kWe**

Table 14.3 **Figures of merit for compound hybrid plants with single-flash [6]**

Wellhead temperature	FM	Wellhead dryness fraction			
		0.10	0.20	0.30	0.40
150°C	F_H	1.032	1.045	1.059	1.074
	F_F	1.034	1.048	1.066	1.087
	F_G	1.612	1.562	1.535	1.517
200°C	F_H	1.066	1.083	1.102	1.123
	F_F	1.072	1.094	1.121	1.153
	F_G	1.759	1.696	1.657	1.631
250°C	F_H	1.095	1.116	1.138	1.163
	F_F	1.109	1.139	1.174	1.216
	F_G	1.745	1.707	1.681	1.663

exchangers. Each combination of wellhead temperature and dryness fraction corresponds to a different reservoir temperature. The results for the overall hybrid, the fossil and geothermal figures of merit are shown in Table 14.3. The F_G values are impressive while the F_H and F_F values are less so, but in all cases synergy exists in the hybrid arrangement.

14.5.4 Gas turbine–topped geothermal flash-steam hybrid plant with superheating

Patterned after the highly efficient combined steam and gas turbine power plants, the gas turbine topping cycle with a geothermal flash plant makes for an interesting hybrid plant [3]. Fig. 14.11 shows a schematic arrangement of such a system. The geosteam from the cyclone separator (CS) first passes through a steam-side recuperator (SR) on its way to being superheated as part of the gas turbine combustion process (SH1). From the high-pressure steam turbine (HPST), the geosteam provides the initial heating of the separated steam in the SR, and then mixes with the low-pressure flashed steam that has been superheated by the gas turbine exhaust (SH2). The merged streams then drive the low-pressure steam turbine (LPST). Other than allowing for the superheating of the separated geosteam, the gas turbine cycle is fairly standard.

The thermodynamic process diagrams are shown in Figs. 14.12 and 14.13 in scale-drawn temperature-entropy coordinates. They are not superimposed in a single diagram owing to the very high temperatures at which the gas turbine cycle operates. All heat transfer and work processes are depicted as arrows.

A set of representative state-points were assigned and a system analysis carried out. This example is not optimized but illustrates the advantages that can be achieved with

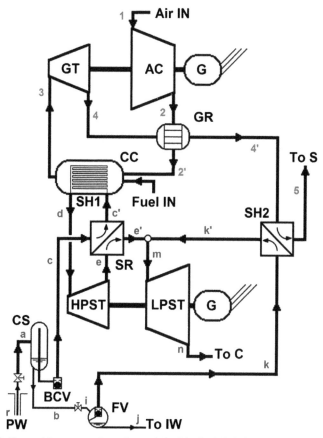

Figure 14.11 Gas turbine—topped geothermal double-flash hybrid system; see Nomenclature.

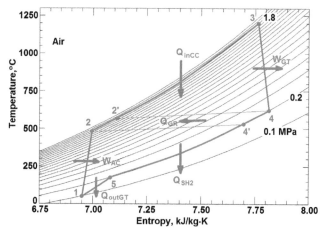

Figure 14.12 Temperature—entropy diagram for gas turbine processes for plant in Fig. 14.11.

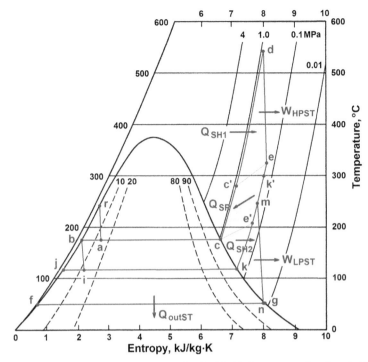

Figure 14.13 Temperature–entropy diagram for double-flash processes for plant in Fig. 14.11.

the hybrid system. Table 14.4 lists the temperature, pressure, enthalpy, and mass flow rates, for an assumed 100 kg/s geofluid flow from the reservoir. The other fixed parameters are shown in **bold numbers** in the table. The following assumptions were made regarding component efficiencies: isentropic efficiency for air compressor = 0.87, for gas turbine = 0.92, for both high- and low-pressure steam turbines = 0.85, and the effectiveness of both the steam recuperator and the gas regenerator = 0.65. There was no need to deploy the Baumann rule for the steam turbines since both of them operated essentially completely dry, as can be seen from Fig. 14.13.

The summary of calculations is given in Table 14.5. It will be seen that the heat rejected from the gas turbine to the surroundings is very small, 1404 kWt, whereas the heat rejected from the steam plant is very large, 57,893 kWt. This is a consequence of the internal heat transfer within the gas regenerator (GR) and from the GT plant to the two-flash plant (SH2). The very high heat rejection per unit of power generated in the geothermal plant is typical of such plants and leads to large cooling towers to absorb the waste heat. Also shown are the exergy inputs to the hybrid plant. Each heat transfer term has an associated exergy found by assuming the heat transfer is used in an ideal Carnot cycle between the highest temperature at which the heat is received and the dead-state temperature (35°C). Thus the exergy equals the heat transfer value multiplied by the Carnot efficiency, namely, $1-(T_0/T_H)$. The thermal efficiency for the GT part of the plant is about 29%, whereas the utilization efficiency

Table 14.4 State-point data for gas turbine–topped geothermal double-flash hybrid power plant

State	Description	T °C	P MPa	h kJ/kg	ṁ kg/s
Gas turbine plant side					
1	Air compressor inlet	**35**	**0.1**	308.52	9.91
2	Air compressor outlet	472.78	1.8	763.90	9.91
2'	Gas regenerator HP outlet	577.92	1.8	879.68	9.91
3	Gas turbine inlet	**1200**	**1.8**	1605.70	9.91
4	Gas turbine outlet	634.34	**0.2**	942.02	9.91
4'	Gas regenerator LP outlet	530.19	0.2	826.24	9.91
5	Superheater 2 outlet	**175**	0.2	450.21	9.91
Geothermal plant side					
r	Reservoir geofluid	**240**	**3.8469**	1037.60	**100**
a	Cyclone separator inlet	**175**	0.8926	1037.60	100
b	Separated liquid outlet	175	0.8926	741.02	85.40
c	Separated vapor outlet	175	0.8926	2772.71	14.60
c'	Steam recuperator HP outlet	275.39	0.8926	3001.93	14.60
d	HP steam turbine inlet	**540**	0.8926	3567.23	14.60
e	HP steam turbine outlet	325.91	0.1692	3125.36	14.60
e'	Steam recuperator LP outlet	211.97	0.1692	2896.13	14.60
i	Flash vessel inlet	**115**	0.1692	741.02	85.40
j	Flash liquid outlet	115	0.1692	482.59	75.44
k	Flash vapor outlet	115	0.1692	2698.58	9.96
k'	Superheater 2 outlet	**300**	0.1692	3072.83	9.96
m	LP steam turbine inlet	247.79	0.1692	2967.80	24.56
n	LP steam turbine outlet	**50**	0.0124	2566.82	24.56
f	Condenser liquid outlet	**50**	0.0124	209.34	24.56
g	Condenser saturated vapor	**50**	0.0124	2591.29	—

is about 37%; the utilization of the hybrid plant is over 57%. This may be compared to 51% utilization efficiency for a pure double-flash plant.

Thus, the hybrid power plant holds a 12% utilization efficiency advantage over a simple geothermal double-flash plant with no superheating, and a 54% advantage

Table 14.5 **Summary of results for gas turbine–topped geothermal double-flash hybrid power plant**

Heat transfer terms	
Rate of heat transfer in combustion chamber	7197 kWt
Rate of heat transfer in superheater 1	8252 kWt
Rate of total heat transfer from fossil fuel	15,449 kWt
Rate of heat transfer inside gas regenerator	1148 kWt
Rate of heat transfer inside steam recuperator	3346 kWt
Rate of heat transfer inside superheater 2	3727 kWt
Rate of heat transfer out of gas turbine cycle	1405 kWt
Rate of heat transfer out of double-flash plant	57,893 kWt
Electrical power terms	
Power output of gas turbine	6579 kWe
Power consumption of air compressor	4514 kWe
Net power developed by gas turbine plant	2065 kWe
Power output of HP steam turbine	6450 kWe
Power output of LP steam turbine	9847 kWe
Total power of steam turbines	16,297 kWe
Total power of hybrid plant	18,362 kWe
Exergy input terms	
Exergy input to gas turbine from heat input in CC	5618 kW
Exergy input to steam from heat input in SH1	4973 kW
Total exergy input from fossil fuel	10,591 kW
Exergy input from geofluid (reservoir state)	21,435 kW
Total exergy input to hybrid plant	32,025 kW
Efficiency terms	
Thermal efficiency of gas turbine plant	0.2869
Utilization efficiency of gas turbine plant	0.3675
Utilization efficiency of hybrid plant	0.5734
Pure double-flash plant	
Power output of HP steam turbine	3477 kW
Power output of LP steam turbine	7430 kW
Total power of steam turbines	10,906 kW
Utilization efficiency of pure double-flash plant	0.5088

over the gas turbine plant with a heat regenerator. These are important gains for the hybrid system and demonstrate high synergy with this arrangement.

14.5.5 Geothermal-biomass hybrid plants

Many geothermal power stations are situated where agricultural or forestry operations exist, creating a supply of vegetation or wood waste that can be considered as fuel for a geothermal-biomass plant. For example, some plants in Larderello, Italy, are being upgraded with biomass. A pair of dry steam plants at Cornia, 2×20 MW, have been redesigned to accommodate a biomass furnace to superheat the geothermal steam. One of the plants has been shut down for several years and will serve as the location for the furnace, while the other plant, only 150 m away, will serve as the power generator. A simplified schematic of the system is shown in Fig. 14.14.

The plant is fed steam from two well areas: area 1 provides dry, slightly superheated steam at 210°C, while area 2 yields two-phase geofluid at the wellhead at 156°C and about 18% dryness fraction. The two streams are mixed at a pressure of 0.55 MPa and fed to the biomass superheater where the temperature is raised to 370°C. The combined mass flow is 30.6 kg/s and produces about 19.2 MW from the turbine, assuming equal steam flows from the two well areas; originally the turbine generated 13.8 MW. The gain of 5.4 MW may be compared to the heat added from the biomass combustion, namely, 12.1 MWt. Thus, the biomass heat has a thermal efficiency of 44.7%, much higher than a typical simple biomass power plant.

New Zealand is another particularly favorable place since numerous large geothermal plants are surrounded by or in close proximity to forest plantations [19]. From among the many possible schemes that can capture these two sources of energy, we show here only two. Fig. 14.15 shows a biomass power plant that is boosted by a geothermal preheat and in which the geosteam is superheated, a compound hybrid arrangement (see Section 14.5.3). Fig. 14.16 shows an integrated geothermal flash-binary plant where a biomass boiler provides superheat to the geothermal steam and includes two binary cycles, one heated by the separated geothermal brine and one by the back-pressure steam exhausted from the geothermal steam turbine.

14.5.6 Geopressure thermal-hydraulic hybrid systems

There are places where reasonably hot geofluids are found under very high pressures and with significant amounts of dissolved petroleum gases. One such region is the Gulf Coast of the United States. If these fluids can be safely brought to the surface via wells, it should be possible to utilize three different forms of energy contained within them. The high pressure can be used to drive a hydraulic turbine, the dissolved gases can be released, captured, and either sold or burned on-site, and the thermal energy used to power a geothermal power plant either flash or binary.

Fig. 14.17 shows one possible arrangement of a geopressured power plant in which the gas, assumed to be methane, is fed to a combustion chamber of a gas turbine after being separated from the exhaust section of a hydraulic turbine. The separated liquid is flashed to generate steam which is superheated by the gas turbine exhaust and then

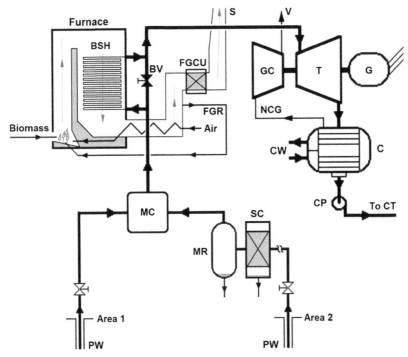

Figure 14.14 Cornia repowered dry-steam/biomass hybrid power plant.
F. Lazzeri, Enel Green Power. [Personal communication]. 15.05.15; see Nomenclature.

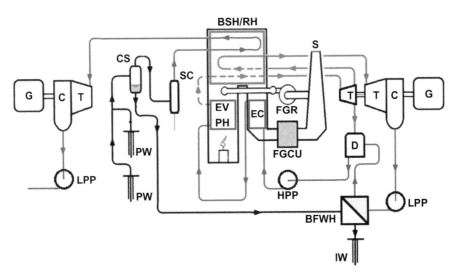

Figure 14.15 Compound hybrid geothermal wood-waste hybrid system [19]; see Nomenclature.

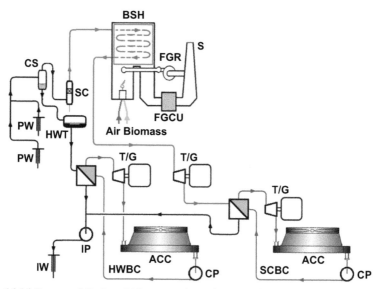

Figure 14.16 Integrated flash and binary geothermal power plant with a biomass superheater; see Nomenclature.

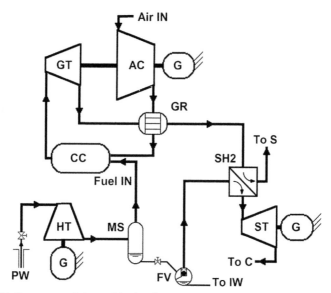

Figure 14.17 Geopressured thermal-hydraulic power system with gas turbine; see Nomenclature.

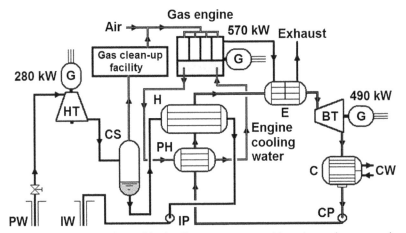

Figure 14.18 Geopressure thermal-hydraulic power system with reciprocating gas engine. Depending on the price of natural gas, the gas engine can be eliminated and the gas sold into a pipeline; see Nomenclature.

expanded in a steam turbine. Many variations on this basic design are possible [20]. Alternatively, a combustion engine may replace the gas turbine, a double-flash system may replace the single-flash, and a binary bottoming cycle may replace the steam turbine. Fig. 14.18 represents one such variation. The U.S. Department of Energy built and operated a pilot plant at Pleasant Bayou in Texas from 1989 to 1990; the plant was modeled after Fig. 14.18 but without the hydraulic turbine [15].

Given the variable thermodynamic characteristics that the geopressured fluid may exhibit, a flexible suite of power system options are needed to exploit the resource. Additionally, the prices of electricity and gas will dictate which option is the thermoeconomic optimum.

Writing in 1981, Khalifa [20] concluded that the levelized cost of electricity from hybrid geopressured plants in the Gulf Coast region would be about 40–50% lower than electricity generated using conventional geothermal flash and binary plants, assuming realistic gas prices and cost escalators. Furthermore, the cost of electricity is strongly dependent on the methane content of the geofluids but only weakly dependent on the wellhead temperature. A more recent study [21] found that near-term conditions as of 2004 were not favorable for exploitation of geopressured resources for electric power.

14.6 Geothermal-solar hybrid systems

The idea of using solar energy to enhance geothermal power plants has a long history [13]. Solar collectors may be used to heat geothermal steam and/or brines to temperatures greater than naturally available from the reservoir. Field experiments have shown that there are advantages from a thermodynamic standpoint. The most important practical consideration is the mismatch between availability of the intermittent solar component and the steady geothermal component. Without a commercial means of storing the solar energy for those periods when the sun is not shining, the geothermal plant must operate under two different modes, one of which will necessarily be nonoptimum. The other important factor is the economics of the hybrid system. Given the high specific capital cost of solar systems relative to geothermal ones, it usually takes special circumstances to justify the addition of a solar system to a geothermal power plant.

Process flow diagrams are presented in Section 14.6.1 for some hybrid systems that incorporate concentrating solar power (CSP) systems with geothermal flash and binary plants. While photovoltaic (PV) systems have been installed at one geothermal plant, the Stillwater binary plant in Nevada, USA, the two energy sources are thermodynamically separate and thus achieve no synergy beyond sharing common infrastructure such as roads and transmission lines; see Section 14.6.2.

14.6.1 Geothermal-concentrating solar power hybrid plants

A parabolic solar collector may be used to superheat the working fluid in a geothermal binary plant as shown schematically in Fig. 14.19. Since binary plants suffer from having a small temperature range across the power cycle, raising the upper temperature in this way will lead to higher thermal efficiency. However, the turbine must deal with saturated vapor when the solar loop is inactive. Thus if the turbine is designed for superheated inlet conditions, it will perform off-design when the sun is not available, or

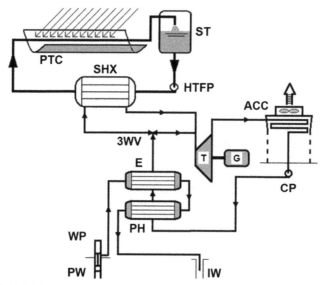

Figure 14.19 Hybrid geothermal-solar system with parabolic trough collector; see Nomenclature.

vice versa. A simple storage system (ST) is shown in the figure to allow for some hours of continued operation in hybrid mode after the sun sets, but when the hot fluid in the storage tank is depleted, the plant reverts back to basic geothermal operation.

A geothermal double-flash plant offers several opportunities for enhancement from solar energy; see Fig. 14.20. The solar heat exchanger (SHX) is shown superheating the low-pressure flashed steam prior to mixing with the steam leaving the first stages of the steam turbine. This will allow the first of the low-pressure stages to see super-heated steam instead of wet steam and to produce a drier expansion line and higher turbine efficiency. Three other options for locating the SHX are shown as *A1*, *A2*, and *A3*. *A1* would preheat the separated brine and allow the flash process to generate more steam; *A2* would superheat the separated steam from the cyclone separator for turbine entry; and *A3* would heat and evaporate a portion of the steam condensate for reuse in the low-pressure turbine section. Recall that there is excess condensate available from the water cooling tower and this would tap into that excess. Field experiments of these systems have been conducted at the Ahuachapán geothermal plant with good results [22,23].

Recently, a proposal [24] has been put forth for a novel system at Cerro Prieto, near Mexicali, Mexico where there is abundant sunlight [26]. Hiriart would like to rejuvenate the now-closed Cerro Prieto I power station by drilling new wells and installing downhole pumps to produce high-pressure brine. The temperature would be fairly low, about 140°C, but would be raised to about 155°C by means of solar collectors during the day. Parabolic collectors would be used to heat a heat transfer medium (oil), as shown in Fig. 14.21, which in turn would impart heat to the brine in a simple tube-in-tube heat exchanger formed by the brine transmission pipeline. The plant would be an air-cooled binary in which the cycle working fluid would enter the turbine

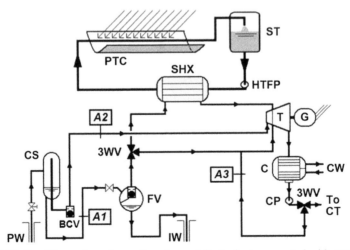

Figure 14.20 Possible configurations of geothermal flash plant augmented with parabolic trough collector; see Nomenclature.

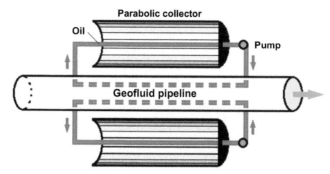

Figure 14.21 Means of superheating geothermal steam traveling from separators to the turbine at Cerro Prieto, Mexico.
After Hiriart Le Bert G. Complejo Híbrido de Energías Renovables con Planta de Almacenamiento en la Frontera Norte [Hybrid Renewable Energy Plant with Storage at the North Border]. Mexico: Academia de Ingeniería A.C.; March 26, 2015 [in Spanish].

as an enhanced superheated vapor when the sun is shining but as essentially a saturated vapor otherwise. The incremental daytime power attributed to the solar input would serve to compensate for the poor heat transfer in the air-cooled condensers which limits power output owing to the hot ambient conditions.

In 2015, a CSP system was added to the Stillwater binary plant in Nevada, USA. The purpose is to raise the geofluid temperature after it is pumped to the surface but before it enters the heat exchangers of the binary cycle. Over the years of operation, the geofluid temperature had declined and the new CSP system restores the plant to its design geofluid temperature. Fig. 14.22, a highly simplified schematic, shows the design; a heat transfer fluid (pressurized water with a corrosion inhibitor) circulates between the solar collector (PTC) and the brine heat exchanger (HXER). The new CSP system cost about US$14.5 million and is expected to add about 2 MWe of power and 3500 MWh of electricity annually to the Stillwater output [25].

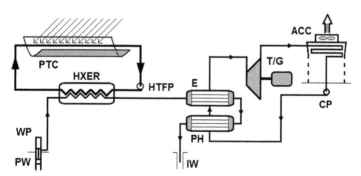

Figure 14.22 Flow diagram for Stillwater CSP-geothermal binary power plant [25]; see Nomenclature.

14.6.2   Geothermal-photovoltaic hybrid plants

One example of a PV-geothermal power plant is at the Stillwater plant in Nevada; see Figs. 14.23 and 14.24. This PV system preceded the CSP system described above. The geothermal binary plant has an installed capacity of 47.2 MW, but has not achieved

Figure 14.23 Aerial view of Stillwater power plant with adjacent PV field. Note: the CSP array lies just north of the PV field.

Figure 14.24 View of Stillwater binary plant showing a few of the 89,000 polycrystalline PV panels.
Photograph by author, July 29, 2012.

that level of output in practice owing to insufficient supply of geofluid from the production wells, lower-than-expected geofluid temperature, and insufficient reinjection capacity, despite the drilling of wells to boost performance. The shortfall of contracted power which the owner is obligated to deliver to the grid is offset using the PV system. This is particularly attractive economically when the price of electricity is highest during hot summer days when sunshine is plentiful at the site. The two power plants share common infrastructure but do not interact in any thermodynamically synergistic manner.

14.7 Conclusions

This chapter has shown that hybrid power plants involving geothermal and other energy sources can be designed and built in a wide variety of ways. Combined geothermal plants of different types are routinely constructed and several examples are described herein. Generally, the goal of these hybrid and combined systems is to achieve some form of synergy, that is, to produce more power from combining the two sources of energy or power systems in a clever way, thereby obtaining more output than could be achieved using two separate, SOTA power plants. The thermodynamic conditions required to achieve this are described for several types of plant. Some successful plants are presented that demonstrate how effective these plants can be. However, the feasibility of hybrid plants turns on many site-specific factors that must be favorable for success, including co-location of the energy sources, prices of energy and electricity, assured availability of energy supplies, environmental permitting, and meeting all regulatory requirements.

Nomenclature

3WV	Three-way valve
AC	Air compressor
ACC	Air-cooled condenser
BCV	Ball check valve
BFP	Boiler feed pump
BFWH	Brine feedwater heater
BHT	Brine holding tank
BSH	Biomass superheater
BT	Binary turbine
BV	Bypass valve
C	Condenser
CC	Combustion chamber
CP	Condensate pump
CS	Cyclone separator
CT	Cooling tower

CW	Cooling water
D, DA	Deaerator
E, EV	Evaporator
EC	Economizer
F, FV	Flasher, flash vessel
FGCU	Flue gas clean-up
FGR	Flue gas recirculator
FSH	Fossil superheater
FWH	Feedwater heater
G	Generator
GC	Gas compressor
GFWH	Geothermal feedwater heater
GR	Geothermal recuperator
GST	Geothermal steam turbine
GT	Gas turbine
H	Heater
HPP, LPP	High-, low-pressure pump
HPS, MPS, LPS	High-, medium-, low-pressure separator
HPST, LPST	High-, low-pressure steam turbine
HPT, IPT, LPT	High-, intermediate-, low-pressure turbine
HT	Hydraulic turbine
HTFP	Heat transfer fluid pump
HWBC	Hot water binary cycle
HWT	Hot water tank
HXER	Heat exchanger
IP	Injection pump
IW	Injection well
MC	Mixing chamber
MR	Moisture remover
MS	Methane separator
NCG	Noncondensable gas
OP	Orifice plate
PH	Preheater
PTC	Parabolic trough collector
PW	Production well
R	Recuperator
RH	Reheater
S	Stack
SC	Scrubber
SCBC	Steam condensed binary cycle
SG	Steam generator
SH	Superheater
SHX	Solar heat exchanger
SR	Steam recuperator
ST	Steam turbine; storage tank
T/G	Turbine/generator
WP	Well pump

References

[1] Janes J, Shurley LA, White L, Deter ER. Evaluation of a superheater enhanced geothermal steam power plant in the geysers area: Final report. Sacramento, CA: California Energy Commission; June 1984. Rep. No. P700-84-003.

[2] DiPippo R. An analysis of an early hybrid fossil-geothermal power plant proposal. Geotherm Energ Mag March 1978;6:31—6.

[3] Kestin J, DiPippo R, Khalifa HE. Hybrid geothermal-fossil power plants. Mech Eng December 1978;100:28—35.

[4] Khalifa HE, DiPippo R, Kestin J. Hybrid fossil-geothermal power plants. In: Proceedings of the 5th Energy Technology Conference, Washington, DC; 1978. p. 960—70.

[5] Khalifa HE, DiPippo R, Kestin J. Geothermal preheating in fossil-fired steam power plants. In: Proceedings of the 13th Intersociety Energy Conversion Engineering Conference, 2; 1978. p. 1068—73.

[6] DiPippo R, Khalifa HE, Correia RJ, Kestin J. Fossil superheating in geothermal steam power plants. Geotherm Energ Mag January 1979;7:17—23.

[7] DiPippo R, Avelar EM. Compound hybrid geothermal-fossil power plants. Geotherm Resour Counc Trans 1979;3:165—8.

[8] DiPippo R. Impact of hybrid combustion-geothermal power plants on the next generation of geothermal power systems. In: Proceedings of the Third Annual Geothermal Conference and Workshop. Electric Power Research Institute; 1979. 6.1—6. Rep. WS-79-166.

[9] Khalifa HE. Hybrid fossil-geothermal systems. Section 4.3. In: Kestin J, DiPippo R, Khalifa HE, Ryley DJ, editors. Sourcebook on the production of electricity from geothermal energy. U.S. Department of Energy, U.S. Government Printing Office; 1980.

[10] DiPippo R, DiPippo EM, Kestin J, Khalifa HE. Compound hybrid geothermal-fossil power plants: thermodynamic analyses and site-specific applications. Trans ASME J Eng Power 1981;103:797—804.

[11] Khalifa HE. Gas-turbine-topped hybrid power plants for the utilization of geopressured geothermal resources. ASME Paper No. 81-Pet-5. In: Proceedings of the Energy-Sources Technology Conference and Exhibition, January 18—22; 1981.

[12] The Ralph M. Parsons Company. System design verification of a hybrid geothermal/coal fired power plant. Prime Contractor: City of Burbank, Project No. 5805. September 1978.

[13] Hiriart Le Bert G, Gutierrez N LCA. Calor del subsuelo para generar electricidad — combinacion solar-geothermia (Heat from the earth to generate electricity—combined solar-geothermal). Ing Civ Mayo 1995;313:13—22 [in Spanish].

[14] Moya R. P, DiPippo R. Miravalles Unit 5 bottoming binary plant: planning, design, performance and impact. Geothermics 2007;36:63—96.

[15] DiPippo R. Geothermal power plants. Principles, applications, case studies and environmental impact. 3rd. Ed. Oxford, England: Butterworth-Heinemann: Elsevier; 2012.

[16] James J. Evaluation of a superheater enhanced geothermal steam power plant in The Geysers area. Rep. P700-84-003. California Energy Commission, Siting and Environmental Div.; 1984.

[17] National Institute of Standards and Technology (NIST), Reference fluid thermodynamic and transport properties, U.S. Dept. of Commerce: http://www.nist.gov/srd/nist23.cfm.

[18] U.S. Energy Research and Development Administration (ERDA). Site-specific analysis of hybrid geothermal/fossil power plants. Prime Contractor — City of Burbank: Div. of Geothermal Energy; June 1977.

[19] Thain I, DiPippo R. Hybrid geothermal-biomass power plants: applications, designs and performance analysis. In: Proceedings of the World Geothermal Congress 2015, Melbourne, Australia; April 2015.

[20] Khalifa HE. Hybrid power plants for geopressured resources. EPRI Project RP 1671-2. United Technologies Research Center; June 22, 1981.

[21] Griggs J. A re-evaluation of geopressured-geothermal aquifers as an energy resource [Masters thesis]. Louisiana State University, Craft and Hawkins Dept. of Petroleum Engineering; August 2004.

[22] Handel S, Alvarenga Y, Recinos M. Geothermal steam production by solar energy. Geotherm Resour Counc Trans 2007;31:503−10.

[23] Alvarenga Y, Handel S, Recinos M. Solar steam booster in the Ahuachapán geothermal field. Geotherm Resour Counc Trans 2008;32:395−9.

[24] Hiriart Le Bert G. Complejo Híbrido de energías renovables con planta de almacenamiento en la frontera norte [[Hybrid renewable energy plant with storage at the north border]]. Mexico: Academia de Ingeniería A.C; March 26, 2015 [in Spanish].

[25] DiPippo R. Geothermal power plants. Principles, applications, case studies and environmental impact. 4th. ed. Oxford, England: Butterworth-Heinemann: Elsevier; 2016.

[26] Hiriart Le Bert G. ENAL: Personal communication. June 01, 2015.

Additional reading

DiPippo R. Geothermal power plants. V. 7, Sect. 7.07. In: Sayigh A, editor. Comprehensive renewable energy. Oxford, England: Elsevier; 2012. p. 207−37.

Lentz I, Almanza R. Geothermal-solar hybrid system in order to increase the steam flow for geothermic cycle in Cerro Prieto, Mexico. Geotherm Resour Counc Trans 2003;27:543−6.

Lentz I, Almanza R. Solar−geothermal hybrid system. Appl Therm Eng 2006a;26:1537−44.

Lentz I, Almanza R. Parabolic troughs to increase the geothermal wells flow enthalpy. Sol Energ 2006b;80:1290−5.

Manente G, Field R, DiPippo R, Tester JW, Paci M, Rossi N. Hybrid solar-geothermal power generation to increase the energy production from a binary geothermal plant. Paper No. IMECE 2011−63665. In: Proceedings of the 2011 ASME International Mechanical Engineering Congress and Exposition; November 2011 [Denver, Colorado].

Part Three

Design and economic considerations

Waste heat rejection methods in geothermal power generation

A. Chiasson

Department of Mechanical and Aerospace Engineering, University of Dayton, Dayton, OH, United States

15.1 Introduction: overview and scope

This chapter focuses on the theoretical and practical considerations of the methods of heat rejection to the environment from geothermal power plants. Thermal energy contained in geothermal fluids that is either reintroduced to the subsurface or otherwise discharged is not included with the waste heat considered here. It is noted, however, that such cascaded uses of geothermal fluids to direct-heat applications, co-generation, and/or bottoming binary cycles can significantly improve utilization efficiencies of geothermal power plants, and therefore also have potential to improve economics. The main barrier to more widespread application of such cascaded uses is lack of proximity of thermal loads to geothermal power stations; most geothermal power plants are located in remote areas, and pipeline costs to convey geothermal fluids over long distances can quickly become prohibitive.

The heat rejected from geothermal power plants (q_{out}) is primarily the heat of condensation of the working fluid exhaust vapor from the turbine. Fundamental to the operation of closed-loop power cycles (eg, binary geothermal power plants) is that a portion of the heat supplied to the cycle must be rejected. Although flash-steam geothermal power plants do not operate in a cycle, the turbine exhaust is condensed back to liquid for practical reasons of returning the geothermal fluid back to the reservoir. In either case, q_{out} can be described with the First Law of Thermodynamics as a function of the thermal efficiency of a power cycle (η) and the net heat input to the working fluid (q_{in}) as:

$$q_{out} = q_{in}(1 - \eta) \tag{15.1}$$

With the source and sink temperature differential being much smaller for geothermal power plants compared with fossil-fuel power plants, the thermal efficiencies of geothermal power plants are correspondingly lower. These lower thermal efficiencies, which are on the order of 10–20%, affect the methods of heat rejection in three main ways: (1) the heat rejection equipment in geothermal power plants will be proportionately larger (and proportionately more costly) than that for conventional power plants, (2) the proportionately larger heat rejection equipment translates to proportionately higher parasitic loads, and (3) additional care is prudent in the design of the heat rejection equipment to provide cooling temperatures as low as

Geothermal Power Generation. http://dx.doi.org/10.1016/B978-0-08-100337-4.00015-2

economically attainable. Thus, methods of heat rejection play an important role in overall geothermal power plant design, operation, and efficiency.

While the methods of heat rejection from geothermal power plants are numerous and many types of specific examples can be found, some generalities can be made. Condensers used in geothermal power plants are generally of the direct-contact or surface type. Methods of heat dissipation to the environment generally consist of wet cooling towers, air-cooled condensers, or a combination of wet-dry (or evaporative) condensers. The actual choice depends on designer preference, and what the geothermal site characteristics dictate. Thus, this chapter is organized into four main subsections. The first subsection discusses condensers in the context of geothermal power plants and the resulting impacts of the condenser on the power generation efficiency. Subsequent sections of the chapter are organized according to method of waste heat rejection: (1) water-cooled condensers, (2) air-cooled condensers, and (3) evaporative (both water- and air-cooled) condensers.

15.2   Condensers in geothermal power plants

15.2.1   General

To obtain good thermal efficiencies of geothermal power generation, the turbine exhaust pressure must be as low as can be economically achieved with the available cooling water or air temperature. Fig. 15.1 shows saturation pressures versus temperature for some common working fluids in geothermal power plants.

A review of the data presented in Fig. 15.1 shows that the exhaust pressure of steam turbines should be well below atmospheric (~ 100 kPa) in order to achieve condensing temperatures reasonably close to the environment temperature. Operating in a vacuum may introduce slight in-leakage of air, which may be a source of noncondensable gases.

Further review of the data presented in Fig. 15.1 shows that binary power plants using the example organic working fluids of ammonia, isobutane, propane, R134a, or R245a have condensing pressures near or well above atmospheric at typical environment temperatures. Thus, air leakage into these systems should not be a problem. It should also be noted that the thermodynamic properties of many organic working fluids used in binary geothermal power plants are such that the turbine exhaust state of the fluid remains in the superheated region. This generally is not the case with geothermal steam turbines.

Fluid temperatures in the condenser as a function of percent heat transfer rate (or relative position in the heat exchanger) are shown schematically in Fig. 15.2. Such plots are sometimes referred to as T-q diagrams. The closer the values of $T_{h,i}$ and $T_{c,i}$, the more effective the heat exchanger, which is typically exponentially proportional to the heat exchanger size and cost. Thus, condenser designers must strike a balance between the heat exchanger size and cost, and acceptable temperature differences between the working fluid and the source/sink fluid. When the turbine exhaust fluid remains superheated, the condenser essentially has three functions: de-superheating, condensing, and sub-cooling.

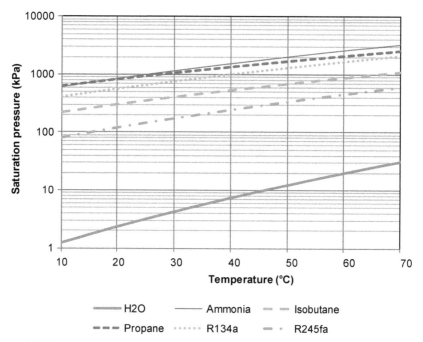

Figure 15.1 Saturation pressures versus temperature for some common working fluids in geothermal power plants.

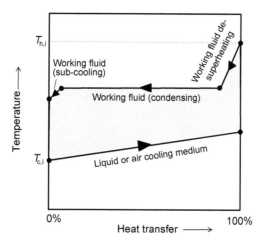

Figure 15.2 Schematic of temperature changes versus percent heat transfer (or T-q diagram) in the condenser of a geothermal power plant. $T_{h,i}$ is the hot inlet temperature and $T_{c,i}$ is the cold inlet temperature.

15.2.2   Impact on thermodynamic efficiency

Waste heat discharged from a geothermal power plant is ultimately absorbed by the Earth's atmosphere, regardless of the method used for heat dissipation. The methods may differ, however, in the lowest effective sink temperature for the plant that can be achieved. The sink temperature is a factor having an important influence on the thermal efficiency of the plant.

In the previous section, it was mentioned that the source and sink temperature differential is much smaller for geothermal power plants compared with fossil-fuel power plants. The impact of the source and sink temperature on the thermodynamic efficiency of a geothermal power plant can be described by the Carnot efficiency of a heat engine, which has the maximum possible efficiency (η_{max}) given by:

$$\eta_{max} = 1 - \frac{T_C}{T_H} \qquad\qquad [15.2]$$

where T_C and T_H are the cold and hot temperature, respectively, in absolute units. By simple mathematical inspection of Eq. [15.2], it is obvious that η_{max} must be less than one since the ratio T_C/T_H is less than unity but never zero. Mathematical inspection of Eq. [15.2] also reveals that the greater T_H is relative to T_C, the greater the maximum possible efficiency. However, due to consequences of the Second Law of Thermodynamics, and as depicted in Fig. 15.2, a condensing temperature of T_C cannot be achieved in practice; thermodynamic irreversibilities exist and heat transfer does not occur perfectly in real world situations.

The performance of an actual heat engine is given by:

$$\eta = 1 - \frac{\dot{q}_C}{\dot{q}_H} = \frac{\dot{W}_{out}}{\dot{q}_{in}} \qquad\qquad [15.3]$$

where $\dot{q}$ is heat transfer rate and $\dot{W}$ is power. Inspection of Eq. [15.3] reveals that decreasing $\dot{q}_C$ or increasing $\dot{q}_H$ increases the actual efficiency. Thus, the heat exchanger effectiveness of the condenser and the heat rejection equipment rejecting $\dot{q}_C$ has a strong influence on the thermodynamic efficiency of the power station.

From further inspection of Eq. [15.1], with geothermal power plant thermal efficiencies of 10–20%, it can be readily seen that the heat rejection from geothermal power plants is 80–90% of the heat input. Thus, per unit of electrical power output, relative rates of heat rejection are higher compared with conventional power plants. This translates to relatively high capital cost of the heat rejection equipment and higher relative parasitic loads.

15.2.3   Noncondensable gases

Working fluids in geothermal power stations contain impurities in the form of gases that do not condense at the same temperature as the working fluid. Thus, when the working fluid condenses, these impurities remain in the gaseous state causing "vapor locks" in the flow channels and are detrimental to system performance, particularly in surface condensers. In flash-steam systems, common noncondensable gases found in geothermal fluids include CO_2 and H_2S. Impurities found in organic fluids used in binary geothermal power plants vary with the working fluid. When first assembled, the piping and equipment will contain gases, usually air and water vapor, and it is important to evacuate the entire system before operation.

For low-pressure working fluids, where the operating pressure of the condenser is less than ambient pressure, even slight leaks can be a continuing source of noncondensables. In such cases, a purge system that automatically and safely expels noncondensable gases may be used, such as vacuum pumps and jet ejectors.

When present, noncondensable gases collect on the high pressure side of the system and raise the condensing pressure above that corresponding to the temperature at which the working fluid is actually condensing. This results in decreased power from the turbine, increased parasitic power consumption, and reduced heat rejection rates. The excess pressure is caused by the partial pressure of the noncondensable gas. These gases form a resistance film over some of the condensing surface, thus lowering the heat transfer coefficient. The actual impact of noncondensable gases is difficult to characterize if they tend to accumulate in confined areas far from the heat transfer surface. In these cases, a relatively large amount of noncondensables may be tolerated. One way to account for noncondensables is to treat them as a fouling resistance to be discussed in the subsections that follow.

15.3   Water-cooled condensers

15.3.1   General

Water-cooled condensers involve a heat rejection loop where condenser heat is dissipated to the environment via water. The most common type of water-cooled condensers in geothermal power plants are direct-contact and surface types. Plate-type heat exchangers (a special kind of surface exchanger) are also possible, but appear to be more common in liquid-to-liquid heat transfer applications. The rate of heat rejection ($\dot{q}$) from the condenser may be calculated by an energy balance on the water stream:

$$\dot{q} = \dot{m}c_p(T_i - T_o) \tag{15.4}$$

where $\dot{m}$ is the mass flow rate of the cooling water, c_p is the specific heat of the water (assumed constant), and T_i and T_o are the inlet and outlet water temperatures, respectively, at the condenser. Heat rejection to the atmosphere is most commonly accomplished with wet cooling towers, although dry fluid coolers are possible. Dry fluid coolers or indirect-contact cooling towers, are discussed in subsections that follow. Once-through systems, using water from a surface water body are also possible, but rare.

15.3.2 Types of condensers

15.3.2.1 Direct-contact

In a typical single-fluid direct-contact condenser, cooling water is sprayed into the turbine exhaust steam and condensation occurs on the water droplets (Fig. 15.3). The temperature difference between the leaving water and the condensing steam temperature could, in theory, be zero, but in actual practice it may be as high as 5—6°C. For a given cooling water inlet temperature, the direct-contact condenser will provide lower turbine back pressures compared with a surface condenser. An energy balance on the condenser yields:

$$\dot{m}_1 h_1 + \dot{m}_2 h_2 = (\dot{m}_1 + \dot{m}_2) h_3 \qquad\qquad [15.5]$$

where subscripts 1, 2, and 3 refer to the turbine exhaust, the cooling water inlet, and the outlet fluid, respectively; h is enthalpy; and $\dot{m}$ is the mass flow rate.

Some advantages of direct-contact condensers are that they tend to be simpler in design than surface condensers and typically have lower initial costs. There are fewer joints and parts, resulting in fewer leakage problems. Spray condensers require little maintenance or cleaning relative to surface condensers, and the heat transfer performance does not deteriorate with time. Space requirements are significantly less for spray condensers than required for surface condensers.

The main disadvantage of direct-contact condensers is that the steam condensate is mixed with the cooling water. The mixed water may require treatment before reuse. This issue of water quality appears to have resulted in development of closed-loop cooling water systems that are air-cooled, known as a "Heller" system. These are discussed in Section 15.4.2.

Direct-contact condensers using two fluids have been studied in the past for use in binary geothermal power plants (Robertson, 1978) and appear to have greater potential in the future. This type of condenser would condense the turbine exhaust vapor by direct contact with cooling water or other fluid that is immiscible with the working fluid.

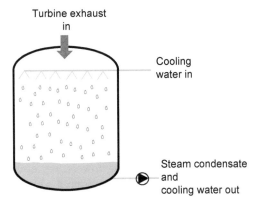

Figure 15.3 Schematic of a direct-contact steam condenser.

15.3.2.2 Shell-and-tube

The shell-and-tube heat exchanger is perhaps the most common type of condenser used in geothermal power plant applications. These types of heat exchangers have numerous applications with detailed discussions found in numerous publications. Specific configurations differ according to the number of shell-and-tube passes. The simplest configuration, which involves a single tube and shell pass, is shown in Fig. 15.4. In geothermal applications, the working fluid is generally on the shell side, and the cooling fluid flows through the tubes in a counterflow arrangement. Baffles are usually installed to improve contact time and to increase the convection heat transfer coefficient of the shell-side fluid by inducing turbulence. Baffles also physically support the tubes. Increasing the number of shell and tube passes increases the heat exchanger effectiveness up to a point of diminishing return.

An essential, and often the most uncertain, part of any heat exchanger analysis is determination of the overall heat transfer coefficient (U_o). Expressed in terms of outside surface area of the tube, the overall heat transfer coefficient may be computed from theoretically calculated or test-derived heat transfer coefficients of the water and working fluid sides, from physical measurements of the condenser tubes, and from a fouling factor on the water side, using the following equation:

$$U_o = \frac{1}{\left(\frac{A_o}{A_i} \frac{1}{h_i}\right) + \left(\frac{A_o}{A_i} R_{f,i}\right) + \left(\frac{A_o}{A_m} \frac{t}{k}\right) + \left(\frac{1}{h_o \eta_s}\right)} \qquad [15.6]$$

where U_o is the overall heat transfer coefficient based on external surface and mean temperature difference between external and internal fluids (W/m^2 K), A_o/A_i is the ratio of external to internal tube surface area, h_i is internal (or water-side) convection coefficient (W/m^2 K), $R_{f,i}$ is a fouling resistance (m^2 K/W), t is the thickness of the tube wall (m), k is the thermal conductivity of the tube material (W/m K), A_o/A_m is the ratio of the external to mean heat transfer surface areas of the tube wall, h_o is tube external (or working-fluid-side) convection coefficient (W/m^2 K), and η_s is the surface fin efficiency (taken as 100% for bare tubes).

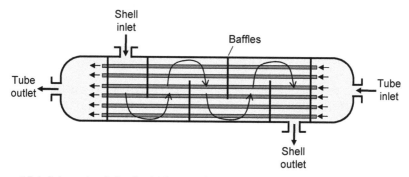

Figure 15.4 Schematic of simple single-pass shell-and-tube condenser.

For internal flows in tubes, the convection coefficient (h_i) is typically correlated to the dimensionless Nusselt number (Nu):

$$h_i = \frac{\mathrm{Nu}\,k}{D_i} \qquad\qquad [15.7]$$

where k is the thermal conductivity of the fluid (W/m K) and D_i is the internal diameter of the tube (m). There are numerous correlations for Nu as a function of the Reynolds number (Re) and Prandtl number (Pr) in the turbulent range. A good example is that of Gnielinski (1976) which is valid over a wide range of $0.5 < \mathrm{Pr} < 2000$ and $3000 < \mathrm{Re} < 5 \times 10^6$:

$$\mathrm{Nu} = \frac{(f/8)(\mathrm{Re} - 1000)\mathrm{Pr}}{1 + 12.7(f/8)^{1/2}(\mathrm{Pr}^{2/3} - 1)} \qquad\qquad [15.8]$$

where f is the Moody (or Darcy) friction factor for smooth pipes for a large range of Reynolds numbers ($3000 < \mathrm{Re} < 5 \times 10^6$):

$$f = (0.790 \ln(\mathrm{Re}) - 1.64)^{-2} \qquad\qquad [15.9]$$

Appropriate correlations for h_o are external convection correlations for horizontal tube geometry with external condensation. If the working fluid enters the condenser in the superheated state, correlations for flow across tube banks are appropriate for the desuperheating section of the condenser. Two correlations are reported by McAdams (1954) for laminar condensing conditions:

$$h_o = 0.951 \left(\frac{k_f^3 \rho_f^2 g}{\mu_f} \frac{L}{\dot{m}} \right)^{1/3} = 0.725 \left(\frac{k_f^3 \rho_f^2 g}{\mu_f} \frac{d_{fg}}{N D_o \Delta T} \right)^{1/4} \qquad [15.10]$$

where k_f is the thermal conductivity of the condensing liquid at the film temperature, ρ_f is the density of the condensing liquid at the film temperature, μ_f is the absolute viscosity of the condensing liquid at the film temperature, g is gravitational acceleration, L is the length of tubes, N is the number of tubes in the vertical direction, D_o is the outside diameter of a tube, $\dot{m}$ is the mass flow rate of the condensate, h_{fg} is the latent heat of the condensing fluid, and ΔT is the temperature difference between the saturation temperature of the condensing fluid and the tube surface.

Performance analysis of shell-and-tube condensers can be modeled using the effectiveness-NTU method. The number of heat transfer units (NTU) may be determined from:

$$\mathrm{NTU} = \frac{UA}{\dot{C}_w} \qquad\qquad [15.11]$$

where UA is the product of the overall heat transfer coefficient (U) and the heat transfer area (A), and $\dot{C}_w$ is the heat capacity rate of the cooling water. Since phase change is occurring in the condenser, the effectiveness (ε) is given by:

$$\varepsilon = 1 - e^{-NTU} \qquad [15.12]$$

The heat transfer rate in the condenser ($\dot{q}$) is then:

$$\dot{q} = \varepsilon \dot{C}_w \left(T_{h,i} - T_{c,i} \right) \qquad [15.13]$$

where $T_{h,i}$ is the hot inlet fluid temperature and $T_{c,i}$ is the temperature of the cold inlet fluid temperature. Eq. [15.13] may be used to a reasonable approximation to describe the average performance of the condenser if $T_{h,i}$ is taken as the condensing temperature of the working fluid.

The pressure drop of the water (or other fluid) is another important factor for designing or selecting water-cooled condensers. In situations where a cooling tower cools the condensing water, water pressure drop through the condenser is generally limited to about 70 kPa (ASHRAE, 2012). If condenser cooling water comes from another source, the pressure drop through the condenser should be lower than the available pressure to allow for pressure fluctuations and additional flow resistance caused by fouling.

The pressure drop through horizontal condensers includes losses through the tubes, tube entrance and exit losses, and losses through the heads or return bends (or both). The effect of tube coiling must be considered in shell-and-coil condensers. Expected pressure drop through tubes can be calculated from a modified Darcy−Weisbach equation (ASHRAE, 2012):

$$\Delta P = N_p \left(K_H + f \frac{L}{D} \right) \frac{\rho v^2}{2} \qquad [15.14]$$

where ΔP is the pressure drop (in Pa), N_p is the number of tube passes, K_H is the entrance and exit flow resistance and flow reversal coefficient, f is the friction factor, L is the tube length (m), D is the inside tube diameter (m), ρ is the fluid density, kg/m^3, and v is the fluid velocity, m/s.

15.3.3 Heat dissipation

15.3.3.1 Once-through cooling systems

These types of cooling systems take water from a plentiful resource, such as a lake or river, pass it through the condenser, and return it to the source. This type of cooling method is quite common in conventional power plants, but rare in geothermal power plants. This is for the obvious reason that geothermal resources are not always located near surface water bodies. Conventional power plants have historically been sited near surface water bodies, as surface water provided an economical means of cooling.

Recent regulations, however, have restricted the use of once-through cooling systems due to environmental impact of "thermal pollution" on surface waters. One example of use of once-through cooling in a geothermal power plant is at Chena Hot Springs, AK, where a creek body is used for cooling water.

15.3.3.2 The cooling tower

A cooling tower cools water by a combination of heat and mass transfer. Operating in a circulating loop, heat is rejected to the atmosphere, thus not requiring water at quantities of once-through systems. Water to be cooled is distributed in the tower by spray nozzles, splash bars, or film-type fill, which exposes a very large water surface area to atmospheric air. Atmospheric air is circulated by fans or by natural convective or wind currents. Thus, cooling towers may be classified into two broad groups: (1) mechanical draft or (2) natural draft. The air and water streams may be configured in either counterflow or cross-flow arrangements.

Temperature relationships between water and air as they pass through a counterflow cooling tower are shown in Fig. 15.5. The curves indicate the drop in water temperature (W_{in} to W_{out}) and the rise in the air wet-bulb temperature (A_{in} to A_{out}), as the streams move through their respective paths in the cooling tower. The temperature difference between the water entering and leaving the cooling tower ($W_{in} - W_{out}$) is the *range*. The cooling tower range is determined by the heat rejection load and water flow rate, not by the size or thermal capability of the cooling tower.

The difference between the leaving water temperature and entering air wet-bulb temperature ($W_{out} - A_{in}$) in Fig. 15.5 is the approach to the wet bulb or simply the *approach* of the cooling tower. The approach is a function of cooling tower capability; a larger cooling tower produces a closer approach (colder leaving water) for a given heat load, flow rate, and entering air condition. Cooling towers are capable of gener-ating approach temperatures as low as 3°C.

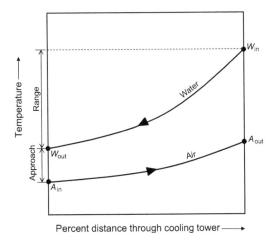

Percent distance through cooling tower ⟶

Figure 15.5 Temperature relationships in a counterflow wet cooling tower.

The term "cooling tower capacity" is ambiguous without a description of the conditions under which the heat rejection occurs. A cooling tower in a dry climate will be a different physical size than one in a humid climate with equal heat rejection rates. Thus, to define cooling tower capacity requires knowledge of the water flow rate through the cooling tower, the entering and leaving water temperature (ie, the cooling tower range), the cooling tower approach, and the ambient wet-bulb temperature. Thermal performance of a cooling tower depends principally on the entering air wet-bulb temperature.

Geothermal power plants commonly use mechanical-draft cooling towers in a counterflow arrangement as shown in Fig. 15.6. Fans may be on the inlet air side (forced-draft) or the exit air side (induced-draft). The type of fan selected, either centrifugal or axial, depends on external pressure needs, permissible sound levels, and energy usage requirements. Towers are typically classified as either factory assembled, where the entire tower or a few large components are factory assembled and shipped to the site for installation, or field erected, where the tower is constructed completely on site. Most factory-assembled towers are of metal construction, usually galvanized steel. Other constructions include stainless steel and fiberglass-reinforced plastic (FRP) towers and components.

Special-purpose towers containing a conventional mechanical draft unit in combination with an air-cooled (finned-tube) heat exchanger are known as *wet-dry towers*. They are used for either vapor plume reduction or water conservation. The hot, moist plumes discharged from cooling towers are especially dense in cooler weather. On some installations, limited abatement of these plumes is required to avoid restricted visibility on roadways, on bridges, and around buildings.

Mathematical modeling of cooling tower performance ranges from the very simple to the complex. Simple models use the cooling tower approach to determine the exiting cooling water temperature (ie, the tower return temperature):

$$T_{\text{return}} = T_{\text{wet bulb}} + T_{\text{approach}} \qquad [15.15]$$

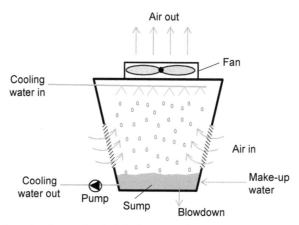

Figure 15.6 Schematic of a mechanical draft, counterflow wet cooling tower.

More complex methods rely on detailed heat and mass transfer theory, but ultimately require some empirical data from cooling tower performance measurements.

An intermediate method is based on the heat exchanger effectiveness concept, where the cooling tower effectiveness is defined as:

$$\varepsilon = \frac{\text{Range}}{\text{Range} + \text{Approach}}$$
[15.16]

Since the air enters the cooling tower at a known state, characterized by a temperature, $T_{a,i}$, and humidity ratio, $\omega_{a,i}$ and exits at $T_{a,o}$ and $\omega_{a,o}$, the limiting exit state for the air would be when the air leaves saturated at a temperature equal to that of the incoming water stream. This condition corresponds to the maximum possible enthalpy of the exit air. In terms of these enthalpies, the air-side heat transfer effectiveness is defined as the ratio of the air enthalpy difference to the maximum possible air enthalpy difference. For a known effectiveness, the heat rejection rate ($\dot{q}$) is then:

$$\dot{q} = \varepsilon \dot{m}_a \left(h_{a,w,i} - h_{a,i} \right)$$
[15.17]

where h is enthalpy, and subscripts a, w, and i refer to air, water, and inlet, respectively.

15.3.4 Operation and maintenance

Aside from the management of noncondensable gases, much of the operation and maintenance of water-cooled systems are related to the cooling tower. Maintenance of wet cooling towers can be significant and mainly consists of water treatment and water consumption management.

With water being exposed to air in the cooling tower, scale, corrosion, and microbiological growth must be managed. Historically, this maintenance item included water treatment with chemicals that are toxic and costly to handle and dispose. More recently, nonchemical methods have emerged for cooling water treatment. The Oregon Institute of Technology, for example, operates two binary geothermal power stations that successfully use a pulse system that employs short, electromagnetic pulses to prevent scale and microbiological growth.

Make-up water must also be provided to cooling towers due to evaporation losses, water losses due to drift, and blowdown. Cooling towers cool water by evaporating about 1% of the water passing through the tower. Drift refers to losses of water droplets that are ejected from the system with the exhaust air. Blowdown refers to removal of water from the cooling tower sump, as evaporation continuously concentrates the cooling water with dissolved minerals. Thus, blowdown is a function of evaporation losses and the so-called cycle of concentration (COC), which is defined as the ratio of dissolved solids in circulating water to the dissolved solids in the makeup water. Thus, blowdown = evaporation loss/(COC − 1).

Cooling towers encounter substantial changes in ambient wet-bulb temperature throughout the year. Accordingly, some form of capacity control is prudent to maintain prescribed condensing temperatures or process conditions. Fan cycling is the simplest method of capacity control on cooling towers and is often used on multiple-unit or multiple-cell installations. Water flow through the condenser can also be modulated. With advances in variable frequency drives (VFDs), VFDs are an economical choice for fan and pump speed control. VFDs save a significant amount energy, given the pump and fan affinity laws, where power is proportional to the cube of the motor speed. VFDs also provide flexible control. For example, the Oregon Institute of Technology does not use antifreeze chemicals in their cooling water. Rather, VFDs are used to control cooling tower fan and cooling water circulating pump speeds to prevent freezing of the cooling tower water.

Nominal cooling tower capacity is based on heat rejection from a chilled water system and is defined as cooling 0.054 L/s from 35°C to 29°C at 26°C wet-bulb temperature (ASHRAE, 2012). This translates to a peak cooling tower flow rate of 0.054 L/s per kW of heat rejection capacity. Power requirements of cooling tower fans are typically on the order of 0.005 kW per kW of heat rejection capacity.

15.4   Air-cooled condensers

15.4.1   General

In air-cooled condensers, condenser heat is rejected directly to the ambient air. Thus, the condensing temperature is a function of the ambient air dry-bulb temperature. The obvious advantage of air-cooled condensers relative to water-cooled condensers is that cooling water is not needed. Consequently, air-cooled condensers are the common choice for heat rejection in geothermal power plants where water is scarce, such as in the desert areas of the western United States. The main disadvantage of air-cooled condensers relative to water-cooled condensers is typically higher capital cost; water is a better heat transfer medium than air, and so larger heat rejection equipment is typically needed for air-cooled power plants. Parasitic losses due to fan power requirements are generally greater for air-cooled condensers than for water-cooled condensers. However, to compare total operating costs of water-cooled versus air-cooled power plants, cooling tower maintenance costs should be considered.

Since the condensing temperature is a function of the dry-bulb ambient air temperature, air-cooled plants perform poorly in hot summer weather. Wind velocity also plays a significant role. The distance between two parallel air-cooled condensers must be considered to avoid thermal influence, as hot discharge air from the windward air condenser may impact the intake air of the leeward air condenser.

15.4.2   Types of condensers

15.4.2.1   Direct-contact

An indirect cooling method using air is known as a *Heller system* or *Heller arrangement* that uses a fluid cooler or natural draft cooling tower. As shown schematically

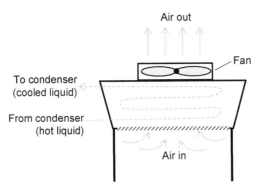

Figure 15.7 Schematic of an air-cooled heat exchanger.

in Fig. 15.7, the Heller system uses an indirect cooling configuration where the vapor is condensed in a direct-contact condenser (as described above in Section 15.3.2.1), and the condensate (working fluid) is pumped through an air-cooled heat exchanger to reject the heat. Thus, the working fluid forms a closed loop, and the liquid is pumped back and forth between the direct-contact condenser and the air-cooled heat exchanger.

The Heller system is typically associated with steam power plants, but, in theory, could be used with other working fluids. In high-temperature plants, the cooling coil can be configured in a spiral pattern and cooled with a natural draft cooling tower. Otherwise, fans may be positioned to either draw or blow the air though the condenser coil. Variants of this design have been proposed where the inlet air is pre-cooled by spray water or other means.

As described by Ashwood and Bharathan (2011), The Heller system has several inherent advantages and disadvantages compared with a conventional air-cooled condenser. Some advantages include: (1) a more simplified distribution arrangement in the air-cooled heat exchanger for the liquid, with potentially "true" countercurrent flow of air and liquid, (2) possibly better control with flexibility to vary the flow rate of the cooling water, (3) a more remote location of the air-cooled heat exchanger.

Determination of the overall heat transfer coefficient (U) can be accomplished by the same means as described above in Section 15.3.2.2. However, there is no condensation on the external surface of the tube; thus, the appropriate convection coefficient is that for external flow over tubes. Churchill and Bernstein (1977) have proposed a single comprehensive correlation for h_0 that covers the entire range of Reynolds and Prandtl numbers:

$$h_{\mathrm{o}} = \frac{\mathrm{Nu}k}{D} = \left(0.3 + \frac{0.62\mathrm{Re}^{1/2}\mathrm{Pr}^{1/3}}{\left[1 + (0.4/\mathrm{Pr})^{2/3}\right]^{1/4}} \left[1 + \left(\frac{\mathrm{Re}}{282000}\right)^{5/8} \right]^{4/5} \right) \frac{k}{D}$$

$$[15.18]$$

where Nu is the Nusselt number, k is the thermal conductivity of the air, D is the outside tube diameter, and all properties are evaluated at the film temperature.

With designs involving free convection over horizontal tubes in a dry cooling tower, Churchill and Chu (1975) have recommended a single correlation for a wide range of Rayleigh number (Ra) values:

$$h_o = \frac{\mathrm{Nu}k}{D} = \left\{ 0.60 + \frac{0.387\mathrm{Ra}^{1/6}}{\left[1 + (0.559/\mathrm{Pr})^{9/16}\right]^{8/27}} \right\}^2 \frac{k}{D} \qquad [15.19]$$

15.4.2.2 Integral-fin

Most air-cooled condensers consist of finned coils, which may be formed to fit various configurations. Coil orientation is generally horizontal, with propeller fans above the coil providing draw-through air circulation (Fig. 15.8). Other orientations include an "A-frame" pattern with the fan positioned underneath to blow air through the coil.

Integral-fin condensers can be made of either copper or aluminum and are made by extruding or forming fins directly from the tube material. Copper tubing can be formed using cold compression to form large fins on the outside of the tubing. Internal fins enhance heat transfer area and promote turbulence. With aluminum, the common method of forming condensers is to rake the surface to form protrusions. Because aluminum is a very soft material, these tubes with the "spiny" fins can be coiled to form complete condensers.

Determination of the overall heat transfer coefficient (U) can be accomplished by the same means as described in Section 15.3.2.2, with h_o determined using Eqs. (15.15)−(15.18). However, with air-cooled condensers, heat is transferred inside the tube by (1) desuperheating, (2) condensing, and (3) subcooling. Conditions within the tube depend strongly on the velocity of the vapor flowing through the tube. As condensation occurs, the mass fraction of vapor (x) decreases along the tube. Condensation begins at the inner tube wall, and if the vapor velocity is small, a pool of liquid forms on the lower part of the tube. The liquid pool is propelled down the length of the tube by shear forces imparted by the flowing vapor. Dobson and Chato (1998)

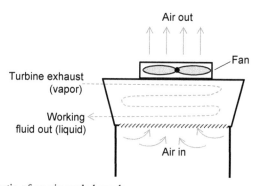

Figure 15.8 Schematic of an air-cooled condenser.

recommend an empirical correlation for a local heat transfer coefficient as a function of the mass fraction of vapor (x). After condensation, the internal convection coefficient (h_i) may be determined by an appropriate correlation for single-phase internal flow.

15.4.3 Operation and maintenance

Condenser coils can be cooled by natural convection or wind or by propeller, centrifugal, and vane-axial fans. Because efficiency increases significantly by increasing air speed across the coil, forced convection with fans predominates in geothermal power plants.

Aside from the management of noncondensable gases, much of the operation and maintenance of air-cooled systems are related to maintaining condensing temperature and pressure within desirable limits. As described earlier with cooling tower operation, fan cycling is the simplest method of capacity control for air. Working fluid pumps are now commonly equipped with variable-speed controls. Advances in variable frequency drives (VFDs) make them an economical choice for both fan and working fluid pump speed modulation.

Unless unusual operating or application conditions exist, fan/motor selections are made by balancing operating and first costs with size and sound requirements. Common air quantities are 80–160 L/s (4.2 kW at 17°C temperature difference) at 2–4 m/s (ASHRAE, 2012). Fan power requirements generally range from 20 to 40 W per kilowatt (thermal).

15.5   Evaporative (water- and air-cooled) condensers

15.5.1 General

As with separate water- and air-cooled condensers, evaporative condensers reject heat from a condensing vapor into the environment. In an evaporative condenser, hot, low-pressure vapor from the turbine exhaust circulates through a condensing coil that is continually wetted on the outside by a recirculating water system. As seen in Fig. 15.9, air is simultaneously blown across the condenser coil, causing a small portion of the recirculated water to evaporate. This evaporation removes heat from the coil, thereby cooling and condensing the vapor.

Evaporative condensers reduce water pumping and chemical treatment requirements associated with cooling towers. In comparison with an air-cooled condenser, an evaporative condenser requires less coil surface and airflow to reject the same heat or, alternatively, greater operating efficiencies can be achieved by operating at a lower condensing temperature.

An evaporative condenser can operate at a lower condensing temperature than an air-cooled condenser because the condensing temperature in an air-cooled condenser is determined by the ambient dry-bulb temperature, whereas in the evaporative condenser, heat rejection is limited by the ambient wet-bulb temperature, which is normally 8–14°C lower than the ambient dry bulb. Also, evaporative condensers

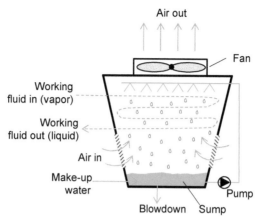

Figure 15.9 Schematic of an evaporative condenser.

typically provide lower condensing temperatures than the cooling tower/water-cooled condenser because the heat and mass transfers (between the working fluid and cooling water, and between the cooling water and the ambient air) are more efficiently combined in a single piece of equipment, allowing minimum sensible heating of the cooling water. Evaporative condensers are, therefore, the most compact for a given capacity. Further, evaporative condensers can be operated in a hybrid or wet/dry configuration; the unit can be operated dry when the ambient dry-bulb temperature is low.

Barriers to further use of evaporative condensers in geothermal power plants are related to operational and maintenance issues. Through field testing, Ormat found that wind speed and direction imposed significant impacts to water-assisted cooling efficiency and to air cooler performance in general (Kaplan et al., 2011). At the Mammoth facility (Mammoth, CA), multiple technologies were tested including misting jets, wetted media, and combinations of both. The use of misting jets caused concern of potential corrosion or deposits resulting from the evaporation of the treated effluent. Wetted media restricted air flow and resulted in significant maintenance to remove mineral deposits (Kaplan et al., 2011). However, subsequent testing at the Galena III Power Plant facility in Reno, NV, showed the potential of an evaporative cooling enhancement system to reduce the air temperature entering the geothermal condenser, thereby increasing power plant performance. In that concept, water is discharged into the atmosphere through spray nozzles that mist into the air entering the condenser tubes.

15.5.2 Heat transfer considerations

Heat transfer in evaporative condensers is similar to that of air-cooled condensers, except for the complexity of mass transfer due to evaporation on the exterior surface of the tube. Mathematical models commonly use a simplifying, single enthalpy driving gradient between air saturated at the temperature of the water-film surface and the enthalpy of air in contact with that surface. Exact formulation of the heat and mass transfer process requires considering the two processes simultaneously. Thus, because evaporative condenser performance cannot be represented solely by a temperature difference or an enthalpy difference, simplified predictive methods can only be used for interpolation of data between test points or between tests of different-sized units, provided that air velocity, water flow, working fluid (eg, a refrigerant) velocity, and tube bundle configuration are comparable. Further work on evaporative condenser performance modeling is under way in the industry (ASHRAE, 2012).

Predictive performance modeling of evaporative condensers is particularly important when designing for variable weather effects. Weather data should ideally be collected prior to determining plant location, and continued collection of weather data is important during operation to monitor changes in weather patterns. Computational fluid dynamics modeling should be calibrated with field testing.

15.5.3   Condenser configuration

The principal components of an evaporative condenser include the condensing coil, fan(s), water circulation pump, water distribution system, cold-water sump, drift eliminators, and water makeup assembly.

Evaporative condensers typically use condensing coils made from bare pipe or tubing without fins; the relatively high rate of energy transfer from the wetted external surface to the air eliminates the need for an extended surface. Bare coils are also less susceptible to fouling and are easier to clean. The high rate of energy transfer from the wetted external surface to the air makes finned coils uneconomical when used exclusively for wet operation. However, partially or wholly finned coils are sometimes used to reduce or eliminate plumes and/or to reduce water consumption by operating the condenser dry when the dry-bulb temperature is favorable. Coils are usually made from steel tubing, copper tubing, iron pipe, or stainless steel tubing.

The water circulation pump circulates water from the cold water sump to the distribution system located above the coil. Water descends through the air circulated by the fan(s), over the coil surface, and ultimately returns to the sump. The distribution system is designed to completely and continuously wet the full coil surface because complete wetting ensures a high rate of heat transfer due to evaporation. A continuously wetted surface also prevents excessive scaling, which is more likely to occur on intermittently or partially wetted surfaces. Scaling is obviously undesirable because it decreases heat transfer efficiency. Water lost through evaporation, drift, and blowdown from the cold water sump is replaced through a water makeup system.

Evaporative condensers are mechanical draft; that is, fans move a controlled flow of air through the unit. These fans may be on the air inlet side (forced-draft) or the air discharge side (induced draft). The type of fan selected, centrifugal or axial, depends on external pressure needs, energy requirements, and permissible sound levels.

15.5.4 Operation and maintenance

Operation and maintenance issues regarding evaporative condensers are similar to those of individual water- and air-cooled condensers. The main difference is that special consideration is needed to avoid scaling and corrosion of the wetted surface. Corrosion of metallic surfaces becomes enhanced on intermittently-wetted surfaces.

The Amedee and Wineagle geothermal power plants in northern California were installed in the 1980s and operated successfully for many decades with evaporative condensers. Some bare steel tubes eventually failed due to corrosion. Other simultaneous operating issues caused these plants to run sporadically since the early 2010s. Since they operate in a remote area, wind-blown dust accumulating on the wetted surfaces is also a maintenance item.

15.6 Concluding summary and future trends

This chapter has reviewed the theoretical and practical considerations of the methods of heat rejection to the environment from geothermal power plants. To obtain good thermal efficiencies of geothermal power generation, the turbine exhaust pressure must be as low as can be economically achieved with the available cooling water or air temperature. Heat rejection methods from geothermal power plants seek to minimize this turbine exhaust pressure by efficiently rejecting heat to the environment. The thermodynamic efficiency of geothermal power generation is strongly influenced by the quantity of heat that can be rejected and at what temperature.

Waste heat discharged from a geothermal power plant is ultimately absorbed by the Earth's atmosphere, regardless of the method used for heat dissipation. Condensers used in geothermal power plants are generally of the direct-contact or surface type. Methods of heat dissipation to the environment generally consist of wet cooling towers, air-cooled condensers, or a combination of wet-dry (or evaporative) condensers. The actual choice depends on designer preference and what the geothermal site characteristics dictate.

With increasing global concern for water shortages, future trends in rejecting heat from both binary and flash-steam geothermal power plants appear to be toward optimizing hybrid methods. In fully air-cooled condensers, the condensing temperature is a function of the ambient air dry-bulb temperature, and consequently, air-cooled plants perform poorly in hot summer weather. Thus, evaporative condensers with minimal water consumption appear to be the main development trend in the near term. However, scaling and corrosion potential of the condenser tubes remain barriers. According to Kaplan et al. (2011), results of the Galena 3 evaporative cooling enhancement tests, where water was misted into the air entering the condenser tubes, will be used in concept and system design of future geothermal power projects.

References

ASHRAE, 2012. HVAC Systems and Equipment Handbook. ASHRAE, Atlanta, GA.

Ashwood, A., Bharathan, D., 2011. Hybrid Cooling Systems for Low-Temperature Geothermal Power Production. Technical Report NREL/TP-5500-48765 March 2011, USDOE Contract DE-AC36-08GO28308.

Churchill, S.W., Bernstein, M., 1977. A correlating equation for forced convection from gases and liquids to a circular cylinder in crossflow. Journal of Heat Transfer, ASME 99, 300−306.

Churchill, S.W., Chu, H.H.S., 1975. Correlating equations for laminar and turbulent free convection from a horizontal cylinder. International Journal of Heat Mass Transfer 18, 1049−1053.

Dobson, M.K., Chato, J.C., 1998. Condensation in smooth horizontal tubes. Journal of Heat Transfer 120 (1), 193−213.

Gnielinski, V., 1976. New equations for heat and mass transfer in turbulent pipe and channel flow. International Chemical Engineering 16 (2), 359−368.

Kaplan, U., Reiss, Z., Sullivan, B., 2011. Evaporative cooling enhancement at the steamboat complex and condenser performance research and development efforts. Geothermal Resources Council Transactions 35, 1315−1317.

McAdams, W.H., 1954. Heat Transmission, third ed. McGraw-Hill, New York.

Robertson, R.R., 1978. Waste Heat Rejection from Geothermal Power Stations. Oak Ridge National Laboratory Report prepared for the U.S. Department of Energy under Contract W-7405-eng-26.

Silica scale control in geothermal plants—historical perspective and current technology[1]

16

P. von Hirtz
Thermochem, Inc., Santa Rosa, CA, United States

16.1 Introduction

Worldwide output from geothermal sources has increased over 40% in the past 10 years. The worldwide installed generation capacity as of January 2015 was 12.8 GW and has been growing at a rate of about 5% annually over the past few years. By 2020, the installed capacity is expected to be as high as 17.6 GW [1]. These geothermal power projects convert the energy contained in hot rock to electricity by using water to absorb heat from the rock and transport it to the earth's surface, where it is converted to electrical energy through turbine generators. It is estimated that more than 97% of current geothermal reservoir production is from magmatically heated reservoirs. Geothermal reservoirs may also develop outside regions of volcanic activity, where deeply penetrating faults allow groundwater to circulate to depths of several kilometers and become heated by the geothermal gradient [2].

More than 90% of exploited fields are "liquid-dominated" under preexploitation conditions with reservoir pressures increasing with depth in response to liquid-phase density. "Vapor-dominated" systems, such as The Geysers in California and Larderello in Italy, have vertical pressure gradients controlled by the density of steam. Even though these dry-steam systems are not the predominant type of geothermal resource, they are relatively large and currently generate about 25% of the worldwide geothermal power production. In vapor-dominated systems, steam is cleaned and then passed directly through condensing steam turbines. Typically, water from "liquid-dominated" reservoirs ($>220°C$) is partially flashed to steam in one or more stages. Heat is converted to mechanical energy by passing steam through steam turbines. About 15% of geothermal generation worldwide is generated using organic Rankine cycle (ORC) technology that involves a heat exchanger and a secondary working fluid to drive turbines. A small percentage of total generation is by combined-cycle technology, which includes a steam turbine and ORC cycle technology in one plant.

The flash, ORC, and combined-cycle plants processing fluids from liquid-dominated fields often have to manage scaling in the separated liquid. Technology

---

[1] Sections 16.3, 16.4, 16.6, 16.7 & 16.8 reproduced by permission of Taylor and Francis Group, LLC, a division of Informa plc. Copyright © 2010 from The Science and Technology of Industrial Water Treatment by Paul von Hirtz & Darrell Gallup/Zahid Amjad.

Geothermal Power Generation. http://dx.doi.org/10.1016/B978-0-08-100337-4.00016-4

development and execution have made it possible to exploit geothermal resources that might not have otherwise been accomplished. A major focus of production chemistry and engineering in the geothermal energy industry has been to control scale deposits from the geothermal fluids. The most ubiquitous and troublesome mineral scale encountered in geothermal power generation and injection facilities is amorphous silica or poorly crystalline silicates. This chapter discusses the occurrence and mitigation of silica scale in geothermal energy production.

16.2   Geochemistry of silica

The concentration of silica (SiO_2) in geothermal brine is usually controlled by the dissolution of quartz and occasionally chalcedony from the geological strata of the reservoirs:

$$SiO_{2(qtz)} + 2H_2O = H_4SiO_4^0{}_{(aq)} \qquad [16.1]$$

It should be noted that produced geothermal waters are termed "brine" by convention in geothermal literature, even though technically this is not the correct geochemical definition for the lower-salinity fluids. The predominant form of dissolved silica is monosilicic acid, $Si(OH)_4$. Silica solubility at reservoir temperatures roughly above about 185°C is usually controlled by the polymorph quartz; solubility at temperatures less than about 185°C may be controlled by the polymorph chalcedony, which is more soluble than quartz. For example, at 185°C the solubility of quartz is 213 ppm as SiO_2, while the solubility of chalcedony is 274 ppm as SiO_2. Therefore, chalcedony can substantially increase the amount of dissolved silica present in lower-temperature produced geothermal fluids above that expected from quartz equilibrium, with a corresponding increase in silica scaling potential.

The solubility of silica deposited within geothermal surface facilities is controlled by amorphous silica and/or metal silicates phases. Quartz or chalcedony is virtually never found in surface production facilities and rarely deposited in or near production wellbores. The deposition of silica scale is primarily controlled by its polymerization and precipitation as amorphous silica or silicates, which are more soluble than quartz and chalcedony. The solubilities of quartz and amorphous silica as a function of temperature in pure water are illustrated in Fig. 16.1. Equations to calculate the solubility of quartz and amorphous silica in pure water are available from various sources [3–5].

Quartz deposition is negligible under most geothermal production conditions due to its slow precipitation kinetics compared to amorphous silica or poorly-crystalline silicates. Precipitation of amorphous silica may occur immediately after cooling or be delayed, depending on the pH and supersaturation ratio.

16.3   Thermodynamics of silica solubility

Although quartz is thermodynamically more stable than amorphous silica, extreme conditions of temperature, pressure, and/or alkalinity are required for the growth of

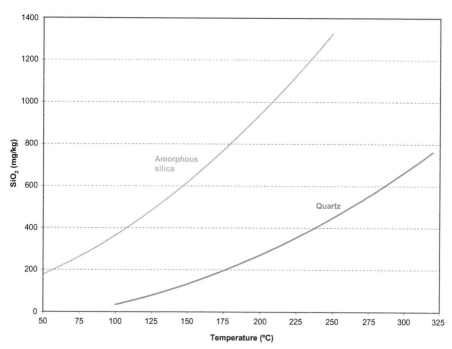

Figure 16.1 Solubility of quartz and amorphous silica as a function of temperature in pure water.

quartz at measurable rates in aqueous solutions. The greater solubility of amorphous silica relative to quartz is a distinct advantage for geothermal resource production facilities because it limits the precipitation of silica from produced waters. The solubility curves of amorphous silica and quartz both increase with temperature to ~300°C and then decrease due to the decreasing density and solvent power of water. The decrease becomes rapid as the critical point of water (374°C) is exceeded.

Dissolved salts and pH also affect the solubility of silica in aqueous solutions. Fournier and Marshall [6] have developed equations to calculate the solubility of amorphous silica at circum-neutral pH from 25 to 300°C using the concept of effective density of water and "salting out" effects of mixed electrolytes. Cations exhibiting elevated "hydration numbers," such as the alkaline-earths, depress the solubility of amorphous silica more than cations exhibiting low "hydration numbers" due to "free" water available for solvation. The effect of salinity and temperature on amorphous silica solubility is shown in Fig. 16.2.

The solubility of silica is substantially independent of pH until the pH increases into the alkaline range. Goto [7] examined the effect of pH on the solubility of amorphous silica from 0 to 200°C and from pH 5.5 to 10.0. As expected, the solubility of amorphous silica increased with increasing temperature, while solubilities remained relatively constant over the pH range of about 5.5–8.5. Above pH 8.5, the solubility increased significantly according to the reactions:

$$H_4SiO_4 = H_3SiO_4^- + H^+ \text{ or } H_4SiO_4 + OH^- = H_3SiO_4^- + H_2O \qquad [16.2]$$

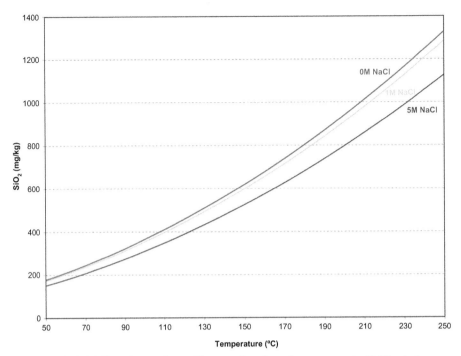

Figure 16.2 Solubility of amorphous silica as a function of temperature in NaCl solutions.

An expression for the relationship between pH and the solubility of monomeric silica is given by Eq. [16.3] [8]:

$$\log\left[\frac{w_s - w_m}{w_m}\right] = \text{pH} + \log K_1 \qquad [16.3]$$

where w_s = total solubility of monomeric silica and w_m = mass fraction of molecular silicic acid (undissociated).

The first ionization constant for molecular silicic acid, K_1, controls the extent of the reaction:

$$H_4SiO_4 = H_3SiO_4^- + H^+ \qquad [16.4]$$

where $K_1 = 10^{-9.7}$ at 25°C.

The ionization constant can be calculated at higher temperatures using Eq. [16.5]:

$$\ln K_1 = -16.76 - \frac{1661}{T} \qquad [16.5]$$

where T = temperature in K.

The effect of pH on the equilibrium solubility of monomeric silica does not generally have practical significance under flashed geothermal brine conditions where pH usually ranges from 5 to 8.5. But above pH 7, monosilicic acid does begin to dissociate according to Eq. [16.4]. This increases the thermodynamic solubility of silica because the silicate ion has a very high solubility. At a fixed concentration of total dissolved silica, the silica saturation ratio or silica saturation index (SSI, the ratio of silica concentration to solubility under the given conditions) decreases with increasing pH. At high pH, silica precipitation can be thermodynamically inhibited.

Nonalkali and alkaline-earth cations present in geothermal fluids may react directly with silicic acid to form metal silicates. These silicates are usually poorly crystalline; their X-ray diffraction patterns exhibit broad humps that are shifted from the normal opal-A hump centered near the diffraction angle of 23 degree 2θ, where 2θ is the angle between the X-ray incident beam and the detector [9−11]. X-ray absorption spectroscopic studies show that the silicates exhibit Si−O−M bonding and are not simply mixtures of silica and metal oxides/hydroxides.

Some examples of iron silicate scales and their formation reactions are

$$Fe(OH)_2 + Si(OH)_4 = FeSiO_3(s) + 3H_2O \qquad [16.6]$$

$$Fe(OH)_3 + Si(OH)_4 + Mg^{2+} = Fe(OH)_3 \cdot SiO_2 \cdot MgO(s) + H_2O + H_2(g) \quad [16.7]$$

$$Fe^{2+} + Si(OH)_4 + H_2O = Fe(OH)_3 \cdot SiO_2(s) + H_2(g) + 2H^+ \qquad [16.8]$$

Depending on the geothermal brine source, silica scaling is often exacerbated by the presence of calcium, magnesium, iron, aluminum, and manganese. Aluminum-rich and iron-rich amorphous silicates are the most common and exhibit significantly lower solubilities than pure amorphous silica. Aluminum can begin to affect amorphous silica solubility at low levels, on the order of only 1 ppm in the brine. Aluminum-rich amorphous silica scale deposits at temperatures that are 25°C or more below that of pure amorphous silica [12], so they can have a profound impact on power cycle and silica scale treatment process design. Increasing a first- or second-stage flash temperature by 25°C (a 4 bar increase from 150 to 175°C flash) could make a geothermal power station economically nonviable.

16.4 Silica precipitation kinetics

While chemical thermodynamics determine the absolute solubility of silica, various factors affect the kinetic rate at which silica polymerizes and precipitates as colloids or by molecular deposition. The kinetics of silica polymerization and molecular deposition are influenced by the degree of supersaturation, temperature, by catalysts (fluoride), and nucleation site availability.

The maximum rate of silica precipitation typically occurs at a temperature approximately 25−50°C below the silica saturation temperature as geothermal brine cools by flashing steam or natural heat loss. As a general rule, once the supersaturation ratio nears 2, silica polymerization and precipitation proceeds rapidly at approximately neutral to alkaline pH (pH 6.6—8.5). Weres et al. [4] postulated that the silica precipitation process consists of the following steps:

1. Formation of silica polymers of less than nucleus size.
2. Nucleation of an amorphous silica phase in the form of colloidal particles.
3. Growth of supercritical amorphous silica particles by further chemical deposition of silicic acid on their surfaces.
4. Coagulation or flocculation of colloidal particles to give either a precipitate or a semisolid material.
5. Cementation of the particles in the deposit by chemical bonding and further molecular deposition of silica.
6. Growth of a secondary phase in the interstices between the amorphous silica particles (occurs rarely).

A solid surface in contact with a supersaturated solution of amorphous silica may have a layer of amorphous silica on it, with continued deposition proceeding by Step 3. If colloidal amorphous silica particles form in the supersaturated solution, these may adhere to the surface (Steps 4 and 5), and Step 6 may follow. The formation of amorphous silica colloidal particles from a supersaturated solution is often referred to as homogeneous nucleation, which is the dominant process at the high initial supersaturation ratios required for rapid polymerization of amorphous silica. Heterogeneous nucleation applies to the deposition of amorphous silica on preexisting colloidal amorphous silica particles, but it is not actually a nucleation process. This relatively slow process resembles homogeneous nucleation and replaces it at low supersaturation. Nucleation by other scale particles can also occur to provide surfaces for amorphous silica deposition [13]. The formation of amorphous silica on Fe^{2+} is most likely, followed by silicates of aluminum, magnesium, and calcium.

The nucleation process frequently includes an induction (delay) time during which the concentration of molecular silicic acid remains constant. After some period of time, the concentration of molecular silicic acid begins to decrease, which indicates that polymerization is occurring. This induction time may be interpreted as the approximate time required for sub-critical clusters of amorphous silica to grow to critical nucleus size. The induction time is longer at lower saturation ratios because the critical nucleus size is larger at these ratios.

Another interpretation of induction time is simply the length of time required for enough particles to nucleate and grow to a point where the concentration of molecular silicic acid measurably decreases. Rapid attainment of steady-state nucleation is implicit in this interpretation. Therefore, an initially slower nucleation rate may be ignored for practical purposes. This interpretation applies to induction times observed for both homogeneous and heterogeneous nucleation. Furthermore, a threshold value for amorphous silica supersaturation may be necessary to achieve nucleation. Similar to dissolution of both quartz and amorphous silica, temperature, degree of supersaturation, pH, dissolved salt concentration, and fluoride ions can affect the rate of silica nucleation [14]. Both Iler [14] and Weres et al. [4], report that the rate of silica

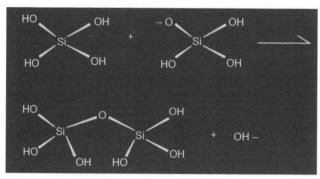

Figure 16.3 Polymerization process of monosilicic acid.

deposition is proportional to the sodium ion concentration; that is, dissolved NaCl or other electrolytes may promote faster solubility equilibrium. They also report a 10-fold increase in deposition rates on increasing the water pH from 5 to 6 and that fluoride catalyzes the silica polymerization reaction. Conversely, the kinetics of silica deposition slows dramatically as the pH is lowered into the acidic range.

Experimental studies have been conducted at the Lawrence Berkeley National Laboratory (LBNL) to determine heterogeneous and homogeneous amorphous silica nucleation rates under specific conditions. Using batch and continuous flow experimental results, Weres et al. [4] developed an empirical equation to calculate the rate of molecular deposition of amorphous silica over a broad range of temperature and molybdate active silica concentrations. A theoretical approach to the formation of colloidal silica particles by homogeneous nucleation was also developed at LBNL and was incorporated into the computer code SILNUC. The molecular deposition model developed at LBNL has been calibrated and modified by von Hirtz and Gallup [15] through numerous field trials measuring actual scale deposition rates in pilot test plants. The new computer modeling code called DEPOSITION also accounts for the effect of aluminum when the process involves precipitation of aluminum-rich silica scale.

The polymerization of monosilicic acid is depicted in Fig. 16.3. At high rates of deposition, the dissolved silica first polymerizes to colloidal silica particles that nucleate in the brine. The rate of homogeneous nucleation depends very strongly on the SSI. Roughly, polymerization is relatively slow when the SSI is below 2 and very fast when the SSI is above 3.

The condensation (polymerization) reaction of silicic acid and the reaction of silicic acid with silicate anions are, respectively:

$$2H_4SiO_4 = (OH)_3Si \cdot O \cdot Si(OH)_3 + H_2O \qquad [16.9]$$

$$H_4SiO_4 + H_3SiO_4^- = (OH)_3Si \cdot O \cdot Si(OH)_3 + OH^- \qquad [16.10]$$

When the SSI is too small for rapid homogeneous nucleation to occur, heterogeneous nucleation and molecular deposition on solid surfaces dominate. Vitreous silica scale formed by molecular deposition alone is sometimes observed at high temperatures. It should be noted that colloidal silica particles also grow by molecular deposition, so in either case molecular deposition is a significant factor in silica scaling processes.

Colloidal silica particles may be coagulated by cations in the brine; calcium and iron are particularly effective in this regard. The same electrostatic forces cause colloidal silica particles to adhere to solid surfaces. Corrosion accelerates the deposition of silica scale on steel because it releases iron ions.

Once attached to a solid surface, colloidal silica is converted to solid scale by molecular deposition of dissolved silica between the particles (earlier step 6). Solid deposits may form at a high rate (centimeters per year) where both colloidal silica and substantially supersaturated dissolved silica are present.

At low pH, molecular deposition and nucleation are catalyzed by traces of hydrogen fluoride. Above pH 4, the rate of molecular deposition is proportional to the negative surface charge on amorphous silica. The rates of nucleation and coagulation also increase with surface charge. Increasing salinity increases surface charge and accelerates all three processes.

16.5   Silica scaling experience in geothermal power production

The historical experience with geothermal fluid injection (disposal of produced brine, sometimes referred to as "reinjection") suggests that silica scaling in injection wells is the greatest concern, more so than thermal breakthrough to production wells. This concern has severely limited the efficiency of early geothermal flash plants due to conservative designs intended to maintain injection fluids at or below the SSI limit. Fig. 16.4 shows the first-stage flash pressure and temperature required to maintain injection fluids at an SSI of 1 (at silica saturation limit), as a function of the reservoir temperature that controls the amount of silica produced based on quartz solubility. At a reservoir temperature of 300°C, the flash temperature is limited to about 200°C in order to prevent silica supersaturation. Over the full range of reservoir temperatures shown, the flash temperature can only be roughly 108°C less than the reservoir temperature to keep the SSI at 1 or less. Many flash plants have been built worldwide that are underusing the resource by maintaining high flash pressures and injecting hot brine that could generate more power if silica scaling were controlled.

The potential injection well impacts are a legitimate concern. Injecting brine that contains colloidal silica or supersaturated dissolved silica risks formation damage; coagulated colloidal silica will accumulate in pores and fractures in the formation. If the brine is hot and contains enough dissolved silica for rapid molecular deposition to occur (>1 mm/y), molecular deposition alone may seal pores and fine fractures in the receiving formation, thereby reducing injectivity. The loss of sufficient injection well capacity may require curtailment of power generation. For most projects, the brine from a geothermal power plant must be injected.

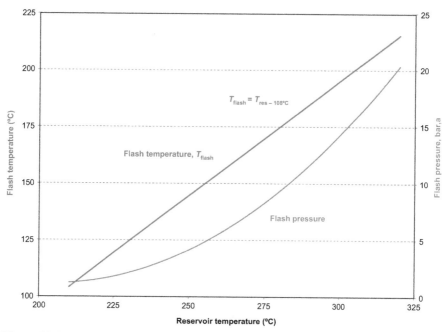

Figure 16.4 First-stage flash pressure and temperature required to maintain an SSI of 1 as a function of the reservoir temperature.

At high levels of supersaturation (>2) and near-neutral pH, massive silica scale deposition may occur in the low-pressure (LP) separators of dual-flash plants and downstream in the injection lines and wells (Fig. 16.5). Typically, amorphous silica scaling does not occur in production well bores. However, silicate scaling has been common in production wellbores tapping high-temperature, high-salinity fluids rich in heavy metals. Current and past examples are the Salton Sea (California), Puna (Hawaii), and Milos (Greece).

Figure 16.5 Massive silica precipitation in a pipeline due to polymerization after a second flash.

The metal silicates behave similar to monomeric deposition forming quickly as hard, glassy solids. Their deposition rate and adherence depend on the metal surface material. For example, it was found that a manganese silicate scale adhered tightly to a variety of steels and corrosion-resistant alloys, an iron silicate scale adhered tightly to carbon steel, but not corrosion-resistant alloys, and a rather pure amorphous silica scale adhered less tightly to carbon steel and not at all to alloys [16].

In the case of the metal silicates, a number of reactions may be responsible for deposition, for example:

$$Fe(OH)_2 + Si(OH)_4 = FeSiO_{3(s)} + 3H_2O \qquad\qquad [16.11]$$

$$Fe(OH)_3 + Si(OH)_4 + Mg^{2+} = Fe(OH)_3 \cdot SiO_2 \cdot MgO_{(s)} + H_2O + H_{2(g)} \quad [16.12]$$

$$Fe^{2+} + Si(OH)_4 + H_2O = Fe(OH)_3 \cdot SiO_{2(s)} + H_{2(g)} + 2H^+ \qquad [16.13]$$

$$2Al(OH)_3{}^0 + Si(OH)_4{}^0 = Al_2O_3 \cdot SiO_{2(s)} + 5H_2O \qquad\qquad [16.14]$$

$$2Al(OH)_4{}^- + Si(OH)_4{}^0 = Al_2O_3 \cdot SiO_{2(s)} + 5H_2O + 2OH^- \qquad [16.15]$$

ORC and combined-cycle geothermal power plants that subcool the brine below silica and silicate saturation temperatures may experience both amorphous silica and silicate scaling in heat exchangers, dramatically lowering the heat transfer, increasing the pressure drop, and reducing power generation.

16.6 Historical techniques for silica/silicate scale inhibition

Silica deposition occurs to some extent in all high-enthalpy, liquid-dominated geothermal operations. Silica scale inhibition/control methods used in geothermal fields have generally been specific to brine chemistry, process conditions, and brine-handling/disposal regimens. The most common methods used for mitigating silica scale deposits in geothermal brine-handling systems include but are not limited to (1) hot brine reinjection at or below amorphous silica saturation conditions, (2) adjustment of brine pH [acid to retard silica polymerization kinetics or base to convert $Si(OH)_4$ to $H_3SiO_4^-$], (3) aging, evaporation, or pond retention to precipitate silica, (4) crystallization—clarification to purposely precipitate silica/silicates, (5) removal of silica from brine by controlled precipitation with metal cations, (6) dilution with fresh water or condensate, (7) reducing agent treatment to control ferric silicate scales, and (8) treatment with "organic" scale inhibitors/dispersants, such as phosphinocarboxylates.

A few geothermal fields dispose of the spent brine at the surface. Some fields located near coastlines or rivers have discharged spent fluids into the ocean or other waterways [17]. This practice has obvious potential environmental impacts due to

Table 16.1 Amorphous silica saturation after cold condensate dilution of hot brine [34]

Condensate fraction[a]	Mixture temperature (°C)	SiO$_2$ concentration (ppm$_w$)	Mixture SiO$_2$ saturation	Salinity (ppm$_w$)	Final pH
0.0	207	1130	1.13	19,000	5.14
0.1	193	1017	1.13	17,100	4.66
0.2	179	904	1.13	15,200	4.47
0.3	164	791	1.12	13,300	4.35
0.4	150	678	1.09	11,400	4.26
0.5	135	565	1.05	9500	4.18

[a]Condensate temperature = 60°C.

the presence of salinity, heavy metals and other toxic species such as boron and ammonia. At least partly because of environmental restrictions, injection of brine, cooling tower water, and steam condensate is now the industry standard. The principal advantage of injection is that the net withdrawal of mass from a geothermal system is greatly reduced. Reservoir pressure is supported so that production well outputs can be maintained for a longer time [18]. The principal disadvantages of injection are scaling and thermal breakthrough where cool brine may flow directly to production wells before it has been in contact with hot rock long enough to reheat, causing a reduction in output from the production wells. This is a common challenge for geothermal energy production. The fractured nature of the rocks in geothermal systems often allows an unpredictable path from injector to producer. This problem is mitigated by increasing the distance between injection and production wells and managed by careful monitoring and interwell chemical tracer testing.

Some geothermal fields use dilution of the silica-supersaturated geothermal brine with fresh water to reduce the silica concentration to below the saturation point. Hot steam condensate or fresh heated water is typically used in these applications. Cold dilution water does not have a significant net effect on silica saturation due to the opposing effects of cooling and dilution. Table 16.1 lists the silica saturation as a function of cold condensate (60°C) dilution of hot brine (200°C) for a combined-cycle geothermal project. A 50% dilution of the brine with condensate results in only a minor decrease in silica supersaturation but causes a drop in pH of about 1 unit. In this case, the reduction in salinity and pH provides a substantial reduction in the silica deposition rate for this fluid.

In some cases, the pH of the flashed brine is high enough (>9) that the solubility of silica is enhanced to the point where amorphous silica deposition is suppressed. The silica becomes thermodynamically inhibited due to the dissociation of silicic acid to soluble silicate ions (Eq. [16.4]). This is more common for very dilute geothermal fluids that are high in alkalinity (Dixie Valley and Beowawe, Nevada, USA, are examples) However, often such high-pH brines deposit aluminum, calcium, and magnesium silicates.

Holding spent brine at near-neutral pH in ponds with net residence time of hours or days can convert dissolved monomeric silica to colloidal silica, reducing supersaturation. This prevents the cementation of weakly adherent deposits and their conversion to solid scale. Deposits from the aged brine tend to be soft and easy to remove. Candelaria [19] postulated that neither adherent gel nor solid scale will form if the silica is in colloidal form. The hypothesis relies on the requirement of monomeric silica to cement silica colloids together. Application of this procedure at the Botong field (Philippines) showed that the colloidal silica (30-nm size range) formed a gelatinous, fluffy precipitate that did not settle in the ponds and was transported into the injection wells. After 3 years, the precipitate had not caused a serious decline in injection capacity. Clearing the settling ponds of deposited silica and allowing passage of the solids to wells resulted in a substantial decline in injection capacity within a week, likely caused by plugging of the reservoir formation. This result demonstrates that the effects of deposition and plugging of colloidal suspensions are related to particle size.

Maintaining a low geothermal brine pH (<6) retards the nucleation, growth, and coagulation of colloidal silica particles and the scaling rate from supersaturated solutions by molecular deposition. The pH of produced brine can be maintained close to the original reservoir pH by simply not flashing the brine, as in an ORC cycle with downhole pumped brine production. Flashing of geothermal brine always results in an increase in pH due to outgasing of CO_2 and disproportionation of bicarbonate. The resulting increase in carbonate consumes hydrogen ions:

$$2HCO_3^-{}_{(aq)} \rightarrow CO_3^-{}_{(aq)} + CO_{2(g)} + H_2O \qquad [16.16]$$

$$H^+ + CO_3^-{}_{(aq)} \rightarrow HCO_3^-{}_{(aq)}$$

The pH of geothermal brine in the reservoir is largely controlled by carbonic acid/bicarbonate equilibria. Carbonic acid is a weak acid at reservoir temperatures ($>200°C$) but becomes a stronger acid at temperatures below $100°C$.

$$H_2CO_3 = H^+ + CO_3^-{}_{(aq)} \qquad [16.17]$$

Therefore, cooling brine without flashing results in a lower pH at the surface than in the reservoir. An ORC geothermal power plant at Blue Mountain (Nevada, USA) is operating without significant silica scale deposition and without treatment even though the SSI is 1.75 at an outlet temperature of about $70°C$. Because of relatively high dissolved CO_2 in the brine and the low heat exchanger outlet temperature, the pH of the brine drops from 6.25 to 5.72 units naturally through this cooling process as a result of the increased acidity of carbonic acid.

When acid is added to brine for silica scale control, the process is termed pH modification (or pH-Mod). This technique has been used at several fields to slow the polymerization and molecular deposition kinetics of silicic acid. The acids that may be used in this application include HCl, H_2SO_4, H_2SO_3, HF, organic acids, and acid precursors such as urea acid adducts or chlorinated hydrocarbons. The selection of the acid must be compatible with the brine to avoid forming byproduct scales [20].

Gallup studied the silica inhibition properties of H_2SO_3 and sulfite salts up to $200°C$. H_2SO_3 inhibits silica scaling by slowing silica polymerization and forming soluble sulfite−silicate complexes, which results in increased amorphous silica solubility. In a comparative test with HCl and H_2SO_3, sulfurous acid (H_2SO_3) maintained the highest silica concentration in solution. The use of H_2SO_4 in brine acidification is undesirable in some cases due to secondary precipitation of alkaline-earth sulfates. The hypersaline brines produced at the Salton Sea require HCl for pH-Mod for this reason. The solubility of alkaline-earth sulfites increases with decreasing pH and the primary product of calcium−bisulfite interactions below pH 5.5 is soluble calcium bisulfite. Consequently, H_2SO_3 can be used to treat most geothermal brines without the formation of byproduct scales. Even though H_2SO_3 is more expensive than HCl or H_2SO_4, the costs can be mitigated by manufacturing H_2SO_3 onsite by incinerating H_2S or elemental sulfur from H_2S abatement systems. Enough H_2S is typically available at most facilities to meet all acid requirements [21].

Hirowatari and Yamauchi [22] reported on the injection of exhausted gases into the brine to decrease the brine pH. The exhausted gases were 70% CO_2 and 2% H_2S. Brine pH decreased with increasing liquid-to-gas ratios as expected, and the gas injection lowered the pH from 7.0 to 5.2, resulting in a 97% reduction in scale deposition.

Gas injection was practiced at the Coso geothermal field in California (USA) for both H_2S emissions and silica scale control for over 12 years and is still used at the Puna field in Hawaii (USA). At the start of the 270-MW Coso project, 100% of the noncondensable gas (NCG) from surface condensers in dual-flash plants was injected at rates of 35,000−45,000 kg/h field-wide. The brine pH was lowered from about pH 8 to 5, effectively controlling silica scaling. Ultimately complications with gas injection such as corrosion due to residual oxygen and gas breakthrough to some production wells led to the installation of conventional H_2S abatement units and pH-Mod systems that injected sulfuric acid to the brine. Condensing turbine plants, even when equipped with surface condensers, have significant levels of oxygen in the off-gas due to inleakage, resulting in severe corrosion if injected with brine and condensate.

The geothermal energy industry has little direct experience using NCG injection for purposes of pH-Mod in silica scale mitigation. Gas injection has historically been performed for H_2S and CO_2 emissions control. NCG injection for pH control alone typically requires much lower injection rates than for emissions control. Given the low gas injection rates usually needed for pH-Mod, gas breakthrough to production wells and "vapor lock" at the injection wells may not be issues.

Gas injection has been used at the Puna plant in Hawaii since start-up, for both emissions control and scale control. The relatively low gas levels at Puna have allowed 100% gas injection without any gas breakthrough to production wells for the life of the project (>20 years). Gas injection at Puna results in an injectate (combined brine/condensate) pH of about 4.5 units and has been very effective in mitigating silica scale at an SSI of over 2.0. The main reason gas injection has not been widely implemented by the industry for pH-control and silica scale mitigation is that only ORC combined-cycle plants with back-pressure vaporizers for steam condensation can provide a suitable source of oxygen-free NCG for injection.

The Salton Sea, CA (USA) geothermal field has iron-rich silica scaling at rates from 0.5 cm/y in production wells to over 50 cm/y in injection wells. Even though the brines contain mostly ferrous (Fe^{2+}) iron, the iron in the scales is primarily ferric (Fe^{3+}). A wide variety of commercial silica inhibitors have proved to be ineffective. Gallup [23] investigated the use of iron-reducing agents to convert the ferric ions to ferrous ions, which are more soluble. Injection of sodium formate at a 2.8 stoichiometric excess achieved up to 50% iron-rich silica scale inhibition. The sodium formate converted 99% of the ferric iron into ferrous iron. Gallup also compared the efficiency of HCl−reducing agent mixtures to HCl alone. In most cases, the acid alone provided adequate scale control, but one benefit of the HCl−reducing agent mixtures was a significant reduction in corrosion [24].

Silica polymer and silicate sludge separated from flashed brine may be recirculated to the primary separators. The particles in the sludge provide a very large surface area for dissolved silica to deposit on, and this prevents large supersaturation and scale deposition. The reaction between freshly flashed brine and silica sludge is completed in a "flash crystallizer" immediately downstream of the separators. Recirculation of sludge to the reaction zone enables silica supersaturation to be quickly reduced. At the Salton Sea, production engineering and chemistry efforts led to the full-scale commercial development of crystallizer−reactor−clarifier (CRC) technology, in which iron silicates are purposely precipitated in surface equipment as sludge to prevent fouling of pipelines and injection wells [25,26].

Threshold-type scale inhibitors have not been effective in preventing amorphous silica scale because these inhibitors are designed to disrupt crystal formation and growth (ie, calcite scale) and have no efficacy on the formation of amorphous solids. Threshold scale inhibitors may be effective on crystalline silicate scales, and there are many commercial inhibitors for this type of scale in cooling water treatment. Amorphous silica scale and silicate scales often get confused in discussions of scale inhibitor treatment for geothermal brine by chemical treatment vendors.

Organic additives were investigated by Gallup [27] and Gallup and Barcelon [28]. They tested over 50 organic additives, whose ingredients included phosphonic acids, phosphonates, phosphinocarboxylic acids, acrylate polymers, polyacrylamides, oxyethylenes, polymaleates, sulfonates, carboxylic acids, polyethyleneimines, caustic soda, and quaternary ammonium compounds. They concluded that organic additives will likely continue to see limited use in the geothermal industry. Organic additives may make silica deposits softer and easier to remove but do not prevent polymerization or increase solubility. This means that the silica exists in solution as suspended solids, which have the potential to cause damage in injection wells. It was found that overdosing of the additives in many cases led to flocculation rather than dispersion. Organic scale inhibitors may just move the scaling problem downhole where it can cause greater damage to an overall geothermal power project. Injection well clean-out and redrilling is much more expensive than cleaning surface equipment and pipelines.

16.7 Current scale control techniques at high supersaturation

A major advancement in the exploitation of liquid-dominated geothermal fields has been the application of "bottoming cycle" heat recovery systems to maximize the heat extraction and power generation from a given resource. An ORC or a second-stage flash added to a single-flash plant is considered a bottoming cycle. In the case of an ORC plant, the heat from single-phase brine is transferred to an organic working fluid. This secondary working fluid is vaporized, passed through a turbine to generate electricity, condensed, and recycled. Many ORC plants are used as the sole source of electricity production (ie, stand-alone plants), in which case the hot brine, using submersible production well pumps, is maintained under pressure through the entire process from production to injection. This process has the advantage of not concentrating silica due to flashing and maintaining the brine pH low by not releasing dissolved CO_2.

In contrast to most stand-alone ORC plants, ORC bottoming plants can result in very high levels of amorphous silica supersaturation, where the SSI may be 2 or greater. Successful processing of high silica fluids through ORC heat exchangers can be attained with pH-Mod and the rapid reduction of temperature through the heat exchangers, which reduces the kinetics of silica deposition. Thermodynamic stability prevents any possibility of scale deposition, while kinetic stability results in an extended period of time before deposition will occur. It is the kinetic stability that allows the successful operation of an ORC bottoming plant by taking advantage of the kinetic effects of both thermal quenching (inherent with an ORC) and low pH (normally augmented by acid injection).

As discussed previously, the nonrelease of the acid-gases (CO_2 and H_2S) in the ORC process maintains the brine pH at a lower level. Upon cooling in an ORC heat exchanger, the brine pH will drop further due to the increased acidity of carbonic acid at the lower temperature. This often provides an advantage in silica scale control for ORC plants since the pH is lower than in a dual-flash process and silica scale is kinetically inhibited to some extent. A comparison of these two processes is depicted in Fig. 16.6.

Whether a dual- or triple-flash system or an ORC system is used as a bottoming cycle, amorphous silica usually becomes highly supersaturated at the outlet temperature of the process. Generally only single-flash plants are operated at or below the silica saturation limit. Historically single-flash geothermal plants were built for this reason, to prevent the possibility of silica scale, but at the expense of maximum resource utilization. Fig. 16.7 again shows the solubility of quartz and amorphous silica as a function of temperature. Also plotted (heavy red line) is the concentration path for brine flashed from single-phase reservoir conditions in equilibrium with quartz at 300°C to a typical single-flash power plant condition of 160°C, then to a second-stage flash to 110°C. The dashed blue line shows the path from first flash to subcooling in an ORC plant as an alternative bottoming process to the second flash. The ORC plant does not further concentrate the silica but usually cools the brine to a lower temperature than the second flash. Note that the concentration of silica in the brine begins to exceed the amorphous silica solubility at less than 200°C.

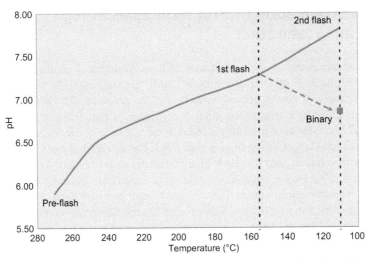

Figure 16.6 Example of brine pH changes as a function of flash and subcooling (binary) [34].

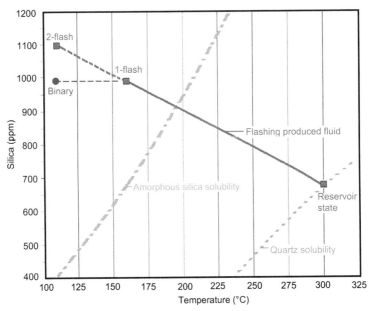

Figure 16.7 Solubility of quartz and amorphous silica as a function of temperature and flash vs. subcooling (binary) processes.

In the multiple-flash processes, steam is generated in stages while brine concentrates and the fluid temperature drops in each stage. As described earlier, the ORC plant does not concentrate the brine but usually extracts more heat resulting in a lower outlet temperature. If amorphous silica or metal silicate saturation is exceeded,

Figure 16.8 Nesjavellir geothermal plant heat exchanger after 1 year at SSI of 4 (55°C) [34].

scaling is thermodynamically possible. Depending on factors such as the fluid temperature, pH, salinity, and concentration of certain cations (in the case of silicates), scale may deposit immediately or be kinetically delayed for a significant amount of time.

As an example of thermal-quenching, the Nesjavellir geothermal plant in Iceland cools geothermal brine in a heat exchanger bottoming cycle to 55°C in order to heat potable water for the city of Reykjavik. In the process, silica in the geothermal fluid becomes supersaturated to an SSI of 4. Due primarily to the rapid cooling and the low outlet temperature of the geothermal brine, scaling is minimal and the heat exchanger tubes are only cleaned once per year (Fig. 16.8). Since the brine is not pH modified, silica polymerization does occur upon aging. The brine is held in ponds and diluted with steam condensate prior to injection [29].

A combined-cycle Puna plant in Hawaii relies on both thermal quenching and pH-Mod to control silica at extremely high levels of supersaturation. The plant flashes 300°C reservoir brine to 200°C where it is slightly undersaturated in silica, then passes that brine through an ORC bottoming cycle where it becomes highly supersaturated (SSI ∼ 3.5) at about 80°C. The natural brine pH is about 6.0 and is reduced to 5.5 after condensate addition. The pH is further reduced to 4.5 with acid injection to control scale through the ORC heat exchangers, which are operating at very high SSI. Fig. 16.9 shows the SSI and predicted scaling rate from DEPOSITION model [15] for the fluids in this plant as a function of temperature. The molecular deposition rate of silica is reduced 80% from a temperature of 140 to 80°C, simply by cooling the brine, due to thermal quenching and kinetic inhibition. The thermal quenching process is only possible with the pH carefully controlled within the target range to prevent any polymerization. This is currently the most extreme example of high-SSI brine being successfully processed with maximum resource utilization.

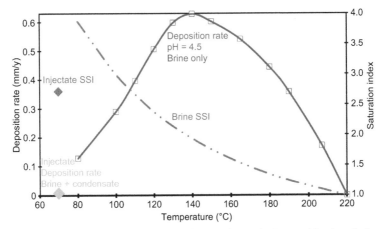

Figure 16.9 The SSI and predicted silica scaling rate in mm/y for combined-cycle bottoming plant with pH-Mod in Hawaii.

16.8 Case study for scale control in a combined-cycle plant design

A combined-cycle single-flash steam turbine with ORC bottoming cycle process was proposed for a geothermal field in Reykjanes, Iceland [30]. The process flow diagram, given as Fig. 16.10, provides a single-flash and bottoming cycle design that will maintain the SSI at or below 1.0 throughout the heat exchangers and injection system. The separated brine would have an SSI of 0.80, the second preheater outlet brine would have an SSI of 1.03, and the brine and condensate injection mixture would have an SSI of 0.96. These saturation indexes include the effects of pH on silicic acid dissociation and salinity on the overall solubility of amorphous silica. The produced brine pH is neutral to acidic and the salinity is moderately high, approximately that of seawater. The wells produce from a single-phase brine reservoir in the range of 290–320°C.

This initial design is conservative in that it will thermodynamically prevent amorphous silica deposition in the power cycle and injection system by maintaining temperatures at or above the silica saturation limits. The preheater outlet brine will be slightly supersaturated (by 3%) but will be immediately diluted with condensate to below saturation (Option 1 in the proposed design).

To recommend the optimum process parameters for the Reykjanes combined-cycle plant, a proposed Option 2 design was developed that would kinetically inhibit silica scale deposition through maximum heat recovery from the brine and acidification with condensate and noncondensable gas. Chemical modeling was performed on the fluids to determine the pH and silica saturation under various process conditions and to determine the relative molecular deposition rates of silica versus temperature and pH.

Fig. 16.11 shows the process flow diagram for the Option 2 configuration. At the flash condition of 220°C (same as for Option 1), silica is undersaturated by 20% and cannot deposit. The brine outlet temperature at the vaporizer is reduced to

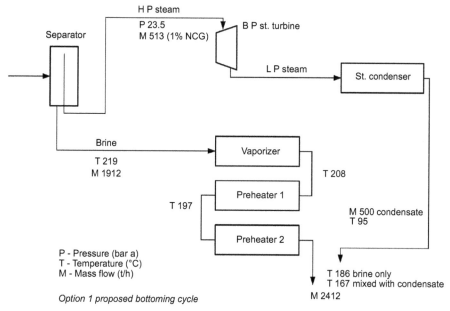

Figure 16.10 Process flow diagram for a conservative combined-cycle design (Option 1); ORC turbine not shown [34].

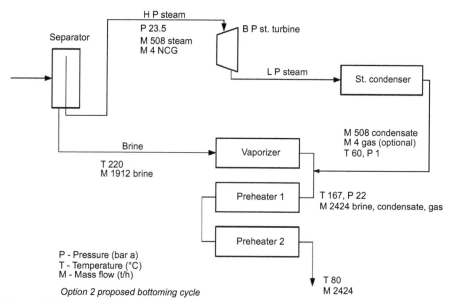

Figure 16.11 Process-flow diagram for an optimized combined-cycle design (Option 2); ORC turbine not shown [34].

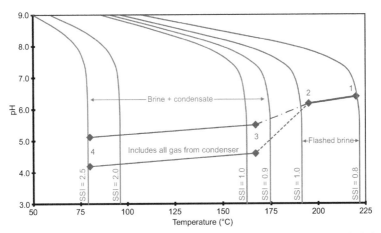

Figure 16.12 Thermal and pH stability map for amorphous silica in the Option 2 design for Reykjanes geothermal plant in Iceland, modified from [34].

195°C, where the silica is undersaturated by 3% and still will not deposit at this temperature. After the brine exits the vaporizer, it is mixed with steam condensate and gas from the condenser. This immediately reduces the temperature to 167°C and the pH to 4.6 units. The silica remains undersaturated (SSI = 0.96) at this point due to dilution. The addition of condensate also reduces the salinity, which helps to increase the solubility of silica and reduce the kinetics of deposition later in the process. The main benefit of condensate and gas addition is the large reduction in pH. Successful processing of this fluid through the preheaters to low temperatures requires that the pH is reduced to below 5.0 units. Addition of the condensate alone without gas will result in a pH of 5.5 units.

In the Option 2 design, the brine—condensate mixture temperature at the outlet of the second preheater would be 80°C or less, to maximize heat recovery and minimize scale. At an outlet temperature of 80°C, the silica saturation will be 2.5. Due to the thermal quenching effect on silica scale kinetics, an even lower temperature would be desirable. At 50°C, the saturation index would be 3.8, but the deposition rate is expected to be less than at 80°C. With acidification to pH 4.5 or less, and cooling of the mixture to 80°C or less, significant silica scaling problems are not predicted to occur in the heat exchangers or injection pipelines over time intervals on the order of 1−2 h.

Fig. 16.12 is a thermal and pH stability map for amorphous silica under the conditions proposed for the Option 2 design. Point 1 represents the separated brine and point 2 the brine after cooling through the vaporizer. Both points are below saturation. The silica is *thermodynamically* stable above 190°C, and at lower temperatures where the pH is high (pH > 8), as indicated by the SSI curves that trend toward lower temperatures at higher pH. Point 3 represents the brine/condensate mixtures at the inlet to the first preheater, which is undersaturated at the lower temperature (167°C) due to dilution. Point 4 represents the mixtures exiting the last preheater, where the saturation is about 2.5. The brine is expected to be *kinetically* stable at temperatures of about 80°C

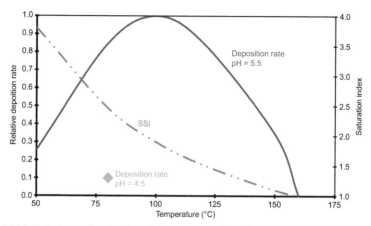

Figure 16.13 Relative molecular deposition rates predicted for Option 2 design for Reykjanes geothermal plant in Iceland, modified from [34].

or less, when the pH is below about 5 units. *Thermodynamic* stability can be predicted precisely and will prevent any possibility of scale deposition. The limits of *kinetic* stability cannot be predicted precisely, but operating in this region does result in an extended period of time before significant deposition will occur. In this design, significant silica scale problems can be prevented in a bottoming cycle by taking advantage of the kinetic effects of thermal quenching and low pH.

Fig. 16.13 is a plot of relative molecular deposition rates calculated for the Reykjanes brine and condensate mixture as a function of temperature, based on the DEPOSITION model [15]. Polymerization of silica can be neglected at these lower pH conditions. The maximum deposition rate occurs at a temperature of 100°C, and drops by 75% from that maximum at 50°C, even though the saturation is over 3.5 at this lower temperature. The rate is dependent on both saturation and temperature, but temperature overwhelms the saturation effect below 100°C. The relative effect of pH is also noted on the figure, where the brine, condensate and gas mixture is plotted at 80°C and pH 4.5, resulting in a deposition rate almost 10 times lower than the brine and condensate mixture at the same temperature.

16.9 Pilot-plant testing for bottoming cycle optimization

Every geothermal combined-cycle or bottoming cycle power plant installed or under construction worldwide, designed for high-SSI with pH-Mod, first had a pilot-plant test conducted to prove that the process would not result in excessive scaling of the surface facilities and injection wells. These sites include the Salton Sea, Coso, Blundell, and Puna (USA); Mak-Ban (Philippines); Kawerau and Nga Awa Purua (New Zealand); and Sarulla (Indonesia).

The most important aspect of these tests is to determine relative scaling rates within the injection well formation. The fouling of injection wells typically has a much greater financial impact on a geothermal project than the scaling of surface equipment. The pH-Mod process is designed to slow the kinetics of silica polymerization and scaling for a time sufficient to allow the brine to migrate away from the wellbore before significant deposition occurs. This process does not stop silica scaling; it simply retards the process long enough to limit precipitation of silica until the brine has penetrated deep enough into the formation where it does not affect injectivity, and/or until the brine is reheated to temperatures above the mineral saturation limits.

A typical pilot test to simulate a combined-cycle plant involves an online pilot testing unit to examine the relative silica scaling impacts within separators, heat exchangers, pipelines, and the injection wells [31]. The test unit takes a side-stream of two-phase fluid from a well test flowline or the main production line to a power plant in the case of a bottoming cycle being added to an existing plant. The test unit separates steam and brine, condenses steam and separates condensate and gas, mixes condensate and brine at a fixed ratio, injects a fixed portion of the gas, or acid is metered into the mixture, and finally passes the mixture through a hold-up vessel (HUV) and packed beds simulating the injection pipeline and wells. The pH, pressure, and temperature are monitored throughout the process.

A critical objective is to predict relative scaling rates for injection wells under different process conditions in addition to surface equipment scaling. The tests are normally run for about 3—6 weeks to obtain sufficient data to evaluate surface and formation scaling under constant process conditions. But the duration of each test depends entirely on the rate of scaling that occurs.

The packed bed consists of drill cuttings from the injection wells intended to be used (or in use) for the plant, sized to between 1 and 3 mm, for a uniform and reproducible test matrix. The use of drill cuttings provides the same geochemical environment to which the actual injection fluids from the plant would be exposed. Relative fouling rates of the packed beds are used to determine the optimum process conditions to minimize scaling in the injection system (separation pressure, injection temperature, and pH). Chemical inhibitors may also be tested in the pilot plant.

Typical test objectives for a combined-cycle plant are summarized as follows:

1. Confirm the quantity of sulfuric acid needed to reduce the injectate pH to the target value based on chemical modeling.
2. Determine the quantity of NCG injection, pressure, and mixing characteristics to reduce the injectate pH to the target value based on chemical modeling.
3. Determine relative simulated injection formation scaling rates for aged brine with no pH-Mod treatment, with sulfuric acid treatment, and/or with gas injection treatment.
4. Determine scaling rates in heat exchanger for untreated brine at target temperature.

Typical process flow diagrams for combined-cycle and ORC bottoming cycle pilot test units are shown in Figs. 16.14 and 16.15, respectively. Depending on the cycle to be simulated, the equipment may include a small two-phase separator, a shell-and-tube steam condenser, NCG pump, a shell-and-tube brine cooler, brine and condensate pumping and metering equipment to combine the brine, condensate and gas streams

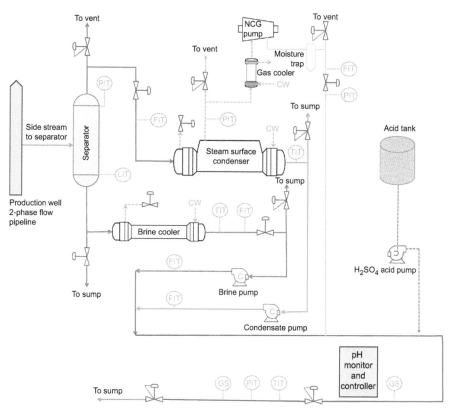

Figure 16.14 Process flow diagram for combined-cycle pilot test unit.

in the desired proportions, an acid-dosing and pH control system, two parallel HUVs to provide controlled residence times similar to the full-scale injection system, two high-pressure (HP) brine metering pumps to force side-streams of the aged brine through the packed-bed test sections at a constant flow rate, and related controls and instrumentation.

The HUV must have a baffled, plug-flow design and stirred sections to keep solid particles suspended in solution. The plug-flow design with mixing is important as it maintains a constant and narrow residence time distribution (RTD) while preventing premature deposition of suspended solids, which would bias scaling and packed-bed fouling rates. Suspended particles of silica and iron can also have a major effect on silica deposition. The HUV is sized for the average hold-up time expected for the injection pipelines and wellbores. Fig. 16.16 shows the distribution of chemical tracer through an HUV designed for a 50-min residence time, comparing an ideal plug-flow simulation model to actual results using chemical tracer.

The fouling rate across the packed beds is monitored by differential pressure at constant flow through the bed to determine the rate of fouling. The packing configuration and flow through the bed is designed to provide accelerated fouling compared to that

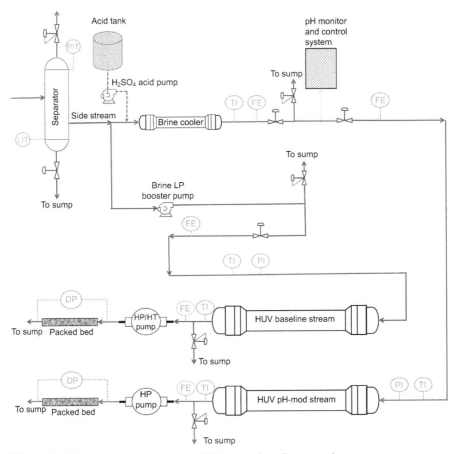

Figure 16.15 Process flow diagram for ORC bottoming pilot test unit.

occurring downhole by maintaining a high interstitial fluid velocity. Normally one HUV train is operated under baseline conditions with no treatment (no gas or acid injection), while the second train is pH modified by sulfuric acid or gas injection. It is important to note that the packed beds provide relative fouling rates only, so it is critical that a baseline stream is tested in parallel.

The primary result derived from these tests is the relative rate of permeability loss over time. Permeability controls the injectivity for a well and determines the pressure needed to dispose of fluid. Permeability of a matrix is defined by Darcy's law as:

$$k = v \, \frac{\mu \, \Delta x}{\Delta P} \qquad\qquad\qquad [16.18]$$

where k = permeability of matrix (m^2), v = superficial fluid flow velocity through matrix (m/s), μ = dynamic viscosity of fluid (Pa s), Δx = length of packed-bed matrix

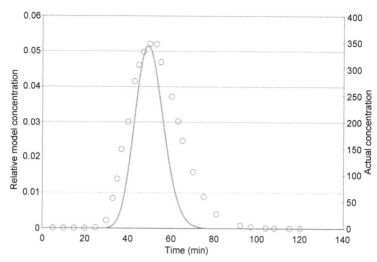

Figure 16.16 RTD through an HUV comparing tracer results (*open circles*) to ideal plug-flow model (*solid line*).

(m), and ΔP = differential pressure (Pa). Since the brine flow rate and viscosity through the packed beds are constant, the permeability is only a function of the differential pressure through the beds.

16.10   Guidelines for optimum pH-mod system design

The only proven technology for the control of amorphous silica scale in both ORC and multiple-flash power cycles at high levels of supersaturation (>1.5) is pH-Mod. The other proven option specifically for flash plants is the CRC process (Salton Sea); see Section 16.6. Although pH-Mod is very effective in controlling silica scale in surface equipment and downhole, it can be difficult to control the pH since the target value is typically at the inflection point of the acid titration curve for the brine (pH 4.5−5.0), where bicarbonate is the dominant alkaline species being neutralized in the process. Poor pH control (±0.5 unit or worse) can result in corrosion of carbon steel pipelines and inadequate scale inhibition. Dissolved CO_2 in the brine (>200 ppm CO_2) smoothes out the pH response curve to acid, resulting in more precise and stable pH control.

Fig. 16.17 shows the pH-Mod dosing curve for an ORC cycle process in Nevada (USA) where the brine is pumped from the wells with no flashing and all the gases remain dissolved (>1000 ppm CO_2). The curve is flat near the target pH region where an 85% change in the total acid-dosing rate results in a pH change of only 1 unit. Note also that the outlet brine pH at 70°C is considerably lower than the inlet brine at 180°C due to the increased acidity at lower temperature of the primary pH-controlling species carbonic acid.

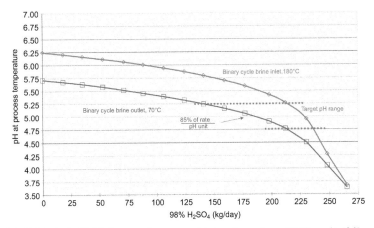

Figure 16.17 Typical pH-Mod curve for pumped brine through an ORC cycle [34].

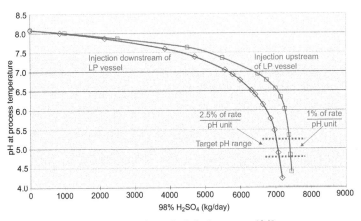

Figure 16.18 Typical pH-Mod curves for a dual-flash process [34].

Fig. 16.18 shows the pH-Mod curves for the dual-flash Kawerau plant in New Zealand [32]. A change of only 1–2.5% in the acid-dosing rate results in a pH change of 1 unit. This is a very difficult process to control and is not possible without stable brine flow. The control is more difficult when acid is injected upstream of the second flash vessel, which results in greater CO_2 loss from the brine than occurs with injection into the single-phase brine after second flash. In many cases, acid must be injected upstream of the second flash to prevent the onset of rapid silica polymerization.

A properly designed pH-Mod system must include careful consideration of the following:

- Brine flow control — *Brine flow stability*
- Acid delivery system — *Acid-dosing response*
- Distributed control system (DCS) — *Acid-dosing response and control*
- Brine flow measurement — *Acid-dosing response and control*
- pH measurement— *Acid-dosing control*
- Acid-mixing and corrosion-resistant materials — *Corrosion control*

16.10.1 Brine flow measurement and control

The pH-Mod process is much easier to control if the incoming brine flow is stable. The primary control variable for acid dosing is the brine flow signal. The brine flow measurement does not need to be accurate but must be linear and reproducible. The absolute accuracy in brine flow is not important since the acid dosing is a proportional response. But brine flow surging and signal noise can result in a substantial increase in the variability of the pH. Fig. 16.19 shows data for brine flow at the outlet of the HP separator at the Kawerau dual-flash plant. The high variability in the brine flow rate measured by an orifice meter was confirmed to be a real phenomenon by using online TFT$^{®}$ measurements (also plotted in Fig. 16.19). Note that the online TFT$^{®}$ measurement is labeled as "TFT CBM" (continuous brine monitor).

The pH control system logic is shown in Fig. 16.20. The acid-dosing pump is proportioned directly to the brine flow signal, and the factor used in that proportional control is updated periodically in a separate control loop that measures the pH of the treated brine downstream. The acid pump control cannot rely on the measured pH alone, as the response time would be far too long. The acid-to-brine ratio needed to

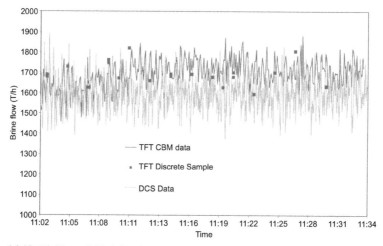

Figure 16.19 Highly variable brine flow at outlet of an HP separator at dual-flash Kawerau plant.

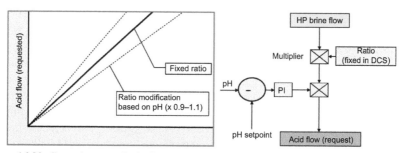

Figure 16.20 Control system logic for pH-Mod process.

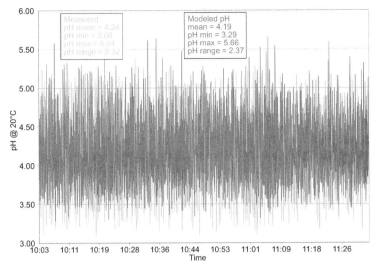

Figure 16.21 pH of highly variable HP brine resulting from unstable brine flow and slow dosing response: set-point = pH 4.8 at 20°C.

reach a given target pH is relatively constant but may change as the brine chemistry varies due to shifts in alkalinity content. The actual dosing ratio also changes due to process variables such as drift in the output of the acid pump at a given setpoint, acid viscosity due to temperature, drift in the brine flowmeter response, etc. All of these errors are compensated for by the secondary pH loop control.

The highly variable brine flow shown in Fig. 16.19 results in highly variable pH values and an unacceptable pH range (Fig. 16.21). The target pH in the example shown is 4.80 units (measured at 20°C), but the actual pH drops to 3 units and has an overall range of 2 pH units. The pH was computer-modeled (plotted in red) based on the brine chemistry, acid-dosing rate and brine flow rate data to demonstrate this was real variability and not pH measurement noise. The wide range in pH caused severe corrosion in the LP two-phase piping and LP separator, requiring a major investigation, redesign and retrofit of the Kawerau plant beginning about a year after it was commissioned [32].

16.10.2 Control system and acid delivery response time

An acid-dosing system can require several seconds to respond to changes in brine flow. Some of this time is consumed by the DCS to command adjustments in the acid pump rate based on brine flow variations. There is an additional electromechanical and/or hydraulic response delay in the acid pump between the DCS command to increase rate and another second or more may be consumed in delivering the acid from the premixing piping to the main brine line. Fig. 16.22 shows an example of a substantial response time off-set between surges in brine flow and the delivery of acid to meet those surges, again for the case where brine flow was highly variable.

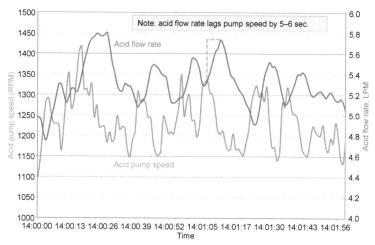

Figure 16.22 Response time offset between surges in brine flow and the delivery of acid.

Efforts must be made in the overall design of a pH-Mod system to reduce the acid delivery response time as much as possible, especially in cases where the inlet brine flow is not stable.

Work has been performed at some plants using a "feed-forward" parameter to predict surges in the inlet brine flow, so that acid dosing could be under a cascade control scheme that includes advance warning of brine flow variations, allowing time for the acid pump to respond appropriately. High correlations ($r^2 = 0.9$) have been identified between the first derivative of the HP separator level and the HP brine flow at the dual-flash plant in New Zealand [32]. The limitation with any feed-forward parameter is that the lead vs. lag time will not remain absolutely constant and will require online statistical analysis of the relationship to maintain synchronization. In this case, fixed time offsets for separator level maintain high correlations only over 10–15 min, and then need to be updated for drift. Any feed-forward approach for a complex two-phase flow system will be difficult and may not produce the required results for acid-dosing synchronization and pH stability.

16.10.3 pH measurement

An online pH monitoring system is required for pH-Mod, with automatic calibration and pH probe cleaning capability, and sample stream flow measurement and control. Accurate pH measurement is critical for both scale and corrosion control in order to maintain the pH within an acceptable range for materials exposed to pH-modified brine, and to provide alarms through the DCS in case of pH excursions outside an acceptable range.

The pH is measured from a small side-stream of cooled brine within about 10-m downstream of the acid-dosing point. The pH at process temperature will be shifted from the measurement temperature, and this must be considered in the system design

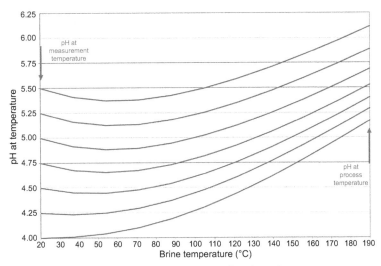

Figure 16.23 Correction curves for HP brine pH as a function of temperature.

and setting of the target pH. Geochemical, process chemistry, and corrosion modeling are always conducted at the actual temperature of the fluid being modeled. Thus, when the target pH for the brine to be treated is first estimated from chemical modeling, the appropriate correction (offset) must be applied to the online measurement.

The pH of geothermal brine in the pH-Mod process is buffered within the range of 4–5 pH units by second dissociation of sulfuric acid, bisulfate ion:

$$1st: \quad H_2SO_4 \rightarrow H^+ + HSO_4^-$$

$$2nd: \quad HSO_4^- \rightarrow H^+ + SO_4^-$$

[16.19]

Bisulfate is a weaker acid at high temperature. At 20°C, only 10% of the total available H^+ ions are associated in the bisulfate ion (HSO_4^-), making bisulfate a fairly strong acid at low temperatures. At 120°C, bisulfate is a much weaker acid, with 75% of the total available H^+ ions still associated. Fig. 16.23 displays the correction curves for high-alkalinity brine that is pH modified, over a range of pH values as a function of temperature. At pH values near 4, the pH is more than 1 full unit lower at measurement temperature than at the HP flash process temperature. At pH 5, it is about 0.5 unit lower at measurement temperature than at the process temperature.

16.10.4 Acid-mixing and corrosion-resistant materials

In the pH-Mod process, the concentrated acid is typically sulfuric and must be diluted immediately prior to injection. Concentrated acid cannot be injected into a hot brine pipeline due to localized "hot-spot" corrosion that may occur, even if high alloys are used. It is not safe or practical to predilute the concentrated sulfuric acid and store it in separate tanks, so it is always diluted online using a side-stream of brine or

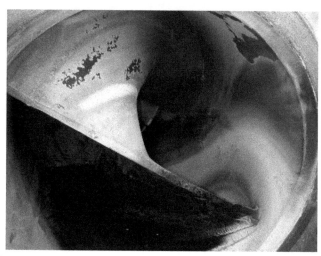

Figure 16.24 A 20-inch Hastelloy C-276 static mixing spool after several years of service.

condensate. Hotwell condensate from direct contact condensers cannot be used for acid dilution as it contains dissolved oxygen on the order of 100 ppb$_w$ or more, which will greatly accelerate corrosion downstream. The acid is diluted to a concentration of about 1%, and then injected into a static mixer in the main brine pipeline to be treated. The dilution water flowrate needed would typically fall in the range of $2-10$ kg/s.

Cold, concentrated sulfuric acid (98%) is not corrosive to carbon steel, but it is typically handled in stainless steel and/or plastic piping and tanks (polypropylene or polyethylene with oxidation-resistant liner). Hot, dilute sulfuric acid (1% range) is extremely corrosive to steels. Below 120°C, dilute sulfuric acid can be handled in Hastelloy C-276 alloy piping and fittings, but above 120°C Hastelloy C-276 alloy will be destroyed quickly. Above 120°C and below 200°C PTFE Teflon-lined pipe and fittings can be used. Above 200°C the only suitable material for hot, dilute sulfuric acid is tantalum.

The diluted acid is injected into a full-line size static mixing spool constructed of Hastelloy C-276 alloy (regardless of brine temperature). A Teflon-lined and jacketed injection quill, a Hastelloy-shrouded solid PTFE quill, or a solid tantalum injection quill is used to deliver the dilute acid to the center of the static mixer inlet. A photo of the inlet to a 20-inch Hastelloy C-276 static mixing spool after several years of service is shown in Fig. 16.24, and Teflon-lined and jacketed injection quills are shown in Fig. 16.25.

After the static mixing spool, the brine piping system may return to carbon steel in cases where the brine flow is stable and pH reasonably well controlled (4.5−5.0 ± 0.2 pH unit). In the case of the Kawerau dual-flash plant discussed earlier (Section 16.10.1), the carbon steel piping between the HP separator and the inlet to the LP separators had to be replaced with super-duplex 2507 alloy piping [32]. This was necessary given the high corrosion rates that had been experienced in the two-phase piping [32] as the result of highly variable pH.

Figure 16.25 Teflon-lined and jacketed injection quill, new and after 6 months of service.

Finally, it is important to ensure there are no possible sources of oxygen intrusion to the pH-modified brine. There may be sources of aerated hotwell condensate passing to the brine through pump seal water flush streams. If noncondensable gas is injected, there should be less than 1 ppm$_v$ in the gas. Air leaks are prevalent in any vacuum extraction system for NCG and the gas from such systems should never be injected into the acidified brine. Even compressors that take gas from a positive-pressure ORC steam condenser may contain oxygen due to air intrusion on the suction side of the compressors. In the complete absence of oxygen, corrosion rates of carbon steel piping and injection well casing exposed to pH-modified brine (4.5–5.0 ± 0.2 pH unit) are typically negligible.

16.11   Summary

There continues to be a huge potential for geothermal energy development worldwide. Renewable geothermal energy is currently used to generate electric power in 24 countries, for a total of 12,800 MWe. Geothermal energy has the potential to be the world's primary source of baseload renewable power [1,2].

Since geothermal energy utilization involves direct heat transfer between water and rock in a reservoir, the silica content of geothermal brine is usually near saturation with respect to crystalline silica minerals (typically quartz or chalcedony). But the kinetics of crystalline silica dissolution/precipitation is slow, so in the time it takes to extract the energy on the surface and reinject the fluids, quartz or chalcedony precipitation does not occur. However, amorphous silica deposition is possible and does occur once the fluid is supersaturated due to cooling. Fortunately, the kinetics of amorphous silica deposition is relatively slow, but scaling still can easily occur within the time-frame of fluid processing and handling.

In early geothermal development, silica scaling was a significant limiting factor in the amount of energy that was extracted from a resource. Recent production engineering and chemistry advances have greatly reduced these barriers by implementing technology to control silica/silicate scaling from geothermal produced fluids. Proven engineering strategies such as combined-cycle plants and pH-Mod are currently the primary means to mitigate silica scale. Many single-flash plants worldwide can achieve an increase of 30% or more in power generation from the same resource by adding an ORC bottoming plant to the existing facility [33], in conjunction with pH-Mod for silica scale control.

References

[1] Geothermal Energy Association (GEA). Annual U.S. & global geothermal power production report. 2015.

[2] Bertani R. Geothermal power generation in the world 2010−2014 update report. In: Proceedings world geothermal congress; 2015.

[3] Arnórsson S. The quartz and Na/K geothermometers. 1. New thermodynamic calibration. In: Proceedings world geothermal congress; 2000. p. 929−34.

[4] Weres O, Yee A, Tsao L. Equations and type curves for predicting the polymerization of amorphous silica in geothermal brines. Soc Petroleum Eng J February 1982:9−16.

[5] Fournier RO. The behavior of silica in hydrothermal solutions. In: Berger BR, Bethke PM, editors. Reviews in economic geology, 2. Society of Economic Geologists; 1985. p. 45−61.

[6] Fournier RO, Marshall WL. Calculation of amorphous silica solubilities at 25 to 300°C and apparent cation hydration numbers in aqueous salt solutions using the concept of effective density of water. Geochimica Cosmochimica Acta 1983;47:587−96.

[7] Goto K. Research on the state of silicic acid in water (Part 2): solubility of amorphous silicic acid. J Chem Soc Jpn Pure Chem Sect 1955;76:1364−6.

[8] Wahl EF. Geothermal energy utilization. New York: John Wiley & Sons; 1977.

[9] Manceau A, Ildefonse Ph, Hazemann JL, Flank AM, Gallup DL. Crystal chemistry of hydrous iron silicate scale deposits at the Salton Sea geothermal field. Clays Clay Miner 1995;43:304−17.

[10] Gallup DL. Aluminum silicate scale formation and inhibition. Geothermics 1997;26: 483−99.

[11] Manceau A, Gallup DL. Nanometer-sized divalent manganese-hydrous silicate domains in geothermal brine precipitates. Am Miner 2005;90:371−81.

[12] Björke J, Mountain B, Seward T. The solubility of amorphous aluminous silica between 100 and 350 C: implications for scaling in geothermal power stations. Proceedings New Zealand geothermal workshop 2012, November 19−21, 2012.

[13] Kindle CH, Mercer BW, Elmore RP, Blair SC, Myers DA. Geothermal injection treatment: process chemistry, field experiences and design options. 1984. PNL-4767, UC-66d.

[14] Iler RK. The chemistry of silica. New York: John Wiley & Sons; 1979.

[15] von Hirtz P, Gallup D. Deposition model, thermochem internal software documentation. 2000.

[16] Gallup DL. Unusual adherence of manganese silicate scale to metal substrates. Geotherm Resour Counc Trans 2004;28:529−32.

[17] Sugiaman F, Sunio E, Molling P, Stimac J. Geochemical response to production of the Tiwi geothermal field, Philippines. Geothermics 2004;33:57—86.

[18] Gallup DL. Advances in geothermal production engineering in recent decades. Geotherm Resour Counc Trans 2007;31:11—5.

[19] Candelaria MNR. Methods of coping with silica deposition — the PNOC experience. Geotherm Resour Counc Trans 1996;20:661—72.

[20] Gallup DL. Brine pH modification scale control technology. Geotherm Resour Counc Trans 1996;20:749—52.

[21] Gallup DL, Kitz K. Low cost silica, calcite and metal sulfide scale control through on-site production of sulfurous acid from H_2S or elemental sulfur. Geotherm Resour Counc Trans 1997;21:399—403.

[22] Hirowatari K, Yamauchi M. Experimental study on a scale prevention method using exhausted gases from geothermal power stations. Geotherm Resour Counc Trans 1990;14: 1599—602.

[23] Gallup D, Sugiaman F, Capuano V, Manceau A. Laboratory investigation of silica removal from geothermal brines to control silica scaling and produce usable silicates. Appl Geochem 2003;18:1597—612.

[24] Gallup DL. The interaction of silicic acid with sulfurous acid scale inhibitor. Geotherm Resour Counc Trans 1997;21:49—53.

[25] Featherstone JL, Van note RH, Pawlowski BS. A cost effective treatment system for the stabilization of spent geothermal brines. Geotherm Resour Counc Trans 1979;3:201—4.

[26] Featherstone J, Butler S, Bonham E. Comparison of crystallizer reactor clarifier and pH mod process technologies used at the Salton Sea geothermal field. In: Proceedings world geothermal congress; 1995. p. 2391—6.

[27] Gallup DL. Investigations of organic inhibitors for silica scale control from geothermal brines. Geothermics 2002;31:415—30.

[28] Gallup DL, Barcelon E. Investigations of organic inhibitors for silica scale control from geothermal brines. II. Geothermics 2005;34:756—71.

[29] Gestur G. Reykjavik energy, personal communication. 2001.

[30] Hirtz P. Thermochem, Inc. internal report to Ormat. 2004.

[31] Compatibility testing of salton sea geothermal brines. In: Featherstone J, Gallup D, von Hirtz P, editors. Proceedings, world geothermal congress; 2015.

[32] Brine silica management at Mighty River power, New Zealand. In: Addison S, von Hirtz P, Gallup D, et al., editors. Proceedings, world geothermal congress; 2015.

[33] Case study to upgrade and optimize an existing geothermal power plant. In: Mulazzani D, editor. Proceedings, Indonesia international geothermal conference and exhibition; 2015.

[34] Gallup D, von Hirtz P. Control of silica scaling in geothermal systems using silica inhibitors, chemical treatment and process engineering, chapter 9, the science and technology of industrial water treatment. CRC Press; 2010.

Environmental benefits and challenges associated with geothermal power generation

17

A.C. de Jesus
First Philippine Holdings Corporation (FPH), Pasig City, The Philippines

17.1 Introduction

"Mega forces" or "mega trends" such as climate change, ecosystem decline, deforestation, and natural resources scarcity, to name a few, have been reported as crucial issues that are facing governments and businesses now and in the next 20 years (KPMG, 2012, 2014; PricewaterhouseCoopers, 2014). The concern for climate change and ecosystem collapse were also assessed as "global risks" in a survey of the World Economic Forum where global risk is defined as an uncertain event or condition that, if it occurs, can cause significant negative impact for several countries or industries within the next 10 years (World Economic Forum, 2015). To address climate change, the UN Intergovernmental Panel on Climate Change cites geothermal as one of the renewable energy options that can provide long-term, secure baseload energy and greenhouse gas (GHG) reductions to minimize climate hazards (Goldstein et al., 2011). According to the World Health Organization, the other major argument for going for renewable energy like geothermal is human health as 50% of air pollution comes from fossil fuels making it the world's largest environmental health risk to man (Health Care with No Harm, 2015). Geothermal is a renewable energy resource that is more than a century old that has environmental and social safeguards in place, ready to be tapped to contribute to the solution on the climate change crisis.

A deeper understanding of the environmental, social, and cultural aspects of geothermal energy can equip stakeholders (governments, communities, civil societies, nongovernmental organizations, academe, project operators, and investors) in addressing a number of sustainability issues. This will lead to an efficient natural resource use, smooth project implementation, and policy formulation for a truly sustainable development process that will unlock this vast and clean renewable energy resource.

17.2 Environmental, social, and cultural benefits and challenges of geothermal power generation

Geothermal was identified in the 1992 UN Conference on Environment and Development also known as the "Earth Summit" as an environmentally advantageous energy

Geothermal Power Generation. http://dx.doi.org/10.1016/B978-0-08-100337-4.00017-6

resource (United Nations, 1992). Like any infrastructure project, geothermal has environmental, social, and cultural concerns. The good news is, these impacts are often localized and manageable and are more likely to be reversible (Arvizu et al., 2011; Gehringer and Loksha, 2012). Positive impacts or benefits also exist that should be given equal importance when considering the feasibility of a geothermal project.

17.2.1 Environmental impacts of geothermal projects

The specific impacts of a geothermal project will depend on the nature of the geothermal resource, location, and extent of human habitation. Hence, only the more common impacts of geothermal projects on air, land, and water and the corresponding management measures are discussed in the following sections. Advancements in geothermal energy technology have allowed for a better management and mitigation of its effects to the environment.

17.2.1.1 Air quality

Generation of power from geothermal sources involves no combustion so there are lower and cleaner emissions compared with other baseload power sources today such as fossil and nuclear fuels (Brophy, 1997; Gehringer and Loksha, 2012; Matek, 2013). On the other hand, binary geothermal plants produce nearly zero emissions (GEA, 2012a). About 75—99.98% of the composition of emissions from geothermal operation is steam and the remaining fraction of 0.02—25% is noncondensable gas (NCG) composed of carbon dioxide (CO_2), hydrogen sulfide (H_2S), ammonia (NH_3), methane (CH_4), and trace amounts of other gases. NCGs are natural constituents of geothermal resources (Chaves, 1996; Odor, 2010; Holm et al., 2012). While geothermal is not exactly carbon free, its CO_2 emissions are significantly lower than that of fossil-based electricity sources.

The average California geothermal carbon dioxide emissions were studied by the California Air Resources Board, California Energy Commission, U.S. Environmental Protection Agency, and Geothermal Energy Association; the results show the significant difference of carbon emissions of geothermal from those of other fuels—1020 kg (2249 lbs) CO_2/MWh for coal, 515 kg (1135 lbs) CO_2/MWh for natural gas, and 82 kg (180 lbs) CO_2/MWh for geothermal (GEA, 2012b). There are also general studies: Using the emission data from Matek (2013) and the U.S. Environmental Protection Agency (2014), the geothermal carbon footprint was computed for major baseload power sources in Table 17.1.

Based on the emission figures shown in Table 17.1, shifting from conventional energy sources to geothermal will lead to net positive benefits to both the environment and the climate change mitigation efforts. Although geothermal energy's NCG emissions are already negligible, they can still potentially alter the air quality (Dickson and Fanelli, 1995; Ozcan and Gokcen, 2009; Holm et al., 2012). For instance, CO_2 is widely identified as a cause of global warming leading to climate change. At the same time, it is needed by vegetation to grow. Thus, even as the amount of its CO_2 emission is low, the Energy Development Corporation (EDC) in the Philippines quantified the carbon storage capacity of the forest in its Leyte geothermal project as early

Table 17.1 Carbon footprint of common baseload power plants

Power plant type	Average emission rate lbs/MWh	Annual carbon footprint tons/MW/year
Coal-fired	1020	9636
Oil-fired	760	7323
Gas-fired	515	3772
Geothermal-steam	82	1736

Emission data are from Matek, B. Promoting geothermal energy: air emissions comparison and externality analysis. http://geoenergy.org/reports/Air%20Emissions%20Comparison% 20and%20Externality%20Analysis_Publication%20May%202013.pdf (accessed 26.02.15.) and U.S. Environmental Protection Agency: Clean energy. http://www.epa.gov/cleanenergy/energy-and-you/ affect/oil.html (for oil emissions) (accessed 27.02.15.).

as the late 1990s. Lasco et al. (2002) found out that the forest can offset the CO_2 emissions from all the geothermal power plants of the company beyond the term of its concession. Also, the company has been reforesting all of its sites since the commissioning of their geothermal power facilities to contribute to the carbon mitigation in its project areas.

In some areas, CO_2 emissions from geothermal may be harnessed for good use. In Kenya, geothermal exhausts are vented to greenhouses to enhance plant growth in the cut flower industry. In Iceland, co-produced CO_2 gases are used not only in greenhouses but are also collected for use in carbonated beverages (Ragnarsson, 2013). Industrial use can also aid in the reported very high carbon emissions from geothermal fields in Turkey due to the affiliation of the resource with carbonate rocks, a unique situation reported for the first time in the geothermal sector (Aksoy et al., 2015).

For H_2S, which is a gas of concern among the components of NCG, air quality modeling is done to determine the projected distribution of the gas in the area and to serve as a guide in the design and location of the power plant, thus avoiding adverse effects to nearby communities and facilities. Being a natural component of geothermal areas, H_2S emissions are regularly monitored and managed during geothermal energy production. For H_2S emissions that are beyond air quality standards, gas abatement methods are available. Olafsdottir et al. (2015) also reported that surface waters and vegetation around the power plants in Iceland are natural sinks for H_2S. There are also experiments on the dissolution of emissions into geothermal waters for subsequent reinjection (Gunnarsson et al., 2015). Once injected, Juliusson et al. (2015) theorized that the water—rock reactions in the geothermal reservoir will mineralize H_2S. Last, a unique biofilm reactor was designed by Contact Energy in New Zealand where pipes were seeded with sulfur oxidizing bacteria for treating the H_2S in cooling waters (Bierre and Fullerton, 2015).

The other areas of concern related to air quality are noise and acid rain. For noise, there are standard equipment noise controls and mufflers that should be used during drilling and construction. New noise barriers being introduced in the market as a reaction of community concerns for various enterprises can be tested in the geothermal industry (Environmental Leader, 2015). During plant operation, noise will come from cooling tower fans and the sounds are found to be negligible (Geothermal Energy Association, 2012a). On the other hand, the concern on acid rain was invalidated by actual sampling around geothermal plants whose rain samples were found to comply with World Health Organization standards (Wetang'ula and Wamalwa, 2015). The same findings were reported in Costa Rica based on a sampling since 1987 where pH has not been altered by geothermal operation (Sequeira, 2015a).

17.2.1.2 Geologic impacts

Slightly more frequent occurrence of localized microseismic activities, hydrothermal steam eruptions, and ground subsidence may be potentially influenced by geothermal power operation. Earthquakes are caused by the rupture of a body of rock, which radiates seismic waves that shake the ground. Because geothermal operations usually take place in areas that are also tectonically active, it is often difficult to distinguish between geothermal-induced and naturally occurring events (GEA, 2009). However, studies have shown that those generated by geothermal projects are usually microseismic activities that are not felt by humans (Kagel et al., 2005). In over 100 years of geothermal deployment, there has been no reported hazardous seismic activity or earthquakes that brought major damage to nearby infrastructure or communities due to their shallow origin (Arvizu et al., 2011; GtV-BV, 2010). Ground subsidence could be addressed by reinjection of geothermal separated waters to maintain the reservoir pressure. Bromley et al. (2015) prescribes injection management as an adaptive tool to limit or control induced stresses to mitigate this effect. Full reinjection at the commissioning of the project is thus the prudent approach.

Conventional geothermal heat and electricity production are developed from near surface hot reservoirs, but regions without volcanoes or "hot spot areas" seek geothermal in deeper layers of the earth. The technology called Enhanced Geothermal Systems (EGS) or "hot dry rock" requires 100°C temperature to be economically viable, thus requiring ultra-deep, usually 5-km, drilling. Sacher and Schiemann (2011) reported the abandonment of the Deep Heat Project in Basel, Switzerland, in 2009 due to tremors that reached the surface as pressurized waters were injected into boreholes to widen fissures to create heated waters. Although there was no major damage that occurred, the public failed to accept the project.

17.2.1.3 Water use and water quality

In the current period of water scarcity due to climate change in many countries, water consumption must be given close attention. The 2011 UN theme of Water-Energy-Food Nexus highlights the interdependence of these three important elements that are derived from nature. They are also most important requirements to sustain humans. All types of power plants require fresh water for cooling. However, geothermal plants cooled by water evaporation can use water derived from geothermal fluids for cooling. Fluids not lost to evaporation are injected back into the geothermal reservoir. In general, geothermal has a smaller water footprint than other power technologies. The total life cycle consumption of water in gallons per kilowatt-hour was estimated per fuel type. These are the estimates in increasing order of water utilization in liters per kilowatt-hour as estimated by GEA (2015): (1) 0.38 for geothermal flash and wind, (2) 0.26−3.98 for solar photovoltaic, (3) 0.30−1.02 for geothermal binary, (4) 0.90−2.72 for natural gas combined cycle, (5) 0.98−5.52 for coal, (6) 1.06−3.72 for nuclear, (7) 1.44−3.70 for natural gas conventional, (8) 2.16−5.79 for coal with carbon capture, and (9) 3.29−4.24 for concentrated solar. On the other hand, water consumption is nil in binary geothermal plants that use air cooling.

The interaction of molten rocks with rain water that percolates through major faults and fractures result in the geothermal brine (Bakane, 2013). If not properly managed, discharge of geothermal fluids through spillage into local surface waterways and leakage into groundwater aquifers may affect the water quality and cause chemical pollution (Goldstein et al., 2011). These incidents may potentially alter the uses of natural water bodies downstream of the geothermal project.

Mitigation measures are already being practiced and have been proved effective in keeping water impacts of geothermal operations within acceptable limits (Dickson and Fanelli, 1995). First on the list is the reinjection of geothermal wastewaters back into its reservoir, which totally removes the pathway of contamination to the environment and allows geothermal operations to remain compliant with water quality standards. Today, the technology is being practiced by almost all geothermal developers. Reinjection also prevents thermal pollution of nearby bodies of water. Another countermeasure is the use of resins as sorption membranes to remove predominant minerals critical to agriculture like boron (Yilmaz et al., 2005). Engineering measures are also put in place to prevent groundwater contamination where production and injection wells are lined with steel casings and cemented to isolate geothermal fluids from the surrounding soil and ground water resources (Odor, 2010).

17.2.1.4 Solid wastes

Solid wastes of geothermal operation in the form of cartons, scrap metals and pipes, plastics, and other domestic wastes come from rig operations, base camps, work areas, and storage yards. There are recyclers for these nontoxic residuals. The solid wastes of concern are the sludges that soak up minerals in the cooling tower as brine is commonly used as an alternative medium to avoid the use of precious freshwater. Sludges from the cooling tower basins may be encapsulated by cement and other materials to make them inert. Wastes can also be managed by chemically reducing solid wastes to soluble forms before reinjection. Biotechnology methods such as the use of bacteria can also dissolve, separate, or immobilize hazardous geothermal sludges (Premuzic and Lin, 1989). There are also waste treaters accredited by environmental bodies to handle specific chemical pollutants. Some minerals may be recovered and recycled for industrial use. For instance, crystallized salts and silica solids that are used in cosmetics production are being collected in the Blue Lagoon of Iceland and are popular with tourists (Odor, 2010).

17.2.1.5   Land use

Geothermal energy operations require relatively little above-ground land (Evans et al., 2009; Geothermal Energy Association, 2012a). An average geothermal power plant is estimated to use only 2.47−19.77 ha of land per megawatt produced; significantly less compared to the land requirements of nuclear (12.36−24.71 ha/MW) and coal (46.94 ha/MW) power plants (U.S. Department of Energy, 2016).

Since minimal above-ground infrastructure is required for geothermal operations, the remaining area of the concession is shared with stakeholders for other activities, for example, as farming, horticulture, and forestry in Mokai and Rotokawa in New Zealand (Koorey and Fernando, 2010) and the national park in Olkaria, Kenya (Odor, 2010; Arvizu et al., 2011). Being found in natural habitats, geothermal locations are often converted to ecotourism sites. In the Philippines, some geothermal concessions are decreed under Executive Order 223 of 1987 to be managed as multiple-use watersheds, facilitating other compatible activities of local stakeholders.

17.2.1.6   Forest ecosystem and biodiversity

Geothermal structures must be carefully designed so that they do not lead to massive clearings that will alter the landscape, permanently damage the environmental functions or ecosystem services of the habitat, close wildlife corridors, and impede natural waterways.

The prohibition of development in national parks and other protected areas has been a major barrier to the geothermal sector. However, international opinion is changing. As early as 1986, the International Union for the Conservation of Nature and Natural Resources (IUCN), an international organization working in the field of nature conservation and sustainable use of natural resources recognized that to maintain services to human settlements, it may require access to protected areas that may be excluded or declared as special zone (MacKinnon and MacKinnon, 1986). Over several decades, IUCN has developed new protocols to address the realities on the ground by considering poverty alleviation and sustainable business in its prescriptions (Edwards, 2015).

Geothermal developers scout for open areas to put up the geothermal facilities but due to the site-specific nature of the geothermal resource sometimes areas need to be cleared subject to sovereign regulations. Tree cutting may potentially cause the loss of biodiversity or lead to ecosystem disturbance (Brunori et al., 2005). An ecosystem, like a forest, is a set of living organisms interacting with its physical environment. If a habitat will be altered in vegetation clearing, the IUCN adopts the principle of "mitigation hierarchy," which is a sequence of steps, starting with the avoidance of the impact, and where avoidance is not possible, minimization of inevitable impacts. Then, if still not possible, it prescribes on-site restoration to reverse the negative impacts to bring an area back to its predisturbance biodiversity state and, finally, biodiversity offset or ecological reconstruction to achieve "no net loss" (NNL) or "net gain" (NG) in biodiversity with respect to species composition and peoples' use of biodiversity (Kate and Crowe, 2014; Pilgrim and Ekstrom, 2014). Biodiversity offsets are quantifiable conservation initiatives to achieve "NNL" and preferably "NG"of biodiversity on the ground (Edwards, 2015).

Experiences in various projects have demonstrated that geothermal has limited negative impacts. Some have even delivered positive effects in forest habitats, especially if the geothermal operator is also assigned to help the government in natural resources management. This was the case of EDC in the Philippines wherein they were tasked by government since 1984 with the watershed administration of their concessions to sustain the water-based geothermal project. In several geothermal project sites, the pristine environments have been maintained allowing successful ecotourism activities such as the waterfalls and bat caves of the Bacon-Manito Geothermal Project in the Philippines, the Hoshino Resort in Japan for geothermal heating, combined safari and geothermal production in Hell's Gate National Park in Kenya, the Surtshellir lava cave and Deildartungaarhver hot springs of Iceland, and the Rotorua geysers of New Zealand (de Jesus, 2007). Costa Rica, which is known to be a leader in nature conservation, also reported the recovery of forests in Miravalles geothermal project since 1973 (Sequeira, 2015b).

Measures are available to limit habitat destruction and avoid loss of biological diversity due to clearings and well testing. These include, but are not limited to (1) integrating biodiversity considerations into the environmental impact assessment (EIA) prior to project implementation; (2) baseline species inventory and the purposeful avoidance of rare, endangered and threatened species during earthworks; (3) minimization of openings; (4) directional drilling that minimizes land openings; and (5) redirecting of the emission during well testing to avoid defoliation (Tuyor et al., 2004; de Jesus, 2007). In cases where there is a constraint in space and the well pad will be close to the forest resulting in its exposure to temporary brine sprays during well testing, EDC in the Philippines has devised a "vertical discharge diffuser." The device will reduce or eliminate the release of geothermal sprays from the well by diverting the geothermal fluids to twin-silencers to separate the brine and the steam. The geothermal fluids are then disposed to a reinjection well.

17.2.2 Social impacts of geothermal projects

Social acceptance by local communities is a prerequisite of a successful geothermal energy development (Arvizu et al., 2011). It is substantially reliant on the ability of the project developers to ensure the enhancement of the positive impacts, the prevention or minimization of the negative impacts, and the creation of benefits and just compensation for resident communities (Rybach, 2010). The positive and negative social impacts are discussed later.

17.2.2.1 Local employment

Positive economic impacts have been identified in geothermal projects. The different stages of geothermal power plant development, from exploration to operation and even decommissioning, create local job opportunities for the resident communities. These help alleviate poverty, particularly in projects located in remote mountain areas where livelihood opportunities are limited (Arvizu et al., 2011; Geothermal Energy Association, 2012b). The total direct, indirect, and induced employment impacts were computed by the Geothermal Energy Association in the United States of America as 4.25 full-time positions and 16 person years per megawatt (Kagel, 2006).

Since not all members of the community can be hired by the company, misunderstanding is avoided by prioritizing qualified residents of communities radiating from the project. Opportunities for livelihoods outside the geothermal facility may be created in other environmental and community projects by the developers. In the Republic of Djibouti, creation of jobs for women was cited as a positive gender impact (Adaweh and Idleh, 2015). The same is the experience in the Philippines, where women are given roles in the community projects provided for host communities by the geothermal developers.

17.2.2.2 Development funds

Regulations in various geothermal countries mandate the payment of royalties for the use of the natural resources of an area. Royalties are also paid to owners or rights holders of the resource (Lund, 2009). For state-owned natural resources, royalties are used for development projects of the communities hosting the geothermal facility. Based on the negotiation with the rights holder for licensing, various benefits may also be committed for the consent. Indigenous peoples (IPs) can also negotiate for development projects and other forms of compensation for the use of geothermal resources within their ancestral domains after a "free and prior informed consent" is secured for the project.

17.2.2.3 Community development assistance

In developing countries, geothermal operators have proven to be appropriate partners for community development. Geothermal companies, whether as part of their Corporate Social Responsibility (CSR) program alone or in partnership with the local government, have been participants in providing social services that may otherwise be inaccessible to rural communities. Zaidi (2015) noted that geothermal power can solve socioeconomic issues in rural communities on unemployment, sustainable power supply, livelihood opportunities, and empowerment. In the Philippines, the typical assistance includes livelihoods, scholarships, roads, school buildings, medical facilities, and other community-shared infrastructures (de Jesus, 2005). Communities in EDC sites (Tongonan and Palinpinon) reported appreciation for their increased economic freedom as well as a positive attitude toward themselves, their community, and the environment as a direct or indirect benefit from the geothermal project. They also learned positive values such as volunteerism, industry, and self-reliance and are now able to build a brighter future for their children (de Jesus and Nieva, 2003).

CSR projects of geothermal developers must be strategic to ensure sustainability such that they benefit both the host community and the company. An example of a strategic project is reforestation. This project recharges the water-based geothermal resource while involving the community to plant and nurture the trees as a means of livelihood which could eventually lead to improvement of the environmental quality in their area. For sustainability, the communities must also be consulted. If they see the benefits of the assistance, they will be willing to devote time and contribute resources to implement the CSR project. The company must include an exit strategy in the design of CSR projects so that communities are being prepared to take over the CSR project from the start of the project (International Finance Corporation, 2010).

17.2.2.4   Resettlement

There are recorded impacts that need more careful planning by developers such as the possible displacement of households due to the site-specific nature of the geothermal resource, although this may be in a smaller scale than other infrastructure projects. In some cases the effects are indirect like the spa owners who reported that their livelihoods are threatened by the possible depletion of hot water resources (Taipei Times, 2008). Key to the agreement of resettlers is early and adequate consultation (Barasa, 2015). If resettlement is unavoidable, prior consultation should be done with the affected households to discuss and reach an acceptable agreement. This usually covers the relocation of affected households to safer locations and the replacement of lost structures, amenities, and livelihoods. Fair compensation should also be given for any damage to property or livelihood such as crops and farm lands. The developers should also maintain an active involvement in helping resettled households regain and maintain their previous living standards (de Jesus, 2007). Operational guidelines in involuntary resettlement are available with multilateral agencies like the World Bank/International Financing Corporation (IFC), Asian Development Bank, and Japan International Cooperation Agency (JICA), as well as private banks around the world that embrace the Equator Principles. Eighty financing institutions in 35 countries have adopted the principles. This is important as these banks own 70% of the international project finance debt (Equator Principles, 2015).

17.2.3   Cultural impacts of geothermal projects

The greatest predicament in terms of culture centers on the potential impacts of geothermal projects to the known way of life of indigenous peoples (IPs). Issues such as encroachment to ancestral lands, desecration of ancestral sites, and limits to their ability to practice their traditional way of life (hunting and gathering, nomadic lifestyle, among others) may all contribute to strong resistance from IPs to accept any development project in their areas. For the interest of the IPs and the geothermal developer to coexist, the geothermal developer must be concerned with protecting the forest and mountain. They must do so to ensure the recharge of their water-based geothermal resource. If the geothermal developer will protect the forest, it will also protect the ancestral domain of the IPs. The geothermal developers and the IPs, therefore, can complement their efforts in protecting the area. It is important to add that this partnership is facilitated when there is trust and respect between the two parties.

The UN Declaration on the Rights of Indigenous Peoples of 2007 highlights the IP right to self-determination, their collective right to the land, territories, and natural resources they occupy, as well as their right to define their development priorities (United Nations, 2008). If geothermal resources are found in IP lands, it is important to abide by the principle of "Free and Prior Informed Consent." It is also important to undertake an ethnographic study of the IPs in order to understand their culture and their perspectives. Ethnography is a systematic study of the people and their cultures observed from the point of view of the subject group. This study is essential in understanding the IP community structure, customary laws and decision-making processes and in engaging them the proper way. Policies and regulations that protect IP rights should also be complied with, along with consultations and discussions throughout the life of the project as any host community.

17.2.4 Summary of benefits and challenges from geothermal projects

Table 17.2 summarizes the common potential positive and negative impacts of geothermal projects and ways to prevent or maintain the status of the environment and the communities in the project area.

17.3 Developing an environmentally sound and socially responsible project

Preventive strategies and mitigating measures for the environmental and social concerns on geothermal projects should be embedded within the project's plans for all stages of operation and are crucial for obtaining public acceptance and the necessary permits from concerned authorities. Even though the process and methods may differ based on a country's environmental laws and policies, the end goal is always the same: to ensure that no harm shall be done to the natural environment and the local communities due to the project's development and operation. The following prescribes the steps to develop a successful geothermal project.

17.3.1 Conduct an environmental impact assessment

The EIA is an internationally recognized legal instrument for assessing the environmental consequences of a project, plan, or program to aid in decision making. At the same time, it is an institutional process governed by a country's local laws and regulation (Ogola, 2009). It involves describing the baseline environment, predicting the environmental impacts of proposed activities, as well as proposing appropriate preventive, mitigating, and compensatory strategies for the identified negative impacts (de Jesus and Nieva, 2003). These together with inputs from concerned stakeholders are taken into consideration by the project proponent in the design of the project. Nowadays, these three levels of consideration—social, environment, and economic—are used as the main decision factors for approving projects in most countries around the world.

Table 17.2 Impacts of geothermal projects and corresponding measures

Potential impact	Enhancement	Management measures
A. Environmental		
1. Reduction in or loss of vegetation during earthworks and clearing activities	**a.** Restoration of vegetation that is net positive to the environment	**a.** Revegetation and/or engineering measures for soil stabilization
2. Soil erosion and siltation		**b.** Proper siting of spoil disposal areas and engineering measures
3. Physical disturbance of waterways		**a.** Proper drainage design
4. Alteration of water quality due to discharge of effluents		**a.** Full reinjection to remove the path of contamination of water bodies
		b. Casing of geothermal boreholes to separate geothermal waters from the watertable
		c. Sorption of minerals by membranes
		d. Water quality monitoring
5. Change in air quality due to NCG emissions (critical components)		
• Carbon dioxide (CO_2)	**a.** Can fertilize agricultural plants and horticultural species	**a.** Vegetation planting to sequester carbon from the environment
		b. Extraction of CO_2 for industrial uses
• Hydrogen sulfide (H_2S)		**a.** Air dispersion modeling to locate facilities away from population
		b. Chemical gas abatement for emissions beyond air quality standards
		c. Dissolution of NCGs in condensates and reinjection to geothermal reservoir
		d. Bacterial treatment
		e. Air quality monitoring

Continued

Table 17.2 Continued

Potential impact	Enhancement	Management measures
6. Elevated noise level		**a.** Noise modeling to locate facilities away from population **b.** Use of silencers to minimize noise during well discharge and during operation of emergency diesel generator set and black start engines **c.** Noise-attenuating equipment barriers **d.** Natural or enhanced vegetation as noise buffer **e.** Noise monitoring
7. Ground subsidence		**a.** Reinjection to maintain reservoir pressure **b.** Subsidence monitoring within the steam field
8. Increase in seismic activities		**a.** Normal geothermal operation only causes microseismic activities which are not felt by man *For EGS, refinement of the method is ongoing.*
9. Generation of solids and other wastes		**a.** Practice of 3Rs (reduce, reuse, and recycle) **b.** Proper waste disposal. Toxic and hazardous wastes are handled by accredited waste treatment and transport facilities
10. Land use alteration	**a.** Adoption of multiple-use management in the area. The limited space used by geothermal facilities allow co-use with local stakeholders **b.** Involvement of the project developer in the management of the area. Due to its interest to protect the geothermal resource, it will ensure minimal alteration of the land use of the area.	**a.** Reduction of openings as a consideration in the design

Table 17.2 **Continued**

Potential impact	Enhancement	Management measures
11. Loss of biodiversity and ecosystem disturbance	**a.** Aim for "no net loss" or "net Gain" of ecosystem services for the environment per protocol of the IUCN **b.** Reforestation beyond the replacement of cut trees but based on lost ecosystem services	**a.** Integration of biodiversity consideration into the EIA of the project prior to implementation **b.** Purposeful avoidance of rare, endangered, and threatened species during earthworks **c.** Adoption of "mitigation hierarchy" approach (from avoidance, minimization, restoration to offset, and ecological reconstruction) **d.** Minimization of openings in the project design and use of directional drilling to minimize pad openings to avoid impairment of ecosystem services, closure of wildlife corridors and impeding natural waterways. **e.** Location of geothermal infrastructure and support facilities outside the environmentally critical zone. **f.** Tree balling to transplant affected trees **g.** Reforestation or replacement planting **h.** Assistance in the government's forest protection to prevent illegal logging and slash and burn farming
12. Migration of wildlife due to disturbance of loss of habitat		**a.** Afforestation by tree species that support and attract back wildlife **b.** Baseline profiling to assess ways for early return of wildlife **c.** Wildlife monitoring and support to biodiversity conservation efforts

Continued

Table 17.2 Continued

Potential impact	Enhancement	Management measures
B. Social		
1. Increase in local employment	**a.** Prioritize populace radiating from the geothermal project **b.** Employment opportunities outside the geothermal operation may be provided in the form of environmental (eg, community reforestation) and other livelihood projects based on available resources in the area and skills of populace	
2. Promotion of gender equity	**a.** Women are given roles in community projects and given equal opportunity to represent the community in monitoring the geothermal project	
3. Increased funds for community development	**a.** Regulatory/mandated (royalties, local taxes, fees, etc.) or voluntary funds (welfare funds, CSR funds, etc.) may be used for community projects	
4. Community development	**a.** Must be strategic or must address the need of the community and the company to ensure its sustainability. For example, reforestation that provides livelihood to the community is good for the water recharge of the geothermal reservoir.	
5. Physical and economic dislocation		**a.** Air dispersion modeling of harmful emissions and to consider alternative location for facilities to avoid or minimize dislocation of residents

Table 17.2 Continued

Potential impact	Enhancement	Management measures
		b. If areas cannot be avoided, undertake resettlement to safer areas and replace lost structures, amenities and livelihoods to maintain or improve their original condition[a]
C. Cultural		
1. Limitation of traditional way of life		**a.** Conduct of ethnographic study to understand the culture and how to engage the IPs the proper way **b.** Respect for IP customs and traditions
2. Encroachment to ancestral domains		**a.** Seek for "free and prior informed consent" for the geothermal project
3. Effects on sacred sites		**b.** Baseline inventory of sacred sites for avoidance

[a]Protocols of multilateral banks and private banks known as Equator Principles that adopt social safeguards.

17.3.2 Develop a comprehensive environmental management plan

On completion of the EIA, an EMP should be developed based on the baseline and impact assessment results. The EMP details the site-specific environmental management practices, as well as preventive and mitigating strategies that will be undertaken throughout the project stages from construction to operation. As a fundamental aspect of the EIA system, the EMP is also a requirement for receiving approval or consent. The Australian government (Department of Infrastructure, Planning and Natural Resources, 2004) purports that an effective EMP should ensure that:

- The best practices in environmental management are applied;
- The conditions of consent or approval from the project EIA are implemented;
- Environmental policies and laws are complied with; and
- The environmental risks associated with the project are managed properly.

The EMP is site and project specific. The contents of the EMPs of geothermal energy projects depend mainly on the location and technology being used.

17.3.3   Conduct a social impact assessment

Parallel to the EIA, and often a component of EIA frameworks in some countries, the SIA describes the baseline socioeconomic and health indicators, scientifically predicts potential project impacts, and recommends appropriate measures both for mitigating negative impacts and enhancing positive ones. Conducting an SIA could be the first step to obtaining social acceptability, as it sets the boundary and identifies the key stakeholder groups that would be under the project's influence. The SIA could also reveal potential problematic areas, allowing the proponents to prepare and improve the construction and operational plans to avoid such issues.

17.3.4   Facilitate social acceptability

The success of a natural resource project like geothermal lies on its cultural/social acceptability based on public perception (Firey, 1960). Social acceptability has been defined by Brunson (1996) as the result of a judgmental process wherein individuals compare their present reality with its known alternatives and decide whether the "real" condition is superior, or sufficiently similar, to the most favorable alternative condition. For a project to be socially acceptable, the alternatives presented or offered by the project proponents should be perceived as superior or equal to their present conditions and that communities are not worse off because of it. This would entail that potential risks are avoided or if unavoidable, they are mitigated or that the compensation schemes are acceptable. Furthermore, social acceptance promotes cooperation, lessening the likelihood of conflicts with other stakeholders. On the side of the project proponents, the process of obtaining social acceptance promotes accountability and responsibility toward their host community.

An important requisite of acceptance is the early consultation of the host community and other stakeholders on how they will be affected by the geothermal project and how these can be prevented or mitigated. Local stakeholders must be empowered with resources and authorities to participate throughout the life of the project.

Social acceptance is dependent on the performance of the developer in terms of translating commitments on environmental and social measures into action. In some countries, an environmental guarantee fund is established as an immediate source for rehabilitation in case of environmental incidents (de Jesus, 2005). It is important to remember that stakeholder analysis, early consultations, and maintaining good relations with local authorities and communities are crucial to the social acceptance of the geothermal project.

17.3.5   Implement a comprehensive monitoring program

A comprehensive monitoring program is often included in the EMP or within the EIA process. The monitoring program puts in place a system for ensuring that the applicable environmental indicators remain within acceptable levels allowed by law throughout the project implementation, taking into account the environmental baseline/indicators established in the EIA. Regular monitoring also ensures that the environmental management measures being implemented remain to be effective and allows for early detection in case adverse environmental impacts are inadvertently being caused by the project operations (de Jesus and Nieva, 2003).

Oftentimes, only water and air parameters are monitored for geothermal projects, but to truly protect the environment and human health, project developers must go beyond these as suggested by Burger and Gochfeld (2012). The proposed coverage of the monitoring includes the traditional media (air, water, and land), ecological impacts (biodiversity, habitat, ecosystems, etc.), and human perceptions for a holistic assessment of an energy project's impacts. Lichens that are sensitive to H_2S and even mosses that are humidity dependent are endorsed as biological monitors (Bargagli et al., 2002; Mutia et al., 2015). EDC in the Philippines has also established long-term forest monitoring stations in all its project sites to determine changes in species structure and composition for biodiversity conservation assessment.

17.4 Geothermal energy in the context of sustainable development

The concept of sustainable development has been discussed many times and characterized in various ways and in different platforms. When one is asked to define sustainable development, the most widely used definition is that of Brundtland Commission, which defines *sustainable development* as: "Development that meets the needs of the present without compromising the ability of future generations to meet their own needs" (World Commission on Environment and Development, 1987). This definition brings to light the importance of including the environment and natural resources concerns into the development agenda. Sustainable development therefore prescribes the need to balance nature, people, and business concerns. Where business does not impair nature, it is feasible. When nature is not impaired, it remains livable by the people. For business to coexist with the communities, they need to share economic values equitably through their CSR program. When this situation is attained, we call this type of development as sustainable.

Chapter 9 of the Rio Declaration of the 1992 UN Conference on Environment and Development declares that geothermal is an environmentally advantageous energy option. In this context of sustainable development, geothermal energy is among the best options because of its high resource potential, renewability of the resource, and low carbon footprint. For the sustainability of geothermal operation, four performance indicators are prescribed: (1) low carbon operation to minimize the effects of climate change, (2) security of the geothermal resource through technological innovations, (3) robust environment which can buffer the geothermal project from extreme weather events, and (4) sharing economic values through the developers' community projects that support the community's appropriate development and to maintain harmonious relations with the host community.

17.5   Conclusions

From the survey of current knowledge and experiences on geothermal development, we draw the following conclusions:

1. Geothermal is a renewable energy option that can provide long-term, secure, baseload energy, and greenhouse gas (GHG) reductions that can contribute to the solution to climate change crisis. It has one of the lowest GHG emissions among the types of baseload power plants today.
2. Like any infrastructure project, geothermal has environmental, social, and cultural concerns. The good news is, these impacts are often localized, manageable, and are more likely to be reversible. Geothermal is a more than a century-old industry that has environmental and social safeguards already in place.
3. Because geothermal developers are dealing with natural resources, they must adopt cautious utilization addressing the effects of their operation following protocol of the "hierarchy of impacts" in decreasing order of importance: avoidance, minimization, restoration, or management of impacts and, as a last resort, compensation or offset to achieve "no net loss" or aim at "net gain" for the environment.
4. The legal and social license of the geothermal developer to operate in the area is facilitated by the:
   a. Participation of host communities, indigenous peoples, and other rights holders early and throughout the life of the project;
   b. Environmental performance and compliance with commitments by developers; and
   c. Sharing of meaningful benefits that support a development that is culturally appropriate to the host community.
5. For an environmentally sound and socially responsible geothermal operation, the following best practices are recommended:
   a. Conduct an EIA prior to project establishment;
   b. Develop a comprehensive EMP;
   c. Conduct an SIA;
   d. Facilitate social acceptability of the geothermal project;
   e. Implement a comprehensive monitoring program; and
   f. Facilitate public awareness.
6. For the sustainability of geothermal operation, the following performance indicators are recommended:
   a. Low carbon operation to contribute to the solution of the global climate change crisis;
   b. Security of geothermal resource through technological innovations;
   c. Maintenance of robust environment to protect geothermal project from natural hazards; and
   d. Sharing of economic values with host communities that supports their appropriate development and to maintain harmonious relations.

References

Adaweh, A.B., Idleh, M.R., 2015. Project evaluation of geothermal resources. In: Proceedings World Geothermal Congress 2015, Melbourne, Australia, April 19–25.
Aksoy, N., Gok, O.S., Mutlu, H., Kilinc, G., 2015. CO_2 emissions from geothermal plants in Turkey. In: Proceedings World Geothermal Congress 2015, Melbourne, Australia, April 19–25.

Arvizu, D.T., Bruckner, H., Chum, O., Edenhofer, S., Estefen, A., Faaij, M., Fischedick, G., et al., 2011. Technical summary. In: Edenhofer, O., Pichs-Madruga, R., Sokona, Y., Seyboth, K., Matschoss, P., Kadner, S., et al. (Eds.), IPCC Special Report on Renewable Energy Sources and Climate Change Mitigation. Cambridge University Press, Cambridge.

Bakane, P.A., 2013. Overview of extraction of minerals/metals with the help of geothermal fluid. In: Proceedings Thirty- Eight Workshop on Geothermal Reservoir Engineering, Stanford University, Stanford, California, U.S.A. February 11−13.

Barasa, J.B., 2015. Public participation in the implementation of 280MW geothermal power projects at Olkaria in Naivasha sub-county, Naukuru County, Kenya. In: Proceedings World Geothermal Congress 2015, Melbourne, Australia, April 19−25.

Bargagli, R., Monaci, F., Borghini, F., Bravi, F., Agnorelli, C., 2002. Mosses and lichens as biomonitors of trace metals. A comparison study of *Hypnum cupressiforme* and *Parmelia caperata* in a former mining district in Italy. Environmental Pollution 116, 279−287.

Bierre, E., Fullerton, R., 2015. Hydrogen removal from geothermal power station cooling water using reactor. In: Proceedings World Geothermal Congress 2015, Melbourne, Australia, April 19−25.

Bromley, C.J., Currie, S., Jolly, S., Mannington, W., 2015. Subsidence: an update on New Zealand geothermal deformation observations and mechanisms. In: Proceedings World Geothermal Congress 2015, Melbourne, Australia, April 19−25.

Brophy, P., 1997. Environmental advantages to the utilization of geothermal energy. Renewable Energy 10 (2−3), 367−377.

Brunori, C.A., Borgia, A., Brunori, A., Boschetti, M., Meroni, M., 2005. Geothermal power plants and forest decline: remote-sensing techniques for impact evaluation of effects of human activity impact in forestal environment. Geophysical Research Abstracts 7. http://meetings.copernicus.org/www.cosis.net/abstracts/EGU05/06776/EGU05-J-06776-1.pdf (accessed 08.02.15.).

Brunson, M.W., 1996. A definition of social acceptability in ecosystem management. In: Brunson, M.W., Kruger, L.E., Tyler, C.B., Schroeder, S.A. (Eds.), Defining Social Acceptability in Ecosystem Management: A Workshop Proceedings, Portland, USA.

Burger, J., Gochfeld, M., 2012. A conceptual framework evaluating ecological footprints and monitoring renewable energy: wind, solar, hydro and geothermal. Energy and Power Engineering 4, 303−314. http://dx.doi.org/10.4236/epe.2012.44040.

Chaves, R.B., 1996. Geothermal Gases as a Source of Commercial CO_2 in Miravalles, Costa Rica and Haedarendi, Iceland. http://www.os.is/gogn/unu-gtp-report/UNU-GTP-1996-03.pdf (accessed 10.04.15.).

de Jesus, A.C., 2005. Social issues raised and measures adopted in Philippine geothermal projects. In: Proceedings of the World Geothermal Congress 2005, Antalya, Turkey. http://www.geothermal-energy.org/pdf/IGAstandard/WGC/2005/0219.pdf (accessed 25.11.14.).

de Jesus, A.C., 2007. Environmental aspects of geothermal utilization − a global perspective. In: A Short Course on Geothermal Development in Central America, San Salvador, El Salvador, November 25 − December 1. http://www.os.is/gogn/unu-gtp-sc/UNU-GTP-SC-04-06.pdf (accessed 15.11.14.).

de Jesus, A.C., Nieva, M.D., 2003. Successful environmental and social initiatives in the Tongonan and Palinpinon geothermal projects, Philippines. In: Proceedings of the 24th Annual PNOC-edc Geothermal Conference. Hotel Intercontinental Manila, Philippines.

Department of Infrastructure, Planning and Natural Resources, 2004. Guidelines for Preparation of Environmental Management Plans. http://www.planning.nsw.gov.au/rdaguidelines/documents/emp_guideline_publication_october.pdf (accessed 23.11.14.).

Dickson, M., Fanelli, M., 1995. Geothermal Energy. John Wiley, Chichester.

Edwards, S., 2015. A Balancing Act. http://www.iucn.org/?18817 (accessed 15.01.15.).

Environmental leader, 2015. Acoustiblok Quiets Water District's Well Motor Noise. http://www.environmentalleader.com/2015/06/03/acoustiblok-quiets-water-districts-well-motor-noise/#ixzz3c5laoRdt (accessed 04.06.15.).

Equator Principles, 2015. Environmental and Social Risk Management for Projects. http://www.equator-principles.com/index.php/about-ep (accessed 07.03.15.).

Evans, A., Strezov, V., Evans, T.J., 2009. Assessment of sustainability indicators for renewable energy technologies. Renewable and Sustainable Energy Reviews 13 (5), 1082−1088.

Firey, W., 1960. Man, Mind and Land. The Free Press, Illinois.

Gehringer, M., Loksha, V., 2012. Geothermal Handbook: Planning and Financing Power Generation. The World Bank, Washington. http://www.esmap.org/sites/esmap.org/files/DocumentLibrary/FINAL_Geothermal%20Handbook_TR002-12_Reduced.pdf (accessed 10.11.14.).

Geothermal Energy Association, 2012a. Geothermal Basics. http://geo-energy.org/geo_basics.aspx (accessed 12.03.15).

Geothermal Energy Association, 2012b. Why Support Geothermal Energy. http://geo-energy.org/pdf/FINALforWEB_WhySupportGeothermal.pdf (accessed 12.03.15.).

Geothermal Energy Association, 2015. Geothermal Energy and Water Consumption. http://geo-energy.org/pdf/Geothermal_Energy_and Water_Consumption_Issue_Brief.pdf (accessed 08.06.15.).

Geothermal Energy Association, 2009. Geothermal Energy and Induced Sesimicity. http://geo-energy.org/pdf/Geothermal_Energy_and_Induced_Seismicity_Issue_Brief.pdf. (accessed 12.03.15.).

Goldstein, B., Hiriart, G., Bertani, R., Bromley, C., Gutiérrez-Negrín, L., Huenges, E., et al., 2011. Geothermal energy. In: Edenhofer, O., Pichs-Madruga, R., Sokona, Y., Seyboth, K., Matschoss, P., Kadner, S., et al. (Eds.), IPCC Special Report on Renewable Energy Sources and Climate Change Mitigation. Cambridge University Press, Cambridge.

GtV-BV, 2010. Induced Seismicity: Position of the German Geothermal Association GtV-bv. http://www.geothermal-energy.org/uploads/media/gtv-bv_position_paper_seismicity_070710.pdf (accessed 08.02.15.).

Gunnarsson, I., Juliusson, B.M., Aradottir, E.S., Sigfusson, B., Arnarson, M., 2015. Pilot scale geothermal gas separation, Hellisheioi power plant, Iceland. In: Proceedings World Geothermal Congress 2015, Melbourne, Australia, April 19−25.

Health Care Without Harm, 2015. World's Public Leaders Call for an End to Coal. https://noharm-global.org/articles/news/global/worlds-public-health-leaders-call-end-coal (accessed 16.02.15.).

Holm, A., Jennejohn, D., Blodgett, L., 2012. Geothermal Energy and Greenhouse Gas Emissions. http://geo-energy.org/reports/GeothermalGreenhouseEmissionsNov2012GEA_web.pdf (accessed 20.04.15.).

International Finance Corporation, 2010. A Quick Guide. Highlights from IFC's Good Practice Handbook. IFC, Washington.

Juliusson, B.M., Gunnarsson, I., Matthiasdottir, K.V., Markusson, S.H., Bjarnason, B., Sveinsson, O.G., Gislason, T., et al., 2015. Taking the challenge of H2S emissions. In: Proceedings World Geothermal Congress 2015, Melbourne, Australia, April 19−25.

Kagel, A., 2006. A Handbook on the Externalities, Employment and Economics of Geothermal Energy. Geothermal Energy Association, Washington.

Kagel, A., Bates, D., Gawell, K., 2005. A Guide to Geothermal Energy and the Environment. Geothermal Energy Association, Washington.

Kate, K., Crowe, M., 2014. Biodiversity Offsets: Policy Options for Governments. IUCN, Gland.

Koorey, K.J., Fernando, A.D., 2010. Concurrent land use in geothermal steamfield developments. In: Proceedings of the World Geothermal Congress. Bali, Indonesia. www.geothermal-energy.org/pdf/IGAstandard/WGC/2010/0207.pdf (accessed 29.11.14.).

KPMG International, 2012. Expect the Unexpected. Building Business Value in a Changing World. KMPG International, London. http://www.kpmg.com/Global/en/IssuesAndInsights/ArticlesPublications/Documents/building-business-value.pdf (accessed 03.04.15.).

KPMG International, 2014. Future State 2030: The Global Megatrends and Shaping Governments. KMPG International, London. http://www.kpmg.com/Global/en/IssuesAndInsights/ArticlesPublications/future-state-government/Documents/future-state-2030-v3.pdf (accessed 06.04.15.).

Lasco, R.D., Lales, J.S., Arnuevo, M.T., Guillermo, I.Q., de Jesus, A.C., Medrano, R.S., Bajar, O., Mendoza, C., 2002. Carbon dioxide (CO_2) storage and sequestration of land cover in Leyte Geothermal Reservation. Renewable Energy 25, 307−315.

Lund, D., 2009. Rent taxation for renewable resources. Annual Review of Resource Economics 1, 287−308. http://dx.doi.org/10.1146/annurev.resource.050708.144216.

MacKinnon, J., MacKinnon, K., 1986. Managing Protected Areas in the Tropics. IUCN, Cambridge.

Matek, B., 2013. Promoting Geothermal Energy: Air Emissions Comparison and Externality Analysis. http://geo-energy.org/reports/Air%20Emissions%20Comparison%20and%20-Externality%20Analysis_Publication%20May%202013.pdf (accessed 26.02.15.).

Mutia, T., Ingibjorg, S.J., Frioriksson, P., 2015. Monitoring protocol for potential hydrogen sulfide effects on moss (*Racomitrium lanuginosum*) around geothermal power plants in Iceland. In: Proceedings World Geothermal Congress 2015, Melbourne, Australia, April 19−25.

Odour, J.A., 2010. Environmental and social considerations in geothermal development. In: Proceedings FIG Congress. Facing the Challenges − Building the Capacity, Sydney, Australia. http://www.fig.net/pub/fig2010/papers/ts01e/ts01e_oduor_3857.pdf (accessed 01.12.14.).

Ogola, P.A., 2009. Environmental impact assessment general procedures. In: A Short Course IV on Exploration for Geothermal Resources, Lake Naivasha, Kenya November 1−22. http://www.os.is/gogn/unu-gtp-sc/UNU-GTP-SC-10-0801.pdf (accessed 01.12.14.).

Olafsdottir, S., Gardarsson, S.M., Andradottir, H.O., Armannsson, H., Oskarsson, F., 2015. Near field sinks and distribution of H_2S from two geothermal power plants in Iceland. In: Proceedings World Geothermal Congress 2015, Melbourne, Australia, April 19−25.

Ozcan, N.Y., Gokcen, G., 2009. Thermodynamic assessment of gas removal systems for single-flash geothermal power plants. Journal of Applied Thermal Engineering 29 (14−15), 3246−3253.

Pilgrim, J.D., Ekstrom, J.M., 2014. Technical Conditions for Positive Outcomes from Biodiversity Offsets. An Input Paper for the IUCN Technical Study Group on Biodiversity Offsets. IUCN, Gland.

Premuzic, E., Lin, M.S., 1989. Advanced Biochemical Processes for Geothermal Brines. Annual Report. Brookhaven National Laboratory, New York.

PricewaterhouseCoopers, 2014. Forces of Change. Sustainability: Building Relationships, Creating Value. www.pwc.com/gx/en/sustainability/index.jhtml (accessed 07.03.15).

Ragnarsson, A., 2013. Geothermal energy use, country update for Iceland. In: Proceedings of the European Geothermal Congress. Pisa Italy. http://www.geothermal-energy.org/pdf/IGAstandard/EGC/2013/EGC2013_CUR-16.pdf (accessed 23.02.15.).

Rybach, L., 2010. Legal and regulatory environment favourable for geothermal development investors. In: Proceedings of the World Geothermal Congress. Bali, Indonesia. www.geothermal-energy.org/pdf/IGAstandard/WGC/2010/0303.pdf (accessed 25.11.14.).

Sacher, H., Schiemann, R., 2011. When Do Drilling Geothermal Wells Make Good Economic Sense? http://www.renewableenergyfocus.com/view/17099/when-do-deep-drilling-geothermal-projects-make-good-economic-sense/ (accessed 22.04.15).

Sequeira, H.G., 2015a. Environmental management at the Miravalles geothermal field after 20 years of exploitation. In: Proceedings World Geothermal Congress 2015, Melbourne, Australia, April 19−25.

Sequeira, H.G., 2015b. Geothermal developments in protected areas: case history from Costa Rica. In: Proceedings World Geothermal Congress 2015, Melbourne, Australia, April 19−25.

Taipei Times, September 21, 2008. Plans for Geothermal Energy Could Put Japanese Spring Resorts in Hot Water. http://www.taipeitimes.com/News/world/archives/2008/09/21/2003423777/1 (accessed 23.02.15.).

Tuyor, J.B., de Jesus, A.C., Medrano, R.S., Garcia, J.R.D., Salinio, S.M., Santos, L.S., 2004. Impacts of geothermal well testing on exposed vegetation in the Northern Negros Geothermal Project, Philippines. Geothermics 34 (2), 252−265.

United Nations, 1992. Agenda 21, UN Convention on Environment and Development. https://sustainabledevelopment.un.org/content/documents/Agenda21.pdf.

United Nations, 2008. UN Declaration on the Rights of Indigenous Peoples. United Nations, Geneva.

U.S. Department of Energy, 2016. Geothermal Power Plants: Minimizing Land Use and Impact. http://www1.eere.energy.gov/geothermal/geopower_landuse.html.

U.S. Environmental Protection Agency, 2014. Clean Energy. http://www.epa.gov/cleanenergy/energy-and-you/affect/oil.html (accessed 27.02.15.).

Wetang'ula, G.N., Wamalwa, H.M., 2015. Trace elements in rainfall collected around Menengai area in Kenya. In: Proceedings World Geothermal Congress 2015, Melbourne, Australia, April 19−25.

World Commission on Environment and Development, 1987. Our Common Future. Oxford University Press, New York. http://www.un-documents.net/our-common-future.pdf.

World Economic Forum, 2015. Global Risks 10th Edition. WEF, Geneva. http://www.weforum.org/reports/global-risks-report-2015.

Yilmaz, I., Kabay, N., Brjyak, M., Yuksel, M., Wolska, J., Koltuniewicz, A., 2005. A submerged membrane-ion exchange hybrid process for boron removal. Desalination 198, 310−315.

Zaidi, S.H., 2015. Economic growth and poverty reduction through geothermal development: Integrated policy measures to rural development and economic sustainability. In: Proceedings World Geothermal Congress 2015, Melbourne, Australia, April 19−25.

Project permitting, finance, and economics for geothermal power generation

18

M.C. Moore
Cornell University, Ithaca, New York, USA

18.1   Introduction

The development and integration of geothermal power into electricity grids are ultimately dependent not only on adequate heat resources but also on adequate capital and financing and securing the contracts necessary to create sustainable projects. This chapter reviews the key elements of project approval and financing for hydrogeothermal projects that are intended to be used for base load electric power.

18.1.1   Chapter structure

This chapter presents a review of the issues surrounding the special financing requirements for geothermal power development. Geothermal power is typically associated with base load power generation, a special category in most electricity grids. In this application, power generated is competitively priced and strategically offered in the base load component of daily generation planning.

The chapter begins with a discussion of the unique financing arrangements that have evolved to address the inherent risks of exploration and development of this technology. Independent investors as well as public agencies or utilities are interested in deploying this technology in both competitive and regulated markets. This level of deployment is best considered at an individual "project" level as opposed to a market or system level.

The process of developing and financing geothermal technologies is complex, initiating and imposing sequential responsibilities and review by public agencies, demanding patience and commitment from developers, financing agencies, and investors.

This chapter includes evidence from new geothermal applications in the most active areas in Australia (Huddleston-Holmes et al., 2014), the United States (Ayers et al., 2004; Matek, 2013), and Europe (IFC, 2013), where we find some of the newest development of geothermal resources. This is followed by a discussion of the nature of costs that can be expected in each phase of project development and follow with a discussion of unique problems and risks involved in finding, permitting, and operating the facilities. The chapter concludes with a brief review of recent projects under development and the potential for the use of renewable energy credits (REC) and avoided

Geothermal Power Generation. http://dx.doi.org/10.1016/B978-0-08-100337-4.00018-8

greenhouse gas (GHG) Credits in carbon markets that may supplement revenue streams over time.

18.1.2 Renewable energy finance

Commercial power generation, no matter what technology is deployed, represents significant, long-term capital and operating commitments. Consequently, given the magnitude and nature of geothermal power facilities, the financing requirements and considerations involve risk at various levels, complicating and making the ultimate financial outcomes — short of actual power delivery and sales — subject to some degree of uncertainty.

Financing as used here represents the set of tools that take that uncertainty and opportunity into account, suggesting arrangements that can be made to evaluate potential, develop a proof of concept, and, finally, deliver power (whether heat or electricity) to a client such as a system operator. The design of financing schemes consequently is highly dependent on the structure of the venture itself. This can include a wide range of variability depending on physical as well as market conditions. For simplicity, in this chapter, it is assumed there are three variants of collaboration, investment, or ownership. These are publicly owned and financed facilities, merchant plants, or public—private joint ventures.

For geothermal facilities, the inherent resource risk can be a significant barrier in financing a project. The resource risk can be defined as the risk associated with locating, evaluating, and drilling a well necessary to test the characteristics of the resource at depth. For electricity projects, this can be assumed to be a temperature gradient $> 35-40°C$ per km with reliable temperature for fluid circulation $> 150°C$ and at drilling depths $< 5-7$ km. The test for direct heat as opposed to electric generation projects is less onerous, as the heat resource can be valuable at temperatures $> 80°C$, but with the caveat that drilling depths must be less ($3-5$ km) and the well system must be within $1-2$ km of the load to be efficient.

The sizable upfront cost[1] and resource uncertainty of geothermal projects may discourage many traditional renewable energy financiers. In developing a geothermal electric project, a developer often must commit their own equity capital (or those from an investment partner) to finance the higher-risk elements of project development. Once the resource has been adequately proved (a subjective milestone), more risk-adverse financiers are likely to invest in the geothermal electric project (Gehringer and Loksha, 2012).

A geothermal project is commonly financed in a series of stages and investor types. Each stage of development attracts a different investor type with appropriate risk and return preferences. The riskiest early to mid stages of project development occur typically financed with equity. As the project proceeds to the construction and operational phases, project risk is reduced and debt financing is increasingly used.

---

[1] The EIA 2013 estimates for overnight capital costs for wind, solar PV, and binary geothermal are US$2213, US$4183, and US$4362/kW, respectively.

18.1.3   Power generation and finance

Power-generating facilities typically support regional or multiregional distribution systems,[2] while high-density customers, groups, or load centers anchor relatively predictable patterns of demand. Most existing grid systems rely heavily on fossil-fired generation with greater, but still not dominant, penetration of renewable generation coming online over time. The nature of fossil fuel systems allows plants to be operated at a distance from fuel sources, if adequate pipes, storage, or transport such as rail lines are available. Electric power systems, once built, are rarely operated intermittently by choice; unless down for maintenance, plant facilities will be operated continuously throughout the year, with minimal ramping and demand response expected from operators.

Financing and building renewable energy projects such as geothermal facilities are based on an ordered series of events that can be viewed independently but that must be sequentially and satisfactorily completed before initiating the following step, or the project will either sit idle or fail.[3] In a sense, this makes the venture riskier than fossil fuel facilities; the experience of the marketplace and grid operators with these thermal facilities reduces uncertainty and further, because there are multiple hearings involved, increases public exposure and subjects the project to changes in taste or need.[4]

Fig. 18.1 shows the financing and development spectrum of a typical geothermal electric plant. The role of sequential permits and permission, implying separate financing arrangements, is apparent in the multiple steps.

The market for renewable energy such as geothermal power is driven by normal considerations for any investor, namely, anticipating a return on investment, managing a commitment or desire to improve environmental quality as society begins a transition away from fossil fuels, or to improve overall grid management.

18.1.3.1   Finance, hedging, and risk management

The issue of financing internalizes known and expected risk, relying on known estimates of project or facility life and power delivery, capacity, and reliability. Geothermal power developers deal with a high degree of risk at the inception of their projects but enjoy the benefits of long project lifespan, high reliability, predictable pay-

---

[2] There are exceptions to this central-station or hub-and-spoke type model. They are referred to here as distributed generation or local distribution systems and they are characterized as remote independent areas or those that have minimized reliance on out-of-region generation or transfers and reliance on high-voltage transmission support. The same thing can be said for some fossil fuel facilities, e.g., coal-fired power plants being phased out because of rule changes, cheaper competition from natural gas.

[3] The literature on renewable energy resources is replete with examples of stalled, idled, or abandoned projects that could not be financed after rule changes, competition from new capacity was installed earlier or costs became excessive.

[4] Both the Bottle Rock and Newberry projects in the western United States demonstrate this characteristic. Without successful stimulation and flow, the project(s) can be canceled or stalled, with any next steps subject to an entirely new round of permits and financing.

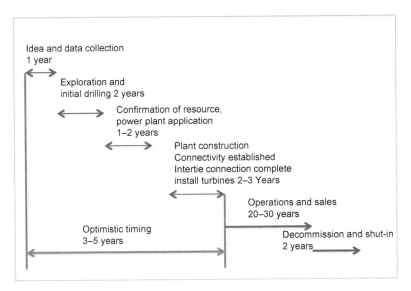

Figure 18.1 Phases and timing of geothermal development.

ment streams, and high-capacity factors. This is not to say there are not uncertain events that must be factored into long-term operations of geothermal facilities, such as water or heat transfer fluid loss, reservoir cooling, or unexpected seismic impacts on wells and reservoirs. In terms of base load generation, however, existing geothermal facilities' performance suggests a highly desirable and predictable power resource for grid and industrial applications.

Tempering this assessment is the fact that the nature of the electric power market has changed relatively rapidly in recent years. Nuclear facilities have begun to reach the end of their economic and, in some cases, physical life, and coal plants are being decommissioned in response to environmental performance standards and rules. Natural gas—fired generation has improved in performance and cost and may be in position to offset nuclear and coal losses by running in a base load configuration with unprecedented ramping and dispatch capability. As a consequence, the attraction for technology like geothermal is challenged and the risk appetite for investors is tempered and more uncertain as a result.

Geothermal power represents a stable technology with predictable long-term performance characteristics, offsetting some of the uncertainty from competitive technologies. Developers may face an alternative type of uncertainty, however, originating in the policy arena. In Australia for instance, a combination of cost concerns and reversal of carbon management targets has effectively diminished both research and commercial development interest in geothermal, especially for EGS-type systems. This type of uncertainty means that changes in policy become unpredictable in terms of both timing and support for the industry.

18.2   Finance background

The role of financing to underwrite the emergence of geothermal power into grid or local industrial operations is critical to success, not only for the developers, shareholders, and owners of facilities but for risk-averse grid operators as well.

Large capital projects must obtain a combination of financing and equity investment from owners or shareholders in order to be built. In addition, most regulatory agencies demand proof of adequate finance capacity to maintain and operate facilities. In broad terms, financing a project entails debt, typically multiple finance partners, and assurances in the form of permits and agreements to allow the project to go forward. Ultimately, all finance arrangements are subject to market confidence in the project characteristics, demand for the technology, and the impact of competitive technologies and market prices.

18.2.1   Financing arrangements

Power generators use a wide range of financing arrangements, combining various types of contributory or participatory agreements. What they share is a focus on long-term, dedicated, and region-specific power generation that is designed to serve specific markets, typically in niches such as base load, load following, or peaking capacity. For simplicity, however, cost comparisons are usually made under one of two sets of financing assumptions — either "merchant" or "public" financing.

Merchant plants *are those built and operated by private investors. These investors pay for their capital through debt and by raising equity and thus pay return on equity and interest on debt throughout their lifetime. These projects must also pay income taxes, both provincial and federal.*

Publicly financed projects *typically are not subject to income taxes or the same constraints on raising finance through issuing debt and equity. However, they are constrained to provide a specified rate of return.*

Public—private partnerships *offer a third option, whereby financing is shared in terms of risk and benefits. Under such an arrangement public financing could be used for construction of a merchant plant that would then be leased over a long period to private operators. A number of other public—private partnership arrangements are also possible.*

18.2.2   Project structure

Electric power—generation projects are conceived, designed, and directed to fill market niches where they can be called on to deliver either electricity or direct heat to a distribution center. The projects may be built in anticipation of growing demand (load) or in response to a call from a utility or system operator. Financing the project depends on an accurate forecast of expected revenue flows based on the technology and design of the facility. Obtaining necessary financial support only occurs when the project is forecast to earn enough to cover all operating and debt-servicing expenses and provide an adequate return on investment over the whole expected life of the project.

Demand or load in electric markets is a relatively predictable characteristic, but varies daily or seasonally and can be extremely volatile in some regions. For this reason, most electric systems require reserve capacity in order to serve load without interruption. Some of the generation capacity will be used periodically for "firming" or backing up intermittent resources such as wind or solar. Agreements to serve load can range from hourly or daily bids called for by the system operator to bilateral agreements and dedicated transmission links for large consumers. Most load growth is forecast in a "needs assessment" routinely developed for the system operator or regional grid and identifies the nature of expected new demand and capacity by technology type.

For regulated utilities, rates of return are determined by a utility *regulator*, who calculates how much the utility and their generators are allowed to invest, what can be charged to consumers, and what profit margin and return on investment are allowed (regulatory compact). Public utility commissions (PUCs) determine a utility's total revenue requirement in what is known as a rate case. The rate "case" is typically heard at intervals of 5 years, where a new basis for charging for expenses and capital expenditures is established. The generalized revenue requirement formula is:

$$\text{Total revenue requirement} = \text{Rate base} \times \text{Allowed rate of return} + \text{Expenses}.$$

where the total revenue expected is equal to the estimated rate base of load or demand multiplied by the set rate of return determined to reflect market conditions plus approved expenses for maintenance, capital replacement, and expansion. The revenue will be expressed in the currency of the country in question in terms of an agreed-on base year to account for inflation.

This practice is conservative in nature, rewarding caution in power technology choices and arguing for prudent maintenance standards and protocols. In return for this conservative approach, the nature of set rates of return ensures that utilities will be able to borrow and raise sufficient capital to make improvements to their infrastructure and provide reliable service to all customers. Since this has the effect of lowering investor risk, the result can be a lower cost of capital than other businesses.

The rate of return expected reflects the cost of paying back its debt holders with interest and the return utilities must provide to their equity shareholders. The unique element of this formula is the calculation of the utility's allowed return on equity (ROE)—that of the revenue received kept as profit. The revenue stream for power generators only results in a profit when average prices exceed marginal costs of production, a factor built into the system average price paid for base load power delivery.

Site location and land use control for new facilities underpin the investment terms that combined with the overnight capital cost estimates will be the basis of either rate recovery at the public utilities commission or the term sheet for investors. This, in turn, will be the basis for establishing the expected financial relationships between investors or the ratepayer, and taxpayer subsidies and support necessary to obtain construction financing. For geothermal plants, the next stage will be to obtain permits for exploration and estimation of production capability and the initiation of environmental and regulatory permission. Only after this phase is complete will the permit process for actual construction and operation begin, necessitating a new round of financing agreements based on more-complete estimates of operating costs and expected returns.

Future financing needs for restimulation and redrilling of wells if necessary will be financed based on operating revenues where debt service can be carried based on current performance within the existing and forecast distribution network.

18.3   Recent evidence in geothermal drilling and construction

Four significant forces have stalled expansion of the installed base of geothermal facilities in recent years. First, a pull-back of funding from public sources for "risky" or novel new technologies has put a higher burden on private investors during a time of public fiscal stress. Second, the emergence of relatively inexpensive, "clean" natural gas reserves has made gas turbine electricity generation more attractive than in recent years. Third, imprecise information about thermal gradients and subsurface performance has introduced an element of uncertainty and increased the cost of risk capital relative to the perceived benefits of geothermal to modern grid operations. Finally, the cost of exploration and proving resource capacity remains high, in part because of a low volume of activity and in part due to the high cost of successive regulatory and permit applications. Some of these costs are shifted to tax and ratepayers through mandatory portfolio standards and "must take" preferential power purchase by systems operators that double as virtual loan performance guarantees after a plant is constructed and operating.

The United States is the world's largest producer of geothermal electricity and has added more geothermal sourced electric capacity than any other country since 2006 (Goldstein and Braccio, 2014). In general, the U.S. government has tended to support geothermal development using loan guarantees that assist during the project development phases when financial risk is at its highest. For instance, the use of a production tax credit (PTC) of up to US$0.019/kWh since 2008 initially encouraged stable investment patterns. State governments, such as those of California and Nevada, have also supported growth in this sector using renewable portfolio standards (a goal of 33% by 2020 in California, 25% by 2025 in Nevada) and performance subsidies through the California Energy Commission (CEC).

In Oregon, difficulty with containing injection water and maintaining flows recently led to reduction in funding for AltaRock, a company engaged in resource development and stimulation. This, in turn, dampened support for a project that is conveniently located near a major transmission facility that was penetrating a high-temperature gradient resource at reasonable depths. The lack of continuing support and subsidy through the exploration and demonstration phase by the U.S. DOE has been seen as a major contributor to diminished investor interest and confidence.

At Cornell University in New York, a project to explore the electric and direct heat potential from relatively low-temperature rock has experienced delays in part due to public concerns over hydraulic fracturing necessary for flow. This can also be traced to the increased attraction for natural gas generation with perceived improved economic performance due to the face of the falling price of natural gas for both power generation and for space heating.

In Canada, research using oil and gas bottomhole temperature measurements (BHT) revealed heat isoquants > 150°C near transmission lines that serve some oil sands operations that could provide power generation for industrial facilities. Associated direct heat resources available could offset part of the cost of developing hydrothermally based electricity with the additional benefit using the offset natural gas demand as a GHG credit (Majorowicz and Moore, 2014).

In Australia, the ARENA expert report (2014) cited the high cost of drilling and exploration, the difficulty in obtaining transmission access, and the cost of drilling test facilities as major reasons to terminate the public support for developing geothermal-derived electric facilities. This reverses a popular policy support program that had been directed toward high-heat resources such as Cooper Basin in the center of the country.

Germany has been very active in supporting the growth of green energy in a strategy known as the Energiewende. The use of generous feed-in tariffs for renewable energy technologies has dramatically changed the overall electricity mix of the country during the past decade. As pointed out in the ARENA (2014) report, Germany's geothermal resources are primarily conductive and, thus,

geologically analogous to the type of geothermal resources available in Australia. As of November 2013, there were a total of 26 operating geothermal projects in Germany, of which six were generating electricity. Of this total, 19 projects provide direct heat, without generating electricity, while four of the projects are combined heat and power, and three are electrical generation only. Several of the facilities include Enhanced (or Engineered) Geothermal System (EGS) type generation for which Germany provides an added feed-in tariff of €50/MWh to the basic geothermal feed-in tariff of €250/MWh. Germany also established the world's first commercial EGS operation at Landau, which has been operating since 2007 without a decline in temperature or output of the plant.

18.4   Cost and financing issues

Obtaining long-term financing is a critical aspect for the development of any power project (European Bank, 2013; Evans and Peck, 2014; Gehringer and Loksha, 2012; IFC, 2013; Matek, 2013). Ultimately, obtaining necessary financing will involve agreements with various lenders that may include private banking or lending institutions, bilateral and multilateral lending agencies, or international and domestic export banks and regional utilities.

18.4.1   Actors and roles

Geothermal resources are widely distributed but not uniformly accessible at reasonable or affordable depths and temperatures; consequently, they are not always conveniently located for grid operations. As a practical matter, the number of potential actors who would participate in geothermal resource development is limited. For purposes of structure, these can be assumed to include a development company, interested shareholders, the ultimate facility operator, power purchaser or system operator, lenders, construction and drilling company, local or regional government, and a higher-order government agency if there are any subsidies involved.

At the core of the project financing structure is the developer, which controls the business plan and initiates the research and site development process. This is likely to be a company whose structure and financing are specially tailored to the site or project and may be a subsidiary of a larger power company in order to provide investment clarity and to isolate risk. Shareholders and investors may include the contractor, operator, and possibly the power purchase agency in some circumstances if guarantees of power delivery were part of their mandate.

Most base load developers will represent private investment with risk offset by long-term power purchase arrangements or contracts and subsidies for performance or to offset the risk of exploration included in the company structure. Private developers, independent power producers (IPPs), or merchant plants represent a potentially attractive and competitive area for private parties (certain utilities may also qualify as quasi-private depending on their status), and some public agencies (such as the Northern California Power Agency (NCPA) at The Geysers in northern California) enter project development in order to ensure that their constituents have power available at affordable rates, as well as taking advantage of financial opportunities.

18.4.2   Exploration phase

Optimal sites for development of geothermal resources would be those located near, but not under, load zones[5] such as cities or industrial centers. In most cases, however, the resources are likely to be found at some distance from load and necessitate remote power interconnections in order to supply grid operations. The exploration and identification of the resource, even when developed at the margins of existing fields, involve the work of a varied team of geophysicists, geochemists, and seismologists and draw heavily on existing data from oil and gas operations.

---

[5] Recent events in Texas, Oklahoma, and Switzerland have made drilling and hydraulic fracturing an area of risk that brings concerns over potential seismic events and property damage to bear on permit approval.

Due to the subsurface nature of the resource, some degree of uncertainty will exist regarding lithographic characteristics and potential heat at depth. The only way to resolve this uncertainty is to drill an exploratory or production well. Most production wells are only completed after a thorough exploration program has yielded physical and seismic data sufficient to reveal the reservoir temperature range and demonstrate adequate flow rates. It can be useful to compare the exploration and drilling phases for geothermal resource with that of the oil and gas industry in the sense that drilling, exploration techniques, and future technology transfers point to areas of cost savings and efficiency.

As the original MIT study (Tester et al., 2006) points out, there is no sure-fire formula for insuring success in geothermal exploration, but in general, it is useful to cite success on the basis of a well's delivered power-generating capacity (MWe). Success is improved by new drilling in an area, where continued drilling in established fields can bring down costs at predictable rates through improvements in the learning curve for drillers and managers. Exploration needs vary on a global and regional scale. "Even within the United States regions such as the Cascades in the Pacific Northwest have undergone less exploration than the Great Basin region and merit a different approach due to geographic differences and other extrinsic factors. As such, varying exploration needs require an open minded approach to future research and development" (Jennejohn, 2009).

The geology of the resource will affect the success of a well, with fields in sedimentary basins, where drilling is above the basement, having the highest success rates. In their 2013 work, IFC found that 16% of the wells analyzed in their worldwide cross section had been redrilled. The average success of those wells was "higher (at 87 percent) than for original wells (77 percent)." They concluded "well depth does not appear to have any bearing on a well's success."

The reference to "success" often implies that lack of success is associated with a "dry hole" for hydrogeothermal operations. The IFC points out, however, "in geothermal fields, a 'dry hole' is a rarity. Almost all geothermal wells flow to some extent... . However, a geothermal well may be deemed unsuccessful for one or more reasons — for example, if:

1. Unexpected mechanical problems are encountered during drilling, and the well is partly filled or bridged by drill cutting and/or casing collapse;
2. It has an inadequate temperature;
3. It has too low a static pressure;
4. It encounters a reservoir that is too "tight" (i.e., the Productivity Index [PI] is low); or
5. it has unacceptable chemical problems (such as gassy, corrosive, or scaling-prone fluids)" (IFC, 2013).

Effectively, a well's success should be determined on the basis of its return on investment (ROI) (IFC, 2013). An illustrative example for success is offered, pointing out that "in addition to the costs of pumping and drilling, the costs of building, operating, and maintaining the attached power plant must also be taken into account, as well as the price of the electricity sold. A 2-MWe or 3-MWe (net) well may be commercially successful if it is, for example, a relatively shallow well costing in the order of $1 million, but such a well may be uneconomic if it is a deep well costing several million dollars. Equally, a system producing 2 or 3 MWe at a relatively shallow depth might be considered successful at an electricity price of $100/MWh, but represents a marginal prospect at a price of $60/MWh".

In the case of very deep wells, new drilling areas, or EGS, this phase of the operation may consume the highest fraction of total costs when measured as a function of final delivered power. This reflects the potential for dry holes, or incomplete fractures, circulation, and connection between wells before construction. There is no rule of thumb for this portion of the operation, and given the potential for failure in this phase, work is often underwritten as research or exploration shared by public agencies.

18.4.3 Construction and operation phases

Proof of the resource and development of the site are coordinated closely with power interconnection facilities and substations. During this phase, the completion of arrangements for sale of power to the grid is perfected, and will typically result in a power purchase agreement (PPA) for both merchant and public power delivery. It will be rare to have spot market sales for geothermal power delivered as base load supplies.

Financing for base load power plant development relies on a high degree of confidence regarding load stability and growth rates. The PPA entered into with a utility or power agency provides the most optimal solution for developers. The agreement is most often between the purchaser "off-taker" such as an independent system operator (ISO) and the plant owner or operator.

These agreements are usually negotiated prior to construction finance and may be a requirement of a power development permit. They secure the payment stream for an independent power plant specifying the delivery and payment terms.

Normally, the power purchaser will agree in the PPA to pay both an energy charge and a capacity charge. The energy charge is designed to reflect and account for actual electricity generated and include the costs of generation. An energy charge represents short-run marginal costs; such as variable operations and maintenance, as well as tolling for the power line interconnect. This is the agreement by which the generator receives sufficient funds to service debt and returns on equity. A typical term for a PPA will range from 10 to 20 years after initial generation. Depending on the service area or adopted rules by the regulatory body, there may also be a capacity charge for users. This is effectively a charge to cover the availability of generation capacity, available for dispatch to meet load. It can also be referred to as a "demand" or load charge.

Arrangements are usually made with the ISO or dispatch agency to plan maintenance and downtime for the facility since they are responsible to ensuring that adequate capacity is available to serve the area. In addition, some of the wells and systems may need to undergo rest periods or recharge periods and can involve redrilling or supplemental reinjection of water such as the wastewater transport system at The Geysers in California. If redrilling or hydraulic fracturing are necessary to improve flow rates, the costs will be included in O&M (operations and maintenance or services required to assure continous functional integrity of the facility) and paid for by system charges through market sales. Redrilling and establishment of replacement (as opposed to new capacity) wells will be issues for both existing and new shareholders and could diminish share value over time.

The rate of restimulation, or the ultimate next step of redrilling existing well fields to improve circulation or maintain heat flows, will vary with the area and lithography. Although geothermal power fuel appears to be "free," acquiring and maintaining the heat resource imply clear and continuing costs. Thus, like wind or solar plants, the life-span of the technology deployed is not infinite and must be periodically updated or refurbished in order to function effectively, a functional reality for every type of energy generation technology.

18.4.4   Decommissioning and site reclamation

Depending on the permitting jurisdiction, new energy facilities, especially those involving wells and underground storage, may incur an end-of-life requirement for the developers. This typically requires cleanup and refurbishment of surface facilities, including remediation to bring the landscape back to a preagreed status and to cap and shut-in existing wells. In the case of geothermal wells, or other wells that have used any hydraulic fracturing, the requirement may specify ongoing seismic monitoring, fluid migration and site control. When the permit authority includes this type of recla-mation, the applicant is typically required to post and maintain a performance bond to cover the estimated cost of compliance.

Since this cost will be borne by shareholders (rather than ratepayers), the commit-ment to maintain a performance bond as well as future monitoring represents a clear long-term association for the project with a public agency, as well as a dilution of the value of the shares held. Some jurisdictions may be persuaded to participate in un-derwriting this obligation if the perceived value of the power or its attributes generated meets a public needs test during its productive life.

18.5   Permitting land use and interconnection

Electricity projects, including generation, transmission, substations, and distribution systems, fall under a variety of regulatory oversight programs. In general, policy-makers delegate the responsibility for permit review, rules enforcement, and rate re-views to utility regulators who may or may not be wholly independent from the policy arm of government. Since energy is produced, sold, and distributed within a "market" and since there are public health and safety concerns involved, *oversight* may involve a wide variety of regulations, from economic such as securities and exchanges, mergers and acquisition, to engineering and safety.

Renewable resources and generation such as geothermal typically need to connect to the wholesale, high-voltage regional grid. This system is often either owned or managed by the utility in the region; thus, one of the requirements for gaining access to transmission capacity is the completion of a utility interconnect study, unique to the specific project at hand.[6] It is possible but not the norm to conduct interconnect studies in parallel with project application reviews. The more likely outcome is to complete the interconnect study, and comply with the consequent recommendations, after project approval is complete. The average time for such a study can range from 12 to 18 months with estimated costs depending on the utility ranging from US$200,000 to US$500,000.

18.5.1 Regulatory permit process

Within the energy arena, there are "regulated" and unregulated utilities; both of these categories imply public oversight and rule enforcement. They differ, however, in terms of the nature of revenue sources and flow. Regulated utilities have a set rate of return to investment while unregulated or market generators have capital and operating assets essentially "at risk." Regulated markets are generally served by vertically integrated utilities that own or control a majority of the flow of electricity from generation to use and metering. By contrast, public utilities (those publicly owned or traded on stock market exchanges) that intend to operate in unregulated markets usually are required to divest ownership in either generation or transmission or both before proceeding. The residual ownership interest for them includes distribution, system operation, bidding, and maintenance from generation to point of interconnect, billing, and responsibility as provider of last resort to consumers.

These investor-owned utilities ("IOUs") are effectively private companies, subject to state regulation but financed by a combination of shareholder equity and bond-holder debt. Most are financially large and may cross state boundaries. Providing or acquiring power on the public side are municipal utilities, consumer-owned utilities, utility districts, and cooperatives that are operated as effective nonprofit organizations with net revenues used to offset consumer fees or to add or replace capacity. Corollaries to this type of organization are public utility districts that are utility-only that are government agencies, governed by a board elected by voters within the service territory (RAP, 2011).

In the past two decades, there has been a widespread effort to deregulate many electricity markets, adding competition in generation and distribution in order to reduce consumer costs while maintaining system integrity. Dispatch functions are similar to fully regulated markets, although diversity of generation type may be different than in deregulated markets. Deregulated markets usually have an ISO or a grid operator/dispatcher that administers wholesale market exchanges in their role, ensuring grid reliability and quality control.

Most regulatory oversight agencies create a periodic needs assessment to forecast capacity and technology demands. Based on the performance and needs of the utilities in meeting demand, the regulator can direct regulated utilities to invest in new

---

[6] For rules of thumb concerning procedure and process, see Energy Trust of Oregon, https://energytrust.org/library/forms/UtilityInter_GD_1108_singles.pdf.

capacity as needed. The utility is rewarded with a guaranteed rate of return and can pass costs on via a periodic general rate review and tariff assignment. The regulator also oversees monopoly and monopsony behavior within their jurisdiction. Most regulators are subject to judicial reviews for fairness by the court system but not by the policy authority.

Publicly regulated utilities have an effective contract with the public to deliver dependable power in a predictable manner and to agreed-on standards over the lifetime of the facility. Merchant plants operate with similar objectives depending on their customers and contracts for service and typically bid in routine competitive auctions to supply power to the grid. Some may have bilateral agreements to serve private load customers within this same area.

18.5.2   Typical project approval process

Geothermal projects are dependent on a complicated series of sequential and often overlapping permits and project approvals. These may extend through various levels of government and public oversight and may vary widely between jurisdictions.

Typically, a proposed project will file for approval before a designated decision-making body or if there are public lands involved with more than one agency where one agency will be assigned lead status and control filing and hearing processes. The process will entail a *prefiling* period where the applicant demonstrates land control and documents need for capacity of this type and the ability of the project proponents to provide adequate financing through the expected permit phase in question.

For most energy projects, this process is inclusive and considers all project characteristics in one hearing process. In the case of geothermal projects, the initial project is likely to be defined as exploratory in nature and will involve test wells, seismic research, and steps to "prove" the resource. The outcome of this phase (including results of hydraulic fracturing for any hydraulic stimulation or for deep EGS projects), water use, and potential power generation determine whether the project can submit a full power generation permit request. The review period for each of these can vary significantly depending on the complexity of the project, proximity to urban areas, the cooperation of the sponsoring utility or ISO, and the interest of other government organizations (see Fig. 18.1).

18.5.3   Involvement of various levels of governance

Most geothermal permits and review will occur before specialized utility regulators. However, since the footprint and physical impacts of the facilities reflect changes in land use, local government at the city or county level may also petition for review and comment, potentially extending the review period and the complexity of information required and ultimately involving a different interested community in the process. As well, specialized agencies concerned with fish, food, and wildlife may petition for involvement no matter whether they are state/provincial or federal in nature.

The consideration of various levels of government will inevitably bring scrutiny on environmental impacts, including any potential subsurface seismic activity induced by the project while circulating hot fluids. As with the case of the Bottle Rock[7] research project sponsored by the U.S. Department of Energy, an impression of sustained or even periodic seismic shaking at low levels was deemed sufficient to terminate the project.

18.6   Long-term economic and financing security

Energy generation projects involve significant capital investments over long periods of time. As such, they commit a level of technology that is matched to unique load, load growth, and policy objectives; they attempt to be flexible enough in design to perform through the full expected lifespan of the unit(s). In this process, the developer, the system operator, and the consumer bear a wide range of risk that must be either mitigated or anticipated in the finance design (Salmon et al., 2011; USDOE, 2010; Gehringer and Loksha, 2012).

18.6.1   Risks and risk indemnification

Power project financing, no matter whether conventional or renewable technology, involves a variety of risks. Most of these are difficult to assign directly to a single portion of the project. In general, though, we can designate them as precommissioning (exploration and test wells fall in this category), commissioning (development, stimulation, and proof of connection between wells, for EGS projects), and postcommissioning (operation and sales). This last phase may include performance failures, unexpected operation and maintenance cost increases, insurance or indemnification charges, grid/transmission system failure or changes in tariffs, regulatory/political rule changes, tax increases, interest rate charges, or exchange rate changes (Novogradac, 2013).

The role and requirements for geothermal finance can be divided into several sequential phases, each with varying levels of uncertainty, risk, and consequent cost changes and information requirements as shown in Table 18.1 (Stefansson, 2001).

18.6.2   Types of expected risk for merchant plants

The majority of geothermal plants are designed to be competitive in serving base load demand, although some may be called on for limited load following. Public agencies generally do not enter into market ownership of geothermal power but may support or even subsidize investments in this sector. They may, however, enter hybrid arrangements with varying degrees of participation with different types of

---

[7] The AltaRock project at The Geysers in California was suspended in 2009 by the Department of Energy. It had financial support from Google.org, Advanced Technology Ventures, and Vulcan Capital, as well as demonstration grants funding from the U.S. Department of Energy (DOE).

Table 18.1 **Phase and risk characteristics: steps and success are sequential**

Phase	Risk characteristics	Information or action needed for next step
Needs determination	Low	Response to call for capacity Interpret demand based on unit retirement, change in technology or environmental standards Forecast cost-benefit assessment
Predrilling resource assessment	Moderate risk, moderate uncertainty, use of equity interest contributions	Develop geoscientific data Interpret and extrapolate oil and gas assessments
Resource assessment and permitting Phase 1	High risk, high uncertainty, use of venture capital	Test well assessment following exploratory or initial production wells Obtaining all regulatory approvals Securing any affected stakeholders' support, the emerging issue of the "social license"
Resource assessment and permitting Phase 2	High risk, high uncertainty	Drilling test injection and production wells and interpreting results, analyzing potential seismic risk (for EGS) Comprehensive cost-benefit analysis
Production and development of well complex, installation of generation facilities, interconnection permitted and built	Moderate risk, moderate uncertainty	Completion of the well complex, after interconnection and testing are complete
Operation	Low risk, low uncertainty	Grid payments and competition from conventional thermal resources such as coal or nuclear
Decommission of facility	Moderate risk, uncertainty over timing and the requirements at time of decommissioning	Agreement with permitting agency on applicable standards and fees

Table 18.2 **Public involvement in geothermal power development**

Characteristic of involvement, ownership, or control	Countries included	Special programs or characteristics
Fully integrated single National public agency	Kenya, Ethiopia, Costa Rica	
Multiple public entities and international country partners	Indonesia, New Zealand, Mexico Mexican OPF model (public works financed by private investment)	Exploration, drilling and field development in the hands of different public agencies Private company construction but ownership and operation by public agency
Government predrills brownfield sites and leases to private entity	Japan, Philippines,* Kenya,* Indonesia,* and Guatemala	Philippines BOT (build-operated-transfer) model Kenya GDC (geothermal development company) strategy
Government may fund exploration and some test drilling	United States, Turkey, New Zealand, Indonesia	
Shared exploration and data collection	United States, Nicaragua, Chile	

*Steam sales separate from power sales.
Adapted from Gehringer, M., Loksha, V., 2012. Geothermal Handbook: planning and financing power generation. Energy Sector Management Assistance Program (ESMAP), World Bank.

power generation depending on the perceived needs. There are a wide range of opportunities for public agencies to participate in regulated as well as unregulated markets as shown in Table 18.2.

The majority of investment in geothermal facilities, however, is undertaken by private or merchant plant developers.

The risks associated with merchant power plants can be broadly grouped into three categories: market risk (which includes the markets for the sale of electricity and competition from other technologies), project development risk (which includes construction, technology, and operating risks), and structural risk (which includes legal/regulatory challenges and the risk of obtaining adequate finance). Some of these risks are more subject to control than others (Goldstein and Braccio, 2014).

18.6.2.1   Risk and market participation

Base load power developers are attracted to serve markets where minimizing impacts of service interruption, lowering costs, and improving reliability are high priorities. Most of the competitive investment in service areas with high load growth takes place in the more volatile categories of load following and peaking generation. Consequently, developers and owners of both regulated and competitive independent base load generators such as installed geothermal capacity typically face limited market risk. However, the advent of significant investments in gas turbine technology is already impacting these arrangements and is likely to continue. Consequently, interest in long-term, large-scale capital facilities such as nuclear, geothermal, hydroelectric, and even advanced coal plants has declined.

In the case of geothermal projects, utilities may find it more difficult to get approval for traditional ratepayer guaranteed scale projects, and merchant providers may have more difficulty in attracting financing without a clear demonstration of resource adequacy (European Bank, 2013; CEC, 2004). Traditional merchant plants providing base load power must carefully examine existing capacity and structure of the market in which geothermal development is proposed in order to successfully compete (Aldas, 1993).

This will include an assessment of current and forecast capacity utilization, technological additions, and changes in demand. For instance, the age and capability of existing stock must be taken into account, as well as the nature and expansion of demand side management and shifts in time of demand. This entails an inventory of the types of technology in use, the life expectancy of the current generation mix, and estimates of any existing coal or other base load facilities. As well, changing regulatory or environmental standards, land access, and even consumer or political/policy preferences may dramatically change the nature of the proposed market.

Structural supply risks can be amplified by transmission or nodal constraints, including changes in transmission pricing policies and available transmission capacity.[8] Furthermore, even if a plant is favorably sited as an initial matter, there is always a risk that the transmission system will change its pricing scheme or invest in increasing the capacity of the grid, thereby eliminating the benefit from congestion derived by the well-sited facility.

An assessment of market risk must forecast the demand for electric power in the region over the life of the project. The evaluation must include demographic estimates and forecasts of growth in large commercial or industrial sites, including fixed high-demand centers such as hospitals, academic institutions, government complexes, or, in the case of hybrid systems, uses such as nurseries. Retail consumers who participate in the market (via changing demand and price signals) are unlikely to dramatically change the nature of base load generation and consequently are not likely to impact contracts, existing PPA's, ISO, or utility-sponsored investments. Industrial users, however, may have a preference for contractual arrangements to indemnify themselves against later price or environmental standards changes.

Given the uncertainty inherent in all forecasts, developers will use all available financial and contractual tools that can help minimize risk exposure. This may include

---

[8] In areas that experience peak load congestion, congestion pricing schemes may make finance and revenue requirements invalid if transmission constraints are not taken into account. Consequently, generator location will reflect the adjustments to hourly prices offered.

dedicated bilateral lines and contracts for uses such as cement plants that must run continuously and be supplied in addition to system base load.

Normally, geothermal centers will not be located adjacent to load centers. However, getting access to transmission capacity that can serve industrial facilities or intense loads such as foundries or steel mills, refineries, or paper mills may increase value and diminish some of the risk financing needed to develop the geothermal project. A good example of this specialized relationship is the siting of aluminum smelters in Iceland, cloud storage, or military and defense installations.

18.6.2.2 Project risk

Project risk associated with plant development, beyond the exploration phase, can be controlled through management oversight and through the use of contractors with project and drilling experience. After test wells have been developed and the field is demonstrating fluid production, injection, and power generation, then maintaining production and power delivery can be described as routine. Projects should plan and budget for maintenance of existing wells and periodic redrilling in existing fields, although the actual timing of these costs will be uncertain. Risks of low productivity holes in established fields are expected to be minimal.

New social concerns constitute risks that must be accounted for as well (Mignan et al., 2015). Majur and Baria (2007) point out that there is a growing public perception of seismic hazards that may be associated with geothermal operations. This can include fluid injection to maintain circulation. Barth et al. (2013) point to the biased "decision threshold due to some ambiguity on hazard and risk estimates" that drive this issue, and Evans et al. (2009) point out the obvious, namely that the uncertainty introduced by this issue poses a severe challenge for developers and financiers, especially given the paucity of real field data to validate or refute the condition.

Most financing sources will be driven by their assessment of technology risk, since the choice and installation of technology will drive costs (but not always revenues) through the life of the project (Mines and Nathwani, 2013).

18.6.2.3 Structural risk

The general category of structural risks includes legal and regulatory challenges. While geothermal projects have been considered benign in the past,[9] an increase in scrutiny by special interest groups, concerned landowners, native and aboriginal populations, and government agencies has intensified examination of project proposals and made them vulnerable to challenge. Even more important, geothermal projects are subject to sequential scrutiny, making the permit process additive and open to criticism or changing regulatory and policy environments. Regulatory

---

[9] As more plants are proposed and built, the attention and focus of special interest groups is likely to increase.

changes including new standards such as seismic risk can delay projects or impose unanticipated conditions and costs, potentially rendering a project noncompetitive.

As well, both utility and merchant electric plants are entirely dependent on the transmission network to reach dispatch nodes. Consequently, rules governing interconnection studies, standards for firm access, or congestion pricing affect overall project economics. Those rules typically reflect the unique conditions of the relevant ISO or dispatch agency as well as regional transmission organizations.

For example, Kriebel et al. (2000) observe that early in 2000

in New England, transmission rules required all new generators seeking interconnection with the transmission grid show, through system impact studies, that the transmission system had the capacity to deliver the planned generator's electrical output to every load-serving entity in the New England Power Pool (NEPOOL), covering a region from Maine to Massachusetts. This had clear negative implications for new generators seeking to sell power into NEPOOL, since, based upon the number and location of proposed new generating plants in the region, transmission "bottlenecks" were projected to occur. After negotiations among affected interests, the ISO adopted a rule whereby a new generator would be required to pay for at least one-half of the system upgrades needed to alleviate the bottlenecks caused by that generator. The U.S. Federal Energy Regulatory Commission (FERC) ruled, however, that the New England ISO must limit its analysis of proposed interconnections to system reliability, stability and operating considerations, while deferring its ruling on how any upgrade costs would be assigned until a transmission congestion management system is developed.

If this were broadly adopted by regulatory agencies, one likely outcome will be more generically based interconnect studies, augmented with an assessment of special cases like geothermal. The upshot would be faster processing, lower cost, and more predictable timetables for developers.

18.6.3   Financing tools

Financiers of merchant power plants are required to analyze a wide range of factors that affect or potentially could affect the cash flow of a particular project (Simhauser and Ariyaratnam, 2013). The key to successfully financing merchant power plants on a "project finance" basis is to craft a financing structure that both anticipates and responds on a real-time basis to dynamic changes in market conditions (Mines and Nathwani, 2013). At risk merchant power plants may require broader financial guarantees and investor contributions than publicly funded and regulated projects.

There are a number of standard financial tools that, with some minor adjustments, will also play a significant role in the financing of merchant power plants. Appropriate use of these tools can build resilience into merchant financing structures, creating more robust capital and economic foundations able to absorb significant swings in market conditions. These tools include substantially higher debt service coverage requirements than are found in traditional debt instruments, including initial equity commitments on the part of the sponsors that are can be as much as twice those of merchant projects of similar size, and full funding of reserves, such as those for debt service and operation and maintenance, earlier and at higher levels than in traditional projects. This financial foundation will also rely on projects that incorporate resiliency in the face of changing market conditions.

18.6.4 Finance vs accounting issues

Developing geothermal projects fundamentally involves long-term finance. This, in turn, generates a need for strict accounting systems to account for expenditures, debt service, and power sales over time. The essence of this type of project finance ensures that there will be sufficient cash flowing into the operation in the future to achieve the goals of the project. Project finance takes into account risk and uncertainty, while the accounting function supports the delivery of financial information for investors as well as regulatory agencies. No long-term project will succeed without a seamless merging of these two functions.

18.6.4.1 Accounting standards

Each country that has the capacity to develop power resources also has a unique taxing and permitting environment. Many of the examples given here are derived from U.S. experience but are broadly descriptive of similar tools used elsewhere in the world.

Finance, per se, is not accounting; however, every energy finance arrangement must adhere to accepted accounting principles. The Financial Accounting Board Standards (FASB) used in the United States provides a good framework with which to view this requirement. For energy projects in general, it is useful in characterizing the nature of risk and reporting for investors generally in subsurface energy generation projects[10].

For instance, geothermal projects that may not be viable initially, either because they have uncertain resource identification or are considered experimental; consequently, they may be the recipients of incentives and stipends to improve their competitive role in the energy mix. They may also participate in policy initiatives such as renewable portfolio standards programs currently in use in states like California.

The support from these programs can enhance the market performance potential for a given project (although the support may have a short and nonrenewable time period in which it can be exercised). In a 2013 comment letter on the topic, a tax energy consultant (Novogradac, 2013) speaking on behalf of the members of the Renewable Energy Tax Credit (RETC) Working Group, submitted responses to the question regarding a proposed accounting standards update dealing with renewable energy

---

[10] FASB Accounting Standards are the source of authoritative generally accepted accounting principles (GAAP) recognized by the FASB to be applied by nongovernmental entities.

investments. The consultant pointed out that two of the most important incentives available for renewable energy were contained within the U.S. Internal Revenue Code (IRC) Sections 45 and 48.

> *IRC Section 45 provides owners or operators of certain electricity generating facilities a production tax credit (PTC) over a 10-year period based on the production of electricity from 'qualified energy sources' that include wind, biomass, geothermal, solar, irrigation, solid waste and hydropower. The PTC rate in effect was 2.3 cents per kilowatt-hour for certain technologies (i.e., wind and closed loop biomass) and half of that rate for other technologies (i.e., open-loop biomass, small irrigation, etc.). IRC Section 48 provides an investment tax credit (ITC) equal to 30 percent of the eligible basis of certain energy property that includes solar (photovoltaic) and other renewable energy property. IRC Section 48 also provides a 10 percent ITC for geothermal, combined heat and power, microturbine equipment and other renewable energy property.*

18.6.4.2   Investment tax credits

Novogradac (2013) also provided an example of a common incentive, the Investment Tax Credit (ITC) in action. They illustrated how "a typical renewable energy project monetizes the value of the ITC. Assume the total cost to develop a project is $11 million, which includes $1 million of costs that are ineligible for the ITC (i.e., land costs, intangible assets and reserves), resulting in ITC eligible basis of $10 million (i.e., major components, labor and related soft costs). Assume the technology utilized in this project qualifies for the 30 percent ITC, meaning the owner(s) of the project is able to claim $3 million of ITCs. As an owner of the project, the tax credit investor earns the majority of its return on investment when it claims the ITC on its tax return."

They went on to point out that tax credit investors typically either purchase the projects outright or contribute equity to a special purpose entity partnership in exchange for an ownership interest in the project's profits, losses, tax credits, and cash flows. The tax credit investor usually derives most of its return on investment from the value of the ITC. Agencies seeking to encourage new or immature technologies often revert to ITCs or other subsidies and timing incentives in order to ensure they can compete in markets where dominant technologies such as fossil fuel generation are well established and may even control access to the grid for certain dispatch operations (USDOE, 2015; Moore and Masri, 1999).

Consequently, most credit equity investors can match their investment based primarily on an instrument such as an ITC, rather than solely on expected project sales revenues over the length of the project. The value of the ITC can be lower after base tax adjustments reducing the taxpayer's *basis* by 50% of the amount of the ITC. Ultimately, in the United States, some of the ITCs can be subject to a 5-year recapture period if certain conditions are not satisfied over a 5-year compliance period.

18.6.4.3 Booking reserves

Formal booking of geothermal energy reserves, for accounting purposes, portfolio management, or annual reporting to shareholders, is not yet a common practice among geothermal companies. In the petroleum industry, booking of oil and gas reserves is routine, and at least two geothermal operators that are subsidiaries of petroleum companies book geothermal reserves. As in the petroleum industry, the reserves are carried in a category reserved for *uncertain* resources. There are three subdivisions that are worth considering for application in this category and that should increase investor confidence in participating in geothermal financing schemes. The traditional category of "proved" is typically equivalent to known resources at the 90th percentile of confidence normally associated with producing fields where further development within the existing formation or subsurface zone is contemplated; the *likely and probable* category is based on geoscientific data or test well evaluation (median or 70th percentile); and the *possible* category is based on extrapolative estimation (equivalent to the 10th percentile).

In the end, volumetric estimation, test drilling, and numerical simulation are the only consistently valid methods of reserve estimation (Lukawski et al., 2014). While numerical simulation is a sophisticated tool for forecasting reservoir performance, estimation of reserves from numerical simulation has a distinctly deterministic character in terms of where to drill, although the application of Monte Carlo methods can account for variability in estimates so the output becomes probabilistic; as such, a simulation forecast represents a single time period reserve estimate. Over time and with the learning curve following continued drilling in a field area, the knowledge of the characteristics of the reserves of heat and the available extent of the resource will become more clearly known and result in more robust developer interest and participation (Thorhallson and Sveinbjornsson, 2012; Tester et al., 2006).

18.6.5 Levelized costs of energy and levelized avoided cost of energy

The decision to invest in any energy technology involves choices in location, technology, market characteristics of demand and growth, and transmission capacity and access. Ultimately, investors will employ standard methods of comparison in order to inform their choices, for instance using net present value (NPV) calculations to compare options. A commonly accepted process is to develop expected and comparable costs between technology options in order to allow for the most appropriate and cost-effective investment strategy. By comparison, in oil and gas operations, the data and history of development of resources are robust and span a long time period with a diverse set of industry investments in a wide variety of locations. Similarly, the number of power-generation technologies used for traditional grid operations is diverse and offers ready comparison for investors.

Since geothermal power is relatively new for most grid operations, the number of installations available to model expected performance is not large; consequently, ready comparison to other power generation technologies is more difficult. Thus, the initial calculations will depend on a combination of levelized costs of energy (LCOE) and levelized avoided cost of energy (LACE) figures for the relevant and comparable technologies.

LCOE represents the per-kWh cost (in real dollars) of building and operating a generating plant over its assumed financial and performance life. The key variables include costs, fuel costs, fixed and variable operations and maintenance (operation and maintenance) costs, financing costs, and some assumed utilization rates to establish a generic base. (Actual plant investment decisions will be affected by the specific technological and regional characteristics of any given project.)

The DOE (2015) points out that

> *since projected utilization rates, the existing resource mix, and capacity values can all vary dramatically across regions where new generation capacity may be needed, the direct comparison of LCOE across technologies is often problematic and can be misleading as a method to assess the economic competitiveness of various generation alternatives. Conceptually, a better assessment of economic competitiveness can be gained through consideration of avoided cost, a measure of what it would cost the grid to generate the electricity that is otherwise displaced by a new generation project, as well as its levelized cost.*

Thus, in addition to LCOE, we can use the concept of LACE to provide an estimate of the cost of generation and capacity displaced by a marginal unit of new capacity to demonstrate the relative value of a given technology in grid operations. For instance, when the LACE of a particular technology exceeds its LCOE at a given time and place, that technology would generally be economically attractive to build.

In their newest calculation of LCOE and benefits, the DOE suggests that

> *a better assessment of the economic competitiveness of a candidate generation project can be gained through joint consideration of its LCOE and its avoided cost, a measure of what it would cost the grid to meet the demand that is otherwise displaced by a new generation project. Avoided cost accounts for both the variation in daily and seasonal electricity demand in the region where a new project is under consideration and the characteristics of the existing generation fleet to which new capacity will be added, thus comparing the prospective new generation resource against the mix of new and existing generation and capacity that it could displace (DOE, 2015).*

The difference between the LACE and LCOE values for the candidate project provides an indication of whether its economic value exceeds its cost, where cost is considered net of the value of any production or investment tax credits provided by federal law. The DOE provides a wide range of comparative data on power plant characteristics on the EIA website, but in the case of geothermal technologies, only an

Table 18.3 **LCOE versus LACE**

	Min	Ave	Max
LCOE			
Conventional natural gas	70.4	75.2	85.5
Advanced natural gas	68.6	72.6	81.7
Geothermal electric	43.8	47.8	52.1
LACE			
Natural gas	65.8	71.4	89.7
Geothermal	70.7	70.9	71
LCOE with/subsidies			
Geothermal	41	44.4	48

US Department of Energy, Energy Information Administration, 2015. Annual Energy Outlook (AEO).

average value for hydrogeothermal is used (see Table 18.3 following). In reality, geothermal plants differ widely based on location and technologies chosen, from binary, to multiflash to hybrid systems.

Generation supply systems are made up of a mix of technologies and fuel types that collectively provide an economic supply of power to meet system demands. The system operator will have choices of technology available to it and in recent times will even be considering gas turbines as an option for traditional base load facilities such as hydroelectric, coal, or nuclear.

A common distinction between generating units is whether they are designed to supply base load or act as peaking units. A base load unit is typically one that is operated to meet part of the minimum load of a power system and consequently is one that produces electricity both continuously and at a constant rate. In contrast, peaking units are those designed to run intermittently in order to meet higher demands during daily, weekly, or seasonal peaks. Consequently, levelized unit cost comparisons will usually be developed not only across technologies but also with different sets of financing assumptions (Ayers et al., 2004; Beckers et al., 2014; Johnston et al., 2011; Mines and Nathwani, 2013).

The comparison of base load generating technologies is typically made using estimates of LCOE. The LCOE is effectively the cost of supply, where the unit cost per MWh or per-kWh represents the amount necessary to recover all costs over the period. By convention, LCOE is determined by finding the price that sets the sum of all future discounted cash flows or NPV to zero.

LCOE is the constant, real wholesale (retail delivery takes place through a utility or local distribution entity) price of electricity necessary to cover debt financing cost, both fixed and variable, income tax and any required cash flow accounting associated with

interim construction operations, and, if required, the decommissioning of the generating plant. Since geothermal wells must periodically be redrilled, this must be
accounted within the expected lifetime of the facility.

The levelized unit electricity cost presents a useful method of comparison between
different technologies. It is especially useful in examining the impact on relative costs
under different sets of assumptions regarding fuel prices, operating characteristics, and
financing assumptions. However, LCOE provides only a partial answer to the
preferred technology for new base load generation. For example, it does not address
exactly when capacity is needed or where quantification based on financial costs fails
to account adequately for some factors. For instance, existing base load facilities may
plan to expand on an existing site or incorporate newer, more efficient technology, or
disruptive technology such as combined cycle gas turbines may present a short-term
challenge for cost-effective dispatch.

The relative competitiveness of different technologies, judged from the perspective
of lowest LCOE, is also a function of the costs and form of financing ultimately available to build different options. Capital-intensive technologies compare more favorably
where lower returns are required or acceptable. This suggests a combination of
public—private partnerships where much of the early risk is reduced for private investment capital. This is most cost-effective prior to widespread establishment of
geothermal technologies for grid scale power generation. Over time, the performance
of the sector is expected to reflect lower marginal costs of delivered electricity with a
corresponding increase in the competitive role of geothermal—literally a positive feedback loop.

By comparison, conditions of very low prices for natural gas, along with highly efficient combined cycle gas turbine (CCGT) plants, may change the competitive attraction for geothermal, even though traditionally CCGT plants would have been excluded
from participating in base load supplies. The construction time for these CCGT facilities is short, and their dispatch and ramp rates are fast. Finally, on a relative basis, the
emissions footprint for gas is significantly lower than other fossil technology such as
coal and may be more attractive to investors simply because of the time and effort
necessary for siting (Kahn, 2000). Here, the relatively low capital cost for gas-fired
generation may mean that despite a relatively high LCOE, some investment capital
will choose the shorter-term gas option over geothermal, especially when overall
dispatch flexibility is taken into account.

18.6.5.1 Limits to LCOE assumptions

The LCOE method of comparison allows examination of a relatively large range of
uncertainties concerning capital costs, operating costs, and other characteristics. Financial analysis based on LCOE provides a basic apples-to-apples comparison between
technologies, with various performance caveats. However, there are important characteristics that must be factored in to develop a traditional pro forma. In terms of normal
operations and dispatch, the NPV or comparison of operating characteristics will not
always take into account the likely concentration of capacity in a dispatch region.
Similarly, it will tend to understate the rates of decline for existing stock, changes

in transmission capacity, or even location impaction (limits to land use expansion). Other externalities, including market financial risks and short-term price volatility, are difficult to portray in the overall LCOE question and comparative analysis. Consequently, the LCOE should be considered as a basic building block in the full project consideration.

18.6.5.2   Merchant vs public projects

For merchant projects, the higher the perceived risk[11] in the marketplace, the higher is the required return. Publicly financed projects may be evaluated on the basis of a given discount rate or may be able to access funds at lower rates, but the risk of cost overruns is implicitly borne by the taxpayer. Merchant-financed plants by nature tend to have levelized costs driven by debt considerations, more so than publicly financed plants. The difference is most easily seen for nuclear generation, the most capital-intensive plants in operation, and consequently those that rely most heavily on debt financing and public insurance.

However, while the LCOE appears to be lower under public financing, all of the risk associated with the construction and operation is implicitly borne by the taxpayer.

Consequently, comparisons between merchant and public financing should be interpreted with care.

There is a third possibility available, namely public—private partnerships. A number of variations and arrangements are possible: for example, public financing of construction and leasing to private owners for operation. All partnership arrangements represent a sharing of risk between the public and private sector.

18.6.5.3   Geothermal siting and financial variables

Geothermal plants are assumed to operate for a period of up to 30 years with adequate maintenance of the well field. By comparison, a realistic operational life for gas-fired units may be considerably less than this (20—30 years), whereas coal and nuclear options could be in operation for 40-plus years. LCOE comparisons must assume a consistent time period (i.e., each option starts and stops generating power at the same time or comparison with other technologies will be invalid).

For geothermal technology, we can assume a capacity factor of 90% (in other words, the generating plant is assumed to produce 90% of the energy it would produce if it were to be run continuously at full power) compared with combined cycle gas turbines (where the highest rated capacity is $\sim 60\%$).

Using geothermal power solely for base load conditions highlights a loss of flexibility for ramping and load responsiveness that is possible with modern turbines. This means that other, more expensive technologies are not offset when meeting load requirements during normal dispatch periods.

---

[11] While not a term of art, the concept of perceived risk may change over time, depending on macroeconomic or even geopolitical trends and conditions.

According to the DOE (2015), estimates of levelized costs expected in 2020 suggest a range of US$41−48/MWh, while the *avoided costs* are estimated to be approximately US$71.00/MWh. In 2013, this difference between LCOE and LACE is then approximately US$26.50/MWh, making the base load attraction of hydrogeothermal resources very high. Table 18.3 shows the relationship between LCOE and LACE as well as adding natural gas values for comparison. Past experience has shown that the role of subsidies, shown in the table, can be powerful in improving competitive costs.

18.6.5.4 Well drilling costs

One of the key variables that influence the cost of geothermal facilities is the cost of wells, from exploratory to production and periodic redrilling (Aldas, 1993; Hammons, 2007; USDOE, 2004). These costs appear significantly at two points in the process, during exploration/proof of resource (the technology is well established and cost curves are known and competitive) and in the site development and operations phase (Thorhallson and Sveinbjornsson, 2012). We can portray the risk(s) involved as a confidence matrix shown in Table 18.4.

Yost et al. (2015) points to the risks of wellbore drilling for any resource extraction − and particularly for EGS − as subject to a high degree of uncertainty, ranging from those related to geology to those related to the drilling procedures (Kahn, 2000). Since well drilling and related subsurface activities for hydro represent a significant portion ((>50%) of the total EGS development cost), assessing the risk is essential (Majorowicz and Moore, 2008).

The range of historical drilling records is limited, in part due to the lack of concentration of existing fields in any geographic area outside existing hydrothermal fields such as The Geysers in California (Matek and Gawell, 2014). Consequently, the majority of cost estimates tend to reflect oil and gas experience and

Table 18.4 Well-cost *risk* in phases of operation

Category	Confidence in outcome	Risk of failure
Exploratory	Low	Moderate to high
Phase 1 Site development	Moderate	Moderate
Phase 1+ Operation	High	Low to moderate
Phase 2 Redrilling in same field	Moderate to high	Low to moderate
Phase 3 Expansion of field into new formation or area	Moderate	Low to moderate

Table 18.5 **Geothermal well costs**

Shallow				Mid range			Deep		
Depth, m	No. of casing strings	Cost, $ million in 2004	Depth, m	No. of casing strings	Cost, $ million in 2004	Depth, m	No. of casing strings	Cost, $ million in 2004
1500	4	2.3	4000	4	5.2	6000	5	9.7
2500	4	3.4	5000	4	7.0	6000	6	12.3
3000	4	4.0	5000	5	8.3	7500	6	14.4
						10,000	6	20.0

From: DOE-EERE, Chapter 6 Drilling Technology Costs, WellCost Lite© figures, 2004

extrapolation. Estimates from 2004 illustrate the relationship of well-drilling activities to depth (but exclude the variable of temperature and flow rates) as shown in Table 18.5.

18.6.5.5 Finance scenarios and timing

Each geothermal project, even those drilling and operating in a related field, will experience slightly different conditions given the nature of drilling, including "trouble[12]" that can be inconsistently apparent and challenges with the actual heat resource being tapped. This, in turn, creates uncertainty in the potential finance community, which can be offset by early success, increased public subsidy, or even confidant equity investor resolve.

As suggested earlier, however, the normal and expected routine of drilling for fossil fuel projects benefits from a consistent and historically established process, well known and followed by most regulatory institutions. The nature of periodic revisiting of well progress and resource discovery subjects geothermal drilling projects to extra scrutiny and potential delays.

As a consequence, the finance scenarios that can be used are illustrative and not definitive, or necessarily transferrable from area to area. They are shown in Table 18.6 and underlie Fig. 18.1 in highlighting the full range of responsibility for developers, from inception to decommissioning, of well sites at the end of competitive generation.

18.6.5.6 Implied costs of regulation

All energy projects, whether traditional fossil fuel–fired combustion or renewable technologies, must undergo regulatory review and, ultimately, permission and

---

[12] "Trouble" is the term used by drilling engineers to describe the unpredictable events that can affect time and costs in downhole operations.

Table 18.6 **Financing scenarios**

Variable	Public finance	Merchant finance
Income tax rates	0%	30%
Discount rate	7% real	Opportunity cost of capital, assume 10%
Debt rate	N/A	10% real
Equity rate	N/A	12–15% real
Debt:equity ratio	N/A	Assume 50:50
Debt life	N/A	20 years plus decommission time
Inflation assumption	2.5% annual	2.5% annual
Sensitivity to change		
Debt rates, equity rates, and debt:equity ratio	6–12% real discount rates	50:50 range to 70:30 debt:equity ratio 6–8% debt rates 12–20% equity rates

oversight from regulatory agencies (Moore and Winter, 2015; RAP, 2011; Huddleston-Holmes et al., 2014). There are expected costs incurred in this exercise, topically known as the cost of entry to a market. To the extent that the requirements, including hearing costs, research, project design, and data collection, fall within a broad average of past experience, then the regulatory process can be deemed an even playing field (Jennejohn, 2009; CEC, 2004; European Bank, 2013).

However, all technologies are not treated equally either by jurisdiction, policy, and regulatory standards or the unpredictable nature of public participation in the regulatory process. Quantifying these phenomena is difficult, in part because each project is different in circumstances, not to mention the experience of the applicants.

In a rule of thumb applied to the issue of "regulatory hurdles" Moore and Winter (2015) found that regulatory hurdles were a cost of business for energy applicants and were best measured as a function of expected time to complete permits. Time for processing is typically cited by regulatory agencies at the commencement of a project, and in many cases is the basis for a performance contract. When time exceeded 130% of the expected value, the issue was not a regulatory hurdle but a more tangible regulatory delay.[13]

---

[13] The majority of the "delays" experienced were attributed to a lack of experience on the part of the applicant, inadequate data presentation, or intervention by other public agencies.

Nevertheless, the cost of such delays can be formidable, not only in terms of direct expenditure but also in terms of the ultimate nature of the mitigations required and the chance that other competitive applications might be approved in the interim (Mignan et al., 2015; Huddleston-Holmes et al., 2014).

18.7 Conclusions

Electricity from hydrogeothermal developments is cost competitive today and provides grid stability and displacement of carbon-intensive technologies. The resource is not ubiquitous but is nevertheless widely distributed and even in distant locations can support and firm grid operations either through interconnections to existing transmission lines or in distributed generation designs. Further, the contribution from geothermal systems offers base load power, some range of load following, and, in some instances, GHG credits or carbon offsets (Majorwicz and Moore, 2008). In situations where coal or other hydrocarbon units are being retired, it is a cost-effective replacement or substitute.

In terms of financing, geothermal projects represent an attractive, proven regional-scale resource where overall performance risk is low, primarily associated with finding and acquiring high-grade hydrothermal resources and identifying sites that can obtain appropriate transmission access. Operational risks are a function of maintaining reservoir quality and proper well maintenance. The primary issues for obtaining finance are tied to site acquisition and control, the nature of the resource itself, the demand market for individual systems operations, and, increasingly, the nature of competition from short-term gas-fired generation (USDOE, 2015).

The base load characteristics of the historic electric grid are profoundly influenced by the availability and flexibility of natural gas turbines. The combination of higher uncertainty and longer siting times makes geothermal projects relatively less competitive, unless carbon contributions are taken into account (Kahn, 2000; Matek, 2013). Changes in carbon markets and accounting are likely to gain currency in the next decade, led by existing structures in Europe and China. GHG and REC markets are increasing in popularity and adoption in Europe but not yet in the United States and not in Australia. If carbon markets become more established in policy and regulatory institutions, then it is possible that geothermal facilities may have more to contribute in terms of carbon offsets than electric generation.

This favorable characteristic, however, can be undone or offset if progress is not made in terms of identifying associated seismic risks and public reaction to the location and in land use commitment that must be made as close as possible to either load or transmission facilities with available capacity.

Ultimately, financing monetizes this litany of costs and benefits, comparing them to other alternatives (Huddlestone-Holmes et al., 2014; Reber et al., 2014). Rates and investment interest may fluctuate over time, but this will fundamentally change only with a predictable power delivery record and a consistent and supportive regulatory environment.

References

Aldas, R., 1993. Survey of Geothermal Costs. California Energy Commission, CE-Pier.

Ayers, M., MacRae, M., Stogran, M., August 2004. Levelised Unit Electricity Cost Comparison of Alternate Technologies for Baseload Generation in Ontario. Canadian Energy Research Institute.

Barth, A., Wenzel, F., Langenbruch, C., 2013. Probability of earthquake occurrence and magnitude estimation in the post shut-in phase of geothermal projects. Journal of Seismology 17 (1), 5−11.

Beckers, K.F., Lukawski, M.Z., Anderson, B.J., Moore, M.C., Tester, J.W., 2014. Levelized costs of electricity and direct-use heat from enhanced geothermal systems. Renewable and Sustainable Energy 6. http://dx.doi.org/10.1063/1.4865575.

California Energy Commission, 2004. New Geothermal Site Identification and Qualification. CE-Pier Report, 500-04-051.

European Bank for Reconstruction and Development, December 10, 2013. Energy Sector Strategy.

Evans and Peck, June 2014. Competitive Role of Geothermal Energy Near Hydrocarbon Fields. A Report to the ARENA Expert Panel, Australia, Brisbane.

Evans, A., Strezov, V., Evans, T.J., 2009. Assessment of sustainability indicators for renewable energy technologies. Renewable and Sustainable Energy Reviews 13, 1082−1088.

Gehringer, M., Loksha, V., 2012. Geothermal Handbook: planning and financing power generation. Energy Sector Management Assistance Program (ESMAP), World Bank.

Goldstein, A.H., Braccio, R., January 2014. 2013 Geothermal Technologies Market Trends Report. U.S. Department of Energy (DOE) Geothermal Technologies Office (GTO) and the U.S. DOE Office of Energy Efficiency and Renewable Energy (EERE).

Hammons, T.J., 2007. Geothermal power generation: global perspectives; USA and Iceland; technology, direct uses, plants and drilling. International Journal of Power and Energy Systems 27.2, 157−172.

Huddlestone-Holmes, C., Grafton, Q., Horne, R.N., Livesay, B., Moore, M.C., Petty, S., May 2014. Looking Forward, Barriers, Risks and Rewards of the Australian Geothermal Sector to 2020 and 2030. Australian Renewable Energy Agency.

International Finance Corporation (IFC), 2013. Success of Geothermal Wells: A Global Study. http://www.ifc.org/wps/wcm/connect/7e5eb4804fe24994b118ff23ff966f85/ifc-drilling-success-report-final.pdf.

Jennejohn, D., December 2009. Research and Development in Geothermal Exploration and Drilling. Geothermal Energy Association.

Johnston, I.W., Narsilio, G.A., Colls, S., 2011. Emerging geothermal energy technologies. KSCE, Journal of Civil Engineering 15 (4), 643−653.

Kahn, R.D., March 2000. Siting struggles: the unique challenge of permitting renewable energy power plants. The Electricity Journal 03/2000.

Kriebel, K.W., Orrick, H., Sutcliff, March 2000. Financing Merchant Power Plants in the United States. https://www.orrick.com/Events-and-Publications/Pages/financing-merchant-power-plants-in-the-united-states-87.aspx.

Lukawski, M.Z., Anderson, B.A., Augustine, C., Capuano Jr., L.E., Beckers, K.F., Livesay, B., Tester, J.W., June 2014. Cost analysis of oil, gas, and geothermal well drilling. Journal of Petroleum Science and Engineering 118, 1−14.

Majorowicz, J., Moore, M.C., 2014. The feasibility and potential of geothermal heat in the deep Alberta foreland basin-Canada for CO_2 savings. Renewable Energy 66, 541−549.

Majorowicz, J., Moore, M.C., 2008. Enhanced Geothermal Systems (EGS) Potential in the Alberta Basin. Calgary, AB, Canada. University of Calgary, Institute for Sustainability, Energy, Environment and Economy.

Matek, B., Gawell, K., February 2014. Report on the State of Geothermal Energy in California. California Energy Commission, Geothermal Energy Association.

Matek, B., September 2013. Geothermal Power: International Market Overview. Geothermal Energy Association.

Majur, E., Baria, R., Stark, M., Oates, S., Bommer, J., Smith, B., Asanuma, H., June 2007. Induced seismicity associated with enhanced geothermal systems. Geothermics 36 (3), 185–222.

Mignan, A., Landtwing, D., Kästli, P., Mena, B., Wiemer, S., January 2015. Induced seismicity risk analysis of the 2006 Basel, Switzerland, enhanced geothermal system project: influence of uncertainties on risk mitigation. Geothermics 53, 133–146.

Mines, G., Nathwani, J., 2013. Estimated power generation costs for EGS. In: Proceedings, Thirty-Eighth Workshop on Geothermal Reservoir Engineering Stanford University, Stanford, California, February 11–13, 2013. SGP-TR-198.

Moore, M.C., Winter, J., 2015. Regulatory Hurdles in Energy Project Review. Forthcoming publication from the School of Public Policy, Calgary, Alberta.

Moore, M.C., Masri, M., January 1999. Renewable energy and electricity restructuring: the California model. IEEE Power Engineering Review 19 (1), 13–14.

Novogradac and Company, LLP, CPA's, EITF-13B, 2013. Comment Letter No.66 to Renewable Energy Tax Credit ("RETC") Working Group.

Regulatory Assistance Project (RAP), 2011. Electricity Regulation in the US: A Guide. www.raponline.org.

Reber, T.J., Beckers, K.F., Tester, J.W., July 2014. The transformative potential of geothermal heating in the U.S. energy market: a regional study of New York and Pennsylvania. Energy Policy 70, 30–44.

Salmon, J.P., Meurice, J., Wobus, N., Stern, F., Duaime, M., March 2011. US Department of Energy, National Renewable Energy Laboratory. Guidebook to Geothermal Power Finance. Subcontract Report NREL/SR-6A20–49391.

Simhauser, P., Ariyaratnam, J., June 2013. What Is Normal Profit for Power Generation. AGL Applied Economic and Policy Research Working Paper No.38 – Normal Profit. http://aglblog.com.au/wp-content/uploads/2013/09/No.38-Normal-Profit.pdf.

Stefansson, V., 2001. Investment cost for geothermal power plants. In: Proceeding of the 5th Inaga Annual Scientific Conference & Exhibitions, Yogyakarta, March 7–10, 2001.

Tester, J.W., Anderson, B.J., Batchelor, A.S., Blackwell, D.D., DiPippo, R., Drake, E.M., Garnish, J., Livesay, B., Moore, M.C., Nichols, K., Petty, S., 2006. The Future of Geothermal Energy. Impact of Enhanced Geothermal Systems (EGS) on the United States in the 21st Century. Massachusetts Institute of Technology, Cambridge, MA.

Thorhallson, S., Sveinbjornsson, B.M., 2012. Geothermal drilling cost and drilling effectiveness. In: Presented at "Short Course on Geothermal Development and Geothermal Wells", Organized by UNU-gtp and LaGeo, in Santa Tecla, El Salvador, March 11–17, 2012.

US Department of Energy, Energy Information Administration, 2015. Annual Energy Outlook (AEO).

US Department of Energy, 2004. Drilling Technology and Costs. http://www1.eere.energy.gov/geothermal/pdfs/egs_chapter_6.pdf.

US Department of Energy, National Renewable Energy Laboratory, May 2010. Geothermal Policymakers' Guidebook, State-by-State Developers' Checklist, and Developers' Financing Handbook.

Yost, K., Valentin, A., Einstein, H.H., 2015. Estimating cost and time of wellbore drilling for engineered geothermal systems (EGS) – considering uncertainties. Geothermics 53, 85–99.

Appendix A

Portrayal of the range and type of geothermal project risk

Stage	Risk level	Project loss	Investor loss	Investor action	Stage	Failure source	Failure mitigation
Initial finance	High	No project	Unsecured cap future market share	Take loss	Initial	Projected return rates	Attract new investors or public subsidy
Public policy and regulatory standards	High	Increased finance costs	Loss of share value	Work to change policy	Mid stage	Inconsistent or changing public policy standards	Sign contract with developers
Lease and land control	High	Equity	Unsecured cap future market share	Take loss do not reinvest	Initial	Private owner refusal public land restriction	Abandon project
Permits	Low	Time equity	Unsecured cap future market share	Take loss do not reinvest	Early	Regulatory denial	Abandon project
Exploration—research	Low	Time	Time	Take loss do not reinvest	Early	Inconsistent or negative results	Secure more data
Characterization and test	High	Time equity	Unsecured cap future market share	Take loss do not reinvest	Early	Inconsistent or negative results	Additional testing
Change design from electric to heat use	Moderate	Time equity	Unsecured cap future market share	Do not invest in alternative design	Early	Inconsistent or negative results	Explore and adapt to alternative market closer to load

Exploration—drilling	High	Time equity	Unsecured cap future market share	Take loss do not reinvest	Early	Inconsistent or negative results	Redrill, add seismic data, move site
Drilling injector and production wells	High	Time equity	Unsecured cap future market share	Take loss, do not reinvest	Early	Inconsistent or negative results	Change location in field area, drill deeper
Stimulation and control	High	Time equity	Unsecured cap future market share	Take loss, do not reinvest	Early	Inconsistent or negative results or micro-seismic fears	Change depth bound field
Install capacity	Low	Equity	Unsecured cap future market share		Early	Unit cost	Adopt alternative technology
Interconnect study	Low	Time	Time		Early	Utility techniques	Submit data seek

Adapted from ARENA Risk Exposure.

Part Four

Case studies

Larderello: 100 years of geothermal power plant evolution in Italy

19

R. Parri, F. Lazzeri
Unione Nazionale Geotermica, Pisa, Italy

Prologue: historical outline on geothermal development in Italy up to 1960, with particular reference to the boraciferous region

R. Cataldi
Unione Geotermica Italiana, Pisa, Italy

From prehistory to the end of eighteenth century

The first human contact in Italy with external manifestations of the Earth's heat and their exploitation for various functions and worship date back more than 5000 years. Since then, the localities with hot springs, fumaroles. and hydrothermal incrustations started to be visited for hot baths or cults, first in Sicily and in the Phlegraean area at the beginning of the second millennium BC (Bronze Age), and afterward in some sectors of central and northeastern Italy during the *Villanovian* and *Venetic* periods (1200—800 BC). In certain cases, they were also chosen as places of permanent settlements. Thermal balneotherapy in Sicily, in particular, had already reached advanced forms of application in the seventeenth century BC, when it was practiced in structures organized with differentiated thermal environments (Fig. 19.1).

Legends, myths, haruspicial rites, popular beliefs, and local traditions related to thermal manifestations started also to grow around the end of the second millennium BC, as had occurred in previous centuries in Anatoly, Greece, and other Mediterranean regions featured with important geothermal phenomena. The myths and legends, in particular, tried to explain in an etiological manner the cause and effect and, in a transcendental way, the formation of surficial manifestations of the Earth's heat. Thus, from then on, information on the peculiar characteristics of the manifestations started to spread to neighboring areas, where they were taken as expressions of chthonian entities of transcendental nature and thus worthy of cult worship.

Beginning in the seventh century BC, the Etruscans first (until the third century BC) and, above all, the Romans from the second century BC through the fifth century AD gave a fundamental boost to multiple uses of the Earth's heat, including thermal baths

Geothermal Power Generation. http://dx.doi.org/10.1016/B978-0-08-100337-4.00019-X

Figure 19.1 The most ancient thermal structure known so far in Italy: San Calogero di Lipari, Sicily (ca. 1600 BC). Entrance to the tholos sudatorium (left), with adjoining pond, probably used as a thermal pool in the open.

and mineral extraction, processing, trading, and transport to distant lands of many geothermal byproducts.[1] The Romans, in particular, developed strongly and extensively, in all their territories, the thermal practice for healing and recreation, not only in localities with geothermal manifestations but also in towns and isolated places with no hot springs by using artificially heated water.

Moreover, as a follow-up of in-depth speculations made by Greek authors in the preceding five or six centuries, starting from the first century BC, many Roman authors contributed notably to the advancement of knowledge of volcanoes, fumaroles, thermal springs, and other surface manifestations of the Earth's heat occurring in Italy and in their territories in the Mediterranean area.

After the fall of the Roman Empire (476 AD), thermal bathing and the extraction and use of most byproducts of the terrestrial heat underwent a strong decline almost everywhere in Italy and in the territories of the old Roman Empire. This decline occurred in a variety of forms and extents and at different times, depending on the historical events then occurring in the peri-Mediterranean area; it lasted for the whole of the Early and most of the High and Late Middle Ages.

From about the beginning of the fifteenth century, balneotherapy in the main Italian thermal spas and the use in many places of hydrothermal minerals and other byproducts of the terrestrial heat started to undergo a new blooming that, while never reaching the level attained in Roman times, nevertheless played an important socioeconomic role at the local and manorial scales.

The new awakening happened mostly in Tuscany, where, as result of the so-called *Guerra delle Allumiere* (War of the Alum Mines) between the communes of Florence and Volterra that ended in 1472, the *boraciferous region* with its rich hydrothermal deposits came under the rule of the Florentine Republic. Since then, the commune of Florence assigned in monopoly the mining of those deposits to its *Corporazione*

---

[1] They are minerals, igneous rocks and fluids formed underground owing to the terrestrial heat, which may reach the surface due to particular geological processes. Among many others, they include the *hydrothermal minerals*, which formed by chemical-physical interactions occurring mostly at depth between circulating hot fluids and their hosting rocks.

della Lana (Wool Guild), which, being so exempted from paying customs to import alum, borates, and other hydrothermal minerals necessary to bleach and process wool and other fibers, became in a few years the most flourishing textile industry in Europe.

It is worth recalling that the borates and other minerals mentioned, as well as the boric acid for pharmaceutical use (then known as *Homberg's sedative salt*), were obtained at those times from tincal, a raw mixture of borates and gangue imported in Europe by land from Persia, India, and China at very high costs. As a consequence, exemption from customs on borates and other hydrothermal minerals, and availability of local and better products, represented for the Florentine textile industry an all-important advantage.

However, the intensive exploitation of the outcropping hydrothermal deposits over two centuries, on the one hand, enabled Florentine industry to reach a peak position in Europe, but on the other hand, resulted in their strong impoverishment. From about the middle of the eighteenth century, the production of alum, borates, and other hydrothermal minerals in the boraciferous region declined rapidly and remained at much-reduced levels for several decades. This notably weakened the Tuscan supremacy of the textile industry in Europe.

The situation began changing again toward the end of the eighteenth century, after Ubert F. Hoefer in 1777 and Paolo Mascagni in 1779 discovered boric acid with high concentrations in the hot waters of the *lagoni*[2] (lagoons), in the outcropping hydrothermal incrustations and in the fumarole steam of the whole boraciferous region.

The chemical industry of Larderello in the nineteenth century

Following the discovery of boric acid, it took some three decades of vain attempts before the successful commercial extraction of boric acid from hot waters of the *lagoni* and hydrothermal deposits of the boraciferous region. From 1818 through the rest of the nineteenth century, the commercial production of boric acid was due mainly to Francesco Larderel and his descendants.

The most significant steps of the industry in question can be summarized as follows.

1816–1827. To produce boric acid, a number of commercial companies were constituted in this period. The mineral was initially extracted from the stagnant brines formed in *natural lagoni*. After a few years it was also obtained from similar brines, which, thanks to leaching by rainwater, were induced to pool in pits dug manually on hydrothermal incrustations, named *artificial lagoni*. The latent heat of evaporation of the brines was in both cases obtained by burning firewood coming from the then plentiful woods of nearby areas.

---

[2] *Lagone* (plural *lagoni*) is the local name of relatively small ponds formed naturally by different geo-lithological causes in areas of active or recent manifestations and hydrothermal incrustations, within which the water issuing from hot springs and fumaroles could gather, evaporate and gradually reach the concentration of a brine of boron and other minerals.

Artificial lagoni represent the first technological innovation applied to produce boric acid of geothermal origin. The most successful of the companies mentioned above was that of Francesco Larderel & Partners, which in the first 10 months of activity in 1818 succeeded in producing as many as 36 tons of boric acid in their relatively small concession areas and to sell it in France at very favorable prices. In the following 10 years this company and others produced in aggregate about 50 tons/year of boric acid from *natural* and *artificial lagoons* of the boraciferous region, for sale in Italy and other European countries.

However, the intensive and prolonged cutting of firewood to evaporate the brines and dry the boric acid resulted in a rapid and notable reduction of the woods all over the boraciferous region and surrounding area, leading to the failure of most companies owing to the increasingly higher cost of the process heat, together with the consequent impossibility to meet supply contracts. The Francesco Larderel & Partners' company, however, concerned about their inability to meet contracts with French and other clients demanding increasing amounts of boric acid, while not wanting to incur the risk of paying high penalties, decided by majority vote to wind up the joint venture in 1827. Francesco Larderel disagreed with that decision because, as the company's technical director, for the evaporation and drying processes of the boric acid, he had started to devise a technology totally different from that of burning firewood. Thus, he decided to buy out his partners and manage the venture alone.

The *lagone coperto (covered lagoon)*, attributable to Francesco Larderel and introduced in 1828, is the second technological innovation used to produce boric acid from geothermal fluids. It consists of a brick hemispheric dome built over *natural* or *artificial lagoons*, within which low-pressure steam could separate from the boiling water of the *lagoon*, thus increasing its saline concentration. The steam formed in the upper part of the dome ($T \approx 100°C$) was channeled outside, first to evaporate the brine that flowed out from the base of the structure and gathered in shallow evaporation tanks and afterward to dry the wet boric gangue. In this way, the natural terrestrial heat was used as process heat in the boric acid plant, thus substituting for the firewood.

1828–1850. Three more important technological innovations were introduced in this period:

- The third (1829) is represented by the so-called *lagoni a cascata (cascading lagoons;* Figs. 19.2 and 19.3). The *natural* or *artificial lagoons* located more or less in the same slope belt in certain hilly areas were connected in series by small open channels enabling the boric water (boron content at the source of 5–10 g/L) to flow slowly by gravity toward the lower *lagoon*, then from this to the following lower one, and so on until reaching the last *lagoon* at the lowest elevation in each area. In this way, the boron-rich waters of the various *lagoons* could enrich gradually in boron content by natural evaporation before being channeled into decantation tanks and shallow evaporation tanks, where they became concentrated boric brines. At this point, they were ready for further evaporation and concentration in stepped tanks and for drying in other tanks by means of steam drawn from the top of the nearest *covered lagoon*, as described above. The combination of the two technologies (*covered lagoon* and *cascading lagoons*) enabled no less than 125 tons of boric acid to be produced in 1829 for sale in Italy, France, and other European countries;
- The fourth innovation (1828–1834) consisted of manual drilling of small diameter (10–12 cm) shallow wells (6–8 m depth), located near the natural *lagoons*. To this end, a three-legged rig with a simple lifting pulley was initially used (Fig. 19.4(a)) by which

Figure 19.2 *Lagone coperto* (covered lagoon).

Figure 19.3 *Lagoni a cascata* (cascading lagoons, A−D); decantation tanks (E−F); stepped evaporation tanks (G−G).

Figure 19.4 Evolution of the drilling equipment used in the boraciferous region to produce boron-rich geothermal fluids from 1828 to 1850 approx. The first two figures are reproductions from lithographs of the time. Left: Three-legged *verga artesiana* ("artesian bar") with lifting pulley only and vertical discharge (1820–1834). Center: Four-legged drilling rig on a wooden platform, with lifting pulley, winch, and vertical discharge (1835–1845); Right: *Ballanzino*, with 2-m high work floor, winch, protective shield for workmen at the winch, walking beam and ropes to facilitate percussion drilling, and *gomito* ("elbow") for lateral discharge of the hot fluid (1845–1855).

production was possible of hot water at a pressure, temperature, quantity, and boric acid content well above the average of the stagnant brines formed in the next natural *lagoons*. These were the first geothermal wells in the world (late 1828).

In a couple of decades, however, first a four-legged drilling rig set on simple wooden platform (Fig. 19.4(b)) and then a much better organized rig named "ballanzino" (Fig. 19.4(c)) were used. Both rigs were devised and constructed in the workshops of Larderel's company. As a consequence, the well depth was increased from the initial 6–8 m to about 20 m in 1834, with diameters up to 15 cm. Subsequently, the depth increased to 30 m in 1842 and to over 50 m in 1850;

• The fifth innovation (1840) was the *caldaia adriana* ("Adrian boiler"), named after the inventor Adrian, who was one of the sons of Francesco Larderel. It consisted of a series of brick conduits, lined internally with lead where a stream of natural steam was allowed to flow below the boric brines that slowly circulated above the steam in counterflow (Fig. 19.5).

Altogether, the three innovations described here resulted in a sizable increase in production of boric acid of geothermal origin, growing from 50 tons/year in 1827 to 125 tons in 1829 and to over 1000 tons/year in 1850.

1850–1900. The first decade of this period was characterized by a further increase in the production of boric acid that grew from 1000 tons/year in 1850–2000 tons/year in 1860. This was the result not only of the intensive exploitation of *natural, artificial*, and *cascading lagoons* and the use of the *adrian boiler* described earlier but also (or perhaps more importantly) of the notable increase in flow rate of the boron-rich fluid produced by wells drilled at gradually increasing depths. Such increased depths, in turn, were possibly the result of the improvement of the drilling technology that enabled an increase from 50 m maximum in 1850 to approximately 300 m in 1900. At the same time, the new wells were sited in areas with no extermal manifestations,

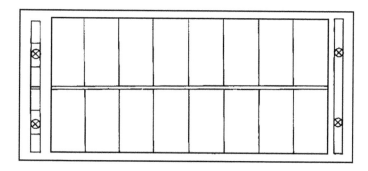

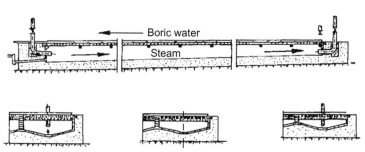

Figure 19.5 Layout and configuration of the "Adrian boiler." Top view (above), longitudinal section (middle), and cross section (below).

at progressively greater distances from the residual hydrothermal incrustations and fumaroles.

In effect, starting from the early 1860s, the incrustations began to run out, the *natural lagoons* to dry up, the fumaroles to vanish, and the hot springs to decrease in flow rate. Consequently, from 1870 to 1880, the only source of boric acid gradually became the geothermal fluid yielded by progressively deeper wells in new areas. Further, a few years later, in the early 1880s, boric acid produced in the United States from rich sedimentary deposits of colemanite [$CaB_3O_4(OH)_3 \cdot H_2O$], rasorite [$Na_2B_4O_6(OH)_2 \cdot 3H_2O$], and ulexite [$NaCaB_5O_6(OH)_6 \cdot 5H_2O$] started to penetrate the European market at favorable prices.

For 10−15 years, owing to its consolidated reputation and the good quality of its product, the Larderello Company with its business solely based on boric acid was able to meet the new market situation and even increase its production, reaching 2550 tons/year in 1900.

But around the end of the century, to face the competition, the company had to devise a new business strategy, which could only materialize at the beginning of the twentieth century when Prince Piero Ginori Conti came on the scene. He was the son-in-law of Count Florestano De Larderel, the last male descendant of Francesco Larderel, the pioneer of the Tuscan boric acid industry. For his important achievements in creating and developing this industry, Francesco had been ennobled by the Grand Duke of Tuscany Leopold II of Lorraine in 1837 as Count of Montecerboli. Moreover, Francesco had been permitted on that occasion to give his family name, Larderello, to the village that had been formed around the family palace and the main industrial establishments.

The chemical and geo-power industries from 1900 to 1960

1900—1930. The new business strategy developed by Piero Ginori Conti from 1903 onward had to consider also that the boraciferous region was located far from the main Tuscan towns and connected to them by unpaved roads. The reorganization involved basic and applied research and development activities in all production phases of the family's industry, including:

*   Geologic surveys and mineralogical/petrographical investigations
*   Field measurements and analyses by a mobile laboratory (Fig. 19.6)
*   Chemicophysical analyses in specialized laboratories on the whole production framework

Figure 19.6 First mobile geothermal laboratory of the world, mounted on a horse-driven cab (1903).

- Applied research on drilling materials and on materials and equipment used in chemical processes
- Development of advanced techniques aimed at drilling large-diameter wells to deeper levels to increase the fluid flow rate as much as possible (Fig. 19.7)
- Diversification of chemical products obtainable from geothermal fluids
- Strict controls to market only high quality products
- Modernization of the mechanical shops to provide services and carry out works in the shortest time possible following requests by the drilling and chemical sectors
- Organize all the work in the most efficient way possible

All of these activities were carried out with the advice and expertise of the best scientists of the time in the geo-mining and chemistry fields, among whom Bernardino Lotti and Raffaello Nasini stand out. It would be too long and out of place here to recount all of the scientific and technological advancements achieved at Larderello in the period 1903–1939 by the encouragement and under the direction of Piero Ginori Conti.

Concerning diversification of boron products, the following were commercialized in the first three decades of the twentieth century: nearly pure boric acid, borax, and

Figure 19.7 The well "Soffionissimo 1" (1931) (depth: 267 m; flow rate: 230 t/h at 4.5 atm; temp. 205°C).

chemical compounds of the same production framework, including sodium perborate, ammonium carbonate, carbonic acid, and talcum powder, all products largely used in pharmacy and in some industrial processes.

Therefore, the aggregate chemical production rose from 2550 tons/year in 1900−4800 tons/year in 1930, a record in the strongly competitive market of boric acid in that year.

In addition, it is worth recalling that following his marriage in 1894 with Countess Adriana De Larderel, daughter of the afore-cited Florestano, after some years of apprenticeship in the business and technical activities of the family's industry, Prince Piero Ginori Conti was appointed general director of the company and could thus develop the idea of using geothermal steam to produce simultaneously boric acid and electric power. With this in mind, following studies and laboratory tests carried out by his consultants at the University of Pisa and after some tests on steam-producing wells of the boraciferous region, the first experiment in the world of geothermal power production was realized under the coordination by Piero Ginori Conti.

A steam well near Larderello was used to this end, by which a piston engine coupled to a 10-kW dynamo supplying five small light bulbs was driven. One of the photographs recalling that event (Fig. 19.8), referred to by some authors as the date proper of the experiment, was actually taken about 3 months earlier, during one of the field tests carried out to investigate the production characteristics of the well chosen for the experiment. Subsequently, in order to check the behavior in time of both the reservoir and the surface equipment, the experiment was repeated several times in different operating conditions.

Figure 19.8 Photograph showing the equipment used by Prince P. Ginori Conti in preparation of the experiment carried out on July 4, 1904, to demonstrate the possibility of producing geothermal power. The prince is observing the shut-in pressure of the well chosen for the experiment: almost 5 atm.
Photograph taken on April 24, 1904.

After that group of experiments, the geo-power generation in the boraciferous region developed jointly with the production of boron chemicals and increased as follows:

1905: First prototype piston engine coupled with a 20-kW dynamo that enabled lighting the family palace and other residential buildings at Larderello for about 10 years;

1908: Second prototype piston engine, different from the one above but still coupled with a 20-kW dynamo. Some chemical plants of Larderello and the nearby production areas were lighted with it;

1913: First turbo-alternator (250 kWe, driven by "pure" steam-indirect-cycle) installed in the power plant called Larderello 1. It enabled the first electrification of all chemical plants of the boraciferous region and then the towns of Saline, Pomarance, and Volterra to the north of Larderello by means of a purpose-built 16-km-long electric trasmission line. After 3 years of regular generation, this unit was replaced by the following ones:

1916: Two turbo-alternators of 3.5 MWe each, still of the indirect-cycle type, both installed in the Larderello 1 power plant. The size of such units was comparable to that of the main hydroelectric and thermal power plants of the period;

1923: First pilot unit (23 kWe) of the direct-cycle type, installed at Serrazzano (western sector of the boraciferous region) to experiment on the behavior of turbines fed directly by natural steam (Fig. 19.9).

1926–1927: Installation at Castelnuovo V.C. of two exhausting-to-atmosphere units (direct-cycle): 600 and 800 kWe, respectively;

1930: Installation in the Larderello 1 power plant of an exhausting-to-atmosphere 3.5 MWe turbo-alternator, next to the two units of indirect-cycle type installed in 1916.

To sum up, 11.9 MWe in total were installed in the boraciferous region by December 1930, of which 7 MWe of the indirect-cycle type produced electrical energy and boron compounds simultaneously, and 4.9 MWe of the exhausting-to-atmosphere (direct-cycle) power plants to generate mainly electricity, with a small contribution also to the production of boric acid from their exhausted steam.

The electricity produced by these power plants was over 80 million kWh/year. Though rather important in economic terms, this represented, in cash terms, only a modest fraction of the total sales of the Larderello Company, which in 1930 was still based on boric acid and other boron compounds derived from geothermal fluids.

Figure 19.9 First exhausting-to-atmosphere geothermal unit (23 kWe) (Serrazzano 1923).

1930–1960. From the early 1930s on, however, a systematic decrease began to occur in content of boric acid in the fluid produced by all wells in the oldest exploitation zones. Although new wells drilled in peripheral areas to progressively deeper depths (up to over 500 m) often yielded sizable flow rates (see, for instance, Fig. 19.7), they were not able to compensate for the lower amounts of boric acid from wells in the old areas.

Consequently, starting from the late 1930s until the summer of 1944 when all industrial activities were halted owing to the final events of World War II that saw the destruction of all wells and power plants, the production of boric acid and other boron compounds from geothermal fluid became gradually less profitable. This happened in part because of the reduced market demand of the period, and in part due to the general decrease in the content of boric acid in the fluid produced by all wells, even the most recent, deep ones.

The decrease continued after the resumption of industrial activities in the late 1940s and subsequent decade, until it was proved definitively that production of sizable quantities of boric acid or other boron compounds from geothermal fluids was economically unsustainable. Therefore, after 1950 no other units were installed to produce electric power and boron chemicals simultaneously. However, some old chemical plants were left to operate for a number of years as a result of importation from abroad of boron minerals from which boric acid and other boron minerals were obtained in the chemical facilities.

The capacity installed in 1930–1960 evolved as follows:

1935–1939: Construction of the new power plant Larderello 2 to host six standard units of 11 MWe each, all of the indirect-cycle type for production of both electric power and boron chemicals.

1940–1943: Installation at Castelnuovo V.C. of four standard units, with the same size and type as the six units above, all to produce power and boron compounds.

Moreover, three exhausting-to-atmosphere units were installed for power generation only: one unit of 3.5 MWe at Sasso Pisano (southern sector of the boraciferous region) and two units of 3.5 MWe at Serrazzano.

As a consequence, taking also into account the substitution or modernization made in this period of some rather old units, the total capacity installed in the boraciferous region as of December 1943 was 126.75 MWe, including 116.25 MWe of the indirect-cycle type and 10.5 MWe of the exhausting-to atmosphere type (direct-cycle).

1944: Annihilation of wells and all power plants due to the final events of World War II. Among the ruins of the technical school of Larderello was found-miraculously intact-the small exhausting-to-atmosphere pilot unit that had been installed at Serrazzano in 1923 to test the behavior of turbines fed directly by natural steam (see Fig. 19.9). The unit had been brought to Larderello toward the end of 1925, where it was used in the training of power plant technicians.

With the tiny amount of electricity produced by this unit and with the firm resolution of the local people, thus began the second and present phase of geothermal development in the boraciferous region of Tuscany in late summer of 1944.

Sept. 1944–1950: Rehabilitation of the few chemical plants and power units that remained less damaged by the war events; restoration of all wells whose shut-in valves had been mined and made to blow out by the retreating armies in July 1944; construction of new power units, some of them of the indirect-cycle type, and others of the condensing or exhausting-to-atmosphere type; and the construction of a new power station, Larderello 3.

Therefore, the installed capacity in the boraciferous region as of December 1950 totaled 258.5 MWe, including 123 MWe of the indirect-cycle type to produce electric power and boron chemicals simultaneously, and 135.5 MWe of the direct-cycle type (118 MWe in the power plant Larderello 3 and 17.5 MWe elsewhere) to generate electricity only.

1950–1960: Completion of post-war reconstruction works, substitution of some indirect-cycle units with condensing units to produce electric energy only, telecommunication system applied to all main power plants in the boraciferous region, and the discovery of a new high-temperature field, Bagnore, in the Mount Amiata region, southern Tuscany.

Thus, the installed capacity in Italy as of December 1960 was 302.7 MWe, with 299.2 MWe in the boraciferous region and 3.5 MWe at Bagnore. The 299.2-MWe capacity installed in the Larderello region included 88-MWe indirect-cycle units to produce both electricity and boron chemicals from geothermal fluids or imported minerals, and the remaining 211.2 MWe (197.2 condensing and 14 exhausting-to-atmosphere) for power generation only.

Concluding remarks

It is clear that the revenue ratio between production of boron chemicals and power generation of the Ginori Conti-De Larderel industry, which was initially 100%, and later (until 1930) about 85–90% based on boric acid, though halved in the early 1940s, still had before World War II a very important role requiring the adoption of synergistic policies regarding the two types of industrial activity. In particular, the synergy justified the use, for power generation, of indirect-cycle units.

However, a systematic decrease of boric acid in the fluid yielded by the wells started in the late 1930s, and the further decrease between late 1940s and through the 1950s, confirmed that production of chemicals at constant levels from geothermal fluids was no longer feasible or economically justified.

The revenue ratio in question reversed completely between 1950 and 1960 in favor of power generation, thus making urgent the substitution of the remaining indirect-cycle units with other units of various types and sizes, optimized for power generation only. These new units are described in the following sections of this chapter. Many aspects of those sections should be read in light of the particular and somewhat peculiar history of geothermal development in Italy as described in this prologue.

Essential references for the prologue

Bianchi MC. Francesco De Larderel, uomo ed imprenditore (a cura di). In: Ciardi M, Cataldi R, editors. Il Calore della Terra. Contributo alla Storia della Geotermia in Italia. Pisa: Ediz. ETS; 2005. p. 209–24. pp. XVI + 344.

Bocci T, Mazzinghi P. I soffioni boraciferi di Larderello. Poggibonsi (Siena): Edit. La Magione; 1994. p. 140.

Burgassi PD. Tecnologie e sviluppo della geotermia nella Regione boracifera (a cura di). In: Ciardi M, Cataldi R, editors. Il Calore della Terra. Contributo alla Storia della Geotermia in Italia. Pisa: Ed. ETS; 2005. p. 195–208. pp. XVI + 344.

Cataldi R. La geotermia nelle antiche civiltà mediterranee (a cura di). In: Ciardi M, Cataldi R, editors. Il Calore della Terra. Contributo alla Storia della Geotermia in Italia. Pisa: Ediz. ETS; 2005a. p. 27–41. pp. XVI + 344.

Cataldi R. Applicazioni della geotermia in Italia dal VI al XV secolo (acura di). In: Ciardi M, Cataldi R, editors. Il Calore della Terra. Contributo alla Storia della Geotermia in Italia. Pisa: Ediz. ETS; 2005b. p. 116–25. pp. XVI + 344.

Cataldi R, Burgassi PD. Le ricerche scientifiche a Larderello dal XVI secolo al 1928 (a cura di). In: Ciardi M, Cataldi R, editors. Il Calore della Terra. Contributo alla Storia della Geotermia in Italia. Pisa: Ed. ETS; 2005. p. 316–30. pp. XVI + 344.

Cataldi R, Chiellini P. Geothermal energy in the Mediterranean before the Middle Ages, a review (a cura di). In: Cataldi R, Hodgson SF, Lund JW, editors. Stories from a heated earth; 1999. p. 165–82. Special Report n. 19 GRC-IGA; Davis, California; pp. XVIII + 569.

Cerruti L. Scienza, industria, estetica. Raffaello Nasini ed i soffioni boraciferi (a cura di). In: Ciardi M, Cataldi R, editors. Il Calore della Terra. Contributo alla Storia della Geotermia in Italia. Pisa: Ed. ETS; 2005. p. 276–92. pp. XVI + 344.

Ciardi M. Da Florestano De Larderel a Piero Ginori Conti: Ferdinando Raynaut ed il primo esperimento di produzione di energia geotermoelettrica (a cura di). In: Ciardi M, Cataldi R, editors. Il Calore della Terra. Contributo alla Storia della Geotermia in Italia. Pisa: Ediz. ETS; 2005. p. 247–74. pp. XVI + 344.

Di Pasquale G. Risorse geotermiche in Etruria (a cura di). In: Ciardi M, Cataldi R, editors. Il Calore della Terra. Contributo alla Storia della Geotermia in Italia. Pisa: Ediz. ETS; 2005. p. 42–52. pp. XVI + 344.

Franceschi F. Vicende della Regione boracifera volterrana nel Basso Medio Evo (a cura di). In: Ciardi M, Cataldi R, editors. Il Calore della Terra. Contributo alla Storia della Geotermia in Italia. Pisa: Ed. ETS; 2005. p. 143–53. pp. XVI + 344.

Fytikas M, Margomenou Leonidopoulou G, Cataldi R. Geothermal energy in ancient Greece: from mythology to late antiquity (3rd Century A.D.) (a cura di). In: Cataldi R, Hodgson SF, Lund JW, editors. Stories from a heated earth; 1999. p. 69–101. Special Report n. 19 GRC-IGA; Davis, California; pp. XVIII + 569.

Giacomelli L, Scandone R. Campi Flegrei-Campania Felix. Liguori, Napoli: Ediz; 1992. p. 121.

Grifoni Cremonesi R. Il rapporto dell'uomo con le manifestazioni geotermiche in Italia dalla Preistoria all'Alto Medioevo (a cura di). In: Ciardi M, Cataldi R, editors. Il Calore della Terra. Contributo alla Storia della Geotermia in Italia. Pisa: Ediz. ETS; 2005. p. 10–26. pp. XVI + 344.

Lungonelli M, Migliorini M. Piero Ginori Conti. Scienza, cultura e innovazione industriale nella Toscana del Novecento. Roma-Bari: Ediz. Laterza; 2003. p. 113.

Mazzoni A. I soffioni boraciferi toscani e gli impianti di Larderello. Seconda edizione aggiornata ed ampliata. Bologna: Editrice Anonima Arti Grafiche; 1951. p. 161.

Nasini R. I soffioni e i lagoni della Toscana e la industria boracifera, Storia, studi, ricerche chimiche e chimico- fisiche eseguite principalmente nell'ultimo ventennio. Roma: Tipografia Editrice Italia; 1930. p. 658.

Papini P. Larderello: Il villaggio e la comunità (a cura di). In: Ciardi M, Cataldi R, editors. Il Calore della Terra. Contributo alla Storia della Geotermia in Italia. Pisa: Ediz. ETS; 2005. p. 293–305. pp. XVI + 344.

Redi F. L'eredità medievale del termalismo romano: Cristiani eIslamici tra ritualità dell'acqua e cura del corpo (a cura di). In: Ciardi M, Cataldi R, editors. Il Calore della Terra. Contributo alla Storia della Geotermia in Italia. Pisa: Ediz. ETS; 2005. p. 82—97. pp. XVI + 344.

Rossi A, Manetti P. La geologia e la geotermia di Bernardino Lotti al passaggio tra il XIX ed il XX secolo (a cura di). In: Ciardi M, Cataldi R, editors. Il Calore della Terra. Contributo alla Storia della Geotermia in Italia. Pisa: Ediz. ETS; 2005. p. 306—15. pp. XVI + 344.

19.1   Introduction: background of geothermal power generation

19.1.1   Situation in Larderello in the late 1800s

The plants in Larderello and the annexed factories produced boric acid and its derivatives, and drilling operations were designed to increase the production of boric waters. The steam retrieved from the "dry" wells, wells without inflowing water, was used for thermal purposes in the evaporation of boric solutions. Starting from the second half of the 1800s, natural steam was used for mechanical energy. This brought boric waters to the surface through ejectors called "pressers" and "pumpers" depending on whether extraction occurred from the well bottom, where there was a certain amount of water, or if the water was to be transferred from one tank to another at higher levels. The principle was as follows: the steam velocity was increased through a reduction of area, the vacuum created in this section attracted the water in the surrounding areas, the steam kinetic energy was transformed into an increase in the steam–water mixture pressure, and this pressure lifted the mixture to the surface in the case for the pressers and up to the upper tank for pumpers.

The method used to create wells [1] during this period should be noted since this conditioned the use of steam as a driving force for some time. The wells were lined with riveted sheet-metal tubes with various diameters between 25 and 10 cm, a standard length of 2 m and a joiner strip. The first tube that was equipped with a toothed ring, which was also used as an expander since the chisels had slightly smaller diameters. The reduction in diameter occurred when the soil encountered was so resistant it did not allow rotation and thus the penetration of the tubes.

The external part of the pipes was not, as such, sealed, and this could lead to the leakage of steam from the lining tubes if the well was under pressure and, consequently, to the loss of control of the well. Accordingly, wells were considered drains for permeable zones rich in steam, useful for transferring boric waters to the surface but could not be considered an artifact capable of transferring the thermodynamic characteristics (pressure and temperature) of the geothermal fluids to the surface. For a long time, the use of the full potential of geothermal fluids was not used due to the precariousness in the construction of the well and the possibility that the steam could choose paths of less resistance and leak from the ground. As a precautionary measure, the maximum pressure used was limited to a few atmospheres. From an energy point of view, this meant that steam was considered only as a heat source for uses with thermal cycles.

19.1.2   First attempts to convert thermal energy into mechanical energy

In the late 1800s, thermal plants used to transfer thermal energy into mechanical energy consisted of a boiler used to evaporate water, with the use of a combustible (coal, wood, etc.) in order to feed a reciprocating engine. These systems were perfected during the 1800s, and an example of these systems is the steam locomotive.

Following this line, in 1895, Ferdinando Raynaut, the director of the Larderello factories [2], designed and had the Pineschi workshops in Pomarance build the first tubular boiler powered by endogenous fluid. The steam produced drove a 9-horsepower (HP) reciprocating engine and machinery such as mills, centrifuges, and stirrers for chemical plants.

This was the first example of the thermal cycle, which used geothermal fluid as a source of heat: a secondary fluid, namely, water, and a reciprocating engine as a processing machine. Reciprocating engines, which were extensively studied and refined during the 1800s, were very reliable. In the late 1800s, turbines were just being discovered (Charles A. Parsons, 1884); the first turbines in Germany date back to 1898. These first attempts to use geothermal fluid were hampered due to the corrosive effect that the geofluid had on the metal parts of the boiler.

At the turn of the century, Raynaut also made an attempt to use kinetic energy to operate a bladed wheel. The energy was acquired from the geothermal fluid emitted by a nozzle from a pressurized well named *Forte,* or "strong" in English. The experiment was a failure due to the short life of the material of the impulse wheel, but it did open the door to the use of pressurized steam. Raynaut's death, which incidentally was an unsolved mystery, put a hold on research, later resumed by Prince Piero Ginori Conti in 1897 when he carried out a similar test on a 5-ft-diameter impeller wheel. The wheel blades were soon covered with boric salts and oxides, which hindered the smooth and prolonged functioning of the wheel.

At the start of the 1900s, experiments were carried out on the direct use of steam in reciprocating engines. These piston engines, which were widely used with pure steam, posed two problems in the geothermal field: the first was connected to the need for pressurized steam, and the second was caused by the chemical attack that the fluid had on the internal parts of the pistons. Raynaut's experiments proved that a pressure of a few atmospheres could be obtained without losing the integrity of the well. In this way, the first reciprocating alternating-piston engines were used. These would operate pumps that transferred boric waters (horse pump) and were used as engines to operate winches used for drilling. With this installation, steam was alternatively conveyed from one side to another of a double-acting piston, causing it to move along its axis. This mechanism, together with a rod and crank system, could produce rotary motion (as happens with steam locomotives) or could transmit the motion to another piston that, in turn, pumped fluids as described above. In fact, from an energy point of view, these were the primary needs of the factory.

The first steam winches replaced the complex and costly system of manual drilling; this provided a major contribution to the increased productivity for such activity and, in fact, this would lead to significant development in the years to come. The boric acid production cycle included the transfer of collected boric waters to evaporators that, up to that time and from an energy point of view, used the pressers and pumpers system as mentioned above. With the introduction of the piston pump, the logistical layout of the plants, which had been based on the transfer of fluids through the use of gravity channels, was much more streamlined and simple. However, the following needs still needed to be fulfilled:

- The energy required during the refining and drying phases (stir mills, filter-press, etc.)
- Lighting (which gave a significant boost to the birth of the electricity industry).

The small thermal cycle, with the boiler and reciprocating engine, tested by Raynaut was not very efficient; the steam produced was low in pressure and the discharge of the reciprocating engine was at atmospheric pressure with a consequent small temperature difference to be exploited.

19.2   1900–1910: first experiments of geo-power generation and initial applications

19.2.1   Rise of Ginori and new company policies

In 1904, Piero Ginori Conti was named general manager of the company, and even though he did not have scientific training, his entrepreneurship was characterized by a strong link to the world of scientific research. Ginori Conti said, "At that time, there were daunting industrial, economic and commercial problems to be resolved and overcome, but, in addition to these, I considered scientific research to be equally important and impressive. If, in fact, technical and commercial problems needed to be resolved, and are still considered difficulties today, I saw research and study as a reality and the power of the future." In 1905, Professor R. Nasini became the scientific consultant for De Larderel, and many of his collaborators either worked in or for the company, including the brothers Plinio and Aldo Bringhenti [3]. The friendship between Ginori Conti and R. Nasini, university professor of chemistry first in Padova then in Pisa, began in the late nineteenth century when Ginori met Prof. Nasini while doing experiments in the Larderello area on radioactive gas. That day began a long history of friendship and cooperation that made Nasini as the technological driving force of the company.

19.2.2   First cycle that converted thermal into electric energy

At the beginning of the twentieth century, Ginori Conti saw the production of electricity as a business to be pursued. For this reason, in 1904, the steam produced by the Forte well after having been separated from the inflowing liquid phase, was sent to a $^3/_4$-HP reciprocating machine. The inlet steam had a gauge pressure of more than 4 atm. In 1904, together with engineer A. Fabaro (director of the Salt Works in Volterra), Ginori Conti coupled the engine to a small dynamo and managed to make it work for a considerable period of time. This was the first use of steam for the production of electricity.

The most important aspect of the event was not the coupling of the engine to the dynamo but rather the pressure reached by the steam and treatment (separation from

Figure 19.10 "CAIL" 40-HP reciprocating-type engine.

the liquid state) of pressurized steam. Piero Ginori Conti, in the famous picture, was not focused on the engine but on the pressure gauge of the separation system (Fig. 19.8, Prologue).

It must be underlined that the installations developed and future ones would have to deal with two aspects: the maximum pressure of the steam and the corrosiveness of the fluid. Through this experiment, a pressure of more than 4 atm, gauge, was achieved and saturated steam from the separation process was used, resulting in steam that was relatively clean.

Following this experience, in 1905 a "CAIL" 40-HP reciprocating-type engine connected to a 20-kW dynamo was installed for the production of electricity in Larderello. This system was used to supply electricity for the lighting in the factory of Larderello and the first electric engines of modest power (Fig. 19.10).

19.2.3   Problems yet to solve

Developments attained after those achieved by Raynaut in 1895 consisted of the direct use of steam in a reciprocating engine. Corrosion problems were reduced because a saturated fluid was used, which released the majority of its salts in the separated liquid phase. This brought a relatively clean fluid to the machine. To increase the overall power produced, the force on the pistons needed to be increased. However, given the low feed pressure, engines with huge cross sections were needed. These would have been slow and, therefore, not easy to use for the production of electricity. Therefore, it was hypothesized to use turbines, the latest technical discovery of that time. In fact, a turbine would make it possible to increase the power while keeping a compact size and to work with a number of revolutions with which the alternator could be directly fed. Therefore, the turbine was the only machine that could produce a significant amount of energy and was therefore preferred to the reciprocating machines that were used up to that time and that demonstrated good results since the CAIL engine remained in operation for about 15 years.

But, there still were problems. The materials available at that time to build the turbines (turbine blades were made of copper alloys) were not compatible with geothermal fluids. To reduce corrosion problems, in the case of direct use, the steam had to be saturated, losing a lot of energy. In order to increase conversion efficiency and the temperature difference, the turbine had to be of the condensing kind, which meant it had to discharge into a vacuum condenser. However, with this solution, the problem of how to extract the noncondensable gases from the condenser had to be resolved. Geothermal steam contained 30 L of gas per 1 kg of steam. Extraction, which at that time was done with reciprocating piston machines, would have required machines with enormous dimensions, whereas by using steam jet ejectors, the energy consumed would have significantly reduced the net electrical production of the condensing cycle.

In the meantime, chemical and physical research was undertaken on the fluids and materials; Professor Nasini was the leading figure in this period of intense research, and Larderello was the field school for his students, especially the brothers Plinio and Aldo Bringhenti. A. Bringhenti published his first and original studies on geyser steam [4].

19.3 1910−1916: first geothermal power plant of the world, experimental generation, and start of geo-power production at the commercial scale

19.3.1 Designing a new plant for the production of electricity

The solution taken some years later (1912) to solve all the problems cited was to copy the indirect cycle of Raynaut but replace the reciprocating engine with a condensing turbine. Subsequently, studies were directed to the use of natural steam as the heating fluid in boilers, which were supplied with pure water in order to produce gas-free steam (Fig. 19.11). The loss of energy was calculated to be less than what would have been required to extract the gas from the condenser. As with all thermal power plants, the extraction of noncondensable gas was limited to the entry of air from the seals, in particular the seal of turbine shaft. To reduce the air in-leakage, they decided to inject steam on the seal. These would "block" the inlet of air. Thus, there was some loss of steam but less air to extract. These systems are still in use today. Feeding good-quality steam, reevaporated from the water, solved the problems of corrosion on turbine materials.

In 1912, work was started for the construction of the first Tosi−Ganz test plant (Fig. 19.12). This had a Tosi-Parsons 350-HP turbine (Fig. 19.13), a Ganz 250-kW three-phase 50-Hz alternator, 4500 V, and was fed by steam produced by four Prache and Bouillon evaporators with a sloping, 120-m^2 surface with a pressure of 0.25-atm, gauge. The geothermal steam acted as a heating medium similar to combustion

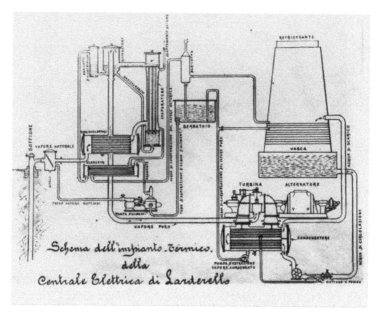

Figure 19.11 Original design of the first project of geothermal cycle.

products in common fuel thermal plants. The clean steam coming out of evaporators that drove the turbine had a relative pressure of only 0.25 atm. Geothermal steam entering the exchangers had a pressure exceeding 1 atm, gauge. The working scheme is described in Ref. [5].

Since the wells had a flow rate that increased by lowering the wellhead pressure and considering that only the latent heat of condensation of the steam was exploited, it was deemed advisable to keep the pressure low to enable a greater flow rate coming from

Figure 19.12 Year 1913 Tosi–Ganz 250-kW unit.

Figure 19.13 Tosi turbine (opened) 250-kW unit.

the evaporators. Consequently, one had to accept a low condensing geothermal steam temperature and a low clean steam temperature and pressure feeding the turbine (only slightly higher than atmospheric). We must remember that the pressure on the wells could not be very high for fear of loss of integrity of wells, and this limited the conditions of exploitation.

Another major difficulty stemmed from the fact that water to cool the condenser was not available in the area. Therefore, research was carried out on the possibility using a surface condenser, where the cooling water was merely steam condensate stored in tank kept cold with a water cooling tower.

So, in 1913 the power station in Larderello became operational and was connected to the electrical distribution line in Volterra and Pomarance on March 10.

19.3.2   Producing and selling electricity becomes more than a side business

The second decade of the twentieth century saw the advent of the sale of electricity. This activity increased in the subsequent years and almost completely replaced chemical production, which had led to the creation of geothermal activities and had generated income for more than a century. The production of electricity and the ease of distribution allowed modification of the drilling rigs once again by equipping them with electric motors to operate the winches. This technological development gave considerable impetus to drilling activities and, after a few years, wells with significant flow rates would be found. This meant an increase in steam production that would, as will be seen later in this chapter, impact the development of electricity cycles.

The Boracifera Company benefited this experience and decided to increase the production of the "new product" electricity, which helped to diversify chemical manufacturing. There was an increase in competition with foreign companies due to a decline in the price of boric acid, and so the company decided to build new groups of units in Larderello similar to the previous ones. The power station was made up of three 5000-HP condensing turbine generators with a low-pressure reaction turbine powered by steam, free of noncondensable gas, from tubular bundle boilers with pressures similar to the first machine, 1.25 atm, absolute. Two units started operating in 1916 (Figs. 19.14 and 19.15) and the third in 1917. Having encountered various problems in the operation of the Parche and Bouillon boilers, the company decided to have Kestner Company build specially designed boilers. The working scheme is described in Ref. [5,11].

Figure 19.14 Larderello 1-5000-HP unit double flux rotor Tosi turbine.

Figure 19.15 Larderello 1 cooling towers.

19.3.3 Numerous plant engineering problems needing solution

The use of these types of plants presented problems related to materials used in contact with geothermal fluid. Although the wisely adopted design choices sought to limit the heat exchanger as the only part of the plant in contact with the geothermal fluid, there were many other problems. The boilers, which were initially only 12, were increased to 18 and then 22 because the deposits reduced the heat transfer coefficient within hours of operation. Moreover, the adoption of aluminum tubes proved to be ineffective since some of the walls were perforated as a result of the action of the fluid. The loss in integrity of heat exchanger tubes led the geothermal fluid to pollute the secondary circuit, thus impacting the turbine blades, which were made of copper alloys and, therefore, also susceptible to attack by the fluid.

These negative experiences led many to conclude that the use of geysers for the production of electricity was unfeasible. Ginori Conti, on the other hand, persisted and established a "task force" to study and find solutions to the problems.

19.4 1917–1930: consolidation of geoelectric power production at the industrial scale and start of a new technology: the direct-cycle geo-power units

19.4.1 Connection with the university

Lago Boracifero was chosen as the training campus to experiment with innovative solutions. The 250-kW unit that was used in Larderello in 1912 was brought there. In 1917, engineer Plinio Bringhenti experimented with various devices that he had created. Among these was the evaporator named after him. With this device, he tried to avoid the formation of condensates on the walls of the exchange surfaces by introducing condensate in mixing plants.

The steam was condensed by mixing it with water (Fig. 19.16). The mixture was heated and the gas that did not dissolve in the liquid phase was discharged into the atmosphere together with a small portion of the saturated steam. The liquid phase rising in height reduced the pressure creating a steam with very little gas, which was then sent to the turbine. Energy performance was a bit lower since a part of the steam was released together with the gas into the atmosphere but, as will be seen later in this chapter, it would be reused in chemical plants.

The successful results of the experiment led to the installation of this equipment in Larderello. This would replace the Kestner boilers, and in 1923 operation of the turbines in Larderello was constant and regular. The working scheme is described in Ref. [5,12].

At this time, other parts that gave rise to operational problems were replaced. Surface condensers on the turbine outlets that were subject to frequent tube failures due to infiltration of geothermal steam in the "pure water" secondary circuit were replaced with direct-contact, jet condensers. The use of direct-contact condensers to eliminate problems with surface exchangers remains to this day.

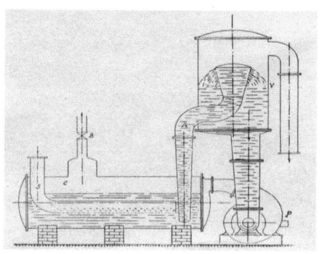

Figure 19.16 Bringhenti's heat exchanger.

19.4.2 Increasing retrieved steam

Enormous advances in drilling operations with greater depths achieved as well as increased productivity of new plants meant that the discovery of steam was increasingly consistent. During those years, steam was mainly used for chemical manufacturing, and only a small part was used for the production of electricity. The electricity market needed new and important customers to justify the substantial investment required to build power plants, which were complex and not easy to manage. This abundance of steam led to the study of systems that produced electricity at a lower investment cost.

19.4.3 New plants to produce electricity are conceived and developed

The first experiments on the direct use of steam in a turbine were carried out in Serrazzano, where, in 1923, a large quantity of steam was found. Technicians of the company Ansaldo designed a turbine directly driven by steam, which was built by the Cerpelli Company (Fig. 19.9, Prologue). It was a turbine with a nominal power of 40 HP driving a dynamo with an effective power of less than 30 kW. The exhausted steam was then released into the atmosphere. This plant had lower efficiency than a condensing turbine, but it also had a considerably lower installation cost. Notably, this marked the first time that a turbine was designed that did not use fluid coming from a boiler or evaporator; instead, it used a fluid mixed with inflowing solids, salts, and various gases.

The steam exhausted from the turbine entered the chemical plants and was first used to evaporate the boric waters and then for its boric acid content. The idea was to fully exploit the steam by sequencing the forms of exploitation, first electricity and then

Figure 19.17 Year 1930: Larderello 1-three condensing units and one atmospheric discharge.

chemical use, in order to minimize required investments and therefore completely use the resource. Economically, it did not make sense to install new and very expensive condensing systems when units could cover the demand for electricity at a lesser lower cost.

Several milestones worth noting include:

In 1926, a 600-kW unit, atmospheric discharge, was installed in Castelnuovo Val di Cecina.
In 1927, two 800-kW turbo-alternators, atmospheric discharge, were installed in Castelnuovo Val di Cecina.
In 1930, the first unit installed in Castelnuovo was brought to Serrazzano.
In 1930, having found very powerful wells in Larderello, the existing power plant was enlarged with a 5000-HP atmospheric discharge turbine.

In summary, in 1930 there were four geo-electric power plants:

- Larderello: four turbines (three condensing and one atmospheric discharge), with an overall power of 12,000 kW (Fig. 19.17)
- Castelnuovo Val di Cecina: two atmospheric discharge turbo-alternators, with an overall power of 1600 kW
- Serrazzano: one 600-kW atmospheric discharge unit
- Lago Boracifero: one 250-kW condensing turbine.

19.5 1930–1943: toward a balanced economic importance of chemical production and geo-power generation

19.5.1 "Complete" use of steam

During this period, earnest attempts were made to work with plants able to completely use the geothermal steam, maximizing both energy and chemical production (Fig. 19.18). Giovanni Ginori Conti, Piero's son, wrote [6], "in order to completely

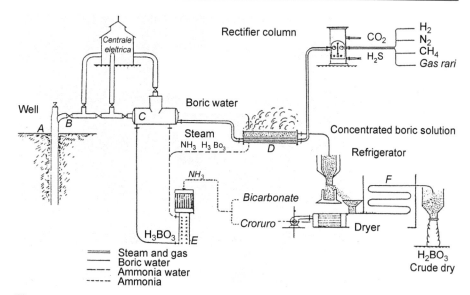

Figure 19.18 Original drawing: "Complete" use of steam maximizing both energy and chemical production.

use the geothermal (geyser) steam, we needed to find equipment that enables us to best exploit its thermal energy without jeopardizing the use of its chemical content."

The aim was to obtain an integrated cycle with complete use in which there was a production of electricity and the production of chemicals, both traditional boric-based and ammonium compounds, and to exploit components of noncondensable gases such as hydrogen, nitrogen, methane, rare gases, and carbon dioxide. Together with the production of electricity, this created a number of important compounds: crude boric acid (95%purity), refined boric acid (99.5%), borax, ammonium bicarbonate, ammonium sulfate and various products such as talcum powder, pink borax (deoxidizing powder), sodium perborate (to use with soap for washing clothes), boric petroleum jelly, as well as sodium borate, calcium borate, manganese borate, ammonium borate, copper borate, sulfate of soda, etc.

19.5.2   New developments in plant engineering

A big steam well erupted in 1931, called "Soffionissimo," the most powerful geothermal well found up to that time; it had a flow rate of 220,000 t/h at a pressure of 4 atm gauge and a temperature of 220°C (Fig. 19.7, Prologue).

Studies on the condensing plants continued (Fig. 19.19). In 1932, a new type of surface steam coil boilers replaced the Bringhenti plants in Larderello (Figs. 19.20 and 19.21). The aim was to increase the energy efficiency of the cycle by returning to surface exchangers while trying to contain the problems experienced with the previous ones, through a project that eliminated the tube sheet, a source of major problems, but mainly to be able to better use the output fluid for the production of chemicals [13].

At that time, for atmospheric discharge plants, it was assumed that the maximum energy obtainable from wells was by maintaining wellhead pressure between 2.5 and 3 atm, gauge.

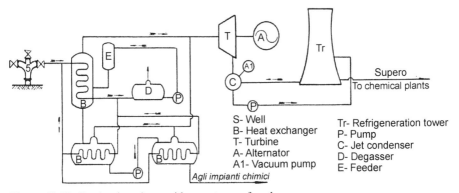

Figure 19.19 Condensing plants with new types of exchangers.

S- Well
B- Heat exchanger
T- Turbine
A- Alternator
A1- Vacuum pump

Tr- Refrigeneration tower
P- Pump
C- Jet condenser
D- Degasser
E- Feeder

Agli impianti chimici

Figure 19.20 Horizontal heat exchanger.

At the start of the century, a system to "wash" the steam was developed. This system injected a minimum quantity of water, with a high concentration of boric acid, into the steam. After the injection, there was a separation that created saturated steam to use for thermal purposes and water with a very high concentration of boric acid to use in the chemical facilities. In these conditions, assuming the use of saturated steam, the specific steam consumption was 28 kg per kWh. The discharge pressure was approximately 0.1 atm, gauge and the temperature was about 105°C. Under these conditions, it was possible to use the steam further for the chemical facilities.

Impulse turbines were used in atmospheric discharge plants because these were the simplest and most economic that existed. For the indirect cycle, the steam entered the coil boilers at the same pressure as they entered the atmospheric discharge units. The thermal performance of the process to create steam with heat exchangers was about 80%, the steam had an effective pressure of 1 atm, gauge, and considering the power absorbed by the auxiliaries, the steam consumption for this cycle was 15 kg per kWh, or roughly half the steam consumption of the atmospheric discharge system.

Figure 19.21 Vertical heat exchanger.

In addition to the turbine and generator, these power plants needed boilers, refrigerants, condensers and other ancillary equipment, such as pumps.

It should, however, be noted that, at this time the availability of steam was considerably greater than the need to produce electricity. An estimate was made during the 1930s that if all available steam were used in condensing power stations there would be available a power of 60,000 kW for a production of 500 million kWh per year, but there was not a demand for so much electricity.

At that time, studies were started on the use, which was defined as "mixed," namely, a condensing cycle with direct utilization of the steam in the turbine, with the vacuum in the condenser maintained by the extraction of noncondensable gases. This is a solution that, as will be seen below, would prevail in the future.

19.5.3 Development of a booming market: electricity for railway transportation

Starting in 1932, Italian State Railways entered into a series of contracts with Boracifera for the supply of electricity. Powering railway lines with the energy produced from the geothermal source was considered strategic for rail transport since it freed itself from the uncertain availability and fluctuating prices of imported coal. On the other hand, it enabled Boracifera to have fairly constant demand that went along well with the continuous and constant availability of geothermal steam.

In 1935, the State Railways made it known that the electrification of the Florence–Rome line would require a greater amount of energy. Two large wells were found in Serrazzano; the first was found in 1933 and the second in 1935, with an availability of almost 200,000 kg/h of steam. The Railways needed 4000 kW. So it was deemed necessary to install a 4000-kW atmospheric discharge unit that would have consumed just half the amount of the newly-won steam. However, the unit was designed in such a way that, with a few changes to the blading and distribution of the steam, it could be transformed into a condensing unit.

In March 1935, the unit was commissioned for construction by Ansaldo. The features were as follows: 5000-HP impulse turbine with a steam pressure of 2.5 atm, gauge, the possibility to transform to condensing operation subject to changes to the blades and distribution nozzles as well as the addition of exhaust section and a 5000-kVA alternator, 0.7 power factor, and output voltage of 4000−4500 V. The entire power plant design was executed by company technicians, and both the shed and crane were built by workshops in Larderello.

Boracifera always had solid profits that enabled it to produce a good dividend for the shareholders and gave it the possibility of self-financing. To develop electricity production, the company also used government grants that were awarded to this extraordinarily innovative company that was making a major national economic contribution, which was particularly felt at that time.

However, the installation of these new units made possible by the exceptional discovery of steam required considerable investments. The relationship with the State Railways became increasingly close. In 1938, the overall production of electricity reached 202 GW-h (GWh), 109 of which were absorbed by the State Railways, 65 by Selt-Valdarno (the electrical energy distribution company that covered the area of Firenze-Arezzo), and 28 directly used by Boracifera or disbursed directly to consumers in the region.

The state considered this business to be increasingly strategic and wanted to promote it, so it negotiated the purchase and used the mining concessions that it had issued to Boracifera as a means of leveraging and promoting geothermal energy. Italian mining laws recognize the state as the owner of the geothermal resource and vest in it the power to grant concessions. However, if the receiver of the concession does not put it to use, the state may revoke the concession, even though in such a case, a refund may be available to the concessionaire. In this way the state would encourage the concessionaire to actually exploit rather than merely hold the resource.

19.5.4   A new company is created

Larderellos. p.a, which subsequently incorporated Boracifera, was founded in 1939. The previous owners, the family of Piero Ginori Conti, who were excluded from the head of the new company were adequately compensated (176 million Lire). However, a decree was passed to transfer mining concessions that included all the wells and the equipment needed to handle and process the geofluids to the State Railways.

The availability of steam and money launched a series of new installations:

A new power station built in Larderello, named "Larderello 2," where three 11,000-kW indirect units were installed in 1938, 1940, and 1941.

Four 11,000-kW indirect units were installed in Castelnuovo in 1942.

Another unit was added to the 3500-kW atmospheric discharge unit in Serrazzano in 1940, and a 3500-kW atmospheric discharge unit was installed in Sasso Pisano in 1944.

Figure 19.22 Larderello 2 power plant.

In summary, in 1944 the Italian geothermal power plants were as follows:

- Larderello 1: two 3000-kW indirect-cycle units
- Larderello 2: six 11,000-kW indirect-cycle units (Fig. 19.22)
- Castelnuovo Val di Cecina: four 11,000-kW indirect-cycle units
- Serrazzano: two 3500-kW atmospheric discharge units
- Sasso Pisano: one 3500-kW atmospheric discharge unit
- Lago: one 250-kW condensation unit.

19.6   1944–1970: destruction, reconstruction, relaunching, and modification of the geo-power system

19.6.1   Situation at the end of World War II

All plants described here earlier (~ 127 MWe) could operate safely for only few months after 1943 before they were completely destroyed in July 1944 during the final events of World War II. The reconstruction started in late 1944 with the rebuilding of most former units, but rehabilitation work lasted almost 2 years. It included also the restoration of all production wells, whose wellheads had been mined by one of the belligerent forces in July 1944, leaving them to flow freely into the atmosphere.

To the existing plants, one atmospheric discharge unit of 3500 kW was added in Sasso Pisano in 1947 and another one in Monterotondo in 1946. The historical Larderello No. 1 power station was shut down in January 1948. Summarizing, indirect cycles were installed in Larderello and in Castelnuovo, while there were direct cycles in Serrazzano, Sasso, and Monterotondo. The start of the reconstruction also saw a relaunch of the projects that the war had put aside.

In 1954, the first successful shutting in of a well was accomplished, namely, the Canteo well in the vicinity of Serrazzano. Until then, in the drilling phase, whenever steam was detected in a well, it was never shut in but left exposed to the atmosphere until being linked to the production network. After an initial failed attempt in 1937 on a well that reached the pressure of 30 atm before damaging the well head, no more attempts were made. The pursuit of new wells meant new labor and financial costs, and the anxieties about losing a well were so great that, until the postwar period, nobody had reattempted the experiment. To facilitate a new attempt an expert commission was established that also included oil industry experts (among others, Dr. C. Sommaruga), who were tasked with defining the course of action. The well was opened and closed several times without any damage or loss of integrity.

19.6.2   Larderello 3 power plant

Indirect cycles made it possible to produce chemical compounds, but the efficiency of transformation into electricity was low due to the complexity of the cycle. Turbines powered directly by steam were already being widely tested with direct cycles, and materials capable of resisting direct contact with steam had already been developed. The efficiency of these plants was equally low, since the turbines discharged into the atmosphere using lower steam enthalpy changes than other stations using vacuum condensation. For some time, researchers had been experimenting with equipment that is able to maintain vacuum in the condenser. They excluded steam jet ejectors whose low efficiency due to high content of non-condensable gases would not have improved plant efficiency.

All of these studies led to the first major postwar project, the Larderello 3 power plant (Figs. 19.23 and 19.24). The result was the first direct condensing cycle, the progenitor of all modern geothermal power plants. For this reason, we are going to illustrate it in detail, in order to appreciate the technological leap that it represented (Fig. 19.23).

Figure 19.23 Larderello 3 power plant (internal view).

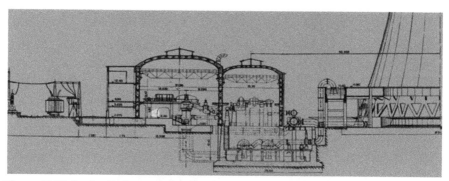

Figure 19.24 Larderello 3 power plant (section).

19.6.3 New machinery to create and maintain vacuum

At this point, studies conducted to find machinery able to maintain vacuum in the condenser by extracting the condensable gas found a solution, namely, the centrifugal extractor (Figs. 19.25 and 19.26). One must understand that up to this time the condenser was kept under vacuum by reciprocating extractors that, due to their limited capabilities, did not allow the use of condensers that could directly receive the geothermal fluid. In fact, this entails a quantity of noncondensables that were variable depending on the region, but at that time fluctuated around 5% by weight of steam.

The Larderello 3 power station started production in 1950 with four 25,000-kW and two 9000-kW units for the auxiliaries. Here, the steam directly entered the turbine, which did not discharge to the atmosphere, as was the case for the previous ones but instead used a vacuum condenser. For the first time, centrifugal compressors devised by Tosi and serving as gas extractors maintained the vacuum.

19.6.4 Optimal vacuum

By decreasing the pressure at the turbine discharge, the power of the turbine increases. However, when this pressure decreases, the extractors work harder and, therefore, absorb more power. The net output power is the difference between the power

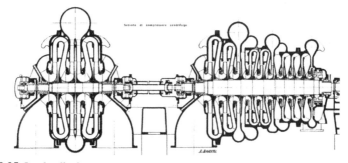

Figure 19.25 Larderello 3 power plant. Gas extractor section.

Figure 19.26 Larderello 3 power plant. Gas extractor internal view.

produced by the turbine minus that absorbed by the compressors. Therefore, we are presented with a problem; we need to know the optimal value of vacuum to maximize the power produced by the plant. Calculations done at that time, which suggested an absolute pressure in the condenser of approximately 80 mbars, do not deviate from the values currently used, despite the fact that these calculations were made more than 50 years ago, without the aid of modern computing technology!

19.6.5   New types of condensers and gas coolants

The power station in Larderello did not only have the first centrifugal gas extractors, but it also was enriched with new components needed to chill down the gas. The geothermal fluid had to be condensed at the exit of the turbine and the gas extractor intake had to be cooled to reduce the power required, thus the condensers were redesigned. This new function had not been needed when the turbine discharged only pure steam to the atmosphere. Many experiments were carried out in order to find optimum operation performance of the equipment. The reduction, for example, of gas temperature, which produces a reduction of the power absorbed by the extractors and consequently increased productivity, was obtained by increasing the water flow rate, that in turn increased the load losses in the condenser, producing less energy from the turbine, which canceled the positive effect mentioned above.

To reduce the power required for the gas compression, the compression process was divided into several stages (three or four) by interposing an intermediate cooling medium that cooled the gas that got heated after each successive compression stage. These gas coolers were designed, just like condenser, as direct-contact type (ie, cold water sprayed into vertical cylinders cooled the gas). The first experiments were not easy since the wet gas attacked all pipes and deposited corrosive products (usually iron sulfides) on the impellers. After a few 1000 h, the impellers needed to be cleaned. To reduce this adverse effect, all tubes that transported damp gas were treated with a hard rubber cover lining or were lead-sealed, and filters were inserted to inhibit inflowing liquids from coming into contact with the impeller intakes. These liquids were responsible for transporting the corrosive products to the impellers.

19.6.6   Power station arrangement

The power station was made up of four 25,000-kW turbines and two 9000-kW auxiliary units that were located in the main bay. The second bay held eight centrifugal gas extractors. The specific steam consumption for this new plant was around 10 kg per kWh produced, the best obtained thus far.

The main double-flow turbines built by the Franco Tosi Company rotated at 3000 revolutions per minute (RPM) and had an inlet pressure of 4.65 bar (abs) with a geosteam temperature of about 190°C. They were connected to 30,000-kVA Ansaldo alternators with an output voltage of 10,500 V and were cooled by a closed-circuit air system; the air was in turn, cooled with the cooling tower water. This measure was taken to prevent the cooling air, which was in contact with all copper parts, from being ambient air because it contains hydrogen sulfide, which is known for attacking copper. The power station had two power units to provide power for auxiliary services (one as back-up for the other), which were composed of a 9000-kW Ansaldo turbine, rotating at 3000 RPM, connected to 12,250-kVA Marelli alternators with an output voltage of 4500 V. These alternators were cooled with an open air system but were designed to be changed to closed-loop air systems. These groups fed the auxiliary services, and thus, the four main units could be started.

Each main turbine had two compressors that worked in parallel. The gas was aspirated by the bottom stage, which was made up of two compression stages; each stage had two parallel impellers, and so the bottom stages were made up of four impellers. The cooled gas entered the intermediate stage, which contained three impellers, arranged sequentially. After the compression, the second coolant was introduced and so the process continued with a second passage through the gas extractor for another compression stage, which was also subdivided into three impellers. Then the third coolant was injected and, finally, the gas passed through the last compression stage, which was divided into four impellers in a row. All impellers rotated at a speed of 4500 RPM and used a 1-kW electric motor. The motor was powered at 4500 V, and revolved at 1460 RPM and with the use of a multiplier drove the shaft on which the impellers were placed. The power absorbed by the compressor was around 630 kW. One of the eight gas extractors that was used for experimental purposes was powered directly by a small 1000-kW turbine, which rotated at 4500 RPM.

In 1951, the power plants consisted of:

- Larderello 2: seven 11,000-kW indirect-cycle units
- Larderello 3: four 25,000-kW and two 9000-kW condensing units, direct-steam
- Castelnuovo Val di Cecina: four 11,000-kW indirect-cycle units and one 2000-kW unit, atmospheric discharge
- Serrazzano: two 35,000-kW units, atmospheric discharge
- Sasso Pisano: two units 3500-kW, atmospheric discharge
- Monterotondo: one unit 3500-kW, atmospheric discharge.

19.6.7 Chemical production decline

Larderello 3 station was the first power station created specifically to maximize the conversion of steam energy into electrical energy. It is completely disconnected from chemical production. In those years, the society was faced with the need to increase electricity production for an advancing country. This need required an increase in the number of plants operating on the basis of this more efficient cycle. Chemical production was, however, the one sector that produced the greatest demand for satellite activities and, therefore, gave work to an area that had always lived off geothermal activities. Atmospheric discharge power plants gave fewer yields from an electricity output point of view. Furthermore, in chemical terms, they sought to use the steam exhausted at the turbine exit point at atmospheric pressure, steam that had a constantly diminishing boron content.

Therefore, these power plants were the first to be converted into direct-steam plants, patterned after the Larderello 3 plant.

In order to stimulate chemical production, the technique of steam washing, already tested in the past, was brought into more intensive use. It consisted of injecting water into the steam manifold, upstream of the power station and, therefore, from an energy point of view, losing the energy from the superheating but obtaining water rich in boric compounds. In-depth research of geothermal fluids discovered fluids containing chlorides, particularly in the area of Lago and Serrazzano. With these fluids, the injection of water caused very acidic condensation, which quickly led to the destruction of many collectors and thus this method was largely abandoned. This experience, however, as will be seen, helped to develop a technique for treating corrosive fluids that remains in use to this day. In the late 1950s, it was decided to build a plant for the production of boric acid from colemanite (a borate calcium hydrate mineral discovered in 1883 in Death Valley, California, U.S.A.) transported from Turkey, in a way that maintains the chemical business and allows all the steam to be used for electricity production, which was becoming increasingly important and profitable.

The last remaining indirect-cycle plants were at Larderello 2 and Castelnuovo. At Larderello 2, the plant conversion began in the 1960s, and would be completed in 1964. Castelnuovo was also converted beginning in the mid-1960s. This was the end of co-production of chemicals and marked the start of electricity-only generation using geothermal fluid.

19.6.8 Travale and Mount Amiata: new areas for geothermal development

Despite the first study on geothermal energy by Professor Paolo Mascagni in 1777 [7] entitled "On the lagoons of Siena and Volterra,"which spoke of the geothermal area of Travale (Siena lagoon), no research projects were launched in this area by Larderello S.p.A. until the 1950s. First, two 3500-kW atmospheric discharge turbo-generators were installed. These units operated for a short time since the steam resource which supplied them went into sudden decline. It was nearly 20 years later that production resumed in that area. It still remains in operation today and is expanding with an installed capacity of 200 MW. This experience is always used as an example of how one should be cautious when expressing final judgements.

In the late 1950s, the first atmospheric discharge unit was also installed in Bagnore in the area of Monte Amiata. In the 1960s, the activity at Mount Amiata was also extended to the area of Piancastagnaio, once again, installing generator units with atmospheric discharge. The fields of Bagnore and Piancastagnaio are still being developed today, with an installed capacity of 120.99 MW [15].

19.6.9 New engineering developments

The negative experience in Travale encouraged the company to develop a plant design with easy installation and a minimum amount of civil engineering works, organized in such a way that it could be installed immediately after the drilling phase. This satisfied the dual purpose of testing the reservoir while producing electrical energy. The project relied on the use of atmospheric discharge units with a capacity of 3—3.5 MW. These units also needed to be operated without staff supervision. They were given the name "monoblocks" because they consisted of a single block (Fig. 19.27). The installations

Figure 19.27 "Monoblocks" view.

served as experience for the first remote control system that in the future would become an integral part of all plant activities in the Larderello area.

By the end of 1960s, the existing plant types were: direct cycle with condensation, such as Larderello 3, and atmospheric discharge units. In converting the plants to the cycle of Larderello 3, it became necessary to adapt turbine sizes to the various types of fluid present in different areas. This led to the proliferation of different sizes of machines with a subsequent increase of spare parts.

Besides the 25-MW turbine size with double reaction flow built by Tosi for Larderello 3 station, 12.5-MW naval turbines by Ansaldo were installed. These naval turbines had two design conditions: cruising (lower capacity) and full-force in which the first stages were bypassed to change the characteristic curve of the turbine and increase the flow rate. The steam enthalpy drop remained almost constant but divided into fewer stages, and therefore the power, which is proportional to the product of flow rate and enthalpy drop, increased.

This was one of the first attempts to adapt the characteristic curve of the machine to the production capacities of the reservoir. Each reservoir provided different steam flows, at different pressures. Therefore, there was an optimum pressure where the product between the flow rate and the specific enthalpy change of the fluid was at its maximum. A turbine that was well-adjusted to the specific reservoir was to be able to work at that optimal condition. Those turbines had two positions in which the turbine performance was relatively good: overhead valves closed with the power of 12.5 MW and open with 18 MW. The adjustment between these two conditions was accomplished by throttling the inlet valve. These intermediate conditions, being realized with a dissipative adjustment process did not yield an optimum output, but would make it possible to pursue optimal overall plant productivity conditions.

Moreover, 15-MW turbines were also built with two bodies: one body with high pressure and single flow and one body with low pressure and double flow. The downwards discharge pressure of the first body was slightly higher than the atmospheric pressure. The high pressure body could then work as an atmospheric discharge turbine for the initial tests of the reservoir, after which the low-pressure body could be added, as well as a condenser and gas extractor, to achieve a complete condensation cycle. It should be recalled that this philosophy was the basis of the design of the first atmospheric discharge unit installed at the Serrazzano station in the mid-1930s. Double flow turbines of the old indirect cycles from Larderello 2 and Castelnuovo were also reused. This involved changing the blades and most critical fixed stages that were originally made of brass with steels containing chromium. In 1985, the fixed parts of one of these turbines from Castelnuovo still featured brass blades!

After 1950, the geothermal capacity increased to 300 MWe in December 1960 and 390 MWe in December 1970, of which 3.5 MWe in 1960 and 25.5 MWe in 1970 were installed in the Mount Amiata region (Table 19.1).

19.6.10   Nationalization of electricity production

In 1962, Italy decided to nationalize the production of electricity. All the companies producing electricity were acquired by Ente Nazionale per la Produzione di Energia Elettrica (National Agency for Electrical Energy) or ENEL. Larderello S.p.A. also

Table 19.1 **Technology development**

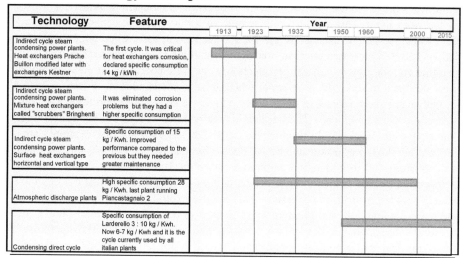

Technology	Feature	Year						
		1913 — 1923 — 1932 — 1950	1960 —				2000	2015
Indirect cycle steam condensing power plants. Heat exchangers Prache Buillon modified later with exchangers Kestner	The first cycle. It was critical for heat exchangers corrosion, declared specific consumption 14 kg / kWh							
Indirect cycle steam condensing power plants. Mixture heat exchangers called "scrubbers" Bringhenti	It was eliminated corrosion problems but they had a higher specific consumption							
Indirect cycle steam condensing power plants. Surface heat exchangers horizontal and vertical type	Specific consumption of 15 kg / Kwh. Improved performance compared to the previous but they needed greater maintenance							
Atmospheric discharge plants	High specific consumption 28 kg / Kwh. last plant running Piancastagnaio 2							
Condensing direct cycle	Specific consumption of Larderello 3 : 10 kg / Kwh. Now 6-7 kg / Kwh and it is the cycle currently used by all italian plants							

became a part of ENEL. However, ENEL's task is only to produce electricity. Therefore, all chemical production activities present in Larderello were separated by creating a new entity that was handed over to a new company called Società Chimica Larderello (Larderello Chemical Company). The conversion of the Castelnuovo station in the late 1960s removed the engineering link between electrical and chemical production. In terms of chemical production, the geothermal resource was used only for thermal purposes to produce boric compounds extracted from colemanite, a mineral which was transported to Larderello by road.

19.7 1970–1990: from reinjection of spent fluids and processing of steam to the renewal of all power units and remote control of the whole generation system

19.7.1 Reinjection of geothermal fluids: initial experiences and results

In the early 1970s, Italy introduced a new regulation on water discharges into the soil. Geothermal water that flowed freely on the surface for years could no longer be conveyed through surface-level watercourses, particularly the condensate water produced by power plants that was now considered the same as water discharged by an industrial plant. All this created the need to re-channel geothermal waters to the basins of origin.

There were dissenting opinions by those who thought that reinjecting water deep under the ground would cool the whole field. With a precautionary approach, in

1974 ENEL technicians began to reinject at the edges of the fields, in relatively marginal wells that were not too hot. Actions were also implemented to assess the effects of the work done. This included the creation of a microseismic network adapted to detect microseismic events, areas and wave epicenters in order to monitor any correlations between increased microseismic activity and the areas of reinjection. The temperature and the flow of adjacent production wells were checked. All these activities were designed to solve the problem of where to send the condensate water.

Only a few more years later did this begin to be viewed as an opportunity. The considerations were as follows. The Larderello area had been subject to intense drilling. The first wells were highly productive, up to 280 t/h. In 1954, the exploited area reached an extraction flow value of 1080 t/h. The majority of wells in this area had a depth from 400 to 600 m and suffered a rather marked decline in production. To compensate for this reduction in flow rate, additional wells were drilled but this ultimately resulted in interference issues. This intense exploitation led to a reduction in the reservoir flow rate and pressure from about 3 to 0.5 MPa [8]. At the same time, the fluid temperature remained nearly constant, and given that these were superheated fluids, the degree of superheating had been increased. This indicated that the rocks still held a large amount of thermal energy. Therefore, the conclusion drawn was to try to reinject right where the level of superheating was at its highest, and right in the surface production levels.

The first test was conducted at Larderello well No. 94, from January 1979 to April 1982, with flow rates in the range 10—50 l/s [9]. The most evident aspect was that the nearby wells were not subject to substantial cooling. However, expectations of an immediate and significant increase in the production went unfulfilled. Reinjected and evaporated water served to increase the quantity of steam present in the reservoir, but the pressure increase could not be immediate owing to the large volume of the reservoir. The flow rate of the wells was linked to reservoir pressure and, therefore, incurred the same time delays. The quantities of water reevaporating in the reservoir were calculated by indirect methods. The variation of the gas—steam ratio and isotopic composition of the fluid extracted before and during reinjection were determined. The reinjected and reevaporated water were free of noncondensable gases. Therefore, any reduction of the content of the gas—steam ratio from producing wells was a consequence of the reinjection and was correlated to the quantity of steam produced by reinjection. Similar considerations were used for isotopic control. Reinjected water is enriched with oxygen 18 and deuterium due to the stripping process in cooling towers. These considerations led to the calculation that the water was almost completely reevaporating and, over time, as this activity intensified, there would be a gradual increase in reservoir pressure and in the flow rate of the wells. This was a very important step toward a more sustainable use of geothermal reservoirs [10].

19.7.2   The oil crises and further development of geothermal research

The 1970s in Italy, and around the world, were marked by the oil crises of 1973 and 1979. This period saw a new impetus toward geothermal investments and, in particular, a fluids research plan was launched, both in new areas and in the reservoirs that were deeper than those traditionally used. The major new areas, developed in joint ventures with Azienda Generale Italiana Petroli (General Italian Oil Company) or AGIP, include areas of Latera, Cesano, Monti Cimini, Lago di Vico in Latium, and Campi Flegrei in Campania.

Even today, after almost 40 years, these areas do not produce any electricity, despite numerous drillings. The thermal resource has almost always been found. The problems come about with the type of fluids encountered. In some areas, they are extremely salty, which makes their economic utilization impossible with commonly known technologies. In others, the fluids are rich in noncondensable gases, in particular those with the presence of hydrogen sulfide (H_2S). Using these fluids with well-known technologies creates problems of odorous and harmful discharges, which ultimately ruled out their use. Further on it will be seen how, from a technological point of view, the issue of the treatment of noncondensables has made some progress.

The Deep Drilling Project relates to traditional areas with greater geothermal capacities and was implemented with wells Sasso 22 and San Pompeo. The traditional and experienced drilling techniques have to cope with unusual temperatures and fluids, even for geothermal fields. From these experiences drilling technology has learned a lot and has evolved considerably.

19.7.3   Standardized production units

The dynamism of the situation at that time, with new fluids and new areas being discovered, clashed with the time frames in which new equipment were supplied and new plants were built. Even the periods of well testing, conducted in order to stabilize the characteristics of fluids and select appropriate equipment dimensions, increased the delay between the drilling phase, where the company invested the largest amount of capital, and the beginning of commercial production, which brought in the first revenues.

For this reason, Italian manufacturers of geothermal machinery (eg, Tosi and Ansaldo) were asked to design new types of systems that met the following requirements:

A turbine, condenser, and gas extractor system were capable of handling fluid flow rates from 55 t/h to 110 t/h with pressures ranging from around 5 to 20 bar,g. The fluid could have from 2% to 12% (wt.) of noncondensables. Conversion performance had to be better than the machine used up until then in all fields of use.

The difficulty was remarkable. The range of turbine inlet volumetric flow rates was significant, almost a factor of nine, from 55 t/h at 20 bar,g to 120 t/h at 5 bar,g. The project, therefore, develops on these lines: a machine body that is equal in all conditions, so that it could be supplied irrespective of the results of the drilling activities. This body is then supplemented with modular components, whose composition is dictated by the specific working conditions. Once the working conditions are known, there is no more need to design and build any components, but only to assemble the premanufactured components. Equally, should the conditions of the fluid change, the structure of the machine could be adjusted in a relatively short time frame, about 2 weeks.

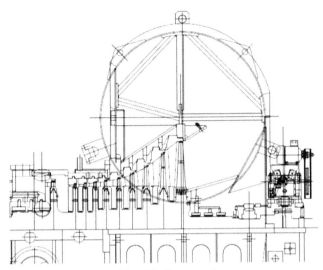

Figure 19.28 A 20-MW Ansaldo turbine. Discharge section.

This was a turning point and a success. Simple machines placed on skids on the ground floor, few civil engineering works, interfaceable equipment between the two suppliers, Tosi and Ansaldo, with the addition of Nuovo Pignone for gas extractors. Depending on the varying fluid characteristics, specific steam consumption was between just under seven and little more than 8 kg per kWh produced. These machines are still in operation and with a fleet distributed across 35 plants; the management of the spare parts stock is extremely economical and effective. The only change made over the years and was to create a discharge turbine with longer blades to reduce discharge losses and thereby further improve performance, as well as to overload the machine, for example during the period of maintenance shutdown of the neighboring units, at flow rates above 110 t/h (Fig. 19.28).

19.7.4   Remote control of power plants

From the times of the first monoblocks, Larderello entered the era of unmanned machines. Gabbro station, with a 15-MW unit built in the late 1960s, was the first condensing unit to be remotely controlled from Larderello (Fig. 19.29). Since then, stations gradually began to be controlled remotely, resulting in considerable savings on operating staff. Manpower was dedicated to both routine and periodic maintenance. For this reason, Larderello still has a maintenance team involved in the direct management of equipment maintenance with a high degree of specialization. The experience gained in maintenance made it possible to perform all the modifications to the equipment, which increased their reliability and greatly extended the intervals between required maintenance.

Initially, remote management was based on sending synthetic data to the remote control room on the proper functioning of the plant and any alarms with two levels

Figure 19.29 In 1960 — Gabbro power plant. First remote controlled.

of urgency. The power plants were still able to secure the machine autonomously. As of the 1990s, all the plants were controlled by a constantly manned single remote control point. For interventions on power stations in case of malfunction, there were staff available on call. Over time, and with the possibility to send a good deal of signals, and there was a gradual shift from remote management to remote diagnostics.

Today, remote management systems have been developed which are able to warn of any abnormal changes in the operating parameters (eg, irregularities in the normal variations of the parameters that are affected by the seasonality). The air temperature and relative humidity (having wet towers) can determine the temperature of the cooling water, which in turn affects the degree of vacuum, which in turn affects the power of the turbine and therefore the electrical production. As such, all parameters are seasonally adjusted to be compared against each other and in order to detect deviations from the optimal operating curve characteristic of each plant, known as the "plant signature."

19.7.5 Steam washing

The technique of injecting water into steam had already been used from the first half of the twentieth century to extract boric acid from the steam. Back in the 1960s, as previously mentioned, with the change of cycles, this has remained a way to effectuate the chemical component, while only slightly reducing the efficiency of conversion into electricity. With the retrieval of fluids containing chlorides, steam condensate became particularly acidic and led to the corrosion of the pipes. To overcome this, it was decided to add a base to the water in order to neutralize the condensate and eliminate the corrosion issue. This has become a method not to produce boric acid but to neutralize the steam acidity. It has laid the foundation for producing electricity using fluids that would otherwise be unusable.

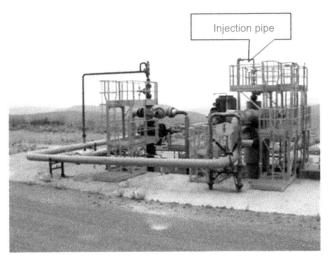

Figure 19.30 Steam treatment inside well.

With the extent of research ever increasing, the fluids discovered become more and more corrosive owing to the presence of chlorides, which makes it necessary to treat them before they reach saturation conditions and forming a particularly acidic condensate. There are therefore plants that treat the steam inside the well (Fig. 19.30), if it is estimated that condensation can happen already inside the casing, or at the wellhead or elsewhere before reaching the station. As such, all the fluids found have become economically usable for the electricity production.

19.8   1990−2014: recent technological advancements, with special regard to the "AMIS Project," new materials, and environmental acceptability

The two main objectives of the activity in this period have been social and environmental acceptability of geo-power production and the efficiency of the generation system. In this light, from the early 1990s (in certain cases even before as a follow-up of previous activities), a number of initiatives have been implemented in the boraciferous region and in the Amiata regions to prevent or mitigate undesirable effects of the geo-power production, together with several technological improvements in the generating units and ancillary equipment.

19.8.1   Reduction of hydrogen sulfide and mercury impact: AMIS Project

Among the activities focused on environment and social acceptance, the main intervention is the so-called "AMIS Project" aimed at eliminating H_2S emissions along

Figure 19.31 AMIS plant.

with its offensive odor and neutralizing the effect of mercury (Hg) contained in the steam. This is a project fully developed by Enel Green Power, a division within ENEL, which has also patented the technology. There are many other processes that can break down the H_2S from geothermal fluids. However, they produce waste and must be closely monitored. Since all Italian geothermal power plants are controlled remotely, ENEL set out to create an extremely reliable system that does not require continuous supervision and does not produce waste. This led to creation of the Hg and H_2S elimination plant. Essentially, the plant consists of an absorber that works to eliminate the traces of mercury found in some fluids, a catalytic reactor to convert H_2S into sulfur dioxide (SO_2) together with an absorber to absorb this gas in water that is then sent to reinjection. The waste products come up to only around 1 m^3/year per facility for the replacement of the mercury absorber once it is exhausted. The raw materials used consist only of sodium hydroxide to neutralize the solution that has absorbed the SO_2. This way, the specific emission of H_2S per kWh produced has been drastically reduced. Today all plants are equipped with this component, which has now become a solid part of best practices for the environment.(Fig. 19.31)

19.8.2 Visual impact

Another aspect that has influenced the design in the recent years is the visual impact. The construction and existence of geothermal wells, piping systems and power stations has an impact on their surrounding land. Drilling pads have been designed to accommodate multiple wells to minimize the number of work sites. Therefore, geothermal goals are achieved with deviated drilling. Piping design techniques have been developed in order to follow the natural boundaries of the woods, perimeters of cultivated areas, and roads, so as to minimize their impact. Also, the use of particularly visible features is avoided and cover plates are colored according to the surrounding area. With regard to the design of station, an attempt is made to try and integrate it into the area while avoiding the classic anonymous, box-like look (Fig. 19.32) The most recently built stations also feature tourist routes to make them a place that can be visited by the general public safely and unaccompanied by plant personnel. The Italian

Figure 19.32 Visual impact.

power plants are located in a tourist area of Tuscany and the design challenge is to transform these industrial facilities into tourist sites where people can connect with and understand the history of this resource and the latest technologies employed.

19.8.3 Improved efficiency of the generation system

The performance of the equipment needs to be enhanced in order to increase the power generated from a geothermal source with a given amount of steam available. Here is presented a description of all measures aimed at obtaining an ever-increasing number of kilowatt-hours from the available resource.

A first measure implemented was to adapt the equipment to make it possible to use fluids considered excessively corrosive. This allowed these fluids to be used for power generation. In the steam washing section, we described the measures adopted and the equipment used.

A second approach is to work on the main equipment to reduce the decline of its efficiency resulting from its operation. Much work was done on materials. The use of materials more resistant to corrosion also made it possible to decrease the need for maintenance and to increase the equipment availability. The equipment in use is of the modular type. This provides a great deal of operational flexibility and allows adapting the installations to the evolution of the geothermal reservoirs. Being able

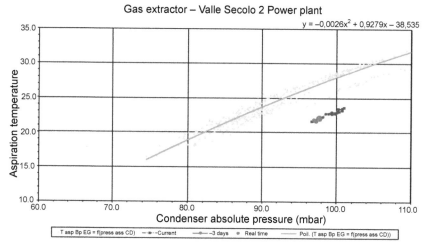

Figure 19.33 Plant signature, a deviation from good operating line: gas extractor operation points (*red points*) compared to the plant signature (*green curve*).

to quickly adapt the equipment configurations to the evolution of the fields, during brief shutdowns, possibly coinciding with normal maintenance shutdowns, makes it possible to keep the power plant constantly attuned to the reservoir's characteristics, in order to optimize the power generated. One of the most significant equipment changes has been to increase the length of the last-stage blades of the turbine to reduce the losses in the form of kinetic energy at the turbine outlet.

A third approach is to increase functionality of the telecontrol system, from a simple remote control tool of the generating units to an advanced diagnostic system of any possible damages or failures, with optimization of the whole plant operation grid.

It is crucial to be able to verify whether the power plant components operate under conditions of maximum efficiency in order to maximize the operational efficiency over time. Since the main operating parameters are affected by the environmental conditions, it is therefore very important to elaborate data to make them independent of season, in order to verify whether the equipment is operating on its characteristic curves or plant signature. This is done with a new telediagnostic system developed in Larderello (Fig. 19.33).

The aggregate capacity installed at the end of 2014 is 915.49 MWe, of which 794.5 MWe is in the boraciferous region, and 12.99 MWe in the Amiata region (Table 19.2).

19.8.4 Hybrid power plant technologies

In the southern area of Larderello, the first hybrid power plant started operation, producing energy from a true integration between geothermal and biomass energy sources. The project was developed through the repowering of preexisting geothermal power plant, Cornia 2.

Table 19.2 **Italian power plants at the end of 2014**

| | | | | Nominal power | |
| | | | | Turbine | Generator |
Name of plant	Prov.	Municipality	Gr.	kW	kVA
Nuova Larderello	PI	Pomarance	1	**20.000**	**23.000**
Farinello	PI	Pomarance	1	**60.000**	**75.000**
Valle Secolo	PI	Pomarance	1	60.000	75.000
			2	60.000	75.000
				120.000	**150.000**
Nuova Castelnuovo	PI	Castelnuovo	1	**14.500**	**23.000**
Nuova Gabbro	PI	Pomarance	1	**20.000**	**23.000**
Nuova Molinetto	PI	Castelnuovo	1	**20.000**	**23.000**
Sesta 1	SI	Radicondoli	1	**20.000**	**23.000**
Age Larderello			8	**274.500**	**340.000**
Pianacce	SI	Radicondoli	1	**20.000**	**23.000**
Rancia	SI	Radicondoli	1	**20.000**	**23.000**
Rancia 2	SI	Radicondoli	1	**20.000**	**23.000**
Travale 3	GR	Montieri	1	**20.000**	**23.000**
Travale 4	GR	Montieri	1	**40.000**	**50.000**
Nuova Radicondoli	SI	Radicondoli	1	**40.000**	**50.000**
	SI	Radicondoli	2	**20.000**	**20.000**
Chiusdino 1	SI	Chiusdino	1	**20.000**	**20.000**
Age Radicondoli			8	**200.000**	**232.000**
Selva 1	PI	Castelnuovo	1	**20.000**	**23.000**
Nuova Lago	GR	Monterotondo	1	**10.000**	**12.000**
Monteverdi 1	PI	Monteverdi	1	**20.000**	**23.000**
Monteverdi 2	PI	Monteverdi	1	**20.000**	**23.000**
Cornia 2	PI	Castelnuovo	1	**20.000**	**23.000**
Nuova Monterotondo	GR	Monterotondo	1	**10.000**	**12.000**
Carboli 1	GR	Monterotondo	1	**20.000**	**23.000**
Carboli 2	GR	Monterotondo	1	**20.000**	**23.000**
Nuova San Martino	GR	Monterotondo	1	**40.000**	**50.000**
Nuova Lagoni Rossi	PI	Pomarance	1	**20.000**	**20.000**

Table 19.2 **Continued**

Name of plant	Prov.	Municipality	Gr.	Nominal power Turbine kW	Generator kVA
Nuova Sasso	PI	Castelnuovo	1	20.000	20.000
Sasso 2	PI	Castelnuovo	1	20.000	20.000
Le Prata	PI	Castelnuovo	1	20.000	23.000
Nuova Serrazzano	PI	Pomarance	1	60.000	75.000
Age Lago			14	320.000	370.000
Bagnore 3	GR	Santa Fiora	1	20.000	23.000
Gruppo Binario Bagnore 3	GR	Santa Fiora	2	990	990
Baqnore 4 gr1	GR	Santa Fiora	1	20.000	22.000
Baqnore 4 gr2	GR	Santa Fiora	2	20.000	22.000
Piancastagnaio 3	SI	Piancastagnaio	1	20.000	22.000
Piancastagnaio 4	SI	Piancastagnaio	1	20.000	22.000
Piancastagnaio 5	SI	Piancastagnaio	1	20.000	22.000

Originally in 1994, Cornia 2 geothermal power plant was a dry steam power plant of the condensing type, located in the territory of Castelnuovo Val di Cecina (PI), south of Larderello. Due to the chloride content of the steam, a caustic scrubbing is carried out upstream from the turbine, so that the inlet steam was saturated vapor.

The main aspect of the project is the superheating of the geothermal fluid from around $160-370°C$ before it is sent into the geothermal turbine. This allows increasing the power output from the original 13 MW up to the actual 19 MW of gross power. The turbine has been modified to extend its range of operation up to a flow rate of 110 t/h, the same as in the original plant but now at a much higher temperature. Further improvement is foreseen involving the dryer system for the wood fuel, using the brine coming from the steam treatments along the geothermal gathering system. See chapter "Combined and hybrid geothermal power systems" for further discussion of hybrid plants.

19.9   Other geothermal areas

Apart from the boraciferous and the Amiata regions, production of geo-electric power had been carried out or started in an experimental way in past years in two other high-temperature geothermal areas of Italy: Ischia island (Campania) and Latera (Latium). Three units were installed, two at Ischia and one at Latera, whose features will be described in this section in relation to the characteristics of the respective feeding fluids.

19.9.1 Ischia

An intensive drilling activity designed to locate geothermal fluids for energy purposes was undertaken on the Island of Ischia from the second half of the 1930s until 1943. The areas where these activities were concentrated were Citara near the area of the Fumarole, San Angelo, and Maronti. In the early years 80 holes had been drilled to depths ranging from 20 to 90 m for a total drilled length of 2160 m. This activity continued after World War II, until the mid-1950s. In total, about 100 wells were [14,16−18,23] drilled.

Concurrent with the search for geofluids, several utilization projects also were started. The idea of using the geothermal springs to power the lifting pumps coincided with the building of the Ischia Island Aqueduct. In that period different cycles were conceived and studied to obtain electric energy from thermal solar panels.

The engineer T. Romagnoli in 1923 [19] and G. Andrei in 1935 published studies on cycles that used sulfur dioxide, methyl chloride, or ethyl chloride vapors. In 1927, the engineer E. Carlevaro formulated for the first time a double-flash system for a geothermal plant. This project consisted of a cycle of thermal waters with three flashes, with steam reaching different velocities during expansion through nozzles placed at increasing distance on the same wheel.

In 1934, the engineer L. D'Amelio designed a plant with an ethyl chloride−powered turbine, the forerunner of the modern binary plant [20−22]. To study the interaction of the materials needed to run an ethyl chloride−powered plant, engineer D'Amelio built a simple plant with a nominal 10-kW power output, which was made up of an evaporator, a turbine, and a condenser at the Istituto di Macchine dell'Università di Napoli. The heat exchangers were designed like normal surface condensers with several pathways, but they were built using different materials to test their performance. The evaporator had iron pipes, the condenser was made of brass, and the preheater was made of copper. The turbine had a single wheel with a single 220-mm-diameter crown of blades.

The problem that needed to be solved involved the large surfaces needed to obtain small temperature differences in the heat exchangers-the large surfaces produced large volumes of secondary fluids and therefore required more expensive setups. To address this problem, engineer D'Amelio experimented with his prototype by using a mixture of nonmiscible fluids, namely water and ethyl chloride. Once preheated to a temperature near evaporation, ethyl chloride was injected by means of a pump into the plant-water in tiny droplets. In this way the large contact surface between the water and the ethyl chloride caused it evaporate while achieving a negligible drop in temperature. The heavier atomic weight and higher vapor pressure of ethyl chloride-compared to water-resulted in the vapor that was produced being almost entirely ethyl chloride. Engineer D'Amelio stated that the experiments proved that the difference in output in the cycle was negligible using this mixture as opposed to using ethyl chloride only.

In 1938, the Società Meridionale di Elettricità (SME) company, together with the Società Romana di Elettricità (SRE) company, founded the Società Anonima Forze Endogene Napoletane (SAFEN), a concern tasked with studying the various possible

Figure 19.34 Citara power plant.

uses of the water and the steam available in the volcanic areas near Naples, for the production of electric energy. SAFEN collaborated with engineer D'Amelio.

And so began the on-site experiments in Ischia, in the Citara area, with an initial 11-kW plant. The excellent results obtained paved the way for the design of a 300-kW plant based on these same principles, which used 60 L of thermal water per second. This took place in 1941−1942. By the end of 1943, the plant was nearing completion, but the construction was interrupted until 1945, the year in which the plant was completed. It seems that even obtaining the ethyl chloride working fluid was made difficult by the events of World War II, and it was only after many negotiations that it was made available from the United States in 1947 (Fig. 19.34).

The power plant became functional in 1955, but the description of the plant made by the Istituto di Giacimenti Minerari e Geologia Applicata dell'Univerita di Roma differed slightly from the original parameters originally proposed by Professor D'Amelio. It was decided that only the steam produced by geothermal wells would be used, by first separating it from the liquid phase. The steam, separated at a pressure of 1.5 atm and at a temperature of 110°C, was then condensed in a mixing condenser using water in a closed system to reduce scaling. The heated water was produced at a temperature of 90°C, and it, in turn, was used to evaporate the ethyl chloride in a surface heat exchanger. The vapor then powered the turbine and was condensed in a condenser cooled by sea water. Although the plant started working in 1955, the available literature does not reveal the power production values and the effective efficiency of the plant.

A second plant was set up in Ischia. In the San Angelo area, two 300-m-deep wells were drilled, the first of which in 1942 produced 7000 kg/h of steam at a pressure of 1.2 atm. The positive results obtained by this first well pushed SAFEN to experiment with machines that used directly low-pressure steam as the motive fluid, as had been done in Larderello. Construction of an electric power plant, which used a steam turbine, was therefore begun. It had a maximum output power of 200 kW. All of these endeavors were halted due to the events of World War II.

Simultaneously with the construction of the power plant, a second well was drilled in 1946 to secure the steam necessary to obtain at least 200 kW of power. The results fell below expectations as only 3500 kg/h of steam were obtained. There is little more technical information on this power plant at this time. What is known is that in 1947, because of irregular flow of wells and deposits, SAFEN begin to doubt being able to supply the island with electricity from geothermal sources. Instead SAFEN began to design a power cable connecting the island to the mainland.

It is clear that these first experiments-particularly those led by the Facoltà di Ingegneria dell'Università di Napoli-on the use of geothermal waters for electricity production on Ischia Island were groundbreaking for their time, the late1930s. Professor D'Amelio in the 1930s had been successful in designing and operating a plant that used ethyl chloride as the working fluid, with the goal of using hot geothermal water.

The events of World War II halted the development of these experiments. In the years, after the war, fluids with greater energy yield were found but the plants were similar to earlier models. The cycle efficiency was very low. The Citara plant had evaporation temperatures lower than 90°C from a 110°C source. San Angelo's atmospheric discharge plant had to operate with a steam pressure difference of only 0.2 bar, less than one-tenth of what was done at similar plants in Larderello. There were no further technological developments that set out to increase yield output. The plants also suffered from the lack of constancy of the flow rate of wells.

By the end of the 1950s, the smaller plants gave way to large-scale petroleum-powered thermoelectric projects. The idea of conveniently producing electric energy from the steam at Ischia gave way to a project for the exploitation of geothermal energy for thermal direct use and tourism.

19.9.2 Latera

The Latera plants consisted first of a 3.5-MWe unit and then two 20-MWe units supplied by flash steam. The project was aimed initially at studying the evolution of the resource in time, and then at exploiting a 150°C water—steam mixture drawn from a CO_2-saturated reservoir. The last unit was installed in 1998 and dismantled in 2000. Even though it proved the technical-economic feasibility of generating geothermal power from this challenging resource, the Latera field had to be closed in late 2000. The olfactory impact of hydrogen sulfide present in the noncondensable gases (AMIS had not yet been developed) was considered unacceptable. The quantities of H_2S present would still have required a plant for its removal incommensurate to the size of the power plant.

Acknowledgments

This work would not have been possible without the support and guidance of Professor Ron DiPippo (Chancellor Professor Emeritus, Mechanical Engineering Department, University of Massachusetts Dartmouth). The authors thank the company ENEL who kindly provided the images of its historical archive and Fabio Sartori who has performed the current pictures of the plants.

References

[1] Burgassi PD. De Larderel and geothermal drilling techniques: studies, experiments and concepts in the XIX and XX century, 1985.

[2] Mazzoni A. Steam vents in Tuscany and the Larderello plant; Arti Grafiche Bologna, 1956.

[3] Conti GG. Geothermal power plant in Larderello. In: Communication during the AEI annual meeting; Florence 25 September; 1932.

[4] Bringhenti A. Chemical and physical research on geyser steam, from Raffaello Nasini I soffioni e I lagoni della Toiscana e la Industria Boracifera Tipografia editrice Italiana. 1930.

[5] Dipippo R. Evolution of geothermal power plants. Perform Assess Geotherm 2015;53: 291−307.

[6] Conti GG. Using boric acid geysers. 1936. Tip.G Cencetti Florence.

[7] Mascagni. On the lagoons of Siena and Volterra. Library of EGP Larderello PI; 1777.

[8] Barelli A, Cappetti G, Stefani G. Optimum exploitation strategy at Larderello − Valle Secolo. In: Proceedings of the World Geothermal Congress; 1995 [Florence].

[9] Cappetti G, Parisi L, Ridolfi A, Stefani G. Fifteen years of reinjection in the Larderello-Valle Secolo area: analysis of the production data. In: Proceedings of the World Geothermal Congress; 1995 [Florence].

[10] Giovannoni A, Allegrini G, Cappetti G. First results of a reinjection experiment at Larderello. In: Proceedings Seventh workshop geothermal reservoir engineering Stanford; December 1981. SGP-TR-55.

[11] Funaioli U. The Larderello natural steam power plant. Engineering (London) May. 1918. p. 568−9. Ginori Conti, P., 1917. L'impianto di Larderello. In: Electrical Engineering Journal and Proceedings, 15−25 September. Italian Assn. of Electrical Engineers [in Italian].

[12] Conti PG. The Larderello natural steam plant. In: Trans., of the first world power Conference. London: Percy Lund Humphries & Co.; 1924. p. 345−65. Ginori Conti, P., 1924, July. The natural steam power plant at Larderello. In: Paper read at the reception given by the Italian Delegation at the World Power Conference, 2nd ed. British Empire Exhibition, Wembly.

[13] Mazzoni A. The tuscan Boracic 'Soffioni' and their development at larderello. Struct Eng 1951;39(4):97−106. April 1951, see also: Discussion, The Structural Engineer, v. 39, Iss. 8, Aug., pp. 227−230.

[14] Sommaruga C. May. Primati Italiani nello Sfruttamento di Risorse Geotermiche di Media e Bassa Temperatura: In Particolare, Pompe di Calore, Cicli Binari, Impianti a Vapore di Flash − (1930−1960). In: Presented at the symposium La Geotermia in Italia Dal 1940 ad Oggi, (Geothermal energy in Italy from 1940 to today). University of Pisa (in Italian); 2010.

[15] Zancani CFA. Summary of thirty years' experience in selecting thermal cycles for geothermal power plants. In: Proceedings: second United Nations symposium on the development and use of geothermal resources, vol. 3; 1975. p. 2069−74. San Francisco, CA, May 20−29.

[16] Penta F. Ricerche e studi sui fenomeni esalativo idrotermali e il problema delle forze endogene. In: Relazione presentala al Convegno dell' Associazione Geofisica italiana, tenuto a Roma il 28−29 Maggio; 1954.

[17] Penta F, Conforto B. Sulle misure di temperatura del sottosuolo nei fori trivellati in presenza di acqua e sui relativi rilievi freatimetrici in regioni idrotermali. Annu Geofis 1951;4: 2−33.

[18] Penta F. Temperature nel sottosuolo della Regione Flegrea. Roma istituto dei Giacimenti Minerari e di Geologia Applicata della Universita' Facolta' di Ingegneria; 15 Agosto 1949.

[19] Carlevaro E. − Il decisivo contributo Italiano allo sfruttamento delle acque termali- Annuatrio della scuola di Ingegneria di Napoli, 1943.

[20] D'amelio L. Le acque termali come fonte di energia- Convegno a Torino. Reale accademia delle scienze di Torino; maggio 1939.

[21] D'amelio L. Utilizzazione dei cascami di energia termica Centrali Termiche azionate da acqua calda. Memoria presentata all'Istituto di macchine della Regia Universita' di Napoli; 13 febbraio 1942.

[22] D'amelio L. La turbina a vapore ed i cicli binari con fluidi diversi dall'acqua fra le isoterme inferiori. Comunicazione tenuta nella sezione della Associazione Elettrotecnica Italiana alla sezione di Napoli il; 23 Maggio 1936.

[23] Sappa M. Prospetto delle ricerche per "Forze Endogene " in varie regioni del mondo durante l'ultimo trentennio. In: Comunicazione presentata all'Assemblea dell'Associazione Geofisica italiana tenuta a Roma il 28−29 Maggio; 1954.

Fifty-five years of commercial power generation at The Geysers geothermal field, California: the lessons learned

S.K. Sanyal[1], S.L. Enedy[2]

[1]Geothermal Resource Group, Palm Desert, California; [2]Calpine Corporation, Santa Rosa, California

20.1 Introduction

The Geysers geothermal field in California has been supplying commercial electric power continuously for the last 55 years (1960–2015). Only two other geothermal fields in the world have had a longer history of power generation, namely, Larderello in Italy (continuously producing since 1948) and Wairakei in New Zealand (continuously producing since 1958). Since the 1970s, the level of power generation at The Geysers has been consistently higher than at either Larderello or Wairakei, or at any other geothermal field in the world. As discussed later, The Geysers field is expected to be capable of supplying several 100 mW of power even decades from now, when the field will have completed a century of commercial power generation. This long production history and the high level of power generation makes this field unique in the geothermal world Sanyal and Enedy (2011).

20.2 Background

The Geysers is a steam field; that is, it produces saturated steam rather than hot water or a steam–water mixture. Furthermore, this field represents an essentially closed system rather than the usual situation of an open geothermal system receiving natural recharge of hot or cold water from outside the reservoir. Steam fields are relatively rare, there being only half a dozen commercially exploited steam fields in the world compared to the numerous geothermal fields producing hot water or two-phase fluids (steam–water mixture). Fig. 20.1 shows a map of this field showing the boundary of the known steam production area, the locations of the power plants, and the "unit area" dedicated to each power plant; the wells for production and injection for a power plant are sited within the unit area dedicated to it. There are 19 unit areas shown on this figure. Many of the unit areas have been cross-tied by steam pipelines. As the electricity price has

Geothermal Power Generation. http://dx.doi.org/10.1016/B978-0-08-100337-4.00020-6

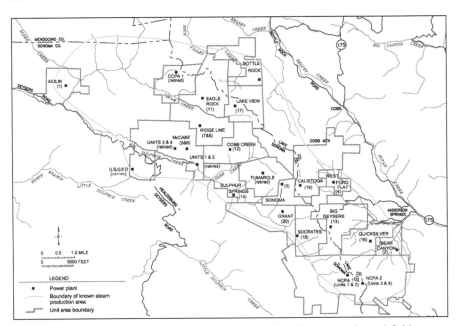

Figure 20.1 Location of power plants and unit areas at The Geysers geothermal field.

escalated, the operators have expanded the use of cross-tied pipelines to better use the existing steam wells. The operators have also retrofitted many of the plants with new steam paths. Introduction of new power conversion technologies has led to lower steam consumption rates per kilowatt and/or lower operating pressures, and lower operating pressures have allowed for higher steam production rates from the existing wells. With these modifications in operation, there are now 16 power plants generating about 800 MW. This field is the largest geothermal field in the world in terms of the productive areal extent (about 100 square kilometers) and has more than 400 active steam wells and 40 injection wells at present.

Fig. 20.2 presents the historical production and injection data from the field during the first five decades of operation. Fig. 20.2 shows the monthly production (uppermost plot) and monthly injection (middle plot) in billion kg, and the percent of the produced mass injected each month (lowermost plot) as a function of time.

As attempted in this chapter, the history of the field can be best described as seven distinct periods of 3—11 years' duration in terms of both field and well behavior and the prevailing socioeconomic forces; these periods are indicated in Fig. 20.2.

Fig. 20.3 presents the cumulative production, injection, and depletion (i.e., production minus injection) history of this field (in trillion kg), while Fig. 20.4 shows the annual steam depletion (billion kg) and the power capacity (MW) versus time. Figs. 20.2 and 20.4 show that this field had increasing steam production between 1960 and 1987, when it peaked at 10 billion kg per month, equivalent to about 1600 MW (net). The installed generation capacity in this field peaked at about

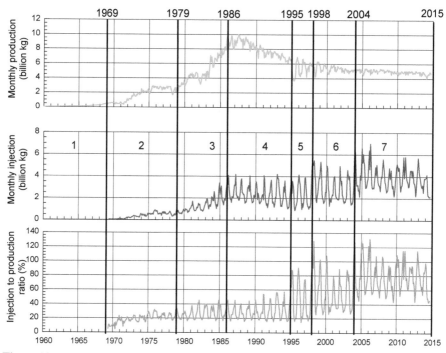

Figure 20.2 Historical production and injection data for The Geysers.

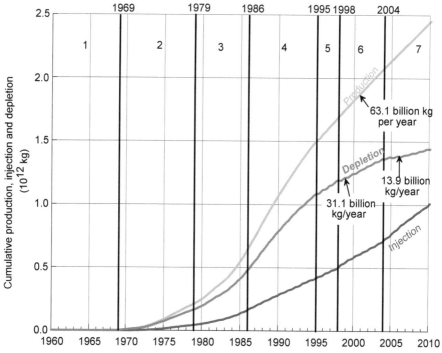

Figure 20.3 Cumulative production, injection, and depletion data for The Geysers.

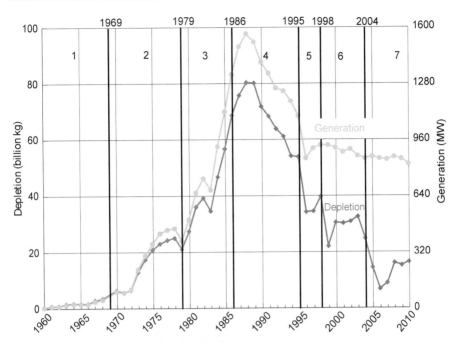

Figure 20.4 Annual depletion data for The Geysers.

2000 MW (net) in 1989. Subsequently, steam production and power generation in the field declined continuously for the next 7−9 years, reaching stability after 1995. Between 1995 and present both steam production and power generation have seen relatively modest declines with time.

Fig. 20.5 presents the monthly steam production rate (1000 kg/month) from the field (on log scale) versus time to illustrate the productivity decline trend in the field since 1987. Fig. 20.6 presents a similar plot of the steam production rate in 1000 pounds per hour (on log scale) versus time since 1980 for a typical well (GDC-12) to illustrate the changes in the productivity decline trend of individual wells.

We believe the long history of this field exemplifies the intricate interrelationship between the socioeconomic forces at play, field development and management practices, and consequent reservoir response. Therefore, in Figs. 20.7−20.9, we present the histories of some indicators of the prevailing socioeconomic forces in the United States over the last half century. Fig. 20.7 shows the history of crude oil and natural gas prices, Fig. 20.8 presents the history of the ratio of inflation rate to interest rate, and Fig. 20.9 presents the history of two leading stock market indicators (Dow Jones Industrial Average and Standard & Poor 500). We will refer to the field behavior trends shown in Figs. 20.2−20.5, an individual well behavior trend shown in Fig. 20.6, and the trends in the socioeconomic forces shown in Figs. 20.7−20.9, in the remainder of this chapter. In light of the above-discussed background, we analyze below the history of this field through each of the seven successive periods.

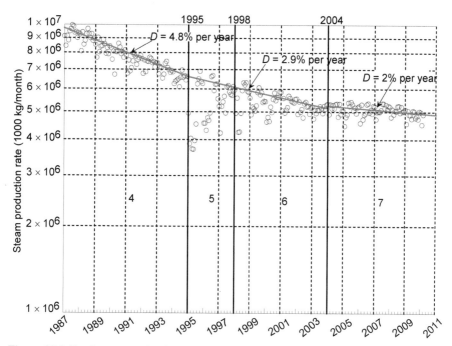

Figure 20.5 Total steam production rate from The Geysers field versus time.

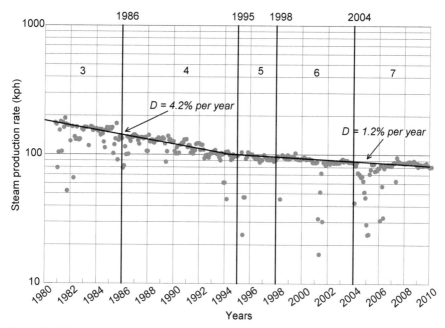

Figure 20.6 Steam production rate of well GDC-12 versus time.

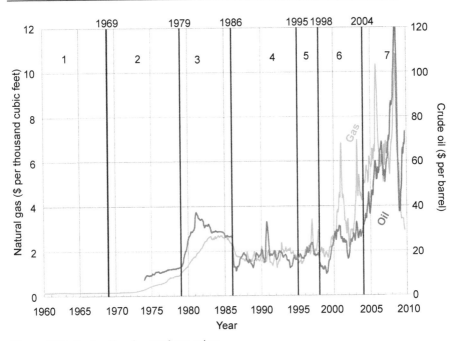

Figure 20.7 Crude oil and natural gas prices.

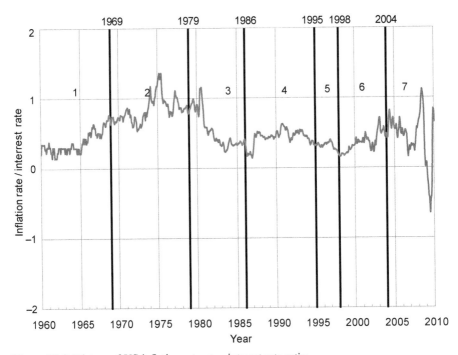

Figure 20.8 History of US inflation rate–to–interest rate ratio.

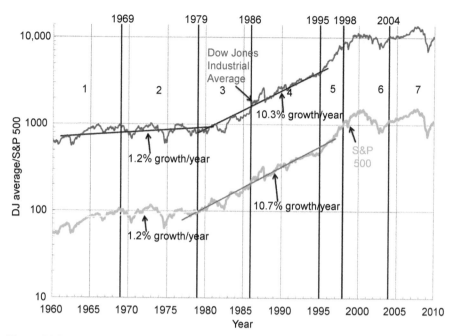

Figure 20.9 Leading stock market indicators.

20.3 The fledgling years (1960–69)

Commercial electric power generation started at The Geysers with the installation of a 12 MW (gross) plant in 1960. From 1960 to 1969, the installed capacity grew very slowly to above 100 MW (Figs. 20.2 and 20.4). During this period, the produced steam was vented to the atmosphere, the steam condensate was disposed of at the surface, and there was no injection. Given the relatively minor rate of production compared to the vast size of the field, the decline rate in well productivity was modest (on the order of 1% per year). In this period, an oil company (Union Oil Company) developed and operated the field. The produced steam was sold to an investor-owned utility (Pacific Gas & Electric Company) that generated power.

20.4 Geothermal comes of age (1969–79)

During 1969 through 1979, generation steadily grew to about 500 MW and injection of the condensed steam (amounting to about 25% of the mass produced) became a practice. The productivity decline trend in wells still remained modest (a few percent per year).

Crude oil and natural gas prices increased slowly (Fig. 20.7) during this period. While the stock market showed modest growth (Fig. 20.9), it was a strongly inflationary environment (Fig. 20.8). The 1973 "oil crisis" causing concerns about future energy supply and the two decades of successful operations by the end of this period began to attract various municipal utilities and independent power producers (IPPs), in addition to oil companies and an investor-owned utility, to the opportunity for development and operation and/or power generation at The Geysers.

20.5   The geothermal rush (1979–86)

During 1979–1986, the oil and natural gas prices skyrocketed, boosting the price of geothermal power, which was then based to a large extent on the price of fossil fuels. By 1981 the oil price reached its peak, followed by the natural gas price in 1984 (Fig. 20.7). This caused a major spurt in the rate of growth in production and installed generation capacity at The Geysers (Fig. 20.4).

During the early 1980s, several incentives for developing geothermal power in the United States became available from the federal and state governments. The first incentive was the Public Utilities Regulatory Policies Act (PURPA), which required a utility to purchase power from any facility of 80 MW or less developed by an IPP at the "avoided cost" for the utility. This ended the near-monopoly of the large investor-owned utilities and provided a guaranteed market for the IPP. The availability of a 10% Business Investment Tax Credit and another 15% Alternative Energy Tax Credit from the federal government during this period allowed up to 25% savings in the capital needed for a new geothermal project. Also during this period, the US Department of Energy introduced a Geothermal Loan Guaranty Program under which up to 75% of a developer's bank loan for a geothermal project could be guaranteed by the federal government. The combination of the Loan Guaranty, Business Investment Tax Credit and Alternative Energy Tax Credit allowed a developer to finance the development of a new geothermal project with a minimal long-term equity investment, while PURPA ensured the developer a market for its power. The wholesale power price offered under PURPA was also on the upswing during this period. In California, the utilities began to offer particularly lucrative power purchase contracts, known as Standard Offer 4, under PURPA.

This combination of an explosive rise in petroleum price, an exceptionally attractive financial incentive package offered by the government, and a guaranteed and lucrative power market drew a whole host of new developers to The Geysers. These included large energy companies capable of bankrolling their projects themselves, and municipal utilities issuing tax-exempt public power revenue bonds, and IPPs securing nonrecourse loans or selling corporate bonds to fund their developments. Hence, the rush for rapid development of new capacity at The Geysers starting in 1979.

The decline rate in reservoir pressure and well productivity began accelerating during this period. This led the operators to augment injection into the reservoir. By 1982, the operators had started augmenting injection, beyond that of condensed steam only, by capturing surface run-off during the relatively brief rainy season and tapping local creeks or aquifers to the maximum extent permitted by the local government. The periodic fluctuations in the monthly injection rate shown in Fig. 20.2 reflect the periodic increases in injection during the annual rainy season in California (October through March).

20.6   The troubled era (1986–95)

This 9-year period stands out as the troubled era in the half century history of The Geysers as production started a steady and steep decline. We discuss next the causes of this unexpected development as presented in Sanyal (2000).

Some dozen operators had developed the power capacity in this field on a strictly competitive, and largely confidential, basis without significant exchange of resource information among them. For some projects, the field developer and the utility were two different, and typically adversarial, entities. There was no single standard steam or power sales contract during this period. Some utilities paid the field developer for the mass of steam supplied, while others paid the developer for steam supply by the kilowatt-hour generated from that steam. Some developers also operated power plants and sold power to the utility based on various pricing formulas. The pricing formula in which the utility paid the field operator by the kilowatt-hour power generation (rather than mass of steam supplied) provided little incentive for the utility to improve power plant efficiency. As such, power generation efficiency varied widely between plants, many being conspicuously inefficient. This diversity of contractual agreements reflected the innovative yet fractious climate of the era. This competitive and, of necessity, secretive spirit of development was precipitated by the checkerboard nature of the myriad federal, state, and private lease blocks that comprised The Geysers field, and the need to win the geothermal development rights through competitive bidding for federal or state leases or through negotiations, often involving hard bargaining, with private land owners.

In this headlong rush, some unpromising, fringe areas of The Geysers got developed, and the plants built on them could not be sustained for long. New developers flocking to The Geysers in this period had little knowledge of the reservoir behavior being encountered by the developers and operators who had preceded them. Therefore, in their optimism, each new developer tended to dedicate less area per megawatts than the ones before. By the end of the 1980s in some parts of the field, the area dedicated per megawatt capacity became alarmingly small. Yet, at least 10 square kilometers of potentially productive area within the field remained undeveloped. Unlike most petroleum or geothermal fields shared by multiple field developers, The Geysers was not unitized; that is, the field developers did not elect one among them to operate the field with full access to the resource information available to all the others. This proliferation of information barriers and paucity of cooperation among the operators soon led to overdevelopment of the field, as evidenced by rapid declines in reservoir pressure and well productivity, and development of superheated steam in some parts of the field (implying local drying out of the reservoir) by 1987–88.

To maintain generation capacity in the face of rapid productivity decline, too many make-up wells were drilled in some parts of the field, which caused excessive pressure interference among wells, further reducing well productivity. This fact plus the unexpected collapse of oil and natural gas prices, and its impact on the steam price for most projects, after 1985 (Fig. 20.7) made make-up well drilling practically uneconomic by 1989. The discounted cash flow rate of return for maintaining generation by make-up well drilling became lower than if no make-up wells were drilled. Therefore, the net generation capacity was allowed to decline from this period on (Fig. 20.4).

During the peak generation year of 1987, the installed capacity was about 1830 MW (gross) or about 1640 MW (net) while generation was in the range of 1500 to 1600 MW; that is, the plant capacity factor was maintained at 91−97% despite forced curtailments. Once the power price dropped and make-up well drilling stopped, the net generation declined steadily from a peak of 1600 MW in 1987 to less than 900 MW by 1995.

Even though the first ominous signs of overdevelopment started appearing by the mid-1980s and the petroleum price as well as the avoided cost of power offered under PURPA had declined precipitously by then, another 500 MW of capacity was installed by 1989. There were several reasons behind this apparent paradox:

1. Some of these late developments had secured unusually lucrative Standard Offer 4 power sales contracts, which guaranteed levelized power rates for a period of 10 years.
2. Most of these projects were too far along to be reconsidered.
3. Mr. Ronald Reagan's presidency had turned around the economic stagnation of the previous decade and ushered in an era of unbridled optimism in the United States. For example, between 1980 and 1983 the inflation rate nose-dived from nearly 15−3% and stabilized at that level. The interest rate plummeted from over 20% in 1982 to 7.5% by 1987. And after 1982, the stock market in the United States took off for the first time in a generation; the Dow Joncs Industrial Average doubled during 1982−88 (Fig. 20.9). This socioeconomic optimism turned project financing easier than ever before.

By the end of the 1980s, signs were unmistakable that the then drilled area of the field had been overdeveloped, perhaps to the extent of the last 500 MW of additional capacity; in other words, the drilled part of the field was overdeveloped by some 25%. In retrospect, there was a twist of irony: while most parts of the field became overdeveloped, some of the last projects to go online proved the most sustainable and profitable.

By 1992 the decline in generation at The Geysers had attracted the attention of the state agency California Energy Commission, which funded a numerical simulation and engineering study of the field to investigate the options available to mitigate the generation decline. The investigation, conducted with the support of the operators, concluded that given the closed nature of the field, augmenting injection into the reservoir was the most effective antidote to pressure and productivity decline. Individual operators had also arrived at the same conclusion several years earlier on their own. However, injection augmentation was a major challenge. By 1995, the percentage of the produced mass injected (condensed steam plus surface run-off) had reached a practical limit of about 35% (Barker et al., 1995). Any augmentation of injection beyond 35% of production, therefore, called for significant operational innovation and capital investment.

It became clear by this time that primarily three critical variables determined the sustainable power capacity at The Geysers: injection strategy, pipeline and gathering system design, and plant design. Throughout this period, the operators had been striving to (1) increase the effective injection rate, (2) reduce pressure drop between the reservoir and the turbine inlet by optimizing the pipeline and gathering system, and (3) reduce the turbine inlet pressure and steam consumption rate by modifying power plants.

The numerical simulation effort funded by the California Energy Commission and conducted by GeothermEx, on behalf of and in collaboration with the operators (Pham and Menzies, 1993), indicated that without further injection augmentation or forced curtailment, generation would continue to decline at a harmonic rate of 9% per year (starting in 1988), which appeared economically unsustainable. This forecast was based on calibration of the numerical model against the production history of over 600 wells spanning 31 years (1960–91); therefore, the model was considered a reasonable representation of the reservoir.

A particularly positive development at The Geysers in this period was the gradual increase in information exchange and co-operation among the operators. For example, Calpine Corporation, Unocal (formerly Union Oil Company) and the Northern California Power Agency jointly implemented some of the injection augmentation strategies developed during this period. A series of acquisitions allowed consolidation of field operation in the hands of fewer and fewer operators; this consolidation finally began to allow optimization and integration of field management and power generation, thus increasing net generation while reducing operating costs.

20.7   The watershed years (1995–98)

During this period (1995–98), the problems at The Geysers eased sharply and unexpectedly because of the cut-back in generation forced by the availability of cheaper hydroelectric power and the sharp increase in injection of the surface run-off (reaching upto 50% of the produced mass), both caused by the unusually high rainfall in California in this period. This production curtailment and increased injection further slowed the depletion rate (Fig. 20.3); this, in turn, reduced the decline rate in well productivity sharply (Figs. 20.5 and 20.6); refer Goyal (2002) for the details.

During this period, oil and gas prices as well the inflation rate continued to remain low but there ensued a bull market in stocks. By the beginning of this period planning had begun for further injection augmentation by piping in and injecting treated municipal effluent from the community of Clear Lake into the reservoir. A 46-km long pipeline from Clear Lake was constructed and 7.8 million gallons per day of effluent injection started by late 1997. The decline rate in well productivity eased further within a year after the start of this effluent injection. The doldrums at The Geysers seemed to be over.

20.8   Stability at last (1998−2004)

This period (1998−2004) confirmed beyond doubt the remarkable stability achieved in this field. By 1998, the injection rate had increased to 60% of the produced mass due to effluent injection even as the production rate remained nearly the same as in the previous period (63.1 billion kg per year); this further reduced the reservoir depletion rate to 31.1 billion kg per year (Fig. 20.3). This reduced depletion rate moderated the well productivity decline to less than 3% per year (Figs. 20.5 and 20.6). In 2003, the Clear Lake pipeline was lengthened to 85 km to connect it to alternative sources of municipal effluent.

Field operation and power generation during this period was consolidated with only two remaining operators and power producers left in the field: an IPP (Calpine Corporation) and a municipal utility (Northern California Power Agency). This consolidation of operation allowed considerable optimization of field operation and power generation, further easing the generation decline rate to about 2% per year (Fig. 20.4).

While the rate of inflation remained low during this period, oil and gas prices started climbing again and the stock market became volatile. Encouraged by the manifest benefit of effluent injection, planning started for installing a larger capacity pipeline for bringing in another 11 million gallons per day of treated effluents from the City of Santa Rosa through a 66-km pipeline to further reduce the reservoir decline rate. This pipeline was completed and put in operation by the end of 2003, thus increasing the effluent injection by 24%.

The numerical model of Pham and Menzies (1993), shown in Fig. 20.10, forecast a production rate of about 11 million lb/h by the end of this period. In actuality, the production rate by 2004 was about 16 million lb/h. This lower decline rate was achieved by augmenting and optimizing injection, improving the pipelines, gathering system and plants, and aided by the curtailment in generation adopted during 1995−98; see Goyal and Conant (2010), Butler and Enedy (2009), and Enedy and Butler (2010) for more details.

20.9   Renewed optimism (2004−15)

Between 2004 and 2015, the injection rate climbed to about 80% of the produced mass as the rate of effluent import from Santa Rosa was increased further to 12.6 million gallons per day, even as production remained the same as in the previous period. The depletion rate was reduced sharply from 31.1 billion kg per year in the previous period to 13.9 billion kg per year (Fig. 20.3) in this period. The decline rate in well productivity as well as generation eased to about 1−2% per year (Figs. 20.4−20.6). Further, in many parts of The Geysers, augmented injection had diluted the noncondensable gases in produced steam (Beall et al., 2007; Pruess et al., 2007), which improved generation efficiency, reduced the discharge of greenhouse gases, and also reduced the cost of customary abatement of hydrogen sulfide gas. The reservoir

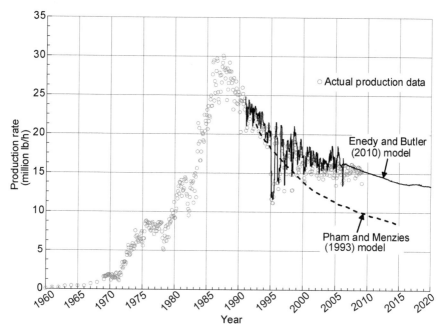

Figure 20.10 Production at The Geysers field — history and future.
Enedy, S.L., Butler, S.J., 2010. Numerical reservoir modeling of 40 years of injectate recovery at The Geysers geothermal field, California, USA. Geothermal Resources Council Transactions 34, 1221−1227.

continues to benefit from the proper siting of injection wells (Enedy, 2014) and augmented injection, especially from the two effluent injection projects (SEGEP or the Southeast Geysers Effluent Pipeline and the Santa Rosa pipeline). A new type of shallow and low-rate injector has been developed to improve heat extraction from the shallower parts of the reservoir.

Fig. 20.2 indicates a sharp decline in injection in 2014. In that year, the water supply from the SEGEP pipeline was reduced by more than 65% due to the prolonged drought in California. Make-up water in the SEGEP is drawn from the vast Clear Lake and this withdrawal rate is based on a specific water level measuring device near the lake. If the lake level does not reach a specified level by the first of May any year, SEGEP cannot draw water from the lake for the following 12-month period. In 2015, Lake County in California received near-normal rainfall such that the lake level was above the minimum required; as such, SEGEP got the right to draw water again from May 2015 through April 2016. Therefore, this fact and the widely accepted belief that the drought in California will break in late 2015 offer the hope that injection water shortfall would be unlikely to be a long-term issue. Based on the historic level readings at Clear Lake over the past 100 years, the present drought condition would occur approximately every 12 years. The drought in 2014 occurred 17 years from the start of augmented injection in 1997. It should be noted that the amount of lake water in the pipeline now is what can be accommodated in the form of future community growth. In other words, when a new connection is made to the SEGEP waste water system it replaces the same amount of water coming from the lake.

Several of the operators have upgraded the power plant's steam path with new rotors, nozzle diaphragms, inner casings, and gland and nozzle packing seal rings (Maedomari and Avery, 2011). This redesign was needed both to fit the reduced steam deliverability from the wells and to take advantage of improved computer-designed blades. After years of power generation with the power conversion efficiency degrading over time, now the redesigned turbine's conversion efficiency does not degrade or even increases over time. High-performance operation of the turbines over the long term has been ensured through deceased main steam pressure and reduced exhaust pressure. Further, generation has benefitted from the reduced parasitic load by shutting down some less-efficient plants.

The oil price increased sharply in this period as did the gas price, but toward the end of this period the natural gas price declined sharply. The stock market was volatile and the US economy suffered a major downturn toward the end of this period. However, these economic uncertainties of the period were largely mitigated in the geothermal industry by a set of already existing and new incentives offered by the federal and state governments as well as the utilities. These incentives, some of which may have only indirectly benefited the operators at The Geysers, consisted of the following:

1. Renewable Portfolio Standard (RPS) enacted by the various states requiring a utility to generate a minimum percentage (20% by 2010 in California) of their power from renewable resources, including geothermal.
2. Production tax credit (PTC), on the order of $0.021 per kilowatt-hour for 10 years, offered by the federal government on geothermal power plants put into service before 2013.
3. In lieu of the PTC, the option of choosing a Federal Investment Tax Credit (ITC) of up to 30% of the initial
4. Rapid increases in the power price offered for renewable power by the utilities seeking to fulfill their RPS obligation.
5. Renewed offer of the Geothermal Loan Guaranty by the federal government.
6. Development of a market for Renewable Energy Credit (REC) and the possibility of availing a carbon credit in the future.
7. A round of "stimulus funding" offered by the federal government in 2009 to encourage the development and demonstration of various new geothermal technologies.
8. More frequent and extensive geothermal lease sales by competitive bidding offered by the US Bureau of Land Management and the California Bureau of Land Management to encourage further geothermal development on government lands.
9. At The Geysers, power generation has historically been a baseload operation. It has been shown that a load-following operation is practicable but may increase operational and maintenance problems. Studies are now under way to establish what additional compensation needs to be provided for the operators to provide the increased flexibility of load-following. This increased generation flexibility is required under the RPS as the California state agencies turn more to renewable power sources, such as geothermal, wind, and solar (Matek, 2015).

In consequence of the economic incentives as well as the demonstrated stability and the minimal rate of well productivity decline established at The Geysers, planning began for new growth in installed power capacity and a new developer, an IPP, entered the field, Western GeoPower Corporation, first acquired by Ram Power Corporation in 2009, and by US Geothermal, Inc. in 2014.

20.10   The future (beyond 2015)

Two major issues characterize the uniqueness of the remarkable history of The Geysers:

1. Fig. 20.3 shows that in 55 years, cumulative steam production has amounted to 2.43 trillion kg, equivalent to 32,000 MW years of electricity. This cumulative generation is equivalent to 1333-MW generation capacity for a typical power plant life of 30 years and a typical plant capacity factor of 90%. No other geothermal field in the world has proved to be so prolific.
2. Fig. 20.10 shows that as of 2010, total steam production rate from this field has been 5.3 million lb/h higher than predicted by Pham and Menzies (1993) before the various operational innovations (such as injection augmentation) were introduced. This higher rate implies that the innovations implemented resulted in 280-MW higher power capacity, or nearly 33% of the power generation level of 850 MW in 2010.

In the foreseeable future, optimization of injection, steam gathering, and power plant operation will be the resource management tools, rather than make-up well drilling. Make-up well drilling is still not attractive because even though the decline rate in well productivity is now minimal, the typical production rate per well is too low (2−3 MW) and the pressure interference between wells too intense to justify the drilling cost. The operator will confront two major challenges in the future: (1) utilization of progressively lower pressure, superheated, and possibly gassier, steam, and (2) further injection augmentation and improvement in injection fluid recovery without causing enthalpy or productivity loss in the wells.

It is likely that additional new plant capacity would be installed at The Geysers. This seeming paradox has a rational explanation. The northwestern boundary of The Geysers reservoir has not been reached by development to date, and there are some yet undeveloped lease blocks with relatively high pressure within and around the currently exploited well field. The steam from these far-flung lease blocks cannot be transported economically to the existing plants; but suitably located new plants should be able to use this steam, if the power price is high enough. Another source of production for new power capacity could be the deeper brine reservoir or ultra-hot rock (approaching the critical point for water in temperature) identified or speculated to exist in several parts of the field (Beall and Wright, 2010), but have not yet been possible to exploit. Hence another energy crisis may trigger another round of new developments at The Geysers. An IPP, US Renewables, had restarted in 2006 a previously shut power plant (Snedaker, 2007) named Bottle Rock. However, this plant was shut down again in 2014 after an 8-year operation because drilling of make-up wells was unsuccessful in maintaining plant capacity at an economic level.

In 2014 US Geothermal, Inc., another IPP, acquired the Western GeoPower Project from Ram Power Corporation and plans to develop a new plant near the southwestern boundary of the field; see Fig. 20.1. At this site, PG&E Unit 15, a 62-MW (gross) power plant, had been built and operated for a decade (1978−88) but was later dismantled; see Sanyal et al. (2010) for details. Calpine Corporation is also expected to add some new capacity in the currently undeveloped north western part of the field. These new installations could perhaps add up to 100 MW in the foreseeable future.

One encouraging fact for the future of The Geysers field is that the operator Calpine Corporation, with the financial support of the US Department of Energy, has successfully conducted a field demonstration of creating and testing an enhanced geothermal system (EGS) involving a production well and an injection well during 2010 through 2013. The project had the following proposed goals:

1. Create an EGS system of 5-MW capacity.
2. Gain the confidence of the local community in the EGS operation.
3. Enhance the permeability of the hot reservoir rock by low-pressure injection of cool water to accomplish thermal fracturing.
4. Lower the noncondensable gas content in the produced steam through injection.

It appears that all of these goals were satisfied; see http://www.geysers,com/egsGeysers.aspx. Finally, after 55 years of supporting conventional geothermal power generation, The Geysers may become available for experimentation, if not exploitation, as the world's largest "enhanced geothermal system"!

20.11   Lessons learned

In addition to optimization of injection, steam gathering system and power plants, and the curtailment issues discussed here, the operators at The Geysers have faced numerous other developmental and operational challenges over the decades; for example, air drilling to depths exceeding 4 km, drilling multileg wells to reduce drilling cost per MW, mitigation of silica scaling associated with superheated steam, handling corrosive steam, utilization of gassy steam, and so on. All these problems have been solved through technical and managerial innovation. For example, some operators gradually adopted effectively "load-following" rather than "baseload" operation to reduce depletion. Guaranteed power sales contracts gave way in many cases to competition in a deregulated power market; the operators quickly adjusted to this market risk and focused on maintaining the "peaking capacity" by reducing the "baseload capacity." In a deregulated power market, a plant is much more profitable if it is operated at the peak capacity for the few hours in the day when power price spikes, the generation being sharply curtailed during the remainder of the day. Therefore, the annual plant capacity factor ceased to be a useful measure of the profitability of a project. Many such technical and managerial innovations pioneered at The Geysers are now commonplace in the geothermal industry.

The important lessons learned from the case history of The Geysers, we believe, are the following:

- Reasonably open exchange of information between field developers and operators is essential when multiple developers and operators exploit the same geothermal field.
- Sustainable generation capacity is determined as much by socioeconomic forces as by resource characteristics, design of surface facilities, or field management strategy.
- Resilience and ingenuity of the operator can overcome the unexpected changes in resource behavior and socioeconomic conditions.
- For steam fields, injection is the most powerful reservoir management tool.
- Augmenting injection by bringing in water from outside the field can be both technically and economically feasible.
- Injection of treated municipal effluents in a steam field offers multiple benefits to society; environmental impact of sewage disposal can be replaced by the generation of relatively cheap renewable energy and reducing the discharge of greenhouse and noxious gases.

Acknowledgments

The authors owe a debt of gratitude to numerous individuals, involved with The Geysers at various times, for sharing their views with them. Of these individuals, Karl Urbank and Keshav Goyal of Calpine, Mark Dellinger of SEGEP, Douglas Glaspie of Western GeoPower/Ram Power, and Steven Butler of GeothermEx are singled out for their invaluable contribution to the development of the perspective presented here. The authors wish to thank Ali Khan of the California Department of Oil, Gas and Geothermal Resources for providing the production and injection database on which this assessment is based. The authors gratefully acknowledge the technical assistance from James Morrow and Amber Thomas of GeothermEx, Inc. in preparation of this chapter.

References

Barker, B.J., Koenig, B.A., Stark, M.A., 1995. Water injection management for resource maximization: observation from 25 years at The Geysers, California. In: Proceedings World Geothermal Congress, Florence, Italy, pp. 1959–1964.

Beall, J.J., Wright, M.C., 2010. Southern extent of the geysers high temperature reservoir based on seismic and geochemical evidence. Geothermal Resources Council Transactions 34, 1199–1202.

Beall, J.J., Wright, M.C., Hulen, J.B., 2007. Pre- and Post-development influences on fieldwide Geysers NCG concentrations. Geothermal Resources Council Transactions 31, 427–434.

Butler, S.J., Enedy, S.L., 2009. Numerical reservoir-wellbore-pipeline simulation model of The Geysers geothermal field, California, USA. In: Proceedings of the Society of Petroleum Engineers, Western Regional Meeting, 24–26 March, San Jose, CA, p. 8.

Enedy, S.L., 2014. Benefit of a typical injection well at the geysers field. Geothermal Resources Council Transactions 38, 985–989.

Enedy, S.L., Butler, S.J., 2010. Numerical reservoir modeling of 40 years of injectate recovery at The Geysers geothermal field, California, USA. Geothermal Resources Council Transactions 34, 1221–1227.

Goyal, K.P., 2002. Reservoir response to curtailments at the geysers. In: Proceedings 27th Workshop on Geothermal Reservoir Engineering, Stanford University, Stanford, California, USA, January 28−30, pp. 39−45.

Goyal, K.P., Conant, T.T., 2010. Performance history of The Geysers steam field, California, USA. Geothermics 39, 321−328.

Maedomari, J., Avery, J., 2011. Turbine upgrade for Geysers geothermal power plant. Geothermal Resources Council Transactions 35, 1325−1329.

Matek, B., 2015. Geothermal Energy Association I: Firm and Flexible Power Services Issue Brief: Firm and Flexible Power Services Available From Geothermal Facilities. Available from: www.geo-energy.org.

Pham, M., Menzies, A.J., 1993. Results from a field-wide numerical model of The Geysers geothermal field, California. Geothermal Resources Council Transactions 17, 259−265.

Pruess, K., Spycher, N., Kneafsey, T.J., 2007. Water injection as a means for reducing non-condensible and corrosive gases in steam produced from vapor-dominated reservoirs. In: Proceedings of the 32nd Workshop on Geothermal Reservoir Engineering. Stanford University, Stanford, CA, pp. 293−300.

Sanyal, S.K., 2000. Forty years of production history at The Geysers geothermal field, California − the lessons learned. Geothermal Resources Council Transactions 24, 313−323.

Sanyal, S.K., Henneberger, R.C., Granados, E.E., Long, M., MacLeod, K., 2010. Expansion of power capacity at The Geysers steam field, California − case history of the western GeoPower Unit 1. In: Proceedings of the 2010 World Geothermal Congress, Bali, Indonesia paper no. 635.

Sanyal, S.K., Enedy, S.L., 2011. Fifty years of power generation at the geysers geothermal field, California − the lessons learned. In: Proceedings of the 36th Workshop on Geothermal Reservoir Engineering, Stanford University, Stanford, California, January 31−February 2, 2011, p. 9.

Snedaker, G., 2007. Re-commissioning non-operating power projects, the bottle rock geothermal case study. Geothermal Resources Council Transactions 31, 535−537.

Indonesia: vast geothermal potential, modest but growing exploitation

21

S. Darma
Indonesia Renewables Energy Society (METI-IRES)/National Research Council of Indonesia (DRN), Jakarta, Indonesia

21.1 Introduction

Indonesia, located between the eastern end of Mediterranean Volcanic Belt and western side of the Circum-Pacific Volcanic Belt, is blessed with abundant geothermal resources. More than 200 volcanoes are located along Sumatra, Java, Bali, and the islands of eastern part of Indonesia, which is known as "The Ring of Fire." The potency is about 29 GWe. Indonesia is also known as having the largest geothermal potential in the world. It gives rise to a large concentration of high-temperature geothermal systems.

This country needs to be balanced in the energy mix to benefit from the clean geothermal energy. The use of renewable geothermal energy would eliminate the dependency on a single source of fossil fuels to generate electricity and meet Indonesia's growing energy demand. Therefore, this geothermal energy of choice will obviously enable Indonesia to export its more portable fuels for much-needed hard currency and increase the energy security of the country.

To speed up geothermal development, a new regulation on the power sector and geothermal development should be introduced. We hope these new regulations will open the opportunities and reduce the challenges and encourage investor to develop geothermal energy in order to fulfill an increasing electricity demand of Indonesia.

In terms of geothermal development and its utilizations, the Government of Indonesia (GOI) was committed to use the biggest geothermal energy resources to be the largest geothermal producer in the world as a leading alternative energy. The increasing use of geothermal will substitute for fossil fuels that may be fulfilling Indonesia's growing demand for electric power and guarantee the security of the energy supply in the long-term National Energy Policy (NEP).

These policies are issued to support having a total geothermal power plant installed capacity of 9500 MWe by 2025 as it is stipulated in the National Energy Mix Policy.

Indonesia is one of the handful of countries to have developed geothermal energy for electricity. The development has proceeded very slowly and is currently facing difficult challenges. Over a span of 40 years, Indonesia has only developed 1438.5 MW or about 4.9% of 29,000 MW of geothermal potential until 2015 and will increase by 390 MW more in 2017. This means the geothermal development is growing, albeit slowly.

Geothermal Power Generation. http://dx.doi.org/10.1016/B978-0-08-100337-4.00021-8

Indonesia started to develop geothermal in 1974 by establishment of the President Decree No. 16 in 1974, strengthened by President Decree No. 22 in 1981, President Decree No. 23 in 1981, by President Decree No. 45 in 1991, and President Decree No. 49 in 1991 to develop geothermal in order to diversify the energy use.

Geothermal Law No. 27/2003 followed by the Government Regulation No. 59 in 2007 is the new regime of geothermal activities established in 2003, to guide the business entities, cooperative bodies, and the government to activate the geothermal business in Indonesia. Since then, unfortunately, the growth of geothermal development is very slow. However, the government has issued so-called President Instruction No. 4 in 2010 to accelerate develop electricity sector by mandating PLN (the public utility) to take almost 4000 MW from geothermal in the 10,000 MW fast-track program phase II. This regulation is in line with the focus of the energy sector to push accelerated use of geothermal energy. The supporting regulation shows that the GOI gives priority to accelerate the use of geothermal in the National Energy Policy. Geothermal is expected to contribute at least 12% of the national electricity needs in 2025. Currently, Indonesia is the world's third largest geothermal electricity producer after the United States and the Philippines.

At the 2010 World Geothermal Congress in Bali, President Susilo Bambang Yudhoyono announced a plan to build 44 new geothermal plants by 2014, more than tripling the capacity to 4000 MW. By 2025, Indonesia aims to produce more than 9000 MW of geothermal power, becoming the world's leading geothermal energy producer. This would account for 5% of Indonesia's total energy needs.

It is also targeted by the government that in 2022, Indonesia will reach 100% national electrification, therefore there exist many business opportunities in achieving the electrification ratio target by 2022. PLN as a state-owned electricity company and independent power producer (IPP) plays a significant role to achieve this target. This information comes from various sources such as Ministry of Finance, PLN, etc.

The government provides the opportunity to the private sector to participate in the development of geothermal power and issued the amendment of the Government Regulation (GR) No. 59 of 2007 to GR No. 70 in 2010 on the geothermal business and operations, the Ministry of Energy Regulation No. 2 in 2011 on the selling price policy, and Regulation No. 22 in 2012 on the Feed in Tariff. In addition, GOI through PIP provides Rp.9 trillion fund facility (Geothermal Fund Facilities (GFF)) taken from national budget period of 2011−16 for the development of a 4840-MW geothermal field and power plant.

Annual growth of energy consumption is 8.65% per year. Table 21.1 shows the annual increase in installed electricity generating capacity (in MW per year) for the period 2012−21. By the end of this period, PLN plans to reach a total at least 70 GW of installed capacity from today's value of 44 GW.

In 2011, the total national generating capacity from PLN (Perusahaan Listrik Negara (state electricity company)), IPPs, and PPU (public−private utility) in Indonesia was of 38.9 GW; in 2015, it is 44 GW, of which approximately 76% is in the Java-Bali region, 13% is in Sumatra, Kalimantan, and the rest is in other islands (Sulawesi, Maluku, NTB-NTT, Papua). In terms of input fuel, coal- and oil-fired plants have the highest share, which amounted to 42% (16.5 GW) and 23% (9 GW),

Table 21.1 National electricity growth projection

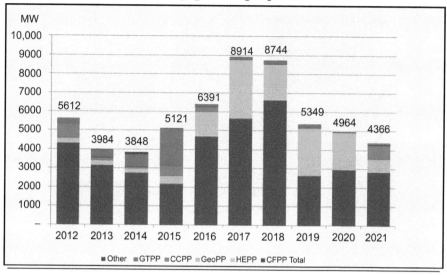

GTPP, gas turbine power plant; *CCPP*, combined-cycle power plant; *GeoPP*, geothermal power plant; *HEPP*, hydro energy power plant; *CFPP*, coal fire power plant.
PLN, 2013. Rencana Umum Penyediaan Tenaga Listrik (RUPTL) - National Electricity General Plan (2012–2021), Jakarta.

respectively, followed by gas-fired plants with a share of around 22% (8.4 GW). The high share of generator fuel is offset by the increasing share of power plants fueled by renewable energy, such as geothermal, with a share approaching 3% (1.2 GW), as well as hydro-based generation in the range of 10% (3.9 GW). In addition, solar power and wind energy have also started operating with a total capacity of 1.6 MW.

21.2 Geological background

Geothermal power in Indonesia is an increasingly significant source of renewable energy. More than 200 volcanoes are located along Sumatra, Java, Bali, and the islands of eastern part of Indonesia, which is known as "The Ring of Fire." It lies between the eastern end of Mediterranean Volcanic Belt and western side of the Circum-Pacific Volcanic Belt and is blessed with abundant geothermal resources. As a result, current calculations indicate that the geothermal potential is approximately 29 GW (Fig. 21.1) and put this country as the largest geothermal energy potential in the world.

The Geological Agency of Indonesia reported that Indonesia is composed of 312 geothermal field locations, of which 58 locations (15,627 MW) of prospective geothermal has been issued the geothermal concession IUP or geothermal working area. These locations are planned to be developed and operated by existing developers with total potency of 10,869 MW and 4834 MW and are mainly composed of

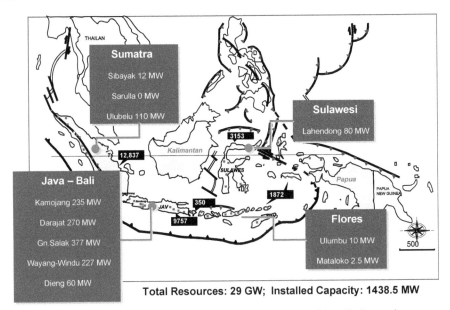

Figure 21.1 Location map of Indonesian geothermal resources and installed capacity.

Pertamina Geothermal Energy (PGE) and its partnerships concession, while 39 locations have been tendered under the geothermal law regimen (4758 MW), from which 19 concessions have been awarded the geothermal permit (IUP).

All the high-temperature geothermal systems are found within the Sumatra, Java, Sulawesi, and Eastern Island Volcanic Zone, which lies over an active subduction zone between the eastern end of Mediterranean Volcanic Belt and western side of Circum-Pacific Volcanic Belt (Fig. 21.1 and Table 21.2).

Lack of infrastructure was the major obstacle in the development of geothermal in Sumatra. Since 2008, PGE began developing this working area, namely, Ulubelu, Lumut Balai, Hululais, and Sungai Penuh in Sumatera. Later in 2011, Supreme Energy started exploring this working area in Muaralaboh, Rajabasa and Rantau Dedap, and it is expected that Sarulla will be developed in 2015 by Medco Power. If all of those projects can be done, in 2016–17 the geothermal installed capacity in Sumatra will increase by 390 MW. But, this will depend on how easy it will be to get permits and on preparation of the infrastructure including the grid.

21.3 Vast geothermal potential

Geological Agency of Ministry of Energy and Mineral Resources of Indonesia (MEMR) data showed that Indonesia's geothermal potential is spread over 312 geothermal locations throughout Indonesia. The potential reserves were calculated using National Standards of Indonesia (SNI) SNI 13-5012-1998.

Table 21.2 Power potential of Indonesian geothermal prospects

| Island | Number of locations | Energy potential (MWe) | | | | | Total | Installed (MW) |
| | | Resources | | Reserves | | | | |
		Speculative	Hypothetical	Possible	Probable	Proven		
Sumatra	93	3183	2469	6790	15	380	12,837	122
Java	71	1672	1826	3786	658	1815	9757	1204
Bali-Nusa Tenggara	33	427	417	1013	0	15	1872	12.5
Kalimantan	12	145	0	0	0	0	145	0
Sulawesi	70	1330	221	1374	150	78	3153	80
Maluku	30	545	76	450	0	0	1071	0
Papua	3	75	0	0	0	0	75	0
Total	312	7377	5009	13,413	823	2288	28,910	1438.5
		12,386		16,524				
		28,910						

Modified from Badan Geologi (2013) and DGEBTKE (2014a,b).

Based on this SNI, Indonesia's geothermal potential can be divided into two classes, namely, resources and reserves, each of which is further divided into subclasses (see Table 21.2).

The resources classification consists of:

1. **Speculative resources**: characterized by the presence of active geothermal manifestations where the vast geothermal reservoir is estimated based on a literature study and preliminary investigations.
2. **Hypothetical resources**: characterized by active geothermal manifestations with a baseline survey of regional geology, geochemistry, and geophysics. Broad prospect area is determined by the advanced preliminary survey.

The reserves classification consists of:

1. **Possible reserves**: evidenced by the data of the drilling geothermal temperature gradient at which the estimated volume and thickness of reservoir, rock, and fluid parameters conducted are based on integrated geoscience data, depicted in the form of a tentative geothermal reservoir model.
2. **Probable reserves**: evidenced by successful exploration wells in which the estimated extent and thickness of reservoir are based on well data and the results of a detailed investigation of integrated geoscience. Parameters of the rock, fluid, and reservoir temperature obtained from direct measurements of the well.
3. **Proven reserves**: evidenced by more than one successful exploration well producing steam and/or hot water, where the estimated volume and thickness of the reservoir are based on well data and the results of a detailed investigation of integrated geoscience. Rock and fluid parameters and reservoir temperature obtained from measured data directly in the well and/or laboratory.

Reserve criterion can be defined as geothermal energy economically and legally mineable, while the criterion of resources is defined as geothermal energy that can technically be mined; resources can be added cumulatively to the reserves in the future once the exploration drilling is conducted. So, the resources and reserves are not static data. As the inventory and exploration activities are carried out either by the government or by the private sector, then the resources and reserves potential of geothermal energy in Indonesia changed from time to time in accordance with the level of investigations that have been carried out. The status of the investigation conducted is based on National Standard Classification Potential of Geothermal Power in Indonesia, referring to the SNI 18-6009-1999. The investigation refers to three geothermal systems in Indonesia today: volcanic, volcano-tectonic, and nonvolcanic geothermal systems.

As can be seen in Table 21.2, from 312 locations, 12.386 GW is considered the resources and 16.524 GW is considered the reserves. But, the proven reserve is only about 2.3 GW and concentrated in Java Island. A best effort is needed to transform energy potential into proven reserves.

Sumatra had the largest geothermal potential, 12.8 GW, or 44% of the potential, is there, but unfortunately only 122-MW geothermal power plant capacities are installed. This installed capacity comes from Ulubelu (110 MW) and Sibayak (12 MW) constructed and operated by PGE.

According to the Geological Agency of Indonesia (GIA), it is indicated that one-third of 29-GW geothermal potential is located in Java and Bali, the most populous islands with the highest demand for electricity. The preliminary surveys for geological data have been carried out on 312 locations, and the geochemical information was taken from 200 locations, while the geophysical surface data for a detailed geothermal survey was done in the 45 locations.

The geothermal development road map released in 2006 was a plan to develop electricity generation of 2000 MW by 2010, 4600 MW by 2016, 6000 MW by 2020, and 9500 MW by 2025 (targeted). Most of the 312 prospects of Indonesia have high temperature geothermal resources. However, the realization of geothermal development is still lower than anticipated by that program. The total geothermal prospects and its potential reserves and resources are tabulated in Table 21.2.

21.4 History of geothermal development in Indonesia

Geothermal exploration in Indonesia is started in 1926 in the Kamojang geothermal field, West Java, Indonesia. The first geothermal well was drilled by the Dutch Colonial at well KMJ-3 at 66 m total depth. The well produced dry steam at wellhead temperature of 140°C and wellhead pressure of 2.5 bar. This well has continued to flow until the present day (Fig. 21.2). From 1926 to 1928, there were five exploration wells drilled, but KMJ-3 was the only well to produce dry steam. The first commercial geothermal energy production was commissioned in 1982 by Pertamina (The National Oil & Gas Company of Indonesia) and PLN (National Electricity Power Company).

Figure 21.2 Well KMJ-3, Kamojang geothermal field, produces dry steam at wellhead temperature of 140°C and wellhead pressure of 2.5 bar and has flowed continuously for 89 years. Photograph taken by author in December 2012.

A 30-MW geothermal power plant was established in 1983 in Kamojang geothermal field. The plant was established after 10 years of exploration, drilling, and construction, which was partially funded by the Government of New Zealand with total funds of US$34 million. Besides the geothermal power plant in Kamojang, Pertamina also established two "monoblock" units with 200-kW and 2-MW total capacities in Kamojang (see Fig. 21.3) and Dieng (Central Java), respectively. Then, the 2-MW "monoblock" plant was relocated to Sibayak Geothermal Field in 1994.

The development by the private party through cooperation with Pertamina in the form Joint Operating Contract (JOC) resulted in the early 1990s with the development of Salak geothermal field by Unocal (which was later taken over by Chevron), Darajat geothermal field by AMOSEAS Inc. (which was later taken over by Chevron), and Wayang Windu by Mandala Nusantara Ltd. (which was then taken over by Deutsche Bank and finally by Star Energy Ltd.). In the early 1990s, geothermal development by private parties was attractive to investors because the price of steam or electricity purchased by PLN at that time was $0.047 and $0.08 per kWh, respectively. GOI did the awarding of 11 contracts for development of geothermal power plants, with a total committed REPELITA VI capacity (GOI's 5-year plans) of 1498 MW in original completion dates between 1998 and 2002 and a commitment contract capacity of 3692 MW. However, as a result of the 1997−98 financial crisis, the government suspended (postponed) 27 IPPs, including 11 from geothermal. As of May 2002, the government successfully resuscitated all geothermal contracts, save one under legal dispute (Karaha Bodas, West Java).

The economic crisis that affected Indonesia in 1997 and started to recover in 2003 has affected the power sector demand and growth in Indonesia. This has resulted in power blackouts in the whole country that occur from time to time and from place to place. Further, changes in the regulatory environment for the geothermal industry

Figure 21.3 A 200-kW "monoblock," the original unit of the first geothermal power plant in Kamojang.

and a firm commitment from the GOI are the major issues affecting the future growth of the Indonesian geothermal industry.

During the energy recovery situation, GOI is setting a strategy to prepare Indonesia to become independent in meeting its energy requirement. GOI has gradually increased the national electricity tariff and assessed a differential electricity tariff structure by region to support electricity infrastructure investment with regard to the needs of each region. In the geothermal sector, GOI continues to encourage investors by offering incentives in exploration activities and lessens the amount of risk by allowing the government to conduct exploration activities.

Recently, the government declared a new Geothermal Law No. 27 in 2003. This law is followed by the Government Regulation No. 59 in 2007 as a guide for companies, cooperative bodies, and the government to activate geothermal business in Indonesia. This regulation is in line with the focus of the energy sector to accelerate the use of geothermal energy. It mainly deregulates the right of regional autonomy, fiscal reform, and sanctity of existing contracts and introduces transparency and levels the playing field, while regulating the geothermal steam field license. The supporting regulation shows that the government will push to accelerate the use of geothermal in the near future. The government blueprint for geothermal development was issued in the Government Regulation No. 5 in 2006 following the issuance of the Geothermal Law No. 27 in 2003. Geothermal was expected to contribute at least 5% of the national energy needs.

These policies were issued to support having a total geothermal power plant installed capacity of 9500 MWe by 2025. The cost of geothermal electricity produced is slightly lower than that generated from fossil fuels such as oil and gas, but still not competitive with that generated from coal-fired plants. It is slightly higher than electricity generated from mine-mouth coal-fired power plants. Thus, the opportunity to develop geothermal power plant in Indonesia is wide open to private companies.

The total capacity of geothermal plants in operation is about 1403.5 MW as can be seen in Fig. 21.4. To restimulate geothermal development in Indonesia, the government under President Joko Widodo inaugurated a 35-MW geothermal power plant, the fifth unit of Kamojang geothermal field, on July 5, 2015.

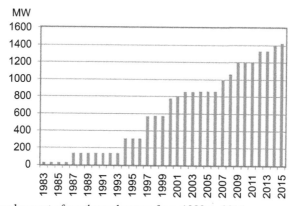

Figure 21.4 Development of geothermal power from 1982 to 2015.

21.5   Geothermal law and other geothermal regulations

By rule, Indonesia started to develop geothermal in 1974 by President Decree No. 16 in 1974, then strengthened by President Decree No. 22 in 1981, President Decree No. 23 in 1981, President Decree No. 45 in 1991, and President Decree No. 49 in 1991 to develop geothermal in order to diversify the energy use (MEMR, 2013). Now, the rule of geothermal development is based on Geothermal Law No. 27 in 2003, which has been amended by Law No. 21 in 2014. Government regulation and regulatory/ministerial regulation, among others, are described next.

21.5.1   Geothermal Law No. 27 in 2003

Geothermal law became the basis of geothermal activities is called Law No. 27 in 2003. This law has been amended and changed by Law No. 21 in 2014. This law allowed the private sector to commercialize geothermal without cooperation with Pertamina, as stated in the previous geothermal business regimen. Granting geothermal permits is done through a tender with the lowest electricity price as the main criterion for the winner.

21.5.2   Government Regulation No. 59 in 2007

Government Regulation No. 59 of 2007 as well as the Minister of Energy and Mineral Resources Regulation No. 11 of 2009 regulate in detail the operation of business activity including the establishment of the geothermal working area, WKP (or Wilayah Kerja Panasbumi). The regulation also allows private parties to do Preliminary Surveys and get a "right to match" at the time of tender of the working area.

21.5.3   Pricing policy

To support the policy, government has issued the Minister of Energy Decree regarding a pricing policy to improve the internal rate of return of the project. In 2008, there was a Decree No. 14 and then revised by the Minister Decree No. 5 in 2009 in order to give the project a fair economic return. Minister of Energy and Mineral Resources Regulation No. 32 in 2009 mandated PLN to buy geothermal power and to set the benchmark price of US$0.0970/kWh. This might increase the use and activate the business entities to develop geothermal even for direct or indirect use of geothermal. Geothermal explorations are just beginning to come back slowly after the government issued a new pricing policy of geothermal energy at the early of 2010.

21.5.4   Presidential Decree No. 4 in 2010

Presidential Decree No. 4 in 2010 as well as the Minister of Energy and Mineral Resources Regulation No. 15 in 2010 commissioned PLN to accelerate power plant development (well known as the second phase of the acceleration of 10,000 MW) includes the construction of 4900 MW of geothermal power that had not been achieved by 2014.

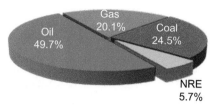

Figure 21.5 National energy condition.
Ministry of Energy and Mineral Resources (MEMR), 2013. Indonesia Energy Outlook 2013, Yearly Publication Book. Center for Data and Information of Ministry and Mineral Resources of Indonesia, Jakarta, Indonesia.

21.6 National energy condition and policy

21.6.1 National energy condition

The use of energy in Indonesia is growing rapidly in line with economic and population growth that has reached over 237 million people in 2014. Energy consumption experienced steady growth of around 7% per annum due to population growth (1.49%) and economic growth ($\sim 6\%$).

Energy supply comes from two sources of energy: fossil energy (oil, natural gas, and coal) and renewable energy such as biomass, hydro power, and geothermal. However, the energy supply is mostly come from fossil energy. Indonesia possesses a variety of energy resources. A substantial portion of the country's energy mix continues to rely on oil (Fig. 21.5). This figure below shows the use of New, Renewable Energy and fossil fuel in the National Energy Mix of Indonesia.

Furthermore, the electricity needs in 2011 is about 40 GW and tend to increase about 90 GW in 2025 and 400 GW in 2050. By using pessimistic estimate assuming conditions into 2025, 2011 shows that by 2025 Indonesia needs 70 GW and by 2050 needs about 200 GW. This might worst to the energy supply security. It is also targeted by the government that in 2022, Indonesia will reach 100% national electrification ratio. Therefore, there will exist many business opportunities in achieving the electrification ratio target by 2022. Geothermal as the biggest potential to generate electricity in the country will play a significant role to achieve this target.

21.6.2 National energy policy

At present, in the early 2014, the government has issued a new National Energy Policy (NEP) following to the Energy Law No. 30 in 2007 in order to enable the coordination and synergy of all stakeholders in energy sector. This NEP regulates the optimizing use of energy resources in Indonesia. The vision of the policy is guaranteeing the sustainable energy supply to support national interest. The targets of NEP are shown in Fig. 21.6.

To reach the energy targets, several directions in energy policy have to be done: (1) availability of energy; (2) priority of energy development; (3) national energy

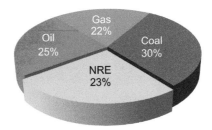

Figure 21.6 New national energy policy (NEP, 2014).

resources utilization; (4) national energy reserve; (5) conservation and diversification; (6) environment and safety; (7) price, subsidy, and incentive; (8) energy infrastructure and industry; (9) energy research and development, and (10) institution and financing.

The new paradigm of energy policy should optimize the use of natural energy resources for economic development of the nation based on "economic value-added process." Further, government should also declare to all the stakeholders that in order to secure the strategic national energy reserves that the fossil energy reserves must be stored as long as possible to guarantee the security of the nation.

21.6.3   Main policy in renewable energy

Firstly, implement energy conservation to improve energy efficiency in supply and demand, among all sectors of industry, transport, household, and commercial. Secondly, implement diversification of energy to increase the share of renewable energy in the national energy mix (supply side), among others.

21.6.4   Power utilities

In the last 5 years (2009–14), Indonesia's GDP 6.2% per year caused the growth of electricity is 8.65% per year and the tariff is about US$0.074/kWh. Capacity growth projection from year 2012–21 is as follows: coal-fired plants, 38 GW; geothermal plants, 6.3 GW; combine cycle plants, 2.5 GW; gas turbine plants: 4 GW; hydroelectric plants: 6.3 GW; and others about 0.28 GW.

In the last 5 years, the electrification of the country has increased from 72% to 80% in 2013. Total installed capacity is 43 GW. In order to have the basic rules in developing power plant, GOI and PLN have issued National Electricity General Planning and General Planning for supplying electricity in 2012–21 (see Fig. 21.7).

21.7   Geothermal energy role in the National Energy Mix

Indonesia currently utilizes geothermal fields mainly for indirect use to generate electricity. However, the direct use of geothermal is also well known in the rural areas

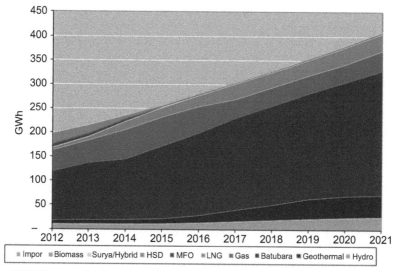

Figure 21.7 The use of primary energy for electricity (in percentage of 2012—21). *HSD*, high speed diesel; *MFO*, mid flat oil; *LNG*, liquefied natural gas. PLN (2013).

since the history of human life. The modern use of geothermal energy is now counted on to support the national energy policy.

Today, 1438.5 MW of geothermal energy has been developed as of 2015. It has increased about 159 MW from an installed capacity of 1187 MW in 2009, and again increased about 90 MW from an installed capacity of 1346 MW as of 2013. The developed geothermal locations are distributed in 11 commercial areas: Kamojang, Darajat, Wayang Windu, Patuha, and Salak in West Java; Dieng in Central Java; Sibayak in North Sumatra; Ulu Belu in South Sumatra; Lahendong in North Sulawesi; and Mataloko and Ulumbu in Flores. Mataloko geothermal field in Flores is a GOI pilot project of 2.5-MW geothermal power plant commissioned by PLN in 2012.

In addition, in the next 2 years there will be an additional 290 MW operated by PGE in 2016 and 2017. There is a 55-MW project under construction to be commissioned in 2016 from Ulu Belu, 55 MW from Lumut Balai, 20 MW from Lahendong, and 30 MW from Karaha geothermal field. There are three other units, totaling 130 MW, that will be commissioned in 2017 in Ulu Belu (55 MW), Lahendong (20 MW), and Lumut Balai (55 MW).

21.8 Geothermal development plan

Indonesia as an industrialized country continues to grow very rapidly. Indonesia's population is undersupplied with electricity; 20% of population is without electricity. To support meeting the energy demand and other GOI policies, local and international companies are now increasing their activities in order to meet the energy demand in the

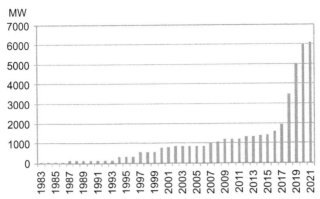

Figure 21.8 Geothermal power history and projected growth in MW, cumulative to 2021.

power sector. PGE with its own activities and its partners are exploring new areas that are not producing yet. Some other new companies are also involved in exploration or even in the development stage to use the geothermal power. All activities are supporting the GOI policy to meet the second stage of 10,000-MW accelerating power plant projects, which are composed of 60% geothermal.

The GOI plans to accelerate the development of electricity in Indonesia by an additional 35 GW between 2015 and 2019. Of this, 10 GW is expected to be developed by PLN and 25 GW by IPPs. Around 7 GW of projects are currently in development.

A number of geothermal growth scenarios were considered in the geothermal development plan, including the overall aspirational development plan of the government, the "Road Map." The most recent geothermal development plan (March 2015) of MEMR shows geothermal generation capacity increasing from the current 1400 MW to around 6000 MW by 2020 and almost 7000 MW by 2025 (refer for details to the Geothermal Development Program — DJEBTKE 2014, see Fig. 21.8 and Table 21.4).

21.9   Geothermal exploitation growth

21.9.1   Installed power and ongoing projects

The geothermal power is generated from 11 areas of high temperature geothermal systems. The total capacity of geothermal power plants installation is about 1438.5 MW. These are operated from Darajat (270 MW), Dieng (60 MW), Kamojang (235 MW), Gunung Salak (377 MW), Sibayak (12 MW), Lahendong (80 MW), Wayang Windu (227 MW), Ulumbu (10 MW), Mataloko (2.5 MW), and Patuha (55 MW). This number accounts for 3.5% of approximately 36,000 MW of total installed electric capacity, of which PLN generates about 21,000 MW, IPPs 1600 MW, and captive power of 13,519 MW. These are operated for electricity generation of 1397 MW by PGE and its joint operation contractors (ie, Chevron, MNL, Geodipa Energy), and PLN (see Tables 21.3 and 21.4).

Table 21.3 **Current installed capacity, location, and developer of Indonesian geothermal plants**

Geothermal concession	Location	Developer/operator	Turbine capacities	Total installed capacity (MW)
Sibayak – Sinabung, Sumut	Sibayak	PT. Pertamina Geothermal Energy	1 × 10 MW; 2 MW monoblok	12.0
Cibeureum – Parabakti, Jabar	Salak	Chevron Geothermal Salak, Ltd.	3 × 60 MW; 3 × 65.6 MW	377.0
Pangalengan, Jabar	Wayang Windu	Star Energy Geothermal Wayang Windu	1 × 110 MW; 1 × 117 MW	227.0
Patuha	PT. Geo Dipa Energi	1 × 55 MW	55.0
Kamojang – Darajat, Jabar	Kamojang	PT. Pertamina Geothermal Energy	1 × 30 MW; 2 × 55 MW; 1 × 60 MW; 1 × 35 MW	235.0
Darajat	Chevron Geothermal Indonesia, Ltd.	1 × 55 MW; 1 × 94 MW; 1 × 121 MW	270.0
Dataran Tinggi Dieng, Jateng	Dieng	PT. Geo Dipa Energi	1 × 60 MW	60.0
Lahendong – Tompaso, Sulut	Lahendong	PT. Pertamina Geothermal Energy	4 × 20 MW	80.0
Waypanas, Lampung	Ulubelu	PT. Pertamina Geothermal Energy	2 × 55 MW	110.0
Ulumbu, NTT	Ulumbu	PT. PLN (Persero)	4 × 2.5 MW	10.0
Mataloko, NTT	Mataloko	PT. PLN (Persero)	1 × 2.5 MW	2.5
Total				**1438.5**

Table 21.4 Installed and planned geothermal power plants in Indonesia (Ministry of Energy, 2014)

No.	Geothermal field	Developer/ operator	Installed capacity by 2015 (MW)	Geothermal development planning to 2025 (MW)
1	Sibayak	PGE	12	12
2	Sarulla	PGE/PLN	0	330
3	Lumut Balai	PGE	0	220
4	Ulu Belu	PGE	110	220
5	Salak	PGE/CGS	377	377
6	Patuha	Geodipa	0	165
7	Wayang Windu	PGE/MNL	227	227
8	Kamojang	PGE	235	235
9	Darajat	PGE/CGI	270	260
10	Karaha	PGE/KBC	0	30
11	Patuha	Geodipa	55	165
12	Dieng	Geodipa	60	170
13	Lahendong	PGE	80	120
13	Bedugul	PGE/BEL	0	10
14	Hulu Lais	PGE	0	110
15	Sungai Penuh	PGE	0	110
16	Kotamobagu	PGE	0	80
17	Iyang-Argopuro	PGE	0	55
18	Ciater	WSS	0	30
19	Ulumbu	PLN	10	10
20	Mataloko	PLN	2.5	5
21	Tulehu	PLN	0	20
22	Cibuni	Yala Teknosa	0	30
23	Jaboi	Sabang Geo Energy	0	10
24	Muaralaboh	Supreme EM	0	220
25	Rantau Dedap	Supreme ERD	0	220
26	Rajabasa	Supreme ER	0	220
	Others[a]		0	3322
		Total	**1438.5**	**6708**

[a]Twenty-two additional areas have been awarded to develop for the 2025 energy mix plan.

Besides, there are also 38 areas that exist and are being developed: (a) Sarulla, Sungai Penuh, Hululais-Tambang Sawah, Lumut Balai, Karaha, Iyang-Argopuro, Bedugul, Tompaso, and Kotamobagu, all being developed by PGE or with its contractors for electricity generation; (b) Kawah Cibuni, Ciater, and Tulehu, three areas outside of PGE activities, which are operated by Yala Teknosa (Cooperative Body), PT Wahana Sambada Sakti and PLN, the National Electricity Company; and (c) of those fields mentioned, the developers are committed and projected to commence geothermal power plants in the new working area (see Table 21.4) with a total capacity of 6703 MW by 2025. Those geothermal fields operate using the existing geothermal rule, that is, PD No. 45 in 1991 and the Geothermal Law No. 21 in 2014.

21.9.2   Geothermal drilling and field development

Since 2010, 58 WKPs (geothermal concessions) and Geothermal Permits have been issued as part of the geothermal acceleration project in the 10,000-MW Fast Track II. The geothermal power contracts expected to reach capacities of 3977 MW, but most projects have not been realized due to the lack of economic viability. Only a few projects are ongoing for exploration and development based on the permit. Three new projects are the new regulation regime and the rest are projects ran by PGE and PLN. Most of the projects were not moving forward due to investment uncertainty as the price of electricity set by GOI is lower than the viability of geothermal development cost. However, through a 10,000-MW acceleration crash program project, GOI has issued a new incentive related to price, tax, and fiscal policy.

From developers' activity, PGE, Chevron, Star Energy, Supreme Energy, and PLN have drilled more than 80 wells in the last 5 years, in addition to 430 wells drilled by PGE, Chevron, MNL, BEL, PLN, GAI, and Yala Teknosa before 2009. From 2009 to 2013, Indonesia has confirmed a total proven and probable reserve of 3101 MW and a total possible reserve of 13,413 MW.

The total wells drilled for electricity generation in each area, both existing and being developed, are: Sibayak (10 wells), Salak (97 wells, 8 wells added from 2009), Wayang Windu (51 wells, 12 wells added from 2009), Kamojang (87 wells, 5 wells more from 2009), Darajat (68 wells, 12 wells more than those in 2009), Lahendong (13 additional wells compared to 23 wells in 2009), and Dieng (52 wells).

The rest of the wells drilled were Sarulla (13 wells), Sungai Penuh (1 well), Hululais-Tambang Sawah (3 wells), Lumut Balai (18 additional from only 3 wells in 2009), Ulu Belu (additional 26 wells from only 9 wells in 2009), Patuha (19 wells), Karaha (7 additional wells in the last 3 years compared to 14 wells in 2009), Bedugul (3 wells), Tompaso (5 wells), Banten (1 well), Muaralabuh (5 wells), Rantau Dedap (2 wells), and Tulehu (1 well).

The other wells drilled are Kawah Cibuni (1 well), Ulumbu (3 wells), Cisolok−Cisukaramai (1 well), Mataloko (6 wells), Sokoria (1 well), and Atadeii (2 wells), which were drilled by Yala Teknosa, PLN, and Geological Agency of Indonesia (GAI), as shown in Table 21.5.

Table 21.5 Numbers of well drilled in Indonesian geothermal area during 1974–2014

No.	Geothermal field/ (contract signed)	Contract size (MW)	Developer/ operator	Total no. of wells drilled (2014)
1	Kamojang (1982, 2012)	230	PGE	87
2	Salak (1982, 1994)	495	PGE/CGS	97
3	Darajat 1,2,3 (1984)	330	PGE/CGI	68
4	Sarulla 1,2,3 (1993)	330	PGE/SOL	13
5	Dieng 1–4 (1994)	400	Geodipa	52
6	Iyang-Argopuro 1	55	PGE	0
7	Karaha 1,2 (1994)	400	PGE	14
8	Patuha (1994)	400	Geodipa	19
9	W. Windu (1994)	400	PGE/MNL	51
10	Bedugul 1,2,3 (1995)	400	BEL	3
11	Cibuni	10	Yala-Teknosa	1
12	Sibayak 1,2 (1996)	40	PGE/DP	10
13	Lahendong (1999)	125	PGE	36
14	Kotamobagu (2009)	40	PGE	3
15	HuluLais 1,2 (2009)	110	PGE	3
16	S. Penuh 1 (2009)	55	PGE	1
17	Ciateur Prahu (1999)	55	WSS	0
18	T. Perahu (2008)	55	Indo Power	0
19	Cisolok (2008)	55	Rekayasa Industri	1
20	Tampomas (2008)	55	Jasa Sarana Jabar	0
21	Jaboi (2009)	20	Sabang Geo Energy	0
22	Sukoria (2009)	20	Sokoria Geo Indonesia	1
23	Jailolo (2009)	20	Star Energy	0
24	Ulu Belu: 1,2,3,4	220	PGE	35
25	Lumut Balai: 1,2,3,4	220	PGE	21
26	Ulumbu 1	50	PLN	3

Table 21.5 **Continued**

No.	Geothermal field/ (contract signed)	Contract size (MW)	Developer/ operator	Total no. of wells drilled (2014)
27	Ata Deii	20	GAI	2
28	Mataloko	40	GAI	6
29	Banten	0	PGE	1
30	Muaralabuh (2012)	220	SEM	5
31	Rantau Dedap (2012)	220	SERD	2
32	Rajabasa (2013)	220	SER	0
33	Tulehu	20	PLN	1
	Total	**5360**		**513**

21.9.2.1 Kamojang (235 MW)

This is the oldest geothermal field developed in Indonesia shown by a 250-kWh mono-block geothermal power plant in 1978. The field is operated by PGE generating 235 MW from five power plant units. The first unit is a 30-MW plant, now in operation for more than 32 years. At present, the gathering facilities are supplied by the total of 87 wells which were drilled for exploration, production, reinjection and monitoring of Units one to five of the Kamojang power plant (Fig. 21.9). The size of the productive area is about 21 km^2 with an estimated electrical potential equal to 270−300 MW.

Figure 21.9 Kamojang geothermal gathering facilities supplied by the total of 87 wells.

Figure 21.10 Sibayak geothermal gathering facilities supply the 2 × 5.65-MW plant.

New exploration activities identified resources sufficient to increase the existing plant by an additional of 60 MW. PGE is completing the engineering-procurement-construction contract for the fifth unit of the geothermal plant with an output capacity of 35 MW. This unit was inaugurated by President Joko Widodo in July 2015.

21.9.2.2 Sibayak (13.3 MW)

Sibayak geothermal field in North Sumatra operates a 13.3-MW plant composed of a 2-MW plant of monoblock manufactured by Westinghouse (USA) in 1942 and has generated electricity since July 1996, and two units of 5.65 MW (Figs. 21.10–21.11) of the Chinese geothermal plant operated by Dizamatra Powerindo. The field is managed by PGE. Although PGE signed an agreement to allow expansion with the next power plant unit, there is no additional unit of power plant invested to increase its capacity generation. In order to increase the capacity of the area, there is a study to use a binary plant to improve the field performance. The current 2-MW geothermal power plant is often damaged due to old age and the 2-MW plant is now retired.

There are some hot springs bathing facilities very near the power plant, but there is no evidence of any effect of the plant (water table drawdown, temperature changes, noise, smell, etc.) on the spas so far.

21.9.2.3 Lahendong (80 MW)

A 4 × 20-MW power plant of 80 MW was commissioned in this area in 2002, 2008, 2009, and 2012. PGE is the developer and plans to develop two more units of 20 MW in the area, mainly in the expansion of Lahendong to the Tompaso geothermal

Figure 21.11 View of Sibayak geothermal field. The power plant building is located in the middle of the right-hand side.
Photograph by author, May 19, 2007.

prospect. The additional 40 MW is planned to commence in 2015 and 2016 using nine production wells drilled in the last 5 years. But, the reality is that those two new units will be commissioned in 2016 and 2017, respectively. The drilling and steam field gathering system is under way (Fig. 21.12). The total number of wells drilled as of 2015 is 36, including exploration, production, reinjection, monitoring, and abandoned wells. Recently, PGE offered using brine technology, ie, an organic Ranking cycle, to optimize the use of Lahendong geothermal fluid to increase the power plant capacity by 7.5 MW.

Figure 21.12 Lahendong geothermal gathering facilities.
Photograph by author, December 18, 2008.

21.9.2.4 Ulu Belu (110 MW)

The Ulu Belu area is located in the Southern Sumatra. It is associated with the volcanic depression of Mt. Sula, Rindingan, and Tanggamus. Since 1993, PGE has drilled three exploration slim holes and 32 development wells in the north block of Ulu Belu area to support four 55-MW power units. Twelve wells were drilled before 2009. The possible and proven reserve is about 540 MW.

The steam field and steam supply is managed by PGE. The geothermal reservoir is a hot liquid-dominated system with temperatures from 240 to 260°C. The first and second units of 55 MW commenced at the end of 2012 using Fuji turbine-generator equipment. Now the field is undergoing construction of the third and fourth 55-MW power units, which are planned to be commissioned in 2016 and 2017.

21.9.2.5 Lumut Balai

The Lumut Balai geothermal system is liquid-dominated. The reservoir temperature varies from 240 to 260°C. It is located in South Sumatra, associated with the Sumatra volcanic belt. The prospect area is predicted to be about 70 km^2. The proven and probable reserve is more than 100 MW. Since 2007 until now, PGE has drilled 21 wells, three of which were drilled before 2009, to support 220 MW of electric capacity. The first unit of 55 MW was proposed to start operating in 2015. But due to various obstacles, the commissioning has been delayed until 2016. The next unit of a 55-MW power plant also is undergoing construction that should commence operation by 2017.

21.9.2.6 Hulu Lais

This field is also located in the southern part of Sumatra. Exploration drilling has been conducted since 2010. Three exploration wells have been completed and encountered a hot liquid-dominated geothermal system. These three wells were originally planned to be drilled at the end of 2009 and to be followed by a development drilling campaign to support two 55 MW power units, anticipated to commence in 2013 and 2014. But, some administrative problem was encountered during the period of drilling preparation. The first and third wells encountered a downhole temperature of 200−210°C, while the second well indicated good production at a downhole temperature of 300°C. The new program for Hulu Lais field has been rescheduled to 2017 and 2019. The potential prospect could be developed to 200 MW from about 500 MW of its potential.

21.9.2.7 Sungai Penuh

PGE prepared the exploration drilling in 2010. The first exploration well has been completed in June 2013 in order to support a first 55-MW unit from total of 200-MW potential. The drilling activities took six months due to a technical problem. This one well might cost double the average well cost drilled by PGE, up to US$6 to 7 million per well.

The main barrier in preparing such activities is a very limited infrastructure at Sungai Penuh and its location in the National Park of Kerinci Seblat. The geothermal prospect is located in the main graben of the Sumatran Fault lying in the middle part of Sumatra. From the first well, a hot water reservoir was encountered at temperature of 211°C.

21.9.2.8   Kotamobagu

Kotamobagu is located 250 km from Manado, North Sulawesi. So far, two exploration wells have been drilled in 2010 and 2011. Both wells are dry and show no indication to produce. It seems that the drilling pad is out of the prospect boundary which located in a conservation forest. The prospect is predicted to have about 230-MW potential as a liquid-dominated resource with temperatures varying from 250 to 290°C. PGE proposed rescheduling the project from 2013 to 2019 and 2020 for the first two 20-MW power units.

21.9.2.9   Tompaso

As part of the Lahendong geothermal area, Tompaso had been expected to host two 20-MW power units managed by PGE to produce electricity by 2012. But, from the nine wells drilled, there were indications that more wells needed to be drilled; thus, the first two units of the plant were rescheduled into 2015 and 2016, respectively. One reinjection well has a good indication of permeability. The probable reserve is about 120 MW from its total potential of about 220 MW. The project is financed using World Bank support. PGE has invested about US$50 million for this project only in the drilling program. It appears that the economy of scale should be recalculated and renegotiated with PLN, the power purchaser.

21.9.2.10   Iyang-Argopuro

This prospect is located in East Java lying mainly in a protective forest and national park. Even though this prospect has 200-MW electric potential and the surface geological and geophysical survey (G&G) has been completed by PGE for resource confirmation, there has been no additional progress in this area. PGE is still waiting for permit clarification from GOI to explore. In this case, and expecting new regulation of geothermal resources, PGE proposes to continue conducting the G&G survey and drilling activities to confirm the reserve and commence the first unit of 55 MW by 2019.

21.9.2.11   Gunung Salak and Darajat

Chevron and its partners, PT PGE and PT Austindo Nusantara Jaya, still operate the Salak and Darajat geothermal fields, both located on the island of Java. Salak is managed by Chevron Geothermal Salak (CGS) and Darajat is operated by Chevron Geothermal Indonesia (CGI).

The Salak geothermal system is a liquid-dominated, fracture-controlled reservoir with benign chemistry and low-to-moderate noncondensable gas content (Stimac et al., 2008). The reservoir is the largest producer of geothermal power in Indonesia (Ibrahim et al., 2005). The Darajat geothermal system is a steam-dominated system. There are nine power plants in the two fields with a combined capacity of 647 MW. Chevron operates five of these plants, with a capacity of 398 MW.

Awibengkok, one part of the Gunung Salak geothermal field, produces steam capable for 377 MW composed of six units, while Darajat produces about 270 MW, composed of three units of 55, 95, and 110 MW. The total of both fields are about 50% of Indonesia's total geothermal power capacity. During the last 5 years, six new wells including one well for reinjection were drilled at Salak to delimit the potential to expand field production. This means there exist 100 wells, of which two wells are for reinjection, two wells for condensate, five wells for monitoring the reservoir, and 23 nonactive wells, including five wells that have been abandoned. Looking to the future, the feasibility of installing additional generating potential has been assessed in both fields. However, it seems hard to predict when any additional power units might be constructed.

If the required steam supply is achieved, then the third unit of Darajat is proposed to up-rate its capacity to 120 MW to produce 280 MW in total. The existing wells at Darajat were supplemented by 12 new wells drilled in 2009, 2010, and 2011 to supply steam for the three Darajat geothermal power units. Currently, 49 wells exist at Darajat, four of which were used for condensate. As a result of these activities total investments in both fields is about US$80 million.

21.9.2.12 Sarulla

The Sarulla area is located 300 km south of Medan in North Sumatra. These projects include Silangkitang, Namora-I-Langit, and Sibualbuali (Darma et al., 2010a,b,c).

Since the project was taken over by a Consortium of Medco Power, Itochu, ORMAT, and Kyushu Electric under Sarulla Operation Ltd. (SOL) regarding of JOC with PGE, there has been no significant progress made in supporting a tight schedule of commissioning originally planned for 2011, 2012, and 2013, as committed for the 10,000-MW crash program. SOL as the operator of the 330-MW project worked to finalize the renegotiation of the project schedule mainly in Power Purchase Agreement (PPA) and price resettlement. For maintaining the 13 wells in three different areas, SOL also worked over some productive wells in Silangkitang. The project is now ongoing and financed by JABIC and ORMAT. Three ORMAT binary units are planned to be installed in 2017.

21.9.2.13 Karaha Bodas

Since Karaha Bodas Company LLC (KBC), a joint venture between Caithness (40.5%) and Florida Power & Light (40.5%), both of the United States, and Tomen of Japan (9%) and a local company Sumarah Daya Sakti (10%), executed the international arbitral award against PT Pertamina and PLN for the postponement of its contract by

Indonesian government, the project now is managed by PGE. In 2008, PGE maintained some wells were drilled by KBC. In the last 3 years, PGE has worked over one well and drilled five new wells to support a 30-MW plant to commence in 2015. This recommissioning date is a new plan of PGE since the original commissioning in 2012 was delayed. PGE has invested at least about US$22−25 million for drilling of new wells. Previously, KBC had drilled about 22 wells of slim holes, standard holes, and large holes.

21.9.2.14   Dieng

Dieng geothermal field, operated by Geo Dipa Energy (Geodipa), the subsidiary company of Pertamina and PLN, produces about 60 MW as a continuance of Himpurna California Energy Limited (HCE). At the beginning, the field is planned to increase its capacity to 180 MW as the resource is greater than that covered by the existing wells. There is no additional investment in the Dieng geothermal field by Geodipa. HCE has invested US$192 million in the 48 exploration and development wells, construction of the steam gathering system, and installation of a 60-MW power plant. Previously, HCE identified the field's potential as 350 MW. Currently, Geodipa has been taken over as geothermal company to focus the Dieng and Patuha geothermal.

21.9.2.15   Patuha

As the third geothermal project identified as a steam-dominated geothermal system in Indonesia by Patuha Power Limited (PPL), Patuha geothermal field is more attractive than Dieng. The project is also operated by Geodipa after the GOI assigned both Dieng and Patuha to the Geodipa for project continuance. Currently, Geodipa proposes to build three units of 55 MW and the first unit was inaugurated in June 2014 using Toshiba power plant. The field is planned to increase its capacity to 110 MW in 2017.

PPL has drilled 17 exploration wells (between 760 and 800 m depth, core holes, six slim holes) and 13 development wells (1000−2172 m depth). PPL identified the field as having a potential reserve is about 400 MW. During operation, the operator HCE has drilled as many as 13 exploration wells, 17 development wells, and six slim holes since 1994.

21.9.2.16   Wayang Windu

This field is operated by Star Energy Geothermal Wayang Windu Limited (SEGWWL) under a Joint Operating Contract with PGE. The field is operating in full capacity of 227 MW. In the initial plan for field expansion, the capacity of the field was to be increased to 400 MW with additional exploration drilling for the expansion unit. A total of 12 wells have been completed in the last 5 years by SEGWWL since the project was taken over to increase its capacity for the next 110-MW units. These additional wells are in addition to 57 wells that now support the Wayang Windu operation. However, the results from the exploration drilling are not encouraging and the expansion program is being reassessed.

Figure 21.13 Production test in Bedugul geothermal field.
Photograph by author, July 2004.

21.9.2.17 Bedugul (Bali)

The Bedugul geothermal field is located on the island of Bali and now was suspended. There is no progress on the project plan. Initially, Bali Energy Limited (BEL) planned to drill about six wells each year starting in early 2008 and ending in 2013; see Fig. 21.13. It was estimated that a total of the 24 production wells and eight injection wells would produce enough steam to supply 175 MW of power generation. BEL would provide eight well pads, ie, three existing pads and five new pads. The government and BEL have agreed to reschedule the project in order to operate the first 10 MW unit by 2019.

21.9.2.18 Ulumbu (10 MW)

Ulumbu geothermal field was initially a bilateral technical cooperation between PLN with the Government of New Zealand to install a small geothermal power unit. The project began in 1989 but the drilling activity commenced in 1994. The field is located at 650 m above sea level, nearly 13 km from the volcano of Anak Ranaka. The first well was drilled in 2003. Drilling resumed later in 2006 with the second and third wells.

The first unit of 5 MW geothermal power plant started operation in November 2011 with 50% of its capacities. Unit 2 of 5 MW at Ulumbu entered a commissioning phase and started in operation in August 2014, sending electricity to the city of Ruteng.

21.9.2.19 Mataloko

Mataloko geothermal field is located in Toda Belu, Flores Island, East Nusa Tenggara, eastern part of Indonesia. Three exploration wells have been drilled since 2000. The field is being developed using a 5-year cooperative research grant between Indonesia and Japan. The subsurface temperature is about 208°C. A 2.5 MW power plant is operated by PLN and started operating in 2013.

21.9.2.20   Tulehu, Maluku

Tulehu geothermal power plant is located in Salahutu, Maluku. The first exploration drilling was by PLN Geothermal, and initially it was expected to have a 10-MW power plant in operation by 2014. The construction will be done after the funding agreement with the Japan International Cooperation Agency is completed. But, the drilling activities have been stopped due to downhole technical problems before reaching total depth. The whole project is now being reassessed.

21.9.2.21   Cibuni

Cibuni is located in West Java, in the vicinity of the Patuha geothermal prospect. The project has been delayed since 1998. The first well was drilled in 1996–97 and the probable reserve is about 50 MW. Cibuni is a steam-dominated geothermal system. The well produced 36 ton/h of steam. The depth of the well is 1600 m. The second and third wells have been canceled due to drilling problems. It is planned to install two units of 5 MW at Cibuni in 2019. The field is operated by Yala Teknosa (Cooperative Body).

21.9.2.21   Ciater

Ciater geothermal field is located in the northern part of Tangkuban Perahu project to install one unit of 30 MW by 2019. Some manifestations in this area have been utilized for Ciater geothermal direct use as a tourist destination, ie, spa and swimming pool. Ciater geothermal project is currently operated by Wahana Sambada Sakti. There are about seven geothermal manifestations of acidic hot springs in the area, precipitating silica sinters and ferro-oxide, aluminum, and phosphate.

21.9.2.22   Muara Labuh

This is one of the three gothermal projects on-going using the Geothermal Law No. 27 in 2003. The two others are Rantau Dedap and Rajabasa in Southern Sumatra. Muara Laboh is located in West Sumatra Province. The concession covers an area of about 62,300 ha at elevations ranging from 450 to 2000 m and is bordered by Taman Nasional Kerinci Seblat (Kerinci National Park) in the west and south. The geothermal prospect is indicated by a long and wide distribution of thermal manifestations, consisting of fumaroles, mud pools, and hot springs, which are located along the graben associated with Great Sumatra Fault zone.

Supreme Energy Muara Labuh (SEML) was awarded in 2010 as the project company to develop Muaralabuh. The PPA was signed in early 2012. The Government Guarantee Letter from the Ministry of Finance of the Republic of Indonesia for the Muara Laboh project was also issued at same time in 2012. The signing of the PPA and issuance of the Guarantee Letter resulted from several months of intensive and constructive negotiations between the parties, driven by the shared goal of accelerating the development of geothermal energy in Indonesia. On September 21, 2012, SEML conducted the first exploration well drilling, ML-A1. This was one of a series of four to six exploration wells in the area, which proved the existence of geothermal resources was sufficient to build a geothermal power plant of 220 MW.

Currently, SEML has completed the exploration drilling program covering six wells (ML-A1, ML-B1, ML-C1, ML-E1, ML-H1, and ML-H2). All of the wells are now being monitored for data evaluation and reserve calculation. The first test of the first well in December 2012 confirmed the existence of a geothermal system and a well with a conservative estimate of 20-MW power capacities.

21.9.2.23 Rajabasa

Rajabasa geothermal project is situated at the southern end of the Sumatra Island and covers an area from sea level to 1280 m high. The project is operated under a partnership between PT Supreme Energy and Sumitomo Corporation. The project has a planned power generation capacity of 2×110 MW and could fuel the complete electricity demand of the province. Supreme Energy expects to start operation of the plant in 2017, providing electricity to the region. The PPA was signed in 2012 and The Government Guarantee Letter from the Ministry of Finance of the Republic of Indonesia for Rajabasa project was also issued in 2012, following the tender award of the Rajabasa concessions to the Supreme Energy Consortium in early 2010.

The delay in exploration drilling is caused by waiting for permits from Minister of Transportation while civil construction activities and other significant tasks have temporarily been suspended since June 2013. Now, the work is again in progress, as the Forest Area Borrow-Use Permit (IPPKH) from Ministry of Forestry was issued in early 2014.

21.9.2.24 Rantau Dedap

The exploration program started in 2011. Rantau Dedap project, South Sumatra, entered into the PPA in 2012. The first exploration drilling RD-B1 was conducted in early February 2014. This well is one of a series of five to seven exploration wells in the area, aimed to prove the existence of geothermal resources sufficient to support building a geothermal power plant with the capacity of 220 MW.

The field is operated by PT Supreme Energy Rantau Dedap (SERD). The SERD project is a joint venture by PT Supreme Energy, GDF Suez and Marubeni Corporation. The project plans to drill five wells of 2400 m depth. Its potential reaches 600 MW.

In the Rantau Dedap power plant project, the consortium is investing around US$3 to 4 million/MW, making the total investment up to $700 million. Half its investment will be spent on the power plants. The electricity that will be generated by Rantau Dedap power plant will be sold to PLN under a 35-year contract.

21.9.2.25   Exploration stage of new geothermal project

The new geothermal project base on Geothermal Law No. 27 in 2003 and all of its derivatives regulation is now slow. Tangkuban Parahu (PT Tangkuban Parahu Geothermal Power), Mount Tampomas (PT Wika Jabar Power), Cisolok-Cisukarame (PT Jabar Rekind Geothermal Power), Jaboi (PT Sabang Geo Energy), Sokoria (PT Sokoria Geo Indonesia), and Jailolo (PT Star Energy Halmahera), all of which were awarded before 2010, are stagnant. It seems no significant progress in these projects is planned. The major issue is caused by pricing of the energy sold, PPA negotiation, location of the projects in conservation forests and or national parks, government guarantee of the project, obligation of PLN to buy the energy from company developer, lack of human resources, etc. However, in the last 5 years, there are a total of 38 new working areas that have been issued the geothermal permit for exploration and exploitation. These fields and the numbers of total capacity install are shown in Table 21.4, while the total numbers of well drilled are seen in Table 21.5. The issuance of the new Geothermal Law No. 21 in 2014 will have attractive terms for the developer.

21.10   Challenges in geothermal development

In the last 5 years, only 212 MW of additional geothermal power plant have been installed and commenced up to 2014. The target of 4000 MW additional to 1189 MW installed capacity in 2014 failed to be achieved. This is likely due to unique attributes of geothermal energy which pose challenges to its development. The following factors are part of the reason for this failure.

21.10.1   The pricing policy

The pricing of energy is the main obstacle to the development of geothermal energy in Indonesia. The high risk of development, fossil subsidy, and the associated electricity tariff required remain core problems in geothermal development. The price needs to be competitive with other energy alternatives, and at the same time offer the producer an attractive rate of return.

For new power plant development, the attractive price for economic viability varies from US$0.09 to 0.12/kWh. To make this happen, GOI issued the Ministry of Energy Regulation No. 22 in 2012, revised from regulation No. 2 in 2011, that mandates PLN to take any power produced from geothermal. A new study from the World Bank (2014) indicated that the price of geothermal energy will vary from US$0.11 to 0.29/kWh based on the avoided cost of coal and the regional use of electricity. The price also depends on the year of plant commissioning from 2014 to 2025. The lowest price is for projects commissioned in the region of Java, Bali, and Sumatra in 2014, and the highest price is for projects in isolated areas to substitute for diesel power plants. This policy will standardize the geothermal price and is expected to raise the geothermal power capacity to more than 7000 MW in 2025.

21.10.2   Long-term benefit

Today, there is no genuine mechanism for considering geothermal long-term benefits for PLN. The low emissions are taken as benefits of the people and to the country. Thus, GOI should support the use geothermal for electricity by PLN. PLN as any other private company, has to survive and compete with other private businesses.

In the long run, Indonesia still has one of the world's most attractive geothermal prospects, but there is a need to look for new development approaches to maximize its potential. There was a study done by West Japan Engineering Consultant in 2009 that showed the economic value of geothermal energy earned by the GOI is about US$0.177/kWh. That value should be taken for issuing some fiscal incentives to attract geothermal development.

In the last 10 years, there is a discussion among geothermal stakeholders to identify the barriers to the growth of the Indonesian geothermal industry, such as competitiveness of geothermal energy price; continuing subsidy of fossil fuel price; lack of political will to intensify geothermal energy utilization; shortage of competence in human resources; absence of technology; lack of renewable incentives; lack of risk appreciation and mitigation efforts; absence of integrated energy planning; lack of information and publicity on indonesia's geothermal potency and benefits; low environmental awareness; geothermal law; absence of government guarantee and mandate for PLN to buy the energy produced by the project company; standard PPA; tendering of geothermal process; and forestry concern. The first action to be taken by GOI to amend the Geothermal Law 2003 has been done in 2014. The draft of the bill has been passed by GOI and agreed to by Parliament to launch a new geothermal law called Law 2014.

In addition, infrastructure is a major concern in the development of geothermal in certain areas such as Sumatra. For example, since 2008 PGE began to develop these working areas in Sumatra: Ulubelu, Lumut Balai, Hululais, and Sungai Penuh. Later on in 2011 Supreme Energy started exploring in Muaralabuh, Rajabasa, and Rantau Dedap, and by 2014, Sarulla was also developed. With so many obstacles of the running project in Sumatra, the geothermal installed capacity in Sumatra will increase by less than 1000 MW in 2017. Further, the capacity building of the human resources should also be improved.

21.10.3   Fiscal incentive

To reduce the price of geothermal power, the government needs to provide fiscal incentives as much as possible. The government should not expect tax revenue from geothermal development; it should be enough to use geothermal as an indispensable source of energy and the reduction of carbon emissions.

At this time, the fiscal incentive provided by the government is that there is an exemption from the value-added tax for exploration activities. However, there is no exemption of exploitation activities.

21.10.4   Feed-in tariff

In 2013, the Minister of Energy and Mineral Resources issued Ministerial Decree No. 22 in 2012 and set a feed-in tariff for geothermal power. But the feed-in tariff has been challenged by several parties, including the government, so it has not been implemented. Although the concept of a feed-in tariff is very good, the value itself was too low and rendered the project uneconomic. This means it is necessary to renegotiate the economic price with PLN, but this process will take a very long time to come to an agreement. The feed-in tariff had fundamental weaknesses in terms of selecting the winning bidder. This can lead to subjectivity that is not transparent. The feed-in tariff system also has the possibility to get a lower rate tariff, which has been set. Hence, the determination of the maximum tariff escalation is the right choice. The maximum rates need to be reviewed from time to time in accordance with the rate of increase in investment costs and others.

Exploration drilling has a high cost. Even an amount as high as US$30 million may not be enough to conduct exploration drilling and obtain sufficient data. It should be remembered that the costs of exploration are not solely the cost of drilling. Infrastructure preparation of access roads, creating drilling platforms and well pads, and others has a very large cost.

21.10.5   Geothermal fund

The government considers that the exploration risk is inhibiting the development of geothermal and leading to high tariffs for geothermal electricity. Therefore, the government provided a revolving fund of 1.2 trillion rupiah in the 2011 budget with the aim to reduce the risk of exploration and accelerate the development of geothermal. These funds will be provided each year and is managed by Pusat Investasi Pemerintah (PIP, the Central Government Investment Agency) through Minister of Finance Regulation No. 286/MK.011/2011. Management procedures and accountability of the revolving fund are set up within the Ministry of Finance Regulation No. 03/PMK.011/2012. Funds may be used through the PIP to get information about the area that will be auctioned by conducting geological, geophysical, and geochemical studies including the drilling of exploration wells. The fund can also be used by holders of geothermal concessions to conduct exploration drilling with the status of a loan as collateral.

Several parties, including the Indonesian Geothermal Association, recommend that the Geothermal Fund should be devoted to conducting a preliminary study, not including exploration drilling. But, the fund is used for exploration drilling only in areas that are not attractive to investors because of their small capacity, and where it is needed to meet the energy needs of the region, for the following reason. Drilling exploration wells is a very high risk operation to be borne by the government, which has limited funds. It is considered that the funds are better used for infrastructure development, which does not contain risks such as road construction and other urgent projects and are not attractive to the private investors.

The best geothermal developers/investors would be willing to take the risk of exploration without the need of government support. But, investors do need a higher tariff to create economic viability for the project.

21.10.6 Government guarantee

The reluctance of the government, the Ministry of Finance, to provide a guarantee for the obligations of PLN against the developer of geothermal power plants is also a factor that has led to this slow development of geothermal power plants.

Until now, almost all commercial banks as well as "multilateral agencies" such as JBIC and ADB still require the government to provide loan guarantees to the geothermal developers as well as other developers (IPPs). Government guarantees are secured as necessary to ensure that the company will meet all its obligations in accordance with the PPA. The provision that the government will offer a guarantee on "the feasibility of PLN" is also contained in the Presidential Decree No. 4 in 2010 about the 10,000-MW Second Phase of the Acceleration Program.

The needs for government guarantees to the obligations of PLN are caused by the following factors:

- Most of the PLN income is still a subsidy from the government because PLN incomes at prices set by the government are still lower than the cost of production. Therefore, the financial viability of PLN is still dependent on ensuring payment of subsidies by the government.
- Investors have still not forgotten the Asian financial crisis in 1998, which resulted in the geothermal electricity price lowered unilaterally by PLN/government to a value far below the price agreed to in the PPA.

In the implementation to obtain the government guarantees issued by the Ministry of Finance, the process requires a very long time because, in fact, the Ministry of Finance is reluctant to give this assurance.

Table 21.6 Road map of future development planning and installation of geothermal plant

System	2013 (MW)*	2014 (MW)**	2015 (MW)***	2020 (MW)****	2025 (MW)****
Java–Bali	1240	1189	35	1867	413
Sumatra	1240	122	0	2270	199
Sulawesi	80	80	0	145	80
Nusa Tenggara	40	12.5	0	45	40
Maluku	20	0	0	20	20
Total	2620	1403.5	35	4347	752
Cumulative	4733	1403.5	1438.5	5946	6638

Notes: *2013 planned in crash program project; **2014 existing installed capacity status; ***2015 on-going projects; ****2015 rescheduled program for power plant installation.

21.11   Future planning of geothermal development

In the last few years, the annual growth rate of electricity demand is about 8.6%. This might create a shortage of power in the near future because the energy sector has not been able to make adequate developments/investments in the power supply capacity to meet its growing electricity demand of about 9% per year. In addition, the government had initial plans to develop geothermal power plants with 2000 MW of capacity in 2008, 3442 MW in 2012, 4600 MW in 2016, and 6000 MW in 2020. By 2025, Indonesia is expected to install a 9500-MW power plant. In fact, there are only a few investors who won tenders, continuing the project as GOI investment plan. Consequently, GOI has revised and released the plan for 2025 geothermal development. The new tariff regulation was also issued, making international standards as a complementary part of MEMR Regulation No. 22 in 2012. This regulation was issued to attract an increase in investment in order to achieve more than 7000 MW from geothermal plants in 2025. The future development and power plant installation expectations are shown in Table 21.6.

21.12   Conclusions

Indonesia is known as having the largest geothermal potential in the world. The geothermal development growth since the 1980s tends to be stable until 2015 and modest in exploitation. The average annual growth is about 45 MW/year during 32 years of exploitation. The total installed capacity now is 1438.5 MW compared to 30 MW in the 1983. Even in the last 5 years, there is a significant increase in geothermal installed capacities as well as a significance use of brine for direct heat use of geothermal. However, the GOI needs to provide clear support for private power developers to minimize risk and uncertainty. In addition, GOI should educate developers and lenders on guaranteeing the viability of the project and to provide a development of 7000 MW electricity from geothermal in the next 10 years, up to 2025.

Other business opportunities in the geothermal sector are geothermal direct heat use, low-temperature geothermal potential, small-scale power plants, and service companies to support the core business of geothermal and human resources for the country.

To achieve the targets, international supports are needed in terms of finance, technology, human resources, and technical assistance. Indonesia, for all its high geothermal potential, has a significant challenge to attract private investors to develop geothermal power.

Acknowledgments

We offer thanks to the Director General of New and Renewable Energy & Energy Conservation of Indonesia, Dr. Rida Mulyana, and PGE for supporting data to publish this report.

References

Badan Geologi, 2013. Sumber Daya Panas Bumi Indonesia: Status dan Potensinya. Indo EBTLKE Conference and Exhibition 2013, Jakarta, Indonesia.

Darma, S., Harsoprayitno, S., Setiawan, B., Hadyanto, Sukhyar, R., Soedibjo, A., Ganefianto, N., Stimac, J., 2010a. Geothermal energy update: geothermal energy development and utilization in Indonesia. In: Proceedings World Geothermal Congress 2010, Bali, Indonesia, 25−30 April, 2010.

Darma, S., Poernomo, A., et al., 2010b. The role of Pertamina geothermal energy (PGE) in completing geothermal power plant in achieving 10,000 MW project in Indonesia. In: Proceeding World Geothermal Congress 2010, Bali, Indonesia.

Darma, S., Dwikorianto, T., Zuhro, A.A., Yani, A., December 2010c. Sustainable development of the kamojang geothermal field. Geothermics 39 (4), 391−399.

Directorate General of New, Renewable Energy and Energy Conservation (DGEBTKE) of Indonesia, 2014a. Statistics Book of New, Renewable Energy and Energy Conservation 2013. Yearly updating data for public report. Ministry of Energy and Mineral Resources of the Republic of Indonesia, Jakarta, Indonesia.

Directorate General of New, Renewable Energy and Energy Conservation of Indonesia, April 2014b. Public Hearing on the Draft of Ministry of Energy and Mineral Resources Regulation Concerning on the Tariff of Steam and Electricity Produced from Geothermal and Mandated PLN to Buy. Jakarta, Indonesia.

Ibrahim, R.F., fauzi, A., Darma, S., 2005. The Progress of Geothermal Energy Resources Activities in Indonesia. World Geothermal Congress, Antalya, Turkey.

Ministry of Energy and Mineral Resources (MEMR), 2013. Indonesia Energy Out Look 2013, Yearly Publication Book. Center for Data and Information of Ministry and Mineral Resources of Indonesia, Jakarta, Indonesia.

Ministry of Energy (MEMR), 2014. Indonesia Energy Out Look 2014, Yearly publication book, Center for Data and Information of Ministry and Mineral Resources of Indonesia, Jakarta − Indonesia.

National Energy Council of Indonesia, 2014. The National Energy Policy (NEP) 2014, Ministry of Energy and Mineral Resources of Indonesia, Jakarta − Indonesia.

PLN, 2013. Rencana Umum Penyediaan Tenaga Listrik (RUPTL) - National Electricity General Plan (2012−2021), Jakarta.

Stimac, J., Nordquist, G., Aquardi, S., Sirad-Azwar, L., 2008. An overview of the Salak geothermal system, Indonesia. Geothermics 37, 300−331.

World Bank, 2014. Geothermal Tariff Methodology, Studi Revisi Tarif Panas Bumi Indonesia, Kementerian Energi dan Sumber Daya Mineral, Jakarta, Indonesia.

Bibliography

Acuna, J.A., Stimac, J., Sirad-Azwar, L., Pasikki, R.G., 2008. Reservoir management at awibengkok geothermal field, West Java, Indonesia. Geothermics 37, 332−346.

APERC, 2007. APEC Energy Demand and Supply Outlook. IEEJ, Japan.

BPPT, 2013. Indonesia Energy Outlook 2013 − Energy Development in Supporting Transportation Sector and Mineral Processing Industry, Yearly Publication Book. Center for Energy Resources Development Technology, Agency for the Assessment and Application of Technology (BPPT), Jakarta, Indonesia.

Darma, S., Tisnaldi, Gunawan, R., 2015. Country update: geothermal energy use and development in Indonesia. In: Proceedings World Geothermal Congress 2015, Melbourne, Australia, 19—24 April, 2015.

Darma, S., Harsoprayitno, S., et al., 2009. Geothermal in Indonesia: government regulations and power utilities, opportunities and challenges of its development. In: Proceedings of the WGC 2010, Bali, Indonesia.

Fauzi, A., Bahri, S., Akuanbatin, H., 2000. In: May 28—June 10, 2000, Geothermal Development in Indonesia: An Overview of Industry Status & Future Growth, World Geothermal Congress, Kyushu, Tohoku, Japan.

Fauzi, A., 1998. Geothermal development in Indonesia: an overview. Geothermia Revista Mexicana de Geoenergia 14 (3), 147—152.

Fauzi, A., Darma, S., Siahaan, E.E., 2005. The role of Pertamina in geothermal development in Indonesia. In: Proceeding World Geothermal Congress 2005, Antalya, Turkey.

Ganefianto, N., Stimac, J., Sirad-Azwar, L., Pasikki, R., Parini, M., Shidartha, E., Joeristanto, A., Nordquist, G., Riedel, K., 2010. Optimizing production at Salak geothermal field, Indonesia, through injection management. In: Proceedings World Geothermal Congress, Bali, Indonesia.

GDF Suez, 2012. Successful Drilling Tests for GDF SUEZ Geothermal Project in Indonesia. GDF Suez online report.

Hadi, J., Ibrahim, R.F., Widiatmoko, Sugiharto, P., March 10—11, 1999. Amsoeas Indonesia's Long-term Commitment to Clean & Efficient Energy Bridges Indonesia's Energy Policy to the Next Millenium. Committee National of Indonesia Energy Council XVII, Jakarta.

National Energy Council of Indonesia, 2014. Final Draft of National Energy Policy. Ministry of Energy and Mineral Resources of Indonesia, Jakarta, Indonesia.

Pertamina, 1994. Indonesia Geothermal Reserves and Resources. Publication of Pertamina Geothermal Division.

Radja, V.T., Saragih, E., 1994. Utilization of small scale geothermal power plants for rural electrification in Indonesia. PLN, Jakarta, Indonesia.

Sanyal, S.K., Morrow, J.W., Jayawardena, M.S., Berrah, N., Li, S.F., darma, S., 2011. Geothermal resource risk in Indonesia — a statistical inquiry. In: Workshop on Geothermal Reservoir Engineering, Stanford University, Stanford, California, January 31—February 2, 2011. SGR-TR-191.

Sitorus, K., Nanlohy, F., Simanjuntak, J., 2001. Drilling activity in the Mataloko geothermal field, Flores, Indonesia. In: Proceedings of the 5th INAGA Annual Scientific Conference & Exhibitions, Yogyakarta, March 7—10, 2001.

Supreme Energy, 2014. Supreme Energy Publish Report. Jakarta, Indonesia.

New Zealand: a geothermal pioneer expands within a competitive electricity marketplace

I.A. Thain[1], P. Brown[2]
[1]Havelock North, New Zealand; [2]Taupo, New Zealand

22.1 Reform of the NZ electricity generation and supply industry

In 1985 the New Zealand electricity power generation and national grid transmission system was a government owned monopoly run by the Electricity Division of the Ministry of Energy. The government held sole right to use rivers and lakes for hydro generation and the right to tap, take, and use geothermal fluid for power generation. The Electricity Division carried out the design, built all new power plants, and operated the power generation plant and national grid including the operation of system control. In the high electricity growth period of the 1950/1960s, this arrangement performed well and was able to satisfy the country's demand for electricity. However, by the 1980s, this arrangement was failing to deliver electricity to the country at the lowest practicable cost—the driving policy of the day. Power generation projects were characterized by commissioning delays, large cost overruns, and electricity production costs well in excess of those predicted. Over investment in generation was being incurred due to over estimation of electricity demand by a business keen to maintain ongoing work projects for its work force, to the detrimental interests of the taxpayer and electricity consumers.

In 1984, New Zealand embarked on an aggressive program of economic restructuring and liberalization aimed at improving the nation's economic efficiency. The chronology of the electricity industry reforms carried out between 1984 and 2012 are detailed in a report [1] prepared by the New Zealand Ministry of Business Innovation and Development. The following account of these reforms outlines the main measures and legislative changes undertaken to create a commercialized electricity industry.

The electricity industry reform of greatest significance began with the corporatization of the Electricity Division of the Ministry of Energy, under the 1986 State Owned Enterprises Act, to form the New Zealand Electricity Corporation (ECNZ). The

Geothermal Power Generation. http://dx.doi.org/10.1016/B978-0-08-100337-4.00022-X

"corporatized" company was subjected to increased commercial discipline and charged to make economic returns on assets used, similar to the financial returns of the best performing privately owned companies. By 1993, this target was being regularly achieved. The design and construction arm of the Electricity Division was made a separated business unit and soon sold into private ownership. Power generation was made open to any person or company wishing to pursue such a development.

On being corporatized, the ECNZ was divided into three operational groups, namely: Production (responsible for all power plant operations), Transpower (responsible for national grid operations and system control), and Marketing (responsible for selling power to industry and commercial retailers).

In 1985, local distribution and retail of electric power were undertaken by local authority-owned power companies that had monopoly rights to provide this service within their designated local area. There were 61 such companies in 1985. Directions were given by the government that there should be consolidation of these local power companies into more efficient and cost-effective units. By 1994, they had consolidated into seven regional distribution and retailing organizations, some owning embedded power generation facilities.

While the government was pleased with the annual financial profits being made by the ECNZ, the lack of competition in power generation and within the local energy retail and distribution sides of the industry made the government set about with more industry reform. These reforms were progressively carried out over the period 1994–1998 and entailed making the following major changes to the industry.

- 1994: Transpower was removed from ECNZ control and made a separate state-owned enterprise. These assets were seen as a natural monopoly, which was required to provide unhindered access to the national grid system by any power generator at fair and transparent cost.
- 1996: Contact Energy was established by carving off around 24% of ECNZ generation assets into a separate company, consisting of a mix of thermal, hydroelectric, and all of the geothermal power stations. It was privatized in 1998 with a controlling stake sold to Mission Energy of the United States, and the remaining shares were sold to New Zealand public and private institutional investors. The sale of Contact Energy earned the government NZ$2.4 billion.
- 1997: ECNZ was broken up into three competing state-owned companies: (1) Meridian Energy, assigned all the remaining South Island hydro generation; (2) Mighty River Power, assigned all hydro generation along the Waikato River in the North Island; and (3) Genesis Energy, based on the large Huntly Thermal Power Station together with Tongoriro Hydro scheme. It was the intention of the government at that time that these new power-generating business units would also eventually be privatized.
- 1997: Wholesale Electricity Market was established to oversee (1) the trading of long-term power sales contracts between generators and large industrial users and energy retail companies and (2) the creation and operation of a time of day spot price market for electricity. Fig. 22.1 provides the outline and principle of operation of the New Zealand electricity wholesale market.
- 1997/98: Legislation was passed forcing the seven regional distribution/retailing companies to choose between becoming a transmission line distribution company or an energy retail company; they would no longer be able to carry out both functions. Most chose to remain as distribution companies and sold their retail customer base to one of the other of the power generating companies.

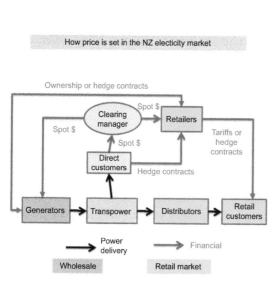

Some characteristics of the NZ wholesale market

• Generators offer to sell electricity into the spot market for each ½ h of each day

• Generation is scheduled and managed in real time by System Operator (Transpower) to ensure that supply matches demand

• All energy is sold at the market clearing price for each ½ h set by the most expensive generator dispatched for each ½ h

• Spot price varies significantly over time and can reach very high levels for sustained periods during times of low hydro inflows

• Retailers and large consumers enter directly into power supply contracts with generators (hedge contracts) in order to reduce the volatility of costs and revenues. (Around 70 to 80% of electricity is sold under hedge contracts)

• Small consumers typically buy electricity from retailers at fixed price, variable volume contracts and are not exposed to the variations in the spot price market

Figure 22.1 Outline and characteristics of the New Zealand wholesale electricity market. NZ Crown.

Under this electricity generation power supply system, customers are able to purchase power from any generation retailer and a distribution company delivers this power within their supply network area. Customers' electricity bill is made up of variable energy cost dependent on amount used plus a fixed price distribution line connection charge.

With a change in government in 1999, further privatization of the three state-owned electricity-generating companies was put in abeyance; it was not until 2014 that the government of the day sold off around 49% of these business units to private and institutional shareholders with the government retaining a controlling interest in these businesses. There was significant public opposition to this move, and it is uncertain if full privatization will eventuate. A large section of the voting public wants these assets retained in government ownership with the profits generated going to sustaining government income.

Twenty years ago, there was no competition in the retail electricity market. Today, residential and commercial customers can choose from 80 retailers. While competition in the electricity retail market is strong, the market is dominated by the ex-ECNZ businesses (Genesis, Meridian, Contact, and Mighty River) that command 80% of the market.

22.2   Geothermal resource management

In New Zealand, environmental and resource management legislation is set out in a single act of parliament, the Resource Management Act 1991 (RMA). This legislation establishes how people can use the land, air, and water (including geothermal fluid). The cornerstone of the RMA is the Maori concept of *kaitiakitanga*, loosely translated as guardianship and requires people to respect the environment and to ensure resources are maintained and handed over to future generations in a healthy condition.

Use the land wisely for the good of all our people and safeguard it for our children and our children's children.

The prime purpose of the RMA is to promote the sustainable and environmentally acceptable management of the natural and physical resources of the country by providing a comprehensive suite of policy, planning, and regulatory rules to manage the effects of using resources. Implementing the RMA has been devolved down from the central government to the regional authority within whose territory the actual resource is located.

With respect to geothermal resource management within the Taupo Volcanic Zone (TVZ), which accounts for around 90% of New Zealand's high-temperature geothermal fields, this function is the responsibility of the Waikato and Bay of Plenty Regional Councils.

The government has devolved to these regional councils the right to grant resource consents for the "take," "use," and "discharge" of geothermal fluid from geothermal resources located within their respective administration boundaries but requires the councils to ensure geothermal resource use is carried out in an efficient and sustainable manner with due regard being given to safeguarding significant natural geothermal features. In New Zealand, there is no financial charge made for the right to tap and use a geothermal resource. The two councils collaborate in managing the TVZ geothermal resource and work with major research organizations in the country in investigating the resource's extent and characteristics. Since 2005, there has been a 637-MWe increase in installed geothermal electricity capacity across eight projects in the TVZ, with most consents being processed within 6 months.

A regulatory perspective of geothermal resource management in the Waikato Region is provided elsewhere [2]. In respect to safeguarding the natural geothermal features, the Waikato Regional Council has applied three classifications to the geothermal systems in the region. These are:

- Protected: The system will remain in its natural state in perpetuity to ensure future generations have opportunity to see these natural unique features.
- Research: These are systems where more research and investigation work required to determine if they can be developed or should remain in their natural state.
- Development: The resource can be exploited for large-scale energy use provided this is carried out in a sustainable and environmentally acceptable manner.

The locations of the main geothermal systems within the TVZ are shown in Fig. 22.2. The systems having a red boundary are those that have been given a protected classification.

Figure 22.2 Map showing location of geothermal systems within the Taupo Volcanic Zone. NZ Crown cadastral regional boundary database with Waikato Regional Council geothermal resource classification overlay.

22.3   Geothermal: a Maori treasure being actively and innovatively used

22.3.1   Traditional involvement by Maori

New Zealand is a relatively young country with the first human habitation at around AD 1000, when it is understood Maori travelled across the ocean from the South Pacific Islands. Geothermal activity has always been regarded as a significant traditional resource among Maori communities of the Bay of Plenty, Rotorua, and Taupo districts. The principal settlements of the tribes of Te Arawa, Ngati Tahu, and Ngati Tuwharetoa were associated with geothermal areas. Outside the Taupo Volcanic Zone, there were numerous other hot springs, which were also highly valued by Maori.

Areas of surface geothermal activity all have some traditions and cultural and historical associations for local tribes. Geothermal resources were used in various ways. Hot pools (*ngawha, puia, waiariki*) provided hot water for cooking and bathing. Hot ground was used for cooking holes and ovens. Mud from some pools had medicinal properties, especially in the treatment of skin infection that was endemic in the Taupo district.

Many hot pools had well-known therapeutic qualities in the treatment of muscular disorders, rheumatic and arthritic ailments, as well as skin conditions. Some had other qualities and were known as *wahi tapu*, such as a place for ritual cleansing after battle, or having other spiritual qualities linked to medicinal or therapeutic use and/or incidents of the past. Some were burial places. Many hot pools are for one reason or another still regarded as *wahi tapu*, sacred places.

Hochstetter, who visited in 1859, provided the first detailed geological description of the surface geothermal activity in the Rotorua Taupo region, "in which hot steam issues forth from the earth at more than a 1000 points and produces all those phenomena of boiling springs, fumaroles, mud pools, and solfataras for which the North Island is so famed."

He also commented, "the natives have (quite rightly) long connected the *ngawha* and *puia* with the centers of volcanic activity that are still active even if they clothe their idea in the form of a strange legend" [3]. Hochstetter was referring to Ngatoroirangi, a Maori high priest and explorer who was freezing in the cold on the mountain Tongariro and called on his sisters in Hawaiki to send him warmth. Hawaiki is an ancient Maori homeland, the places from which they migrated to New Zealand [3].

22.3.2   The British arrive

Some 800 years after Maori settled in New Zealand, the British arrived, and a little later the Treaty of Waitangi was crafted. Regarded as New Zealand's founding document, the Treaty of Waitangi has been a source of much debate and controversy since 1840. The differences between the English and Maori language versions of the treaty are at the heart of this debate, along with the fact that Maori language was predominantly spoken rather than written.

In 1952, the Crown (the New Zealand parliament) bequeathed to itself the sole right to take, tap, and use all geothermal resources when it introduced the Geothermal Energy Act. The reason for controlling rights to geothermal was to facilitate government industrial development at Wairakei (electricity generation) and Kawerau (newspaper production). Maori consider that they have rights to geothermal and that the geothermal energy legislation is in breach of the Treaty of Waitangi — the issue remains unresolved.

22.3.3   Commercial use of geothermal

While holding fast to their view that Maori have ownership of New Zealand's geothermal resources, they have adopted a pragmatic approach to the commercial use of geothermal. In the interim, Maori are going ahead with geothermal business in accordance with the government "rule book," the RMA.

Today, Maori engage in (and benefit from) participation in the geothermal industry. Participation ranges from 100% development and ownership of geothermal enterprises to access rights, akin to an energy royalty.

The 1980—90 electricity reforms enabled Maori to participate in geothermal electricity generation projects. One of the earliest geothermal generation projects was a 55-MWe development that is 100% owned by a Maori Land Trust. This development has now increased to 113 MWe. The land trust has about 2000 owners made up of seven Hapu (subtribes); it is a wonderful testament to the owners and trustees that they continue to work together in harmony.

The commercial use of geothermal by Maori is not limited to electricity generation. Maori have led the way in the direct use of geothermal energy for heating and drying. High-temperature geothermal is used to produce milk powder, tissue paper, and timber. Lower-temperature geothermal provides heat for climate-controlled glasshouses and hot pools.

In 2005, Ngati Tuwharetoa Geothermal Ltd. (a subsidiary of the Maori Tribe Tuwharetoa ki Kawerau) acquired from the government the steam supply business that provides direct geothermal heat to Kawerau industry. The steam supply business commenced in 1956 and continues to be the largest geothermal steam supply business in the world.

22.3.4   Sustainability

With an engagement in geothermal of 1000 years plus, Maori continue to do their best to secure geothermal energy for future generations by careful and wise management of the commercial use of geothermal. The ongoing employment of people, present and future, continues to be a priority.

22.4   Geothermal developments—2000 to 2015

New Zealand has seen a period of rapid growth in the utilization of geothermal energy over the past 15 years, particularly for electricity generation. There has also been a resurgence in direct use over the past 5 years, which has offset a large direct use loss when one of the paper-making facilities at the Kawerau plant was shut down. The availability of good high-temperature geothermal resources within the Taupo Volcanic Zone has seen geothermal energy being the lowest-cost electricity generation facilities to build and operate, compared with other renewable energy or fossil fuel options. In 2000, the installed capacity of geothermal power generation on the New Zealand Electricity supply system was 428 MWe; by 2015 the installed capacity had increased to 1004 MWe. This represents a 135% increase in geothermal capacity over the past 15 years. The growth in installed geothermal capacity since the commissioning of the Wairakei Power Station in 1958 until 2015 is illustrated in Fig. 22.3.

In 2014, total New Zealand electricity generation was 43,200 GWh; geothermal supplied 7000 GWh, which represented 16% of the electric power needs of the country. Fig. 22.4 shows 2014 generation source breakdown, of which renewable generation accounts for 83% of the total generation. The government has set a target for the country to achieve 90% renewable electricity generation by 2025.

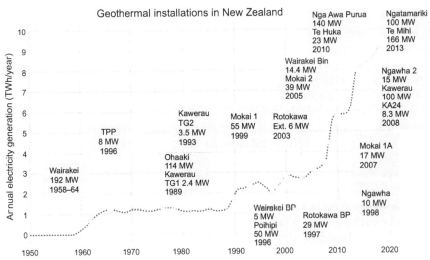

Figure 22.3 Graph showing growth in New Zealand geothermal installed capacity and electricity generation since commissioning of Wairakei Power Station in 1958 until 2015. B. White, NZGA.

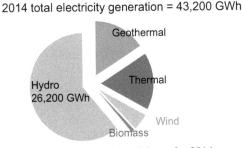

Figure 22.4 New Zealand generation source breakdown for 2014. Data sourced from WGC2015 NZ country update report.

22.5   Field review of geothermal power, tourism, and direct use developments

The following account of geothermal developments in the country relies heavily on the NZ Country Update Report [4] presented at the WGC 2015 in Melbourne, Australia.

22.5.1   Wairakei

The geothermal resource at Wairakei is now used in four separate plants: the original Wairakei power station (commissioned 1958/63), the Wairakei Organic Rankine Cycle (ORC, 14 MWe) binary plant added in 2005, Poihipi (55 MWe) power station built in 1996, and the recently commissioned (2014) Te Mihi (166 MWe) plant. Total installed generating capacity of the combined Wairakei stations is 357 MWe, with all the plants owned and operated by Contact Energy.

The Poihipi Road plant, when commissioned in 1996, was the first geothermal nongovernment power plant to be built in the country. The plant initially tapped a low-pressure (LP), shallow, dry steam cap created over the western part of the Wairakei field as a result of deep liquid pressure drawdown that occurred between 1960 and 1980. It was a joint venture (JV) between a local landowner whose property encroached over a small portion of the field containing the shallow dry steam cap and Mercury Energy, a distribution/retail company serving the Auckland area. The JV obtained resource consents for a 30-MWe power development and purchased a surplus unused Fuji 55-MW turbine generator, which had initially been bought for installing at The Geysers field in the United States. This unit was successfully reengineered to operate at 50 Hz, the New Zealand system electric power frequency. To comply with its 30-MWe resource consent, the unit was operated at full load for 14 h per day and at no load overnight. Problems within the JV led to the Poihipi Road plant being sold to Contact Energy in 2000. Using part of the resource consents held for the Wairakei Power Station, Contact Energy diverted steam from the Wairakei steam field to enable the Poihipi Road plant to operate in a base load mode at its full design output of 55 MWe.

The Te Mihi power station, which was commissioned in 2014, is the most recent development at Wairakei and consists of two 83 MWe Toshiba dual-pressure (5.2 bar,a and 1.3 bar,a) turbine/generators; see Fig. 22.5. Construction and design information on this project are contained elsewhere [5], including details of the project from inception to execution. An aerial view of Te Mihi is shown in Fig. 22.5, taken during initial commissioning of the power station. The plant was conceived as a staged replacement for the existing Wairakei A and B power stations. Space was left for a third unit to be added when the aged Wairakei plant will be retired; its existing resource consents expire in 2026. The double-flash system will produce 25% more power from the same amount of geothermal fluid that is currently used at Wairakei. The power plant is located close to the main steam production area, at a higher elevation relative to the existing Wairakei Power Station, an added advantage for the gravity-fed reinjection system operated on the field. A unique feature of the Te Mihi design is the incorporation of a high-temperature acid dosing system to prevent precipitation of silica in the LP separated geothermal fluid as the waste fluid flows under gravity head to the reinjection wells.

The original Wairakei power station has been in continuous operation for over 55 years, but operation of the plant has been altered to suit a changing steam supply.

Figure 22.5 Photograph of new Te Mihi Power Station.
Contact Energy.

The plant now has only sufficient steam flow to fully load 122 MWe of the original
157 MWe of plant still in operation. The previously supplied steam is currently being
diverted to the more modern and efficient Te Mihi (166-MWe) plant. In 2007, resource
consents for the Wairakei Power Station were granted that extended the operational life
of the plant until 2026, by which time the original Wairakei power plant will have been
in continuous operation for 70 years. These consents, however, were conditional on
(1) annual maximum allowable geofluid take from the field capped at a level sufficient
to support its lower operational capability; (2) a stepped reduction in total geothermal
fluid discharges to the Waikato River; and (3) limit set on discharge of hydrogen sulfide to the Waikato River.

To achieve this latter condition, a novel bioreactor has been incorporated into the
original Wairakei direct-contact condenser cooling water discharge to the Waikato
River. This facility reduces the hydrogen sulfide contaminant in the discharge to the
river by about 95%. The bioreactor consists of an array of 1890 parallel 100-mm-
diameter HDPE pipes, each 200 m long, buried underground, through which the mixed
condensate and river cooling water flow before being discharged back to the river. In
total, 378 km of pipe is used in the bioreactor bed. The research work undertaken in the
conception of this facility is described elsewhere [6]. A schematic cross section of the
bioreactor is shown in Fig. 22.6 along with an aerial view of the facility (Fig. 22.7).
The tubes provide a surface habitat for on which naturally occurring bacteria can
grow. The bacteria consume the hydrogen sulfide in the condensate, converting it to
sulfate in the 4-min transit time through the bioreactor.

At Wairakei, separated geothermal water discharged from the binary plant is used to
heat fresh water to around 30°C for use in a 6-ha prawn aquaculture facility. Separated
geothermal fluid is also made available to create a novel silica terrace feature at a
nearby geothermal themed tourist facility, which is illustrated in Fig. 22.8.

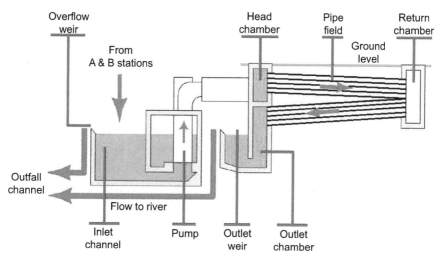

Figure 22.6 Schematic cross section of Wairakei Bioreactor (one of five banks).
After Bierre E, Fullerton B. Hydrogen sulphide removal from geothermal PS cooling water using a bio film reactor. In: Proceedings WGC 2015, Melbourne, Australia, 19−25 April, 2015; 2015. Courtesy: Contact Energy.

Figure 22.7 Aerial view of Wairakei bioreactor.
After Bierre E, Fullerton B. Hydrogen sulphide removal from geothermal PS cooling water using a bio film reactor. In: Proceedings WGC 2015, Melbourne, Australia, 19−25 April, 2015; 2015. Courtesy: Rob Fullerton & Milly Bierre, Contact Energy, taken from Wairakei Bioreactor, WGC, 2015, and BECCA, constructors and photographers of the bioreactor.

Figure 22.8 Wairakei Silica Terraces created with waste geothermal fluid from Wairakei Power Station.
Raewyn Hill, Taupo.

22.5.2   Tauhara

The Tauhara geothermal system is located adjacent to the Taupo township and is hydrologically connected to the Wairakei system. Field investigations were undertaken by the government in 1970s with one of the test bores used to provide geothermal heating to a plant propagating nursery. Interest in exploiting the field was renewed in the 1990s when Contact Energy was granted Resource Consents for a 30-MWe power plant. Part of this consent was used in 2005 to provide a 20-MWth direct use energy source, using two-phase heat exchangers, to supply hot water for timber-drying kilns. From 2005 to 2009, further exploratory drilling was carried out that confirmed the field was larger and hotter than previously thought. Several of these new wells were used by Contact Energy to fully use the Resource Consents already held for the Te Huka 23-MWe binary power plant, commissioned in 2010 and later expanded to 26 MWe in 2011 when increased fluid production became available from the field.

Due to the proximity of the field to the Taupo township, serious concerns were raised about geothermal-induced subsidence, affecting household properties and possibly linked to the Wairakei operations. As part of Contact Energy's application to further develop the Tauhara system, it undertook an extensive scientific subsidence investigation involving a large network of benchmarks monitored with regular repeat surveying to determine the extent of the subsidence problem. Reinjection at Wairakei has assisted in supporting field pressure at depth at Tauhara, alleviating some of local concerns over possible subsidence effects on properties should major geothermal development take place on the field. Extensive scientific investigations have also been carried out by Contact Energy to gain new knowledge on subsidence mechanisms. Contact Energy hold geothermal Resource Consents for the take and use of geothermal fluid from the field sufficient for supplying a 250-MWe power plant.

Construction of this plant has been put on hold in preference to the development of the Te Mihi Power Plant on the Wairakei field.

Overlying the Tauhara system are two shallow hot water aquifers that are tapped for domestic use with downhole heat exchangers and submersible hot water pumps to supply the local bathing pool complex and provide space heating and hot water services to the Taupo Hospital and nearby convalescent home.

22.5.3 Ohaaki

The Ohaaki power plant, commissioned in 1989, has been the least successful of New Zealand's geothermal power plants. It entered commercial service with around 25% excess steam capacity, but by 1991 this had been used to compensate for steam depletion, and from then on faced a continuing decline in steam production. This decline in steam production was due to a cool subsurface aquifer perched over the field invading the deeper production levels and quenching the deep production fluid. The draining of the upper cool aquifer led to significant field subsidence. Replacement well drilling on the eastern production field has stabilized steam production but at a level sufficient to supply only 57 MWe of the installed plant. The remaining 57 MWe of generating plant has been retired from operational service. Ohaaki plant is owned and operated by Contact Energy.

Approximately 4% of spent brine from the Ohaaki field, equivalent to a thermal energy supply of around 10 MWth, is supplied to a small timber-drying plant producing thermally dried firewood.

22.5.4 Rotokawa

Exploratory drilling of the Rotokawa system was carried out by the government from 1965 to 1986. This exploratory work identified a large high-temperature resource of about 28 km^2 with a power potential assessed as 300 MWe. Production is around 2.0−2.5 km, with temperatures up to 320°C.

Utilization of the resource commenced in 1997 with the commissioning Rotokawa I power plant, an Ormat combined cycle power plant of 24 MWe. In 2000, the Rotokawa I project was procured by two companies: Rotokawa JV, a 50:50 partnership between Mighty River Power and Tauhara North No 2 Trust, which own the steam field, and Mighty River Power, which own the generating plant. In 2003, due to achieving greater production well output, the Rotokawa I plant was expanded to 34 MWe.

Mighty River Power and Tauhara North No 2 Trust again partnered together as the Nga Awa Purua Joint Venture in a second development on the Rotokawa field and obtained resource consents in 2007 for the 138-MWe Nga Awa Purua power station, which entered commercial service in 2010. The 138-MWe plant was supplied by Fuji Electric and when commissioned was the largest single-shaft triple-flash (23.5, 8.4, and 2.3 bar,a) geothermal turbine in the world. The design and construction details of the Nga Awa Purua power plant may be found elsewhere [7]. A key feature of the design is the concept of acid dosing of the high-pressure (HP) brine with sulfuric acid in order to inhibit silica polymerization within the steam flash and reinjection systems by maintaining LP line brine at pH 5.0. Mighty River Power operates both power plants and the steam field at both Rotokawa facilities.

22.5.5 Mokai

Initial drilling of six geothermal wells on the Mokai field was undertaken by the government in the early 1980s on Maori land owned by the Tuaropaki Trust. This work indicated the Mokai potential to be around 150 MWe with one well deemed capable of generating 25 MWe. The field has some of the hottest geothermal wells in the country with recorded downhole temperatures of up to 325°C. These wells were acquired by the Tuaropaki Trust in 1987; these assets were placed in the Tuaropaki Power Company, which was formed in 1988 as a business unit of the trust. With these productive geothermal well assets in hand, a design/build contract was entered into with Ormat for the construction of a 55-MWe combined cycle power plant that entered commercial service in 1999.

The stated intent of the Tuaropaki Trust is to "act as a beacon of hope and prosperity for our people," and in acting on this intent over the past 15 years, by using the Mokai geothermal resource: (1) a 12-ha greenhouse complex started in 2002; (2) a second combined cycle power plant of 39 MWe commissioned in 2005; (3) they carried out a reengineering of the initial Mokai power plant to increase its output by 17 MWe that was commissioned in 2007, and (4) they developed the Miraka milk treatment plant. The reengineering of the Mokai I power plant became necessary as the production fluid enthalpy was higher than initial design expectations, resulting in less HP brine being available to power the two HP brine-fed binary units. To overcome this problem, the 32-MWe back-pressure turbine was replaced with a 45-MWe unit plus an additional 8-MWe LP steam unit. Sufficient HP brine was also made available to fully load the initial two HP 6-MWe binary units. The combined output of the Mokai Power plants is 113 MWe.

The latest business venture of the Tuaropaki Trust is the development of the Miraka milk processing facility in 2013 that uses a high-temperature geothermal energy reboiler system to produce clean process steam for dehydrating milk and to make milk powder; see Fig. 22.9. Taylor [8] provides design details of the clean steam-generating facility installed at the Miraka plant. The milk is delivered from nearby dairy farms including those of the Tuaropaki Trust. The geothermal reboiler system consisting of $2 \times 100\%$ duty heat exchangers and standby streams capable of generating up to a maximum of 30 ton/h of clean steam at pressures between 20.5 and 22.5 bar,g. The milk dehydration plant operational steam flow varies between 2 and 22 ton/h. The geothermal condensate from the reboiler plant is passed through a flash stripping process to be cleaned up for use as feed water on the clean steam side of the reboiler heat exchangers. This clean make-up water complements the condensate return from the milk treatment plant and maximizes the use of on-site water without having to resort to outside potable water supplies. Switching the reboiler heat exchangers from operation to standby can be carried out in under 10 min without interruption to clean steam delivery pressure or flow rate. The reboiler system used at Mokai was designed and manufactured in New Zealand.

Figure 22.9 Miraka milk-processing plant.
Miraka Ltd.

Geothermal fluid, at enthalpies varying between 1625 and 2050 kJ/kg, is supplied to the clean steam plant from two existing Mokai Power Station production wells. Approximately 10% of production capacity of either one of the connected wells is required to generate the 30 ton/h maximum clean steam flow. The clean steam plant is remotely controlled and monitored by an SCADA system linked to the Mokai Power Station control room.

The Miraka milk-processing facility is the first plant in New Zealand, and likely in the world, to use geothermal energy to produce clean steam for such a purpose and to use geothermal steam condensate as the feed water source for the clean steam. This is a highly efficient and sustainable use of geothermal energy and should place it at the head of the Lindal diagram of geothermal energy uses, ahead of electricity production.

22.5.6 Kawerau

Direct geothermal energy use commenced at Kawerau in 1957 with geothermal steam supplied to the Tasman Pulp and Paper Mill. The geothermal energy was used in a reboiler system to produce clean steam for the paper-making process. By the 1970s, three paper-making machines were being supplied with clean steam from the Kawerau resource together with an 8-MWe back-pressure turbine generator used to supply some in-house electricity for the mill. In 1990, this supply was cut back to service two machines and, in 2012, to a single machine. The current owner of the paper-making plant is Norse Skog Tasman (NST).

This drop-off in direct use of the Kawerau geothermal resource has been compensated by its increased use for power generation and for the production of a 24 ton/h clean steam supply, using reboiler technology, for a paper tissue plant operated by Syenska Celluloas Aktiebolaget. A schematic diagram of the clean steam plant is shown in Fig. 22.10. Since 2010, geothermal energy has replaced the natural gas fired steam generation process previously used by the tissue mill.

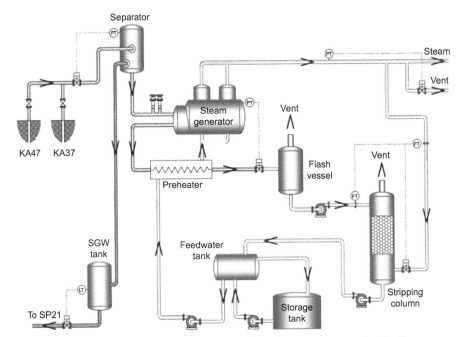

Figure 22.10 Schematic diagram of paper tissue plant clean steam-generating facility.
G. Moore, Dobbie Eng.

Bloomer [9] provides details of the Kawerau clean steam plant including details of chemistry relating to stripping the condensed geothermal steam to produce clean feed water from which clean steam is produced for the tissue plant. The Kawerau geothermal energy reboiler plant is shown in Fig. 22.11; a similar installation is used at the Miraka dairy plant at Mokai.

Timber drying is another long-term use of the Kawerau resource; 10 bar,g steam at 185°C is used to provide the thermal energy needs for 100 m^3 fast-drying kilns

Figure 22.11 Geothermal reboiler heat exchangers for production of clean steam for Kawerau tissue paper plant.
Ngati Tuwharetoa Geothermal Ltd.

Figure 22.12 Geothermal steam supply pipeline blow through at commissioning of Sequal Lumber new timber-drying kilns at Kawerau.
Sequal Lumber Ltd.

operated by Carter Holt Harvey Wood Products Ltd. The actual kiln temperature ranges from 80 to 140°C depending on drying regimen. Typically, the wet timber moisture content is reduced to 10% in 20 h [9].

In 2015, a second timber-drying facility was commissioned on the Kawerau field for Sequal Lumber Ltd. This facility consists of two large-capacity kilns; however, the energy delivery system was designed to be capable of delivering 30 ton/h of geothermal steam at 7 bar,g because two similar sized kilns are planned for future installation on the site. The timber-drying capacity of the initial facility is reported to be 30,000 m³ per year. Fig. 22.12 shows the geothermal steam blow-through during commissioning of the Sequal lumber plant.

On-field power generation is currently provided by a 3.5-MWe binary plant operating on separated geothermal brine. Commissioned in 2008, an 8.3-MWe binary plant, named KA24, was supplied with geothermal fluid from well KA24, from which it derives its name. In 2010, a 100-MWe double-flash Fuji Electric steam plant power plant (pressures: 12.0 and 1.8 bar,a), supplied by the Sumitomo Corporation, was built for Mighty River Power Company; see Fig. 22.13. The design and construction details of the 100-MWe Kawerau power plant may be found elsewhere [7]. On commissioning, the net output achieved by the plant was 106 MWe. The latest power plant, built by NST, was a 23-MWe Ormat binary plant operating on steam and brine from the paper mill geothermal fluid supply. As mentioned in Section 22.3.3, in 2005 Ngati Tuwharetoa Geothermal Ltd. acquired the steam supply business for provision of direct geothermal heat to Kawerau industries.

22.5.7 Ngatamariki

Four exploratory wells were drilled by the government to test the energy potential of the Ngatamariki system in the mid 1980s, with two of these exploration wells discovering a high permeability resource, with a temperature greater than 280°C. In 2004,

Figure 22.13 Kawerau 100-MWe double-flash steam power station.
Mighty River Power.

Mighty River Power acquired access to the field and further investigation of the system
was carried out from 2005 to 2010, with the drilling of three more wells that resulted in
a resource consent being granted in 2010 for supply of 60,000 ton/day of geothermal
fluid. Construction began in 2011 of an 83-MWe power plant consisting of four Ormat
binary units. The heat exchangers are fed with separated steam and brine at 192°C and
14 bar,a from three production wells. The plant, shown in Fig. 22.14, began commercial service in 2013 and has a net electrical output of 83 MWe. It is the largest Ormat
binary plant in the world. The Ngatamariki plant was purpose-built as a binary power
plant for high temperature, high-enthalpy use; design details are provided elsewhere
[10]. The plant is owned and operated by Mighty River Power using Resource Consents they jointly hold with Tauhara North No 2 Trust.

22.5.8 Ngawha

Ngawha, in Northland, is the only high temperature geothermal system located outside
the TVZ. It was investigated by the government between 1977 and 1983 and was
intended to be the next geothermal power plant to be built after the completion of
Ohaaki. These investigations resulted in the drilling of 13 production bores. Restructuring of the Electricity Division of the Ministry of Energy in 1987 curtailed this development. Power development utilizing some of these production bores was carried out
in 1998 by a joint venture between Tai Tokerau Maori Trust Board and the local electricity supply company Top Energy which resulted in the construction of a 10-MWe
binary power plant. Another 15-MWe binary plant was commissioned by the joint venture company in 2008. This 25-MWe facility produces around 70% of the electricity
demand for the far north of the country.

Figure 22.14 Ngatamariki binary power plant.
Mighty River Power.

22.6   Geothermal outlook

Geothermal energy has the potential to make a significant contribution to New Zealand's economy in the energy (electricity generation and direct heat use) and tourism sectors. The current estimated future development potential, which can be realized from existing geothermal fields, is around 1000 MWe, or if used directly around 6500 MWth. For the foreseeable future (up to 2020) the electricity load growth in New Zealand is expected to remain static; consequently, no new major geothermal power stations are likely to be built until after 2020. In future energy outlook scenarios (beyond 2020) the geothermal energy share of electricity generation is predicted to rise from its current level of 16% to between 21% and 29% by 2040. An insight into New Zealand's future electricity needs can be found elsewhere [11].

With the current curtailment in geothermal power station development, this is an opportune time for New Zealand to focus more of its geothermal energy resource development toward direct heat use, the recovery of valuable mineral products contained in the geothermal fluids, and tourism. Opportunities to use geothermal beyond power generation include:

- Extraction of minerals from geothermal brine − a pilot plant to extract silica indicates that commercial operations are likely
- Drying timber from the abundant supply of logs from plantation forests
- Heating and cooling of commercial and residential properties
- Geothermal tourism aspects of geothermal − natural features, hot pools, and spas

There is also the opportunity to enhance the efficiency of geothermal power generation and take advantage of synergy by using heat energy from biomass to produce superheated steam [12]. Superheated steam is a colorless, odorless gas with the ability to carry and transfer large amounts of thermal energy due to its high enthalpy. This high enthalpy makes geothermal steam an excellent candidate for a wide variety of applications in heat treatment, sterilization, pharmaceuticals, energy production, drying, and many more such applications. This is an area of development where New Zealand is favorably positioned because its main high-temperature geothermal resources are located in close proximity to vast sustainable plantation forests and thus have access to large amounts of waste material from processing wood (bark, sawdust, wood off-cuts) and material left on the forest floor after log harvesting. This power generation concept is covered in Chapter "Combined and hybrid geothermal power systems" of this publication.

These endeavors will call for visionary entrepreneurship that looks beyond the low-risk, economically viable application of geothermal energy resources for power production.

The delay in further power generation projects provides a welcome opportunity to apply increased focus on the management of existing geothermal reservoirs to ensure their sustainability for future generations.

References

[1] NZ Ministry of Business Innovation and Employment, Communications Branch. Chronology of New Zealand electricity reform. July 2012. MBIE-MAKO — 3727675, http://www.med.govt.nz/sectors-industries/energy/electricity/industry/chronology-of-new-zealand-electricity-reform.

[2] Luketina K. New Zealand geothermal resource management- a regulatory perspective. In: Proceedings WGC2000, Kyushu, Tohoku, Japan, May 28—June 10, 2000; 2000.

[3] Evelyn S. The legacy of Ngatoroirangi — Maori customary use of geothermal resources. Department of Geography University of Waikato; October 2000.

[4] Carey B, et al. New Zealand country update. In: Proceedings WGC2015, Melbourne, Australia, 19—25 April, 2015; 2015.

[5] Hudson R, et al. Te Mihi geothermal power station project-from inception to excecution. IPENZ Trans 2014;41. ISSN:1179 9293.

[6] Bierre E, Fullerton B. Hydrogen sulphide removal from geothermal PS cooling water using a bio film reactor. In: Proceedings WGC 2015, Melbourne, Australia, 19—25 April, 2015; 2015.

[7] Horie T. Kawerau and Nga Awa Purua geothermal power station projects New Zealand. Fuji Rev 2009;55(3).

[8] Taylor TAC. Tuaropaki clean steam supply for Miraka. In: Proceedings NZ geothermal workshop 2011, Auckland, New Zealand, 21—23 November, 2011; 2011.

[9] Bloomer A. Kawerau direct heat use: historical patterns and recent developments. In: NZGA workshop proceedings, Auckland, New Zealand, 21—23 November; 2011.

[10] Legmann H. The 100 MW Ngatamariki geothermal power station — a purpose built plant for high temperature, high enthalpy use. In: Proceedings WGC 2015, Melbourne, Australia, 19—25 April, 2015; 2015.

[11] NZ Ministry of Business Innovation and Employment. New Zealand energy outlook. Electricity Insight 2012. ISSN:1179 4011, http://www.med.govt.nz/sectors-industries/ energy/energy-modelling/modelling/pdf-docs-library/electricity-insight/electricity-insight.pdf.

[12] Thain I, DiPippo R. Hybrid geothermal-biomass power plants: applications, designs and performance analysis. In: Proceedings WGC 2015, Melbourne, Australia, 19–25 April, 2015; 2015.

Central and South America: significant but constrained potential for geothermal power generation

P. Moya Rojas

West Japan Engineering Consultants, Inc. (West JEC), Liberia, Guanacaste, Costa Rica

23.1 Central America

23.1.1 General information

The following Central American countries with geothermal potential will be studied in this chapter: Costa Rica, El Salvador, Guatemala, Honduras, Nicaragua, and Panama. While Belize is a part of Central America, it has no geothermal resources. For context, the geographic and economic conditions of these countries are presented in Table 23.1.

Geothermal energy in Central America has been studied and developed for decades. These countries were able to cultivate their geothermal developments, taking advantage of soft loans from U.S. and Japanese banks to the countries (sovereign guarantees) when these banks routinely lent money for geothermal projects. Most recently, some Japanese banks are providing loans for these projects again, such as JICA funding for future geothermal developments in Costa Rica (Las Pailas 2, Borinquen 1 and 2),

Table 23.1 **Selected data for geothermally active countries in Central America [1]**

Country	Land area (km²)	Population (2015)	GDP, 10⁹ US$ (2014)	Installed electric power, GW (year)	Electricity consumed, GWh (2011)
Costa Rica	51,060	4,814,144	70.97	2.944 (2011)	8792
El Salvador	20,721	6,141,350	50.94	1.507 (2011)	5412
Guatemala	107,159	14,918,999	119.10	2.813 (2011)	8143
Honduras	111,890	8,746,673	39.08	1.815 (2011)	5091
Nicaragua	119,990	5,907,881	29.47	1.275 (2013)	2777
Panama	74,340	3,657,024	76.42	2.391 (2012)	6626

Geothermal Power Generation. http://dx.doi.org/10.1016/B978-0-08-100337-4.00023-1

Bolivia (Sol de Mañana, Laguna Colorada), and perhaps even in Guatemala and Nicaragua in the near future.

Owing to the large number of agencies, government and private, involved with geothermal developments in the region, we present a listing of them with their acronyms in Table 23.2.

Table 23.2 Names and acronyms for agencies, banks, institutions, and methods involved in geothermal development in Central and South America

Acronym	Agency/bank/institution
CABEI	Central American Bank of Economic Integration
CEL	Hydroelectric Commission of Lempa River in El Salvador
CELEC EP	Public Corporation for Electricity Generation in Ecuador
CFG	French Geothermal Company
CHEC	Hydro-Electric Central of Caldas (Colombia)
CNRH	National Council for Water Resources in Ecuador
CODELCO	National Copper Corporation of Chile
CONELEC	National Council for Electricity in Ecuador
CORFO	Chilean Development Corporation
DGE-MEM	General Direction of Electricity from the Ministry of Energy and Mines in Peru
EBI	European Bank of Investments
ENAP	National Company of Oil in Chile
ENDE	National Company of Electricity of Bolivia
ENEE	National Company of Electric Energy of Honduras
ENEL GP	ENEL Green Power from Italy
ENG	National Geothermal Company
ESPE	Polytechnic University of the Army in Ecuador
GEA	Geothermal Energy Association
GESA	Andina Geoenergy (Colombia)
GQP-SAC	Geotérmica Quellaapacheta Peru SAC
IAEA	International Atomic Energy Agency
ICC	International Chamber of Commerce
ICE	Costa Rican Institute of Electricity
ICEL	Colombian Institute of Electrification

Table 23.2 **Continued**

Acronym	Agency/bank/institution
IDB	Inter-American Development Bank
ICSID	International Center for the Solution of Investment Disputes
IGA	International Geothermal Association
IIE	Electrical Investigations Institute (Mexico)
INDE	National Institution for Electrification in Nicaragua
INECEL	Ecuadorian Institute of Electrification (now defunct)
INER	National Institute for Energy Efficiency and Renewable Energy in Ecuador
INIE	National Institute of Statistics and Computer Science from Peru
IPSE	Planning and Promotion of Energetic Solutions Institute in Colombia
IRENA	International Renewable Energy Agency
IRHE	Institute for Hydraulic Resources and Electrification in Ecuador
ISAGEN	ISA Colombian Generation
JBIC	Japan Bank for International Cooperation
JETRO	Japan External Trade Organization
JICA	Japan International Cooperation Agency
LaGeo	Geothermal Company from El Salvador, subsidiary of CEL
LANL	Los Alamos National Laboratory
MAE	Ministry of the Environment in Ecuador
MEER	Ministry of Electricity and Renewable Energy in Ecuador
MICSE	Ministry for the Coordination of Strategic Resources in Ecuador
MINEM	Ministry of Energy and Mines from Peru
MT	Magnetotelluric, geophysical exploration method
NGO	Nongovernment organization
NRDC	Natural Resources Defense Council
OLADE	Latin American Energy Organization
PCA	Paris Court of Appeals
SERNAGEOMIN	National Geological and Mining Survey in Chile
SGC	Geological Colombian Survey (previously known as INGEOMINAS)
SIGET	Institution for Electrical Regulation in El Salvador
SYR	Servicios y Remediación S.A. (private company from Ecuador)

Continued

Table 23.2 **Continued**

Acronym	Agency/bank/institution
TDEM	Time-domain electromagnetics, geophysical exploration method
UNDP	United Nations Development Program
UNFSTD	United Nations Fund for Scientific and Technological Development
UPME	Mining and Energy Planning Unit in Colombia
West JEC	West Japan Engineering Consultants, Inc.

Countries that began their geothermal efforts earliest have already developed two fields in each country. The first geothermal plants in Central America started production at El Salvador's Ahuachapán geothermal field (Unit 1, 30 MW) in 1975, in Nicaragua at the Momotombo geothermal field (Unit 1, 35 MW) in 1983, at Costa Rica's Miravalles geothermal field in 1994 (Unit 1, 55 MW), and at Guatemala's Amatitlán geothermal field in 1998 (Unit 1, 5 MW). Each of these four countries now has two fields in operation: Costa Rica — Miravalles and Las Pailas; El Salvador — Ahuachapán and Berlin; Guatemala — Amatitlán and Zunil; and Nicaragua — Momotombo and San Jacinto Tizate.

The other two countries in Central America, Honduras and Panamá, are not producing geothermal energy yet but are expected to produce geothermal in the coming years. In fact, there is already a geothermal project currently under development in Honduras at the Platanares geothermal field.

There has been extensive geothermal development since late 1970s, and the latest information on these geothermal fields will be found in the following sections. However, in order to properly introduce the Central American region, tectonic plates, the seismic hazard map, and the distribution of the volcanoes in the area are briefly discussed first.

23.1.2 Central America geologic information

23.1.2.1 Central America location

The countries of Central America are shown in Fig. 23.1.

23.1.2.2 Tectonic plates

As described in detail in Chapter "Geology of geothermal resources," the theory of plate tectonics describes the large-scale motion of Earth's lithosphere. These tectonic plates can be seen in Fig. 23.2.

The main tectonic plates for the Central America region are the Cocos Plate and the Caribbean Plate; see Fig. 23.3. The Cocos Plate continuously subducts under the Caribbean Plate, forming what is known as the Mesoamerican trench. This led to the formation of volcanoes throughout Central America, which run almost parallel

Figure 23.1 Map of central America [2].

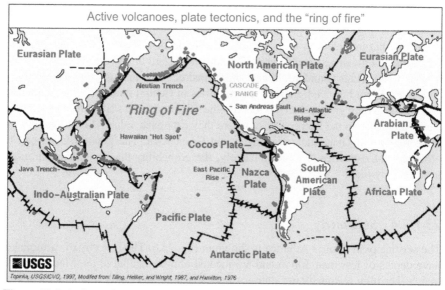

Figure 23.2 Active volcanoes, plate tectonics, and the "ring of fire" [3].

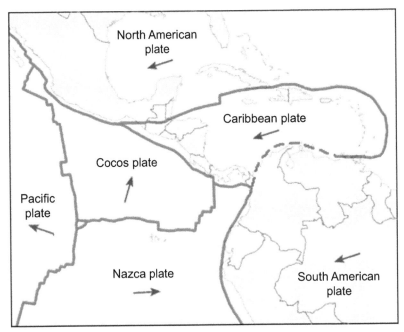

Figure 23.3 Main tectonic plates for the Central America region [4].

to the Mesoamerican trench. This chain of volcanoes is conducive to the formation of geothermal fields in the region.

23.1.2.3 Seismic hazard

A seismic hazard refers to the statistical likelihood of a seismic event (earthquake) occurring in a given geographic area. The seismic hazard of any area is used to assess the risk to buildings (standard, larger, and infrastructure), land use, and overall insurance rates. From Fig. 23.4, most of the countries in Central America with geothermal potential also rank greater than 4.0 m/s^2 on the seismic hazard scale. The accelerations are greater close to the Pacific Coast, where they can reach values close to 4.8 m/s^2 in some regions of Guatemala, Nicaragua and Costa Rica. The greatest value can be observed in Costa Rica (9.8 m/s^2). Clearly, the probability that an earthquake might occur is very high.

23.1.2.4 Seismicity

The seismicity in Central America is shown in Fig. 23.5. The majority of earthquakes have depths of less than 300 m and are located almost parallel to the Mesoamerican trench. The high seismicity in Central America is an indirect indication that the region has a high potential to develop geothermal resources.

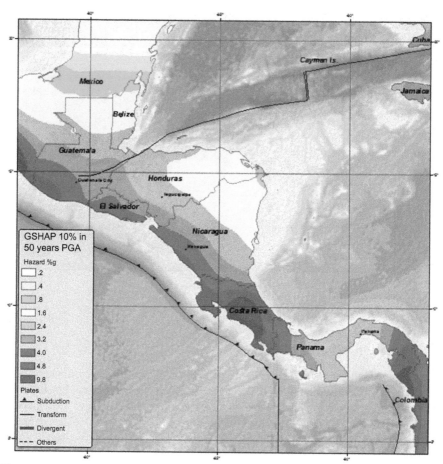

Figure 23.4 Earthquake hazard map for Central America [5].

23.1.2.5 Volcanoes

There is a high density of volcanos in this geographical region. The names of the 29 most notable one are listed in Fig. 23.6. These volcanoes are just the main ones, and there are more volcanoes in each country, further indication that there is a significant potential for geothermal power generation in Central America.

23.1.3 Costa Rica

In the early 1970s, Costa Rica satisfied its electricity needs using hydro and thermal energy sources (70% and 30%, respectively). The continuous rise in oil prices, especially during the 1973 crisis, motivated the authorities of ICE to study the possibility of using alternative energy sources for generating electricity, including geothermal energy [8].

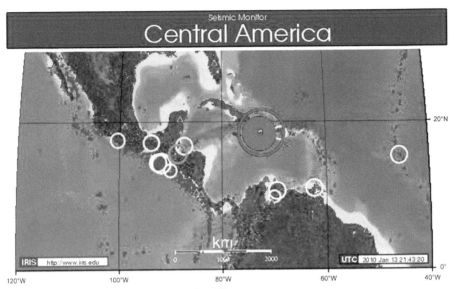

Figure 23.5 Seismicity in Central America [6].

23.1.3.1 Introduction

The first evaluations of the geothermal resources of Costa Rica were undertaken in 1963—64 when, at the request of ICE, a team of experts from the United Nations recommended a detailed study be carried out at Las Pailas (on the slope of the Rincón de la Vieja volcano) and at Las Hornillas (on the slope of the Miravalles volcano), both in the Guanacaste Mountains (Fig. 23.6). These two volcanoes are very close together; the lineal distance between their craters is less than 70 km.

Thus, ICE began to collect geological, hydrologic, and geochemical data over a region of more than 500 km^2 between the volcanoes (Rincón de la Vieja, Miravalles, and Tenorio volcanoes) and the Inter-American Highway. Preliminary exploratory studies of the geothermal areas in the Cordillera Volcánica de Guanacaste were performed in 1975, and the first technical report (a set of prefeasibility studies) on the possibility of exploiting geothermal resources for generating electricity within this area was completed in 1976 [9]. The positive outcome of this work allowed ICE to apply for loans from IDB, loans that facilitated the development of the Miravalles geothermal field.

Since 1976, many investigations, prefeasibility studies, and feasibility studies have been carried out, primarily at the Miravalles geothermal field but also at the Tenorio and Rincón de la Vieja volcanoes, where areas of geothermal interest have been found.

A huge quantity of scientific documents resulted, including volcanological, geophysical, geological, fluid geochemistry, soil geochemistry, petrography, and

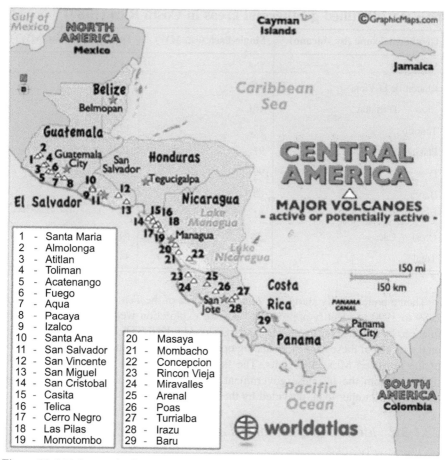

Figure 23.6 Volcanoes in Central America [7].

mineralogical studies. In addition, temperature gradient wells, as well as exploratory, production, and injection wells, have been drilled in the main geothermal areas of Miravalles, Tenorio, and Rincón de la Vieja.

Once the financing was approved for the construction of the first 55-MW unit in Miravalles, ICE carried out a national geothermal reconnaissance study and a geothermal prefeasibility study using its own funds and those of the Italian government. This was done in two phases. First, a reconnaissance study of the geothermal resources in Costa Rica was carried out from November 1987 to January 1989 to characterize the country into different zones of geothermal interest, delineate the most favorable areas, and select one or two of the preferred areas in which a prefeasibility study would be carried out in the second phase [10]. Table 23.3 shows the results.

Table 23.3 **Identified geothermal areas in Costa Rica (1989)**

Geothermal zone (by volcano)	Single-flash (est. MW)	Double-flash (est. MW)
Miravalles	164	213
Rincón de la Vieja	137	177
Irazú − Turrialba	101	130
Tenorio	97	123
Platanar	97	122
Poas	90	116
Barva	85	109
Fortuna	61	77
Orosi − Cacao	33	41
Total	865	1108

Then a prefeasibility study was done in the area of the Tenorio volcano from mid-1989 to 1990 to identify appropriate sites for exploration wells. These studies were financed by the Italian government and managed by UNFSTD through UNDP. Based on the results of these studies, financing options were investigated for the construction of the proposed Miravalles units. The first unit went online in March 1994, with financing from the Japanese government and the IDB loans. The second unit was financed by another loan, provided by the IDB.

23.1.3.2 Identified geothermal areas

Rincón de la Vieja

The second priority to develop the geothermal resources in Costa Rica was the Rincón de la Vieja volcano. In this geothermal zone, ICE has already installed a binary plant that has a capacity of 35 MW in the Las Pailas geothermal field. Currently, ICE is developing a new geothermal project there (Unit 2) intended to generate an additional 55 MW based on a prefeasibility study that was completed in 2012.

Besides the Las Pailas geothermal sector, ICE has identified another geothermal sector, Borinquen, in the Rincón de la Vieja zone. Here, ICE plans to develop two more units to complete the geothermal development at Rincón de la Vieja. The prefeasibility study for Units 1 and 2 in Borinquen was finished in 2012.

Other geothermal areas

ICE drilled exploratory wells in the Tenorio Volcano area, but the wells showed no permeability and no production. ICE abandoned the site mainly because of the poor results, but also because the area of Tenorio had by that time become a national park. The National Park Law in Costa Rica forbids any geothermal development inside any national park.

A similar situation is happening with the remaining identified geothermal zones in Table 23.3: Irazú-Turrialba, Platanar, Poas, Barva, Fortuna, and Orosi-Cacao. The majority of these geothermal zones correspond to volcanoes and are situated within various national parks in Costa Rica. Due to this constraint, ICE has decided to investigate other zones such as Mundo Nuevo (next to Las Pailas), the northern sector of the Rincón de la Vieja volcano, the northern sector of the Tenorio volcano, Arenal volcano-Pocosol area, and the Platanar-Porvenir zone. These are also protected zones but do not fall within in the national park category. Some independent studies have been carried out, but most of them require more studies in several disciplines to identify a possible geothermal anomaly.

23.1.3.3 Geothermal plants in operation

Current installed capacity

Table 23.4 shows the current geothermal units in Costa Rica. Since the construction of Unit 2 was under way and some of the production wells were already drilled, ICE made an agreement with CFE to produce more electricity at Miravalles by using wellhead units coming from CFE. Thus, the first wellhead unit WHU-1 (5 MW) in the Miravalles geothermal field came online by the end of 1994. Once Unit 2 (Fig. 23.7) was ready to be commissioned, the production wells used for WHU-2 and WHU-3 were switched for use in Unit 2. Units WHU-2 and WHU-3 were dismantled and sent back to CFE.

While drilling the wells for Unit 2, some extra wells were drilled to supply steam to Unit 3, which began operations in March 2000. Unit 3 was a build-operate-transfer (BOT) project for 15 years and in March 2015 became ICE property. Units 1 to 3 are all single-flash units (Figs. 23.7 and 23.8).

Table 23.4 Geothermal plants in Costa Rica

Plant name	Power (MW)	Owner	Start-up	Shut-down
Miravalles				
Unit 1	55	ICE	3/1994	
WHU-1	5	ICE	11/1994	
WHU-2	5	CFE	9/1996	4/1999
WHU-3	5	CFE	2/1997	4/1998
Unit 2	55	ICE	8/1998	
Unit 3	29	ICE	3/2000	
Unit 5	19	ICE	1/2004	
Las Pailas				
Unit 1	35	ICE	7/2011	

Figure 23.7 Miravalles Units 1 and 2 (two single-flash units of 55 MW each).
Photograph by Paul Moya Rojas.

Figure 23.8 Las Pailas Unit 1 (one 35-MW binary plant).
Photograph by Paul Moya Rojas.

ICE installed a binary plant (Unit 5) at Miravalles in January 2004 to take advantage of extra energy from the brine. Unit 5 currently has a capacity of 19 MW.

ICE installed the Las Pailas Unit 1 binary plant (Fig. 23.8) using its own financial resources with aid from BCIE. This unit belongs to BCIE, but after 12 years of monthly payments, ICE will get the first opportunity to buy the geothermal unit. Costa Rica currently has an installed capacity of 163 MW in Miravalles and 35 MW in Las Pailas, for a total of 198 MW.

23.1.3.4 Future geothermal development

Future geothermal projects are shown in Table 23.5. ICE has already applied for soft loans to continue the geothermal development in Las Pailas and Borinquen. Law 9254 was published in August 2014 in which JICA as well as EBI will provide a loan to build these three units. At Miravalles, ICE has been able to install a 163-MW (single-flash) plant, very close to the original estimation of 164 MW in the reconnaissance study. However, for the time being, ICE does not plan to increase the current capacity at Miravalles.

23.1.3.5 Constrained potential for geothermal growth

The majority of the geothermal sites are associated with volcanoes that are located within national parks where current National Park Law does not allow any geothermal development. This constraint has already had negative consequences for Unit 1 at Las Pailas, such as lack of access to better permeability zones inside the national park, and might have similar consequences for the next units ICE is planning to install at Las Pailas and Borinquen geothermal fields.

In 2006, a bill called "Regulated Law for the Production of Geothermal Energy in National Parks" [11] was submitted to the Costa Rican congress in order to permit ICE to develop geothermal projects inside national parks, which included the environmental aspects involved in this type of development. However, this bill was tabled some time later. After this bill, several others have been presented, but they have also been tabled. Unlike the first bill written in 2006, the subsequent bills had major errors that make approval impossible. No more bills have been submitted, and geothermal development apparently is not a topic of interest for the current

Table 23.5 Future geothermal development in Costa Rica

Plant name	Power (MW)	Owner	Planned start-up
Las Pailas			
Unit 2	55	ICE	2019
Borinquen			
Unit 1	55	ICE	2023
Unit 2	55	ICE	2024

government (2014—18). Without changes to the present law, electrical generation from geothermal resources will be severely hampered.

23.1.4   El Salvador

23.1.4.1   Introduction

The initial exploration studies in El Salvador were undertaken in 1958, with the first geothermal exploration well being drilled in 1968 in Ahuachapán. These efforts were completed because the country was looking for ways to develop its available geothermal resources, as well as to reduce its dependency on the fossil fuels. The program was conducted under UNDP and was very important in terms of exploratory work, leading to 28 geothermal discoveries. Of these, 24 sites were found to have excellent potential: 12 systems with temperatures >150°C and 12 more with temperatures >90 and <150°C [12].

Initially, geothermal development in El Salvador was done by a public institution, CEL that sought soft loans for exploration work and later was able to install the majority of the units in Ahuachapán and Berlin geothermal fields. CEL carried out all the geothermal development until 1999, when the government of El Salvador created LaGeo, a public/private company, to take charge of the developments. LaGeo found a strategic technical and financial partner in Italy's EGP in 2002 to continue and accelerate the geothermal development in the country and the region. Between 2008 and 2014, CEL and EGP entered into a dispute over EGP's right to finance LaGeo's expansion projects in exchange for new shares in the company. The dispute was heard in the arbitration tribunals of the ICC, the Paris Court of Appeals, the ICSID, and local Salvadorian courts. In December 2014, the dispute was resolved when CEL agreed to buy back EGP's shares in LaGeo, and EGP pulled out of El Salvador.

23.1.4.2   Geothermal plants in operation

The geothermal plants in El Salvador are shown in Table 23.6. At Ahuachapán, Units 1 and 2 are single-flash, and Unit 3 is double-flash.

In the Berlin geothermal field, two back-pressure wellhead units were installed and ran for some years to produce electricity while the field was being developed. Later, it was decided to change from wellhead units to single-flash units, and increase the production capacity of the field. The wellhead units were dismantled and two units of 28 MW were activated in 1999 with another unit of 44.2 MW activated in 2007. Also in 2007, a binary plant of 9.2 MW began production at Berlin.

The current installed capacity in El Salvador is 95 MW in Ahuachapán and 109.4 MW in Berlin, for a total of 204.4 MW.

23.1.4.3   Future geothermal development in El Salvador

LaGeo envisions improvements to the power generation in each of the fields in operation. In Ahuachapán, plans are to increase the capacity of Unit 3 by an extra 5 MW, and in Berlin, it is planned to install a new unit of 28 MW and a binary plant of 5.7 MW [13].

Table 23.6 **Geothermal plants in El Salvador [12]**

Plant name	Power (MW)	Owner	Start-up	Shut-down
Ahuachapán				
Unit 1	30	LaGeo	1975	
Unit 2	30	LaGeo	1976	
Unit 3	35	LaGeo	1981	
Berlin				
WHU-1	5	LaGeo	1992	1999
WHU-2	5	LaGeo	1993	1999
Unit 1	28	LaGeo	1999	
Unit 2	28	LaGeo	1999	
Unit 3	44.2	LaGeo	2007	
Unit 4	9.2	LaGeo	2007−08	

SIGET used to be in charge of awarding the geothermal concessions within the country, but in 2013 legislation was changed and now the National Congress of El Salvador awards geothermal concessions [13].

The main identified high-enthalpy geothermal areas in El Salvador that do not have power plants installed are shown in Table 23.7. In all these areas, some reconnaissance studies have been carried out, but more geothermal studies (prefeasibility and feasibility studies) are needed to verify if the estimated potentials are correct.

Table 23.7 **Identified high-enthalpy geothermal areas in El Salvador [14]**

Geothermal zone	Est. potential (MW)
Chinameca	76
San Vicente	117
Caluco	15
Coatepeque	70
Chambala	26
Chilanguera	11
Olomeca	11
Conchagua	13

LaGeo Branch San Vicente 7 Inc. is the owner of two geothermal concessions, San Vicente and Chinameca. Three exploratory wells were drilled in San Vicente during 2004 and 2007, with no important results (temperatures between 150 and 250°C). Between 2012 and 2013, two more wells were drilled and after testing them, it was found that one of them had the capability to produce 7 MW, which encouraged continued drilling in order to install a unit of about 30–40 MW. In Chinameca, two exploratory wells were drilled during 2009 and 2010, one of which could produce 7 MW. Recently a new well was drilled, but even though the temperature found was around 240°C, the permeability was poor [13].

23.1.4.4 Constrained potential for geothermal growth

Even though there are many geothermal areas identified in El Salvador, only six exploratory wells have been drilled: three in Chinameca and another three in San Vicente. Feasibility studies shall be carried out in order to understand the geothermal characteristics of these zones.

The lack of studies in El Salvador is due in part to the dispute between the government of El Salvador and EGP mentioned earlier. When EGP entered into LaGeo as strategic partner, part of their obligation was to prepare feasibility studies of two geothermal areas, including exploratory wells. The feasibility studies led to EGP installing Unit 3 in Berlin geothermal field and increasing production at Ahuachapán by ~20 MW.

Fearing that additional investment by EGP would put in jeopardy the status of LaGeo as a majority government-owned entity, subsequent Salvadoran governments rejected EGP's offer of continued financing of the expansion program in exchange for more shares. Thus in 2008, there began a series of lawsuits and countersuits that resulted in a final settlement in 2014 when CEL and EGP agreed that CEL would buy back all EGP's shares in LaGeo for $284 million, and EGP dropped its lawsuit in ICSID.

Geothermal generation in El Salvador stalled at around 1420 GWh/year between 2008 and 2015 and is unlikely to increase in the next 2 years. Recent drilling campaigns in Ahuachapán, San Vicente, and Chinameca have not yielded additional steam production.

23.1.5 Guatemala

23.1.5.1 Introduction

Similar to previous countries in this chapter, the development of geothermal resources in Guatemala began in the mid-1970s as a consequence of increasing oil prices. In that decade, there were initial studies in some of the geothermal areas such as Moyuta, Zunil, and Amatitlán. In Moyuta, a prefeasibility study was done in the 1970s, and in the 1980s a national geothermal inventory was carried out with the collaboration of OLADE. Prefeasibility studies in Zunil and Amatitlán were also completed, as well as initial studies in Tecuamburro. A prefeasibility study was completed in Zunil 2 (1989−92) and identified the drill sites to continue the development in this field [15].

In the 1990s, the feasibility study in Zunil was finished and the first private geothermal plant was installed. This unit is a 24-MW binary plant, which came online in 1999. In Amatitlán, the feasibility study was also finished and a wellhead unit of 5 MW was installed, while at the same time a prefeasibility study in San Marcos was completed. In addition, preliminary studies were carried out in another area called Totonicapán [15].

In 2007, a private company installed a 22-MW binary plant in the Amatitlán geothermal field and more studies in the Tecuamburro geothermal zone were conducted [15].

23.1.5.2 Geothermal plants in operation in Guatemala

The geothermal plants in operation in Guatemala are listed in Table 23.8. Production began in 1998 with the installation of a wellhead unit in Amatitlán. Wellhead Unit 1 was in operation until 2001, when it was dismantled and replaced by another wellhead unit that was installed and operated by INDE from 2002 to 2006. Wellhead Unit 2 belonged to CFE and had operated at the Miravalles geothermal field (Costa Rica) for a brief period; see Table 23.4. The 22-MW binary plant at Amatitlán belongs to Ortitlán under a Build-Operate-Transfer (BOT) contract until 2027. Expansion of the field is expected depending on its ultimate capacity. At Zunil, the 31.5-MW binary plant is comprised of seven 4.5-MW modular units, where Orzunil initially maintained a 25-year BOT contract that was later extended by 10 years. The current installed capacity in Guatemala is 31.5 MW in Zunil and 22 MW in Amatitlán, for a total of 53.1 MW.

23.1.5.3 Future geothermal development in Guatemala

INDE is planning to develop another sector of the Zunil geothermal field, called Zunil 2, and located in the Quetzaltenango region, around 230 km west of Guatemala City and only a few kilometers east of the Zunil 1 plant. Although tectonically the two fields are located inside the Quetzaltenango Caldera, they are two different reservoirs,

Table 23.8 Geothermal plants in Guatemala

Plant name	Power (MW)	Owner	Start-up	Shut-down
Zunil 1	31.5	Orzunil, BOT until 2034	1999	
Amatitlán: Wellhead 1	5	Ingenieros Civiles Asociados, BOT	1998	2001
Amatitlán: Wellhead 2	5	INDE	2002	2006
Amatitlán 1	22	Ortitlán, BOT until 2027	2007	

Table 23.9 Identified undeveloped geothermal areas in Guatemala [15]

Geothermal area	Est. potential (MW)	Concession holder
Zunil 2	50	
Tecuamburro	50	
San Marcos	24	
Moyuta	30	
Joaquina	–	Centram Geothermal Inc.
Atitlán	–	Centram Geothermal Inc.
Gloria	–	Recursos del Golfo, S.A.
La Chinita	–	Recursos del Golfo, S.A.

divided by a fault that runs parallel to the Samalá River. Prefeasibility studies were conducted in this area (1989–92), as well as scouting for the location of the possible drilling targets. So far, three exploratory wells (slim holes) have been drilled, which confirmed a reservoir with temperatures of around 300°C, and allowed for an initial estimation of the geothermal potential in this zone of 50 MW. In 2003, a deep (1928 m) geothermal well was drilled, but the well lacked permeability. Two more deep wells were scheduled to be drilled, but due to social problems with the neighbors, geothermal exploration in this zone was stopped [15].

The identified geothermal areas in Guatemala currently without power plants are shown in Table 23.9. Some studies have been carried out in Zunil 2, San Marcos, Moyuta, and Tecuamburro, but feasibility studies must be completed to estimate the potential in each area.

23.1.5.4 Constrained potential for geothermal growth

Between 1970 and 1990, there was strong support from the government for the exploration of geothermal resources of the country. This important support was lost once the government of Guatemala decided to privatize some of its institutions and/or allow private companies to participate in the development of geothermal resources. Companies such as Orzunil in 1999 and Ortitlán in 2007 have been in control of the development of geothermal resources in Guatemala.

INDE is exploring options in order to continue geothermal development in Guatemala. On April 28, 2015 the Forum in Geothermal Energy was held by INDE to show to some of the government institutions, as well as NGOs, the great benefits of geothermal energy. INDE is definitely interested in developing Zunil 2, but to do so INDE will have to settle the social issues in the region, as well as in the preparation of its personnel to continue the development of the geothermal resources in the country.

23.1.6 Honduras

23.1.6.1 Introduction

Geothermal exploration in Honduras began in the late 1970s, when Geonomics, Inc. began preliminary investigations. These investigations pointed to the areas of Pavana and Choluteca as possible areas of geothermal interest. The UNDP provided the necessary funds for Honduras to carry out more geothermal studies. These studies identified: Pavana, Sambo Creek, Azacualpa, Platanares, El Olivar, and San Ignacio as geothermal zones. These areas were later reviewed by LANL from the United States, together with ENEE. Simultaneously, the UNDP provided more funds so that Honduras could continue its investigation into its geothermal potential. More studies were carried out in El Olivar, San Ignacio, and Azacualpa, all of which are located in the central region of the country. Three exploratory wells were drilled in Platanares, two in Azacualpa and one in San Ignacio [16].

23.1.6.2 Identified geothermal areas

The identified geothermal areas in Honduras are shown in Table 23.10.

In 1988, LANL carried out geological and geochemical studies, which aided in estimating geothermal potential. In Platanares, Azacualpa, and Pavana, the prefeasibility studies have been completed, whereas in San Ignacio and Sambo Creek, only reconnaissance studies were done. Nevertheless, more prefeasibility and feasibility studies are needed to establish the real potential in all these areas [16].

23.1.6.3 Future geothermal development in Honduras

GeoPower, the owner of the concession in Platanares (as well as other geothermal zones), has decided to carry out more preliminary studies in order to identify the geothermal anomaly at Platanares. In 2014, GeoPower made a 5-year BOT agreement with Ormat to develop the geothermal field. By early January 2015, four geothermal wells had been drilled, with one well producing and three that lacked permeability. Thus, more studies have to be carried out in order to identify the geothermal anomaly before drilling the next exploration well. The initial plans anticipated an 18-MW plant by 2016 and a 17-MW plant by 2017, but these plans assumed a successful initial drilling phase.

Table 23.10 **Identified geothermal areas in Honduras [16]**

Geothermal area	Est. potential (MW)	Concession holder
Platanares	48	GeoPower
San Ignacio	20	–
Azacualpa	22	GeoPower
Pavana	11	GeoPower
Sambo Creek	15	–

23.1.6.4   Constrained potential for geothermal growth

There are a number of obstacles in regards to Honduras' geothermal development. From Table 23.10, the geothermal resources in Honduras are very small in comparison to the identified fields in neighboring countries. The resource in Honduras is unlike the volcanic systems found in Guatemala, El Salvador, and Nicaragua and more like the basin and range province in the western United States. Thus, the government of Honduras decided against searching for funds to develop their geothermal developments from international geothermal companies. The government hopes that geothermal development of the country can be done by private companies since the government itself does not have the interest, the funds, or the personnel to develop their own geothermal fields. So far only GeoPower, a private company, has been involved with carrying out the rest of the geothermal studies in Honduras and hopes to produce geothermal energy in the coming years.

23.1.7   Nicaragua

Nicaragua is highly active geothermally, mainly due to the volcanic mountain chain that runs parallel to Pacific coast. This chain is composed of 16 active volcanoes, lagoons, residual volcanic structures, volcanic calderas, and many areas with hydrothermal activity, which are very active, indicating the presence of magmatic bodies with high geothermal potential [17].

23.1.7.1   Introduction

Early in the 1980s, OLADE promoted the execution of new geothermal investigations in Nicaragua, which were done by Geotérmica Italiana. This company completed a new study called "Reconnaissance of the Geothermal Resources of the Republic of Nicaragua," and geothermal studies were conducted in the Maribios mountain chain, primarily in the area between the Momotombo and Telica volcanoes. From 1999 to 2001, a study was carried out on the geothermal resources in the Maribios mountain chain in order to identify the high-temperature fields and evaluate their potential utilizing volumetric methods. The results were published in the Nicaraguan Master Plan in 2001 [17].

23.1.7.2   Geothermal plants in operation

The geothermal plants in operation in Nicaragua can be seen in Table 23.11.

The first two units at Momotombo were single-flash plants. In late 2002, two binary units were installed in the Momotombo geothermal field. The San Jacinto-Tizate geothermal field initially had two wellhead units installed (2005−11) to prove the existence of a commercially exploitable geothermal reservoir while the construction of two single-flash units of 36 MW were built.

The current installed capacity in Nicaragua is 78 MW (in Momotombo) and 72 MW (in San Jacinto-Tizate), for a total of 150 MW.

Table 23.11 Geothermal plants in operation in Nicaragua [17]

Plant name	Power (MW)	Concession holder	Start-up year
Momotombo			
Unit 1	35	Momotombo Power Company	1983
Unit 2	35	Momotombo Power Company	1989
Binary Unit 1	4	Momotombo Power Company	2002
Binary Unit 2	4	Momotombo Power Company	2002
San Jacinto-Tizate			
WHU-1	5	Polaris Energy de Nicaragua, S.A. (PENSA) (Ram Power Corp.) (Ram Power Corporation)	2005
WHU-2	5	Polaris Energy de Nicaragua, S.A. (PENSA) (Ram Power Corp.) (Ram Power Corporation)	2005
Unit 1	36	Polaris Energy de Nicaragua, S.A. (PENSA) (Ram Power Corp.)	2012
Unit 2	36	Polaris Energy de Nicaragua, S.A. (PENSA) (Ram Power Corp.)	2013

23.1.7.3 Future geothermal development

The geothermal areas that were identified from the OLADE reconnaissance study in 1980, as well as from the Nicaraguan Master Plan in 2001, and that do not have power plants are shown in Table 23.12.

Numerous geothermal areas have been identified in Nicaragua. At the very least, reconnaissance studies have been conducted in each of these areas. More prefeasibility and feasibility studies need to be completed in order to verify the real potential in these geothermal zones.

The government of Nicaragua decided some years ago to allow private companies to continue the development of the country's geothermal resources. This is evident in the geothermal concessions that the government has approved for the two developed geothermal fields (Momotombo and San Jacinto-Tizate). Currently the company having the concession at Momotombo is Momotombo Power, but before this company, Ormat Momotombo Technologies S.A. had the concession for over 5 years. The concession holder for the San Jacinto-Tizate field is Polaris Energy Nicaragua, S.A. (PENSA), a subsidiary of RAM Power Corporation.

Three more concessions have been awarded in other geothermal areas and they are summarized in Table 23.13.

Geotérmica de Nicaragua (Geo-Nica) was a consortium formed by EGP (60%) and LaGeo (40%). However, this company was liquidated following the settlement of the ENEL-CEL dispute; see Section 23.1.4.1.

Table 23.12 **Identified geothermal areas in Nicaragua [17]**

Geothermal area	Est. potential (MW)	Completed study
El Hoyo-Monte-Galán	159	Prefeasibility
Managua-Chiltepe	111	Prefeasibility
Casita-San Cristóbal	225	Reconnaissance
Volcán Cosigüina	106	Reconnaissance
Volcán Telica-El Ñajo	78	Reconnaissance
Tipitapa	9	Reconnaissance
Caldera de Masaya	99	Reconnaissance
Caldera de Apoyo	153	Reconnaissance
Volcán Mombacho	111	Reconnaissance
Isla Ometepe	146	Reconnaissance

Table 23.13 **Geothermal concessions in Nicaragua [17]**

Geothermal area	Concession holder	Est. potential (MW)
El Hoyo-Monte Galán	Geotérmica de Nicaragua (Geo-Nica)	159
Managua-Chiltepe	Geotérmica de Nicaragua (Geo-Nica)	111
Casita-San Cristóbal	Consorcio Cerro Colorado Power	225

Two commercial wells in El Hoyo-Monte Galán and one slim hole in Managua-Chiltepe geothermal areas were drilled [22], but these wells did not identify the geothermal anomaly since they lacked permeability. Both concessions were renounced by Geo-Nica in 2011, and Mangua-Chiltepe was subsequently awarded to Alba Geotermia, an ALBA Group company.

In Casita-San Cristóbal, Cerro Colorado Power drilled a slim hole to a depth of 850 m and found the steam cap of a geothermal system. This confirmed the commercial-size potential for a single-flash development in the field [17].

Besides the development that has been taking place in the concession areas, such as drilling of both commercial and slim hole wells, the Nicaraguan government has recently accepted a new geothermal study from JICA in the Caldera de Apoyo and Volcán Mombacho geothermal areas, that will provide a better understanding of the potential of these two areas.

23.1.7.4 Constrained potential for geothermal growth

Due to the devastating 1983 earthquake and the civil war, the government of Nicaragua could not continue the development of their geothermal fields and decided to turn to private concessions. Therefore, concessions for the two producing fields

(Momotombo and San Jacinto-Tizate) as well as concessions for the development of El Hoyo-Monte Galán, Managua-Chiltepe, and Casita San Cristóbal were awarded.

The government soon realized that the private sector was not developing the geothermal resources in Nicaragua as quickly as expected and began looking for options to mitigate drilling risk in the exploration phase, as well as attempting to obtain financial resources to carry out more studies that could provide more information to the private sector about the possibilities of development. Unfortunately for the government of Nicaragua, these actions have failed thus far to attract more private companies to invest in the Nicaraguan geothermal resources.

Nevertheless, the initial studies carried out so far by all means show that Nicaragua has a huge geothermal potential.

23.1.8 Panama

23.1.8.1 Introduction

In 1971, the government of the Republic of Panama initiated studies into the thermal manifestations in the country, but there was no methodology in place and no data with which to compare. In 1981, IRHE, together with OLADE, carried out an evaluation of the existing data and developed programs of Geoscientific Studies with Geothermal Criteria. In March 1983, contracts were signed between the government of Panama and the IDB to carry out these programs, which were completed almost 15 years later.

From that point on, there have been several efforts to develop geothermal energy with the goals of reducing fossil fuel imports, avoiding increasing oil prices, and producing clean energy for the population. In 2000, ETESA with the support from SEN-ACYT and the cooperation of OIEA began studies to characterize geothermal resources with the application of new technologies (isotopic hydrology), representing the most recent efforts to evaluate the geothermal resources of Panama [18].

23.1.8.2 Identified geothermal areas

Several estimations of Panama's geothermal potential have been made by institutions, international experts, and international organizations such as IGA and GEA. In August 2006, West JEC, while carrying out consulting services for JBIC under the Plan Puebla—Panama program, presented the most recent estimation of the geothermal potential in Panama (Table 23.14). Fig. 23.9 shows the location of the identified geothermal areas in Panama.

23.1.8.3 Future geothermal development

Since the prefeasibility studies in the identified areas have not been completed, the government of Panama is preparing the Terms of Reference to carry out a study that will better identify their geothermal potential, as well as a prefeasibility study for one geothermal area. In the Barú-Colorado area, Centram Geothermal Inc. plans a development of 5 MW [18].

Table 23.14 Identified geothermal areas in Panama [18]

Geothermal area	Est. potential (MW)	Type of study
Barú-Colorado	24	Geoscientific studies above ground
El Valle de Antón	18	Geoscientific studies above ground
Chitra − Calobre	−	Geoscientific studies above ground
Isla de Coiba	−	−
Tonosí	−	−

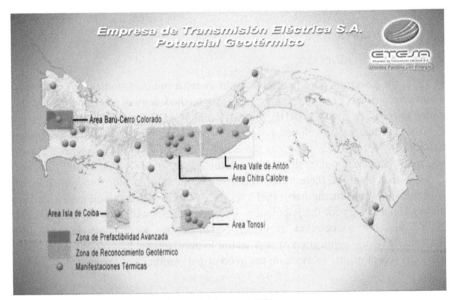

Figure 23.9 Identified geothermal areas in Panama [18]

23.1.8.4 Constrained potential for geothermal growth

The main obstacle to develop Panama's geothermal resources is the perception that they do not seem expansive enough to justify investment by the government or private companies. The high risk in the initial phases of geothermal energy development in terms of finances and human resources is such that neither party is prepared to take.

23.2 South America

23.2.1 South America general information

A map of South America is shown in Fig. 23.10.

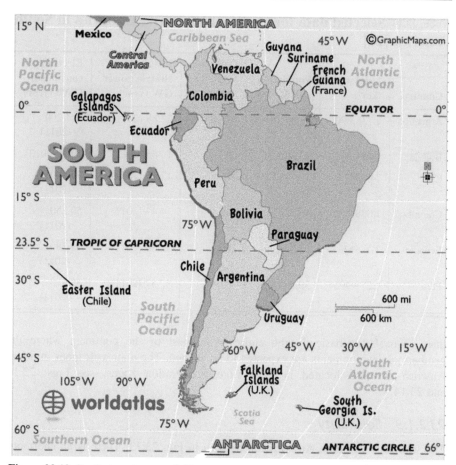

Figure 23.10 South America map [19].

The countries near and parallel to the western coast along the Pacific Ocean are those with the possibility to develop geothermal resources. These countries are Argentina, Bolivia, Chile, Colombia, Ecuador, and Peru; see Fig. 23.10.

Table 23.15 gives some pertinent geographic and economic information on the countries that hold geothermal potential.

23.2.1.1 Tectonic plates

The most important tectonic plates in South America are the Nazca and South American plates; see Fig. 23.11. The interaction of these two plates led to the formation of the volcanoes in South America, principally in three defined sectors.

23.2.1.2 Seismic hazard

The seismic hazard in the southern Pacific Coast area is shown in Fig. 23.12. It is important to note that the greater ground accelerations range between 3.2 and more

Table 23.15 Selected data for geothermally active countries in South America [20]

Country	Land area (km^2)	Population (2015)	GDP 10^9, US$ (2014)	Installed electric power, GW (year)	Electricity consumed, GWh (year)
Argentina	2,736,690	43,431,886	947.6	32.88 (2011)	114,200 (2011)
Bolivia	1,083,301	10,800,882	69.9	1.365 (2012)	6944 (2012)
Chile	743,812	17,508,260	409.3	17.95 (2011)	57,890 (2011)
Colombia	1,038,700	46,736,728	640.1	14.47 (2011)	50,250 (2011)
Ecuador	276,841	15,868,396	180.2	5.336 (2011)	19,380 (2011)
Peru	1,279,996	30,444,999	371.3	8.557 (2011)	35,710 (2011)

than 4.8 m/s^2 for basically the entire west coast of the continent where the geothermal developments are expected to be located. The main volcanoes in South America are also located in these high acceleration zones; see Figs. 23.13 and 23.14.

23.2.1.3 Seismicity

The seismicity in South America can be seen in Fig. 23.13. There is an enormous amount of seismicity, with depths ranging from 0 to 800 m below sea level, mainly along the west coast of the continent.

23.2.1.4 Volcanoes

There are 199 volcanoes in South America, and several of them may be related to future geothermal developments. A breakdown of the number of volcanoes in each country and/or region is given in Table 23.16.

As mentioned, the volcanoes are located in three main sectors, northern, central, and southern; these sectors may be seen in Fig. 23.14.

23.2.2 Argentina

23.2.2.1 Introduction

Argentina is the only country in South America that once had an operating power plant. The area where it was located, Copahue, is one of the main areas under consideration for further development. Since 1995, a number of reports have been published

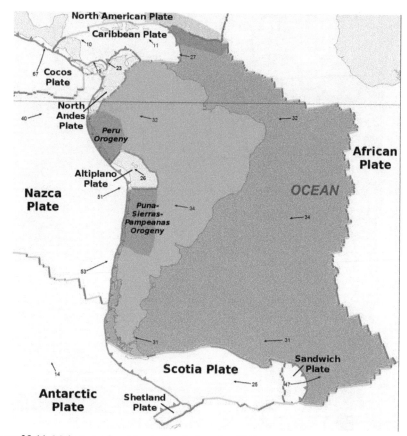

Figure 23.11 Main tectonic plates in South America [21].

on the National Geothermal Plan [26–29]; the objectives are the usual ones for early exploration:

1. To expand knowledge of the thermal resources of the country,
2. To develop the thermal cartography of Argentina
3. To systematize the information by means of catalogs and directories, and
4. To establish an adequate legal framework, in order to generate an economic takeoff.

23.2.2.2 Identified geothermal areas

The main identified geothermal areas were reported to be the Puna and Cordillera Principal [28]. In these locations, the following geothermal areas were identified:

Copahue geothermal area

This is located at the western border of Argentina, near Chile, in the Neuquén Province. Characteristics include vapor-dominant reservoir at 1200-m depth; temperature

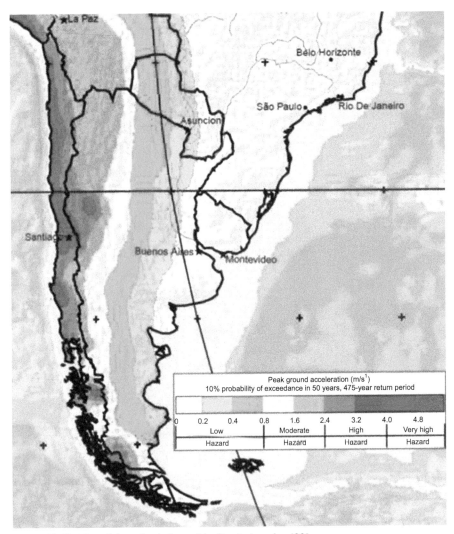

Figure 23.12 Map of the seismic hazard in South America [22].

of the reservoir at commercial depth between 230°C and 240°C; and area could support a 30-MW geothermal power plant for 30 years. The government called for a public bidding for private investors who would be interested in building and operating a future plant of 30 MW called "Las Mellizas of Copahue" with an investment of about US$70−100 million.

Geothermal One, a Canadian company, was awarded with the tender. However, according to the secretary of energy [29] in August 2013, for various reasons, the company decided to back out.

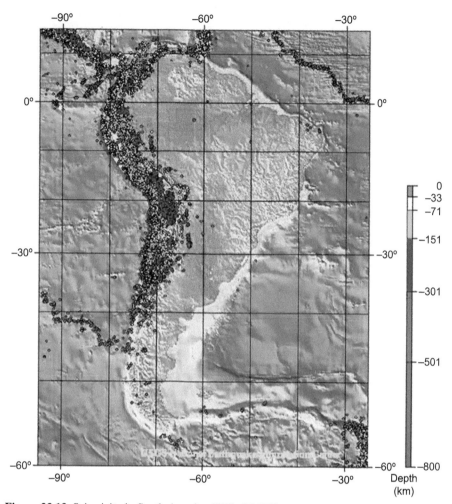

Figure 23.13 Seismicity in South America 1975–95 [23].

Domuyo geothermal area

This is located to the north of the Neuquén Province. After several geoscientific studies, the temperatures ranged from 214 to 233°C. The next step would be drilling exploratory wells [28]. A bid process was made, but there have been no positive results thus far [29].

Tuzgle-Tocomar geothermal area

This is located in the Central Puna, inside the Altiplano Salteño-Jujeño. Tocomar sector has the highest temperatures, between 132°C and 142°C. No significant progress has been made in the development of this geothermal area [29].

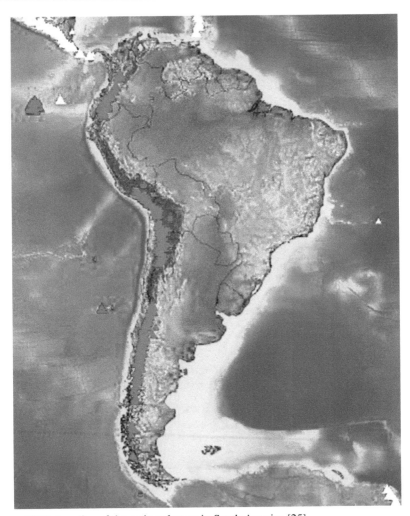

Figure 23.14 Location of the main volcanos in South America [25].

Table 23.16 **Volcanoes in South America [24]**

Country or region	Number of volcanos
Colombia	18
Ecuador	27
Peru	16
Northern part of Chile, Bolivia, and Argentina	63
Central part of Chile and Argentina	45
Sothern part of Chile and Argentina	30

Termas de Río Hondo geothermal area

This is located on the border of the central-western province called Santiago del Estero, where a strong anomaly of heat was found. Some geoscientific studies have already been done to site the location of a deep, $\sim$2000-m exploratory well [29].

23.2.2.3 Future geothermal development in Argentina

The experimental plant Copahue I was completed in 1988 with Japanese capital and produced 0.67 MW. Currently, this plant is out of service. The construction of a 100-MW geothermal plant called Copahue II (Neuquén) is being considered by the government, but several financial and technical issues remain to be defined. It was estimated that the plant could provide electricity for some 15,000 inhabitants and that the cost would be around US$600/kW [29].

23.2.2.4 Constrained potential for geothermal growth

Argentina, as with most South American countries, has encountered several problems in developing their geothermal resources. The main obstacles are financial and technical issues, but politics also comes into play when defining the role of the government in such developments. Thus, no major geothermal developments are expected in Argentina since the geothermal potentials of the known geothermal areas are technically unknown and not attractive to private companies.

23.2.3 Bolivia

23.2.3.1 Introduction

Since the 1970s, ENDE has investigated the possibility of developing geothermal energy in Bolivia by carrying out several studies; their topics and the year in which they were prepared are:

- Geothermal Manifestations Reconnaissance Study (1971–79)
- Drilling of five geothermal wells (1987–89)
- Laguna Colorada Prefeasibility Study done by ENEL GP (1990–91)
- Estimation of the geothermal potential by the Federal Commission of Electricity (1996–97)
- Geological, geochemical, and geophysical studies at the Sol de Mañana geothermal area done by JETRO (ENDE-JICA)
- Environmental analysis as well as economic viability to develop the Sol de Mañana geothermal area done by West JEC (ENDE-JICA).

Based on these studies, the government of Japan, through JICA, offered Bolivia a loan to develop the Sol de Mañana geothermal field (Laguna Colorada).

23.2.3.2 Identified geothermal areas

Bolivia has devoted much effort to geoscientific studies aimed at developing the Sol de Mañana area (near Laguna Verde, close to the southern border between Bolivia

Figure 23.15 Sol de Mañana geothermal area (geothermal well).
Photograph by Paul Moya Rojas.

and Chile) and is planning to develop a geothermal project here in the near future
(Fig. 23.15).

Besides Sol de Mañana (Fig. 23.15), Bolivia has identified two other geothermal
areas. The first is Sajama, located close to Oruro city, and the second is Empexa,
Department of Potosí. But no major scientific studies have been carried out.

23.2.3.3 Future geothermal development

The government of Bolivia submitted its application to obtain a loan from JICA, in
order to develop the Sol de Mañana resource. This loan was awarded by JICA during
the second half of 2014. The government of Bolivia has decided to initially install
100 MW in two stages of 50 MW each, although the geothermal potential is probably
greater than 100 MW.

23.2.3.4 Constrained potential for geothermal growth

Bolivia has the potential to develop geothermal energy, at least at Sol de Mañana, but it
has been unable to carry out any geothermal developments for several reasons:

- Lack of financial resources and technical knowledge
- Until recently, lack of commitment from the government in the development of the
  geothermal energy
- There has not been a geothermal department until recently, so that this group could have a
  person in charge of the geothermal development in the country, as well as guide the preparation of technical personnel in several geothermal disciplines

- Definition of property rights regarding geothermal developments
- Low ambient temperatures, ~6°C in summer time, and high altitude, ~4900 masl, at Sol de Mañana, and
- The need to develop infrastructure near the geothermal areas, including roads, work camps, electricity, telephone, Internet, source of water to drill geothermal wells, etc.

23.2.4 Chile

23.2.4.1 Introduction

Interest in geothermal energy in Chile goes back to the beginning of the last century. In 1908, members of the Italian colony at Antofagasta city created a private society, the Preliminary Community of El Tatio. This society carried out the first geothermal exploration program in the country. During 1921–22, an Italian technical group from Larderello drilled two wells of about 70–80 m depth [30].

Systematic exploration in the northernmost region of the country was summarized by the end of 1968 as result of a project funded by CORFO and UNDP. Geologic and geochemical reconnaissance studies of many hot-spring areas and detailed geological, geophysical, and geochemical surveys in selected areas such as Suriri, Puchuldiza, and El Tatio geothermal areas were performed from 1968 to 1976. These were followed by drilling a number of exploratory wells and prefeasibility studies for power generation at El Tatio and Puchuldiza [31].

Up to 2000, numerous volcanological, geochemical, and geophysical studies were conducted at many sites by numerous entities, including the University of Chile, the National Geological Survey, ENAP, UNOCAL, CFG, and others (eg, Refs [32–34]). One outcome was an exploratory well in the Nevados de Chillán geothermal area that encountered wet steam with a temperature of 198°C [35].

In January 2000, a geothermal law was enacted that provided the framework for the exploration and development of geothermal energy that establishes exploration and exploitation concessions, which are granted by the Ministry of Mines [36].

From 2000 to 2003, a 4-year geothermal research project was conducted by the Geology Department of the University of Chile, in association with ENAP, in the south central part of the country to characterize and assess the geothermal resources and their possibilities for power generation and direct applications. Several sites were selected as promising for electrical generation, including Nevados de Chillán and Puyehue-Cordón Caulle. During the same period, a joint venture between ENAP and COLDECO explored northern Chile (as well as some of southern Chile). There was no follow-up to all these investigations due to lack of support and commitment from any lead institution, research center, or permanent program [37].

During 2005–10, there was some support from the government encouraging the development of geothermal energy. SERNAGEOMIN carried out detailed geological studies in those geothermal areas for which geothermal maps were not available. A cofinancing tool allows for subsidies of up to 50% (not exceeding US$60,000) of the cost of studies and up to 2% of the total estimated investment [36].

23.2.4.2   Identified geothermal areas

Country-wide geological and geochemical reconnaissance together with detailed geological, geochemical, and geophysical surveys in selected areas yield a preliminary assessment of the geothermal potential of about 16,000 MW for 50 years [32]:

Puchuldiza geothermal area

This is located in the Tarapacá Region of northern Chile, 160 km northeast of the city of Iquique. Results include five slim exploration wells drilled between 1975 and 1977 to depths from 428 to 1013 m and encountered feed zones shallower than 600-m depth with temperatures greater than 175°C.

Apacheta geothermal area

This is located in the Antofagasta Region in northern Chile, 120 km east of Calama city. Results include being accidentally discovered by CODELCO [35]; surface manifestations are two superheated fumaroles atop Cerro Apacheta; gas geothermometry gives reservoir temperatures >250°C; and magnetotelluric (MT) and time domain electromagnetic (TDEM) show a low resistivity boundary (<10 Ω m) extending over an area of 5 km^2 [38].

El Tatio geothermal area

This is located near the Bolivian border, 95-km east of Calama city and the copper mine of Chuquicamata. Results include that it lies in a north-south down-faulted block (El Tatio Graben) [40]; Schlumberger surveys show a low resistivity anomaly (<10 Ω m) reaching over 30 km^2 and extending farther southeast beyond the limits of the graben; and 13 slim wells drilled from 1969 to 1974 reached temperatures of 252–256°C in some of the wells. A technical and economic feasibility study for a 15–17-MW power plant was made in 1975 by ELC-Electoconsult [46].

Calabozos geothermal area

This is located 240-km southeast of Santiago, in the central-southern zone of Chile, near the border with Argentina. Results include promising possibilities pointed out [47] and [48], based on large volumes of erupted silicic magma, high flow rates of the thermal springs (>240 L/s), and the estimated subsurface temperatures >250°C; hot springs range from chloride and chloride-sulfate to bicarbonate types, suggesting a liquid-dominant geothermal system; and MT-TDEM geophysical surveys, plus water and gas geothermometry, indicate subsurface temperatures 235–300°C [35].

Nevados de Chillán geothermal area

This is located 76-km southeast of the city of Chillán, in central south Chile. Results include being associated with 13-km long, northwest-trending volcanic chain composed of polygenetic and flank volcanoes and calderas [49]; existence of a shallow bicarbonate aquifer is supported by a 274-m-deep exploratory well drilled by CFG and ENAP in La Termas that tapped a wet steam feed zone at 198°C; inferred chemistry of the parent fluid was a mixed chloride-bicarbonate aquifer with total dissolved solids (TDS) <1000 ppm; and solute geothermometry suggests subsurface temperatures >200°C for the shallow aquifer [35].

Cordón Caulle geothermal area

This is located 65-km east of the city of Osorno, in southern Chile. Results include being associated with a 15-km-long, 5-km-wide volcanic graben bordered by northwest-trending faults [37]; Na-K and pH-corrected silica temperatures range from 170 to 180°C, interpreted as subsurface temperatures of a secondary steam-heated aquifer overlying a main vapor-dominant system; and gas geothermometry suggests temperatures of 225−300°C for the deep reservoir [50].

23.2.4.3 Future geothermal development

By the end of 2009, 38 concessions were under regional reconnaissance with the objective to narrow the focus and identify areas of potential interest [44]. In the northern volcanic-geothermal zone, six geothermal prospects are under exploration by ENG (ENAP-ENEL), Energía Andina, a joint venture between ENAP and Antofagasta Minerals S.A., and by other companies related with cooper mining. The areas being explored were Suriri, Puchuldiza, Lirima, Apacheta, and El Tatio. At Apacheta and El Tatio geothermal areas, four geothermal wells were drilled to depths close to 1700 m [36].

By 2015, in the northern Chile geothermal zone, about 90 hot-spring areas had been identified [43] and 45 exploration concessions were being surveyed. In this region, the most advanced exploration programs were Colpitas, Apacheta, Pampa, Lirima, and El Tatio-La Torta geothermal prospects [38−40]. Exploratory wells were drilled in all four of these prospects.

In the central-southern volcanic-geothermal zone, seven geothermal prospects were under exploration by ENG, the University of Chile, and some private companies. The exploration programs were undertaken in Tinguiririca, Calabozos, Laguna del Maule, Chillán, Tolhuaca, Sierra Nevada, and Puyehue-Cordón Caulle. Exploration slim holes have been drilled at Calabozos, Laguna del Maule, Chillán, and Tolhuaca [36].

By 2015, in the central-southern Chile geothermal zone, more than 200 geothermal areas were identified [34] and 31 exploration concessions were under way. The most advanced exploration surveys have been completed at Tinguiririca, Calabozos, Laguna del Maule, Chillán, and Tolhuaca geothermal areas [41−44]. Exploratory wells were drilled in all these prospects [45].

Exploration concessions were granted for Laguna del Maule (Mariposa sector) and Tolhuaca (San Gregorio), where production size wells were drilled. An environmental impact assessment was submitted to the authorities for approval to install a 70-MW power plant at Tolhuaca, where well Tol-4 has an output of 12 MW [45].

Detailed exploration studies are being carried out by 14 private companies in 76 geothermal concessions areas (2015). Exploration drilling has been conducted in at least nine of these areas. Eight exploitation concessions have been awarded, and in Apacheta and Tolhuaca, the results of the environmental impact studies have been submitted to the national authorities to further develop this field with power plants.

Exploitation concessions have been granted for Apacheta and El Tatio geothermal fields, and the environmental assessment for the installation of a 50-MW power plant has been approved for Apacheta. El Tatio geothermal project is the most advanced, but recently the concession has been canceled due to the fact that company failed to comply with environmental and safety requirements. The company Geotérmica del Norte, a joint venture of CODELCO, the State Copper Company, ENAP, the State Oil Company, and ENEL, has indefinitely suspended work at this site (from Think GeoEnergy website).

23.2.4.4 Constrained potential for geothermal growth

A study done by the Natural Resources Defense Council in 2013 indicated that in order to develop the geothermal potential of Chile, it is necessary to consider the barriers and find political solutions so as to:

1. Attract investment
2. Reduce the regulatory and legal issues, and
3. Take under consideration institutional restrictions.

The following actions should be considered:

1. Actions by the government:
   a. Provide a clear direction for the geothermal development and reevaluate the support and institutional structure for the geothermal development.
   b. Investigate and provide support to the establishment of incentives or financial mechanisms to encourage the geothermal development.
   c. Reevaluate the laws and the rules on geothermal energy, mainly those related to the impact on and the participation in local communities, the concession system, and the time-table and requirements for studies on the environmental impact.
2. Actions by the developers:
   a. Guarantee transparency with the communities which are located next to the geothermal projects.
   b. Develop general standards of the industry for geothermal developments next to established communities, mainly the ones located in the northern arid regions.
   c. Demonstrate a willingness to participate in preliminary environmental studies with greater detail than the current ones.

23.2.5 Colombia

23.2.5.1 Introduction

The possibilities to develop the geothermal energy in Colombia started in the late 1970s. Colombian companies such as CHEC and GESA, as well as entities such as OLADE, IPSE, SGC (previously known as INGEOMINAS), and UPME, conducted many studies to explore the geothermal potential in Colombia. The main studies, with their names, who conducted them, and the period in which they were done, are found in the following list [51]:

- Reconnaissance Study of geothermal fields in Colombia and Ecuador (OLADE, AQUATER, BRGM, and Geotérmica Italiana, 1979–82),

- Prefeasibility studies for geothermal development in the Tufiño-Chiles-Cerro Negro area (INECEL-OLADE, 1982; OLADE-ICEL, 1986—87),
- Prefeasibility studies for geothermal development in the Nevado del Ruíz Volcano area (CHEC, 1983; Geoconsul, 1992; GESA, 1997. Drilling Nereidas-1 Geothermal Well),
- Research studies of the geothermal systems in Azufral and Cumbal volcanos areas (INGEO-MINAS, 1998—99, 2008—09; INGEOMINAS-National University of Colombia, 2006),
- Research studies of the geothermal systems of the Paipa and Iza areas (INGEOMINAS, 2005, 2008—09),
- Additional researches of the geothermal resource in the Tufiño area and the geothermal development plan in Ecuador (MEER, 2008—09),
- Feasibility studies for the generation of geothermal energy in Colombia (ISAGEN-USTDA-BPC-INGEOMINAS, 2008—09),
- Report on well PGT-1, drilled in Aguas Hediondas (MEER, 2010),
- Strategic program for modeling of the hydrothermal-magmatic system of the Nevado del Ruíz volcano (ISAGEN-UNAL-INGEOMINAS, ISAGEN, COLCIENCIAS, 2010—12),
- Prefeasibility studies to develop the Tufiño-Chiles-Cerro Negro Binational Geothermal Project (ISAGEN-CELEC EP, 2010—12),
- Modeling of the resistive structure by MT studies (ISAGEN-INGEOMINAS-CIF-UNAM-COLCIENCIAS, 2011—12),
- Prefeasibility studies of the Macizo Volcánico del Ruíz Volcanic; drilling of three TGW, between 174 and 300 m in depth (ISAGEN-BID-NIPPON KOEI-GEOTHERMAL-INTE-GRAL, 2011—12),
- Catalytic investments for geothermal energy; complementation of a resistive model, advice and support during exploratory drilling stage (ISAGEN-BID/GEF, 2011—14).

Despite all these studies, as of 2015 there is no geothermal plant installed in the country, but ISAGEN is committed to develop the geothermal resources in Colombia in the coming years.

23.2.5.2   Identified geothermal areas in Colombia

The principal identified geothermal areas are Chiles, Cerro Negro, Cumbal, Azufral, Galeras, Doña Juana, Sotará, Puracé, Nevado del Huila, Nevado del Ruíz, and Nevado del Tolima volcanoes. These are quaternary volcanos with hot springs, fumaroles, surficial hydrothermal alteration, and other thermal features, which might be evidence of a geothermal resource, possibly with adequate characteristics to support geothermal power plants [51]. These prospects together with other nonvolcanic areas that could have geothermal potential sum to an estimated geothermal potential of 2210 MW [52].

23.2.5.3   Future geothermal development in Colombia

Colombia is planning to develop renewable energy projects because they are cleaner and environmentally friendly. For this reason, Colombia has set up a National Energy Plan that contains the following objectives:

- Expand and warrant the energy provision
- Promote regional and local development
- Introduce new sources and technologies of energy generation
- Contribute to reduce the greenhouse emissions and climate change
- Promote the use of renewable energy sources.

Figure 23.16 Fumaroles at Nevado del Ruíz Geothermal Area.
Photograph by Paul Moya Rojas.

ISAGEN is studying the possibility to develop two of the identified geothermal areas, the Macizo Volcánico del Ruíz (Fig. 23.16) and the bilateral project Tufiño-Chiles-Cerro Negro complex. In both geothermal areas, environmental impact assessments as well as prefeasibility studies are being carried out. The presidents of Colombia and Ecuador signed a bilateral agreement to jointly develop the Tufiño-Chiles-Cerro Negro geothermal area. The joint effort will be between ISAGEN and CELEC EP (Fig. 23.16).

23.2.5.4 Constrained potential for geothermal growth

In Colombia, as in other South American countries, the main constraints in developing the geothermal developments are:

• The lack of financial sources and technical knowledge
• The initial institution in charge of the geothermal development was closed, forcing subsequent institutions to restart the geothermal process
• There has not been a geothermal group or unit until recently, to be in charge of geothermal development and guide the preparation of the technical personnel in several geothermal disciplines
• The need to develop infrastructure nearby the geothermal areas.

23.2.6 Ecuador

23.2.6.1 Exploration studies

Geothermal resource explorations were carried out from the mid-1970s through the early 1990s by Ecuadorian government institutions (the now defunct INECEL and

INE in Ecuador) with the aid of foreign technical assistance programs (mainly OLADE, IAEA, UNDP). Today, there are numerous agencies for geothermal energy development and regulation, including MICSE, MEER, CONELEC, CELEC EP, INER, CNRH, and MAE. Public funding has been allocated already through MEER for geothermal exploration of both high and low-temperature resources. In accordance with the constitution (2008), joint ventures are favored in government-to-government agreements where the State of Ecuador owns at least 51% [53].

The Reconnaissance Study of the Geothermal Resources of Ecuador, carried out from 1979 to 1980, began geothermal exploration in Ecuador. It was aimed to find high-temperature hydrothermal systems along the Andes in the areas of recent volcanism, following the methodology recommended by OLADE (1978). The report (INECEL/OLADE 1980) produced by INECEL and OLADE, together with AQUATER (Italy) and BRGM (France), summarized the areas of interest in two main groups: Group A, high-temperature; Tufiño, Chachimbiro, and Chalupas, and Group B, low-temperature; Ilaló, Chimborazo, and Cuenca. Follow-up prefeasibility studies were done between 1981 and 1992, first at Tufiño with OLADE and ICEL and later at Chachimbiro and Chalupas.

The Tufiño prospect gained the first priority for exploration, and the results of advanced geological, geochemical, and geophysical surveys indicated a high-temperature resource underneath Chiles volcano (INECEL-OLADE-AQUATER, 1987). Results of scientific surface studies at Chachimbiro and Chalupas indicated the existence at depth of high-temperature resources. INECEL'S Geothermal Project was shut down in 1993 due to political reasons and financial cutbacks [54], and since then, no further government-run geothermal exploration has been done in Ecuador. A privately funded MT survey in Tufiño (Tecniseguros, 1994) confirmed the presence of a deep, high-temperature resource, but the number of soundings was insufficient to fully define the size of the reservoir.

Geothermal exploration was interrupted again in 2002 by Ecuador's internal financial crisis, but in 2007, the need to diversify the country's energy matrix became a national policy. Consequently, attention was put again on geothermal energy due to its high-capacity factor [54].

In 2008, the government, through MEER, restarted geothermal exploration, aiming to develop the former INECEL geothermal prospects for power generation. In 2010, 11 geothermal prospects countrywide were ranked: Chachimbiro, Chalpatán, Chacana-Jamanco, Chalupas, Guapán, Chacana-Cachiyacu, Tufiño, Chimborazo, Chacana-Oyacachi, Baños de Cuenca, and Alcedo.

Based on the expertise and experience that Iceland has in geothermal energy exploration and use, the government of Ecuador, through MEER, signed a memorandum of understanding with the Ministry of Industry and Tourism of Iceland in 2009 that established an institutional relationship to promote bilateral technical cooperation in matters of geothermal development between Ecuador and Iceland [54].

During 2010—15, the following actions took place: prefeasibility studies up to predrilling stage were put out to bid and contracted by CELEC EP for prospects Chachimbiro, Chacana-Jamanco, and Chacana-Cachiyacu; the bids were won by an Ecuadorean company SYR in early 2011 and final reports were handed in January

2012 for Chachimbiro and in May 2012 for both the Chacana-Cachiyacu and Chacana-Jamanco prospects. A fourth geothermal prospect, Chalpatán, was given by INP, under an agreement with CELEC EP and MEER, to CGS, a Spanish consulting company, which submitted the final report in April 2013. By the end of 2013, SYR won the bid to perform complementary exploration work in the Bi-National Tufiño-Chiles geothermal project to establish a geothermal model and eventually drill the first deep exploration well to culminate the prefeasibility stage of the project [53].

23.2.6.2 Identified geothermal areas in Ecuador

Tufiño-Chiles-Cerro Negro geothermal area
This is located in the CO (Cordillera Occidental or Western Cordillera), 35-km west of the city of Tulcán, 7-km west of the villages of Tufiño and Chiles, in the province of Carchi (Ecuador) and Nariño department (Colombia). The development area lies across the Ecuador–Colombia border. In 2009, US$1 million of state funds was allocated to the Tufiño prospect to carry out four shallow (500 m), small-diameter (76 mm) gradient bore holes, which were operated by MEER. Well PGT 1, the first geothermal hole to be drilled in Ecuador, reached a total depth of 554 m in May 2009 [55].

Many geoscientific studies have been carried out in specific areas of the Tufiño-Chiles-Cerro Negro prospect. Nevertheless, the prospect has not yet been studied integrally. Therefore, additional geological and geochemical studies are required to enhance the conceptual models of the prospect. Complementary MT and TDEM surveys will also provide a better understanding of the resistivity anomaly in the main area of the prospect. Reanalysis of geological, geochemical, and geophysical surface exploration data were endorsed to a private consulting group that is currently executing field activities. If these complementary studies are positive and a high-temperature resource is proved, feasibility studies must be undertaken to prove the resource's production capacity [54]. Beate [56] gives an estimate of 138 MW for the Tufiño prospect based on surface data geology presented by Almeida [57].

Chachimbiro geothermal area
This is located on the east slopes of the CO (Western Cordillera), about 20-km west of the city of Ibarra in the province of Imbabura. Results: estimated potential varies from 81 MW [54] to 113 MW [57]; in 2009, US$1 million of state funds was allocated for exploration activities, mainly geophysics (MT) to be operated by ESPE as a research project [55]; preliminary geoscientific studies concluded in 2012 showed a 65% probability of success. Next steps include drilling of shallow exploration wells to quantify and evaluate reservoir; a low-cost 1500-m-deep slim hole is recommended. If the results from exploration wells are positive, advanced feasibility studies must be done. The project is currently undergoing environmental impact assessment. JICA has showed interest in financing the feasibility stage.

Chacana geothermal area
This is located related to the silicic Chacana caldera (60-km east of Quito) on the CR (Eastern Cordillera) in the province of Napo; the prospect mostly located in

environmentally sensitive territory, the Antisana and Cayambe-Coca ecological reserves [53].

Exploration studies carried out between July 2011 and April 2012, covered the southern half of the Chacana caldera, and were conducted by the consulting firm SYR under a contract with CELEC EP. The studies, condensed here [58], encompassed a full scope of work in the areas of geology, geochemistry, and geophysics and culminated in the preparation of conceptual models of the geothermal resource in four areas of interest: Cachiyacu, Jamanco, Chimbaurcu, and Plaza de Armas [63].

Chacana has a good potential for hosting a geothermal reservoir at a shallow depth due to the geological conditions and rhyolite volcanic properties. The next stage consists of drilling two exploratory slim holes to depths of 600 and 900 m in hopes of intersecting the main faults inside the caldera and to reach the reservoirs in Cachiyacu and Jamanco. The project is currently undergoing environmental impact assessment. The potential expected in Jamanco is 13 MW, and that of Cachiyacu is 39 MW [54].

Chalpatán geothermal area

Due to its proximity to Tufiño—Chiles-Cerro Negro, the Chalpatán caldera was also studied by OLADE, INECEL, and ICEL from 1982 to 1987. Prefeasibility studies were completed in 2013. These studies included the use of state-of-the-art technologies, such as satellite and airborne infrared thermal imagery, audio-MT, and magnetometry. Preliminary results indicate temperatures below 120°C, making the resource more suitable for direct use rather than electricity production [54].

Chalupas geothermal area

This is located 70 km SSE of Quito, at the crest of the CR (Eastern Cordillera), in the province of Napo. Latacunga is the nearest load center at a distance of 30 km [59]. Although prefeasibility studies were carried out in Chalupas, additional research activities are required to complete the geothermal conceptual model presented by INECEL in 1983. The project has been temporarily delayed by CELEC EP and will resume once the feasibility studies are finalized in Chachimbiro. Almeida [57] determined an estimated potential of 283 MW based on surface data geology [54].

23.2.6.3 Future geothermal development in Ecuador

Recently, the Tufiño-Chiles-Cerro Negro prospect regained its status of Bi-National Geothermal Project and is managed jointly by ISAGEN of Colombia and CELEC EP of Ecuador, the former being the operator. Public funding for geothermal exploration at Tufiño-Chiles is given in equal parts by each government. Reassessment of available geological, geochemical, and geophysical information by ISAGEN-CELEC EP recommended complementary exploration surveys to properly locate the resource and to see whether a commercial resource might exist. An international contract was won by SYR in late 2013, and field work on geology and geochemistry started February 2014 on the Ecuadorean side [63].

23.2.6.4 Constrained potential for geothermal growth

An Ecuadoran delegation to a multistakeholder workshop organized by IRENA and OLADE held in Lima, Peru, in November 2013 reported that the following commitments were agreed on [54]:

1. Technical assistance from the IDB to develop a regulatory framework based on existing regulations,
2. Legal assessment provided by the National Energy Authority of Iceland in the development of new policies and regulations for a geothermal law in Ecuador; and
3. Assistance from IRENA to connect financial resources from bilateral and multilateral organizations to support the development of geothermal regulatory framework.

Taking in consideration the above recommendations, the main constraints to develop the geothermal development in Ecuador are:

- More detailed geothermal information is needed to incentivize the geothermal development for both public and private sectors.
- The lack of financial sources, technical knowledge, legislation (regulatory framework), and political decisions have delayed development.
- Some of the public institutions that had initiated the preliminary studies were closed by political decisions.
- There is no geothermal law and the government is not preparing a geothermal bill.
- Political decisions have prioritized hydroelectric and has subsidized thermal (fossil) energy, hurting the development of geothermal.
- There has not been a geothermal group or unit in charge of the geothermal development in the country.

23.2.7 Peru

23.2.7.1 Exploration studies

The first geothermal studies in Peru began in the 1970s with the first inventory of mineral and thermal springs carried out by INGEMMET. Since then, several studies have been conducted [60]:

1. In 1975, MINERO PERU carried out preliminary studies of exploration of geothermal manifestations in Calacoa and Salinas zones in Moquegua.
2. In 1976, Geothermal Energy Research of Japan conducted preliminary exploration work on Vilcanota basin in Cusco.
3. In 1977, the INIE carried out the first inventory of geothermal manifestations.
4. In 1978, INGEMMET compiled an inventory and geographic clustering of geothermal regions.
5. In 1979 and 1980, INGEMMET and AQUATER (Italy) conducted geothermal reconnaissance studies in Region V, where the geothermal region Eje Volcánico Sur is located.
6. In 1980, Geothermal Energy System Ltd. conducted geothermal studies in Calacoa and Salinas zones, located in Moquegua region and Tutupaca zone in Tacna region.
7. Between 1983 and 1985, British Geological Survey and INGEMMET made a partial inventory of geothermal manifestations of Region VI, located in the Cusco and Puno regions.

8. Between 1983 and 1986, ELECTROPERU and CESEN carried out geothermal reconnaissance studies in geothermal Regions I to IV.
9. In 1986, ELECTROPERU with technical assistance from OIEA and United Nations conducted geochemical investigations in Region V.
10. In 1994, INGEMMET, commissioned by ELECTROPERU, performed a geovolcanologic study and a systematic inventory of the geothermal manifestations in Tutupaca.
11. In 1995, INGEMMET (requested by the Especial Tacna Project) carried out the evaluation of the hydrothermal zones in Pampas de Kallapuma and its surroundings [61].
12. In 1996, CENERGIA and IIE carried out the study called "Analysis of the Geochemical Information of the geothermal zones in the Southeast of Peru" [61].
13. In 1997−2003, INGEMMET carried out the National Inventory of Thermal and Mineral Waters from Peru.

Most of these studies have been focused on the southern-volcanic-axis region (Eje Volcánico Sur), located in southern Peru where several zones with different geothermal importance have been recognized. All of this information, together with field work, allowed a new map of geothermal resources that was elaborated on by Cruz and Vargas [60]. All these elements [60] redefined the boundaries of six geothermal regions, presenting their findings in WGC 2010 in Bali.

In addition, in 2007 the Ministry of Energy and Mines of Peru through JICA managed to get the technical support of the Japanese government to develop geothermal energy in the country. Two Japanese banks financed the prefeasibility studies in Borateras and Calientes. The study was done by West JEC and resulted in the geothermal potential for the two sites that was estimated at 50 MW and 100 MW, respectively [60].

On December 28, 2009, MINEM and the government of Japan signed an agreement of technical cooperation with the idea to develop the master plan for the development of the geothermal energy in Peru [62]. As part of this project, 13 geothermal zones were evaluated (including field work), most of them located in the southern portion of the country. The master plan estimated the geothermal potential of the country at around 3000 MW corresponding to 61 geothermal fields, which are allocated along all the country, with the most promising fields located in Eje Volcánico Sur. Furthermore, as part of this project, geoscientific studies were carried out in two areas, Ancocollo and Tutupaca, with estimated geothermal potentials of 90 and 105 MW, respectively; these areas are located in the Tacna region along with Calientes and Borateras [60].

23.2.7.2 Identified geothermal areas in Peru

So far the exploration studies in Peru are at the reconnaissance and geoscientific level. Drilling of exploratory wells for geothermal exploration has never been conducted. Therefore, present data and information of any geothermal areas in Peru are not sufficient to develop a particular geothermal area [62].

Only in the two fields, Calientes and Borateras in Tacna Region, have prefeasibility studies, including the resource assessment with an MT resistivity survey, been conducted for planning of geothermal power development. The estimated geothermal power potential for the two fields was Calientes, 100 MW (80% probability), and Borateras, 50 MW (70% probability), using the volumetric method with Monte Carlo analysis [62].

23.2.7.3   Future geothermal development in Peru

The legal framework, begun in 2010, for geothermal resource development in Peru has been established that developments will be basically carried out by the private sector. Thus for development to be enhanced, the Peruvian government needs to proclaim the policy and strategy for the promotion of geothermal developments [62].

According to DGE-MEM, more than 30 geothermal authorizations for exploration have been granted, but no company has drilled any geothermal wells or exploratory or slim holes [60].

The most promising geothermal areas identified in the Master Plan for Development of Geothermal Energy in Peru [62] are listed in Table 23.17.

GQP-SAC, EDC Energía Verde S.A., and ENEL GP are in the first stage to find and develop geothermal resources. Superficial exploration campaigns (geological, geochemical, and geophysical) are mainly being carried out in southern Peru. One of the three companies, GQP-SAC, is moving into the "phase 2" of the exploration authorization, as indicated by DGE-MEM at the beginning of 2015. GQP-SAC filed their intention to continue with exploration activities and plan to drill the three wells needed by current regulations [60].

23.2.7.4   Constrained potential for geothermal growth

The main constraints to develop the geothermal development in Peru are:

- Lack of detailed geothermal information, financial sources, technical knowledge and legislation (regulatory framework), and political decisions.
- There has not been a geothermal group or unit in charge of the geothermal development.
- The Master Plan for Development of Geothermal Energy in Peru [62] provides ideas for the direction that the country should follow to develop their geothermal resources, but so far the government has not implemented any of the suggested actions, other than the ones already established by law.

Table 23.17 Most promising (priority A) geothermal areas in Peru

Region	Geothermal area	Estimated power (MW)
Tacna	Tutupaca	105
Moquegua-Tacna	Crucero	70
Moquegua	Quellaapacheta	100
Puno	Pinaya	35
Ayacucho	Puquio	30

- The government expects that the private sector will take the risks and develop the geothermal resources in the country, but at the same time, the private sector is hesitant to do so with the available geothermal information.
- For the drilling phase, the definition of the environmental instruments (studies, regulations, processes) takes too much time, probably due to the lack of knowledge related to the environmental instruments that apply to geothermal developments.
- The lack of government involvement in the risky drilling phase.
- Some geothermal areas are located inside national parks or protected areas, and others require consultation with indigenous peoples before development can take place.

23.3 Final remarks

Geothermal developments in Central and South America are radically different. While in Central America there are already six geothermal fields in operation, in South America there are none. This difference is based in the need of countries to produce a cheaper form of energy. In the 1970s, the Central American countries were forced to look for cheaper types of energy due to the expense of importing fossil fuels, while the South American countries were not in that position. This situation has not been favorable to the development of the geothermal resources in South America, since other types of energy can be cheaper to produce in some South American countries and there is no need for high investments with high risks as in the initial phase of geothermal energy development.

Fig. 23.17 describes the nature of risk and investment with time in geothermal development. The highest risk occurs in the early phases, which discourages possible geothermal development for both public institutions and private companies.

The countries in Central America were able to pass this critical initial stage because they received loans to advance geothermal developments. In this way, the countries prepared their geothermal personnel as well as their legal infrastructure to produce geothermal energy. That is, strong support from the government was necessary in order to carry out the initial geothermal development.

South American countries have not had this support from their own governments, and their geothermal resources have not developed as fast as expected. When obtaining concessions from the different countries, in many cases, private companies felt that the existing regulatory framework was not suitable for geothermal developments, and projects were halted.

Nevertheless in South America, IRENA and OLADE have launched an initiative to improve access to geothermal energy in the Andean Region. This initiative, supported with expertise from Iceland, Mexico, New Zealand, France, and the IGA, aims to contribute to the development of the vast geothermal potential in this region. Five countries are participating in this initiative: Bolivia, Chile, Colombia, Ecuador, and Peru [54].

The history of geothermal development in the Central and South America has demonstrated that government needs to play the most important position in the development of its resources. Governments interested in their geothermal development should

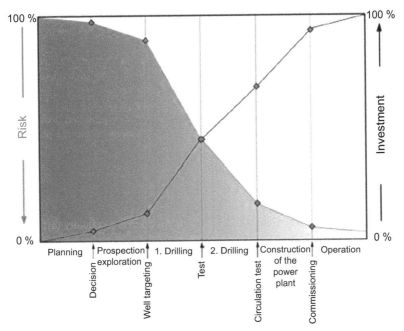

Figure 23.17 Risk of a geothermal project versus time.
GeoThermal Engineering.

invest in the critical early stage to reach the prefeasibility and feasibility stages of the
project. The support of international lending agencies is also often crucial at this stage.
Thereafter, the government can decide to do the development itself or to give it to a private
company through a concession. For this last option, the regulatory framework must
be clear and with incentives that will encourage geothermal energy development.

Acknowledgments

Special thanks to Mr. José Antonio Rodríguez (from El Salvador) for his technical revision of the
sections on El Salvador and Nicaragua. Also many thanks to my acquaintances from institutions
that provided me with geothermal information for this chapter: Enrique Lima (West JEC), Cristian
León Cruz (from Bolivia), Javier Alonso Méndez S., Julián Echeverri Ocampo and Claudia
Alfaro (from Colombia), and Gonzalo Guerrón (from Ecuador).

References

[1] Central Intelligence Agency (US) https://cia.gov.
[2] http://www.infoplease.com/atlas/centralamerica.html.
[3] http://es.justmaps.org/maps/images/thematics/volcanoes-map.gif   USGS,   Topinka,
    USGSICVO, 1997, Modified from: Tilling, Heliker and Wright, 1987, and Hamilton, 1976.

[4] http://geology.com/volcanoes/santa-maria/.
[5] http://geology.about.com/od/seishazardmaps/ss/World-Seismic-Hazard-Maps.htm#step6.
[6] http://earthquake.usgs.gov/earthquakes/world/seismicity_maps/.
[7] http://www.worldatlas.com/webimage/countrys/camvolc.htm.
[8] Moya P. Costa Rican geothermal energy development 1994–2006. In: Workshop for decision makers on geothermal projects in Central America, organized by UNU-GTP and LaGeo, San Salvador, El Salvador, 26 November–2 December; 2006.
[9] ICE. Guanacaste geothermal project, technical prefeasibility report. Rogers Engineering Co., Inc. and GeothermEx, Inc.; December 1976. 68 pages [prepared for Instituto Costarricense de Electricidad].
[10] ICE. Evaluación del Potential Geotérmico de Costa Rica. Internal Report. San José: Instituto Costarricense de Electricidad; Noviembre 1991. p. 70.
[11] Rodríguez EM, Moya P, Mainieri A. Law for the regulation of geothermal energy exploitation in national parks. In: Geothermal bill for the Costa Rican Congress; 2006.
[12] Montalvo F, Guidos J. Estado Actual y Desarrollo de los Recursos Geotérmicos en Centroamérica. San Salvador, Pisa, San José: Instituto Italo-Latino Americano (IILA); Abril 2009–Mayo 2010.
[13] Burgos J, Montalvo F, Gutiérrez H. El Salvador country update. In: Proceedings World Geothermal Congress 2015, Melbourne, Australia, 19–25 April; 2015.
[14] Campos T. The geothermal resources of El Salvador: characteristics and preliminary assessment. In: United nations workshop on the development and exploitation of geothermal energy in developing countries, Pisa, Italy; 1987.
[15] Ortiz V. Estado Actual del Desarrollo de los Recursos Geotérmicos en Guatemala. San Salvador, Pisa, San José: Instituto Italo-Latino Americano (IILA); Abril 2009–Mayo 2010.
[16] Lagos C. Estado Actual del Desarrollo de los Recursos Geotérmicos en Honduras. San Salvador, Pisa, San José: Instituto Italo-Latino Americano (IILA); Abril 2009–Mayo 2010.
[17] González M. Estado Actual del Desarrollo de los Recursos Geotérmicos en Nicaragua. San Salvador, Pisa, San José: Instituto Italo-Latino Americano (IILA); Abril 2009–Mayo 2010.
[18] http://www.etesa.com.pa/estudios.php?act=geotermico.
[19] http://www.worldatlas.com/webimage/countrys/sa.htm.
[20] https://www.cia.gov/library/publications/resources/the-world-factbook/wfbExt/region_soa.html.
[21] http://3.bp.blogspot.com/_pAEitDiZvOw/TAF5gj51fEI/AAAAAAAAODg/aK5lwy8cxJw/s640/terremoto4.jpg USGS, Topinka, USGSICVO, 1999, Modified from: Tilling, Heliker and Wright, 1987, and Hamilton, 1976.
[22] http://geology.about.com/od/seishazardmaps/ss/World-Seismic-Hazard-Maps.htm#step8.
[23] http://epod.usra.edu/blog/2001/06/seismicity-in-south-america.html.
[24] http://www.volcanodiscovery.com/volcanoes.html.
[25] http://www.argentinasmountains.com.ar/images/Contenido/Volcanoes%20of%20South%20America.png.
[26] Pesce AH. Argentina country update. In: Proceedings World Geothermal Congress 2000, Kyushu – Tohoku, Japan, May 28–June 10; 2000.
[27] Pesce AH. Argentina country update. In: Proceedings World Geothermal Congress 2005, Antalya, Turkey, 24–29 April; 2005.
[28] Pesce AH. Argentina country update. In: Proceedings World Geothermal Congress 2010, Bali, Indonesia, 25–29 April; 2010.

[29] Pesce AH. Argentina country update. In: Proceedings World Geothermal Congress 2015, Australia; May 2015.

[30] Tocchi E. "Il Tatio", Ufficio Geológico Larderello SpA. 1923. Unpublished Report.

[31] Lahsen A. Geothermal exploration in northern Chile — summaryHalbouty MT, Maher JC, Lian HM, editors. Circum-pacific energy and mineral resources. Am Assoc Petrol Geol Mem 1976;25:169−75.

[32] Lahsen A. Origen y potential de energía geotérmica en los Andes de Chile. In: Frutos J, Oyarzún R, Pincheira M, editors. Geología y Recursos Minerales de Chile. Chile: Univ. De Concepción; 1986. p. 423−38.

[33] Lahsen A. Chilean geothermal resources and their possible utilization. Geothermics 1988; 17:401−10.

[34] Hauser A. Catastro y caracterización de las fuentes minerales y termales de Chile, vol. 50. Servicio Nacional de Geología y Minería, Subdirección Nacional de Geología, Boletín; 1997.

[35] Salgado G, Raasch G. Recent geothermal industry activity and the market for electric power in Chile. Geotherm Resour Counc Annu Meet Trans 2002;26:55−8.

[36] Lahsen A, Muñoz N, Parada MA. Geothermal development in Chile. In: Proceedings World Geothermal Congress 2010, Bali, Indonesia, 25−29 April; 2010. 7 p.

[37] Lahsen A, Sepulveda F, Rojas J, Palacios C. Present status of geothermal exploration in Chile. In: Proceedings, World Geothermal Congress 2005, Antalya, Turkey, 24−25 April; 2005. 9 p.

[38] Urzua L, Powell T, Cumming WB, Dobson P. Apacheta, a new geothermal prospect in northern Chile. Geotherm Resour Counc Trans 2002;26:65−9.

[39] Aguirre I, Clavero J, Simmons S, Giavelli A, Mayorga C, Soffia JM. Colpitas—a new geothermal project in Chile. Geotherm Resour Counc Trans 2011;35:1141−5.

[40] Soffia J, Clavero J. Doing geothermal exploration business in Chile, Energía Andina experience. Geotherm Resour Counc Trans 2010;34:637−41.

[41] Clavero J, Pineda G, Mayorga C, Giavelli A, Aguirre I, Simmons S, et al. Geological, geochemical, geophysical and first drilling data from Tinguiririca geothermal area, central Chile. Geotherm Resour Counc Trans 2011;35:731−4.

[42] Melosh G, Cumming W, Benoit D, Wilmarth M, Colvin A, Winick J, et al. Exploration results and resource conceptual model of the Tolhuaca geothermal field, Chile. In: Proceedings, World Geothermal Congress 2010, Bali, Indonesia, 25−29 April; 2010. 7 p.

[43] Melosh G, Moore J, Stacey R. Natural reservoir evolution in the Tolhuaca geothermal field, southern Chile. In: Proceedings, 36th workshop on geothermal reservoir engineering. Stanford, CA: Stanford University; 2012. SGP-TR-194.

[44] Hickson CJ, Ferraris F, Rodriguez C, Sielfeld G, Henriquez R, Gislason T, et al. The Mariposa geothermal system, Chile. Geotherm Resour Counc Trans 2011;35:817−25.

[45] Lahsen A, Rojas J, Morata D, Aravena D. Geothermal exploration in Chile: country update. In: Proceedings World Geothermal Congress 2015, Melbourne, Australia, 19−25 April; 2015.

[46] ELC − Electroconsult.: aprovechamiento del Campo. Geotérmico de El Tatio, en el Norte de Chile. United Nations; 1975. Unpublished Report.

[47] Hildreth W, Grunder A, Drake R. The Loma Seca Tuff and the Calabozos Caldera: a major ash-flow and caldera complex in southern Andes of central Chile. Geol Soc Am Bull 1984; 95:45−54.

[48] Grunder AL, Thompson JM, Hildreth W. The hydrothermal system of the Calabozos Caldera, central Chilean Andes. J Volcanol Geotherm Res 1987;32:71−81.

[49] Dixon HJ, Murphy MD, Sparks SJ, Chávez R, Naranjo J, Dunkley JA, et al. The Geology of Nevados de Chillán Volcano, Chile. Rev Geol Chile 1999;26(2):227−53.

[50] Sepulveda F, Lahsen A, Dorsch K, Palacios C, Bender S. Geothermal exploration in the cordón caulle region, southern Chile. In: Proceedings, World Geothermal Congress 2005, Antalya, Turkey, 24−25 April; 2005.

[51] Mejía E, Rayo L, Méndez J, Echeverri J. Geothermal development in Colombia. In: Short course VI on utilization of low- and medium-enthalpy geothermal resources and financial aspects of utilization, organized by UNU-GTP and LaGeo, in Santa Tecla, El Salvador, March 23−29; 2014.

[52] Battocletti L. Geothermal resources in Latin America and the Caribbean. BoB Lawrence & Associates, Inc.; 1999. 214 pp. Website: http://bl-a.com/ecb/PDFFiles/GeoResLAC.pdf.

[53] Beate B, Urquizo M. Geothermal country update for Ecuador: 2010−2015. In: Proceedings World Geothermal Congress 2015, Melbourne, Australia, 19−25 April; 2015.

[54] Lloret A, Labus J. Geothermal development in Ecuador: history, current status and future. In: Short course VI on utilization of low- and medium-enthalpy geothermal resources and financial aspects of utilization, organized by UNU-GTP and LaGeo, in Santa Tecla, El Salvador, March 23−29; 2014.

[55] Beate B, Salgado R. Geothermal country update for Ecuador, 2005−2010. In: Proceedings World Geothermal Congress 2010, Bali, Indonesia; 2010. p. 16.

[56] Beate B. Plan para el aprovechamiento de los Recursos Geotérmicos en el Ecuador. 2010. p. 177. Quito.

[57] Almeida E, Sandoval G, Panichi C, Noto P, Bellucci L. Modelo geotérmico preliminar de áreas volcánicas del Ecuador a partir de estudios químicos e isotópicos de manifestaciones termales (Geothermal investigations with isotope and geochemical techniques in Latin America). In: Proceedings of the final research co-ordination meeting, held in San José, Costa Rica; 1990.

[58] CELEC EP/SYR. Initial prefeasibility study to develop a conceptual model for the Chacana geothermal project. 2012. Unp.Tech. Report prepared by SYR on behalf of CELEC EP. (Directors: B. Beate-Project, M. Hall-Geology, S. Inguaggiato-Geochemistry; E. Anderson-Geophysics; G. Smith-Environmental). Quito. Final Report.

[59] Beate B, Salgado R. Geothermal country update for Ecuador, 2000−2005. In: Proceedings World Geothermal Congress 2005, Antalya, Turkey, 24−29 April; 2005.

[60] Cruz V, Vargas V. Geothermal country update for Peru, 2010−2014. In: Proceedings World Geothermal Congress 2015, Melbourne, Australia, 19−25 April; 2015.

[61] Vega LA. IV Mesa de Trabajo Técnica, sobre Recursos Geotérmicos en la Macro Región Sur. Plan Maestro para el Desarrollo de la Geotermia en Perú, Power Point Presentation, Moquegua. Octubre 11, 2013.

[62] (JICA-West JEC), Final Report The master plan for development of geothermal energy in Peru. February 2012.

[63] Beate B, Urquizo M. Geothermal country update for Ecuador, 2010−2015. In: Proceedings World Geothermal Congress 2015, Melbourne, Australia, 19−25 April; 2015.

Mexico: thirty-three years of production in the Los Azufres geothermal field

24

L.C.A. Gutiérrez Negrín[1], M.J. Lippmann[2]
[1]Geoconsul, S.A. de C.V., Morelia, Michoacán, Mexico; [2]LBNL, Berkeley, CA, United States

24.1   Geothermal power in Mexico

Geothermal power in Mexico has a long tradition, whose beginnings can be traced back to 1937 when the Mexican geothermal pioneer Luis F. de Anda started to work for the then recently formed Comisión Federal de Electricidad (CFE), the Mexican electric utility owned by the federal government. De Anda promoted the use of geothermal energy for generating electricity among the CFE's authorities and published a seminal study of the geothermal potential of the Ixtlán de los Hervores zone, located in the western state of Michoacán (Fig. 24.1) (Quijano-León and Gutiérrez-Negrín, 2003).

The Geothermal Energy Commission (CEG, or Comisión de Energía Geotérmica) was created in 1955, headed by De Anda. The exploration studies were focused in the

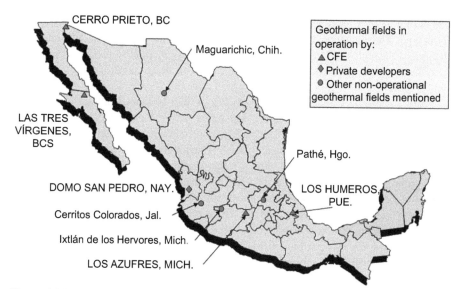

Figure 24.1 Current geothermal fields and historic geothermal zones in Mexico. Prepared by the authors.

Geothermal Power Generation. http://dx.doi.org/10.1016/B978-0-08-100337-4.00024-3

Pathé geothermal field, located in the state of Hidalgo in central Mexico (Fig. 24.1). The first production well (Pathé-1) started to produce steam in January 1956. In 1958, an exploratory well in the Ixtlán de los Hervores zone produced steam as well, and the first exploration studies started at the Cerro Prieto field in northern Mexico (Fig. 24.1).

Also in 1958, CFE acquired a geothermal power plant of 3.5-MW capacity built in Italy, which was installed in Pathé and officially commissioned on November 20, 1959, thus becoming the first commercial geothermal power plant in the Americas. The plant never could be operated at its nameplate capacity due to insufficient steam but was in operation until 1973, when it was dismantled (Quijano-León and Gutiérrez-Negrín, 2003). Years later, the plant was reassembled and mounted in the Los Azufres geothermal field, which is currently displayed as a historic piece.

Exploration continued in the Cerro Prieto geothermal field whose large and powerful surface manifestations were known since the Colonial time. By 1968 around 14 production wells had been drilled in the field, and CFE, supported by CEG, decided to start the construction of the first two power units of 37.5 MW each. The CEG was dissolved and all its personnel and assets were transferred to the CFE in 1971 (Quijano-León and Gutiérrez-Negrín, 2003). Since then and up to 2015, CFE has been in charge of all the geothermal development in Mexico.

CFE commissioned the first 37.5-MW power unit at Cerro Prieto in April 1973 and the second unit followed in October 1973. Since then, another 11 power units have been installed and commissioned in this field up to 2000, reaching an installed capacity of 720 MW. In other words, there are four condensing single-flash plants of 37.5 MW, one low-pressure plant of 30 MW, four condensing double-flash, double-flow turbine plants of 110 MW each, and the most recent four 25-MW power units of the single-flash type. The first four 37.5-MW units, installed in 1973 and 1979 and known as the units 1 through 4 of Cerro Prieto I were put off service in 2011 and 2012; nevertheless, they are still installed in the field. The current running capacity at Cerro Prieto is only 570 MW.

From a geologic view point, the Cerro Prieto field is located in a pull-apart basin formed as result of two active strike-slip faults of the San Andreas Fault System. The heat source seems to be a basic intrusion producing a thermal anomaly related to the local thinning of the continental crust. A sequence of sedimentary rocks (sandstones interbedded with shales) with a mean thickness of 2400 m are hosting the geothermal fluids. During the last 50 years, almost 450 geothermal wells have been drilled at Cerro Prieto, with an average depth of 2430 m and a maximum depth of 4400 m (Gutiérrez-Negrín et al., 2015). On average, 159 production wells were in operation during 2014 producing 35.6 million metric tons of steam, at an annual mean rate of 4063 tons per hour (t/h). In 2014, the electricity generation in the field was 3957 GWh (Flores-Armenta, 2015), at an annual average specific steam consumption (SSC) of 9 t/h of steam per megawatt effectively produced (ie, 2.5 kg/s per MW). That year, the annual mean production rate per well at Cerro Prieto was 25.6 t/h. There were also 16 injection wells to partially dispose of the separated brine together with the solar evaporation pond.

Los Azufres was the second Mexican geothermal field to be developed, with the first plants starting to operate in 1982. Its main features are presented in detail in the following sections since it is the central subject of this chapter.

Following Los Azufres, the Los Humeros field was developed, its first two power plants being commissioned in April and September 1990. The field is located in the central Mexico at the eastern section of the Mexican Volcanic Belt (Fig. 24.1), within the Quaternary volcanic caldera of Los Potreros, which in turn is nested in the larger Los Humeros Caldera. The geothermal fluids are found in andesites overlying a complex basement composed of metamorphic, sedimentary, and intrusive rocks. The heat source is the magma chamber that produced both calderas; the last volcanic activity occurred 20,000 years ago (Gutiérrez-Negrín et al., 2015).

The first units in Los Humeros were two back-pressure, wellhead plants of 5 MW each. After their commissioning in 1990, they were followed by five other similar plants installed in 1991, 1992, and 1994, raising the total power installed to 35 MW. In 2003, one of these units was replaced by a similar one moved from Los Azufres, and in 2008, one more back-pressure unit was installed. With this additional unit, the installed capacity increased to 40 MW. In 2012 and 2013, two condensing, single-flash power plants of 25 MW (26.8 MW gross) each were installed in Los Humeros, and thus currently the total is 93.6 MW. However, five 5-MW units remain in standby and are used only when the main flash units must go out of service. Therefore, the running or operative capacity in Los Humeros is 68.6 MW, that is, 2×26.8 MW plus 3×5 MW (Gutiérrez-Negrín et al., 2015).

During 2014, there were 22 production and two injection wells in operation at Los Humeros. The steam produced was 5.38 million metric tons at an annual average rate 614 t/h (Flores-Armenta, 2015). Each of the 22 production wells produced an average of 27.9 t/h of steam. With that amount of steam, the power units operating in Los Humeros produced 321 GWh (Flores-Armenta, 2015). Currently, another 25-MW (net) power plant is under construction, which is expected to be operating in 2016.

Simultaneously with the Los Azufres and Los Humeros field, CFE was also developing the Cerritos Colorados field (previously known as La Primavera). This field is also in central Mexico, at the western section of the Mexican Volcanic Belt near the city of Guadalajara (Fig. 24.1). It is located inside a Quaternary volcanic caldera and near the confluence of three major continental structural elements: the N—S oriented Colima Graben, the E—W Chapala Graben, and the NW—SE Tepic Graben. The heat source is the magma chamber that produced the caldera around 95,000 years ago and whose last eruptive activity is about 20,000 years old. The high-temperature geothermal fluids are encountered in a lithological unit mainly composed of Tertiary andesites with some tuffs and rhyolites, whose permeability is basically secondary produced by local and regional faults and fractures. This unit rests on a low-permeability granodiorite basement (Gutiérrez-Negrín, 2015).

During the 1980s, CFE drilled 13 exploration/production geothermal wells in Cerritos Colorados. Several of them were properly evaluated producing fluids with an average temperature of ~300°C. At request of the local government CFE suspended its developing activities in 1989 and the field remained on standby since then. Thus, no power plant has been installed so far, but the steam produced by six of the 13 wells was considered enough to feed a 25-MW power plant. CFE has estimated that the Cerritos Colorados field has a minimum potential of 75 MW (Gutiérrez-Negrín, 2015).

During the 1980s, CFE started to explore another geothermal area, the Las Tres Vírgenes field located about half-way down the Baja California Peninsula (Fig. 24.1). The field is within a Quaternary volcanic complex, composed of three N−S aligned volcanoes that are related to the NW spreading of the Gulf of California and the drifting of the peninsula. Las Tres Vírgenes volcanic complex is the most recent one in the area, the others being La Reforma caldera and El Aguajito. The field lies within an NW−SE trending Pliocene to Quaternary depression known as the Santa Rosalía Basin that seems to be the western limit of a deformation zone related to the opening of the Gulf of California. The geothermal system is structurally controlled and located near the northern edge of the volcanic complex, which, in turn, is emplaced into a system of right strike-slip faults related to a tension zone and some left strike-slip lateral faults (Macías Vázquez and Jiménez Salgado, 2012).

The geothermal fluids at Las Tres Vírgenes are hosted by the granodioritic basement, part of the California Batholith (Macías Vázquez and Jiménez Salgado, 2012), whose top is found at 900−1000 m depth. Temperatures range from 250 to 275°C. In 2001 CFE commissioned two condensing, single-flash 5-MW plants, currently fed by four production wells. In 2014 these wells produced 667,774 tons of steam (Flores-Armenta, 2015) at an average rate of 76.2 t/h. The power units generated 51 GWh with a mean SSC of 13.1 t/h of steam per megawatt (3.6 kg/s per MW).

Since the late 1980s, CFE has conducted several exploration studies in the Piedras de Lumbre geothermal zone near the village of Maguarichic, located in the northern state of Chihuahua, high in the Sierra Tarahumara (Fig. 24.1). This geothermal zone lies inside the Basin and Range tectonic province and present hot springs with sodium-chloride waters at 42−93°C, fumaroles, and hydrothermal alteration zones composed mainly of kaolin. CFE drilled a shallow ($\sim$ 300 m depth) well that produced around 35 t/h of 135°C water and decided install a small power unit to provide electric energy to Maguarichic. The village, of about 600 inhabitants then, was located around 75 km from the nearest power transmission lines of the national grid (Quijano-León and Gutiérrez-Negrin, 2003). The settlers operated a small 150-kW diesel generator only 3 h a day (from 7 to 10 PM) at a very high cost since the liquid fuel had to be transported via an unpaved road in a 4-h drive, and the road usually had to be closed during the winter.

Partially financed by federal and state programs for rural electrification, CFE installed a 300-kW binary cycle power unit manufactured by Ormat that included a modular cooling tower with eight fans and a 480-V synchronous generator; the unit used isopentane as working fluid. A 6.5-km-long, 34.5-kV transmission power line was also constructed from the geothermal unit to the village. The Maguarichic plant started to operate in April 2001. CFE provided training to villagers for start-up, reset alarms, and repair of minor problems, including restarting the plant when it shut down. For major problems, CFE used to send its technical personnel. The plant was operated satisfactorily up to 2007 when the expansion of the national grid reached the village. Eventually the plant was dismantled, but during those 6 years it was the only geothermal power plant operating off the grid in Latin America.

No new geothermal fields came into operation after Maguarichic and Las Tres Vírgenes until early 2015, when two back-pressure power units of 5 MW each were commissioned in the Domo San Pedro geothermal field (Del Valle, 2015). This geothermal field is located in the state of Nayarit in the western tip of the Mexican Volcanic Belt (Fig. 24.1). It is related to a couple of massive Quaternary ($\sim$0.1 million years old) dacitic domes whose magma chamber seems to be the heat source of the geothermal system. Along with other volcanic structures, the San Pedro domes were emplaced in the northwestern edge of the Tepic Graben, which is considered as a pre-rifting regional structure. The about 280°C geothermal fluids are found in Tertiary andesitic rocks and Cretaceous granitic rocks.

The Domo San Pedro is the first geothermal field in Mexico developed and operated by a private company, thanks to a self-supply permit issued by the Mexican Energy Regulatory Commission (CRE, Comisión Reguladora de Energía) under the former energy regulations. Several production and injection wells have been drilled in the field to feed the two power units whose electric energy is currently used by partners of the private developer. This could change when the new regulatory electric framework is fully implemented in January 2016. In any case, a condensing, single-flash plant of 25 MW (net) is under construction and is expected to be operating in 2016 (Del Valle, 2015).

It is worth mentioning that in late 2013 the Mexican federal and state congresses passed amendments of three constitutional articles in what was called the energy reform that involved a deep transformation in the oil, gas, and electric industries and particularly in the two state companies Petróleos Mexicanos (PEMEX) and CFE. Although still owned by the state, both companies are being transformed into operative, finance, and management autonomous companies that will compete with private companies in free power and oil national markets (Gutiérrez-Negrín et al., 2015).

The energy reform involved a series of 21 legislative initiatives composed of nine new laws and 21 reforms to preexisting laws in the energy sector, as well as in related areas like federal financing, income and taxes, and the structure of the federal administration. Two of the new laws are the Electric Industry Act (LIE, Ley de la Industria Eléctrica) and the Geothermal Act (LEG, Ley de Energía Geotérmica) (SENER, 2014) that were approved in August 2014.

The main change contained in the new LIE is the creation of a free power market in Mexico. Although the old law permitted the construction and operation of power units by private investors (ie, independent power producers (IPP)), they had to sell all the energy to CFE through long-term contracts. That was a fair deal for IPPs, but they were not allowed to make other arrangements with private consumers — or offer their output in a public market that did not exist at that time. Now, the LIE defines the main characteristics of a wholesale public power market that will be managed and controlled by an autonomous entity. CFE and private generators and commercial participants will take part in the spot and/or the auction markets. CFE and private investors will operate and maintain the national transmission and distribution grid and CFE will continue providing the final service to domestic and small consumers (Gutiérrez-Negrín et al., 2015).

On the other hand, the new Geothermal Act (LEG) divides the process of geothermal development into three successive stages: reconnaissance, exploration, and exploitation, for whose implementation it is necessary first to obtain a registration, permit, or concession, respectively, issued by the Ministry of Energy (SENER: Secretaría de Energía). The exploration permit shall be issued for 3 years and may be extended for another 3 years but only once. The licensee shall drill and complete at least one exploration well per every $30\,km^2$ of area granted, up to a maximum of $150\,km^2$. The holder of the exploration permit can apply for an exploitation concession in the area, if results are successful. The exploitation concession will be valid for 30 years and can be extended by SENER. Where geothermal resources are extended to other geothermal areas and both are subject to different concession holders, a joint operation may be agreed to with prior authorization from SENER. Neither the exploration permit nor the exploitation concession can be sold, this being a cause for revocation.

CFE retains its four geothermal fields under operation, as well as a set of 13 geothermal zones still undeveloped but previously explored by CFE. In any case, CFE is allowed to form public—private partnerships with private entities — which was de facto prohibited before. In all private geothermal projects that started before the act was passed, the developers have the preferential right to request the proper permit or concession. This is the case of the Domo San Pedro field and the Ceboruco Volcano zone —in the latter, a private joint venture has been carrying out exploration studies.

Altogether the four geothermal fields in operation in Mexico during 2014 accounted for an installed capacity of 1018 MW, the running or effective capacity was 840 MW (Table 24.1). That year the electric output in those four fields was 6000 GWh according to statistics from the electricity sector (SENER, 2015). Thus, in 2014 the national average capacity factor of the geothermal plants operating in Mexico in 2014 was 81.8%.

Table 24.1 Geothermal fields, installed and running capacity in Mexico in 2014 and 2015

Geothermal field	Owner and operator	Capacity (MW) in 2014		Capacity (MW) in 2015	
		Installed	Running	Installed	Running
Cerro Prieto, Baja California	CFE	720	570	720	570
Los Azufres, Michoacán	CFE	194.4	191.4	227.4	224.4
Los Humeros, Puebla	CFE	93.6	68.6	93.6	68.6
Las Tres Vírgenes, Baja California Sur	CFE	10	10	10	10
Domo San Pedro, Nayarit	Grupo Dragón	0	0	10	10
Total		**1018**	**840**	**1061**	**883**

The installed capacity in the geothermal fields represented 1.9% of the electric power capacity installed in Mexico in 2014, considering the power plants dispatched by CFE for public electricity service, which amounted to 54,372 MW (SENER, 2015). The share of geothermal power to the national electricity generation in Mexico was a bit higher (2.3%) considering that the national total was 258,256 GWh (SENER, 2015). This indicates that the geothermal plants operated at a higher average capacity factor than the average of the rest of the plants connected to the grid.

As of 2015, the geothermal installed capacity in Mexico has increased to 1061 MW, thanks to a new power unit of 50 MW (net) commissioned in Los Azufres in February, and the two back-pressure units installed in Domo San Pedro in February and March. In consequence, the effective or running capacity increased to 883 MW, as shown in Table 24.1.

24.2   Main features of the Los Azufres field

As mentioned before, Los Azufres is located in the central part of the country (Fig. 24.1) in the Mexican Volcanic Belt province at an average elevation of 2850 masl. The field occurs in a complex Plio-Pleistocene succession of basalts, andesites, dacites, and rhyolites that represent three probable volcanic cycles. The second cycle started with basalts and continued with basaltic andesites, porphyritic and microcrystalline andesites, dacites, and rhyolites, each rock type accompanied by its pyroclastic equivalents. The first cycle is represented only by scarce dacites at the bottom of some of the deepest wells, stratigraphically below thick andesitic sequences. The third, final volcanic cycle is represented by some young basaltic and diabase dikes stratigraphically and radiometrically younger than the youngest rhyolites, which are the final, acid stage of the second volcanic cycle (Gutiérrez-Negrín and Aumento, 1982). The geothermal fluids are found in the middle and lower parts of the second, more complete cycle, mainly in andesites.

In the 1990s, it was postulated that the field was located in the southern portion of a large (about 80 km diameter) caldera developed inside the Morelia-Acambay rift (Ferrari et al., 1991). However, more recent studies did not found any evidences for such caldera (Pérez-Esquivias et al., 2010). In any case, this caldera would not be related to the current geothermal system whose heat source seems to be the magma chamber of the San Andrés volcano, the highest peak in the area (Gutiérrez-Negrín, 2015).

The volcanic rocks un-conformably overlie metamorphic and sedimentary rocks of Late Mesozoic to Oligocene age. This prevolcanic basement has not been cut by the geothermal wells but is supposed to consist of gently folded shales, sandstones, and conglomerates. The oldest volcanic activity reported in the area began with andesite flows and some subordinated pyroclastic flows of about 18 million years of age (Ma), belonging to the Mil Cumbres Unit (Fig. 24.2), that represent the local, more than 2700-m-thick, basement. It was followed by micro-granular 5.9 ± 0.6 Ma up to 1.4 Ma andesites, highly fractured and faulted with layering resembling in places that of sedimentary rocks (Gutiérrez-Negrín and Aumento, 1982; Dobson and Mahood, 1985; Pérez-Esquivias et al., 2010). Covering these Miocene-Pliocene

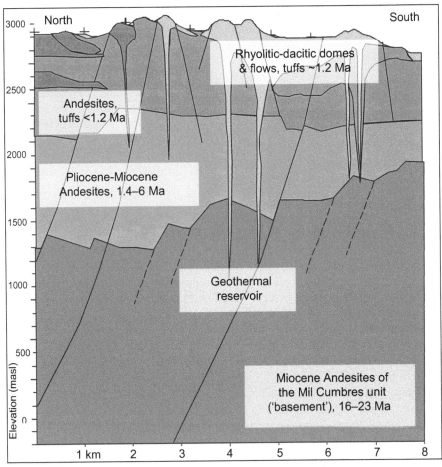

Figure 24.2 Schematic conceptual model of the geothermal reservoir.
Adapted from Pérez Esquivias, H., Macías Vázquez, J.L., Garduño Monroy, V.H., Arce
Saldaña, J.L., García Tenorio, F., Castro Govea, R., Layer, P., Saucedo Girón, R., Martínez, C.,
Jiménez Haro, A., Valdés, G., Meriggi, L., Hernández, R., 2010. Estudio vulcanológico y
estructural de la secuencia estratigráfica Mil Cumbres y del campo geotérmico de Los Azufres,
Mich. Geotermia 23 (2), 51−63.

andesites, there are Quaternary andesites, basaltic andesites, and pyroclastic rocks less
than 1.2 Ma in age, belonging to the Mexican Volcanic Belt (area of Michoacán-
Guanajuato volcanic field). The geothermal fluids are found in the middle and lower
portions of the local "basement" (Fig. 24.2).

The andesites are overlain by flow-structured 1.2 ± 0.4 Ma rhyolites, which in
some sites include obsidian flows and perlite structures. The rhyolites are also fractured and present wide areas covered by superficial kaolin resulting from the hydrothermal alteration by surface discharging geothermal fluids. The silicic volcanism
also includes rhyodacites and dacites with ages as recent as 0.15 Ma and thickness

of up to 1000 m, with five different units: the Agua Fría rhyolite, Tejamaniles dacite, Cerro Mozo and San Andrés dacites, and Yerbabuena rhyolite. They form domes and short lava flows with glassy structures and are generally fractured on the surface. Close to hydrothermal manifestations, they show strong alteration, characterized by kaolinitization and silicification. The more recent outcropping rocks include porphyritic andesites, glassy and pumicitic rhyolites, and basalt flows and cinder cones on the western part of the Los Azufres field, product of the youngest volcanic activity in the area and belonging to the Michoacán-Guanajuato Volcanic Field (Gutiérrez-Negrín and Aumento, 1982; Dobson and Mahood, 1985).

In the field and its surroundings, three main fractures trends of NNW-SSE, NE–SW, and E–W orientation have been identified (Fig. 24.3). The first one corresponds to a Miocene deformation with semivertical geometry affecting the basement. The two other trends were formed as part of the Mexican Volcanic Belt and show semivertical and subhorizontal geometry in regionally affected Miocene basement rocks and in Quaternary rocks outcropping in the field (Pérez Esquivias et al., 2010).

Hydrothermal surface manifestations are represented by hot springs, some with up to 90°C fluids (especially those ascending through faults that penetrate the vadose zone), fumaroles, steaming soil, mud pools, and even small thermal lakes. The thermal features are mostly located along faults or lineaments suggesting a fracture control (Fig. 24.3) (Viggiano-Guerra and Gutiérrez-Negrín, 1995).

Most rocks in the field show several grades of hydrothermal alteration. Primary minerals and rock matrix, as well as vesicle and fracture filling, present alterations ranging from incipient to complete. The subsurface hydrothermal activity can be divided into two types: acid sulfate (near surface) and alkali chloride below the water table, which is located at about 400-m depth. Separation of steam from the boiling alkali-chloride water provides steam to some relatively shallow wells in the southern part of the field (see later) (Viggiano-Guerra and Gutiérrez-Negrín, 1995).

At least three main calc-silicate alteration zones have been identified at Los Azufres. The zeolite zone is the shallowest, located below 400 m depth and composed of calcite $\pm$ anhydrite + pyrite + smectite + chlorite + quartz (chalcedony) and zeolites (heulandite and laumontite). Below it is the epidote zone, composed of epidote + wairakita + chlorite (penninite) + quartz + illite/smectite + illite + calcite + pyrite $\pm$ prehnite. Generally it is located between 400 and 2000 m depth and is the most important one because it contains the producing reservoir with temperatures between 250 and 285°C. Finally is the amphibole zone, presenting amphibole + epidote $\pm$ wairakite + biotite $\pm$ illite $\pm$ chlorite $\pm$ garnet $\pm$ diopside. This zone is located below 2200 m depth with temperatures up to 285°C and pressures above 170 bars. However, its porosity is less than 3% with anhydrous minerals appearing; thus the thermal regimen is forced convective or almost conductive. No well drilled into this zone is a producer; therefore, it has been assumed this zone acts as an aquitard with alternating regimens from conductive to convective or vice versa but also channels fluids to the upper reservoir (Viggiano-Guerra and Gutiérrez-Negrín, 1995). This deep zone seems to be a good candidate area for developing an enhanced geothermal system (EGS).

The natural geothermal waters are rich in sodium chlorides with high CO_2 contents and pH around 7.5. The average composition of the separated water in

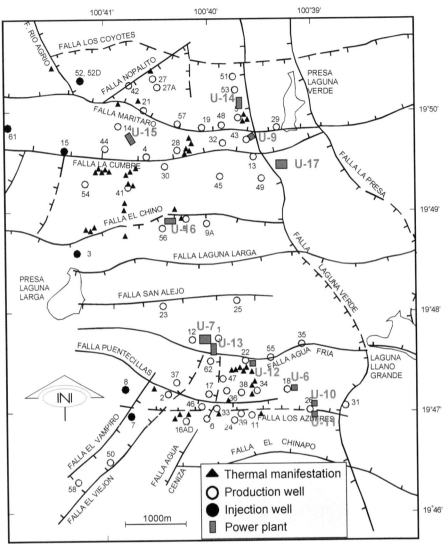

Figure 24.3 Wells, faults, superficial manifestations, and power plants in Los Azufres.
Adapted from Arellano G.V.M., Ramírez M.M., Barragán R.R.M., Paredes S.A., Aragón A.A,
López B.S., Casimiro E.E., 2015. Condiciones termodinámicas de los fluidos del yacimiento de
Los Azufres (México) y su evolución en respuesta a la explotación (1979−2011). Proceedings of
the 22nd Annual Congress of the Mexican Geothermal Association (AGM), Cuernavaca, Mor.
Mexico, 10−11 March 2015 (in Spanish).

the northern part of the field (see below) includes 2870 parts per million (ppm) of Cl,
1580 ppm of Na, 1070 ppm of SiO_2, and around 240 ppm of boron, with a pH of 6.3.
In the separated waters of the southern zone of the field, the concentration of chlorides is 5000 ppm, Na is 2700 ppm, SiO_2 is about 880 ppm, boron is 400 ppm,
and the pH is 7.

Although fluid temperatures can be as high as 320°C, the usual range is 240–280°C. Geochemical studies have shown that chemical reactions between the volcanic rocks and geothermal fluids are close to equilibrium (Verma et al., 1989).

Since the exploration stage, the field was divided into two distinct zones, Maritaro in the north and Tejamaniles in the south, with a separation of a few kilometers between them. Both seem to join at depth to form a single geothermal reservoir, but in between the reservoir seems to deepen, as suggested by a high-resistivity anomaly related to the two normal E–W trending faults and to a higher thickness of the Agua Fría rhyolites.

Analysis of production, chemical, and isotopic ($\delta^{18}O$, δD) data of the northern zone of the field have shown the influence of the reinjected fluids in some wells. In the southern zone, the long-term response is more complex; the following patterns were identified: pressure and mass flow-rate drop, boiling, cooling, increased steam production, and in some wells affected by injection, there is an increase in both pressure and mass flow rate. The isotopic composition of fluids from southern zone wells showed the occurrence of two processes. The first, with a positive slope, indicates in some wells a mixing of reservoir and injection fluids. The second, with a negative slope, results from the reservoir steam separation and partial condensation (Arellano et al., 2004, 2005).

In general, production wells have shown a decrease in the fluid temperatures and an increase in the reservoir steam fraction. Main noncondensable gases (NCGs) in the separated steam are CO_2 (94% by volume), H_2S (2.5% by volume), and minor concentrations of H_2, CH_4, N_2, and NH_3 (combined 3.5% by volume) (Barragán-Reyes et al., 2012).

24.3 Geothermal production

The first exploration studies in Los Azufres were carried out by CFE in the 1970s, including geologic, geophysical, and geochemical surveys. According to the geophysical results, that is, low-resistivity anomalies separated by a high-resistivity zone in the central part, the field was divided, as mentioned earlier, into the north (Marítaro) and south (Tejamaniles) zones.

The first exploration well was drilled by CFE in 1976, and by 1977 the well results confirmed the existence of an important geothermal resource. More exploration wells were successfully drilled in the following years, and by 1980 there was an important amount (about 170 t/h) of steam had become available. Considering that the construction of a condensing, flash power plant could take more additional time, CFE decided to buy and install a number of smaller wellhead, back-pressure units, in order to quickly begin generating power and, at the same time, learn more about the response of the reservoir under actual exploitation conditions. CFE was aware that these units with atmospheric discharge were less efficient than condensing turbines, but their construction and installation were less expensive and faster. Additionally, any individual unit could be moved to other wellpads if production of the original well dropped significantly or if another area of the field needed to be tested.

In 1982, the year when the first five back-pressure plants started operating, the steam produced was almost 3.1 million tons at an average production rate of 351.6 t/h. The steam was accompanied by 3.1 million tons of water and the total production was roughly 6.2 million tons, with a steam fraction of about 50% (Fig. 24.4). That was the only year when the amount of steam and water produced in the field was approximately equal. From 1983 onward, production of steam has been always larger than that of water, yielding a steam fraction higher than 50%. Variations go from 51% in 1986 to 79.1% in 2007, with a mean historic value of 67%, as explained later. Based on this and other evidence, it became clear that the geothermal reservoir in Los Azufres is vapor dominant.

Part of the steam sent to the power units comes from production wells located in the south zone of the field that have been producing almost dry steam from a shallow (at 600- to 700-m depth) portion of the reservoir. However, most of the wells produce from deeper portions of the reservoir (from about 2000-m depth) that present a minor liquid phase.

In 1986, another 5-MW back-pressure plant was commissioned but production remained almost steady up to 1988, when the first 50-MW condensing plant was installed. In that year the steam production increased to 7.1 million tons at an average rate of 810 t/h. In 1989 it reached 8.3 million tons at 947 t/h in average. That year the separated water also increased to 4.65 million tons (Fig. 24.4). Further increases in

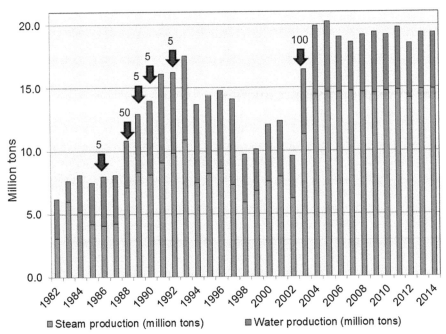

Figure 24.4 History of geothermal production in Los Azufres. Arrows indicate the commissioning year of new power plants and their capacity in MW. Does not include the 50-MW plant installed in February 2015.
Prepared by the authors.

production occurred in 1990 through 1992, when another three back-pressure plants were put into commercial operation. By 1993, the annual steam production in Los Azufres surpassed 10 million tons, at an average rate of 1239 t/h. Water production increased correspondingly to 6.65 million tons (Fig. 24.4).

The reduction in steam production between 1994 and 1998 was due to problems in the performance of Unit 7 that forced frequent shutdowns. Additionally, between 1996 and 1998, a couple of back-pressure plants were dismantled and moved to a geothermal field in Guatemala to comply with an international services contract of CFE. One of these plants returned to Los Azufres in 1999 and the other (Unit 1) remained in Guatemala and was finally sold by CFE in 2002. As a result steam extraction plummeted to 5.9 million tons in 1998, with water production falling to 3.8 million, the lowest figure since 1987 (Fig. 24.4).

In the following years, steam production and power generation at Los Azufres experienced some ups and downs up to 2003 when another back-pressure plant (Unit 8) was dismantled and moved to the Los Humeros field in February of that year. The Los Azufres II project, composed of four 25-MW plants, was commissioned between January and July 2003. The steam production recovered again and reached 11.26 million tons that year. In 2004, it reached its current level of 14.5—14.9 million tons (1650—1700 t/h). Water production also increased to 5.15 million tons in 2003 and up to a historic peak of 5.59 million tons in 2005 (Fig. 24.4).

The steam production in 2014 was 14.9 million tons (Flores-Armenta, 2015), the highest historic production (Fig. 24.4), at an annual average rate of 1700 t/h ($\sim$470 kg/s), accompanied by around 4.4 million tons of water. It is estimated that between 1976 and 2014 the field has produced 482.8 million metric tons of geothermal fluids, of which 323.7 are steam and 159.1 are water. Based on these data, the historic average steam fraction of 67%, mentioned earlier, was obtained.

In February 2015, the 50-MW Unit 17 started to operate in Los Azufres. Even though four of the oldest back-pressure plants were put out of service at that time, it is expected that steam production will increase this year and in the future.

During 2014, there were 39 production and six injection wells in operation in Los Azufres (Flores-Armenta, 2015). Most of these wells are located in the north zone, including injection wells Az-3, Az-15, Az-52D and Az-61; the two remaining injectors, wells Az-7 and Az-8, are operating in the south zone (Fig. 24.3). Taking into account that the average rate of steam production in 2014 was 1700 t/h, it follows that the average well output was 43.6 t/h (or $\sim$12 kg/s and $\sim$5.6 MW), the highest of all the Mexican fields and comparable to other geothermal fields in the world. The peak steam production per well in Los Azufres was reached in 2004, with an annual average of 47.2 t/h ($\sim$13 kg/s or $\sim$6.3 MW).

As of December 2014, the steam in Los Azufres is separated at the wellpads and transported through a pipeline network whose total length is approximately 28 km: 15.2 km (54.3%) in the northern zone and 12.8 km (45.7%) in the southern zone. The steam pipes have diameters ranging between 0.25 and 1.07 m (10 in. and 42 in.) and are covered by thermal insulation. The steam transportation network is a distributed system with interconnections to provide an adequate steam supply to the different power plants. However, since there are no production wells in the central section of the field, there is no interconnection between the pipelines in the northern and southern zones, that is, the network is composed of two independent networks (García-Gutiérrez et al., 2010).

The separated water is cooled in ponds generally located at the same wellpad. It is then transported by flexible pipelines made of high-density polyethylene running along the roads up to the injection wells where gravity supplies the needed head for injection. Since 2003, when the four 25-MW power plants started to operate, the excess of condensate in the cooling towers is also injected along with the brine. Thus, currently the volume of injected water is higher than the volume of liquid water extracted; it is hoped that this will enhance the reservoir recharge. For instance, in 2011 the separated water was 4.97 million tons and the injected water was around 7.6 million tons—more than 50% higher.

Unlike other Mexican fields, such as Cerro Prieto, where the chemical contents of the separated water gives place to deposition of silica, calcite, or other minerals provoking scaling problems in the wells and/or the superficial installations, in Los Azufres, scaling problems are not relevant.

A recent study investigated the response of the Los Azufres reservoir to exploitation between 1979 and 2011 by gathering and processing historical chemical, isotopic, and production data from 33 wells (17 from the southern zone and 16 from the northern zone). The reservoir thermodynamic conditions (pressure, temperature, enthalpy, flow rate, and steam quality) were obtained by using a well simulator and wellhead monitoring parameters as input. Besides the simulator results, reservoir temperatures were estimated by using a cationic geothermometer in two-phase wells and a gas geothermometer (FT-HSH2) in dry steam wells. Subsequently, the rates of declining of pressures, mass flow rates, and temperatures were estimated for individual wells, for wells related to specific faults, and for every zone of the field by using the linear and the harmonic models (Arellano et al., 2015).

Results of that study indicate that in the southern zone, the average declining rates of pressure and temperature are -0.39 bar/year and $-0.78°C$/year, respectively, while the rate of enthalpy increase was 18.7 (kJ/kg)/year. Such rate changes were considered to be rather moderate and are attributed to the natural and artificial recharge that seems to be adequate to compensate for the fluid extraction. For the northern zone, the average rates of change of pressure, enthalpy and temperature obtained by the linear method were -0.53 bar/year, 20.3 (kJ/kg)/year and $-0.73°C$/year, respectively. Such rates were also considered to be very moderate.

Therefore, from 1979 to 2011, the pressure of south zone wells decreased about 12 bar, the enthalpy increased 598 (kJ/kg) and temperature decreased 25°C. On the other hand, the pressure of the northern zone wells decreased about 17 bar, the enthalpy increased 650 (kJ/kg) and the temperature decreased 23°C (Arellano et al., 2015).

24.4 Power plants and output

The first power plants at Los Azufres began operation in 1982 (Table 24.2), Unit 1 in July and the others in August. Units 1 to 5 were 5-MW back-pressure, wellhead type,

Table 24.2 Some historic and current data on the Los Azufres power units

Power unit	Type	Capacity/MW		Manufacturer	Status	Dates*
		Net	Gross			
1	Wellhead, back-pressure, single-flash	5		Mitsubishi	Dismantled	1982 (1986)
2		5				1982 (2015)
3		5				
4		5				
5		5				
6		5	5	Toshiba	Operating	1986
7	Condensing, single-flash	50	NA	General Electric	Operating	1988
8	Wellhead, back-pressure, single-flash	5		Ansaldo/Makrotek	Dismantled	1989 (2003)
9		5	5		Operating	1990
10		5	5		Operating	1992
11	Binary cycle (ORC)	1.45	1.5	Ormat	Out of service	1993 (2006)
12		1.45	1.5			1993 (2006)
13	Condensing, single-flash	25	26.6	Alstom	Operating	2003
14		25	26.6		Operating	2003
15		25	26.6		Operating	2003
16		25	26.6		Operating	2003
17		50	53	Mitsubishi	Operating	2015
Current total		**217.9**	**227.4**			

NA, data not available.
* Commissioning date (and decommissioning).

built by Mitsubishi. Every turbine was composed of one horizontal cylinder with five reaction or impulse stages running at 3600 revolutions per minute (rpm). The steam entered to the turbine at 8 bar,a (8.16 kg/cm^2) and 170.4°C and was discharged to the atmosphere through a silencer at 0.7 bar,a (0.714 kg/cm^2) and 91°C. The units did not need a condenser, cooling system and system for removal of NCGs (Fig. 24.5). The electric generator, also manufactured by Mitsubishi, rotated at 3600 rpm and worked at 4160 V at a power factor of 80% (Table 24.3). All these units had a low efficiency; when new, each one required 59 t/h (16.4 kg/s) of steam at full load, or around 11.8 t/h (3.28 kg/s) per MW, but in the last years of operation, the specific steam consumption (SSC) rose to more than 12 t/h per MW. Unit 1 was dismantled and moved to Los Humeros in 1996 and the other four units were dismantled in early 2015 when the new Unit 17 started to operate (Table 24.2).

Unit 6 was installed in 1986. It is also a wellhead and back-pressure unit, similar in features of the first five units, but in this case fabricated by Toshiba (Table 24.2). The main differences with the Mitsubishi's units were the number of stages of the turbine (four instead of five), the rotating speed (6088 rpm instead of 3600) of the turbine, a speed-reducing gear box, and the SSC that was 7% higher (3.51 kg/s per MW instead of 3.28). Despite its high SSC, this unit is still in operation (Table 24.2) but is destined to be taken out of operation when a new plant is installed.

The last three back-pressure plants were installed in Los Azufres in 1989 (Unit 8), 1990 (Unit 9), and 1992 (Unit 10) (Tables 24.2 and 24.3). They were manufactured by Ansaldo-Makrotek, a joint venture between a Mexican and an Italian company that produced several similar plants for CFE. These units are more similar to the Toshiba design than the Mitsubishi one, but the speed of the turbine is between both. As in all

Figure 24.5 General view of the back-pressure Unit 10, constructed by Ansaldo-Makrotek. The feeding well (Az-26, see Fig. 24.3) is on the left side and the power house is on the right. Photograph by Alfredo Mañón-Mercado.

Table 24.3 Main features of the wellhead, back-pressure power units in Los Azufres

Plant	Units 1, 2, 3, 4, 5	Unit 6	Units 8, 9, 10
General data			
• Manufacturer	Mitsubishi	Toshiba	Ansaldo/Makrotek
• Start-up year	1982	1986	1989, 1990 & 1992
• Rating (MW net)	5	5	5
Turbine			
• Cylinders	1	1	1
• Flows	1	1	1
• Stages per flow	5	4	4
• Inlet pressure (bar,a)	8	8	8
• Inlet temperature (°C)	170	170	170
• Exhaust pressure (bar,a)	0.7 (Atmospheric)	0.83 (Atmospheric)	0.73 (Atmospheric)
• Steam mass flow (kg/s)	16.4	17.6	17.6
• Speed (rpm)	3600	6088	5600
Generator			
• Output (kVA)	6250	6250	6250
• Voltage (kV)	4.16	4.16	4.16
• Frequency (Hz)	60	60	60
• Power factor	0.8	0.8	0.8
• Cooling medium	Air	Air	Air
Condenser	None	None	None
Cooling system	None	None	None
NCG system	None	None	None
Plant performance			
• SSC (kg/s per MW)	3.28	3.51	3.51

Notes: *NCG*, noncondensable gas; *SSC*, steam-specific consumption. Units 1, 2, 3, 4, 5, and 8 are currently dismantled, as indicated in Table 24.2.

back-pressure plants, there is no condenser, cooling system, or system for removal of NCGs. Unit 8 was dismantled and moved to the Los Humeros field in 2003, while Units 9 and 10 remain operating (Table 24.2).

Unit 7, the first large condensing plant in Los Azufres, was synchronized to the grid in November 1988 and officially inaugurated on June 27, 1989. It is a 50-MW (net) single-flash power plant manufactured by General Electric (GE) that was supposed to be installed in New Mexico, USA but ended up in Los Azufres. The turbine has a single cylinder with two flows. Steam enters into the turbine via the central plenum and one half flows in one direction while the other half flows in the opposite direction; each flow exits the turbine to the condenser through two ducts, one at each end of the turbine. The pattern of blades is symmetric with the same number of stages (six) and the same size blades in each side (Fig. 24.6); the rotor turns at 3600 rpm. The steam enters at 8 bar,a and 169°C at a rate of 385 t/h (107 kg/s) at full load and it is exhausted to the condenser at 0.135 bar,a. The electric generator, also made by GE, works at 13,800 V with a power factor of 0.9 and is cooled by hydrogen. The cooling system is composed of an eight-cell, counterflow water cooling tower with induced draft and top-mounted fans (Table 24.4). The original SSC was around 7.7 t/h (2.14 kg/s) per MW, which was very efficient for that time (mid-1980s). But the plant had been originally designed for different reservoir conditions and it was difficult for CFE to make it work effectively under the actual conditions of the Los Azufres reservoir. Thus, the plant had frequent shutdowns and periods of de-rated operation, until CFE made a major refurbishment in the mid-1990s. Since then the plant has been working regularly.

Two small binary cycle units, described later, were installed in 1993, but it was about one decade before a new unit would be installed in Los Azufres. Unit 13 started to operate in January 2003, soon accompanied by Unit 14 in February, Unit 15 in May, and Unit 16 in July of the same year (Table 24.2), all of them being part of the

Figure 24.6 Turbogenerator group of the condensing flash Unit 17 of 50 MW, constructed by Mitsubishi.
Photograph by Emigdio Casimiro-Espinoza.

Table 24.4 Main features of the condensing power units at Los Azufres

Plant	Unit 7	Units 13, 14, 15, 16	Unit 17
General data			
• Manufacturer	General Electric	Alstom	Mitsubishi
• Start-up year	1988	2003	2015
• Rating (MW net)	50	25	50
Turbine			
• Cylinders	1	1	1
• Flows	2	1	1
• Stages per flow	6 for each side	7	5
• Inlet pressure (bar,a)	8	8.63	8
• Inlet temperature (°C)	169	173.6	170
• Exhaust pressure (bar,a)	0.135	0.13	0.095
• Steam mass flow (kg/s)	107	55.6	101.9
• Speed (rpm)	3600	3600	3600
• Last stage blade height (mm)	508	NA	749.3
Generator			
• Output (kVA)	64,000	44,200	60,222
• Voltage (kV)	13.8	13.8	13.8
• Frequency (Hz)	60	60	60
• Power factor	0.9	0.9	0.9
• Cooling medium	Hydrogen	Air	Water
Condenser			
• Type	Double-flow, surface, horizontal	Low-level, spray-jet, direct contact	Low-level, spray-jet, direct contact
• Vacuum pressure (bar,a)	0.135	0.12	0.095
• Cooling water flow (kg/s)	4040	1300	2024

Continued

Table 24.4 Continued

Plant	Unit 7	Units 13, 14, 15, 16	Unit 17
• Cooling water temperatures			
Inlet (°C)	21.1	26.9	22.9
Outlet (°C)	35.5	49.9	40.8
Cooling system			
• Type	Counterflow, induced draft cooling tower	Counterflow, induced draft cooling tower	Counterflow, induced draft cooling tower
• No. of cells	8	2	4
• Water flow rate (kg/s)	4543	1485	2351
• Fan power (kW)	746	298	744
NCG system			
• Steam-jet ejector	Yes	Yes	Yes
• Stages	2	2	2
• Steam flow (kg/s)	9.7	2.22	4.12
• Vacuum pump power requirement (kW)	NA	210	582 (Annual average)
Plant performance			
• SSC (kg/s per MWh)	2.14	2.22	2.04

Notes: *NA*, not available; *NCG*, noncondensable gas; *SSC*, steam-specific consumption. All units are in operation.

Los Azufres II project. The plants were constructed, installed, and put in operation by Alstom through an engineering, procurement, and construction (EPC) contract signed with CFE as a result of an international tender.

Units 13 to 16, each with 25 MW of net capacity, have a seven-stage turbine spinning at 3600 rpm where the first five stages are of impulse design with the two last stages using specialized three-dimensional blade forms with aerodynamically optimized surface contours for increased efficiency prior to the exhaust. Each unit has a hybrid gas extraction system consisting of ejectors and vacuum pumps to remove NCGs and prevent them from building up inside the condenser and reducing performance. The condenser is of direct contact type, almost entirely made of stainless steel. Cooling is provided by a standard wooden and fiberglass counterflow cooling tower with two cells and PVC packing. The turbines were manufactured by the Alstom's local facility in Morelia, nearby the Los Azufres field (Martínez-Toledo, 2011). The units consume an average of 200 t/h (55.6 kg/s) of steam at 8 bar,a and 173.6°C. Therefore, they have a nominal nameplate SSC of 8 t/h (2.22 kg/s) per MW, which is slightly higher (worse) than the nominal 2.14 kg/s/MW of the older Unit 7 (Table 24.4). All units are operating properly.

The more recent addition to the Los Azufres fleet is Unit 17. It is also a condensing, single-flash power plant constructed, installed and put in operation by Mitsubishi through an EPC contract awarded by CFE. The plant has 50-MW net capacity and was successfully synchronized to the electric grid of CFE in February 2015, after a long span of 12 years with no new plants added to the field. The turbine is composed of only five stages, which results in a relatively large last stage of 749 mm in height. An average of 367 t/h (101.9 kg/s) of steam enters to the turbine at 8 bar,a and 170°C, driving it at 3600 rpm and is extracted by the condenser at a very low absolute pressure of 0.095 bar,a. The condenser is also of direct contact type and spray-jet, and the condensate is cooled by a tower of counterflow and induced draft type with four cells and fans. Unit 17 presents the best efficiency with a nominal nameplate SSC of 7.34 t/h (2.04 kg/s) per MW (Table 24.4).

The only binary cycle plants currently at Los Azufres, and in all of Mexico, are Units 11 and 12. These were manufactured and installed by Ormat in July and October 1993, being officially inaugurated in February 1994. Each unit had an installed capacity of 1.5 MW (Table 24.5), with the aim to take advantage of part of the separated

Table 24.5 **Main features of the binary cycle power units at Los Azufres**

Plant	Units 11, 12
General data	
• Manufacturer	Ormat
• Start-up year	1993
• Rating (MW net)	1.45
Turbine	
• Cylinders	1
• Flows	1
• Stages per flow	2
• Inlet pressure of the working fluid (bar,a)	12.64
• Inlet temperature (°C)	139
• Exhaust pressure (bar,a)	1.31
• Speed (rpm)	3600
• Last stage blade height (mm)	NA

Continued

Table 24.5 Continued

Plant	Units 11, 12
Generator	
• Output (kVA)	1875
• Voltage (kV)	4.16
• Frequency (Hz)	60
• Power factor	0.8
• Cooling medium	Air
Condenser	
• Type	Air-cooled
• Vacuum pressure (bar,a)	1.31
Cooling system	
• Type	Air-cooled
NCG system	None
Working fluid	
• Type	Isopentane
• Boiling temperature (°C)	36
• Flammability in air limit (%)	1.4 to 8.3
• Flow to the preheater (kg/s)	39.8
• Storage capacity (m^3)	10.6
Hot water	
• Total flow (kg/s)	43
• Inlet temperature to vaporizer (°C)	175
• Inlet temperature to preheater (°C)	139
• Outlet temperature from preheater (°C)	110
Plant performance	
• SWC (kg/s per MWh)	28.7

Notes: *NA*, not available; *NCG*, noncondensable gas; *SWC*, specific water consumption. Note that both units are currently out of service and partially dismantled, as indicated in Table 24.2.

water to generate electricity while learning about their performance and possible associated problems. Both were of organic Rankine cycles (ORC) using isopentane as the organic working fluid. They were typical ORC units composed of a turbogenerator, air-cooled condenser, two heat exchangers (one a preheater, the other a vaporizer), a tank for storage of the isopentane and a pump to move it in liquid phase from the condenser to the heat exchangers. From there, the working fluid in gas phase passed through to the turbine and the condenser, where it returned to the liquid state and the cycle repeated in a closed loop.

The turbines had only two stages that spun at 3600 rpm propelled by gaseous isopentane at 12.64 bar,a and approximately 139°C. Each turbogenerator system demanded around 155 t/h (43 kg/s) of hot water that reached the vaporizer at 175°C. After passing through this first heat exchanger, the water temperature dropped to 139°C and it entered the second heat exchanger (the preheater). At the exit of the preheater the temperature had decreased to 110°C, and the water was sent to the injection system (Table 24.5). As far as is known, the operation of the plants never reached expectations since their electric output was a fraction of their nominal capacity. Eventually, in December 2006, CFE took them out of operation.

As of 2015, there are 11 power plants installed at Los Azufres with a total capacity of 218 MW (net) and 227.4 MW (gross), as shown in Table 24.2. Two of those 11 plants (the binary plants) remain installed in the field but not in operation; thus the running or operative capacity is 224.4 MW (gross) as reported also in Table 24.1.

As of December 2014, the installed and running capacity in the field was 194.4 MW and 191.4 MW, respectively, as indicated in Table 24.1. As mentioned before, those figures come from the fact that Unit 17 had not yet entered service and the back-pressure Units 2 to 5 were still in operation. Thus, at that time, there were 14 power plants installed and 12 in operation.

The electricity output at Los Azufres was 1538 GWh in 2014 (Flores-Armenta, 2015). Taking into account the gross installed capacity in that year, the mean capacity factor in the field was 90.3%, the highest of the four operating Mexican geothermal fields. However, the peak generation in Los Azufres was reached in 2011 with 1576 GWh (Gutiérrez-Negrín, 2012), with capacity factor of 92.5%.

24.5 Perspectives

In general, the outlook for the geothermal market in Mexico looks encouraging since the new energy framework makes it possible for private investors to participate in developing new, greenfield geothermal projects in the country. Recent estimates indicate that the geothermal potential in Mexico for conventional hydrothermal resources at temperatures $\geq 150°C$ is approximately 2310 MW, composed of 125 MW of proven reserves, 245 MW of probable reserves, 75 MW of measured resources, 655 MW of indicated resources, and 1310 MW of inferred resources, according to the definitions of the Australian geothermal code (Gutiérrez-Negrín, 2012). The potential for nonconventional, hot rocks exploitable using EGS technologies is still under assessment, but a preliminary estimate shows more than 5200 MW (electric). In addition, the potential development for low- to medium-enthalpy geothermal resources for direct uses is even larger since they are underdeveloped in Mexico (Gutiérrez-Negrín et al., 2015). Thus, the combination of private investment and an important potential to be tapped, allows one to expect high growth rates for Mexico's geothermal market.

One supportive element in the geothermal panorama in Mexico is the creation of the Mexican Center for Innovation in Geothermal Energy (CeMIE-Geo: Centro Mexicano de Innovación en Energía Geotérmica) in February 2014. The center is a consortium composed of 22 entities headed by the Center for Scientific Research and High Studies of Ensenada. Twelve of its members are public research institutes and universities, nine are private companies, and the other entity is CFE. A steering group (Grupo Directivo) was formed, composed of three representatives from private companies, six representatives from the academia, and one from CFE. The CeMIE-Geo is not intended to have any physical offices, labs, or administrative structure but uses the infrastructure of its 22 parties (Gutiérrez-Negrín et al., 2015).

The CeMIE-Geo is conducting 30 research and innovation geothermal projects between 2014 and 2017, which encompass improvements in exploration tools and methods, evaluation of Mexico's EGS potential, development and testing of direct-use prototypes, improvement of steam and brine pipelines and turbine materials, and specific studies in geothermal areas. All the 30 projects have been defined, described, programmed, scheduled, and budgeted for about US$87 million, $74 million of which (88%) is provided by the federal government and the rest by the private companies as in-kind contributions (Gutiérrez-Negrín et al., 2015).

Some geothermal plants are currently under construction in the Los Humeros and Domo San Pedro fields. In both cases, they are 25-MW condensing plants to be completed in early and late 2016, respectively. The EPC contract for Los Humeros plant is being developed by Alstom and the EPC for Domo San Pedro by Mitsubishi.

In the particular case of Los Azufres, in 2015 CFE is launching an international tender to award the EPC contract for another 25-MW condensing, single-flash plant, scheduled to be commissioned in 2018. This new and more efficient plant will replace the last three 5-MW back-pressure plants (Units 6, 9, and 10; Table 24.2), while using almost the same amount of steam. The back-pressure units are slated to be dismantled and removed from the field. With this new plant (Unit 18), the net addition will be 10 MW, raising the total installed capacity in the field to 237 MW. However, the number of plants will reduce to nine, including the two idle 1.5-MW binary plants.

The total power-generating capacity of the geothermal reservoir in Los Azufres seems to be larger than the current level of exploitation. Some volumetric models applied in Los Azufres estimated 230 MW in the south zone and 480 MW in the north zone, for a total of 710 MW. According to that, the current capacity could be easily doubled. Stricter and history-matched mathematical simulations have estimated up to 260 MW in both zones which would barely support one more 25 MW plant, besides the future Unit 18. However, even if the conventional resource is approaching the field's limit, the hot rock resource seems to extend toward the eastern portion. In the next decade, this area could be commercially exploited by using EGS technologies, widening the extent of the exploitable resource. Thus, the geothermal perspectives seem to be as promising in Los Azufres as in the rest of Mexico.

Acknowledgments

The authors of this chapter want to thank Leonardo Huazano-Arredondo, Carlos A. Cuervo-Padilla, Emigdio Casimiro-Espinoza, and Héctor Pérez-Esquivias from CFE, as well as Alfredo Mañón-Mercado for the information, photographs, and support provided.

References

Arellano, G.V.M., Torres, R.M.A., Barragán, R.R.M., Sandoval, M.F., 2004. Respuesta a la explotación (1982−2003) del yacimiento geotérmico de Los Azufres, Mich. (México). Parte I: Zona Norte. Geotermia 17 (1), 10−20.

Arellano, G.V.M., Torres, R.M.A., Barragán, R.R.M., Sandoval, M.F., 2005. Respuesta a la explotación (1982−2003) del yacimiento geotérmico de Los Azufres, Mich. (México). Parte II: Zona Sur. Geotermia 18 (1), 18−25.

Arellano, G.V.M., Ramírez, M.M., Barragán, R.R.M., Paredes, S.A., Aragón, A.A., López, B.S., Casimiro, E.E., 2015. Condiciones termodinámicas de los fluidos del yacimiento de Los Azufres (México) y su evolución en respuesta a la explotación (1979−2011). In: Proceedings of the 22nd Annual Congress of the Mexican Geothermal Association (AGM), Cuernavaca, Mor. Mexico, 10−11 March 2015 (in Spanish).

Barragán Reyes, R.M., Arellano Gómez, V.M., Mendoza, A., Reyes, L., 2012. Variación de la composición del vapor en pozos del campo geotérmico de Los Azufres, México, por efecto de la reinyección. Geotermia 25 (1), 3−9.

Del Valle, J.L., 2015. Perspectivas de desarrollo privado con la nueva Ley de Energía Geotérmica. In: Presentation at the 22nd Annual Congress of the Mexican Geothermal Association (AGM), Cuernavaca, Mor. Mexico, 10−11 March 2015 (in Spanish).

Dobson, P.F., Mahood, G.A., 1985. Volcanic stratigraphy of the Los Azufres geothermal area, Mexico. Journal of Volcanology and Geothermal Research 25, 273−287.

Ferrari, L., Garduño, V.H., Pasquaré, G., Tibaldi, A., 1991. Geology of Los Azufres Caldera, Mexico, and its relationships with regional tectonics. Journal of Volcanology and Geothermal Research 47, 129−148.

Flores-Armenta, M., 2015. Perspectivas de la nueva empresa productiva del estado. In: Presentation at the 22nd Annual Congress of the Mexican Geothermal Association (AGM), Cuernavaca, Mor. Mexico, 10−11 March 2015 (in Spanish).

García-Gutiérrez, A., Martínez-Estrella, J.I., Hernández-Ochoa, A.F., Canchola-Félix, I., Mendoza-Covarrubias, A., 2010. Numerical modeling of complex geothermal steam transportation networks: the cases of Cerro Prieto and Los Azufres, Mexico. In: Proceedings World Geothermal Congress 2010, Bali, Indonesia, 25−29 April 2010 paper 2506, 12 pp.

Gutiérrez-Negrín, L.C.A., 2012. Update of the geothermal electric potential in Mexico. Geothermal Resources Council Transactions 36, 671−677.

Gutiérrez-Negrín, L.C.A., 2015. Mexican geothermal plays. In: Proceedings World Geothermal Congress 2015, Melbourne, Australia, 19−25 April 2015 paper 11078, 9 pp.

Gutiérrez-Negrín, A., Aumento, F., 1982. The Los Azufres, Michoacán, Mexico, geothermal field. Journal of Hydrology 56, 137−162.

Gutiérrez-Negrín, L.C.A., Maya-González, R., Quijano-León, J.L., 2015. Present situation and perspectives of geothermal in Mexico. In: Proceedings World Geothermal Congress 2015, Melbourne, Australia, 19−25 April 2015 paper 01002, 10 pp.

Macías Vázquez, J.L., Jiménez Salgado, E., 2012. Actualización vulcanológica del complejo de Las Tres Vírgenes, BCS. In: Proceedings of the 20th Annual Congress of the Mexican Geothermal Association (AGM), Morelia, Mich., Mexico, 26–28 September 2012 (in Spanish).

Martínez-Toledo, L.A., 2011. Geothermal turning up the heat at the Los Humeros geothermal field, Puebla, Mexico. In: Proceedings of the 19th Annual Congress of the Asociación Geotérmica Mexicana (AGM), Los Humeros, Pue., Mexico, September, 2011.

Pérez Esquivias, H., Macías Vázquez, J.L., Garduño Monroy, V.H., Arce Saldaña, J.L., García Tenorio, F., Castro Govea, R., Layer, P., Saucedo Girón, R., Martínez, C., Jiménez Haro, A., Valdés, G., Meriggi, L., Hernández, R., 2010. Estudio vulcanológico y estructural de la secuencia estratigráfica Mil Cumbres y del campo geotérmico de Los Azufres, Mich. Geotermia 23 (2), 51–63.

Quijano-León, J.L., Gutiérrez-Negrín, L.C.A., September–October 2003. Mexican geothermal development, an unfinished journey. Geothermal Resources Council Bulletin 5, 198–205.

SENER, 2014. Iniciativas de leyes secundarias. Electricidad. Information available at. http://www.energia.gob.mx/webSener/leyes_Secundarias/Seg_nivel_Elect.html. consulted May 12, 2014.

SENER, 2015. Statistics of the Electricity Sector. Information available at: http://egob2.energia.gob.mx/portal/electricidad.html. consulted June 15, 2015.

Verma, M.P., Nieva, D., Quijano, L., Santoyo, E., Barragán, R.M., Portugal, E., 1989. A hydrothermal model of Los Azufres geothermal field. In: Miles, D.L. (Ed.), Proceedings of the 6th Symposium on Water-Rock Interaction (WRI-6). Malvem, UK, 3–8 August 1989, pp. 723–726.

Viggiano-Guerra, J.C., Gutiérrez-Negrín, L.C.A., 1995. Comparison between two contrasting geothermal fields in Mexico: Los Azufres and Los Humeros. In: Proceedings World Geothermal Congress 1995, Florence, Italy, 18–31 May 1995, pp. 1575–1579.

Enhanced geothermal systems: review and status of research and development

E. Huenges
Helmholtz Centre Potsdam, GFZ German Research Centre for Geosciences, Potsdam, Germany

25.1 Introduction

Today, most of the large geothermal power plants in the world use steam and hot water from volcanically active regions to generate electric power. Geothermal reservoirs available in nonvolcanic areas are hot, either deep water—bearing systems (hydrothermal systems) or systems without or with limited water (petrothermal systems) (Fig. 25.1) (Huenges, 2010). Hydrothermal systems are deep water—bearing layers (aquifers) with naturally sufficient hydraulic permeability. The permeability of continental crustal rocks is defined by the capacity of the geological medium to transmit fluid. Crucial for the economic operation of hydrothermal systems is a sufficient temperature of the aquifer, and a hot water production rate of at least 100 m³/h has to be achieved. The development of hydrothermal systems in general entails an exploration risk. While a certain temperature is reached at a depth nearly everywhere, the number of economically interesting locations is restricted considerably by the second condition due to low permeability and thus too low thermal water production. Petrothermal systems are not bound to water-bearing formations in the subsurface. They use the existing heat in the rock which increases with an average of 3 K per 100-m depth.

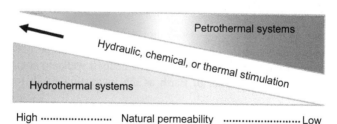

Figure 25.1 Definition of hydrothermal and petrothermal geological systems by their natural permeability and the EGS concept, which includes artificial improvement of the hydraulic performance of a reservoir with the goal to use it for an economical provision of heat or electric energy. The enhancement challenge is based on several nonconventional methods for exploring, developing, and exploiting geothermal resources that are not economically viable by conventional methods.

Geothermal Power Generation. http://dx.doi.org/10.1016/B978-0-08-100337-4.00025-5

Many of the hydrothermal and all petrothermal systems can be developed to an economic state by means of the so-called enhanced (or engineered) geothermal systems (EGS) concept. EGS are geothermal reservoirs in which technologies enable economic utilization of low permeability conductive dry rocks or low productivity convective water-bearing systems by creating fluid connectivity through hydraulic, thermal, or chemical stimulation methods or advanced well configurations. Hydraulic or acid stimulation treatment technologies are available to help generate that artificially higher hydraulic permeability in initially low-permeability rocks. Therefore, the concept of EGS is designed to make geothermal energy utilization feasible in most environments and thus offers an enormous untapped potential. EGS technologies represent the sum of the engineering measures that are required for the transfer of heat and to optimize the exploitation of the reservoir.

Hydraulic EGS treatments are generally intended to improve productivity (or injectivity) of a geothermal reservoir by increasing the overall transmissivity of the reservoir rocks. This goal can be achieved by the various stimulation methods that are dependent on the geological system comprising the rocks, the rock structures, the tectonic situation, as well as the stress field. Besides the controlled enhancement of the reservoir, the sustainable operation of the system presents a challenge, as newly opened fractures may close again with reduced reservoir pressure and because of chemical interaction with the fluid, which may lead to neoprecipitation and reduced permeability. Another challenge lies in the side effects of hydraulic treatments, as the high fluid pressures applied and the large fluid volumes injected in such treatments sometimes induce seismic events that can, in some cases, be felt at the surface and jeopardize the public acceptance of a project.

This chapter reviews issues of EGS based on a general characterization of geothermal energy systems, the critical parameters of EGS, followed by a presentation of potential treatments and their interaction with the environment. The sustainability of operating EGS will be addressed and an outlook given.

25.2   Characterization of geothermal energy systems

Geothermal energy is related to the temperature at depth and to the pathways to extract the heat from the Earth with a heat carrier, usually water. The property "permeability" is required to characterize pathways in Earth. The deepest wells do not exceed 13 km. We consider the uppermost crust of Earth as the interesting depth for commercial geothermal heat recovery.

The temperature gradient that has been measured in deep wells on average is 30 K/km. The continental Earth crust reaches depths of about 30 km, where the boundary between the crust and the mantle is located. In geologically old continental crustal shields (eg, Canada, India, South Africa), lower temperature gradients can be observed (eg, 10 K/km). In contrast, much higher gradients are measured in tectonically active young crustal areas, such as at the boundaries of lithosphere plates (eg, in Iceland, in Larderello in Italy (approximately 200 K/km)) or in rift regions (eg, up to 100 K/km in the Rhine rift).

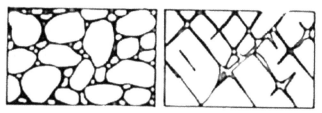

Figure 25.2 Potential pathways for geothermal fluids: left, porous media; right, fractured media.

The permeability of the continental crust is the essential critical parameter for the definition of the geothermal reservoir as it plays a fundamental role in heat and mass transfer.

Permeability is related to two basic properties of the rocks:

1. The porosity is the ratio of pore volume to the total volume (Fig. 25.2). The *intrinsic permeability* is the prerequisite of the fluid flow through the pore network of the rock and can be correlated to the porosity. Porosity and permeability are directly linked to the packing of the minerals within the rocks, which is a result of the size sorting of the minerals and elements and of the compaction and diagenetic history of the rocks. Sedimentary rocks such as limestone, sandstone, or conglomerate are generally porous and can store large quantities of fluids within their pore network. They constitute natural reservoirs in the crust for all kind of fluids. The intrinsic permeability parameter is the primary control on fluid flow as it will vary from 10^{-23} m^2 in intact crystalline rocks to 10^{-7} m^2 in porous sediments, meaning a 16-order of magnitude variation.
2. The *fracture permeability* is linked to the discontinuities that are present within the rock along which fluid circulation is possible. This type of permeability is generally well developed in crystalline massifs but can also be found in deep sediments. Thus, although granite is an impermeable rock, a granitic massif will be considered as a permeable massif as a whole because of fluid circulating along the fracture network. Implicitly, such permeability will be well developed in the vicinity of large fracture systems, whether active or fossil. Because of the discontinuous character of the fracture and their geometrical complexity, the intrinsic permeability of such system is more difficult to evaluate compared to stratified permeable layers.

The knowledge of underground physical conditions, especially the magnitude and the stress direction at site, is important for reliable drilling into EGS reservoirs. Awareness of the stress conditions is also a prerequisite for starting hydraulic fracturing treatment which is addressed in the following section.

25.3 Reservoir types applicable for EGS development

The natural occurrences of formation pathways rely on the 3D kinematic evolution of fault systems and fracture networks. Favorable settings generally involve subvertical conduits of highly fractured rock along fault zones oriented approximately perpendicular to the least principal stress (Fig. 25.3), which is caused by the tectonic regime. These pathways follow tensile fracture behavior of the rocks. Conjugated systems, which have a shearing potential, are oriented with an angle of about 30 degrees from the direction of the maximum principal stress and provide additional permeability. Knowledge of

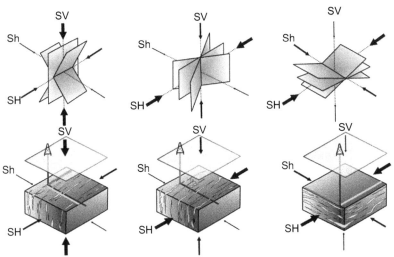

Figure 25.3 Geometrical relation between stress axes, stress regimes, and fracture planes. Upper part: shear fractures (brown); tensile fractures (blue). Stress regimes from left to right: normal (SH > SV > Sh) faulting, strike-slip faulting, and reverse faulting. Lower part: From left to right, orientation of tensile fractures in the three faulting regimes. Red drill path is least stable; green drill path is most stable. In strike-slip regimes, the most stable drill path depends on the stress ratios of SV and SH (Moeck in).

Bruhn, D., Manzella, A., Vuataz, F., Faulds, J., Moeck, I., Erbas, K., 2010. Exploration methods. In: Huenges, E. (Ed.), Geothermal Energy Systems: Exploration, Development and Utilization. Wiley-VCH, Berlin, pp. 37—111.

the regional stress field is required (Holl and Barton, 2015), usually determined preferably from deep wells (ie, borehole breakouts or drilling induced fractures). Other, off-well methods to determine the stress field using seismic or structural geological observations serve as first-order stress indicators.

However, permeability development is the result of a time-dependent process as fluid—rock interactions can result in permanent dissolution and recrystallization that can modify the permeability network. The intensity and orientation of the stress field exert a direct control on this process by determining zones of compression and extension in relation to the relative position of the main stress axis and resulting strain.

All necessary system components for applying the EGS concept are available, but there is still potential for improvement in terms of reliability and efficiency of the technologies. Techniques and experiences from EGS sites are described in the following section, providing a set of methods available for addressing the goal of increasing well productivity.

25.4   Treatments to enhance productivity of a priori low-permeable rocks

In many cases, drilling operations will not open up a geothermal reservoir under such conditions that an extraction of geothermal energy is economically viable without any further measures. Geothermal wells often have to be stimulated in order to increase well productivity. Different stimulation concepts have been developed to enhance the productivity of geothermal wells. Formally, stimulation techniques can be subdivided with respect to their radius of influence. Techniques to improve the near wellbore region up to a distance of a few tens of meters are chemical treatments and thermal fracturing. Hydraulic fracturing is the only demonstrated stimulation method with the potential to improve the far field, up to several hundreds of meters away from the borehole.

25.4.1   Thermal stimulation

Thermal stimulation treatments are performed in order to increase the productivity or injectivity of a well. Thermal stimulation can enhance the near well permeability, which may have been reduced by drilling operations (drill cuttings or mud clogging feed zones), or may open hydraulic connections to naturally permeable zones that were not intersected by the well path. Cold water injection has an effect on the heterogeneous contraction of the minerals of the hot rocks. This leads to local stresses. In connection with the outer stress field, fractures will be generated in preferred directions. In addition, the parts of rocks in the neighborhood of preferred fluid pathways will get a greater temperature decrease with cold water injection and thus preferred further opening. This self-reinforcing effect either leads to reopening of existing, possibly sealed fractures, or creates new fractures through thermal or additional hydraulic stresses.

25.4.2   Chemical stimulation

Chemical injection with specific chemicals that are able to dissolve hydrothermal minerals during long-term circulation can also improve the injectivity with time. Acid treatment is a preferred method to enhance flow performance near wells. Productivity index and injectivity index may be significantly enhanced by up to 30% to a distance of up to 1 m from the borehole, depending on the lithology and the preexisting permeability—porosity field.

25.4.3   Hydromechanical stimulation

An alternative method is the generation and propping of small tensile fractures or inducing self-propping shear fractures by hydraulic stimulation. This method is routinely used in tight-gas formations, for example in the Netherlands and northern Europe and—to our knowledge—has never produced unplanned seismicity by itself, but possible reactivated unknown faults nearby (Westaway and Younger, 2014). Using this method, the Productivity Index (PI) and the Injectivity Index (II) (ie, production/injection rate divided by the driving pressure gradient) can be enhanced by up to 100—180%, with fracture lengths ranging from 25 to 100 m, respectively. The technologies of massive water injection and hydraulic-proppant treatments are well known and described next.

25.4.3.1 Massive water injection treatments

Massive water injection treatments are applied in low-permeable or impermeable rocks by injecting large amounts of water to produce long fractures with low width compared with the following treatments (see Fig. 25.4). Parameters to control the treatment itself are the rate of injection, which affects the pressure in the reservoir and properties of the injection fluid such as its viscosity. This hydraulic fracturing technology in geothermal reservoirs is still in the trial and error phase, and standards for treatments and the required completions of the well such as well heads and frac strings are not yet fixed. The special treatment design is site dependent and is usually based on expertise about the target sections. In order to get an optimum result from different depth sections in the same well with distinguishable properties, it is recommended to treat the sections individually with hydraulic seals in between.

The lessons learned applying massive water injection treatment can be summarized as follows: It is recommended to use water as the fracture fluid—best taken from a shallow well—with a viscosity of $1-10$ cP. A small proppant concentration of $c = 50-200$ g/L seems to support the stimulation goals. A fracture length can be determined by matching the pressure history using a modeling program that addresses fracture generation. See, for example, in Soultz-sous-Forêts, Jung (2013) or a review by Schill et al. (2015); in Habanero, Hogarth et al. (2013) and Mills and Humphreys (2013); and in Groß Schönebeck, see Legarth et al. (2005) and Zimmermann et al. (2009, 2010, 2011).

In general, massive water injection treatments with injection rates up to 150 L/s produce long fractures in the range of a few 100 m with low apertures of approximately 1 mm and hence low conductivity. The success of the treatment depends on the self-propping of the rock and on the potential of shear displacement (Fig. 25.5). Costs of this treatment are significantly lower compared to hydraulic-proppant fractures, which will be described later.

Additional measures to enhance the success of the treatment design include adding some abrasive agent in the fluid during the high injection rates, such as sand or proppants. This will help conductivity into the fractures to be created and using a proppant suspending agent that gives the proppant mechanical suspension while traveling

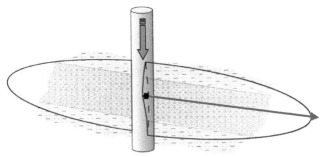

Figure 25.4 Artificial fracture with fracture width w_f and fracture length x_f generated by the massive water injection technique in a normal faulting or strike-slip faulting tectonic regime.

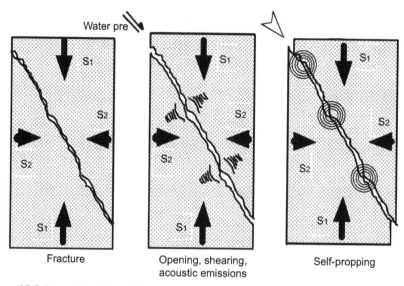

Figure 25.5 Potential self-propping mechanism after water fracture treatment.
From Jung, R. 1999. Erschließung Permeabler Risszonen für die Gewinnung Geothermischer Energie aus Heißen Tiefengesteinen. Final report, project 0326690A funded by the German Ministry for Education and Research (BMBF), 82 pp.

through the artificial fracture will increase the height of the fracture that will be generated and allow the proppant to travel to the end of the fracture. It can be considered using a friction-reducing agent in the fluid as opposed to using a guar-based gel in case pH values do not correspond between the fluids injected and the cross-linked gel.

25.4.3.2 Hydraulic-proppant treatments

Hydraulic-proppant treatments are used to stimulate reservoirs with cross-linked gels in conjunction with proppants of a certain mesh size. Parameters to control the treatment are, in addition to the rate of injection, the injection fluid viscosity, the opening or closure pressure for fractures, and the amount and characteristic of proppants (ie, well-sorted, small, high-strength spheres with a diameter between 0.4 and 1 mm). Knowledge of reservoir parameters, such as the tortuosity (ie, the winding of the fluid pathways), and the potential for friction during the fluid flow is crucial. Hydraulic-proppant treatments can be applied in a wide range of formations with varying permeabilities and a good control of stimulation parameters. The produced fractures have a short length of about 50−100 m, but a higher aperture of up to 10 mm compared to massive water injections. Hydraulic-proppant treatments are especially used to bypass the wellbore skin in high-permeable environments. In general, this kind of treatment is more expensive than a massive water injection treatment.

Typically, the hydraulic-proppant treatments start with an injection test to obtain information about friction and tortuosity of the perforated interval. In this specific experiment, one first pumps an uncross-linked gel, and the measured hydraulic data give an indication of any near-wellbore problems that could potentially adversely affect the result of the fracture treatment. This is followed by pumping a cross-linked fluid, which gives an idea of leak-off (ie, lateral fluid losses) as well as helping to predict closure pressures, the fracture geometry, and whether there is any indication of pressure-dependent leak-off.

The main treatment after such pretesting is an injection of gel-proppants with a step-wise increase of proppant concentration with a highly viscous cross-linked gel into the fracture. The result of the treatment (ie, the propagation of the fracture) mainly depends on the slurry rate and the concentration of proppants added and their variation as a function of time (Fig. 25.6).

The lessons learned applying the hydraulic-proppant treatment several times in Groß Schönebeck (Zimmermann and Reinicke, 2010) can be summarized as follows: A fracture fluid (gel) with a viscosity of 100−1000 cP is recommended. The gel, a cross-linked polymer, must be able to change its high viscosity into transportable conditions in the reservoir after sometime. The temperature of the reservoir or injected acids usually accelerates the destruction of the cross-linked polymer. For a sustainable fracture width, a proppant concentration of 200−2000 g/L is needed. A fracture length was determined by matching the pressure history using a modeling program, which is usually based on empirical experiences and addresses fracture generation.

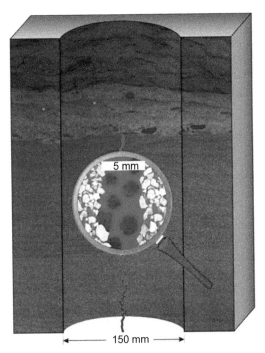

Figure 25.6 Sketch of artificial fracture generated by the hydraulic-proppant treatment with fracture (blue) and proppants (*dark dots*) in a normal faulting or strike-slip faulting tectonic regime.

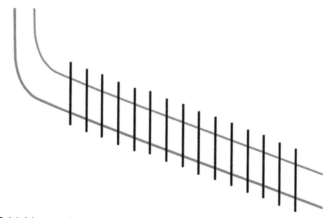

Figure 25.7 Multistage stimulation; *blue and red lines* show the boreholes, and the *vertical lines* are the foreseen generated fractures.
Yoon, J.-S., Zimmermann, G., Zang, A., 2015. Discrete element modeling of cyclic rate fluid injection at multiple locations in naturally fractured reservoirs. International Journal of Rock Mechanics and Mining Sciences 74, 15–23.

An adjustment during the treatment is possible and often necessary to avoid a screen-out of the well, that is, to fast sedimentation of proppants within the well, and preventing any further transport into the fractured reservoir. One can adjust the treatment by varying the flow rate and the proppant concentration in case the pressure progression indicates a failure of the treatment.

25.4.4 Multistage treatment

Multistage treatments (Jung, 2013) are based on an efficient method for sectioning and successfully stimulating individual sections of a well (Fig. 25.7). Boreholes are deviated from vertical and steered to intersect a maximum number of natural fissures. The injection/production borehole will be sectioned with a completion string either in an open hole or with special sectioning methods. Each section will be stimulated independently by injecting water.

The multistage system has an advantage over the single-stage system because the increase in the costs of generating a multistage system compared with a single-stage system is compensated by higher production rates. This is valid even if a majority of the stages is not successfully stimulated.

25.5 Environmental impact of EGS treatments

For existing sites requiring treatment, the focus is on the transfer of know-how from hydrocarbon exploitation, as well as from improving drinking water recovery. Special attention has to be paid to environmentally safe application. A risk assessment requires a deep understanding of the ongoing processes in order to identify the complex fluid–rock interaction and the protection of drinking water reservoirs.

Beyond the specific technical goals, the overall ambition is to show that geothermal utilization can be managed at low specific costs and low risks with high-level performance and with an overall public acceptability.

25.5.1 Induced seismicity

The process of high-pressure injection of cold water into hot rock, which is the preferred EGS method of stimulating fractures to enhance fluid circulation, dramatically changes the stress field in the immediate vicinity of the injection point (Majer et al., 2007). The reduction in effective normal stress unlocks preexisting faults in rocks that are tectonically near-critically stressed, typically causing seismic events. The variability in occurring seismicity depends on regional and local stress history (Evans et al., 2012; Ellsworth, 2013). Occasionally, seismic events large enough to be felt at the surface have occurred, causing annoyance for the population and occasionally nonstructural damage to nearby buildings.

Human-made earthquakes related to energy technologies in the subsurface appear in news headlines and are receiving increasing public attention. Seismic hazard by acid treatment is not expected. However, for hydraulic fracturing in oil and gas operations, CO_2 storage, and deep geothermal energy, each technology faces challenges, including seismic risks. Researchers of the EU-funded Geothermal Engineering Integrating Mitigation of Induced Seismicity in Reservoirs project (GEISER) from seven European countries (France, Germany, Italy, Iceland, Norway, Switzerland, the Netherlands) turned their attention to the analysis of seismicity related to the exploitation of geothermal reservoirs, in particular, EGS (Zang et al., 2014) (Fig. 25.8).

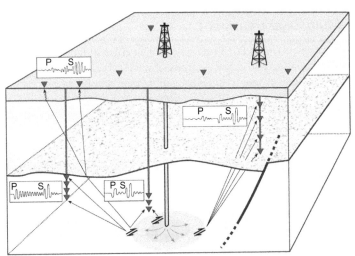

Figure 25.8 Sketch of potential seismic networks to monitor seismicity related to a hydraulic stimulation at depth. Schematic waveform examples are shown for four different scenarios: closely placed downhole sensors show the highest quality P- and S-wave onsets, whereas for surface seismic sensors the observed amplitudes may have attenuation, high impedance contrasts that introduce secondary or converted phase arrivals and higher noise level.
Zang, A., Oye, V., Jousset, P., Deichmann, N., Gritto, R., McGarr, A., Majer, E., Bruhn, D., 2014. Analysis of induced seismicity in geothermal reservoirs – an overview. Geothermics 52, 6–21.

It is remarkable that, based on observations of some geothermal projects and many other underground uses in middle Europe, seismicity induced by geothermal operations seems to be lowest (Grünthal, 2014). The starting point for seismic risk assessment is an improved understanding of failure of faults with respect to the in situ stress directions and magnitude. The orientation and strength of preexisting faults explain why such faults are close to failure in some cases, while in other areas faults are stable, far from failure, and require very high-pressure changes to induce seismicity. Since the differential stresses increase generally with depth, the depth of the injection is also a critical parameter that determines the level of the induced seismicity.

The understanding of the key parameters that control induced seismicity at a given injection site in response to an injection serves as input for the necessary seismic risk assessment of future EGS projects, that is, estimating the likelihood to induce felt and potentially damaging seismic events during reservoir creation and operation. Such events do not only pose a nuisance to the local population; they can also substantially lower the efficiency of the engineered underground heat exchanger by creating "hyperpermeable" pathways as occurred in Rosemanowes Quarry (Jung, 2013). Fluids will circulate preferentially along these shortcuts, thus cooling out rapidly resulting in a nonsustainable system.

The decrease of the number of seismic events with increasing event magnitude seems to be higher in areas with low natural seismicity (eg, Groningen/the Netherlands and Groß Schönebeck and Horstberg/North Germany) than observed at sites with some naturally occurring background seismicity (eg, the Rhine valley). A lower seismic hazard in aseismic environments is assumed not only for natural but also for induced events. In some areas such as Groß Schönebeck, induced seismic events were so weak due to the low regional stress magnitude that they could not be detected at the surface. In addition, the thick sedimentary cover above the reservoir attenuates the signals that were only be detected in a nearby well (80 events) (Kwiatek et al., 2010).

An important factor affecting seismic hazard versus risk probability is related to the volume of fluid involved. Geothermal applications, especially in systems with production and injection, are volume balanced and differ from exploited systems such as gas and coal. The exploitation of gas and coal causes changes in the mass balance in the underground. Such changes can lead to subsidence and the likelihood for seismicity is higher than in volume-balanced systems like geothermal.

Ground-motion prediction equations as specifically derived for geothermal sites for the first time by Douglas et al. (2013) become important. Ground-motion prediction equations forecast the level of shaking expected at the surface for a given seismic event. They are a key ingredient in any seismic risk assessment. At geothermal sites, small magnitudes, high frequencies, and short distances are much more relevant than for normal tectonic hazard assessment.

25.5.2 Measures to mitigate seismic events

25.5.2.1 Traffic light system

The traffic light system of Bommer et al. (2006) is based on the following rules, starting with the safest levels.

- The green zone is characterized by levels of ground motion that are either below the threshold of general detectability or, at higher ground-motion levels, at occurrence rates lower than the already established background activity level in the area. Injection operations proceed as planned.
- The amber zone is defined by ground-motion levels at which people would be aware of the seismic activity associated with the hydraulic stimulation, but damage would be unlikely. Pumping has to proceed with caution, possibly at reduced flow rates, and observations are intensified.
- The lower magnitude bound of the red zone is the level of ground shaking at which damage to buildings in the area is expected to set in. Injection is suspended immediately.

Best practice guidelines for safe and reliable EGS operations, centered on a probabilistic framework for the assessment of the risk posed by induced seismicity, are required during all phases of a project. The cornerstone of this framework is a well-tested, forward-looking traffic light system to be implemented in real-time in future EGS applications. Such a dynamic forecasting framework should be able to predict the expected seismicity in the next hours and days. It is based initially on prior information, such as the proximity to faults, the subsurface stress conditions, etc., but is then updated on the fly with real-time measurements of the observed induced seismicity (see Fig. 25.8) and downhole pressure conditions.

25.5.2.2 Cyclic treatments

The flow rate during massive water injection treatments can be constant during the whole treatment or vary in a cyclic manner with several high flow rates followed by low stages. Simulations have shown that the impact of high flow rates for the fracture performance is better, even if the intervals are limited in time, compared with a constant flow rate (Yoon et al., 2015).

Monitoring of cyclic stimulation treatments (ie, massive water injection with varying flow rates inducing different pressure levels above fracture opening state) shows the potential to control the fracture propagation and simultaneously reduce the risks of unwanted seismic events (Zimmermann et al., 2010; Kwiatek et al., 2010). This is a chance to remain below a given threshold of intensity of seismicity in order to minimize the risk on the vulnerability and exposure of people, buildings, and infrastructure. Cyclic stimulation was already performed in Groß Schönebeck, resulting in low-magnitude and low-frequency seismicity (Kwiatek et al., 2010). The layout of cyclic stimulation treatments depends on the geological setting such as the recent stress field, preexisting fractures and fault zones, and the properties of the different geological units involved. Fig. 25.9 shows the seismicity of continuous and cyclic stimulation treatments where both stimulation treatments were modeled with the same total injection reaching a similar final productivity. Seismicity was validated based on seismic activity assessments from Soultz-sous-Forêts data (Zang et al., 2013; Yoon et al., 2015). It can be concluded that cyclic stimulation induces seismic events with lower frequency and lower magnitude at lower seismic energy.

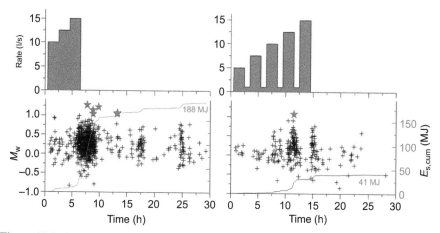

Figure 25.9 Comparison of continuous stimulation (left) and cyclic stimulation (right) with (beneath) the magnitude (M_w) of induced seismic events and accumulated seismic energy $E_{s,cum}$ (*red curve*) (Zang et al., 2013; Yoon et al., 2015). *Red stars* indicate measured maximal seismic magnitudes.

Frick, S., Regenspurg, S., Kranz, S., Milsch, H., Saadat, A., Francke, H., Brandt, W., Huenges, E., 2011. Geochemical and process engineering challenges for geothermal power generation. Chemie Ingenieur Technik 83 (12), 2093−2104.

25.5.3 Processes with treatment fluids

Hydraulic fracturing of geothermal reservoirs uses treatment or fracture fluids (ie, in most cases, pure water). Additives are required for following technical reasons (usually <1% of volume of injected fluids):

- For mitigating iron reactions, light acids such as citric acid are added. Citric acids are used also as food additives. For chemical stimulation stronger acidization is performed. The acids react with rock minerals to yield salt and water. Therefore, acidization is not a special issue of environmental concern.
- Proppant treatments require gel for preventing early sedimentation within the borehole while injecting the fluid−proppant mixture. For this purpose, cross-linker or guar gum are used, components known as laundry detergents or as food additives. After being placed in the fractures it is necessary to reduce the viscosity of the fracture fluid by breaking the cross-linked polymers. The breaking process is initiated by the temperature in the reservoir and accelerated by special breakers like ammonium persulfate which is used for bleaching hair. In some cases friction reducers are used such as polyacrylamide known from water and soil treatments.
- To prevent biologically induced or catalyzed chemical reactions that are likely to occur within a temperature window between 30°C and 80°C, biocides may be added such as glutaraldehyde which is used also as dental disinfectant.
- Corrosion and scaling is a general effect in any thermal water loop. Finding suitable inhibitors is still a matter of research although some products are on the market. A special challenge is the compatibility of inhibitor-bearing fluids in the reinjection reservoir.

The use of all these additives must be approved by the mining authorities. Therefore material safety data sheets exist either for handling the components during operation or for ensuring environmental safety.

25.5.4 Connecting drinking water horizons

Any induced hydraulic connection to drinking water horizons must be excluded. Salinity of deep water horizons increases with depth. Therefore, drinking water horizons are expected near surface usually at a distance from the reservoir to be treated. The generation of artificial fractures depends on the stress regime, the properties of the stressed rocks and the fracture fluid injection procedure. A risk assessment, which is a prerequisite of any EGS-treatment, requires this information. Vertical fracture growth in normal faulting or strike-slip faulting tectonic regimes has to be considered (see Fig. 25.3), whereas horizontal fracture growth is expected in reverse faulting regimes. In a regime in which the vertical is the lowest stress direction, ie, a reverse faulting regime, horizontal fractures are more probable coming closer to the surface and to drinking water horizons. Another aspect is that fracture propagation usually stops at sheet silicate-bearing rocks which act as barriers. Many of such layers can be found in any geological setting. Taking into account that geothermal reservoirs are explored and developed at greater depth, it can be concluded that the artificial connection to drinking water horizons is not probable.

25.6 Sustainable operation

25.6.1 Auxiliary energy

Most of EGS reservoirs are non-artesian and require a downhole pump for the production of thermal water. The installation requires the operation of a drill rig to hoist the weights and organize a reliable anchoring of the pump. Fig. 25.10 shows a pump solution with the motor downhole. Another solution uses shaft pumps with the motor at the surface. The advantage is an easier maintenance of the motor. The disadvantage is to keep the shaft from the surface to the pump running reliably with low friction, which demands lubrication. In any case, the shaft pump is restricted to shallow installation of a few hundreds of meters in maximum.

In operation the production borehole thermal water is pumped from the reservoir to the surface, leading through the surface equipment, such as a heat exchanger, and reinjecting the cooled fluid into a second well. Most of the auxiliary energy for the downhole pumping is required for overcoming the friction within the reservoir and within the tubes (casing and connecting lines). A successful stimulation may reduce the friction in the reservoir significantly. However, this is valid so long as a short circuit of the downhole hydraulic connection is excluded.

25.6.2 Flow control, monitoring systems

For extracting the heat from the natural or stimulated reservoir, it is required to close the thermal water loop as drawn in Fig. 25.11. The temperature of the water remains almost at the reservoir temperature level after a starting phase. However, the pressure decreases as shown in Fig. 25.10 and increases again within the injection well.

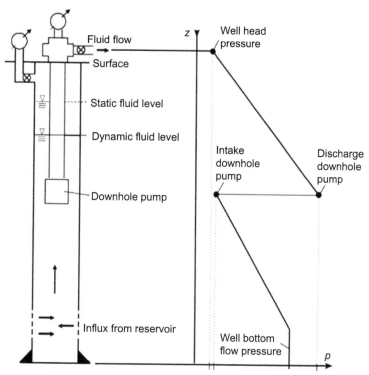

Figure 25.10 Scheme of a production well showing downhole pump, static fluid level, and dynamic fluid level (left) and the pressure curve during fluid production (right).
Frick, S., Regenspurg, S., Kranz, S., Milsch, H., Saadat, A., Francke, H., Brandt, W., Huenges, E., 2011. Geochemical and process engineering challenges for geothermal power generation. Chemie Ingenieur Technik 83 (12), 2093–2104.

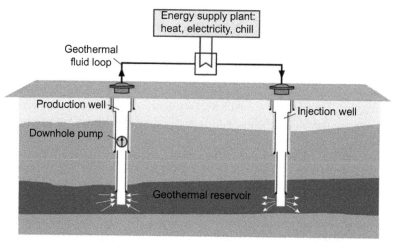

Figure 25.11 Schematic drawing of a well doublet with geothermal fluid loop and heat extraction (Frick et al., 2011).

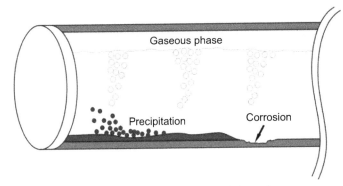

Figure 25.12 Probable processes during the circulation of thermal water.

Therefore, pressure-induced thermodynamic processes have to be considered. Modeling instruments to calculate the effect of such processes are available based on changing conditions and varying fluid parameters within the borehole during production, such as that offered by Francke (2014).

One important aspect of EGS in operation is that involved fluids are not in equilibrium with the given surrounding and a deep understanding of chemical processes is important. Therefore, a detailed knowledge of the physicochemical properties of the fluids is crucial (Regenspurg et al., 2015). Processes like corrosion, precipitation, and generation of gas phases may occur as shown in Fig. 25.12. Therefore, corrosion and scaling should be monitored within the boreholes by wireline logging, determining deviations from the nominal casing diameter using caliper measurements (Feldbusch et al., 2013). All these processes have been observed at the Groß Schönebeck sites where combined scaling and corrosion resulted in clogging the production well (Regenspurg et al., 2015 or Blöcher et al., 2015). Fluid flow rate and phase composition can be determined using production logging tools for measurement of pressure, temperature, and fluid velocity, as well as using novel techniques like distributed fiber-optic sensing (Henninges et al., 2012). Such measurements are also diagnostic for changes in reservoir behavior (eg, productivity).

Safe operations can be ensured by an operation regime derived from the full understanding of potential involved processes. The measures may include the use of inhibitors or taking into account temperature and pressure levels or the choice of nonreacting material in the technical system components. Continuous monitoring is recommended, possibly with recently developed instruments (eg, Milsch et al., 2013).

25.7 Outlook

The EGS concept expands the geothermal resource base and supports utilization options from domestic geothermal resources enabling renewable heat provision at any site close to customers. Fig. 25.13 shows the economic effect of applying stimulation

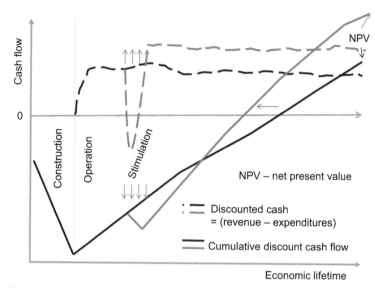

Figure 25.13 Cash flow and net present value of a geothermal project. The discounted cash is calculated by revenue minus capital and operational expenditure. In red, the changes are shown induced by the demonstrated treatment (stimulation), which can be done any time during the operating phase. Please notice the increase in the slope and the earlier achievement of a positive discount cash flow.

treatments on potential cash flow. That way, a huge energy market can be reached due to the high potential of geothermal resources. As a consequence, these activities will boost the development of new geothermal projects providing a key contribution for the transition to a low-carbon society.

A better knowledge of EGS will be useful for both the development of new sites and for improving the sustainability of sites in operation. The performance improvement is expected to be of the order of a factor of 2 for permeable sedimentary rocks and can be around an order of magnitude for tight impermeable rocks. For operations with damaged reservoir formations, EGS methods provide ways to regain productivity.

It is important to note in this context that the utilization of geothermal energy provides a domestic energy supply. Therefore, application of EGS methods will have an impact on domestic industry because the basic knowledge and the availability of services and instruments require short distances. Geothermal energy systems have a strong research need perspective for domestic energy provision (Huenges et al., 2013).

The ambitious plans for the development of geothermal energy are based on successful testing and installation of EGS in recent years, and now the geothermal industry has a unique record of industrial exploitation of this technology, particularly in Europe. It is known that many sites are suitable for further development of geothermal energy in basement/sedimentary environments suitable for EGS. Geothermal energy's main bottlenecks for a more widespread use are the high initial investment costs and the uncertainty of exploitation of the geothermal reservoir. EGS addresses this latter bottleneck by improving and developing productivity and injectivity of geothermal reservoirs to be used. Progress in this field will significantly enhance the potential of geothermal energy in the worldwide energy mix.

References

Blöcher, G., Reinsch, T., Henninges, J., Milsch, H., Regenspurg, S., Kummerow, J., Francke, H., Kranz, S., Saadat, A., Zimmermann, G., Huenges, E., 2015. Hydraulic history and current state of the deep geothermal reservoir Groß Schönebeck. Geothermics (online).

Bommer, J., Oates, S., Mauricio Cepeda, J., Lindholm, C., Bird, J., Torres, R., Marroquı, G., Rivas, J., 2006. Control of hazard due to seismicity induced by a hot fractured rock geothermal project. Engineering Geology 83, 287–306.

Bruhn, D., Manzella, A., Vuataz, F., Faulds, J., Moeck, I., Erbas, K., 2010. Exploration methods. In: Huenges, E. (Ed.), Geothermal Energy Systems: Exploration, Development and Utilization. Wiley-VCH, Berlin, pp. 37–111.

Douglas, J., Edwards, B., Convertito, V., Sharma, N., Tramelli, A., Kraaijpoel, D., Cabrera, B.M., Maercklin, N., Troise, C., 2013. Predicting ground motion from induced earthquakes in geothermal areas. Bulletin of the Seismological Society of America 103 (3), 1875–1897.

Ellsworth, W.L., 2013. Injection-induced earthquakes. Science 341 (6142).

Evans, K., Zappone, A., Kraft, T., Deichmann, N., Moia, F., 2012. A survey of the induced seismic responses to fluid injection in geothermal and CO_2 reservoirs in Europe. Geothermics 41, 30–54.

Feldbusch, E., Regenspurg, S., Banks, J., Milsch, H., Saadat, A., 2013. Alteration of fluid properties during the initial operation of a geothermal plant: results from in situ measurements in Groß Schönebeck. Environmental Earth Sciences 70 (8), 3447–3458.

Francke, H., 2014. Thermo-hydraulic Model of the Two-Phase Flow in the Brine Circuit of a Geothermal Power Plant (Ph.D. thesis). Technical University, Berlin. https://opus4.kobv. de/opus4-tuberlin/frontdoor/index/index/docId/4712.

Frick, S., Regenspurg, S., Kranz, S., Milsch, H., Saadat, A., Francke, H., Brandt, W., Huenges, E., 2011. Geochemical and process engineering challenges for geothermal power generation. Chemie Ingenieur Technik 83 (12), 2093–2104.

Grünthal, G., 2014. Induced seismicity related to geothermal projects versus natural tectonic earthquakes and other types of induced seismic events in Central Europe. Geothermics 52, 22–35.

Henninges, J., Brandt, W., Erbas, K., Moeck, I., Saadat, A., Reinsch, T., Zimmermann, G., 2012. Downhole monitoring during hydraulic experiments at the in-situ geothermal lab Groß Schönebeck. In: Proceedings, 37th Workshop on Geothermal Reservoir Engineering (CD-ROM), Edited, pp. 51–56 (Stanford, USA).

Hogarth, R., Holl, H., McMahon, A., 2013. Flow testing results from Habanero EGS project. In: Proceedings Australian Geothermal Energy Conferences 2013 Brisbane, Australia.

Holl, H.G., Barton, C., 2015. Habanero field - structure and state of stress. In: Proceedings World Geothermal Congress 2015 Melbourne, Australia.

Huenges, E. (Ed.), 2010. Geothermal Energy Systems: Exploration, Development, and Utilization. Wiley-VCH, Berlin.

Huenges, E., Kohl, T., Kolditz, O., Bremer, J., Scheck-Wenderoth, M., Vienken, T., 2013. Geothermal energy systems: research perspective for domestic energy provision. Environmental Earth Sciences 70 (8), 3927–3933.

Jung, R., 2013. EGS — Goodbye or back to the future 95. In: Bunger, A.P., McLennan, J., Jeffrey, R. (Eds.), Effective and Sustainable Hydraulic Fracturing, ISBN 978-953-51-1137-5.

Jung, R., 1999. Erschließung Permeabler Risszonen für die Gewinnung Geothermischer Energie aus Heißen Tiefengesteinen. Final report. project 0326690A funded by the German Ministry for Education and Research (BMBF), 82 pp.

Kwiatek, G., Bohnhoff, M., Dresen, G., Schulze, A., Schulte, T., Zimmermann, G., Huenges, E., 2010. Microseismicity induced during fluid-injection: a case study from the geothermal site at Groß Schönebeck, North German Basin. Acta Geophysica 58 (6), 995—1020.

Legarth, B.A., Huenges, E., Zimmermann, G., 2005. Hydraulic fracturing in a sedimentary geothermal reservoir: results and implications. International Journal of Rock Mechanics and Mining Sciences 42 (7—8), 1028—1041.

Majer, E.L., Baria, R., Stark, M., Oates, S., Bommer, J., Smith, B., Asanuma, H., June 2007. Induced seismicity associated with enhanced geothermal systems. Geothermics 36 (3), 185—222.

Mills, T., Humphreys, B., 2013. Habanero pilot project — Australia's first EGS power plant. In: Proceedings Australian Geothermal Energy Conferences 2013 Brisbane, Australia.

Milsch, H., Giese, R., Poser, M., Kranz, S., Feldbusch, E., Regenspurg, S., 2013. Technical paper: FluMo - a mobile fluid - chemical monitoring unit for geothermal plants. Environmental Earth Sciences 70 (8), 3459—3463.

Regenspurg, S., Feldbusch, E., Byrne, J., Deon, F., Driba, D.L., Henninges, J., Kappler, A., Naumann, R., Reinsch, T., Schubert, C., 2015. Mineral precipitation during production of geothermal fluid from a Permian Rotliegend reservoir. Geothermics 54, 122—135.

Schill, E., Cuenot, N., Genter, A., Kohl, T., 2015. Review of the hydraulic development in the multi-reservoir/Multi-well EGS project of Soultz-sous-Forêts. In: Proceedings World Geothermal Congress 2015 Melbourne, Australia.

Westaway, R., Younger, P.L., 2014. Quantification of potential macroseismic effects of the induced seismicity that might result from hydraulic fracturing for shale gas exploitation in the UK. Quarterly Journal of Engineering Geology and Hydrogeology 47, 333—350.

Yoon, J.-S., Zimmermann, G., Zang, A., 2015. Discrete element modeling of cyclic rate fluid injection at multiple locations in naturally fractured reservoirs. International Journal of Rock Mechanics and Mining Sciences 74, 15—23.

Zang, A., Oye, V., Jousset, P., Deichmann, N., Gritto, R., McGarr, A., Majer, E., Bruhn, D., 2014. Analysis of induced seismicity in geothermal reservoirs — an overview. Geothermics 52, 6—21.

Zang, A., Yoon, J.S., Stephansson, O., Heidbach, O., 2013. Fatigue hydraulic fracturing by cyclic reservoir treatment enhances permeability and reduces induced seismicity. Geophysical Journal International 195 (2), 1282—1287.

Zimmermann, G., Blöcher, G., Reinicke, A., Brandt, W., 2011. Rock specific hydraulic fracturing and matrix acidizing to enhance a geothermal system - concepts and field results. Tectonophysics 503 (1—2), 146—154.

Zimmermann, G., Moeck, I., Blöcher, G., 2010. Cyclic waterfrac stimulation to develop an enhanced geothermal system (EGS): conceptual design and experimental results. Geothermics 39 (1), 59—69.

Zimmermann, G., Reinicke, A., 2010. Hydraulic stimulation of a deep sandstone reservoir to develop an enhanced geothermal system: laboratory and field experiments. Geothermics 39 (1), 70—77.

Zimmermann, G., Tischner, T., Legarth, B., Huenges, E., 2009. Pressure dependent production efficiency of an enhanced geothermal system (EGS): stimulation results and implications for hydraulic fracture treatments. Pure and Applied Geophysics 166 (5—7), 1089—1106.

Geothermal energy in the framework of international environmental law

E. Rodríguez Arias
Independent Consultant, Costa Rica

26.1 Introduction

As a reader, you might find the title of this chapter unusually dry and consider the task of learning about the finer points of environmental legislation a daunting proposition, perhaps even a journey you would rather avoid. My aim, however, is to guide you gently through to the final paragraph so that not only will you have a clear appreciation of the close, almost intimate, relationship between geothermal science and law, but you will be glad to have taken the trip.

So to begin, let us go through some general terms to familiarize ourselves with the topic. Have you noticed that every time you take a taxi, a bus, or a plane, you're actually part of an agreement? You pay a fee, and in exchange you are taken to a certain place. In the same way, every time you buy something at the supermarket you're actually part of a purchase and sale agreement. You pay a certain amount and the other person allows you to leave the premises with products. Traffic signals and corresponding laws cause you stop at red lights. These and other commonplace activities are supported by laws.

Let us think about it. Life: birth, marriage, divorce, death. Are there really laws governing even those very basic human activities? Well, yes, there are. Laws deal with all the significant aspects of life, which is nourished by them. There are laws for when you are born (citizenship), or even before birth, when a human fetus already has rights that protect him/her while in the mother's womb.

In this context, and to start being specific, will energy be an important matter for the law, just as agreements, homicides, labor relationships, and inheritances are?

It is a fact; the use of energy is absolutely vital for human development. Energy is associated with comfort and quality of life, advances in medicine, technology, and the economy; moreover, humanity is interested (because of the survival instinct) in fostering clean energy use. This is where geothermal science steps in. However, should it be allowable for geothermal science to stomp on a few details of lesser importance just for the sake of fostering clean energy? Let me ask this in a different way: Should polluting a river, losing flora and fauna species, or increasing noise pollution be acceptable collateral damage in the quest to realize a geothermal project? This leads some of us to the question: Does the law always protect the environment? Well, yes, it does.

Geothermal Power Generation. http://dx.doi.org/10.1016/B978-0-08-100337-4.00026-7

Humans have come to the understanding that we all live in a common place called Earth and that by destroying it, we are destroying our only available habitat. Thus, the great tree known as the law is interested in this. This is why alongside the branches of family law, criminal law, labor law, and commercial law, there is also the branch called environmental law. Furthermore, the topic is so important that it has gone beyond any one country's governmental jurisdiction, and this branch is now properly called international environmental law. This is the branch of the law that we want to present to you in this chapter so that you will understand its relationship to geothermal energy.

26.1.1 Definitions and environmental components

Since it is inevitable that all human activities are going to have an impact on our shared environment, protecting the environment is a way of ensuring human survival; it is a way of protecting ourselves. Currently, the large number of problems affecting the planet has brought to light the absolute necessity of using a powerful tool like the law to legally protect the environment. So, to have a better understanding of this issue, we will review some essential definitions to put into use later when we focus on the principal international treaties that have been created as a legal warranty to protect the environment.

26.1.1.1 Environment

It seems there are as many definitions for the word "environment" as there are authors. So, I will try to define it as simply as possible. According to the *Didactic Ecology Dictionary*, the word "environment" means "not only the environment per se, but it also includes/covers the circumstantial conditions surrounding individuals or things." Those circumstances can be physical (cold, heat, humidity, dryness, illumination, noise) or social and psychological (happiness, sadness, ignorance, misery, wealth); they could also be biological or natural (tropics, mountains, deserts) and anthropogenic (urban, industrial, rural) [1]. In other words, environment is everything surrounding or affecting humans; it comprises natural, physical, biological, artificial, social, and psychological elements, and all those elements interact with each other.

The Highest Constitutional Court of Costa Rica defined environment as follows: "it is a potential area of development to be properly and wholly recognized in its natural, sociocultural, technological and political dimensions, because in each specific case, its productivity for today and tomorrow can be harmed, and the estate assets of future generations can be endangered (...)." The main purpose of its use and protection is to obtain a favorable human outcome, and the environment is not less important than health, food, work, housing, education, etc. Essentially, we must understand that if humans have the right to use the environment for their own development, then they have also the right, and obligation to protect and preserve it for the use of present and future generations [2].

The legitimate exercise or execution of a law has two essential limitations: on one hand, rights must be equal for everybody; and on the other hand, there is the rational exercise or execution and the useful benefit of the law in and of itself. Environmental quality is a fundamental criterion of quality of life. As we can see, the definition of "environment" is very broad. To sum up, we can agree that the environment makes all activities of human life possible. That is why it is necessary that the quality of the environment is taken into account as a parameter of human activity; in other words, a polluted environment is not beneficial for the activities of human life. Consequently, while it is a goal that every human being has the "right" to live in a healthy environment, it is also unavoidably true that humanity has the duty of preserving that environment not only for the current population but also for future generations.

26.1.1.2 Environmental components

So, we can agree that the environment is an incredibly all-encompassing entity. However, we may find it easier to handle if we break it apart into two subcategories.

The natural environment

This is what is usually meant by the term "environment." It consists of the climate, soil, water, flora, fauna, minerals, and energy found in a given area. These components in turn are "natural resources." which humans use to survive. Natural resources are goods provided by nature, which are exploited to satisfy the needs of human society. They have been classified in different ways, but the most often-used classification system is the one that divides them into "renewable" and "nonrenewable" resources. Renewable resources are the ones that can replenish themselves over time, as long as they are exploited in a way that allows for natural replenishment, in other words, as long as the level left for replenishment is higher than the level of consumption. Nonrenewable resources are slowly formed, they cannot be replaced in a natural way or during a human scale of time. For example, minerals and fossils are nonrenewable because they require millions of years to be formed. From a purely human perspective, resources are not renewable when the consumption rate is higher than the replenishment rate. In the context of geothermal energy, we need to ask ourselves: Is geothermal vapor renewable or nonrenewable? This is a valid question, as it appears the inappropriate exploitation of this resource can make it become nonrenewable [3].

The social environment

The social environment refers to the human beings themselves and the human activities, which use natural resources. One example is the creation of infrastructure. The Declaration of the United Nations Conference on Human Environment presented in Stockholm in 1972 gives a more complete definition: "Man is both creature and molder of his environment, which gives him physical sustenance and affords him the opportunity for intellectual, moral, social and spiritual growth. In the long and tortuous evolution of the human race on this planet, a stage has been reached when, through the rapid acceleration of science and technology, man has acquired the power to transform his environment in countless ways and on an unprecedented scale. Both aspects of man's environment, the natural and the man-made, are essential to his well-being (…)." [3].

26.1.2 What is sustainable development?

No effort on the preservation of the environment can be carried out without a central axis on sustainable development.

26.1.2.1 Context and definition

Before defining "sustainable development," it is necessary to define the context from which it came into being. We can start back with the economic growth that took place during the Industrial Revolution in the beginnings of the eighteenth century, starting with the mechanization of the sewing machine. This is when industrial production started to substitute the traditional handmade (and homemade) manufacturing processes. The resulting textile industries were developed near places with coal deposits, the energy source.

Subsequently, the invention of the steam engine was a huge success for the agricultural industry, reducing the need for a human workforce at home even further, and consequently leading rural people to flee to the city, which greatly affected the lifestyles of most countries in Europe. By the end of the nineteenth century, a huge growth in industry and world commerce was realized. The steam-powered railways, boats, and ships reduced the time and price of raw material and finished good transportation.

In the beginning of the twentieth century, the development of new technologies like electricity, the gasoline-powered automobile, and the transformation of petroleum products, started what historians call the Second Industrial Revolution [4]. However, all this growth was causing severe damage to the environment. Beginning in the 1980s, the ecologists started to loudly voice their concern about the environmental damage caused by industrial society. First they talked about industrial pollution, coal residues in the sea, strip-mining, excessive logging, poor air quality, and the like, and the conversation never stopped as the major disasters of Chernobyl, Bhopal, Exxon Valdez, and Fukushima occurred. Then, people started to become familiar with terms such as "global warming" to the increasing concentration of "greenhouse gases" and other related terms that gave humanity the vocabulary to discuss worldwide environmental issues. In countries around the world, people were becoming conscious that this human industry and development based on nonrenewable natural resources, despite often bringing social and economic improvements, could ultimately destroy the planet.

Consequently, the United Nations convened a Conference on Environment and Development in 1992, also known as "The Earth Summit," in Rio de Janeiro, Brazil, led by Mrs. Gro Harlem Brundtland, First Lady of Norway at that time. The report from this conference ended up in a document called "Our Common Future" also known as "The Brundtland Report." It says, in part: "Sustainable Development is the development that meets the needs of the present without compromising the ability of future generations to meet their own needs."

26.1.2.2 Sustainable development as a goal

From this very brief summary of recent human history, we can say that before that moment in Brazil, the world commonly understood "development" just as economic and social progress. The Brundtland Report clearly shows that for the understanding to be complete, we all need to consider the environmental impact as well. So now, the goal for every country is to say "yes" to development as long as it is in harmony with the environment. Increasing the environmental awareness and sensitivity of citizens and governments is a key path in right direction. Programs to spread environmental education or boost the use of renewable energies play an important part in achieving that goal.

If we make a short evaluation, we will notice that the environment is implicit in every aspect of our daily life. To mention a few examples (you can think of some others): humans need to eat, have a place to live, communicate with each other, produce goods and services that enhance the quality of life, and we need spaces for recreation, etc. All of these aspects of living presuppose a relationship with the environment [5].

The need to eat: Agriculture and water, for example, are greatly impacted by agro-chemicals; desertification; river and aquifer pollution and depletion; droughts; floods; rich countries with access to food and poor countries going through starvation.

A place to live: The rise in global population has increased the urbanization process; what used to be more the habitat of flora and fauna, or farms, is now covered with big cement buildings, asphalt streets, and homes seeming to require a TV, a washing machine, a stove and a lot of other appliances that need electricity. Real estate development not only affects the land being built on, but the acquisition of needed materials: wood, cement, clay, iron, and plastic, etc. all have an environmental cost.

Communications: To be able to communicate a country needs commerce and defense; to have an infrastructure able to satisfy needs such as bridges, highways, airports, ports, vehicles, and fuels. All of these impact the environment in many ways, mentioning a few: rubble, plastic packaging, dust, chemical products, etc.

Production of goods and services: Industrial production and electronic manufacturing produce waste, electricity use, smoke pollution from factories, etc.

Recreational spaces: Tourism represents an income source for countries that have a lot of attractive places. It is an activity that promotes the use of transportation, generates increased waste, and could damage the biodiversity of natural spaces. It also implies the development of infrastructure for the construction of hotels, restaurants, airports, marinas, golf courses, and swimming pools. As we see, all of these can cause great damage to the environment if they are not properly managed.

Thus, it is undeniable that a country getting income from tourism, with an industry that enables it to export products, with modern ways of communications, etc. might be a developed or at least developing country. But definitely in measuring the successful development of a country, we should not just be focused on the social and economic benefits, because this will compromise the needs of future generations. The unsustainability of cutting down a whole forest goes beyond the loss of plant and animal species and beyond the loss of the wood as a future resource; it impacts air and water quality, soil stability, etc. well into the future. In other words, the environmental benefits of development must be an essential measure of any country's progress, for without it we would be destroying our habitat.

26.1.2.3 Rules of sustainable development management

It is clear that the environment is an essential component of development. But now we have another question: How should natural resources be managed so that they will be sustainable?

Fortunately, there are a lot of authors who agree on the key foundation of sustainable development policy: No renewable resource will be used at a greater pace than its generation. If we start from the fact that a resource is renewable if it is obtained from natural sources which are almost endless, either because of the great quantity or because it can replenish itself naturally, then we can ask if geothermal energy is or is not a renewable energy resource. Also, as we have said before, if a resource is used with greater speed than what it takes it to replenish itself, overexploitation will use up that resource. Can this happen with geothermal energy? Some think that it is possible, and further that this is what makes geothermal energy "partially renewable." Whichever the case, it is clear that if a geothermal field is not well preserved and it is overexploited or improperly managed, this field is liable to cool off or become useless. Thus, it needs to not be used at a greater speed/rate than its regeneration.

No pollutant should be produced at a greater rate than what it takes to be recycled, neutralized, or absorbed by the environment. Nature has a great self-purifying power, as long as its capacity is not exceeded. Microorganisms in the ocean can decompose oil, but if the leak is of tons of raw oil, it surpasses the capacity of the sea to purify itself.

Human interference in the environment needs to be reasonable, according to what it takes the environment to react and to stabilize itself. For example, massive overfishing of a single species does not allow enough time for the species to reproduce and for the young to reach maturity; thus it will become extinct.

26.1.2.4 Sustainable development challenges and achievements

Since the environment has been put on the agendas of organizations and governments, achievements have increased but enormous challenges remain. Since governments and nongovernment organizations (NGOs), businesses and consumers globally have become conscious of the importance of environmental conservation as an imperative to survival on this planet, the goals of a country are no longer measured only in terms of gross domestic product (GDP).

Now we are going to look at the most important challenges and goals for certain countries by region. These are only examples for we know there are many countries and organizations working efficiently to create better ways to preserve the environment and increase sustainability. Here, we are only going to mention a few that help illustrate our point. And even though we have mentioned before that sustainable development comprises environmental, social, and economic factors, we will keep our focus on the environmental aspects.

26.1.2.5 Sustainable development in the European Union

Europe has a long history as the leader on environmental conservation. The European Union's (EU) environmental policy is carried out through environmental action programs, each a 10-year strategy to solve a certain problem in a certain region. The objectives are not considered in a separate way, rather, they are modified according to the development of the environmental problems. The first of this type of program began in 1973, just as global environmental awareness started.

The EU has faced multiple problems. We can mention some: creation of financial funds for the environment and the need to have environmental policies embedded within all public policy in the countries of the EU. Environmental issues as global warming, use and management of natural resources, waste management, soil deterioration, and so much more have been discussed.

Currently, the "Seventh Program" has been in effect since January 17, 2014, and its motto is: "Living well, within the limits of our planet." This program will guide European environmental policy until December 31, 2020. It identifies three key objectives:

- To protect, to conserve, and to enhance the Union's natural capital.
- To turn the Union into a resource-efficient, green, and competitive low-carbon economy.
- To safeguard the Union's citizens from environment-related pressures and risks to health and well-being.

One topic that has always been part of these programs is renewable energy, owing to ever-increasing demand. More recently, a new technology is being studied: carbon capture and storage (CCS). CCS has been recognized at the highest political level in the EU as part of the action framework in terms of climate and energy until 2030, and as part of the European Strategy for competitive, sustainable, and secure energy [6].

26.1.2.6 Sustainable development in Latin America and the Caribbean

In general, Latin American and Caribbean countries face similar challenges since they are all developing countries. Some of those challenges include:

- To include environmental issues in economic planning budgets.
- To create financing mechanisms for environmental programs.
- To create and to support strong environmental institutions.
- To establish and operate effective environmental tracking systems.
- To strengthen the environmental education of children and adults.
- To increase modern technological access.
- To implement international treaties in internal environmental legislation, policies, and institutions. In other words, to have a strong legal framework.
- To promote the use of renewable energies to decrease the dependence on fossil fuel.

Even though these challenges would become the goals, this region is very rich—rich but with great inequality [7].

Here are some of the efforts that are being done by some countries of this region to try having a balanced environment. And again, these are just some examples for illustration since there are many more. Here, we will not focus solely on geothermal energy examples as we will see those in a later section. However, we will refer to geothermal when we think it is necessary.

Argentina

The Sustainable Development and Environmental State Ministry is the department in charge of environmental issues. The sustainable energy growth in this ministry has brought up the challenge to achieve a fuel and energy grid that will allow the Argentinean economy to grow in a sustainable way. In 2008, a study called "Framework for a Strategy to Promote Renewable Energies and Energy Efficiency" served as a scientific foundation for the creation of law. This report shows that the geothermal potential of Argentina in 2010 was 30 MW and that the promotion of this type of energy is limited by available technologies, regulation limitations, and the need to create ways of measuring renewable energy capacity. It also shows that there are approximately 90 geothermal energy projects of low enthalpy dedicated for thermal uses. In 2004, Argentina started an incentive-based project of rational energy use [8].

Brazil

One of the most interesting environmental strategies developed in Brazil is the "Sustainable Cities" program, focused on the reduction of high environmental impact materials, the reuse of materials, waste management, air quality, the promotion of public transportation, and the construction of urban green areas. In 2002, PROINFA was created as an incentive program to encourage alternative sources for electric energy [9].

Chile

In 2003, a law was passed to promote nonconventional renewable energies and aimed to give licenses for geothermal energy development. Its motivation is to diversify the natural energy complex, to reduce the dependence on external fuels, and to reduce the electric system's vulnerability caused by concentrating electric energy in hydroelectric energy and natural gas [10].

Colombia

One of the greatest environmental achievements in Colombia has been the strengthening of environmental justice through the creation of a natural resources criminal unit. Complaints of crimes against nature can be filed through this legal entity. Another achievement is the National Environmental System (SINA) creating a legal platform for scientific and technological support for the formulation of environmental policies [11].

Costa Rica

Costa Rica has a long history of environmental protection. In the 1970s when the world was just creating environmental awareness, Costa Rica was already creating a network of national parks. The National System of Preservation Areas (SICAC) is formed by nine different categories of environmental management: national parks, biological reserves, absolute national reservation areas, national monuments, protected areas/zones, forest reservation areas, wildlife refuges, wetlands, and indigenous territories. These nine categories represent almost 26% of the national territory with a total of 169 protected areas. This small country, only 51,100 km^2 in size, is home to a total of 4% of the world's species, which puts it among the 20 countries with the greatest biodiversity in the world. Costa Rica has also developed multiple types of renewable energies: hydroelectric energy, geothermal energy, wind power, and, in less quantity, solar energy, and biomass. Costa Rica plans to be the world's first carbon neutral country by 2021 [12].

Mexico

This country has placed great importance on sustainable tourism. All environmental aspects of marine recreation and golf course use are managed according to "Good Practices" environmental manuals. It has also designed a Guide for Promoting Renewable Energies program. In the same way, it has programs for support and subsidies for managing waste and for environmental education projects [13].

The Organization of Eastern Caribbean States (OECS)

This is a regional entity that promotes the technical cooperation and sustainable development of Antigua and Barbados, Dominica, Grenada, Saint Lucia, Saint Kitts and Nevis, Saint Vincent and the Grenadines, Montserrat, Anguilla, and the British Virgin Islands. The OECS was created on June 18, 1981, with the Treaty of Basseterre, which was named after the capital city of St. Kitts and Nevis. The Environment and Sustainable Development Unit in this organization is responsible for providing the environmental management framework for its members [14].

Panama

The government's Environmental Authority rules in all matters related to natural resources and the environment. Its mission is to ensure the fulfillment and application of government environmental laws and policies. One of the greatest environmental achievements is Law No. 028 of October 13, 2014, which mandates environmental education for primary and secondary schools. This is very important because raising awareness on environmental protection should start at an early age as they will be the future generations of environmentally responsible citizens.

When we talk about Panama, we also need to talk about the Panama Canal, a monumental piece of civil engineering that is a significant international means of communication and transportation. It is managed by the governmental Authority of the Panama Canal (ACP) and its most significant task is to protect the hydrological resource, indispensable for its labor. The catchment area is not only vital for the canal operations, but it also is the main source of fresh water to satisfy two-thirds of the country's population. The Interinstitutional Commission of the Catchment Area of the Panama Canal (CICH), also part of the ACP, is in charge of implementing the environmental strategy. It carries out programs of vegetation coverage surveillance, soil use, reforestation, protection, control, and tracking of the hydrological resource [15].

Peru

This is another country concerned about environmental issues. Through the Environmental Ministry, Peru has started ambitious environmental projects. For example, government universities are researching environmental issues. In the same way, the Peruvian Amazonia Research Institute (IIAP) was created to protect the environment specifically in such a rich area of natural resources. There is also a citizen service where people can complain about environmental damage. The National Policy for Environmental Education was implemented so that all the population will feel responsible for protecting the environment. It works with the education and environment sectors. Japan is helping Peru to preserve their forests. The Environmental Evaluation and Auditing Organization (OEFA) is also in charge of controlling and tracking the correct use of natural resources [16].

26.1.2.7 Sustainable development in North America

In 1993, the United States, Mexico, and Canada signed the North American Agreement on Environmental Cooperation (NAAEC) from which the Commission for Environmental Cooperation was created. This is an intergovernmental group that supports cooperation in environmental matters among the members of the North American Free Trade Agreement (NAFTA). The conservation, protection, and enhancement of the environment in their territories is developed in the context of increasing economic, trade, and social links between Canada, Mexico, and the United States. This commission was created in 1994 to address environmental issues of continental concern, to prevent possible environmental conflicts from commercial relationships, and to promote effective environmental legislation in all three countries. The basic programs ran by this commission are:

- Environment, economy, and commerce
- Diversity preservation
- Pollution and health
- Environmental policies and legislation [17]

26.1.2.8 Origin and development of environmental international law

All human activities impact the environment, either positively or negatively. Negative impacts include climate change, pollution, a hole in the Earth's ozone layer, and the loss of biodiversity. These problems are not only happening in Europe, Africa, Asia, Oceania, or the Americas; these issues are happening all around the world.

That is why a common and binding legal instrument is needed to regulate different human activities, to control and establish a common guide. In other words, all these countries reports should go beyond stating the problems and concerns; there should be a coercive mechanism to measure all activities. Such a mechanism is the law. It regulates human conduct; it is perfectly suited to establish norms to control how human activities affect the environment.

However, the aim of the first international environmental agreements was not always environmental protection in itself. It was the protection of some natural resources for economic reasons or interests. As an example, at the end of the nineteenth and the beginning of the twentieth centuries, using feathers from certain species of birds in the making of hats and accessories became a fashion trend for ladies as a way to show wealth and social status. The acquisition of feathers was on such a large scale that some species nearly became extinct. In 1868, a group of farmers asked the Austro-Hungarian Emperor Francisco Jose to protect certain bird species that were helpful for the agricultural industry but were being excessively hunted for their feathers. As we see, the farmers' interest was not the welfare of all the affected bird species exactly, but just those birds that were beneficial for their harvests, their economic interests. So in 1902, the Paris Convention was held to protect birds "that eat worms and are useful for the harvest." But the protection was only given because they were useful for agriculture. It was so specific that the document had an appendix called "non-useful birds," thus making hunting these birds legal.

Many events along the way contributed to the formation of what we now know as International Environmental Law. In the middle of the twentieth century, the Trail Smelter Arbitration case between the United States and Canada was known because of the pollution in the state of Washington caused by the Consolidated Mining and Smelting Company in Canada, which processed lead and zinc. The arbitration tribunal concluded that no country has the right to use its territory in a way that causes harm in another country's territory. This decision established a basis for modern international environmental law.

Later, the use of oil tankers caused the first ecological catastrophes. The oil spill of tons of Torrey Canyon oil that happened in England in 1967 affected the coastal areas of several countries and was the reason for the United Nation to call for the Conference on the Human Environment in Stockholm in 1972. Genuine measures to protect the environment were carried out after this conference.

26.1.2.9 Main international treaties on international environmental law

In this section, we will look at the main international treaties of international environmental law. Though the list is very long, we will only look at one that made an important start of an international environmental law.

UN Conference on the Human Environment in Stockholm in 1972
This conference was attended by 114 countries and many international NGOs. This one was the first of many aimed at solving global environmental issues and was

when the United Nations Environment Program (UNEP) was created. The main result of this conference was the contribution and development of legal and institutional mechanisms to protect the environment. Treaties were created, such as the MARPOL Convention for the prevention of pollution from ships, the CITES Convention on international trade in endangered species of wild fauna and flora in 1973, the Convention of the World Cultural and Natural Heritage (1972), and the United Nations Convention on the Law of the Sea (UNCLOS). The Stockholm Conference led the way for the creation of regional regulations for the environment in the EU. The World Bank and regional development banks also included environmental matters when processing loans. Another important consequence of this conference was the creation of the World Commission on Environment and Development also known as the Brundtland Commission, which created the Brundtland Report: "Our Common Future," where the concept of Sustainable Development was introduced.

UN Conference on Environment and Development
The Earth Summit resulted in the following documents:

- Framework Convention on Climate Change (UNFCCC) of NU
- Convention on Biological Diversity
- Rio Declaration on Environment and Development Sustainable
- Forest Principles
- Agenda 21 Program
- Creation of the United Nations Commission on Sustainable Development (CSD)

26.1.2.10 Second Earth Summit

The Second Earth Summit convened in New York City, USA, in 1997 to review progress since Rio. Even though major accomplishments were not realized, they issued a plan of action called Agenda 21 for the next 5 years.

26.1.2.11 Johannesburg Summit

This took place in 2002, in Johannesburg, South Africa. It is also known as "The Rio+10 Summit." The main achievements of this summit were the commitment by the countries to reduce the number of people without access to clean water and that more countries would follow the Kyoto Protocol.

26.1.2.12 International environmental law definition

From all that is mentioned above, we could conclude that the environment has stirred up the world. We can say that there is no area that has not been affected: economy, medicine, engineering, law, politics, and government. The environment influence in different areas of human activities has been strengthened by complex legal norms regulating environmental matters created as a consequence of environmental international treaties and environmental national regulations from different countries. When we refer to international environmental law, we refer to all the treaties, international agreements, conventions, pacts, covenants that are obligatory, and to all different declarations, resolutions, and recommendations that are not legally obligatory but that are politically important. Environmental law refers to all national legislation to defend a country's natural resources. Environmental law, either from a national or international perspective, penetrates the whole judicial system of a country; that is why it has become a branch or a specialized area inside the law.

Without trying to be exhaustive, we could try using all of these ideas to define international environmental law. We could define it as the international law branch that comprises all international juridical norms aiming to protect the environment so that future generations will be assured of enjoying it. Some might include the bilateral or multilateral relationships among countries and international organisms to protect the environment. However, for the purposes of this report, our definition only tries to help us understand the topic.

26.2 Environmental international law and geothermal energy

We list here some of the environmental impacts caused by geothermal projects. This is not a comprehensive list because it will depend on the specific characteristics of the site where the project is developed. We will also cite the main international treaties related to each topic. So, when developing a geothermal project, one will have to figure out which treaties and which specific environmental legislation apply within the host country.

This list will provide a clearer idea of the environmental impacts, of the factors that are being impacted, and of the treaties that are applicable in each case. Some of the environmental impacts of a geothermal project are described later [18].

26.2.1 Land subsidence

Can affect:

- Air quality due to contamination from emissions from the machinery used
- Changes in the overflow of surface waters
- Surface or ground water pollution by fuel or lubricant leaks
- Land degradation
- Increase of erosion and sedimentation.
- Certain species will experience habitat disruption or loss, corresponding feeding spaces may be reduced.
- Interruption of fauna's crossing
- Landscape modifications
- Archeological patrimony alteration; in the excavation process necessary for the remediation effort, historically or culturally important artifacts may be uncovered, or disturbed.

Regulating legal framework:

- Kyoto Protocol (December 11, 1997)
- Vienna Convention for the Protection of the Ozone Layer No. 7228. Vienna, Austria (March 22, 1985)

- Montreal Protocol on Substances that Deplete the Ozone Layer Montreal, Canada, 1989
- The Convention for the Protection of Flora, Fauna, and Natural Scenic Beauty of the Americas, Washington, DC, USA (October 24, 1940)
- Convention on the Prevention of Marine Pollution by Dumping of Wastes and Other Matter, 1972

26.2.2   Well drilling

Can affect:

- Air quality because of emissions from machines and field operations
- Air quality by the H_2S causing acid rain
- Air quality because of emissions, especially H_2S, which can affect people's health.
- Dust and waste production

Regulating legal framework:

- Kyoto Protocol (December 11, 1997)
- Vienna Convention for the Protection of the Ozone Layer, March 22, 1985
- Montreal Protocol on Substances that Deplete the Ozone Layer Montreal, 1989.

26.2.3   Natural surrounding sounds

Can be affected by:

- Machine, equipment, and vehicular noise
- Geothermal plants and field activities noise
- Noise can affect people's health in residential areas, or affect nearby tourism industries

Applicable legal framework:

- Working Environment (Air Pollution, Noise, and Vibration) Convention 148 from the International Labor Organization Convention. Geneva, Switzerland (June 20, 1977)

26.2.4   Surface water

Can be affected by:

- Changes in the overflow of surface waters
- Surface water pollution by fuel or lubricant leaks
- Pollution by leaks from piping systems or holding pond rupture; if geothermal water comes into contact with surface water, it can become polluted owing to the difference in their chemical compositions.
- Water quality reduction by solid waste pollution
- Water pollution by chemical spills

Applicable legal framework:

- Agenda 21, Chapter 18.
- Water Framework Directive (Directive 2000/60/EC of the European Parliament and of the Council of October 23, 2000—establishing a framework for community action in the field of water policy)

26.2.5 Soils

Can be affected by:

- Soil compaction would be reduced, caused by earth movements (slopes) when preparing the roads, drilling sites, and the field and plant facilities.
- Hydraulic conductivity of the soil would be reduced in specific areas when compacting for the construction of the different facilities in the field and in the plant site.
- Soil's capacity for water filtration would be modified, caused by earth movements (slopes) when building the different facilities in the field and in the plant site.
- Dangerous substance spills
- Construction waste (cement, metal, and paint)

  Applicable legal framework:

- European Soil Charter (May 30, 1972)
- World Soil Charter Food and Agriculture Organization of the United Nations (FAO), 1981
- World Soils Policy United Nations Environment Program (UNEP), 1982
- Europe Recommendation on Soil Protection (Council of Europe, 1992)
- United Nations Convention to combat desertification in countries seriously affected by drought, particularly in Africa. Paris (June 17, 1994)

26.2.6 Flora

Can be affected by:

- Forest coverage reduction
- Physiological and reproductive processes can be altered by the accumulation of dust or geothermal spray on leaves, flowers, and fruits.

  Applicable legal framework:

- Convention on International Trade in Endangered Species of Wild Flora and Fauna, No. 5605, Washington, DC, USA (March 3, 1973)
- Convention on Biological Diversity, Rio de Janeiro, Brazil (June 13, 1992)

26.2.7 Fauna

Can be affected by:

- Aquatic fauna can be affected by the mud.
- Aquatic fauna can be affected by pollution from dangerous substances introduced to rivers.
- Wildlife habitat can be affected by the loss of forest coverage.
- Increased risk to wildlife of being run over by vehicles.
- Animals' ability to move within normal territorial range would be interrupted during construction.
- Animals' ability to move within normal territorial range could be permanently interrupted by the tubes from the geothermal facilities.
- Some animals' diets would be changed by eating some of the waste from the residue places.
- Rodent or plague increase
- H_2S nontolerant mammals will have to be relocated.
- Light pollution can affect the biological cycles of some mammals.

Applicable legal framework:

- Convention on International Trade in Endangered Species of Wild Flora and Fauna, Washington, DC, USA (March 3, 1973)

26.2.8 Landscape

Can be affected by:

- Landscape quality would be affected by buildings, construction, and electric transmission lines and towers

Regulating legal framework:

- Convention for the Protection of Flora, Fauna, and Natural Scenic Beauty of the Americas, No. 3763, Washington, DC, USA (October 4, 1940)

26.2.9 Archeological patrimony

Can be affected by:

- Earth movements because of construction works

Legal framework:

- Convention on the Protection of the Archeological, Historical, and Artistic Heritage of the American Nations, No. 6360 (Convention of San Salvador) (approved on June 16, 1976)
- Convention concerning the Protection of the World Cultural and Natural Heritage, No. 5980. Paris (November 23, 1972)

26.2.10 Social environment/context

Can be affected by:

- Increase in vehicular traffic
- Loss of ability to conduct business if public access is disrupted
- Health and tranquility of the surrounding communities
- Tourism can be temporarily or permanently affected by the changes in the landscape, noise, or bad odors.
- Respiratory disease outbreak or aggravation, such as an increase in asthma-related hospitalizations
- Cattle grazing can also be affected.

Regulating legal framework:

- Indigenous and Tribal Peoples Convention, No. 169, from the International Labor Organization Convention, Geneva, Switzerland (June 7, 1989)
- The United Nations Conference on the Human Environment Stockholm, Sweden (June 16, 1972)
- United Nations Declaration on Environment and Development (UNCED), Rio de Janeiro, Brazil (June 14, 1992)
- World Charter for Nature, New York, USA (October 28, 1992)

In brief, the main areas of possible environmental impact from geothermal projects are air, surficial water, and subsoil pollution; land alterations; noise pollution; cultural and archeological artifact disturbance; socioeconomic issues with surrounding communities; and dangerous solid waste production. Fortunately, proper environmental management in the planning of a project will significantly reduce negative impact and the possibility of legal action.

26.2.11 Advantages and disadvantages of geothermal energy

The geothermal energy is considered a renewable energy, and therefore, it has many advantages, but also, it could have some few disadvantages.

Advantages:

- It is more widely available than oil, which is found only in specific places.
- It is abundant; it is believed that there are more geothermal energy deposits in the world than oil deposits.
- It is a national energy.
- A geothermal energy plant costs less to build than a nuclear plant.
- The commodity is not subject to international prices fluctuations.
- Geothermal energy plants produce low levels of pollution.
- Water extracted and separated on the surface is "reinjected" to the subsoil so that geothermal water (brine water) will not stay on the surface thus avoiding pollution.
- It is always available throughout the whole year.
- It can be used to produce electricity (high enthalpy) and for other uses (low enthalpy) such as heating and cooling, industrial drying of fruit and vegetables, tourist use for swimming pools and spas.

Disadvantages:

- Some geothermal deposits are located in protected zones where current legislation prohibits the extraction of the resource, thus even though it is clean energy, it cannot be accessed
- If geothermal deposits are not well managed, they can be lost.
- In the case of direct use, everything is only used locally because there is still no technology able to transport that steam from one place to another.
- Sulfur can produce an unpleasant smell.
- It is available only in certain places.
- Some geothermal projects can produce a negative visual impact, it is important to use camouflage on the facilities, for example in the piping.

26.3 Environmental features in public and private companies developing geothermal projects; green sells

26.3.1 Introduction

When initially taking into account the cost increases inherent in the management of these environmental issues, some companies will try to fulfill only the minimum required by law—and even that may be economically challenging for a company. In other words, taking care of the environment is not always seen as something beneficial but as an additional and unwelcome expense. Luckily nowadays, environmental sensitivity and awareness play such an important role that they have become part of the business strategy of more and more companies. Geothermal projects are developed by public and private companies which hope to "sell" their product—electricity. Responsible clients and buyers want to know the conditions in which their "product" is developed.

26.3.2   Traditional environmental management and current environmental management in companies

Years ago, profit was the main goal for companies. It did not matter to the soap manufacturer if soap waste ended up in a nearby river, for example. Environmental protection and control was nonexistent or minimal or was not really taken seriously because fines were not that high. Furthermore, consumers were not interested in knowing if a particular company was polluting. In brief, we can state that many companies used to (1) not have environmental plans at all (for waste management, measures to save water or electricity, etc.), (2) look at laws and fines only as obstacles to business activity, and (3) cooperated with environmental goals at a minimum level, only doing what the law required when compelled.

Nowadays, many companies around the world have environmental programs and advertise their products as "environment friendly," but has this change happened only because there is increased "environmental awareness"? In a way, yes, it has. Humans have grown much more aware of the importance of protecting the environment. However, these societal changes are not enough for companies, they still need to make a profit out of their investment, or they cease to exist.

Therefore, if companies have taken environmental issues into account as an important part of their business plans, it is also because "green sells"—it is a plus to a company's bottom line. In today's marketplace, environmental sensitivity makes the company attractive to consumers and/or it transmits confidence to clients and investors, all of which add up to increased profits. Consequently, we can state that modern-day environmental management in companies is characterized by preventing pollution and natural resource exploitation in all its myriad forms, and we can say that management of environmental factors is an inherent part of business strategy because they have real economic value [19].

26.3.3   Important motives for companies to take environmental factors into account

If various aspects of environmental protection have been taken into account because they produce profit, let us see the details of the motivation from a company's point of view [20].

26.3.3.1   Environmental factors

If we take into account the environmental factor, it is easy to notice that companies will be benefited by environmental conservation. Raw materials needed by many companies come from natural resources, and many of these depend on the quality of the air, soil, and water supply of a region. Businesses can suffer the negative consequences of the greenhouse effect or the ozone layer damage just as individuals can. Thus, the future livelihood of a company depends greatly on environmental preservation.

26.3.3.2 Legal factor

Legal sanctions for environmental harm done by companies have existed for years, but now they are stricter. What has always been true is that companies avoid legal problems. Breaking an environmental law brings legal sanctions, anything from fines or the cost of reparations to going to jail, depending on the severity of the crime.

26.3.3.3 Social factor

People's opinions are very important today, without a doubt. NGOs, protest groups, ecologists, neighbors, media (newspapers, TV news, the Internet), providers, consumer associations, insurance companies, workers, clients, etc. … their opinions are essential to companies, so essential that they have enough power to even close a company which is polluting.

26.3.3.4 Technical factor

The technical factor comes into action because if there are not correct techniques in the processes of the geothermal plant, then it is more likely to pollute the environment.

26.3.3.5 Economic factor

Environmental preservation also plays an important role when it comes to making a profit. Reduced use of energy, water, and raw materials is good for profit. Even the commercialization of waste has become recognized as beneficial, for example, golf courses paying less to use "gray water" for irrigation.

26.3.3.6 The case of geothermal energy

All geothermal projects are either public or private companies selling electricity to their clients. As with any company all the above factors are applicable to it:

- *Environmental factor*: Since it needs natural resources as raw materials, for example water and steam to produce electricity. Whenever the steam is gone, then the facility will be useless.
- *Legal factor*: If a geothermal project pollutes the environment in any way, there will be legal consequences.
- *Social factor*: Since the people's voice has a strong influence when it comes to shutting down a project when it is damaging the environment, respecting the environment will encourage people's approval of the project.
- *Technical factor*: Working with up-to-date technology will bring more efficiency, will optimize the processes and will be reflected in the costs.
- *Economic factor*: The viability of the company will depend on protecting the natural resources they need to be able to continue.

26.3.4 Environmental management tools: Environmental Impact Assessment

As protecting the environment has become more commonplace, various methods have been used to assist companies in meeting legal requirements. One essential tool is the Environmental Impact Assessment (EIA), legally recognized in all countries.

EIAs are judicial-administrative procedures to predict and identify the environmental impacts caused by a project in order to prevent, correct, reduce, value, and/or compensate for those impacts. EIAs were created in the 1960s to begin to measure and control negative human influence on the environment in the United States. After that, EIAs started to be implemented in other countries. By 1973, EIAs were being put into practice in countries like Canada, and in 1976 in France, and by 1985, they were included in the laws governing the EU. Today, EIAs are pretty much used worldwide, even though each country might implement its own variation. They are used in both public and private projects.

26.3.5 Environmental management mechanisms

Environmental procedures are characterized by an obligatory sense when the law says so. However, nowadays companies want to take a step forward protecting the environment and do not want to limit themselves by just fulfilling only what the law says. There are other mechanisms that can guarantee an adequate and respectful environmental management. We will cite some of those mechanisms.

26.3.6 International Organization for Standardization (ISO)

The ISO was created in 1974 with its main branch located in Geneva, Switzerland. It is a federation of national organizations from different countries. Its main objective is to establish manufacturing norms to provide ease in the exchange of goods and services among nations. These are nonobligatory international norms. So, whereas EIAs are almost always a legal requirement before starting a project, ISOs are completely voluntary; companies decide if they want to put them into practice. If they decide to apply the ISO norms, then they will have to fulfill even stricter requisites then the ones demanded by other laws. Companies that apply for ISO certification go through a process of changes, corrections, and improvements in their projects' plans, so that after an audit (to make sure they actually meet the criteria) they will be able to be certified in an ISO-normed aspect, like environmental management [21].

ISO norms exist:

- To optimize production procedures
- To use a common language for the exchange of goods and services
- To standardize different practices
- To ensure the client that the product used was made under standards that ensure quality and respect the environment
- To improve the company's reputation among clients and public administration, which will eventually turn into an economic benefit

26.3.7 Eco-management and Audit Scheme (EMAS)

EMAS is an optional regulation among the EU countries that certifies companies that have environmental policies, have implemented environmental management procedures, and have assumed a commitment to continual improvement. Audits will verify that all norms are being followed, just as happens with ISO norms. ISO and EMAS norms aim to evaluate, improve, and make known a companies' commitment to the environment. Geothermal companies can also opt for these certifications as a tool to improve their reputation with clients, neighbors, organized groups, environmentalists, and others. Additionally, these norms can optimize electricity production processes creating supplies and materials savings, which will become long-term benefits for the project [22].

26.4   Global interest in geothermal energy

26.4.1   Early international conferences

As we have said before, geothermal energy is relatively new to the energy industry. When we talk about renewable energies, most people only think about solar energy, wind power, and hydropower. Geothermal energy and tidal power are still little known. Despite that, interest in geothermal science and interest in taking advantage of its benefits increase every day.

In this context, it is important to know where and when global interest in geothermal energy started. It all began with the First World Power Conference in 1924 held in London, UK, as part of the British Empire Exhibition at Wembley. It was attended by 1700 delegates from 40 countries. Prince Piero Ginori Conti read his paper describing the natural steam plant at Larderello, Italy. No other country had such power plants and a considerable amount of interest was expressed. During the discussion session, one attendee mentioned that New Zealand had similar geothermal resources but that they were more difficult to utilize than dry steam since the natural fluids contained both hot water and steam.

The next important international meeting was the UN Conference on New Sources of Energy (Solar Energy, Wind Power, Geothermal Energy) held in Rome, Italy in August of 1961. Its aim was to study those energy sources that, back then, were just starting to be developed. The conference topic was so important and relevant that the Pope John XXIII was invited for the opening speech. Here is a quote from the Pope:

> Our Creator has sown abundant energy in the world, and man's wisdom throughout centuries has caught and used it for his needs. But nowadays, in what we can call the age of human technology, the possibilities of using this energy is growing at a rapid pace, not only "classic" energy but also the ones that come from sources still a little unused until now, like the sun, the wind, water, and steam hidden in Earth's core: solar energy, wind power, and geothermal science.

There was also another fact that contributed to the interest in these clean sources of energy. The decade after that conference in Rome, the 1970s, was characterized by oil crises and consequently the worry about consistent energy supply. The Rome Conference gave the first look at the topic.

About a decade later, a UN conference was devoted solely to geothermal energy, the first of its kind: the UN Symposium on the Development and Utilization of Geothermal Resources, held in Pisa, Italy, from September 22 to October 1, 1970. The proceedings were published in two parts as a special issue of the journal *Geothermics*. There were a total of 198 technical papers presented in 11 categories covering the full spectrum of activities related to geothermal exploration and exploitation.

This was followed by the Second United Nations Symposium on the Development and Use of Geothermal Resources held in San Francisco, California, on May 20–29, 1975. It took three large hard-bound volumes to present all the technical papers that were delivered there. And from this conference began the worldwide quinquennial geothermal congresses that continue to this day.

Yet another UN Conference on New and Renewable Sources of Energy was held in Nairobi, Kenya, on August 10–21, 1981. Representatives from 125 countries came to this conference, including Prime Minister Gandhi of India. Nairobi adopted the Nairobi Program of Action for the Development and Utilization of New and Renewable Sources of Energy as a blueprint for national and international action. In 1984, an assessment meeting was held as a complement of the Conference in Nairobi. This meeting was held in New York on June 27–29 and resulted in the Report on the Consultative Meeting for the Promotion of Latin American and Caribbean Projects in the Area of New and Renewable Sources of Energy, including geothermal energy. In regard to the latter, the following agreements were approved:

*   Reconnaissance and prefeasibility studies in the Eastern Caribbean.
*   Regional project for the exploration and development of geothermal resources
*   Implementation of an isotopic laboratory
*   Installation and start-up of wellhead generators, and training in this area
*   Evaluation of the geothermal potential in Central America
*   Regional low- and medium-enthalpy projects

26.4.2 International organizations related to energy issues

As a result of this surge in international interest in renewable energy sources, including geothermal energy, there have arisen many organizations dedicated to energy, environment, and especially geothermal energy. Here are a few of the most important ones.

United Nations Environment Program (UNEP): This is the most important international organization related to the environment. Located in Nairobi, Kenya, it is a UN program to coordinate all activities related to the environment. It was created as a recommendation in the UN Conference on Human Development held in Stockholm in 1972 [23].
American Council of Renewable Energies (IRENA): IRENA is an intergovernmental entity to promote renewable energies around the world. It was created in 2009 by Germany, Spain, and Denmark to assess the politics of clean energies and help to transfer technologies and now includes 160 countries [24].
Development Bank of Latin America (CAF): Located in Caracas, Venezuela, it was created in 1970, formed by 19 countries (17 from Latin America and the Caribbean, plus Spain and

Portugal) and 14 private banks from the region. It promotes a Sustainable Development model through credit operations, nonrefundable resources, and support for technical and financial organization for private and public projects in Latin America [25].

Geothermal Development Fund (GDF): The GDF, formed by CAF, is an initiative to promote the relief or mitigation of global warming in the Latin American region by using geothermal energy.

Latin American Energy Organization (OLADE): Founded in 1973 by 27 countries in Latin America and the Caribbean through the Lima Convention, this organization's goal is to develop energy resources of the region and to monitor factors related to its rational and efficient use [26].

United Nations University Geothermal Training Program: The United Nations University was established in 1975. It created and has operated the Geothermal Training Program (GTP) since 1979 in collaboration with the government of Iceland. Its main purpose is assisting developing countries having geothermal potential and to train geothermal specialists [27].

26.5 Conclusion

The relation between humans and nature is so close that we depend on it to survive. The expression "Mother Nature" is not frivolous. She gives us what we need to have a whole development and our duty is to protect her.

At first, a country's development was measured in economic terms. This led to greater industrialization and better technologies. The protection of the environment was not a priority. When the first ecological disasters started to happen, the world realized that the environment needed to be protected if we wanted to survive. Environmental problems do not distinguish countries and do not respect borders.

The problem is a global concern and needs a global solution. And there is no more efficient mechanism to order human conduct than the law.

Thus, environmental law was created as a mechanism for social coercion to set limits, to establish sanctions, restrictions and to normalize conduct to know what to do, what not to do, and how far to go. It sets rights, obligations, and sanctions. Countries have issued international treaties on environmental protection that later became part of their national jurisdiction and that have become part of their legal system.

Environmental law is sometimes criticized as a soft law because it is perceived as not very effective. This criticism is based on the idea that the toughest law is criminal law because it is believed that people will not commit crimes because of severe punishment. And yet jails are full of people and crimes are still being committed. Laws are made to regulate human beings, not angels; thus, they will always be broken. The environmental law is *an answer*; it is a tool to efficiently protect the environment.

And geothermal energy, with everything it implies in all stages, has multiple interactions with natural resources protected by international and national legislation. Consequently, all geothermal project developers need to know that laws regulate everything they do.

Hopefully, all that you have read so far will increase your interest to know which environment treaties have been implemented in your country. Also, you might have an interest on getting acquainted with the environmental legislation in the country where your geothermal project will be developed. The legal framework applicable for a geothermal development is something that must be taken into account; ignorance of this can lead to disaster. Respecting the law and taking all care toward protecting the environment in a geothermal project will warrant the continued operation of your project.

References

[1] Mata A, Quevedo F. Diccionario Didáctico de Ecología. Chap. 1. In: Salazar R, 1rst CMV, editors. Investigación, Análisis y Desarrollo del Derecho Ambiental. San José, Costa Rica: Editorama; 2004.

[2] Verdict 3705-93 of Constitutional Court of Costa Rica. Chap. 1. In: Salazar R, 1rst CMV, editors. Investigación, Análisis y Desarrollo del Derecho Ambiental. San José, Costa Rica: Editorama; 2004.

[3] Declaration of the United Nations Conference on Human Environment. 1972. Stockholm.

[4] Haywood J. In: Atlas histórico del Mundo. Bonner Strasse, Colonia: Könemann Verlagsgesellschaft; 2000.

[5] Ludevid M. Un vivir distinto. Cómo el medio ambiente cambiará nuestra vida. 1st ed. Nivola, libros y ediciones, S.L.; 2003.

[6] http://eur-lex.europa.eu/homepage.html.

[7] http://www.cepal.org/es/publicaciones/desarrollo-sostenible-en-america-latina-y-el-caribe.

[8] www.ambiente.gov.ar.

[9] www.ministeriodomeioambiente.gov.br.

[10] http://portal.mma.gob.cl.

[11] https://www.minambiente.gov.co.

[12] www.minae.go.cr.

[13] http://www.semarnat.gob.mx.

[14] www.occs.org.

[15] http://miambiente.gob.pa.

[16] www.minam.gob.pe.

[17] http://www.state.gov/e/oes.

[18] Personal communication with Mr. Rogelio Zeledón Ureña, Environmentalist (Costa Rican Institute of Electricity). Emilia Rodríguez Arias, invited speaker in short con Geothermal Development in Central America, Resource Assessment and Environmental Management. Environmental legislation applicable to geothermal developments. San Salvador, El Salvador, November 25—December 1 in 2007.

[19] (coord) Conde J. Empresa y Medio Ambiente, hacia la gestión sostenible. 1st ed. S.L. España: Nivola libros y ediciones; 2003.

[20] Conesa Ripoll V. In: Los instrumentos de la gestión ambiental de la Empresa. Madrid, España: Mundi-Prensa; 1997.

[21] www.iso.org.

[22] http://ec.europa.eu/environment/emas/index_en.htm.

[23] www.unep.org.

[24] www.irena.org.

[25] www.caf.com.

[26] www.olade.org.

[27] http://unu.edu.

Index

Ingram Content Group UK Ltd.
Milton Keynes UK
UKHW051611250423
420674UK00001BA/23